Lecture Notes in Electrical Engineering

Volume 802

The book series *Lecture Notes in Electrical Engineering* (LNEE) publishes the latest developments in Electrical Engineering - quickly, informally and in high quality. While original research reported in proceedings and monographs has traditionally formed the core of LNEE, we also encourage authors to submit books devoted to supporting student education and professional training in the various fields and applications areas of electrical engineering. The series cover classical and emerging topics concerning:

- Communication Engineering, Information Theory and Networks
- Electronics Engineering and Microelectronics
- Signal, Image and Speech Processing
- Wireless and Mobile Communication
- Circuits and Systems
- Energy Systems, Power Electronics and Electrical Machines
- Electro-optical Engineering
- Instrumentation Engineering
- Avionics Engineering
- Control Systems
- Internet-of-Things and Cybersecurity
- Biomedical Devices, MEMS and NEMS

For general information about this book series, comments or suggestions, please contact leontina.dicecco@springer.com.

To submit a proposal or request further information, please contact the Publishing Editor in your country:

China

Jasmine Dou, Editor (jasmine.dou@springer.com)

India, Japan, Rest of Asia

Swati Meherishi, Editorial Director (Swati.Meherishi@springer.com)

Southeast Asia, Australia, New Zealand

Ramesh Nath Premnath, Editor (ramesh.premnath@springernature.com)

USA, Canada:

Michael Luby, Senior Editor (michael.luby@springer.com)

All other Countries:

Leontina Di Cecco, Senior Editor (leontina.dicecco@springer.com)

** **This series is indexed by EI Compendex and Scopus databases.** **

More information about this series at https://link.springer.com/bookseries/7818

Hasmat Malik · Md. Waseem Ahmad · D. P. Kothari
Editors

Intelligent Data Analytics for Power and Energy Systems

Springer

Editors
Hasmat Malik
BEARS
National University of Singapore (NUS)
Singapore, Singapore

D. P. Kothari
Chairman, Board of Governors
THDC Institute of Hydropower
Engineering and Technology
Tehri, Uttarakhand, India

Md. Waseem Ahmad
Department of Electrical and Electronics
Engineering
National Institute of Technology Karnataka
Surathkal, Karnataka, India

ISSN 1876-1100 ISSN 1876-1119 (electronic)
Lecture Notes in Electrical Engineering
ISBN 978-981-16-6080-1 ISBN 978-981-16-6081-8 (eBook)
https://doi.org/10.1007/978-981-16-6081-8

This Springer imprint is published by the registered company Springer Nature Singapore Pte Ltd.
The registered company address is: 152 Beach Road, #21-01/04 Gateway East, Singapore 189721, Singapore

Preface

Intelligent data analytics for power system is an evolving field of research at the crossroad of power system with the information science. The efficient management of power and energy system encompasses systems planning, maintenance, operation, and scheduling. The increased penetration of renewable sources, enabling of conventional consumers as prosumers and digitization have increased the volume of data received from system components and operations. The application of data analytics to the power system brings tremendous opportunities and challenges due to its availability for developing algorithms for the efficient management of power system networks. Improving demand response, disaggregation of consumers, better forecasting, detection of fault outage, and monitoring of equipment are some application areas where the data analytics have been successfully applied.

The source of data in power and energy system can be from SCADA, synchrophasor, consumption data from the smart meter, financial data, weather data, etc. Model-based analysis has been the traditional methodology for power system analysis, whereas the IT sector approach has been data-driven. The fusion of model and data-based approaches seems to be the most effective solution for handling complex power and energy issues. To process and visualize these data, various tools include but are not limited to are data mining and machine learnings, statistical analysis, natural language process, artificial intelligence, and predictive analytics.

The goal of this book is to collect state-of-the-art contributions that discuss recent developments in the application and tools of data analytics for power and energy system. Some chapters are overview-oriented, while others describe applications of data analytics for improved management of power system.

Chapter "Intelligent Approach for Analysis and Diagnosis of Attack, Fault, and Load Variation in SCADA Systems: A Power System Application" introduces the problems that SCADA-based automated power station faces in distinguishing the normal and faulty operation of the system. Different machine learning algorithms are analyzed for accuracy and complexity in classifying the normal, faulty, and load varying conditions in the electrical system.

Chapter "Intelligent Approach for Distributed Generation Planning and Distribution Network Reconfiguration Using Metaheuristic Technique" discusses the metaheuristic technique for optimal sizing and location of DGs in power system. The multi-objective function considers active and reactive power loss, voltage deviation, and system reliability and tests the efficacy of adaptive particle swarm optimization algorithm on different configuration of system with placed DGs.

Chapter "Intelligent Modelling and Analysis of Direct Torque Control-Space Vector Modulation of Doubly Fed Induction Generator (DFIG)" discusses the modeling and design of direct torque control (DTC) of a doubly fed induction generator (DFIG). Detail mathematical model of DFIG and space vector modulation (SVM) of the voltage source inverter is presented. Simulink model of DTC-SVM for DFIG is illustrated, and the results under different operating condition are presented to show its fast dynamic response and good steady-state performance.

In Chapter "Modeling and Analysis of an Intelligent Approach for Load Frequency Control in a Deregulated Power System: A Case Study Based on Different Control Schemes," different control schemes such as model predictive control (MPC), proportional integral derivative (PID), linear quadratic Gaussian (LQG), and fuzzy cascaded PID for load frequency control of power system are presented. The gains for the PID controller are selected using the big bang big crunch optimization algorithm. Different case studies are performed to show that MPC control outperforms the control scheme.

Chapter "Mathematical Approach-Based Power System Analysis: A Review of Short-Term Hydro-Scheduling" provides a comprehensive analysis of the various methodological strategies for modeling and solving the unit-based short-term hydro-scheduling (STH'S) issue. The mathematical formulation for different STH'S and the different optimization algorithms for solving such issues are discussed.

In Chapter "Novel Approach for Power System Stability Analysis and Improvement: A Case Study Based on UPFC Application," the issue and identification of the most sensitive voltage are discussed. A MATLAB-based Simulink and simulation toolbox, which utilizes L-index method for voltage stability analysis and sensitive nodes determination, is developed. It is suggested that UPFC should be connected between these buses for the overall effective improvement of the voltage profile.

Chapter "Intelligent Modelling and Analysis of *P-Q* Control Technique for SPV Plant Supplying Power to Grid" discusses the P-Q control approach for solar photovoltaic (SPV) generators connected to the grid. The algorithm is designed to extract maximum power from the SPV, and the active and reactive power control is decoupled to independently control the power fed to the grid. Performance under different irradiation and temperature conditions is presented to show its performance.

Chapter "Intelligent Approach for Fuel-Constrained Economic Emission Dispatch Analysis Using Multi-objective Differential Evolution" investigates the utility of the multi-objective differential evolution (MODE) to address the fuel constrained economic emission dispatch (FCEED) issue of the thermal generation unit. Comparison between strength pare to evolutionary algorithm 2 (SPEA2) and DE algorithms is made to conclude that the use of MODE gives better results compared to SPEA2.

In Chapter "Intelligent Adaptive Critic Scheme Implementation and Investigation Using Policy Iteration Technique: A Case Study of Multi-area Automatic Generation Control (AGC) Problem," a comparative study of policy iteration (PI) control and a linear quadratic regulator (LQR)-based control for addressing frequency issues in AGC system is carried out. Different case studies in AGC system are simulated to show the effective performance of the policy iteration-based optimal control.

Chapter "Intelligent Method for Installation and Investigation of VSC-HVDC Converter Using Metaheuristic Algorithm: A Case Study of Unified Optimal Power Flow Problem" discusses a methodology to find the optimal location of the converter station for high-voltage direct current (HVDC) networks. The unified optimal power flow model is developed for a hybrid network and solved using the PSO algorithm to satisfy operational and economic constraints. The result shows the proper replacement of HVDC lines can improve the economic and technical performance of the overall system.

Chapter "Implementation and Analysis of TID Controller for Power Apparatus Applications Using Flower Pollination Algorithm" applies flower pollination algorithm (FPA) for the design of tilt integral derivative (TID) controller for speed control of a DC motor, which is one of the power apparatuses. A comparative status of TID with other controllers like PID and fractional-order PID (FOPID) has been carried out. It is shown that tuning TID using FPA is better in performance than tuning with genetic algorithm (GA) and firefly algorithm (FA).

Chapter "Intelligent Technique for Eccentricity Fault Diagnosis of Power Apparatus Using Signal Processing Method" provides a health monitoring approach for eccentricity fault detection in an induction motor. The EMD and wavelet algorithm has been used jointly for efficient health monitoring purpose for inverter-fed induction motor machines. It has been observed that the hybrid technique has given encouraging results over FFT technique.

Chapter "Intelligent Methods for Power System Analysis: Advancement in Optimization and Its Application" gives an overview of multi-objective optimization methods used in power system analysis. Performance indices for accessing the efficiency of MOO algorithms and various mathematical test problems are summarized.

In Chapter "Investigation and Analysis of Harmonics in Different Control Techniques of PMSM Drives: An Application of Power System Health Monitoring," the permanent magnet synchronous motor (PMSM) drive scheme has been analyzed. A comparative analysis of SVPWM and hysteresis current control technique of PMSM control drive in simulation is done for assessing the performance of the drive.

Chapter "Design and Analysis of Artificial Intelligence Method for MPPT Control: An Application of Grid-Connected PV System" proposes fuzzy logic (FL) controller for extracting maximum power from the designed PV array. Comparative analysis of FL and P&O in terms of the efficiency, convergence speed, and reduced overshoot is done through the Simulink model. It is concluded that FL-based MPPT is better than traditional methods and is able to minimize the THD injected into the grid.

Chapter "Intelligent Tools and Techniques for Data Analytics of SPV Systems: An Experimental Case Study" discusses the use of simulation software solar PVGIS, PVsyst for the planning and forecasting of SPV plan. A case study has been undertaken to show the use of different modules of the software. Performance of software packages: PVGIS, PVsyst, and SCADA has been done for the case study to highlight the various pros and cons of the software.

Chapter "Intelligent Tools and Techniques for Renewable Energy Analytics: A Review on PV Generation Prediction" provides a summary of the various PV power prediction methodologies in the literature. An overview of tools and techniques used in different fields for PV energy forecasting has been discussed. This chapter also highlights the research gap in PV forecasting.

Chapter "Design and Performance Analysis of Grid-Connected Solar PV System Using PV Syst Software" discusses different steps involved in the design of a solar PV rooftop grid-connected system using PVsyst software. The result obtained for a case study shows that the GCS PV system plan will be profitable to the installer due to its operational advantage.

Chapter "Intelligent Approach-Based Maximum Power Point Tracking for Renewable Energy System: A Review" provides an insight of latest development and advancement of AI-based MPPT algorithms suitable for solar power system. Different AI algorithm for MPPT operation has been discussed along with their merits and demerits. Simulation has been carried out for AI-based MPPT techniques to investigate their dynamic behavior.

Chapter "A Novel Approach for Estimating and Analyzing the Environmental Parameters: A Case Study for Renewable Energy Prospective" introduces an energy model-based approach to estimate the energy generated from a solar array and wind with a complex profile. A simple model to calculate the efficiency loss is also provided.

Chapter "Novel Application of Data-Driven Intelligent Approaches to Estimate Parameters of Photovoltaic Module for Condition Monitoring in Renewable Energy Systems" introduces the use of extreme learning machines and multi-layer feed-forward networks with Levenberg–Marquardt algorithm for estimation of the PV parameters. An experimental database of solar radiation, air and back surface module temperatures, and electrical parameters of PV module is created by developing an experimental setup.

Chapter "Intelligent Modeling and Analysis of a Transformerless Inverter System for Renewable Energy System" models and simulates a single-phase two-stage SPV inverter system in MATLAB. The MPPT controllers and the converter controllers are implemented in the Simulink environment. The system has been tested at varying load and climatic conditions to show its performance.

In Chapter "Intelligent Approach for Performance Investigation of Direct–Drive Generator-Based Wind Energy Conversion System Under Variable Speed Operation," the dynamic model of direct-drive generator WECS is developed in MATLAB/Simulink environment. A discussion on the MPPT model based on TSC, ANN, GA, and PSO is provided. Simulation results show the better performance of the MPPT integrated converter.

Chapter "Performance Analysis of PV Module Using Pyramid Surface Texturing Approach" analyzes the performance of thin-film solar modules using the pyramid texturing technique. TCAD and Pvsyst software are used to analyse surface texture and temperature-dependent loss in PV module. It is concluded that by surface texturing of PV modules by inserting random pyramid structures, significant performance can be achieved.

Chapter "Analysis and Modelling of Basic Wireless Power Transfer Compensation Topology: A Review" discusses different topologies of the wireless power transfer system. Standard parameters in wireless power transfer operation are also presented.

The emergence of microgrids and its importance in light of renewable energy system are discussed in Chapter "Comprehensive Data Analysis of Power and Energy System: A Review of Microgrid Applications and Status." Various data analytics related with microgrids besides power and energy systems are discussed.

Chapter "Role of Blockchain in IoT Enabled Power and Energy Related Healthcare System Platform for the Development of IoT Security" investigate and proposed a modified power and energy-efficient architecture to integrate IoT and blockchain technology in digital healthcare system. Various benefits, challenges, opportunities, and security challenges in IoT-enabled healthcare with a blockchain system are discussed.

Chapter "Security Challenges in Internet of Things (IoT) Integrated Power and Energy (PaE) Systems" discusses security challenges in the Internet of things (IoT)-enabled power and energy systems. Different recommended solutions and their comparative advantages and disadvantages are discussed.

Chapter "Intelligent Analysis of a Hybrid Energy System with Telecom Load" provides an analysis of a hybrid energy system with a telecom load. A case study using hybrid optimization model for electric renewable (HOMER) software examines the practical financial possibilities for grid-connected solar PV systems. It is inferred that a hybrid power system could be an intelligent solution for supply power to telecommunication load.

Chapter "Comparative Analysis of Nature-Inspired Optimization Techniques for Data Analytics" provides a comparative analysis of nature-inspired algorithms for data analytics. The MATLAB implementation of some nature-inspired algorithms is provided for understanding the convergence issue and effect of parameters such as objective function, numbers of variables taken, numbers of iterations considered, lower bound, and upper bound.

First of all, we are thankful to the contributors of this edited book. We are indebted to more than 70 authors experts in the field of intelligent data analytics for power and energy systems and its advances in models and applications, who have authored and co-authored their original research findings.

We are sincerely thankful to the Intelligent Prognostic Private Limited India to provide all types of technical and non-technical facilities, cooperation, and support in each stage to make this book in reality.

We wish to thank our colleagues and friends for their insight and helpful discussion during the production of this edited book. We would like to highlight the motivation and contribution of Prof. Atif Iqbal, Qatar University, Qatar; Prof. Majid Jamil, Jamia Millia Islamia Delhi, India; Prof. Sukumar Mishra, IIT Delhi, India; Prof. Imtiaz Ashraf, Aligarh Muslim University, India; Prof. Kouzou Abdellah, Djelfa University, Algeria; Prof. Jaroslaw Guzinski, Gdansk University of Technology; Prof. Irfan Khan, Texas A&M University at USA; Prof. A. P. Mittal, NSUT Delhi, India; Prof R. K Jarial, NIT Hamirpur (HP), India; Prof. Rajesh Kumar, GGSIPU, India; Prof. Anand Parey, IIT Indore, India; Prof. M. Rizwan, Delhi Technological University, India; Prof. Narender Kumar, Delhi Technological University, India; Prof. Y. R. Sood, NIT Hamirpur (HP); Prof. O. P. Rahi, NIT Hamirpur (HP); Prof. A. Chandel, NIT Hamirpur (HP); Prof. Majed A. Alotaibi, King Saud University, Saudi Arabia; and Prof. Abdulaziz Almutairi, Majmaah University, Saudi Arabia.

We further would like to express our love and affection to our family members for their intense feeling of deep affection.

Woodlands, Singapore/New Delhi, India — Dr. Hasmat Malik
Surathkal, India — Dr. Md. Waseem Ahmad
Nagpur, India — Prof. (Dr.) D. P. Kothari

Contents

Editors and Contributors

About the Editors

Dr. Hasmat Malik (Senior Member IEEE) received M.Tech. degree in electrical engineering from the National Institute of Technology (NIT) Hamirpur, Himachal Pradesh, India, and the Ph.D. degree in Electrical Engineering from Indian Institute of Technology (IIT), Delhi. He has served as an assistant professor for more than five years at the Division of Instrumentation and Control Engineering, Netaji Subhas Institute of Technology (NSIT), Dwarka, Delhi, India. He is currently a chartered engineer (CEng) and a professional engineer (PEng). He is also a research fellow with Berkeley Education Alliance for Research in Singapore (BEARS), a research centre of the University of California, Berkeley, University Town, National University of Singapore (NUS), Singapore, since January, 2019. He has published widely in international journals and conferences his research findings related to intelligent data analytics, artificial intelligence, and machine learning applications in power system, power apparatus, smart building and automation, smart grid, forecasting, prediction and renewable energy sources. Dr. Hasmat has authored/co-authored more than 100 research papers and eight books and thirteen chapters in nine other books, published by IEEE, Springer, and Elsevier. He has supervised 23 PG students. His principal area of research interests is artificial intelligence, machine learning and big data analytics for renewable energy, smart building and automation, condition monitoring and online fault detection and diagnosis (FDD). Dr. Malik is also a member of the Computer Science Teachers Association (CSTA), the Association for Computing Machinery (ACM) EIG, the Institution of Engineering and Technology (IET), UK, and Mir Labs, Asia, a life member of the Indian Society for Technical Education (ISTE), the Institution of Engineers (IEI), India, and the International Society for Research and Development (ISRD), London, and a fellow of the Institution of Electronics and Telecommunication Engineering (IETE). He received the POSOCO Power System Award (PPSA-2017) for his Ph.D. work for research and innovation in the area of power systems. He also received the Best Research Papers Awards from IEEE INDICON-2015 and the Full Registration Fee Award from IEEE SSD-2012, Germany.

Dr. Md. Waseem Ahmad (Member, IEEE) received the B.Tech. and M.Tech. degrees in electrical engineering from Aligarh Muslim University, Aligarh, India, in 2008 and 2011, respectively, and the Ph.D. degree in electrical engineering from the Indian Institute of Technology Kanpur, Kanpur, India, in 2018. He worked as a research fellow with the Department of Electrical and Computer Engineering, National University of Singapore, Singapore, and a graduate trainee engineer with Siemens Ltd., India. He is currently an assistant professor with the National Institute of Technology Karnataka, Surathkal, India. His research interests include fault diagnostics and condition monitoring of power electronic converters.

Prof. (Dr.) D. P. Kothari obtained his BE (Electrical) in 1967, ME (Power Systems) in 1969, and Ph.D. in 1975 from BITS, Pilani, Rajasthan. From 1969 to 1977, he was involved in teaching and development of several courses at BITS Pilani. Earlier Dr. Kothari served as vice chancellor, VIT, Vellore, director in-charge and deputy director (Administration) as well as head in the Centre of Energy Studies at Indian Institute of Technology, Delhi, and as a principal, VRCE, Nagpur. He was a visiting professor at the Royal Melbourne Institute of Technology, Melbourne, Australia, during 1982-83 and 1989, for two years. He was also NSF Fellow at Perdue

University, USA, in 1992. He also taught at Melbourne University Australia for one semester in 1989.

Dr. Kothari, who is a recipient of the most Active Researcher Award, has published and presented 830 research papers in various national as well as international journals, conferences, guided 56 Ph.D. scholars and 68 M.Tech. students, and authored 67 books in various allied areas. He has delivered several keynote addresses, 100 plus Webinars and invited lectures at both national and international conferences. He has also delivered 42 video lectures on YouTube with maximum of 60,000 hits!

Dr. Kothari is a fellow of the National Academy of Engineering (FNAE), fellow of Indian National Academy of Science (FNASc), fellow of Institution of Engineers (FIE), fellow IEEE, honourable fellow ISTE, and fellow IETE. Having received 76 awards till now, his many awards include the National Khosla Award for Lifetime Achievements in Engineering (2005) from IIT, Roorkee. The University Grants Commission (UGC), Government of India, has bestowed the UGC National Swami Pranavandana Saraswati Award (2005) in the field of education for his outstanding scholarly contributions. He is also the recipient of the Lifetime Achievement Award (2009) conferred by the World Management Congress, New Delhi, for his contribution to the areas of educational planning and administration. Recently, he received Excellent Academic Award at IIT Guwahati by NPSC-2014. He has received six Life Time Achievement awards by various agencies on 19 February, 4 March, 11 March, 18 March, 20 March, and 25 March 2016, respectively. On 20th April 2016, he received 'Living Legend' Award in Chennai Conference. Recently, he received Malviya Award for Excellence in Power System at IIT Gandhinagar on 17 December 2020. Dr. Kothari is also distinguished Emeritus professor and adjunct professor at several institutes such as Charutar University, Gujrat, and Wainganga College of Engineering, Nagpur. Currently, Dr. Kothari is with S. B. Jain Institute of Management, Research, and Technology, Nagpur, serving as a director research and professor. He is also a chairman board of governors of THDC Institute of Hydropower Engineering and Technology, Tehri, and a visiting professor at Wainganga College of Engineering, Nagpur.

Wikipedia Link: http://en.wikipedia.org/wiki/D.P. Kothari

Contributors

Sunilkumar Agrawal Department of Electrical Engineering, SVNIT-SURAT, Surat, Gujarat, India

Rafik Ahmad Electrical Engineering Department, BBDITM, Lucknow, Uttar Pradesh, India

Tausif Akram Department of Electrical Engineering, Z.H.C.E.T., Aligarh Muslim University, Aligarh, India

Bilal Alam Department of Electrical Engineering, Z.H.C.E.T., Aligarh Muslim University, Aligarh, India

Mohammad Irfan Alam VIT Bhopal University, Sehore, India

Asim Rahman Ansari FET, MJP Rohilkhand University, Bareilly, India

Naladi Ram Babu Chalapathi Institute of Engineering and Technology, Guntur, Andhra Pradesh, India

Farhad Ilahi Bakhsh Department of Electrical Engineering, National Institute of Technology Srinagar, Hazratbal, Srinagar (Jammu and Kashmir), India

Rahul Kumar Bansal Department of Electrical Engineering, Dr. Radhakrishnan Institute of Technology, Jaipur, Rajasthan, India

Devansh Bhatnagar Ajay Kumar Garg Engineering College, Ghaziabad, India

Aashish Kumar Bohre Electrical Engineering Department, NIT Durgapur, Durgapur, West Bengal, India

Tirumalasetty Chiranjeevi EED, REC Sonbhadra, Churk, Uttar Pradesh, India

Pallavi Choudekars Electrical Engineering Department, Amity University, Noida, Uttar Pradesh, India

Ramesh Devarapalli EED, BIT Sindri, Dhanbad, Jharkhand, India

Maitreyee Dutta National Institute of Technical Teachers Training and Research, Chandigarh, India

Kulsoom Fatima Department of Electrical Engineering, Integral University, Lucknow, India

Ushakiran Huiningsumbam Department of EEE, Amity University, Noida, India

Arshad Ali Hurra Department of Electrical Engineering, SSM College of Engineering, Baramulla (Jammu & Kashmir), India

Mohammed Aslam Husain Department of Electrical Engineering, REC, Ambedkar Nagar, Ambedkar Nagar, India

Tanzeel Imtiyaz Electrical Engineering Department, Amity University Noida, Noida, Uttar Pradesh, India

Noorul Islam Department of Electrical Engineering, ZHCET, AMU, Aligarh, India

Anjali Jain Department of EEE, Amity University, Noida, Uttar Pradesh, India

C. Jena School of Electrical Engineering, KIIT University, Bhubaneswar, India

B. P. Joshi Department of Applied Science, Seemant Institute of Technology Pithoragarh, Pithoragarh, Uttarakhand, India

Kamaldeep National Institute of Technical Teachers Training and Research, Chandigarh, India

Balmukund Kumar Electrical Engineering Department, NIT Durgapur, Durgapur, West Bengal, India

Gaurav Kumar BGSB University, Rajouri, Jammu and Kashmir, India

Jitendra Kumar Department of Electrical and Electronics Engineering, SRM Institute of Science and Technology, Delhi NCR, Uttar Pradesh, India

Munish Kumar Integral University, Lucknow, India

Nagendra Kumar Department of Electrical and Electronics Engineering, G.L. Bajaj Institute of Technology & Management Greater Noida, Greater Noida, India

Pradeep Kumar Electrical and Electronics Engineering Department, National Institute of Technology Sikkim, Ravangla, Sikkim, India

Prasanta Kundu Department of Electrical Engineering, SVNIT-SURAT, Surat, Gujarat, India

Niranjan Lal CSE, SET, Mody University, Lachamangarh, Sikar, Rajasthan, India

Hasmat Malik BEARS, University Town, NUS Campus, Singapore, Singapore

Manisha Malik National Institute of Technical Teachers Training and Research, Chandigarh, India

Ashish Mani Department of EEE, Amity University, Noida, India

Parul Mathuria Malaviya National Institute of Technology, Jaipur, India

Ahmad Faiz Minai Department of Electrical Engineering, Integral University, Lucknow, India

Ahmad Khubaib Minai Department of Electrical Engineering, Integral University, Lucknow, India

Sheikh Fareed Mohammad Department of Electrical Engineering, SSM College of Engineering, Baramulla (Jammu & Kashmir), India

Sheikh Suhail Mohammad Department of Electrical Engineering, National Institute of Technology Srinagar, Hazratbal, Srinagar (Jammu and Kashmir), India

Shafqat Nabi Mughal BGSB University, Rajouri, Jammu and Kashmir, India

Prahlad Mundotiya Malaviya National Institute of Technology, Jaipur, India

Safwan Mustafa Department of Electrical Engineering, Z.H.C.E.T., Aligarh Muslim University, Aligarh, India

Alina Naaz Department of Electrical Engineering, Z.H.C.E.T., Aligarh Muslim University, Aligarh, India

Lipika Nanda School of Electrical Engineering, KIIT University, Bhubaneswar, India

Babita Panda School of Electrical Engineering, KIIT University, Bhubaneswar, India

Amjad Ali Pasha King Abdul-Aziz University, Jeddah, Saudi Arabia

Amruta Pattnaik Dr. Akhilesh Das Gupta Institute of Technology and Management, Delhi, India

Arjyadhara Pradhan School of Electrical Engineering, KIIT University, Bhubaneswar, India

Dinanath Prasad Ajay Kumar Garg Engineering College, Ghaziabad, India;
Delhi Technological University, Delhi, India

Md Atiqur Rahman Department of Electrical Engineering, Government Engineering College, Vaishali, India

Abhishek Rajan National Institute of Technology Sikkim, Ravangla, Sikkim, India

Rajesh Ranjan Electrical and Electronics Engineering Department, National Institute of Technology Sikkim, Ravangla, Sikkim, India

Pulivarthi Nageswara Rao GITAM (Deemed to be University), Visakhapatnam, Andhra Pradesh, India

Anjan Kumar Ray Electrical and Electronics Engineering Department, National Institute of Technology Sikkim, Ravangla, Barfung Block, South Sikkim, India

Ruchira Electrical Engineering Department, Amity University, Noida, Uttar Pradesh, India

Sarita Samal School of Electrical Engineering, KIIT University, Bhubaneswar, India

Mohammad Sami Integral University, Lucknow, India

Sriram Sankaran Center for Cybersecurity Systems and Networks, Amrita Vishwa Vidyapeetham, Amritapuri, India

Mohammad Sarfraz Department of Electrical Engineering, ZHCET, AMU, Aligarh, India

Bikash Kumar Saw Electrical Engineering Department, NIT Durgapur, Durgapur, West Bengal, India

Vishal Sharma CSE, SET, Mody University, Lachamangarh, Sikar, Rajasthan, India

Khadim Moin Siddiqui Electrical Engineering Department, SRMGPC, Lucknow, Uttar Pradesh, India

Mohd. Khursheed Siddiqui Integral University, Lucknow, India

Akhilesh Singh Department of Electrical and Electronics Engineering, Seemant Institute of Technology Pithoragarh, Pithoragarh, Uttarakhand, India

Alok Kumar Singh Department of Electrical Engineering, Nirwan University, Jaipur, Rajasthan, India

B. K. Singh Department of Electrical Engineering, Bipin Tripathi Kumaun Institute of Technology Almora, Dwarahat, Uttarakhand, India

Omkar Singh Electrical and Electronics Engineering, Raghu Engineering College (A), Visakhapatnam, India;
Electrical and Electronics Engineering Department, National Institute of Technology Sikkim, Ravangla, Barfung Block, South Sikkim, India

S. P. Singh Department of Electrical Engineering, REC, Ambedkar Nagar, Ambedkar Nagar, India

Satyendra Singh School of Electrical Engineering, Bhartiya Skill Development University, Jaipur, Rajasthan, India

Nikhil Kumar Sinha B.I.T. Sindri, Dhanbad, Jharkhand, India

Ibra Subhan Department of Electrical Engineering, ZHCET, AMU, Aligarh, India

Vamshi Sunku Mohan Center for Cybersecurity Systems and Networks, Amrita Vishwa Vidyapeetham, Amritapuri, India

Md. Tabrez Department of Electrical and Electronics Engineering, Motihari College of Engineering, Motihari, India

Ritu Tak School of Electrical Engineering, Bhartiya Skill Development University, Jaipur, India

Mohd Tariq Department of Electrical Engineering, Z.H.C.E.T., Aligarh Muslim University, Aligarh, India

Harpal Tiwari Malaviya National Institute of Technology, Jaipur, India

Anuradha Tomar Netaji Subhas University of Technology, Delhi, India

T. Usmani Department of Electronics and Communication Engineering, Integral University, Lucknow, India

Sami Uz Zaman Department of Electrical Engineering, Integral University, Lucknow, India

Kiran Babu Vakkapatla EED, Lingayas Institute of Management and Technology, Madalavarigudem, Andhra Pradesh, India

Nasir Ul Islam Wani Electrical Engineering Department, Amity University, Noida, Uttar Pradesh, India

Md. Waseem Ahmad Department of Electrical Engineering, National Institute of Technology Karnataka, Surathkal, India

Amit Kumar Yadav Electrical and Electronics Engineering Department, National Institute of Technology Sikkim, Ravangla, Barfung Block, South Sikkim, India

Intelligent Approach for Analysis and Diagnosis of Attack, Fault and Load Variation in SCADA Systems: A Power System Application

Vamshi Sunku Mohan and Sriram Sankaran

Abstract Supervisory Control And Data Acquisition (SCADA) automates power station processes such as component switching, and load shedding. SCADA collects, processes and displays the electric parameters recorded by the end components at the control panels. In case of malfunctions, SCADA neither distinguishes between faults, attacks and maintenance events nor identifies their type or location of occurrence.A distinction between them would avoid component damage and power system blackout caused due to resulting fault current. Hence, there is a need to develop efficient models to classify between faults, attacks and maintenance events based on the type and location of occurrence. In this work, we propose classification models to enable speedy attack detection and distinguish between faults and attacks using machine learning algorithms. We compute the classification accuracy, energy consumed and execution time of the algorithms to analyze the trade-offs in choosing between lightweight and accurate models. While, k-NN, SVM and Naive Bayes classify the faults with lower accuracy and energy, algorithms such as OneR, Adaboost and JRip are more accurate and incur higher energy consumption. Further, we classify the faults and attacks based on the type and location of occurrence, sequence currents and phase voltage constraints. Evaluation of proposed approach using cyber-attack datasets on Industrial Control System (ICS) shows that the models classify between faults and attacks with 97.25% accuracy. In addition, depending on the type and location of occurrence, faults, attacks and load variations observed during normal operation have been classified with accuracies of 96.55%, 97.22% and 98.98% respectively.

Keywords SCADA · Attack classification · Common electric faults · Lightweight algorithms · Power relays

V. Sunku Mohan (✉) · S. Sankaran
Center for Cybersecurity Systems and Networks, Amrita Vishwa Vidyapeetham, Amritapuri, India
e-mail: vamshis@am.amrita.edu

S. Sankaran
e-mail: srirams@am.amrita.edu

H. Malik et al. (eds.), *Intelligent Data Analytics for Power and Energy Systems*,
Lecture Notes in Electrical Engineering 802,
https://doi.org/10.1007/978-981-16-6081-8_1

1 Introduction

Supervisory Control And Data Acquisition (SCADA) [1] is a control system architecture comprising controllers, networked data communications and GUIs for high-level process supervisory management. Two important components of SCADA are Remote Terminal Units (RTU) and Programmable Logic Controllers (PLC). RTUs are microprocessor-based devices installed at the end components to monitor and control their activity. PLCs operating at control panels process and display the end component parameters to enable automated tracking of power statistics such as power transfer, frequency stability, and load shedding. This function provides the flexibility to the operator to enable and disable the relays either manually or remotely.

RTUs located at remote locations are prone to malfunction caused either by cyber-attacks or internal faults. Therefore, SCADA is not immune to cyber-attacks [2] due to varying state-wise security guidelines, infrastructure and design specification to provide more protection at the control panel compared to the end devices. Cyber-attacks disrupt power supply by either injecting false data or changing the relay operating conditions. False data injection cause sequential overloads leading to power outage or imbalance. Internal faults are caused due to natural calamities such as heavy wind and rain, wear-out of component insulation or irregular maintenance. These faults introduce out of phase voltages which, when not monitored, may cause the power to go out of phase resulting in malfunction of generators and transformers.

Under such malfunctions, SCADA displays electric parameters under fault, but fails to decide the type and severity of malfunction. Hence, a need arises to distinguish normal operation from malfunctions. In order to determine the component affected, malfunctions are to be further classified into attacks and faults and normal operation into maintenance cases and load variations due to power imbalance. Maintenance tests are introduced by power station to check the working and responsiveness of components. These events when classified and recorded helps to carry out the system maintenance on a regular basis. System imbalance is caused due to sudden increase or decrease in power generation and load variation. This phenomenon when ignored produces noise harmonics which eventually damages electric components. Considering the classification requirements discussed above, we develop machine learning algorithms and propose a trade-off to choose between high-efficiency and lightweight alternatives. High efficiency models such as OneR, JRip and Adaboost require a longer period to execute in addition to requiring a powerful GPU, extra memory and register files, thereby extracting additional energy and resources from power-hungry SCADA [3]. Hence, they can be used in situations such as power system blackout and load shedding to precisely detect the component under fault. Whereas, energy-efficient algorithms such as SVM and Naive Bayes are easy to implement, require less time and memory to execute and are scalable [4, 5]. Therefore, these models provide a time and energy-efficient alternatives.

Figure 1 shows the power station sub-system drawn based on the parameters defined in datasets, accessible from [6], developed by University of Alabama, Huntsville in collaboration with Oak Ridge National Laboratories (ORNL). Each

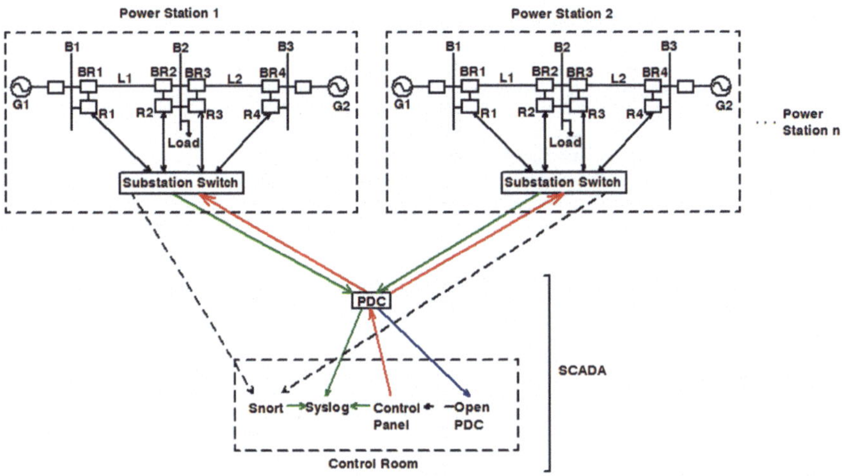

Fig. 1 Line diagram of power station

of the power stations comprises three buses B_1, B_2 and B_3 with generators G_1 and G_2 connected to the side buses B_1 and B_3. The middle bus B_2 is connected to the load. The buses are connected by transmission lines L_1 and L_2 which in turn are protected by breakers: L_1 by BR_1, BR_2 and L_2 by BR_3, BR_4. The breakers are connected to the relays R_1, R_2, R_3 and R_4 respectively. The relays of each of the power stations are connected to Power Distribution Center (PDC) and control room through their respective Substation switch. The control room houses the Syslog containing snort log, relay log and control log whose operation states along with the magnitude of line current determines whether the malfunction is a fault, an attack or a maintenance event. The breakers along with the relays and substation switch are a part of power station, whereas, the PDC and the control room encompass the SCADA. RTU recorded electric parameters such as phase and sequence voltages and currents, impedance and switch status of the relays in addition to the frequency and rate of change of frequency form the SCADA generated dataset.

In this paper, we propose to develop machine learning algorithms to classify malfunctions based on the type and the location of occurrence. Our contributions include,

- Classifying the data into faults, attacks and normal operation.
- Defining classification rules based on line voltages and currents to detect the components under faults and attacks.
- Determining their location with high accuracy and at the shortest time possible.
- Categorizing normal operation into maintenance cases and system imbalances due to load variation.

Further, classification accuracy, energy consumed and execution time of the algorithms are analyzed and a trade-off is proposed to select between lightweight and

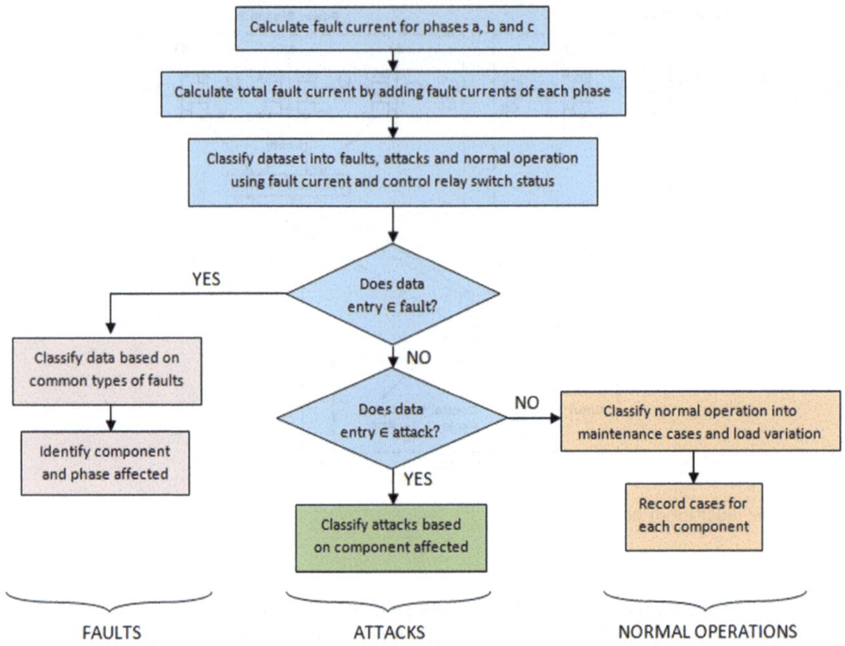

Fig. 2 Procedure overview

accurate models. An overview of the process is shown in Fig. 2. Our approach classifies the attacks from faults and normal operating conditions with 97.25% accuracy. Based on the type and location, faults, attacks and maintenance events are classified with accuracies of 96.55%, 97.22% and 98.98% respectively. As classification rules are defined based on phase and line voltages and currents, our model is generic and that can be implemented on any bus system.

2 Motivation

SCADA automates power transmission, distribution and load shedding, thus minimizing human supervision. To enable these functions, SCADA uses Ethernet for wired and IEEE 802.x for wireless communication [2]. As security is given to central host rather than end components, RTUs are more prone to cyber-attacks. Faults may occur when RTUs are prone to heavy winds and rain, component malfunction or irregular maintenance. During such malfunctions, SCADA classifies them from normal operation, but it neither identifies the type or exact location of maintenance events, faults or attacks nor distinguishes between them [7]. A clear distinction would determine the cause and assess its severity, in order to initiate end component protection, load shedding or damaged component replacement.

3 Problem Statement

Existing approaches propose the use of high accuracy algorithms such as OneR, Adaboost and XGboost [8] to classify malfunctions. These algorithms require a higher execution time along with a powerful GPU with high processing speed, large memory and register files [3] thereby consuming additional energy from power-hungry SCADA. This necessitates the replacement of hardware components demanding a huge infrastructure, rendering the process infeasible. Hence, they may not be suitable under all cases, thus limiting their usage to emergency situations such as load shedding during power system blackout.

This necessitates the development of lightweight algorithms to assist SCADA under normal operation or when only a few casualties are detected. Existing approaches [9–16] work on efficient detection of malfunctions, but do not focus on classifying them further into faults, attacks and maintenance events and determining their exact number based on the type, severity and location of occurrences. Such a classification would help in initiating precautionary measures in time to avoid worsening of malfunctions. Hence, we propose a trade-off to choose between high accuracy and energy-efficient models while addressing the goals elucidated below.
Classify the attacks from normal operation and faults.
Classify the faults, attacks and maintenance events further based on the type, phase and system frequency.

Our model satisfies both the goals. To achieve the first objective, magnitude of fault current and switch status of control relay are fed to various classifiers [8] resulting in an accuracy of 97.25%. The second goal is achieved by classifying the faults, attacks and maintenance events based on phase and sequence voltage conditions with 96.55%, 97.22% and 98.98% accuracy respectively.

4 Related Work

Existing papers propose a classification of casualties from normal operation. The works range from creating a Bayesian graph, predicting a malfunction based on the sequence of previous system states using SVM, forecasting the future casualties based on present states using hybrid classification algorithms.

Idaho National Laboratory's report [7] discusses the causes of cyber-attacks, probability of occurrence and challenges faced. It further proposes the security guidelines and legal help to be provided to citizens. In view of these requirements, [9–11] discuss various machine learning models to optimize power loss and distribution. Iqbal et al. [12] use meta-heuristics to minimize usage of electrical components. Pan et al. [13] have created a Bayesian network to predict the possible malfunctions based on the existing system status. Similar approach is used in [14, 15] to detect malfunctions in electrical systems and forecast power requirements. Pan et al. [16] have utilized the rate of change of frequency of the relay and its operating time to decide efficacy of relays to decide severity of fault. Extension of this work is proposed by Pan et al. [17], where, the time-defined state paths classify electric faults using common path algorithm. Borges et al. [8] have used algorithms such as OneR, JRip and NNge to classify the malfunctions scenarios such as short circuit attack, and data injection, with accuracy plotted against precision and recall to determine the best performing algorithm. A similar approach has been used in [18, 19] to classify the malfunctions from normal operation using supervisory learning. Yang et al. [20] has proposed a technique to identify traffic patterns in SCADA using CNN with which the time windows having network traffic are predicted. Sayegh et al. [21] test the severity and effectiveness of common cyber-attacks such as replay and cryptographic attack, on PLC and HMI.

In contrast to existing works, we define classification rules based on the end component parameters such as voltage, current and impedance. These rules are used to propose simple classification techniques using machine learning algorithms to classify attacks from faults. Further, sequence current conditions are used to determine the number of faults, attacks and maintenance events based on the severity and location of occurrence.

5 Background of the Models

In this section, we provide a brief mathematical background of the models used for evaluation.

5.1 K-NN

k-NN selects k number of nearest points to the given sample. The distance between the query instance and remaining samples is calculated using Euclidean, Manhattan or Hamming distance. Most commonly used is the Euclidean distance given by the Eq. 1. The values thus obtained are arranged in ascending order. The top k number of rows are chosen from the sorted array. The test point is then sorted to determine the nearest neighbors based on the most frequent class in the rows selected.

$$d(p,q) = \sqrt{\sum_{i=1}^{n}(q_i - p_i)^2} \tag{1}$$

5.2 SVM

SVM deals with constructing a maximum marginal hyperplane given by, $H : w^T(x) + b = 0$ to classify the labeled test data iteratively to minimize the error in categorizing new test points. This is done using kernels to convert feature vector $\phi(x)$ to a high-dimensional space $\phi(x) : R^D \longmapsto R^M$ to provide accurate classification. The data points segregated by the hyperplane are given by Eq. 2.

$$y_n[w^T\phi(x_n) + b] \leq 0 \ \forall n \tag{2}$$

Hyperplane is defined based on C (precision factor) and gamma (curvature factor). A low C gives high precision results with a distorted curve whereas a high C results in a smooth curve. Gamma determines the curvature and is used when Gaussian kernel is implemented. A high gamma outputs a hyperplane with a high curvature whereas a lower gamma results in straighter curve.

5.3 Naive Bayes

Naive Bayes calculates the posterior probability of the dependent variable based on the probability of independent variable. The posterior probability is adjusted continuously with the incoming data. The algorithm works under the assumption that all data attributes are independent and given equal weight. Posterior probability of class variable C_k with respect to the feature vector $X = (x_1, x_2, \ldots, x_n)$ is given by Eq. 3. Considering the naive independence condition and that $P(X)$ is constant for a given input, the output and its probability is given by Eqs. 4 and 5.

$$P(C_k|X) = \frac{P(X|C_k)P(C_k)}{P(X)} \tag{3}$$

$$C_k = argmax_{C_k} P(C_k) \prod_{i=1}^{n} P(x_i|C_k) \tag{4}$$

$$P(C_k|X) \propto P(C_k) \prod_{i=1}^{n} P(x_i|C_k) \tag{5}$$

5.4 State-Path Algorithm

The algorithm extracts data patterns from the time-stamped RTU measurements. Such patterns are called states which are given by, $S_i = \{q(s_1), q(s_2), \ldots, t_i\}$. The states thus recorded from various RTUs are arranged based on the time they were observed in an increasing order to define a path $P = (S_1, S_2, \ldots, S_n)$. The paths are searched for in the data to find similar patterns [17].

5.5 OneR

OneR learns a one-level decision tree by generating a set of rules for each attribute, based on which, each branch assigns the most frequent class. Error rate is calculated for the data instances not belonging to the majority class. The feature (p_i) with the lowest error rate given by Eq. 6 is chosen to be the decision branch.

$$\text{Error\_rate} = 1 - \sum_i {p_i}^2 \tag{6}$$

5.6 JRip

JRip works based on repeated incremental pruning to reduce the error rate by integrating the association rules with decision tree pruning. The algorithm divides the dataset into growing sets and pruning set. It then generates rules for a subset of the training samples and removes all samples covered by the rules in the training set. The rule with highest gain given by Eq. 7 is selected.

$$\text{Gain} = \text{Entropy(dataset\_before\_split)} - \sum_j^K \text{Entropy(subset\_}j\text{\_after\_split)} \tag{7}$$

5.7 Adaboost

Adaboost combines weak classifiers by retraining them iteratively until a predetermined number of weak learners have been created or no further improvement can be made on the training dataset. Training set is chosen based on accuracy of previous training as given by Eq. 8.

$$H(x) = \text{sign}(\sum_{t=1}^T (\alpha_t h_t(x)) \tag{8}$$

where, $\alpha_t = 0.5 * ln \frac{1-\text{error}_t}{\text{error}_t}$.

The weight of each training output (D_t) given by Eq. 9 is updated to achieve high accuracy.

$$D_{t+1}(i) = \frac{D_t(i)\exp(-\alpha_t y_i h_t(x_i))}{\text{Sum\_of\_weights}} \tag{9}$$

6 Proposed Approach

We propose to develop a model which addresses the limitations of the existing papers and satisfies the goals as discussed under Sect. 2.2. To meet the design objectives, we consider the magnitude of fault current and the switching status of the relay log, snort log and control log to classify the system states into attacks, faults and maintenance events. As seen in Fig. 3, the magnitude of fault current varies from 0 to 900 A. In order to distinguish the malfunctions from normal operation, we classify the fault current into four categories based on its magnitude. These current categories and relay states are used to classify malfunctions as faults, attacks or normal operation.

To classify the attacks and faults further based on the common electric faults (symmetrical and unsymmetrical faults), voltage and current constraints are defined. Some faults may occur due to the deterioration of insulation on the transmission lines, whose location is measured from the bus connected to the generator in terms of percentage, e.g., 10% fault on L1 means fault occurs on line L1 at 10% of distance from breaker BR1. These faults can be identified based on phase voltage and impedance constraints. Attacks on power system components are commonly caused due to relay state manipulation and aurora effect [22]. These conditions can be identified using fault current pattern observed during relay switching.

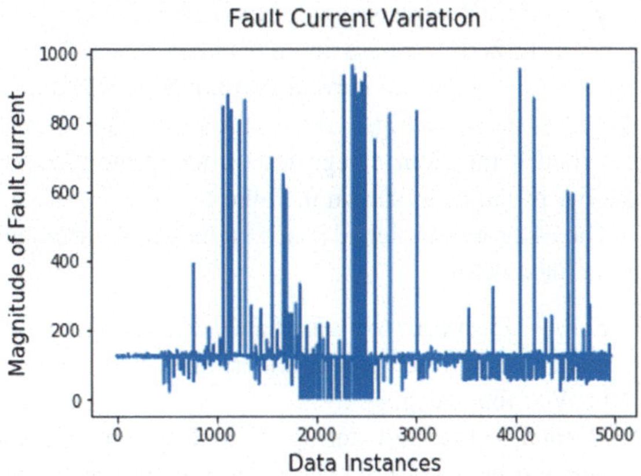

Fig. 3 Variation of Fault current magnitude

Table 1 Categorization of Fault Current

Category	Current magnitude (A)
Zero	0–50
Normal	50–200
Warning	200–400
High	>400

6.1 Classification of System States

Classification rules are defined based on the fault current category and the switch status of relay log, snort log and control log. These rules along with the relay currents $R_1(I_{a_f})$, $R_2(I_{a_f})$, $R_3(I_{a_f})$, $R_4(I_{a_f})$ are given as input to the machine learning algorithms to classify the SCADA events as faults, attacks or normal operation. The procedure to calculate the fault current and classify system states is described as follows.

6.1.1 Calculation and Categorization of Fault Current

In an unsymmetrical fault, the fault current [23] is calculated as shown in Eq. 10.

$$I_f = I_0 + \alpha I_1 + \alpha^2 I_2 \tag{10}$$

where,

I_0, I_1, I_2 =Zero, positive and negative sequence components of fault current (A),
$\alpha = 1\angle 120°$ = phase angle between I_0 and I_1,
$\alpha^2 = 1\angle 240°$ = phase angle between I_0 and I_2.

Using the sequence current components [6], fault current (I_f) is calculated and plotted as shown in Fig. 3. As the fault current has a wide magnitude range, we have classified it into four categories to determine present state of the system and the severity of faults, as shown in Table 1.

The categories assigned based on the fault current magnitude described in Table 1 are defined below.

1. *Zero*—Occurs due to maintenance delay, ripping of cables, wear out of component insulation or when precautionary measures are not taken when power system is in 'Warning category'.
2. *Normal*—Observed during normal system operation and maintenance events, where transmission line voltages and currents do not exceed permissible limits.
3. *Warning*—Occurs due to sudden change in state of control relays or due to strong wind and rain when the cable and relay impedance may experience a slight reduction.

Table 2 Switch states of relays and operations performed

Logs	Relay state	Meaning
Control	0	Abnormal operation
Log	1	Normal operation
Snort	0	Control log/relay operated
Log	1	Control log/relay not operated
Relay	0	Breaker not open
Log	1	Breaker open

Table 3 Declaration of system states based on log status and fault current category

Logs			Fault current category			
Control log	Snort log	Relay log	Zero	Normal	Warning	High
0	0	0	Attack	Attack	Attack	Attack
0	0	1	Fault	Fault	Attack	Attack
0	1	0	Attack	Attack	Attack	Attack
0	1	1	Attack	Attack	Attack	Attack
1	0	0	No Event	Attack	No Event	No Event
1	0	1	Attack	No Event	Attack	Attack
1	1	0	Attack	Fault	Attack	Attack
1	1	1	Fault	Fault	Fault	Fault

4. *High*—Occurs either due to natural disasters or when load is not shed when power system is in 'Zero' status. Any delay in implementing precautionary measures may result in power system blackout [7].

6.1.2 Classification Outcomes

Output states are assigned by defining the classification rules derived based on the fault current category and the relay switch status. The functions performed in power station based on the position of relays is described in Table 2. Table 3 shows the declaration of output states. The states thus assigned are defined as follows.

1. *No-event*—This state comprises the normal operation and maintenance events.
2. *Fault*—Defines malfunctions due to internal fault or irregular maintenance.
3. *Attack*—Includes malfunctions caused due to unauthorized external (cyber) attack.

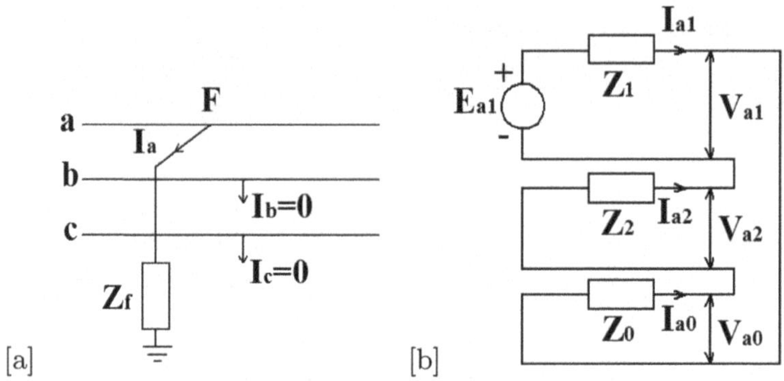

Fig. 4 Single line to ground fault, **a** circuit diagram, **b** equivalent diagram

6.2 Description of Common Electric Faults

Faults recorded using the classification rules are further classified based on some of the common electric faults observed in power stations. Equivalent circuits used to analyze the faults are derived based on [23]. Electric parameters observed under these cases are described as follows.

6.2.1 Single Line to Ground Fault (1LG)

The fault occurs when a conductor comes in contact with ground or neutral wire. The fault and equivalent circuits are shown in Fig. 4. As seen in the circuit diagram, phase *a* is grounded and *b* and *c* are open circuited. Therefore, $I_b = I_c = 0$ and fault current $I_f = I_a$.

The terms defined in the equivalent circuit are described as follows.

I_a, I_b, I_c = Currents in phase *a*, *b* and *c* (A)
$I_{a_0}, I_{a_1}, I_{a_2}$ = Zero, positive, negative sequence currents (A)
V_a, V_b, V_c = Voltages in phase *a*, *b* and *c* (V)
$V_{a_0}, V_{a_1}, V_{a_2}$ = Zero, positive, negative sequence voltage (V)
Z_0, Z_1, Z_2 = Zero, positive, negative sequence impedance (Ω)
Z_f = Fault impedance (Ω)
E_a = Voltage across generator (V)

6.2.2 Line to Line Fault (LL)

Two phases *b* and *c* are short circuited and phase *a* is unfaulted. Therefore, $I_a = 0$, $I_b = -I_c$ and $V_b = V_c$. Fault and equivalent circuits are shown in Fig. 5.

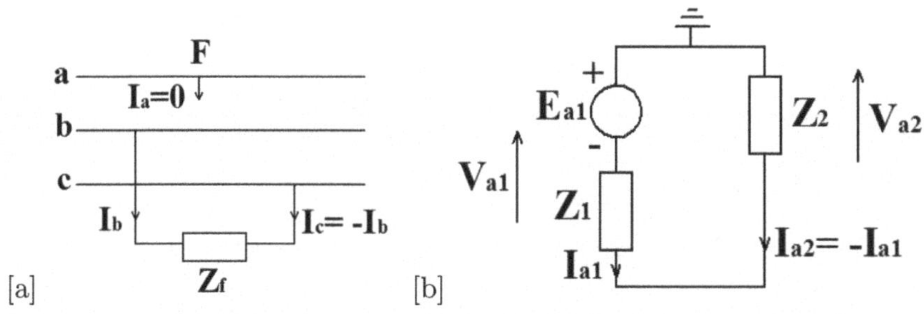

Fig. 5 Line to line fault, **a** circuit diagram, **b** equivalent diagram

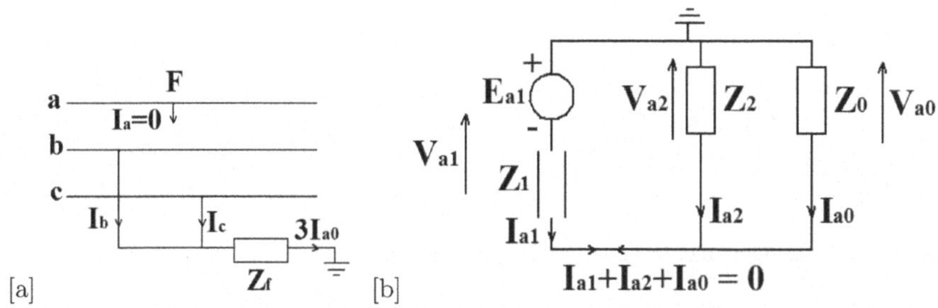

Fig. 6 Double line to ground fault, **a** circuit diagram, **b** equivalent diagram

6.2.3 Double Line to Ground Fault (2LG)

Two phases are accidentally connected to the ground resulting in $I_a = 0$ and $V_b = V_c = 0$. The fault and equivalent circuits are shown in Fig. 6.

6.2.4 Triple Line to Ground Fault (3LG)

This occurs between the three phases and the ground resulting in $V_a = V_b = V_c = 0$ and $I_a + I_b + I_c = 0$, with the sequence current $I_{a_1} = I_a = E_{a_1}/Z_1$. The fault and equivalent circuits are shown in Fig. 7. The probability of occurrence of the fault is 2–3% [23].

6.2.5 Triple Line Fault (3L)

It is the most dangerous fault which occurs when all the three phases are short circuited. Probability of occurrence is 1–2% [23]. Fault parameters are $V_a = V_b = V_c$. Sequence currents and voltages are $I_{a_1} = E/Z_1$, $I_{a_2} = I_{a_0} = 0$, $V_{a_0} = V_{a_1} = V_{a_2} = V_a/3$. The fault and equivalent circuits are shown in Fig. 8.

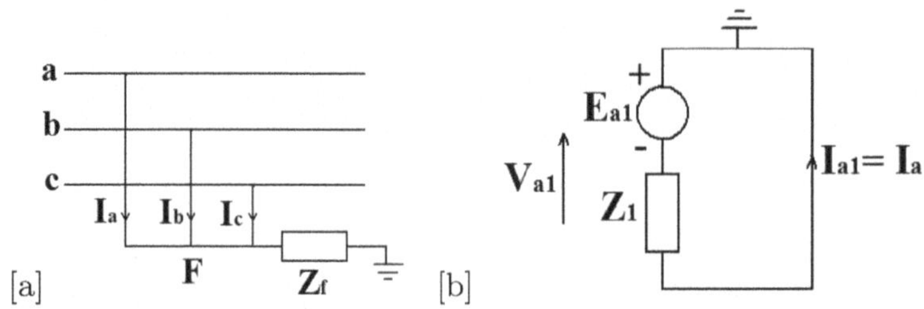

Fig. 7 Triple line to ground fault, **a** circuit diagram, **b** equivalent diagram

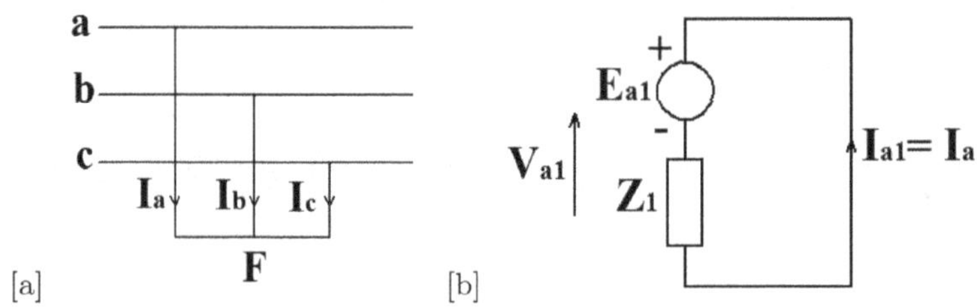

Fig. 8 Triple line fault, **a** circuit diagram, **b** equivalent diagram

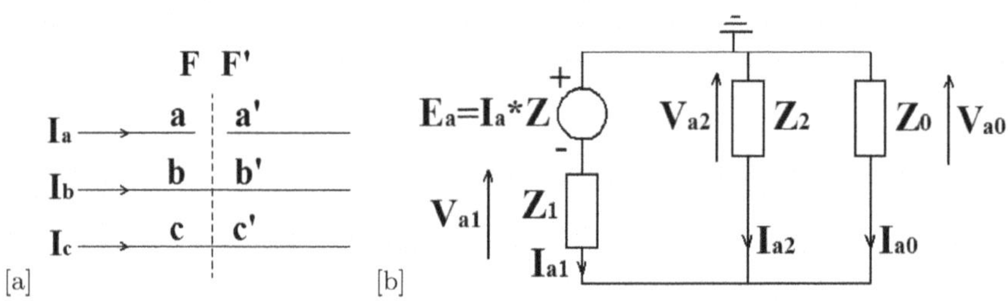

Fig. 9 One conductor open fault, **a** circuit diagram, **b** equivalent diagram

6.2.6 One Conductor Open Fault

It is a symmetrical fault with one conductor open. The fault and equivalent circuits are shown in Fig. 9. As phase *a* open circuited, $I_a = 0$ and $I_{a_0} + I_{a_1} + I_{a_2} = 0$. As phases *b* and *c* are closed, $V_b = V_c = 0$.

6.2.7 Two Conductor Open Fault

It is a symmetrical fault where two conductors are open. The fault circuit and equivalent circuit are shown in Fig. 10. As phases *b* and *c* open circuited, $I_b = I_c = 0$ and $V_{a_0} + V_{a_1} + V_{a_2} = 0$. Phase *a* is closed resulting in $V_a = 0$ and $V_{a_0} = V_{a_1} = V_{a_2} = 0$.

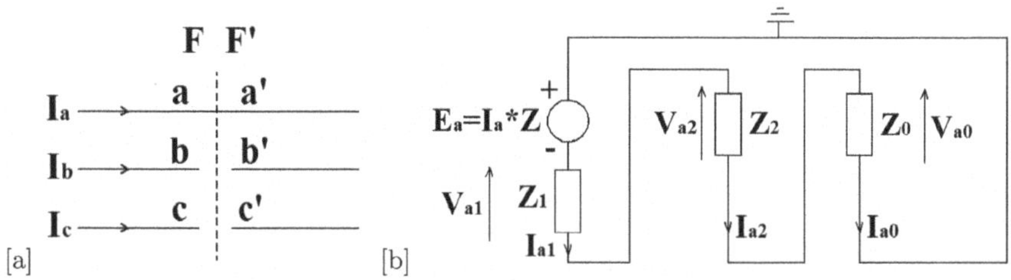

Fig. 10 Two conductor open fault, **a** circuit diagram, **b** equivalent diagram

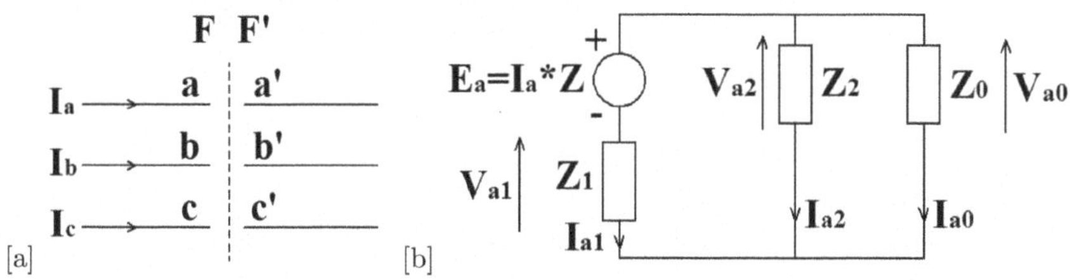

Fig. 11 Three conductor open fault, **a** circuit diagram, **b** equivalent diagram

6.2.8 Three Conductor Open Fault

The fault occurs when all the three conductors are open. Hence, $I_a + I_b + I_c = 0$, $I_{a_1} = I_{a_2} = I_{a_0} = 0$ and $V_{a_0} = V_{a_2} = 0$. Fault voltage is given by $V_f = V_{a_1}$. The fault and equivalent circuits are shown in Fig. 11.

6.2.9 Bolted Fault

All the conductors are physically held together resulting in zero impedance. The protective devices are tripped when powered, resulting in high short-circuit current which is balanced equally within the three phases.

6.2.10 Ground Fault

The fault is caused due to damaged wiring, usage of old appliances or overloaded circuit leading to unintended connection of conductors with ground. When the fault occurs, the pathway to ground has zero resistance resulting in a sudden rise in current flow and energizing the equipment to a dangerous voltage.

Table 4 Voltage and current conditions to identify fault types

Fault	Fault location	Conditions
1LG	Phase a	$V_a = Z_f * I_a$, I_b =0 , I_c =0
L-L	Lines b, c	I_a =0 , $I_b = -I_c$
2LG	Lines a, b	$V_{a_1} = V_{a_2}$, $I_{a_1} = I_{a_0} + I_{a_2}$, $V_a = V_b$
3L	All lines	$I_a = I_b = I_c$
3LG	All lines	$V_a = V_b = V_c$, $I_a + I_b + I_c$ =0 , I_{a_0} =0
1 conductor open	Phase a	I_a =0, $V_b = V_c$ =0
2 conductors open	Lines b, c	V_a =0, $I_b = I_c$ =0
3 conductors open	All lines	$I_a + I_b + I_c$ =0, $I_{a_1} = I_{a_2} = I_{a_0}$ =0
Bolted	All lines	Z_f =0, low I_f
Ground	any phase	Z_f =0, very high V_f and I_f
Arcing	any two phases	very high Z_f and V_f

6.2.11 Arc Fault

Fault occurs when non-connected conductors come close enough such that current jumps across the gap creating an arc resulting in ions to create a plasma. This causes increased power flow resulting in an arc flash or arc blast. The arc fault has high ground path impedance and low current magnitude limited by the resistance of the arc.

6.3 *Classification of Faults*

6.3.1 Classification Based on Targeted Relays and Common Electric Faults

The faults obtained using the classification rules described in Sect. 5.1 are classified further based on some of the common electric faults. The voltage and current constraints defining these faults are summarized in Table 4.

6.3.2 Classification of Transmission Line Faults

Some faults occur along the length of the transmission lines due to erosion of insulation. They have varying attack potency based on location of occurrence on transmission lines. Location is calculated in terms of percentage starting from the generator-connected bus. This is followed by the calculation of fault impedance (Z_f) using the equivalent circuit of power system as described in [23], phase current (I_a) and fault voltage (V_f). Formulae to find Z_f are given in Table 5. I_a and V_f are calculated as

Table 5 Formulae to calculate fault impedance at various percentage locations

Line	Fault Location (%)	Approximate Location (%)	Fault impedance (Z_f)
L1	10–19	10	$0.1 * Z_{L_1}$
	10–49	50	$\frac{(Z_1+0.5*Z_{L_1})(Z_2+Z_3+Z_4+0.5*Z_{L_1}+Z_{L_2})}{Z_1+Z_2+Z_3+Z_4+Z_{L_1}+Z_{L_2}}$
	50–79	75	$\frac{(Z_1+0.75*Z_{L_1})(Z_2+Z_3+Z_4+0.25*Z_{L_1}+Z_{L_2})}{Z_1+Z_2+Z_3+Z_4+Z_{L_1}+Z_{L_2}}$
	80–90	90	$\frac{(Z_1+Z_2+0.9*Z_{L_1})(Z_3+Z_4+0.1*Z_{L_1}+Z_{L_2})}{Z_1+Z_2+Z_3+Z_4+Z_{L_1}+Z_{L_2}}$
L2	10–19	10	$0.1 * Z_{L_2}$
	10–49	50	$\frac{(Z_4+0.5*Z_{L_2})(Z_1+Z_2+Z_3+0.5*Z_{L_2}+Z_{L_1})}{Z_1+Z_2+Z_3+Z_4+Z_{L_1}+Z_{L_2}}$
	50–79	75	$\frac{(Z_4+0.75*Z_{L_2})(Z_1+Z_2+Z_3+0.25*Z_{L_2}+Z_{L_1})}{Z_1+Z_2+Z_3+Z_4+Z_{L_1}+Z_{L_2}}$
	80–90	90	$\frac{(Z_3+Z_4+0.9*Z_{L_2})(Z_1+Z_2+0.1*Z_{L_2}+Z_{L_1})}{Z_1+Z_2+Z_3+Z_4+Z_{L_1}+Z_{L_2}}$

given in Eqs. 11 and 12 respectively.

$$I_a = R_1(I_{a_f}) + R_2(I_{a_f}) + R_3(I_{a_f}) + R_4(I_{a_f}) \tag{11}$$

$$V_f = Z_f * I_a + \frac{I_a}{3} * (Z_{a_0} + Z_{a_1} + Z_{a_2}) \tag{12}$$

where,

Z_1, Z_2, Z_3, Z_4 = Impedance of breakers BR_1, BR_2, BR_3 and BR_4 respectively (Ω)
Z_{L_1}, Z_{L_2} = Impedance of transmission lines $L_1, L_2(\Omega)$

In 1LG faults, the positive (V_1) and negative sequence (V_2) voltages are close to zero. Therefore, zero sequence voltage (V_0) is considered for calculations. Table 6 shows the voltage constraints to determine the number of faults at various locations for each relay.

6.4 *Classification of Attacks*

Based on the data recorded by the RTUs at the end components, some of the attacks that can be observed are due to aurora effect [22] and forceful disabling of relays [24] resulting in cascading failures.

Table 6 Voltage conditions to calculate transmission line faults

Line	Relay	Location (%)	Condition
L1	R_1	10	abs$[R_1(V_{a_0}) - V_f$ at 10%$] \leq 0.2 * R_1(V_a)$
		50	abs$[R_1(V_{a_0}) - V_f$ at 50%$] \leq 0.2 * R_1(V_a)$
		75	abs$[R_1(V_{a_0}) - V_f$ at 75%$] \leq 0.2 * R_1(V_a)$
		90	abs$[R_1(V_{a_0}) - V_f$ at 90%$] \leq 0.2 * R_1(V_a)$
	R_2	10	abs$[R_2(V_{a_0}) - V_f$ at 10%$] \leq 0.2 * R_2(V_a)$
		50	abs$[R_2(V_{a_0}) - V_f$ at 50%$] \leq 0.2 * R_2(V_a)$
		75	abs$[R_2(V_{a_0}) - V_f$ at 75%$] \leq 0.2 * R_2(V_a)$
		90	abs$[R_2(V_{a_0}) - V_f$ at 90%$] \leq 0.2 * R_2(V_a)$
L2	R_3	10	abs$[R_3(V_{a_0}) - V_f$ at 10%$] \leq 0.2 * R_3(V_a)$
		50	abs$[R_3(V_{a_0}) - V_f$ at 50%$] \leq 0.2 * R_3(V_a)$
		75	abs$[R_3(V_{a_0}) - V_f$ at 75%$] \leq 0.2 * R_3(V_a)$
		90	abs$[R_3(V_{a_0}) - V_f$ at 90%$] \leq 0.2 * R_3(V_a)$
	R_4	10	abs$[R_4(V_{a_0}) - V_f$ at 10%$] \leq 0.2 * R_4(V_a)$
		50	abs$[R_4(V_{a_0}) - V_f$ at 50%$] \leq 0.2 * R_4(V_a)$
		75	abs$[R_4(V_{a_0}) - V_f$ at 75%$] \leq 0.2 * R_4(V_a)$
		90	abs$[R_4(V_{a_0}) - V_f$ at 90%$] \leq 0.2 * R_4(V_a)$

6.4.1 Classification of Attacks Due to Cascading Failures

When a relay is to be turned off by a power station, the breakers connected to the transmission lines and their corresponding relays are disabled in an order [16] starting from the generator connected bus followed by the breaker located at the other end of transmission line. This process ensures a steady voltage flow. In case of an attack, the relays are forcefully disabled resulting in a voltage spike [24]. The line current rises to a high value and reduces to zero [25] causing the breakers to get opened. These steps occur at an interval of 1 s giving rise to current patterns called states. The states along with the timestamps form state-paths [17], which is utilized to analyze the dataset to determine similar attacks.

6.4.2 Classification of Aurora Attacks

Aurora attacks occur when circuit breakers or relays are operated randomly and simultaneously resulting in an out-of-phase connection of electric components [22]. This generates a synchronous noise and asynchronous power signal. When left unattended, it results in overheating of cables and end components and can damage alternating current (AC) equipments such as generators, motors and transformers connected to the electric grid.

6.5 *Classification of No-Event States in Power System*

The RTU observations classified as No-event using the classification rules described in Sect. 5.1 comprise normal system operation and maintenance events. Maintenance conditions [26] are described as intentional faults introduced periodically by the power station to test the responsiveness and efficient working of components. No-event states are defined by a line current of 50–200 A with control relays set to OFF state.

An efficient classification of maintenance events would help in keeping a record of the components inspected and date of maintenance which could be used in to determine the period for future service. Classification of normal operation would identify the abnormal currents and large load variations. This could be used to take precautionary measures such as load shedding or component replacement to ensure stability of the power grid.

6.5.1 Classification of Maintenance Conditions

Some of the maintenance tests carried out in power station are transmission line maintenance along with failure test and open test [26] on relays and breakers. Transmission line maintenance involves disabling of breakers connected to it in order to physically inspect the lines, relays and breakers and then energizing them in the reverse order. Breaker failure test involves overloading the power grid with end components or by connecting a faulty appliance to ensure the resilience of the breaker. Breaker open test is conducted by initiating short circuit under normal current to examine the breaker opening time.

6.5.2 Classification of Normal System Operation Conditions

Normal operating conditions do not have an impact on power generation. But there are certain cases such as accidental load variations during which precautionary measures need to be taken to prevent the power system from going to 'Warning' state. Load variations at power grid are caused due to sudden addition or deletion of load, loss of generators and malfunctioning of motors. This condition introduces asynchronous noise and synchronous power leading to the destabilization of frequency [27], but it does not effect line current, voltage and control relay status. Hence, they are to be identified only by observing the 'Rate of Change of Frequency' (ROCOF or df/dt). This would help to initiate load shedding or additional power generation before abnormal frequency initiates under-frequency relays or unit-auxiliary components to shut down the power grid. Positive df/dt indicates loss of load or power generation, whereas a negative value suggests peak load demand. The frequency variations observed on Relay 1 (R_1) is plotted in Fig. 12. The formulae to calculate ROCOF and the corresponding change in phase angle of voltage are given by Eqs. 13

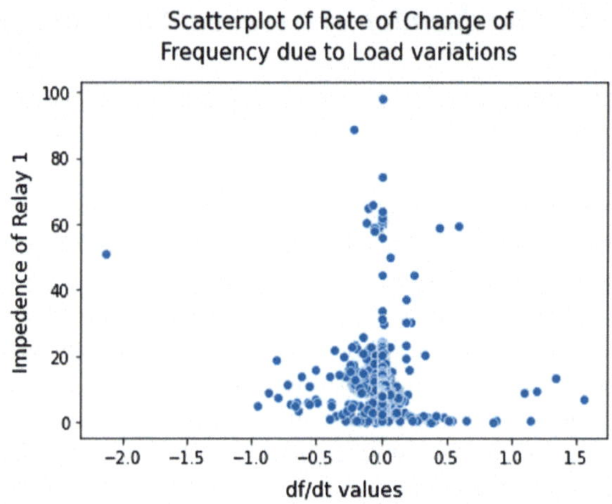

Fig. 12 Rate of change of frequency of Relay-1 observed at various instances

and 14 respectively. The angle change is used to determine the power imbalance due to load change [28]. The terms used in the equations are defined below.

$$\frac{\mathrm{d}f}{\mathrm{d}t} = -\frac{P_L - P_G}{2HS_{GN}} f_r \tag{13}$$

$$\Delta\delta = \pi \frac{\mathrm{d}f}{\mathrm{d}t} \Delta T^2 \tag{14}$$

P_L, P_G = Load and generator power (W)
H = Inertia constant of the generating device
S_{GN} = Rated capacity of the generating device (W)
f_r = Rated frequency (Hz)
$\Delta\delta$ = angle change due to frequency change (deg)
ΔT = phase angle calculation interval (s) assumed to be 1s

7 Evaluation

The algorithms were run on various datasets [6] containing the RTU observations of end components such as phase and sequence voltages and currents, system frequency and relay switch status. The accuracies obtained by classifying attacks from faults based on targeted relays and location were observed to be ranging from 91 to 98% across all the datasets. Hence, we include the results obtained from one of them in our paper.

Table 7 Execution time, energy consumed and classification accuracy for various algorithms

Algorithm	Execution Time (s)	Energy Consumed (J)	Classification Accuracy (%)
k-NN	0.303	0.084	93.43
Naive Bayes	0.003	0.020	91.52
SVM	0.002	0.023	91.14
OneR	0.034	0.092	96.30
JRip	0.025	0.075	96.17
Adaboost	0.025	0.082	97.25

7.1 Classification of Attacks from No-Event and Faults

The relay currents $R_1(I_{a_f})$, $R_2(I_{a_f})$, $R_3(I_{a_f})$, $R_4(I_{a_f})$ and switch states of snort log, relay log and control log were given as input and the system states defined using classification rules were given as the output to the AI models. Classification accuracy, energy consumed and execution time for the models hence obtained are recorded in Table 7. Based on the classification accuracy, energy and execution time calculated, we analyze the trade-offs in choosing between lightweight and accurate model alternatives. Naive Bayes and SVM classify the attacks with a slightly lower accuracy. Whereas, k-NN records the lowest percentage of attacks, but, classifies them with higher accuracy and execution time. These models require less energy to execute. Considering the performance factors, Naive Bayes provides an ideal lightweight solution. OneR, JRip and Adaboost classify the malfunctions with high accuracy, but they incur additional energy consumption and execution time. Considering these factors, Adaboost proves to be a high accuracy alternative.

7.2 Classification of Faults Based on Common Electric Faults

Based on the voltage and current conditions mentioned in Table 4, faults are classified based on the targeted relays. The faults thus obtained are classified based on the unsymmetrical and symmetrical faults. The faults are further classified based on the phases (a, b and c) for 1LG fault and lines (a–b, b–c and c–a) for L-L and 2LG faults. The classified unsymmetrical faults are detailed in Table 8. The symmetrical faults classified are listed in Table 9. The faults thus classified, sum up to an integer less than the total number of faults. These unclassified faults can be termed as ‘Unknown Faults’. They comprise 10.84% of total number of faults recorded. These faults occur due to breakage of transmission lines, wear-out of component insulation, etc. Given the component specification unique to each power station, the ‘Unknown Faults’ can be further categorized. The accuracy of fault classification based on unsymmetrical and symmetrical faults are tabulated as a confusion matrix as shown in Table 10

Table 8 Classification of unsymmetrical faults based on location, type and phase

Target relays		Number of faults	Type of faults					
			1LG	L-L	2LG	3L	3LG	Unknown
R_1, R_2, R_3, R_4	R_1	436	a=10,b=10, c=10	ab=10,bc=10, ca=10	0	10	10	146
	R_2		$a = 4, b = 4, c = 4$	$ab = 4, bc = 4, ca = 4$	$ab = 3, bc = 3, ca = 3$	4	4	
	R_3		$a = 10, b = 10, c = 10$	$ab = 10, bc = 10, ca = 10$	$ab = 3, bc = 3, ca = 3$	3	10	
	R_4		$a = 12, b = 12, c = 12$	$ab = 12, bc = 12, ca = 12$	0	3	12	
R_1, R_2, R_3	R_1	40	$a = 1, b = 1, c = 1$	$ab = 1, bc = 1, ca = 1$	$ab = 1, bc = 1, ca = 1$	1	1	0
	R_2		$a = 1, b = 1, c = 1$	$ab = 1, bc = 1, ca = 1$	$ab = 1, bc = 1, ca = 1$	1	1	
	R_3		$a = 2, b = 2, c = 2$	$ab = 2, bc = 2, ca = 2$	$ab = 1, bc = 1, ca = 1$	1	2	
R_2, R_3, R_4	R_2	44	$a = 1, b = 1, c = 1$	$ab = 1, bc = 1, ca = 1$	0	1	1	13
	R_3		0	0	0	10	12	
	R_4		0	0	0	1	0	
R_3, R_4, R_1	R_1	88	$a = 1, b = 1, c = 1$	$ab = 1, bc = 1, ca = 1$	0	1	1	0
	R_3		0	$ab = 16, bc = 16, ca = 0$	0	0	0	
	R_4		$a = 16, b = 16, c = 16$	0	0	0	0	
R_4, R_1, R_2	R_1	44	0	0	0	0	0	44
	R_2		0	0	0	0	0	
	R_4		0	0	0	0	0	
R_1, R_2	R_1	54	0	0	0	0	1	39
	R_2		0	0	0	8	6	
R_2, R_3	R_2	1250	$a = 55, b = 55, c = 55$	$ab = 55, bc = 55, ca = 55$	$ab = 55, bc = 55, ca = 55$	55	55	40
	R_3		$a = 55, b = 55, c = 55$	$ab = 55, bc = 55, ca = 55$	$ab = 55, bc = 55, ca = 55$	55	55	
R_2, R_4	R_2	83	0	0	0	0	13	56
	R_4		0	0	0	5	9	
R_3, R_4	R_3	83	0	0	0	0	0	83
	R_4		0	0	0	0	0	

Table 9 Classification of symmetrical faults based on targeted relays

Target Relays	Types of faults					
	1 conductor open	2 conductors open	3 conductors open	Bolted	Ground	Arc
R_1	0	0	10	0	1	8
R_2	3	3	4	0	1	9
R_3	3	3	10	0	3	1
R_4	0	0	12	0	6	6

Table 10 Confusion matrix for unsymmetrical faults

Fault	1LG	LL	2LG	3L	3LG
1LG	495	0	8	0	1
LL	0	485	0	1	2
2LG	4	0	352	0	1
3L	0	3	0	145	0
3LG	0	2	6	0	176

Table 11 Confusion matrix for symmetrical faults

Faults	1 conductor open	2 conductor open	3 conductor open	Bolted	Ground	Arc
1 conductor open	6	0	0	0	0	0
2 conductors open	0	6	0	0	0	0
3 conductors open	0	0	36	0	0	0
Bolted	0	0	0	0	0	0
Ground	0	0	0	0	11	0
Arc	0	0	0	0	0	24

and Table 11 respectively. Faults occurring on transmission lines due to wear-out of insulation discussed under Sect. 6.3.2 are classified based on various locations in terms of percentage on lines L_1 and L_2. The classification accuracy of these faults is represented as a confusion matrix in Table 12.

7.3 Classification of Attacks Based on Targeted Relays

Attacks were classified using state-path algorithm, SVM and k-NN. Accuracies thus obtained are given in Table 13. Results show that state-path algorithm has the highest classification accuracy. Hence, state-path algorithm has been used to classify the attacks further based on cascading failures and aurora effect. The number of attacks thus observed for each of the targeted relays is recorded in Tables 14 and 15 respectively.

7.4 Classification of Maintenance Events from Normal Operating Conditions

The No-event states recorded using the classification rules described in Sect. 5.2 are classified to determine the number of maintenance events and load imbalances. The number of line maintenance, breaker failure test, breaker open test cases and load variations observed using ROCOF are recorded in Table 16. Accuracy with which these events are classified at each of the targeted relays is recorded as a confusion matrix in Table 17.

8 Conclusion and Future Work

In this work, we develop a machine learning-based approach to identify the electric components under malfunction and thus determine their cause of occurrence, i.e., internal faults or cyber attacks. Our model classifies data based on rules defined using line voltages and currents thereby making it suitable for implementation on any bus system. We show that our approach is effective in providing a trade-off to choose between energy-efficient and accurate algorithms by classifying faults, attacks and maintenance cases with accuracies of 96.55%, 97.22% and 98.98% respectively. Hence, it is ideal for implementation during light load and peak load hours. A further work in this regard can be done by recording power station location and specifications of relays, breakers, transmission lines unique to each power station along with the communication history of IP layer in SCADA software. These parameters help to decide the malfunctions occurring due to wear and tear, ill-maintenance and weather conditions specific to a particular region and cyber-attacks due to DoS, relay, fragmentation attacks, etc., thus providing fine-grained classification.

Table 12 Confusion matrix for transmission line faults

Faults			Line L1								Line L2							
			R_1				R_2				R_3				R_4			
			10 %	50 %	75 %	90 %	10 %	50 %	75 %	90 %	10 %	50 %	75 %	90 %	10 %	50 %	75 %	90 %
Line L1	R_1	10%	638	0	0	0	0	0	0	0	0	0	0	0	0	0	0	0
		50%	0	638	0	0	0	0	0	0	0	0	0	0	0	0	0	0
		75%	0	0	636	0	0	0	2	0	0	0	0	0	0	0	0	0
		90%	0	0	0	638	0	0	0	0	0	0	0	0	0	0	0	0
	R_2	10%	0	0	0	0	637	0	0	0	1	0	0	0	0	0	0	0
		50%	0	0	0	0	0	638	0	0	0	0	0	0	0	0	0	0
		75%	0	0	0	0	0	0	602	0	0	0	0	0	0	0	0	0
		90%	0	0	0	0	0	0	0	638	0	0	0	0	0	0	0	0
Line L2	R_3	10%	0	0	0	0	0	0	0	0	635	0	4	0	0	0	0	0
		50%	0	0	0	0	0	0	0	0	0	639	0	0	0	0	0	0
		75%	0	0	0	0	0	0	0	0	0	0	639	0	0	0	0	0
		90%	0	0	0	0	0	0	0	0	0	0	0	639	0	0	0	0
	R_4	10%	0	0	0	0	0	0	0	0	0	0	0	0	628	0	3	0
		50%	0	0	0	0	0	0	0	0	0	0	0	0	0	631	0	0
		75%	0	0	0	0	0	0	0	0	0	0	0	0	5	0	590	0
		90%	0	0	0	0	0	0	0	0	0	0	0	0	0	0	0	638

Table 13 Classification accuracy of attacks

Algorithm	Classification accuracy (%)
k-NN	93.43
SVM	91.14
State-path	97.22

Table 14 Classification of attacks caused due to cascading failures

Location	Attacks	Location	Attacks
R_1	87	R_1, R_3	112
R_2	85	R_1, R_4	117
R_3	68	R_2, R_3	112
R_4	72	R_3, R_4	241
R_1, R_2	334	R_1, R_2, R_3	1271

Table 15 Classification of attacks based on aurora effect

Location	Attacks	Location	Attacks	Location	Attacks
R_1	1	R_3	1	R_1, R_2	2
R_2	1	R_4	1	R_3, R_4	2

Table 16 Number of maintenance events

Component tested	Number of events			
	Line maintenance	Breaker open test	Breaker failure test	Load variations
R_1	0	18	23	158
R_2	0	73	27	147
R_3	0	73	23	145
R_4	0	15	26	157
L_1	0	0	0	0
L_2	0	0	0	0

Table 17 Classification accuracy of maintenance events

Relays	R_1	R_2	R_3	R_4
R_1	190	4	5	0
R_2	0	247	0	0
R_3	0	0	241	0
R_4	0	0	0	198

References

1. M. Ahmed, W.L. Soo, Supervisory Control and Data Acquisition System (SCADA) based customized Remote Terminal Unit (RTU) for distribution automation system (2009), pp. 1655–1660. https://doi.org/10.1109/PECON.2008.4762744
2. B. Zhu, A. Joseph, S. Sastry, A taxonomy of cyber attacks on SCADA systems, in *2011 International Conference on Internet of Things and 4th International Conference on Cyber, Physical and Social Computing*, Dalian, China, 2011, pp. 380–388. https://doi.org/10.1109/iThings/CPSCom.2011.34.
3. W. Feng, Y. Cao, D. Patnaik, N. Ramakrishnan, Temporal Data Mining for Neuroscience (Chapter 15), in *Applications of GPU Computing Series, GPU Computing Gems Emerald Edition*, ed. by W.W. Hwu, Morgan Kaufmann, 2011, pp. 211–227. ISBN 9780123849885. https://doi.org/10.1016/B978-0-12-384988-5.00015-2
4. Advantages of Support Vector Machines (SVM). Available (Online): https://iq.opengenus.org/advantages-of-svm/
5. Naive Bayes. Available (Online): https://towardsdatascience.com/all-about-naive-bayes-8e13cef044cf
6. Industrial Control System (ICS) Cyber Attack Datasets. Available (Online): https://sites.google.com/a/uah.edu/tommy-morris-uah/ics-data-sets
7. Cyber Threat and Vulnerability Analysis of the U.S. Electric Sector: Mission Support Center Analysis Report, Idaho National Laboratory, United States. Department of Energy. Office of Scientific and Technical, United States. Department of Energy, 2016, Length 62 pages
8. R. Borges, J. Beaver, M. Buckner, T. Morris, U. Adhikari, S. Pan, Machine Learning for Power System Disturbance and Cyber-attack Discrimination (2014). https://doi.org/10.1109/ISRCS.2014.6900095
9. J.A. Alzubi, *AI and Machine Learning Paradigms for Health Monitoring System: Intelligent Data Analytics*, 1E (Springer Nature, Part of the Studies in Big Data, vol. 86, p. 513, 2020). ISBN: 978-981-33-4412-9. https://doi.org/10.1007/978-981-33-4412-9
10. A. Iqbal, et al., *Renewable Power for Sustainable Growth*, 1E (Springer Nature, Part of the Lecture Notes in Electrical Engineering, vol. 723, 805 p). ISBN: 978-981-33-4080-0. https://doi.org/10.1007/978-981-33-4080-0
11. S. Srivastava, et al., Applications of Artificial Intelligence Techniques in Engineering, vol. 2, 1E (Springer Nature, Part of the Advances in Intelligent Systems and Computing, vol. 697, 647 p, 2018). ISBN 978-981-13-1822-1. https://doi.org/10.1007/978-981-13-1822-1
12. A. Iqbal, et al., *Meta Heuristic and Evolutionary Computation: Algorithms and Applications*, 1E (Springer Nature, Part of the Studies in Computational Intelligence, vol. 916, p. 849, 2020). ISBN 978-981-15-7571-6. https://doi.org/10.1007/978-981-15-7571-6
13. S. Pan, T. Morris, U. Adhikari, A specification-based intrusion detection framework for cyber-physical environment in electric power system. Int. J. Network Secur. **17**, 174–188 (2015)
14. A.K. Yadav, et al., *Soft Computing in Condition Monitoring and Diagnostics of Electrical and Mechanical Systems*, 1E (Springer Nature, Part of the Advances in Intelligent Systems and Computing, vol. 1096, p. 496, 2020). ISBN 978-981-15-1532-3. https://doi.org/10.1007/978-981-15-1532-3
15. N. Fatema, et al., *Intelligent Data-Analytics for Condition Monitoring: Smart Grid Applications*, 1E (Academic Press, 2021). ISBN: 978-0-323-85510-5. https://doi.org/10.1016/C2020-0-02173-0
16. S. Pan, T. Morris, U. Adhikari, Developing a hybrid intrusion detection system using data mining for power systems. IEEE Trans. Smart Grid. **6** (2015). https://doi.org/10.1109/TSG.2015.2409775
17. S. Pan, T. Morris, U. Adhikari, Classification of disturbances and cyber-attacks in power systems using heterogeneous time-synchronized data. IEEE Trans. Ind. Inform. **11**, 1 (2015). https://doi.org/10.1109/TII.2015.2420951

18. L. Maglaras, J. Jiang, Intrusion detection in SCADA systems using machine learning techniques, in *Proceedings of 2014 Science and Information Conference, SAI 2014*. (2014) https://doi.org/10.1109/SAI.2014.6918252
19. S. Srivastava, et al., *Applications of Artificial Intelligence Techniques in Engineering*, vol. 1, 1E (Springer Nature, Part of the Advances in Intelligent Systems and Computing, vol. 698, 643 p, 2018). ISBN 978-981-13-1819-1. https://doi.org/10.1007/978-981-13-1819-1
20. H. Yang, L. Cheng, M.C. Chuah, Deep-Learning-Based Network Intrusion Detection for SCADA Systems, pp. 1–7 (2019). https://doi.org/10.1109/CNS.2019.8802785
21. N. Sayegh, A. Chehab, I. Elhajj, A. Kayssi, Internal security attacks on SCADA systems, in *2013 3rd International Conference on Communications and Information Technology, ICCIT 2013*, pp. 22–27 (2013). https://doi.org/10.1109/ICCITechnology.2013.6579516
22. M.F.M. Arani, A. Abiri Jahromi, D. Kundur, M. Kassouf, Modeling and simulation of the aurora attack on microgrid point of common coupling, in *7th Workshop on Modeling and Simulation of Cyber-Physical Energy Systems (MSCPES)*. Montreal, QC, Canada, vol. 2019, pp. 1–6 (2019). https://doi.org/10.1109/MSCPES.2019.8738801
23. Analysis of Unsymmetrical Faults. Available (Online): https://www.eeeguide.com/analysis-of-unsymmetrical-faults/
24. J. Zhang, Y. Dong, Cyber attacks on remote relays in smart grid, in *2017 IEEE Conference on Communications and Network Security (CNS)*, Las Vegas, NV, USA, 2017, pp. 1–9. https://doi.org/10.1109/CNS.2017.8228637
25. E.O. Schweitzer, B. Kasztenny, Distance protection: Why have we started with a circle, does it matter, and what else is out there? in *71st Annual Conference for Protective Relay Engineers (CPRE)*. College Station, TX, USA, vol. 2018 (2018), pp. 1–19. https://doi.org/10.1109/CPRE.2018.8349791
26. P. Velayutham, F. Alnaimi, A review on power plant maintenance and operational performance. MATEC Web Conf. **225**, 05003 (2018). https://doi.org/10.1051/matecconf/201822505003
27. H. Bevrani, G. Ledwich, J. Ford, On the use of df/dt in power system emergency control, in *2009 IEEE/PES Power Systems Conference and Exposition*. PSCE, vol. 2009, pp. 1–6 (2009). https://doi.org/10.1109/PSCE.2009.4840173
28. J. Xing, L. Mu, A new passive islanding detection solution based on accumulated phase angle drift. Appl. Sci. **8**, 1340 (2018). https://doi.org/10.3390/app8081340

Intelligent Approach for Distributed Generation Planning and Distribution Network Reconfiguration Using Metaheuristic Technique

Bikash Kumar Saw, Balmukund Kumar, and Aashish Kumar Bohre

Abstract The intelligent planning approach of distributed generations (DGs) and the distribution network reconfiguration is very challenging but it offers lot of advantages and efficient to improve system techno-economic parameters. Therefore, this work pays attention on mentioned issue to improve the system voltage profile and reliability with reduced power losses. The IEEE 118-bus test system with different combination of DG and reconfiguration using the adaptive particle swarm optimization (APSO) technique is examined in this study. The objective function is minimized using APSO which consists of different objective parameters of system like active and reactive power losses, voltage stability and reliability parameter. The backward–forward with BIBC and BCBV matrices has been used for load flow analysis due to its effectiveness to solve power flow analysis then other method. The different system case configurations are implemented under with and without new work reconfiguration condition including DGs. Afterward, the achieved results are scrutinized and compared which express the system case Configuration-6 including both reconfigurations and multi-DGs perform better than other cases to improve the overall system performances.

Keywords Distribution network reconfiguration · Distributed generation (DG) · Active/real and reactive power loss · Voltage deviation · System reliability · APSO technique

B. K. Saw · B. Kumar · A. K. Bohre (✉)
Electrical Engineering Department, NIT Durgapur, Durgapur, West Bengal, India
e-mail: aashishkumar.bohre@ee.nitdgp.ac.in

B. K. Saw
e-mail: bks.18ee1105@phd.nitdgp.ac.in

H. Malik et al. (eds.), *Intelligent Data Analytics for Power and Energy Systems*,
Lecture Notes in Electrical Engineering 802,
https://doi.org/10.1007/978-981-16-6081-8_2

1 Introduction

Generally, the distribution system associated with some issues like more losses, worst reliability and under voltage of system. The planning of distributed generations (DGs) and distribution network reconfiguration offers many benefits in today's load growth environment of system. The optimal location of DG capacitor based on optimal network capacity in IEEE 33-bus radial system with different techniques is proposed in [1]. The main objective is to reduce the distribution system losses and voltage profile improvement including released loading of lines. The multi-objective function-based problem solution through weed optimization, crowding distance and non-dominated sorting technique for IEEE 33-bus radial and 84-bus Taiwan distribution network is investigated in [2]. The distribution network reconfiguration with objective to reduce loss and enhance the reliability of system by imperialist competitive approach is described in [3]. The details of adaptive particle swarm optimization (APSO) technique with the required equations to update the velocity and position are discussed in [4]. The allocation of renewable solar PV distributed generation and the optimal capacitor placement for IEEE 118-bus test system is reported in [5, 6]. Additionally, the same system is also planned using different heuristic optimization techniques and algorithms like genetic algorithm (GA), harmony search algorithm (HSA), artificial bee colony (ABC), cuckoo search algorithm, shark smell optimization, particle swarm optimization (PSO), biogeography-based optimization (BBO), etc. The optimal allocation of different type of DGs considering the distinct practical load models like residential, commercial and industrial loads using genetic algorithm (GA), particle swarm optimization (PSO) and butterfly particle swarm optimization (BF-PSO) is given in [7–13]. The multi-objective approach for simultaneous DERs to maximize yearly savings by means of increasing the annual system parameters benefits using improved PSO and guided search algorithm (GSA) is suggested in [14]. The planning of hybrid renewable DGs such as solar PV, wind and energy storage devices based on the reliability assessment parameters [15] techno-enviro-economic approach is explained in [16, 17]. The optimal allocation of DGs in radial system using comprehensive technique [18] and novel stochastic fractal search algorithm [19] to improve the system performance parameters is explored. The simultaneous planning of DGs and network reconfiguration with performance parameter-based objective function using PSO technique is given in [20, 21]. Using scalable enumeration method for large distribution network, optimal DG allocation in distribution system [22] and decomposition method for primary and secondary distribution network [23] using optimization techniques are presented to improve the various system performances. Literature [24] said about integrated DG planning and smart grid monitoring. In [25–28], application of soft computing, artificial intelligence and metaheuristic algorithms is presented. The graphical abstract is represented in Fig. 1.

In this work, following major work are carried out as follows:

- Intelligent planning of distributed generations (DGs) and the distribution network reconfiguration are implemented.

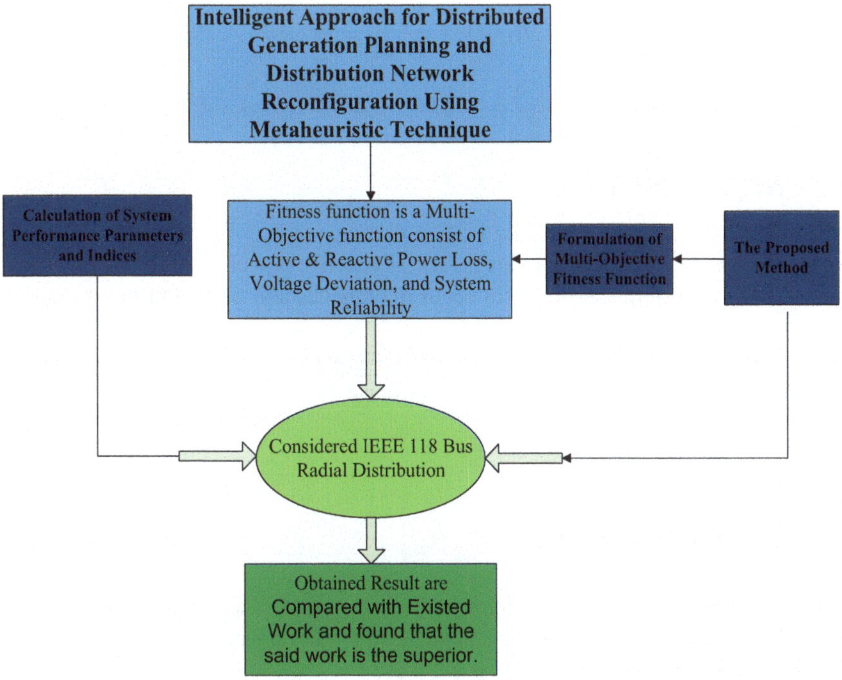

Fig. 1 Graphical abstract

- Novel multi-objective-based fitness function problem including system parameters such as active/real and reactive power loss, deviation of voltage, and system reliability is considered.
- The multi-objective fitness function problem is minimized using metaheuristic technique, namely APSO technique which is also compared with other techniques for IEEE 118-bus test system.
- The different proposed system configurations are analyzed and compared based on the achieved results for various system parameters such as voltage profile, active and reactive power losses, and system reliability parameters.

2 System Performance Parameters and Index

2.1 AP_{LOSS} Index

Real power loss (AP_{LOSS}) index is ratio of real power losses for the proposed system to power losses for base system. These losses occur in branches and tie-line of the system.

$$I_{\text{PL}}(\alpha) = \frac{P_l(\text{case-}\alpha)}{P_l(\text{case-}1)} \tag{1}$$

2.2 *RP*

Reactive power loss (RP_{LOSS}) index is ratio of reactive power losses in the proposed system to base system. These losses occur in branches and tie-line of the system.

$$I_{\text{QL}}(\alpha) = \frac{Q_l(\text{case-}\alpha)}{Q_l(\text{case-}1)} \tag{2}$$

2.3 $\mathbf{V}_{DEVIATION}$ *Index*

Voltage deviation ($V_{\text{DEVIATION}}$) index is the ratio of difference of slack or reference bus voltage to the minimum calculated voltage of system at buses to reference voltage, where reference voltage magnitude is 1.1 p.u.

$$I_{\text{VD}} = \max\left(\frac{V(n)}{v_{\text{ref}}}\right) \tag{3}$$

2.4 R *Index*

Reliability (R) index is the ratio of RI_{SUM} of modified system to standard basic system.

$$I_R(\alpha) = \frac{RI_{\text{SUM}}(\text{case-}\alpha)}{RI_{\text{SUM}}(\text{case-}1)} \tag{4}$$

Calculation of RI_{SUM} which is the summation of SAIFI, SAIDI, CAIDI, and AENS of a different system.

2.5 *The System Average Interruption Frequency Index (SAIFI)*

$$\text{SAIFI} = \frac{\sum \text{Interrupted customers}}{\text{customer serve}} \quad (5)$$

2.6 *The Energy Not Supplied (ENS)*

$$\text{ENS} = \sum \text{energy not supplied at load point} \quad (6)$$

2.7 *The System Average Interruption Duration Index (SAIDI)*

$$\text{SAIDI} = \frac{\sum \text{Duration for customers interrupted}}{\text{customers serve}} \quad (7)$$

2.8 *The Customer Average Interruption Duration Index (CAIDI)*

$$\text{CAIDI} = \sum \text{total number of customers interrupted for specified duration} \quad (8)$$

3 The Proposed Method

3.1 *Formulation of Multi-objective Fitness Function*

This section introduces a fitness function that consists of the multiple objective-based problem design for optimal size and location of multiple DGs, including network reconfiguration through PSO optimization.

$$\text{Multi-objective-Function} = \eta_1 * I_{\text{PL}} + \eta_2 * I_{\text{VD}} + \eta_3 * I_{\text{QL}} + \eta_4 * I_R \quad (9)$$

where η_1, η_2, η_3, and η_4 are weight factors with values of 0.40, 0.20, 0.25, and 0.15 respectively which is decided based on the priority. I_{PL}, I_{VD}, I_{QL} and I_R are AP_{LOSS} index, $V_{\text{DEVIATION}}$ index, RP_{LOSS} index, and R index.

3.2 Backward–Forward Sweep Load Flow Based on BIBC-BCBV Method

This study utilizes backward–forward sweep based on BIBC and BCBV method, which is effective than basic load flow.

The backward–forward sweep based on BIBC and BCBV process to update the buses voltage and branches current is as follows:

Step (1) In system for load flow analysis in first iteration, each bus voltage is equal to slack bus voltage. For second and so on iteration, bus voltage is updated value by using forward sweep.
For kth iteration, load current of each bus is:

$$\mathrm{LC}(t)^k = \left(\frac{\mathrm{PL}(t) + \mathrm{QL}(t)}{V^{t-1}}\right)^* \tag{10}$$

Step (2) Calculation of branch current for load flow analysis. We are using KCL for each bus from open end of the system. This is calculated using backward sweep.

- The BIBC matrix in terms of branch current is:

$$[\mathrm{BC}] = [\mathrm{BIBC}][\mathrm{LC}] \tag{11}$$

- The forward sweep voltage updates as:

$$V_i = V_j - \mathrm{BC}(j-1) \times Z_{ji} \tag{12}$$

Step (3) The VD is calculated as:

$$[\Delta V] = [\mathrm{BCBV}][\mathrm{BC}] \tag{13}$$

Step (4) Power losses in system are calculated through basic theorem like multiplication of difference of voltage magnitude of each branch node and conjugate of that branch current.

- The losses for apparent power of the system are calculated as,

$$S_{\mathrm{LOSS}} = \pi r^2 = \sum_{j=1}^{\mathrm{NB}} [\Delta V][\mathrm{BC}]^* \tag{14}$$

- The losses for active and reactive power are considered as,

$$\mathrm{AP}_{\mathrm{LOSS}} = \sum [\mathrm{BC}].*[\mathrm{BC}].*[R_{\mathrm{BR}}] \tag{15}$$

$$\mathrm{RP_{LOSS}} = \sum[\mathrm{BC}].*[\mathrm{BC}].*[X_{\mathrm{BR}}] \tag{16}$$

3.3 Metaheuristic Technique

The used metaheuristic technique for said work is APSO technique. The PSO is initially proposed by Eberhart and Kennedy in 1995. PSO is a technique for optimization based on population [1, 2]. For every associated iteration of the PSO in its search zone, every particle follows a defined inertia and speed (velocity). The is developed by Zhan et al. [4] in 2009 based on the some modification in PSO parameter, inertia weight, and additional incorporating constriction factor (ξ). The updating PSO to APSO increases the searching ability and convergence of optimization in APSO technique. The modified equation for APSO is given in Eqs. 17 and 18.

$$v(k+1) = \xi[w(k)*v(k) + c_1*\mathrm{rand}*(Y_{\mathrm{locbest}}(k) - Y(k)) \\ +c_2*\mathrm{rand}*\left(Y_{\mathrm{gbest}}(k) - Y(k)\right)] \tag{17}$$

$$Y(k+1) = Y(k) + v(k+1) \tag{18}$$

$$w = w_{\max} - (w_{\max} - w_{\min})\frac{\mathrm{Itr}}{\mathrm{Itr}_{\max}} \tag{19}$$

$$\xi = \frac{2}{\left|2 - \psi - \sqrt{\psi^2 - 4\psi}\right|} \tag{20}$$

$$\psi = c_1 + c_2 = 4.1 \tag{21}$$

Assume total population is N, where "k" denotes variation of P and its value is from 1, 2, 3, …, N and individual velocity is "v." For kth iteration, the velocity and population are $v(k+1)$ and $Y(k+1)$ separately. The v and Y indicate velocity and position of past generation interaction. Two equations of velocity and position are given to update the same for $(k+1)$th generation as in [1–4].

Where inertia weight is "w," Itr and $\mathrm{Itr}_{\max}$ are iteration and maximum iteration. The minimum and maximum weights are $w_{\min}$ and $w_{\max}$ with set value of 0.9 and 0.4, respectively [4]. ξ is the constriction factor with 0.729 value and both c_1 and c_2 having set value of 2.05. $Y_{\mathrm{localbest}}$ and Y_{gbest} are the local and global best populations, respectively. Velocity, swarm population for APSO are denoted by "v," Y. "i" is the ith iteration, rand resulted into random number in between 0 and 1, respectively.

Figure 2 shows a flowchart to obtain the optimal site and size for the proposed system with reconfiguration. In the flow diagram, we have considered five switches

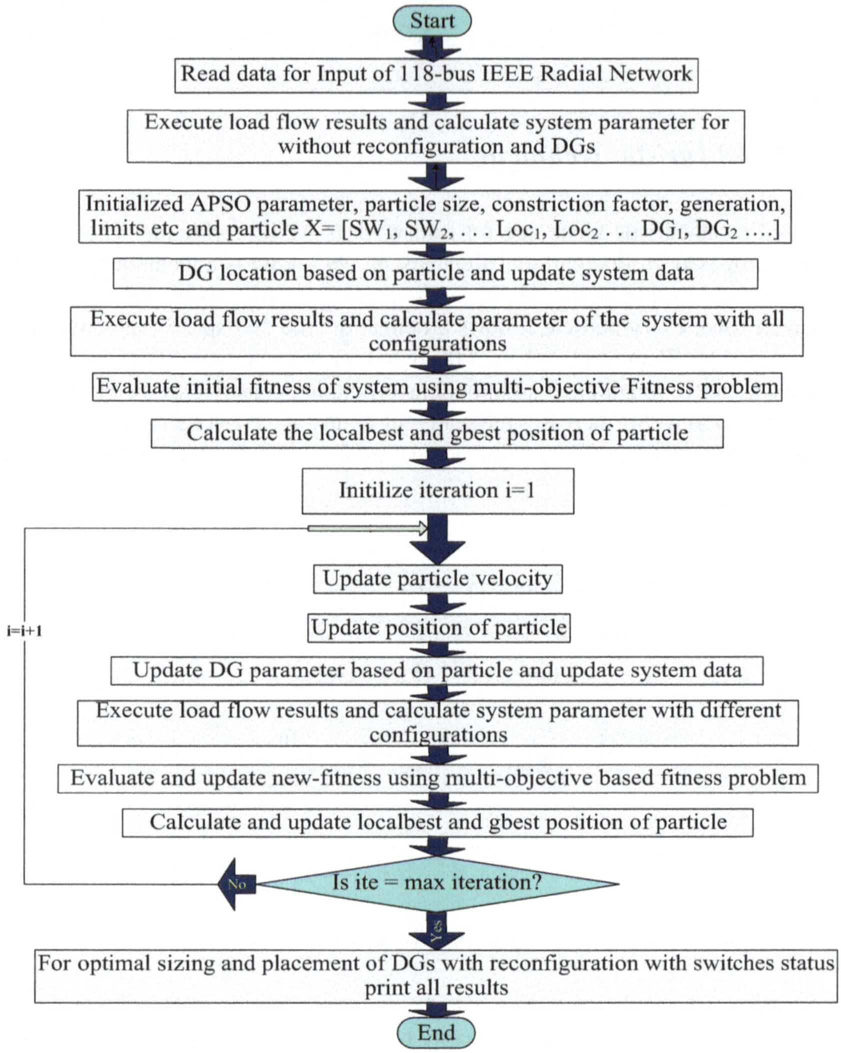

Fig. 2 Process flowchart of particle swarm optimization technique

(SW_1, SW_2, …, SW_5) location (Loc_1, Loc_2,…) and sizing of DGs, respectively. All the switches, DG locations, and DGs are transformed and signified as particle.

4 Test Systems

In the proposed work, six configurations are analyzed, which are:

Configuration-1: The base case system.
Configuration-2: The system with only reconfiguration.
Configuration-3: The system with only single DG.
Configuration-4: The system with only multi-DGs.
Configuration-5: The system with both reconfiguration and single DG.
Configuration-6: The system with both reconfiguration and multi-DGs.

5 Results and Discussions

In Configuration-1, we have considered a basic test system without reconfiguration and DG. We noticed the losses (active and reactive) are 1.2981 MW (AP_{LOSS}) and 0.9787 MVAr (RP_{LOSS}); also, RI sum, reliability (R) index, and voltage deviation ($V_{DEVIATION}$) of the system as 1.8223, 0.9839, and 0.2312 p.u. $V_{DEVIATION}$ is as compared to rated p.u voltage. Configuration-2, only reconfiguration of 118-bus test network is reported. The system losses (active and reactive) are 0.4232 MW and 0.7785 MVAr; also, RI sum is 1.3694073 and the maximum $V_{DEVIATION}$ of the system is 0.1428 p.u with fitness value of 0.3425 as illustrated in Tables 1, 2, and 3.

In the same fashion Configuration-3, present only-single DGs using APSO, which shows active and reactive power losses, (AP_{LOSS} and RP_{LOSS}) reduced by 21.59%

Table 1 IEEE 118-bus radial system losses for all different configurations

Configurations	System losses			$V_{DEVIATION}$
	AP_{LOSS}	RP_{LOSS}	S_{LOSS}	
Configuration-1	1.2981	0.9787	1.6257	0.2312
Configuration-2	0.4232	0.2873	0.5115	0.1428
Configuration-3	1.0178	0.7785	1.2814	0.1947
Configuration-4	0.9414	0.7391	1.1969	0.1947
Configuration-5	0.3706	0.2404	0.4418	0.1304
Configuration-6	0.3606	0.2201	0.4225	0.1275

Table 2 Proposed system REL index for all configurations

Configurations	Indices of system reliability				RI SUM	% Reliability
	SAIFI	SAIDI	CAIDI	AENS		
Configuration-1	0.4993	0.2996	0.6	0.4234	1.8223	98.39
Configuration-2	0.3160	0.1896	0.6	0.2638	1.3694	98.79
Configuration-3	0.4263	0.2558	0.6	0.3969	1.679	98.52
Configuration-4	0.3253	0.1952	0.6	0.3564	1.4769	98.7
Configuration-5	0.2750	0.1650	0.6	0.2488	1.2888	98.86
Configuration-6	0.2204	0.1322	0.6	0.2234	1.176	98.97

Table 3 Fitness function and indices value of the proposed system for all configurations

Configurations	Multi-objective fitness function indices				Fitness function
	I_{PL}	I_{VD}	I_{QL}	I_R	
Configuration-1	1	0.2102	1	1	0.842
Configuration-2	0.3260	0.1298	0.2935	0.7515	0.3425
Configuration-3	0.7840	0.1770	0.7954	0.9213	0.6860
Configuration-4	0.7252	0.1770	0.7552	0.8105	0.6359
Configuration-5	0.2855	0.1186	0.2456	0.7072	0.3054
Configuration-6	0.2778	0.1159	0.2498	0.6453	0.2936

and 20.46%, and also, the $V_{DEVIATION}$ and RI sum reduced by 15.78% and 7.86% as shown in Figs. 3, 4, 5, and 6 related to Configuration-1.

In Configuration-4, only multi-DG considered in which the AP_{LOSS} and RP_{LOSS} are reduced to 72.53% and 75.52%; $V_{DEVIATION}$ and RI sum decreased to 84.22% and 81.05%; and also the reliability is improved up to 98.7% as shown in Table 2 and Fig. 8 as compared to base Configuration-1.

After that, implementation of system reconfiguration including single and multi-DGs is reported. A better reduction in all indices is noticed in Configuration-5 and 6. The Configuration-5 includes the implementation of system reconfiguration with single DG. The active and reactive power losses are decreased by 71.45% and 75.44% as noticed from the results. The $V_{DEVIATION}$ and RI-sum decreased to 43.59% and 29.28% than Configuration-1.

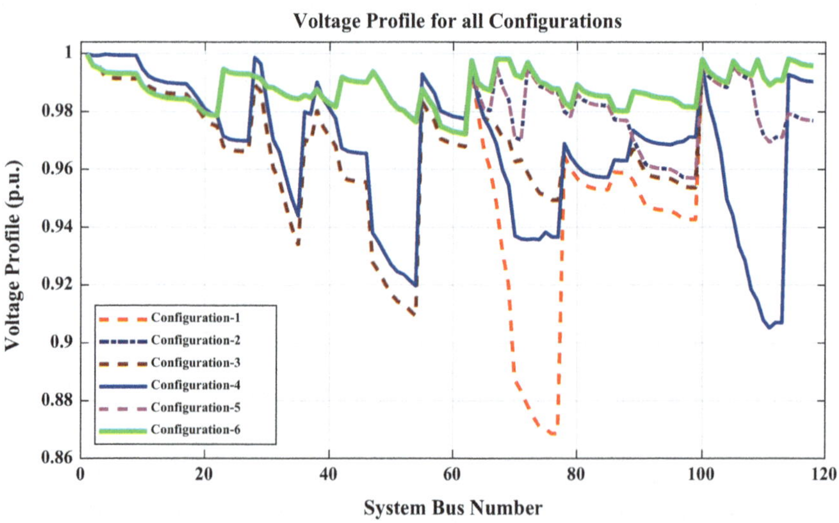

Fig. 3 118-bus radial system voltage profile for all configurations

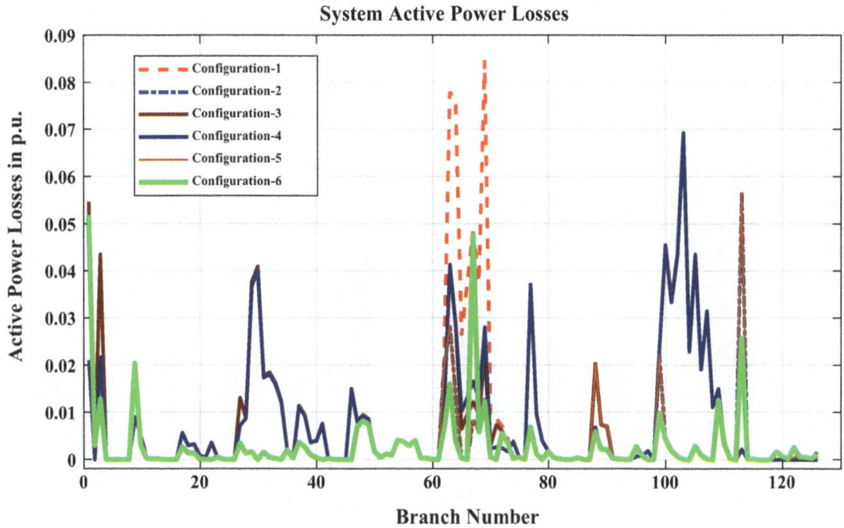

Fig. 4 Proposed system active power loss for all configurations

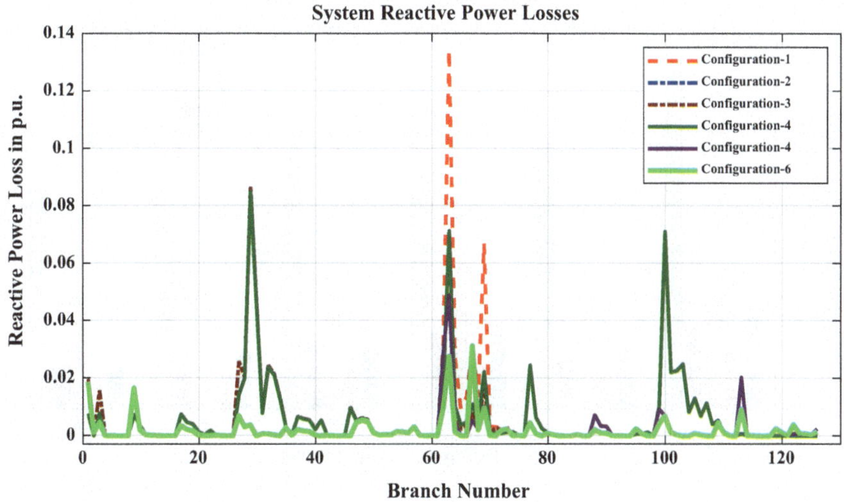

Fig. 5 Proposed system reactive power loss for all configurations

In the final stage (i.e. Configuration-6), we have implemented reconfiguration with multiple DGs using APSO. A considerable decrement in multi-objective function is observed. The AP_{LOSS} and RP_{LOSS} are decreased up to 72.22% and 77.51% as given in Table 1 and Figs. 4, 5, and 6. The reduction in $V_{DEVIATION}$ and RI-sum up to 44.85% and 90.34% are achieved as shown in Figs. 2 and 7. Along with 0.2936 p.u,

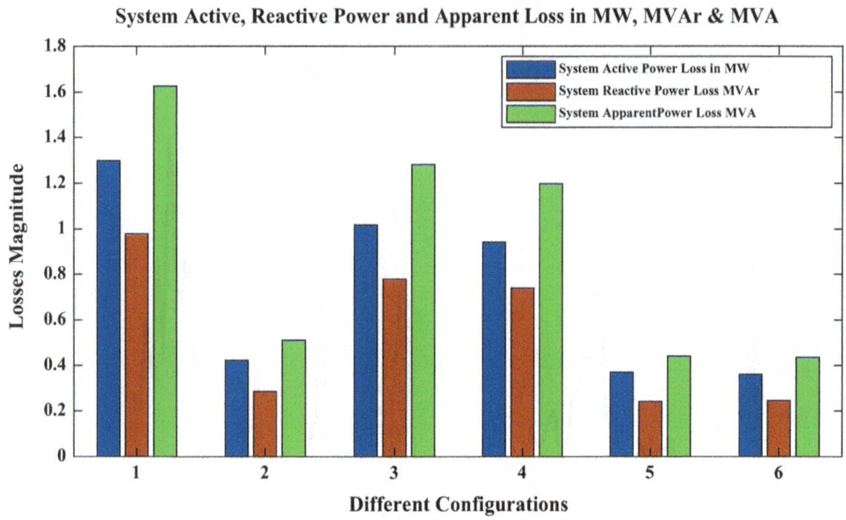

Fig. 6 Proposed system losses for all configurations

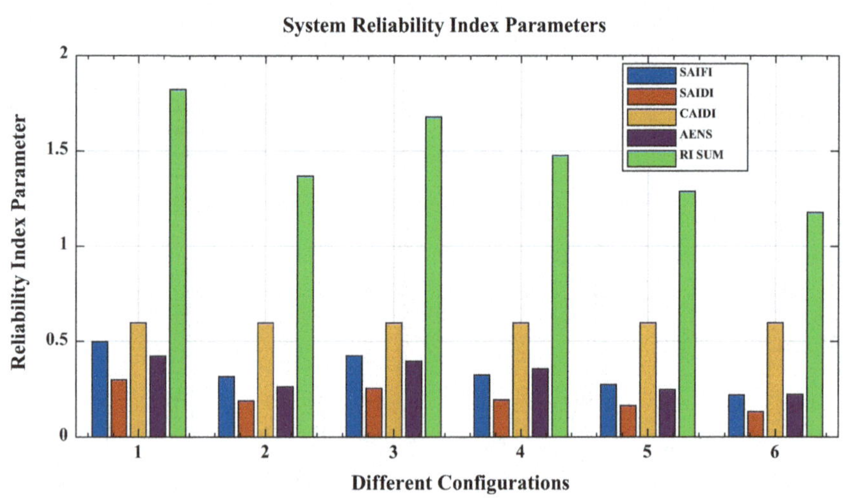

Fig. 7 Proposed system reliability index for different configurations

fitness function value is minimized by 65.13% as presented in Table 3 and Fig. 10. The reliability is improved up to 98.97%, as tabulated in Table 2, and the related bar chart is shown in Fig. 8.

The result analysis concludes that intelligent optimal site and size of multiple renewable-based DG with reconfiguration are the efficient methods using APSO to discover an optimal or minimum solution for multi-objective function optimization. The data for the optimal solution is presented in Table 3 and Figs. 9 and 10. Finally,

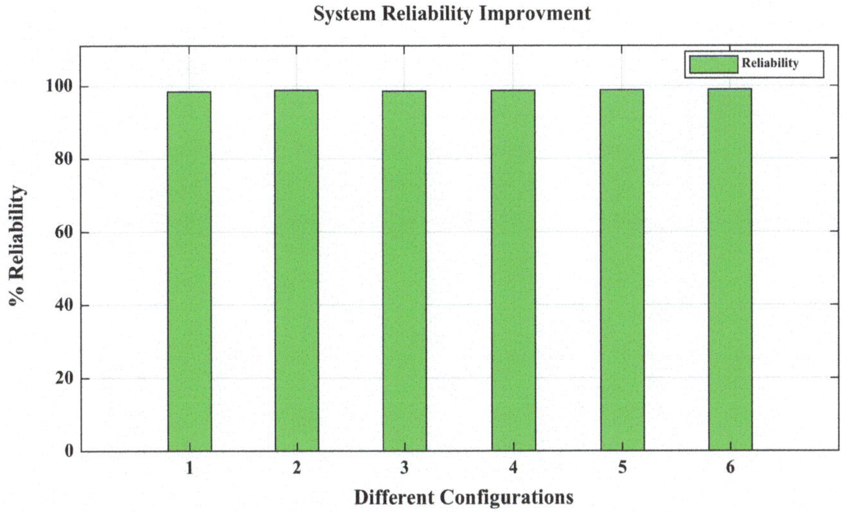

Fig. 8 Proposed system reliability for all configurations

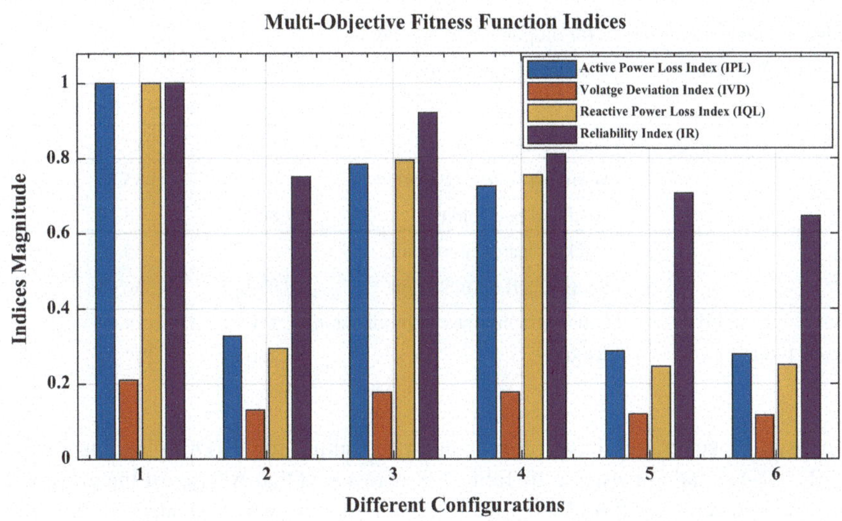

Fig. 9 Value of proposed system fitness function indices for all configurations

in the proposed work, i.e., Configuration-6, the AP_{LOSS} is decreased up to 72.22%, and reliability improved up to 98.97%, and the voltage profile improved as revealed in Tables 1 and 2 and figures from Figs. 3, 4, 5, 6, 7, 8, 9, and 10.

The comparison Table 4 tabulated the data of optimal capacitor allocation by genetic algorithm, harmony search algorithm, artificial bee colony, cuckoo search algorithm, and shark smell optimization as given in [5]. While Bawazir et al. [5]

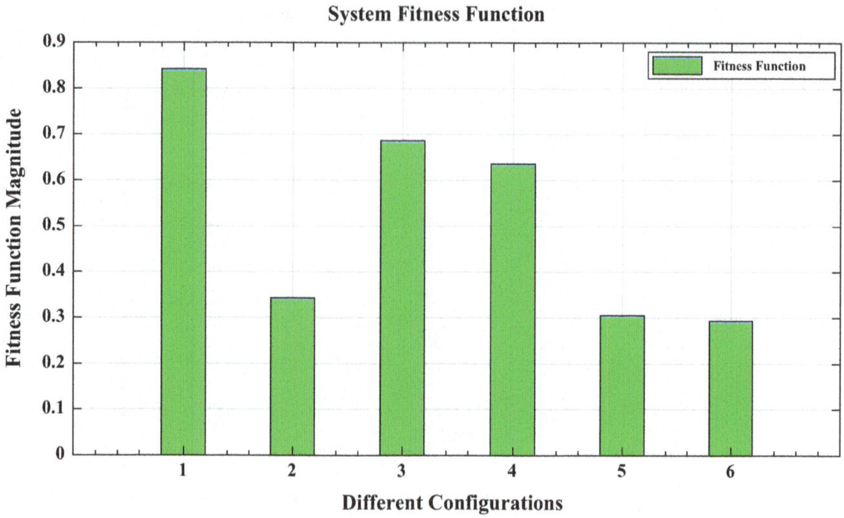

Fig. 10 Value of multi-objective fitness function for all configurations

Table 4 Comparative results for proposed and existing work

Existed/Work proposed	Method of optimization	Loss in kW	
		Magnitude	% Reduction
Hartono et al. [6]	Genetic algorithm	1120.9	13.65
	Harmony search algorithm	833.51	35.79
	Artificial bee colony	862.03	33.59
	Cuckoo search algorithm	837.37	35.49
	Shark smell optimization	830.15	36.05
Bawazir et al. [5]	Biogeography-based optimization	1108.1	14.64
Work Proposed	APSO	360.6	72.22

proposed biogeography-based optimization technique for PVDG planning in the IEEE 118-bus radial network. In Table 4, it is revealed that AP_{LOSS} of the proposed work is reduced to 360.6 kW that is 72.22% reduction which is superior than the other compared existed works.

6 Conclusions

The DGs optimal sizing and location with reconfiguration through intelligent approach using metaheuristic technique, including multiple objective based on system parameters in the fitness function, are proposed. This consists of system parameter indices like voltage deviation, real/active and reactive power loss, and R index of IEEE 118-bus standard system using APSO metaheuristic technique. Author analyzed all parameters of fitness function for different configurations. The results analysis confirms that the proposed method effectively reduces power loss (apparent, real and reactive power loss) as well as decreases $V_{\mathrm{DEVIATION}}$ and also improves the R index which constitutes with SAIFI, SAIDI, CAIDI, and AENS. The 118-bus radial system result shows that in Configuration-1, i.e., base configuration, $\mathrm{AP_{LOSS}}$ and $\mathrm{RP_{LOSS}}$ are 1.2981 MW 0.9787 MVAr. The $\mathrm{V_{DEVIATION}}$ and reliability index parameter (RI sum) are 0.2312 p.u and 1.8223. In Configuration-3, after implementing single DG, the $\mathrm{AP_{LOSS}}$ and $\mathrm{RP_{LOSS}}$ are 78.41% and 79.54% using APSO. The $V_{\mathrm{DEVIATION}}$ and RI sum for this configuration are 84.22%, and 92.14% compared with base Configuration-1. In Configuration-4 after implementing multiple DGs, the $\mathrm{AP_{LOSS}}$ and $\mathrm{RP_{LOSS}}$ reduced to 72.53%, 75.52%, and $V_{\mathrm{DEVIATION}}$ and RI sum reduced to 84.22% and 84.22% using APSO than Configuration-1. Similarly, the other configurations can be analyzed, which shows the superiority of the proposed methodology. Finally, we have implemented multiple DGs with reconfiguration using PSO; it is observed that considerable decrement in $\mathrm{AP_{LOSS}}$ and $\mathrm{RP_{LOSS}}$ reduced to 27.78% and 22.51%. The $V_{\mathrm{DEVIATION}}$ and RI sum decrease to 55.15% and 9.66% as illustrated in Tables 1 and 2. Similarly, reliability improved up to 98.97%. Correspondingly, the system voltage profile improves. All the above data in graph and bar are shown in Figs. 3, 4, 5, 6, 7, 8, 9, and 10, than the base configuration.

References

1. S. Gopiya Naik, D.K. Khatod, M.P. Sharma, Optimal allocation of combined DG and capacitor for real power loss minimization in distribution networks. Int. J. Electr. Power Energy Syst. **53**, 967–973 (2013)
2. D. Sudha Rani, N. Subrahmanyam, M. Sydulu, Multi-objective invasive weed optimization—an application to optimal network reconfiguration in radial distribution networks. Int. J. Electr. Power Energy Syst. **73**, 932–942 (2015)
3. M. Sedighizadeh, M. Esmaili, M.M. Mahmoodi, Reconfiguration of distribution networks to improve reliability and reduce power losses using imperialist competitive algorithm. Iranian J. Electr. Electron. Eng. **13**(3), 287–302 (2017)
4. Z.-H. Zhan et al., Adaptive particle swarm optimization. IEEE Trans. Syst. Man Cybern. Part B (Cybernetics) **39**(6), 1362–1381 (2009)
5. R. Bawazir, N. Çetin, M. Mosbah, S. Arif, Improvement of the voltage profile of the distribution network by optimal integration of PV based decentralised source, in *2020 International Conference on Electrical Engineering (ICEE)* (IEEE, 2020), pp. 1–4
6. H. Hartono, M. Azis, Y. Muharni, Optimal capacitor placement for IEEE 118 bus system by using genetic algorithm, in *2019 2nd International Conference on High Voltage Engineering and Power Systems (ICHVEPS)* (IEEE, 2019)
7. A.K. Bohre, G. Agnihotri, M. Dubey, Optimal sizing and sitting of DG with load models using soft computing techniques in practical distribution system. IET Gener. Transm. Distrib. **10**(11), 2606–2621 (2016)
8. A.K. Bohre, G. Agnihotri, M. Dubey, S. Kalambe, Impacts of the load models on optimal planning of distributed generation in distribution system. Adv. Artif. Intell. **2015**, 1–6 (2015)

9. A.K. Bohre, G. Agnihotri, M. Dubey, The optimal distributed generation placement and sizing using novel optimization technique. Middle-East J. Sci. Res. **10**, 1228–1236 (2015)
10. A.K. Bohre, G. Agnihotri, M. Dubey, S. Kalambe, Impacts of the load models on optimal sizing and siting of distributed generation in distribution system. World Appl. Sci. J. **33**(7), 1197–1205 (2015)
11. A.K. Bohre, G. Agnihotri, M. Dubey, Distributed generation planning including load models using different optimization techniques. J. Electr. Eng. **2016**(1), 1–12 (2016)
12. A.K. Bohre, G. Agnihotri, M. Dubey, S. Kalambe, Assessment of intricate DG planning with practical load models by using PSO. Electr. Comput. Eng. Int. J. (ECIJ) **4**(2), 15–22 (2015)
13. A.K. Bohre, G. Agnihotri, M. Dubey, The optimal planning of distributed generation using OPF and butterfly-PSO (BF-PSO) technique. J. Electr. Eng. **2015**(3), 1–12 (2015)
14. N. Kanwar, N. Gupta, K.R. Niazi, A. Swarnkar, R.C. Bansal, Simultaneous allocation of distributed energy resource using improved particle swarm optimization. Appl. Energy **185**, 1684–1693 (2017)
15. T. Adefarati, R.C. Bansal, Reliability assessment of distribution system with the integration of renewable distributed generation. Appl. Energy **185**, 158–171 (2017)
16. S.S. Tanwar, D.K. Khatod, Techno-economic and environmental approach for optimal placement and sizing of renewable DGs in distribution system. Energy **127**, 52–67 (2017)
17. A. Kumar, M. Rizwan, U. Nangia, Optimal sizing of renewable energy resources in a microgrid for a distributed generation system, in *2019 International Symposium on Advanced Electrical and Communication Technologies (ISAECT)* (IEEE, 2019), pp. 1–6
18. I.A. Quadri, S. Bhowmick, D. Joshi, A comprehensive technique for optimal allocation of distributed energy resources in radial distribution systems. Appl. Energy **211**, 1245–1260 (2018)
19. T.P. Nguyen, D.N. Vo, A novel stochastic fractal search algorithm for optimal allocation of distributed generators in radial distribution systems. Appl. Soft Comput. **70**, 773–796 (2018)
20. B. Kumar, B.K. Saw, A.K. Bohre, Planning of multi-renewable DG with distribution network reconfiguration based on MOFF approach using PSO technique, in *Recent Advances in Power Electronics and Drives* (Springer, Singapore, 2021), pp. 167–179
21. B. Kumar, B.K. Saw, A.K. Bohre,Optimal distribution network reconfiguration to improve the system performances using PSO with multiple-objectives, in *2020 International Conference on Computational Intelligence for Smart Power System and Sustainable Energy (CISPSSE)* (IEEE, 2020), pp. 1–6
22. Y. Takenobu, N. Yasuda, S.I. Minato, Y. Hayashi, Scalable enumeration approach for maximizing hosting capacity of distributed generation. Int. J. Electr. Power Energy Syst. **105**, 867–876 (2019)
23. J.R. Mantovani, A decomposition approach for integrated planning of primary and secondary distribution networks considering distributed generation. Int. J. Electr. Power Energy Syst. **106**, 146–157 (2019)
24. H. Malik, N. Fatema, A. Iqbal, *Intelligent Data-Analytics for Condition Monitoring: Smart Grid Applications* (Elsevier, 2021)
25. H. Malik, A. Iqbal, A.K. Yadav, Soft computing in condition monitoring and diagnostics of electrical and mechanical systems, in *Novel Methods for Condition Monitoring and Diagnostics* (Springer, Berlin, 2020), p. 499
26. H. Malik, *Metaheuristic and Evolutionary Computation* (Springer Nature, Berlin, 2020)
27. S. Srivastava et al., *Applications of Artificial Intelligence Techniques in Engineering*, vol. 1, 1E, Advances in Intelligent Systems and Computing, vol. 698 (Springer Nature, Berlin, 2018), 643 p. ISBN 978-981-13-1819-1
28. S. Srivastava et al., *Applications of Artificial Intelligence Techniques in Engineering*, vol. 2, 1E, Advances in Intelligent Systems and Computing, vol. 697 (Springer Nature, Berlin, 2018), 647 p. ISBN 978-981-13-1822-1

Intelligent Modelling and Analysis of Direct Torque Control-Space Vector Modulation of Doubly Fed Induction Generator (DFIG)

Tanzeel Imtiyaz, Farhad Ilahi Bakhsh, and Anjali Jain

Abstract Intelligent modelling and analysis of direct torque control-space vector modulation of doubly fed induction generator (DFIG) done in MATLAB/Simulink is presented in this paper. DTC was a control technique for squirrel cage motors and now is applied to induction generators with wound rotor by means of a rotor connected voltage side inverter. First, the rotor flux and electromagnetic torque of DFIG are estimated on the basis of rotor currents and voltages measurement, and then space vector modulation is implemented for generating the vectors of reference voltage for the windings of rotor. Control is done for rotor side which is aimed at controlling the speed and the electromagnetic torque of DFIG with the help of hysteresis controllers. The designed control strategy helps to generate suitable voltage vector which have varying magnitude and angle. DTC-SVM is used to keep switching frequency constant and to reduce the flux and torque ripples. DTC-SVM has other advantages like simple coordinate transformation and fast dynamic response. Results of MATLAB/Simulink of DTC-SVM of DFIG are illustrated at various operating conditions, and it is verified that the speed and electromagnetic torque of DFIG are controlled directly.

Keywords Doubly fed induction generator (DFIG) · Space vector modulation (SVM) · Direct torque control (DTC) · Voltage vectors · *d–q* reference frame · Hysteresis controller

T. Imtiyaz (✉) · A. Jain
Electrical Engineering Department, Amity University Noida, Noida, Uttar Pradesh 201313, India

A. Jain
e-mail: ajain5@amity.edu

F. I. Bakhsh
Electrical Engineering Department, NIT Srinagar, Srinagar, Jammu and Kashmir 190006, India

H. Malik et al. (eds.), *Intelligent Data Analytics for Power and Energy Systems*,
Lecture Notes in Electrical Engineering 802,
https://doi.org/10.1007/978-981-16-6081-8_3

Nomenclature

$V_{s,r(dq)}$	Voltage of stator and rotor winding (*d–q* components)
$I_{s,r(dq)}$	Current of stator and rotor winding (*d–q* components)
$R_{s,r}$	Resistance of stator and rotor winding
$L_{s,r}$	Inductance of stator and rotor winding
$L_{(s,r)l}$	Leakage inductance of stator and rotor winding
L_m	Mutual inductance
$\Psi_{r(dq)}$	Flux linkage of rotor winding (*d–q* components)
$\Psi_{s(dq)}$	Flux linkage of stator winding (*d–q* components)
T_e	Electromagnetic torque of DFIG
p	Number of poles

1 Introduction

Nowadays in wind power generation systems, a lot of attention is drawn to doubly fed induction generators (DFIGs) because of advantages like varying speed operation, less power losses, low converter cost as compared to other generators [1]. The basis for controlling the DFIGs that are connected to the grid is stator voltage [2] or rotor/stator flux oriented [3] called vector control (VC). Such techniques decouple the currents of the rotor into two perpendicular components that are along d-axis and q-axis frame [4] which are in phase with the flux and the electromagnetic torque. Then proportional-integral (PI) controllers are utilized for control of speed and electromagnetic torque by varying the current components of the rotor. These control techniques are not reliable as they highly depend on the machine parameters like stator and rotor resistances and inductances and on the PI controller tuning. Thus, when the used values in control system are different from actual machine values, performance of machine is affected.

In [5] and [6], an alternative for vector control (VC) of machines like induction motor and DFIG is given by direct torque control (DTC) because of voltage source inverters (VSI) discrete nature. In DTC, it is possible to regulate the torque directly that minimizes the machine parameters usability and also reduces the VC strategy complexity. In DTC, the speed and the electromagnetic torque can be controlled by selection of voltage vectors through a predefined look-up table that is built on the information of flux and torque [7]. Torque/flux hysteresis controllers are used for the control mechanism as they overcome the drawbacks of PI controllers. One of the drawbacks of hysteresis controller is that the switching frequency of controller is dependent on its bandwidth which results in pulsation of torque. To overcome this disadvantage, DTC-SPM is implemented in this paper [8]. In a fixed switching time period, two successive active voltage vectors followed by a zero vector are applied to the rotor. The order of these three vectors is predefined in a look-up table to maintain switching frequency constant. Based on the minimization cost function of flux and torque ripple, each active vector's time duration is calculated. Although there is minimization of ripples, the time duration calculations are complex.

Comparison between various control techniques like vector control (VC), direct torque control (DTC), maximum power point tracking (MPPT) control and direct power control (DPC) is presented in [9]. In [10] and [11], the performance of DTC of DFIG is improved by injection of three active vectors. One more scheme is sliding-mode control (SMC) used in electric drive systems [12], and another one is SCM-DTC control of induction motor with PI controller that is presented in [13] and [14]. Sliding-mode control is applied to high power DFIG in [15, 16]. It is concluded in [17] that DTC-SVM offers constant switching frequency, fast dynamic response and low switching losses.

This paper is divided in seven sections. The mathematical model for doubly fed induction generator is given in Sect. 2. Section 3 presents the space vector modulation technique for DFIG. Section 4 depicts the direct torque control proposed model. Section 5 shows the MATLAB/simulation models of DTC-SVM for DFIG. Section 6 demonstrates the results obtained using direct torque control strategy. At last, Sect. 7 gives the conclusion of the paper.

2 Mathematical Model for DFIG

Figure 1 gives the mathematical DFIG modelling in synchronous *d–q* axis reference frame [18]. This model is used to find the equations of the voltages, flux linkages and electromagnetic torque in *d–q* axis reference frame for stator and rotor side, and these equations are used for modelling the DTC-SVM for DFIG.

The stator and rotor winding voltage equations written in *d–q* axis reference frame are shown as under

$$V_{sd} = R_s I_{sd} - \Psi_{sq}\omega_s + \frac{d\Psi_{sd}}{dt} \tag{1}$$

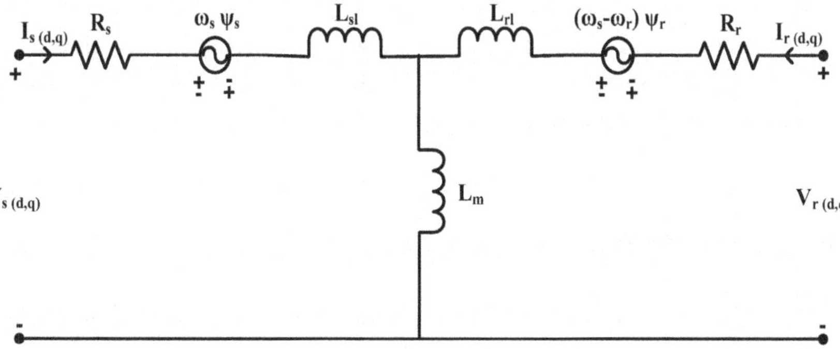

Fig. 1 Mathematical model of DTC-SVM for DFIG

$$V_{sq} = R_s I_{sq} + \Psi_{sd}\omega_s + \frac{d\Psi_{sq}}{dt} \quad (2)$$

$$V_{rd} = R_r I_{rd} - \Psi_{rq}\omega_r + \frac{d\Psi_{rd}}{dt} \quad (3)$$

$$V_{rq} = R_r I_{rq} + \Psi_{rd}\omega_r + \frac{d\Psi_{rq}}{dt} \quad (4)$$

The flux linkages developed in the DFIG are as below:

$$\Psi_{sd} = L_m I_{rd} + L_s I_{sd} \quad (5)$$

$$\Psi_{rd} = L_m I_{sd} + L_r I_{rd} \quad (6)$$

$$\Psi_{sq} = L_m I_{rq} + L_s I_{sq} \quad (7)$$

$$\Psi_{rq} = L_m I_{sq} + L_r I_{rq} \quad (8)$$

$$\Psi_r = \sqrt{\Psi_{rd}^2 + \Psi_{rq}^2} \quad (9)$$

where $L_s = L_{sl} + L_m$ and $L_r = L_{rl} + L_m$.

The electromagnetic torque generated in DFIG is as under

$$T_e = -\frac{3}{2} p \frac{L_m}{L_s} (\Psi_{sd} I_{rq} - \Psi_{sq} I_{rd}) \quad (10)$$

3 Space Vector Modulation

A widespread PWM technique designed for voltage source inverters (VSI) is space vector modulation (SVM) that is used in control strategies for permanent magnet synchronous machines and ac induction machines.

3.1 Basic Principle

Figure 2a shows the construction of a three-phase VSI. The voltage (line-to-neutral) vector and the switching state vector are related by Eq. (11)

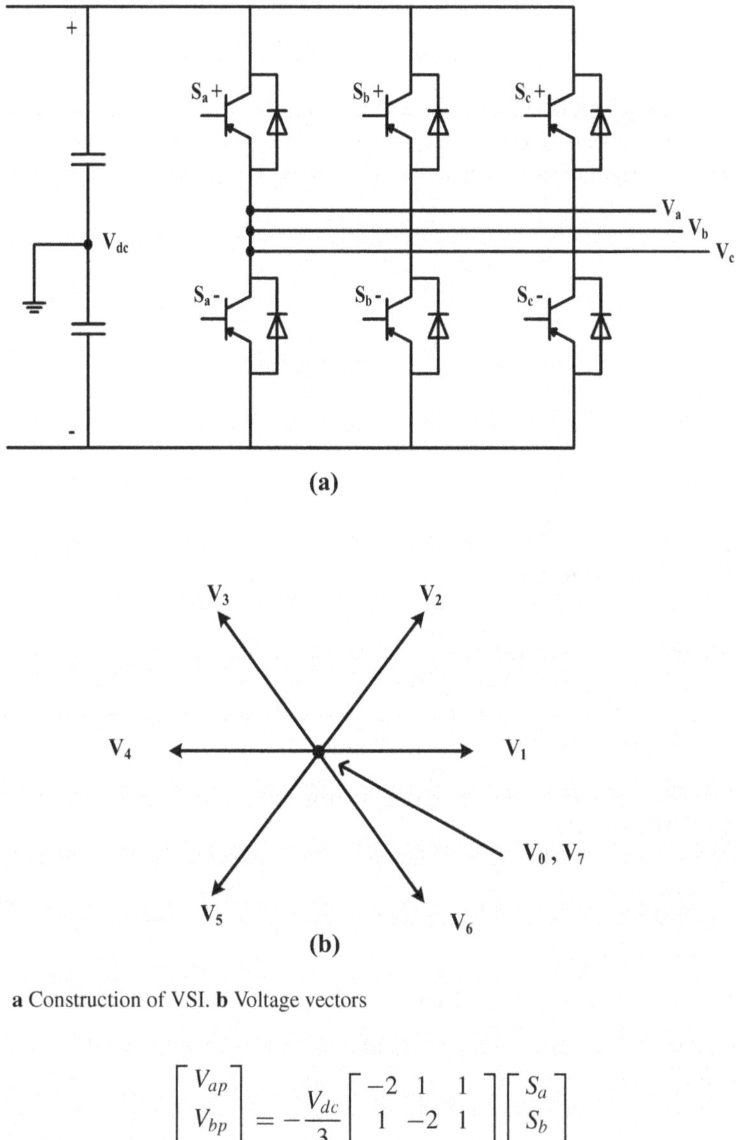

Fig. 2 **a** Construction of VSI. **b** Voltage vectors

$$\begin{bmatrix} V_{ap} \\ V_{bp} \\ V_{cp} \end{bmatrix} = -\frac{V_{dc}}{3}\begin{bmatrix} -2 & 1 & 1 \\ 1 & -2 & 1 \\ 1 & 1 & -2 \end{bmatrix}\begin{bmatrix} S_a \\ S_b \\ S_c \end{bmatrix} \tag{11}$$

For the three switches of the upper wing, there are total eight combinations of on/off states, and the results of these are eight vectors known as basic space vectors consists of two zero vectors and six nonzero vectors depicted in Fig. 2b. Any two adjacent nonzero vectors are 60° apart. A voltage of magnitude zero is applied to the load by zero vectors. The main aim of this technique is generation of an arbitrary voltage vector by combination of the basic space vectors [19]. For realization of this, the output voltage is given as

$$V_{\text{out}} = \frac{1}{t_{\text{PWM}}}(t_1 V_n + t_2 V_n \pm 1 + t_0 V_{0/7}) \tag{12}$$

where V_n and $V_{n\pm1}$ represent any two nonzero voltage vectors that are adjacent, and their time durations are denoted by t_1 and t_2, respectively. The algebraic sum of t_1 and t_2 must be equal to or less than the total PWM time period. The time durations can be calculated as

$$t_{0/7} = t_{\text{PWM}} - (t_1 + t_2) \tag{13}$$

$$t_1 = \sqrt{2} V_{\text{out}} \cos(\delta + 30) t_{\text{PWM}} \tag{14}$$

$$t_2 = \sqrt{2} V_{\text{out}} \sin(\delta) t_{\text{PWM}} \tag{15}$$

3.2 Switching Patterns

The total number of arrangements possible in each PWM time period for V_n, $V_{n\pm1}$ and $V_{0/7}$ are two. In [20], two switching patterns are implemented on TI's C2000 digital signal processor (DSP) known as hardware and software-implemented switching.

For switching pattern determined by software, the characteristic features are [20]:

- With the help of a program, the order of switching is fixed for each sector of the three PWM channels.
- Every time period of PWM will start as well as end with V_0. Time period of V_0 and V_7 is same.
- In every PWM time period, each PWM channel is switched twice except for duty cycle value 0 and 1.

Features of switching pattern determined by PWM hardware module are [20]:

- The order of basic space vectors should be such so that minimal switching is required.
- For the entire time period of PWM, one of the upper switches is always constant.
- There is no dead band effect for the switch which is constant for whole time period. Because of this, the three-phase outputs are uneven and harmonics are generated in the output of the inverter.

Figure 3 represents the switching pattern for software and hardware technique. The switching pattern determined by software technique has better performance.

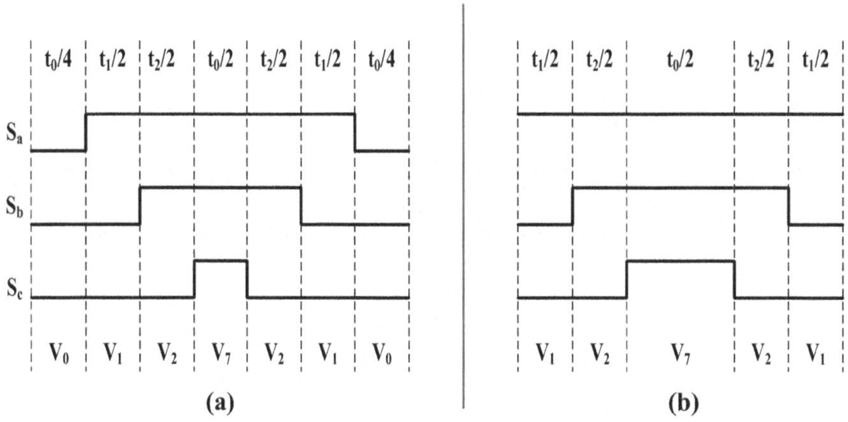

Fig. 3 Switching patterns determined by **a** software **b** hardware

4 Direct Torque Control Proposed Model

In DFIG model, the windings of stator have direct connection with the grid, and the stator flux magnitude is kept constant. The windings of rotor have connection with the grid via back-to-back power converters [21]. For this paper, voltage of DC link is kept at a constant value, and control is done by rotor side converter (RSC). Figure 4 depicts the complete diagram of DTC-SVM for DFIG.

Equation 16 gives the modified equation of electromagnetic torque of DFIG. It is evident from the equation that the torque is dependent on the parameters of machine and the magnitude and angle of stator and rotor linkage fluxes. The magnitude of stator flux is ideally kept constant and the rotor flux will be controlled by RSC. Thus, the electromagnetic torque of DFIG can be assumed to be dependent on the angle between the flux linkages of the stator and rotor vector "α".

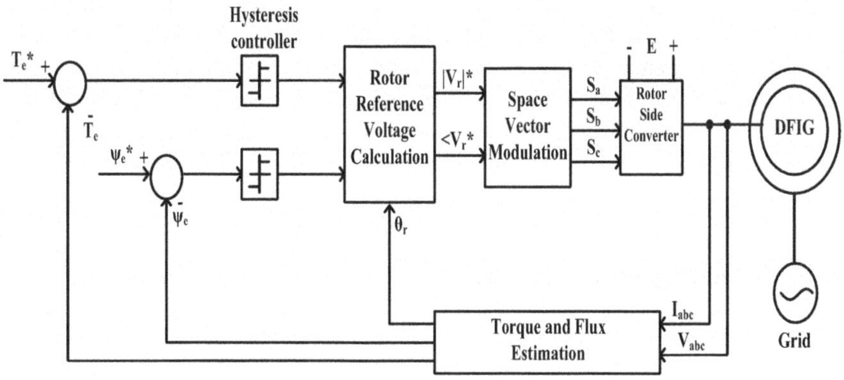

Fig. 4 DTC-SVM for DFIG

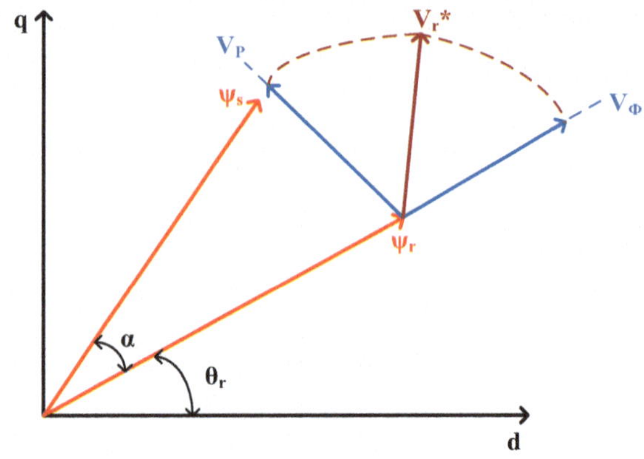

Fig. 5 Reference voltage vector of rotor

$$T_e = \frac{L_m}{\sigma L_s L_r}\left|\vec{\Psi}_s\right|\left|\vec{\Psi}_r\right| \sin\alpha \tag{16}$$

To explain the mechanism, introduction of two voltage vectors V_ϕ and V_P is needed, where V_ϕ is the voltage vector that aligns with the rotor flux and V_P is the voltage vector perpendicular to the rotor flux. The magnitude of rotor flux is controlled by V_ϕ and V_P controls the angle α. Thus, for controlling the flux, speed and torque, the resultant of V_ϕ and V_P is taken as the rotor reference voltage vector. Selection of reference vector is shown in Fig. 5.

Hysteresis controllers are used for control with zero band detect variation of flux and torque. Table 1 gives the angle of reference vector that depends on the output of hysteresis controllers. The presented controller will be able to work in both motoring and generating modes.

The magnitude of reference rotor voltage can also be modified since SVM has the ability to generate any arbitrary voltage vector. Thus, for reference voltage vector magnitude, the controller utilizes two different values. When the flux and torque errors are small in magnitude in the steady-state operation, ripples can be reduced by providing a voltage vector whose value is less than one per unit (pu). Hence, SVM is directed for generating voltage vector magnitude of 0.5 pu when the flux and torque errors are in maximum permissible values. In other conditions, voltage vector magnitude is kept at 1 pu to have excellent dynamic response [7]. Therefore, when the presented controller generates reference rotor voltage with appropriate magnitude and angle, then SVM model will generate gate pulses for voltage source inverter.

Table 1 Selection of reference voltage vector angle

	T_e		
		−1	1
Ψ_r	−1	$\theta_r + 3\pi/4$	$\theta_r + 5\pi/4$
	1	$\theta_r + \pi/4$	$\theta_r + 7\pi/4$

5 Simulation Model of DTC-SVM for DFIG

Figure 6 gives the simulation model of direct torque control-space vector modulation (DTC-SVM) for DFIG. The simulation model comprises of gird which supplies three-phase voltages, two *V*–*I* measurement blocks for measuring the currents and voltages, DFIG model, DTC block, and inverter block connected to the ground. The voltages coming from the inverter are fed to one of the *V*–*I* measurement blocks to give voltage to the DFIG rotor side. The DFIG input is the load torque, and electromagnetic torque and rotor angle are the outputs taken from the DFIG block. The inputs given to the DTC block are the speed and the flux. For speed reference, stair-case wave is used.

Direct torque control subsystem is depicted in Fig. 7. The inputs to this subsystem are reference speed, DFIG speed and the reference flux. The first subsystem in DTC is the flux, torque and sector estimator block shown in Fig. 8. Currents are taken from

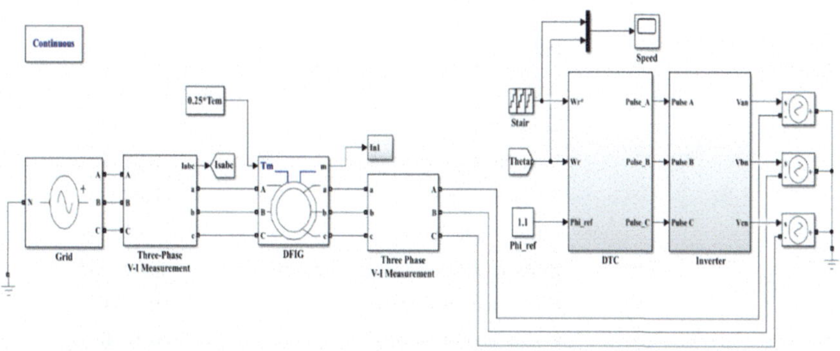

Fig. 6 Simulation model of DTC-SVM for DFIG

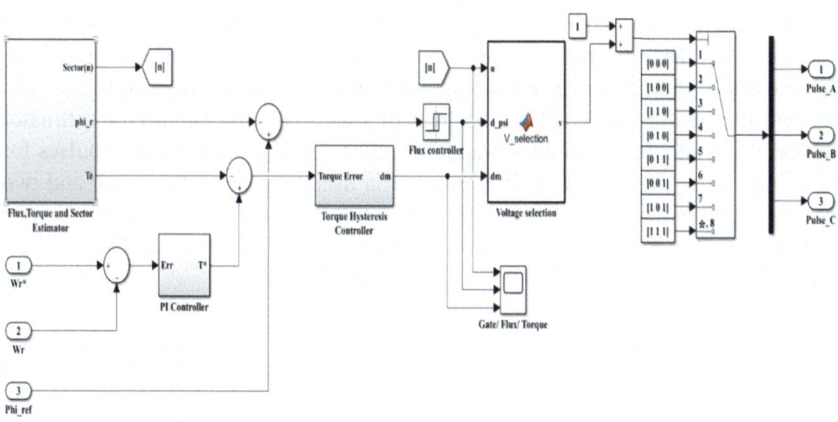

Fig. 7 Direct torque control subsystem

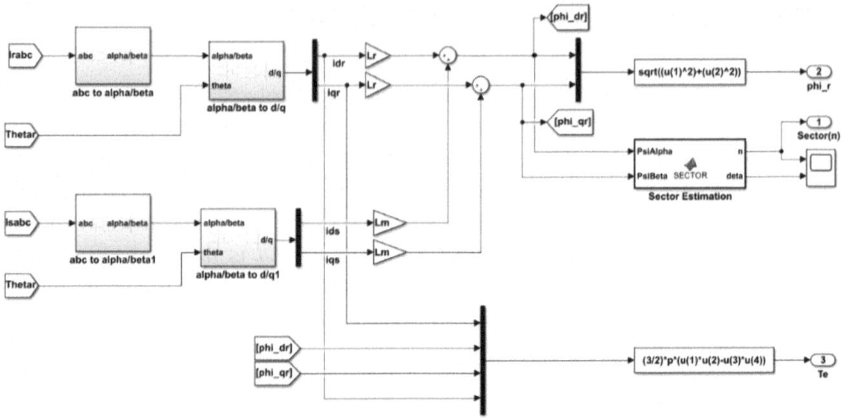

Fig. 8 Flux, torque and sector estimator block

Table 2 Simulation parameters

Parameters	Values
Frequency	50 Hz
Grid voltage	385 V
Resistance of rotor	0.38 Ω
Resistance of stator	0.48 Ω
Mutual inductance	0.0258 H
Rotor and stator leakage inductance	0.0156 H
Coefficient of friction	0.0114
Coefficient of inertia	0.1390
Number of poles	4

the rotor and the stator windings, and angle used for this transformation is taken from the rotor. Then Clarke and Park transformation [22, 23] is done to transform the currents from *abc* to α/β reference frame and α/β to d/q reference frame. The flux and torque are generated by the help of Eqs. (5)–(10). The sector estimator block consists of a predefined look-up table and helps in generating gate pulses for the voltage selector block. One PI controller is used to control the speed, and two hysteresis controllers are used for controlling the flux and electromagnetic torque of DFIG [24, 25]. Table 2 defines the simulation parameters used in DTC-SVM for DFIG.

6 Results Demonstration and Discussion

The simulation results of DTC-SVM for DFIG are presented in this section. The simulation model is run for a time period of $t = 18$ s. Figure 9 shows the waveform of stator voltages, and it is evident from the figure that the voltages are purely sinusoidal.

Figure 10 represents the gate pulses and the flux and torque controller's output. The gate pulses are generated from the section estimator block and are given to the voltage

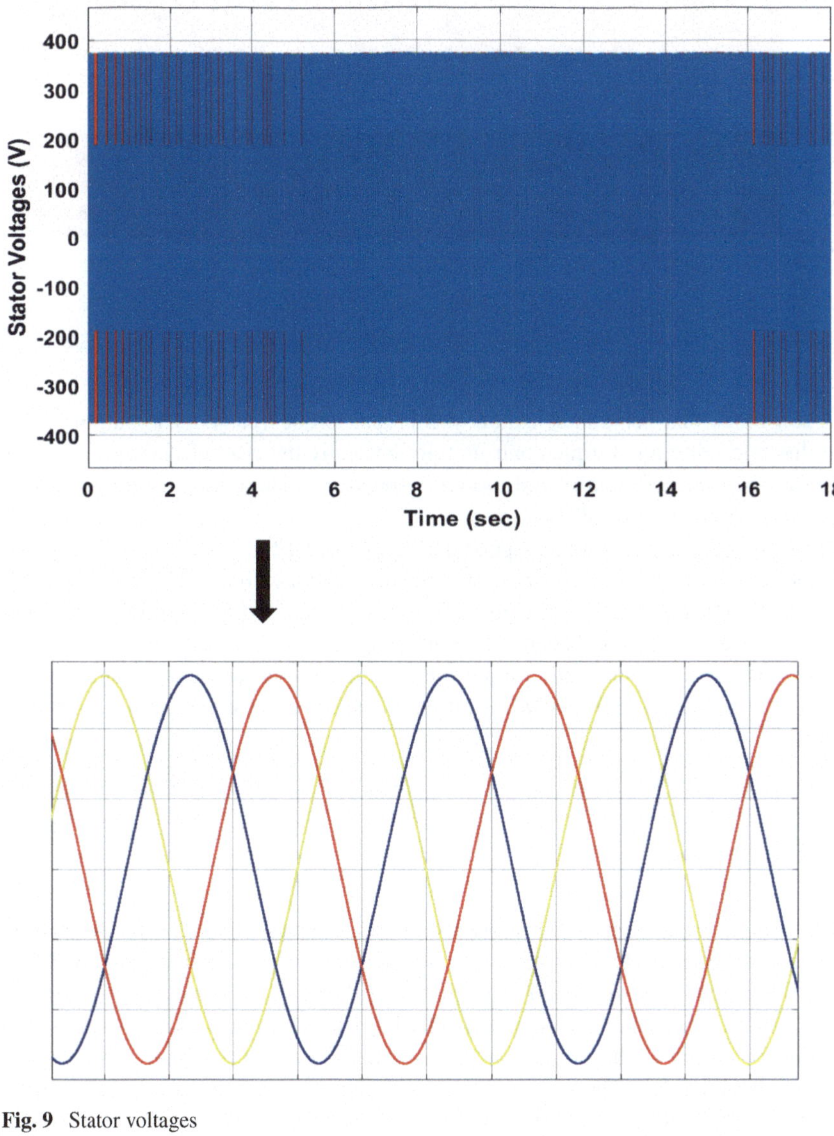

Fig. 9 Stator voltages

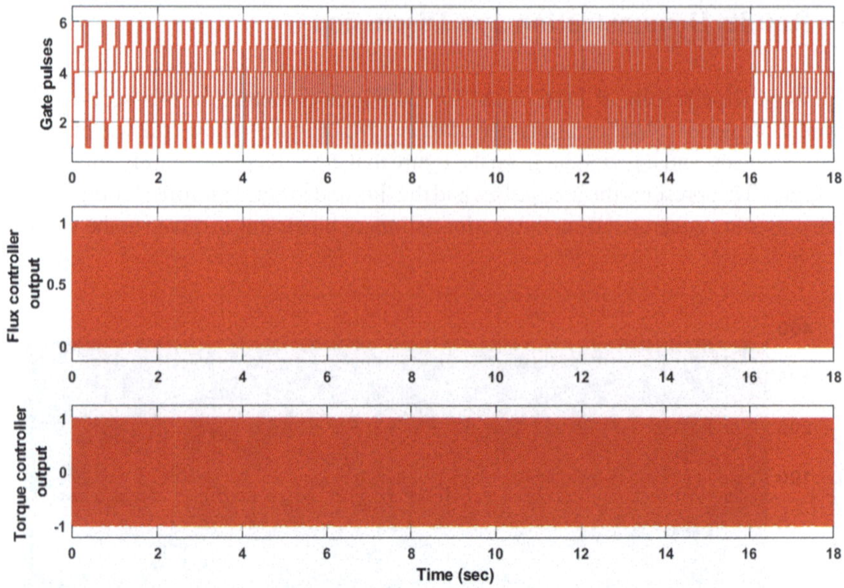

Fig. 10 Gate pulses, flux controller output and torque controller output

selector block that helps in selection of flux and torque vectors. The outputs from the flux controller and torque controller are basically the errors between the actual and desired value of the flux and torque, respectively. These outputs are generated from the hysteresis controllers.

Figures 11 and 12 depict the speed and torque of DFIG, respectively. In Fig. 11 for the reference speed, a stair-case waveform is taken who is periodic in the time interval $t = 0$ s to $t = 16$ s. It is evident from Fig. 11 that the speed of DFIG follows the reference signal in an acceptable manner with minute disturbances. Figure 12 shows that the electromagnetic torque follows the load torque quickly. Thus, it can be concluded that by DTC-SVM for DFIG, the speed and torque can be controlled directly.

7 Conclusion

In this paper, direct torque control-space vector modulation of DFIG is proposed. For analysis purposes, the DFIG's stator side is connected to the grid directly, and DFIG's rotor side is connected via power converters to grid. Direct torque control is modelled with the help of fundamental equations of DFIG. The selection of rotor voltage vector is done according to the outputs of the hysteresis controllers which have varying magnitude and angle. Then space vector modulation technique is implemented to generate voltage vectors for the rotor side with the help of a voltage source inverter.

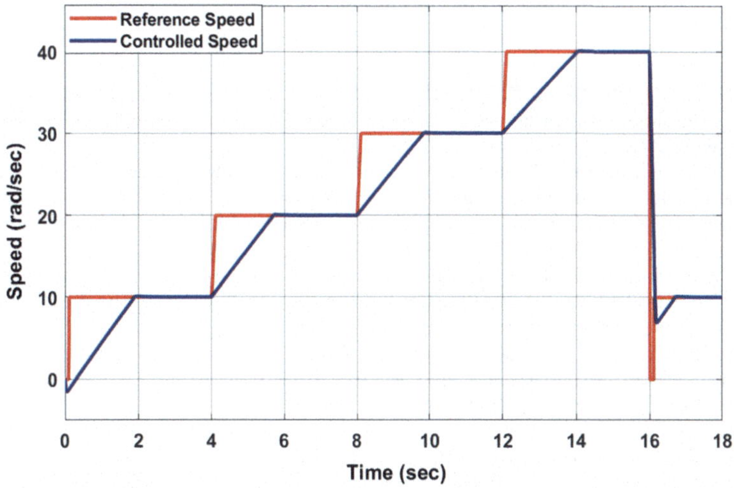

Fig. 11 Speed of DFIG

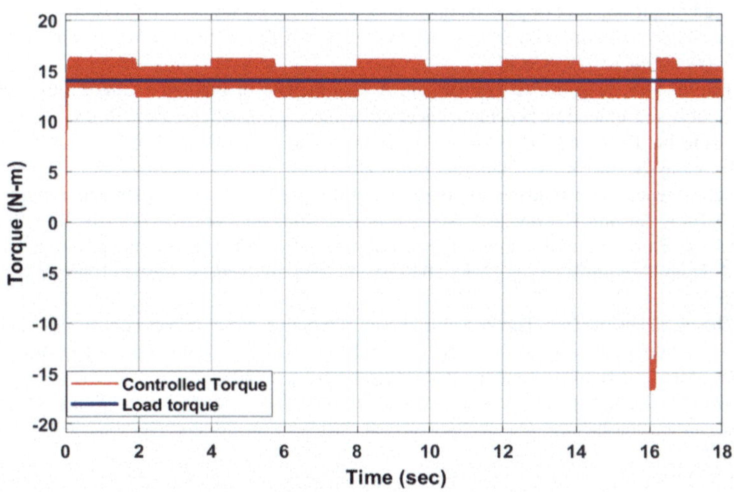

Fig. 12 Electromagnetic torque of DFIG

DTC is an effective technique to achieve direct control of speed and torque without any complicated techniques. With the fact that no PI controllers for flux and torque control and no speed sensors are used in this method, this method also has fast dynamic and steady-state performance. This model can also be used for different operating conditions and hence capable of working at different wind speeds making it suitable for wind range of wind power applications.

References

1. H. Malik, A. Iqbal, P. Joshi, S. Agrawal, F.I. Bakhsh, Meta heuristic and evolutionary computation: algorithms and applications, in *Studies in Computational Intelligence*, vol. 1096 (Springer, Singapore, 2020). https://doi.org/10.1007/978-981-15-7571-6
2. H. Sheikh, F.I. Bakhsh, Maximum power extraction and monitoring from wind power generation system using intelligent controllers, in *Soft Computing in Condition Monitoring and Diagnostics of Electrical and Mechanical Systems* (Springer, Singapore, 2020), pp. 345–362. https://doi.org/10.1007/978-981-15-1532-3_15
3. W. Ayrir, M. Ourahou, B. El Hassouni, A. Haddi, Direct torque control improvement of a variable speed DFIG based on a fuzzy inference system. Math. Comput. Simul. **167**, 308–324 (2020). https://doi.org/10.1016/j.matcom.2018.05.014
4. C.J. O'Rourk, M.M. Qasim, M.R. Overlin, J.L. Kirtley, A geometric interpretation of reference frames and transformations: dq0, Clarke, and Park. IEEE Trans. Energy Convers. **34**(4), 2070–2083 (2019). https://doi.org/10.1109/tec.2019.2941175
5. H.A. Wiam, Direct torque control-based power factor control of a DFIG. Energy Procedia **162**, 296–305 (2019). https://doi.org/10.1016/j.egypro.2019.04.031
6. Y. Sahri, S. Tamalouzt, S.L. Belaid, Direct torque control of DFIG driven by wind turbine system connected to the grid, in *International Conference on Wind Energy and Applications in Algeria (ICWEAA)* (2018), pp. 1–6. https://doi.org/10.1109/icweaa.2018.8605083
7. M.R.A. Kashkooli, S.M. Madani, R. Sadeghi, Improved direct torque control of DFIG with reduced torque and flux ripples at constant switching frequency, in *7th Power Electronics and Drive Systems Technologies Conference (PEDSTC)* (IEEE, 2016), pp. 58–63. https://doi.org/10.1109/pedstc.2016.7556838
8. A. Izanlo, M.V. Kazemi, A. Gholamian, A new method of predictive direct torque control for doubly fed induction generator under unbalanced grid voltage. Rev. Roum. Sci. Techn.-Electrotechn. Energ. **63**(3), 332–337 (2018). https://doi.org/10.1002/9781119172093.ch10
9. E. Tremblay, S. Atayde, A. Chandra, Comparative study of control strategies for the doubly fed induction generator in wind energy conversion systems: a DSP-based implementation approach. IEEE Trans. Sustain. Energy **2**(3), 288–299 (2011). https://doi.org/10.1109/tste.2011.2113381
10. X. Liu, X. Kong, Nonlinear model predictive control for DFIG-based wind power generation. IEEE Trans. Autom. Sci. Eng. **11**(4), 1046–1055 (2014). https://doi.org/10.1109/tase.2013.2284066
11. Y. Zhang, J. Hu, J. Zhu, Three-vectors-based predictive direct power control of the doubly fed induction generator for wind energy applications. IEEE Trans. Power Electron. **29**(7), 3485–3500 (2014). https://doi.org/10.1109/tpel.2013.2282405
12. T. Azar, Q. Zhu (eds.), *Advances and Applications in Sliding Mode Control Systems* (Springer International Publishing, Cham, Switzerland, 2015). https://doi.org/10.1007/978-3-319-11173-5
13. X. Liu, X. Su, X., P. Shi, C. Shen, Y. Peng, Event-triggered sliding mode control of nonlinear dynamic systems. Automatica **112**, 108738 (2020). https://doi.org/10.1016/j.automatica.2019.108738
14. C. Lascu, I. Boldea, F. Blaabjerg, Direct torque control of sensor less induction motor drives: a sliding-mode approach. IEEE Trans. Ind. Appl. **40**(2), 582–590 (2004). https://doi.org/10.1109/tia.2004.824441
15. C. Lascu, I. Boldea, F. Blaabjerg, A class of speed-sensor less sliding-mode observers for high-performance induction motor drives. IEEE Trans. Industr. Electron. **56**(9), 3394–3403 (2009). https://doi.org/10.1109/tie.2009.2022518
16. Y.T. Weng, Y.Y. Hsu, Sliding mode regulator for maximum power tracking and copper loss minimisation of a doubly fed induction generator. IET Renew. Power Gener. **9**(4), 297–305 (2015). https://doi.org/10.1049/iet-rpg.2014.0125
17. J.B. Alaya, A. Khedher, M.F. Mimouni, DTC, DPC and nonlinear vector control strategies applied to the DFIG operated at variable speed. J. Electr. Eng. JEE **6**(11), 744–753 (2011). https://doi.org/10.1109/ssd.2011.5767498

18. G. Abad, J. Lopez, M. Rodriguez, L. Marroyo, G. Iwanski, *Doubly Fed Induction Machine: Modelling and Control for Wind Energy Generation* (Wiley, Hoboken, 2011), p. 85. https://doi.org/10.1002/9781118104965.ch4
19. A. Benachour, E.M. Berkouk, M.O. Mahmoudi, DTC-SVM control of induction machine fed by three level NPC matrix converter, in *8th International Conference on Modelling, Identification and Control (ICMIC)* (2016), pp. 628–633. https://doi.org/10.1109/icmic.2016.7804188
20. Z. Yu, Space-vector PWM with TMS320C24x/F24x using hardware and software determined switching patterns. Texas Instruments Application Report SPRA524 (1999). https://doi.org/10.1049/ip-epa:19990417
21. G.H. Li, B.H. Zhang, Z.G. Hao, J. Wang, Z.Q. Bo, D. Writer, T. Yip, Modelling of DFIG based wind generator and transient characteristics analysis, in *10th International Conference on Environment and Electrical Engineering* (2011), pp. 1–4. https://doi.org/10.1109/eeeic.2011.5874588
22. S. Chattopadhyay, M. Mitra, S. Sengupta, Clarke and park transform, in *Electric Power Quality* (Springer, Dordrecht, 2011), pp. 89–96. https://doi.org/10.1007/978-94-007-0635-4_12
23. J.A. Alzubi, *AI and Machine Learning Paradigms for Health Monitoring System: Intelligent Data Analytics*, 1E, Studies in Big Data, vol. 86 (Springer Nature, Berlin, 2020), 513 p. ISBN: 978-981-33-4412-9. https://doi.org/10.1007/978-981-33-4412-9
24. A. Iqbal et al., *Renewable Power for Sustainable Growth*, 1E, Lecture Notes in Electrical Engineering, vol. 723 (Springer Nature, Berlin), 805 p. ISBN: 978-981-33-4080-0. https://doi.org/10.1007/978-981-33-4080-0
25. H. Malik et al., *Intelligent Data-Analytics for Condition Monitoring: Smart Grid Applications*, 1E (Academic Press, 2021). ISBN: 978-0-323-85510-5. https://doi.org/10.1016/C2020-0-02173-0

Modeling and Analysis of an Intelligent Approach for Load Frequency Control in a Deregulated Power System: A Case Study Based on Different Control Schemes

Nagendra Kumar, Akhilesh Singh, Jitendra Kumar, and Hasmat Malik

Abstract For a power system network to operate reliably, load frequency control (LFC) is required. LFC's main job is to manage the generator's output in response to changes in network frequency and tie-line power, such as restoring the scheduled frequency and power exchange with other locations within prescribed limits. This chapter carries out an analysis of the performances of model predictive control (MPC), proportional integral derivative (PID), linear quadratic Gaussian (LQG), and Fuzzy cascaded PID as load frequency control schemes for a two-area non-reheat thermal system. In both areas, the disco participation matrix (DPM) is taken into account for various power exchanges between Genco and Discos. To generate optimal gains for PID controllers, a well-known optimization approach called big bang big crunch (BBBC) is applied. To test the responsiveness of controllers, three scenarios were considered: load fluctuation in one region only, load disturbance in both areas, and additional demand requested by Disco beyond the contract. In terms of overshoot, undershoot, and settling time, a comparative analysis of the results has been produced. Apart from certain peaky oscillations in replies, it is clear that MPC outperforms alternative controllers in terms of settling time in all circumstances. The MPC controller's robustness is also superior to that of other controllers.

Keywords ACE · AGC · FLC · LFC · LQG · MPC · PID

N. Kumar (✉)
Department of Electrical and Electronics Engineering, G.L. Bajaj Institute of Technology & Management, Greater Noida 201306, India

A. Singh
Department of Electrical and Electronics Engineering, Seemant Institute of Technology Pithoragarh, Pithoragarh, Uttarakhand 262501, India

J. Kumar
Department of Electrical and Electronics Engineering, SRM Institute of Science and Technology, Delhi NCR, UP 201204, India

H. Malik
BEARS, University Town, NUS Campus, Singapore, Singapore

H. Malik et al. (eds.), *Intelligent Data Analytics for Power and Energy Systems*, Lecture Notes in Electrical Engineering 802,
https://doi.org/10.1007/978-981-16-6081-8_4

1 Introduction

Load frequency control (LFC) is used in power networks if there is an imbalance between generation and load demand occurs. The LFC job is to keep the scheduled frequency and tie-line power stable under normal conditions. Primary control and secondary control are two loops used by the LFC. Automatic generation control (AGC) is a mechanism in an electrical system that handles secondary frequency regulation. It keeps a predetermined level of generation load in a certain area and compensates for any frequency and power exchange mismatches induced by load variations [1]. The power system is a vast network, therefore, to have a better control it is divided into control areas. One control area is connected to another using tie-line to exchange the power and to support one another in case of emergency conditions. The structure of power system is now becoming complex, consequently, the growth of more specialized entities like Genco, Transco, and Disco, independent system operator (ISO), under this scenario has become conceivable. The estimation of LFC in a deregulated environment has been perused in [2]. In addition, a detailed review of the subject and the simulation of an LFC system after deregulation are presented. Generally, regulation of frequency in electrical grid is obtained by power balance between generation and demand through load pattern [3]. The Gencos regulate their output power as the load demand varies to keep power system at balance. If the Gencos are at their full capacity an additional load demand will fluctuate the frequency. It is seen from the literature survey that power system performance depends on control structure and fitness function [4]. Therefore, it is always welcome to propose and apply new high-performance optimization algorithms to real-world issues. In LFC, to maintain equilibrium, the controller plays an important role. Therefore, many methods have been used in literature, such as classical, adaptive, optimal, nonlinear, modern, etc. Additionally, control systems are also used to implement LFC controllers based on various soft computing methods such GA, PSO, HS, etc. [5–7]. A conventional system comprising thermal/hydro or mixture of both has been studied in the past. The controller like LQG, i.e., a regulator which uses state space model of the system to determine optimal feedback gain [8], I, PI, and PID [9] are popular among researchers due to their merits like simple construction, easy implementation, good performance, etc. The effectiveness of these control approaches has successfully been seen in many areas from power system to other system. However, it has been shown that conventional technique is definitely not preferred when the number of parameters to be optimized is high. Now days, power system is being changed from conventional to deregulated environment because of merits of deregulated ones. However, the stability of power system in this deregulated environment is a point of prime concern. This increases the role and responsibility of the control approaches are being used in power system [10, 11]. The changed environment requires other control schemes. Zadeh [12] presented the theory of fuzzy sets and Mamdani and Assilian [13] implemented the first fuzzy logic control algorithm on a steam engine. Fuzzy logic controller performance is based on the experience and knowledge of a human operator. Model predictive control (MPC) predicts future output of the system and produces optimal control. It also has found its application in frequency control of power systems composed of wind turbines and SMES [14, 15]. The various AI-based techniques have also been used in the literature for same and other applications [16–24]. Since different control approaches have their merits and limitations, therefore, in this study authors have carried out a comparison of performances of the approaches, i.e. LQG, PID, Fuzzy cascaded PID, and MPC on a two area non-reheat thermal power system in a deregulated environment.

2 Highlight and Graphical Abstract

The main contributions of this chapter are:

- Two area deregulated thermal power system has been taken for the case study.
- LQG, PID, Fuzzy cascaded PID, and MPC control schemes have been designed for an LFC.
- A state space and a simulation models are utilized to design control system.
- All control approaches have been tested for performance and efficacy in a variety of disturbance conditions, including step load changes and parameter adjustments.

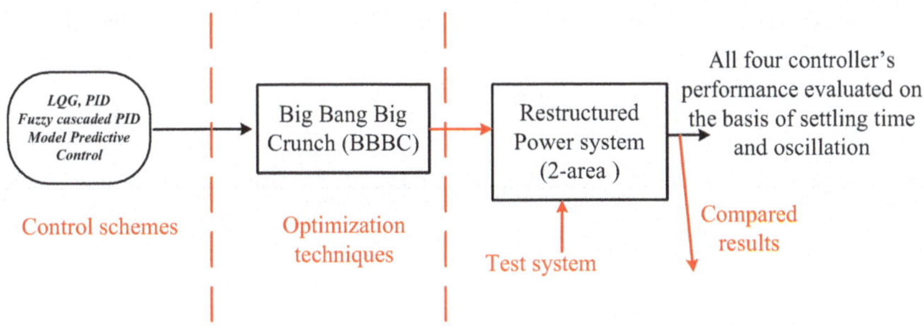

There are five sections to this chapter. The modeling of two-area power systems is introduced in Sect. 3. Section 4 explains all of the control methods. In Sects. 5 and 6, the findings achieved utilizing all control systems are reported and contrasted. Finally, in Sect. 7, the complete analysis has been performed.

3 System for LFC

Whenever a load imbalance occurs, the speed of prime mover changes to compensate that effect. The frequency also gets affected from that load imbalance that change in speed is controlled by the governor of generators. Once the speed change eliminates, the governor speed is made constant. Due to a varying nature of load governor speed is also not fix to a set point, therefore, secondary control mechanism with integrator is used to remove offset and bring the system to its normal value called reset point. A two equal area thermal power system has been considered as a test system to check the performance of the designed controllers. Both areas have 2 Gencos and 2 Discos each. The scheduled tie-line power of 0 and −0.05 (area-2 to area-1) pu has also been taken to simulate all three cases. As it is known that with the introduction of deregulation, conventional power system is being through many reforms. The task of frequency control becomes more important in this scenario. Nowadays, a Disco is free to have power contract from Genco of its area or from another area, and this way of power transaction is known as bilateral trading which is implemented by disco participation matrix (DPM). The block diagram representation of ith area power system is shown in Fig. 1. It can be seen that apart from Genco and Disco, several bilateral power signals are also play role in deregulated power system. These signals have been implemented using DPM which shows the relation between Genco and Disco. Now the controller has to take care the generator's output, various power transaction signals as per the load requirement. Any unbalance between power generation and load gives rise to area control error (ACE). A controller is used to eliminate area control error given in (1).

$$\mathrm{ACE}_i = B_i \Delta F_i + \Delta P_{\mathrm{tie}_i} \tag{1}$$

In deregulated environment, different transactions like Poolco, bilateral, and combination of both are being used for power transactions. Bilateral transactions have an effect on tie-line power and are modify as (2).

$$\Delta P_{\mathrm{tie}_{i\text{-new}}} = \Delta P_{\mathrm{tie}_i} + \sum_{\substack{j=1 \\ i \neq j}}^{n} D_{ij} - \sum_{\substack{j=1 \\ i \neq j}}^{n} D_{ji} \tag{2}$$

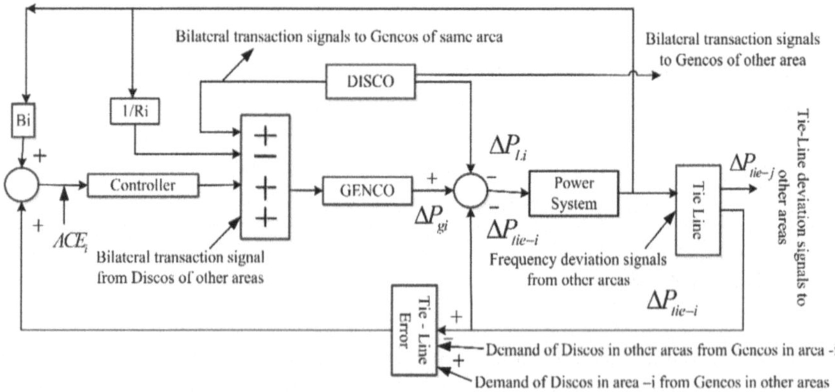

Fig. 1 ith area deregulated power system

where n = areas, D_{ij} = area-j demand to the area-i generation, D_{ji} = area-i demand to the area-j generation, ΔP_{tie_i} = tie-line variation. The existing ACE is modified by this modification in tie-line flow (3),

$$\text{ACE}_i = B_i \Delta F_i + \Delta P_{\text{tie}_{i\text{-new}}} \tag{3}$$

As per the load deviation, the total generation of Gencos can be calculated using (4),

$$\Delta P_{g_i} = \sum_j \text{cpf}_{i_j} \Delta P_{D_j} \tag{4}$$

where P_D is the total load demand.

4 Control Approaches

Many approaches have been used in past, however, in this work four controllers, namely LQG, PID, Fuzzy cascaded PID, and MPC have been designed and tested for the system given in Fig. 1.

4.1 LQG Control Scheme

Designing of LQG regulator requires the state space model. The system given in Fig. 1 has been modeled in state space by choosing following state variables:

$$x = \begin{pmatrix} \Delta w_1, \Delta w_2, \Delta P_{\text{GV}_1}, \Delta P_{\text{GV}_2}, \Delta P_{\text{GV}_3}, \Delta P_{M_1}, \Delta P_{M_2} \\ \Delta P_{M_3}, \Delta P_{M_4}, \int \text{ACE}_1, \int \text{ACE}_2, \Delta P_{\text{tie}1-2} \end{pmatrix}$$

Power demands vector: $P_L = \begin{bmatrix} \Delta P_{L_1} & \Delta P_{L_2} & \Delta P_{L_3} & \Delta P_{L_4} \end{bmatrix}^{\text{T}}$. Control inputs: $u = \begin{bmatrix} u_1 & u_2 & u_3 & u_4 \end{bmatrix}^{\text{T}}$. Uncontracted demands vector: $P_{\text{uc}} = \begin{bmatrix} \Delta P_{\text{uc}_1} & \Delta P_{\text{uc}_2} & \Delta P_{\text{uc}_3} & \Delta P_{\text{uc}_4} \end{bmatrix}^{\text{T}}$.

The state space description of the system shown in Fig. 1 is

$$\dot{x} = Ax + Bu + FP_L + \Gamma P_{\text{UC}} \tag{5}$$

where A is the parameter matrix, B is the control, F and Γ is the disturbances matrices. The structure of all these matrices is shown above.

To design a controller, eigenvalues of the closed loop system can be calculated from (6),

$$\dot{x} = (A - BK)x \tag{6}$$

These Eigen values can be placed at desired locations on satisfying the objective given in (7).

$$J = \frac{1}{2}\int x^{\mathrm{T}}Qx + u^{\mathrm{T}}Ru \tag{7}$$

where Q is a positive semidefinite and R is a positive definite matrix.

The optimal control minimizes (7) is given by

$$u = -Kx, \quad \text{where } K = R^{-1}B^{\mathrm{T}}P \tag{8}$$

P is calculated from Ricatti equation

$$A^{\mathrm{T}}P + PA - PBR^{-1}B^{\mathrm{T}}P + Q = 0 \tag{9}$$

Once this feedback matrix is known, the control law will be obtained to use as an input to the system to take care of the deviations in load and to keep load frequency in the limit.

$$F = \begin{bmatrix} -\frac{K_{P_1}}{T_{P_1}} & -\frac{K_{P_1}}{T_{P_1}} & 0 & 0 \\ 0 & 0 & -\frac{K_{P_2}}{T_{P_2}} & -\frac{K_{P_2}}{T_{P_2}} \\ 0 & 0 & 0 & 0 \\ 0 & 0 & 0 & 0 \\ 0 & 0 & 0 & 0 \\ 0 & 0 & 0 & 0 \\ \frac{\mathrm{cpf}_{11}}{T_{G_1}} & \frac{\mathrm{cpf}_{12}}{T_{G_1}} & \frac{\mathrm{cpf}_{13}}{T_{G_1}} & \frac{\mathrm{cpf}_{14}}{T_{G_1}} \\ \frac{\mathrm{cpf}_{21}}{T_{G_2}} & \frac{\mathrm{cpf}_{22}}{T_{G_2}} & \frac{\mathrm{cpf}_{23}}{T_{G_2}} & \frac{\mathrm{cpf}_{24}}{T_{G_2}} \\ \frac{\mathrm{cpf}_{31}}{T_{G_3}} & \frac{\mathrm{cpf}_{32}}{T_{G_3}} & \frac{\mathrm{cpf}_{33}}{T_{G_3}} & \frac{\mathrm{cpf}_{34}}{T_{G_3}} \\ \frac{\mathrm{cpf}_{41}}{T_{G_4}} & \frac{\mathrm{cpf}_{42}}{T_{G_4}} & \frac{\mathrm{cpf}_{43}}{T_{G_4}} & \frac{\mathrm{cpf}_{44}}{T_{G_4}} \\ \mathrm{cpf}_{31} + \mathrm{cpf}_{41} & \mathrm{cpf}_{32} + \mathrm{cpf}_{42} & -(\mathrm{cpf}_{13} + \mathrm{cpf}_{23}) & -(\mathrm{cpf}_{14} + \mathrm{cpf}_{24}) \\ -(\mathrm{cpf}_{31} + \mathrm{cpf}_{41}) & -(\mathrm{cpf}_{32} + \mathrm{cpf}_{42}) & \mathrm{cpf}_{13} + \mathrm{cpf}_{23} & \mathrm{cpf}_{14} + \mathrm{cpf}_{24} \\ 0 & 0 & 0 & 0 \end{bmatrix}$$

$$B = \begin{bmatrix} 0\,0\,0\,0\,0\,0 & \frac{\mathrm{apf}_1}{T_{G_1}} & 0 & 0 & 0 & 0\,0\,0 \\ 0\,0\,0\,0\,0\,0 & 0 & \frac{\mathrm{apf}_2}{T_{G_2}} & 0 & 0 & 0\,0\,0 \\ 0\,0\,0\,0\,0\,0 & 0 & 0 & \frac{\mathrm{apf}_3}{T_{G_3}} & 0 & 0\,0\,0 \\ 0\,0\,0\,0\,0\,0 & 0 & 0 & 0 & \frac{\mathrm{apf}_4}{T_{G_4}} & 0\,0\,0 \end{bmatrix}^{\mathrm{T}}$$

$$\Gamma = \begin{bmatrix} -\frac{K_{P_1}}{T_{P_1}} & 0 & 0 & 0 & 0 & 0 & 0 & 0 & 0 & 0 & 0 & 0 & 0 \\ \frac{K_{P_2}}{T_{P_2}} & 0 & 0 & 0 & 0 & 0 & 0 & 0 & 0 & 0 & 0 & 0 & 0 \\ 0 & -\frac{K_{P_3}}{T_{P_3}} & 0 & 0 & 0 & 0 & 0 & 0 & 0 & 0 & 0 & 0 & 0 \\ 0 & \frac{K_{P_4}}{T_{P_4}} & 0 & 0 & 0 & 0 & 0 & 0 & 0 & 0 & 0 & 0 & 0 \end{bmatrix}^{\mathrm{T}}$$

$$A = \begin{bmatrix} -\frac{1}{T_{P_1}} & 0 & \frac{K_{P_1}}{T_{P_1}} & \frac{K_{P_1}}{T_{P_1}} & 0 & 0 & 0 & 0 & 0 & 0 & 0 & 0 & -\frac{K_{P_1}}{T_{P_1}} \\ 0 & -\frac{1}{T_{P_2}} & 0 & \frac{K_{P_2}}{T_{P_2}} & \frac{K_{P_2}}{T_{P_2}} & 0 & 0 & 0 & 0 & 0 & 0 & 0 & -a_{12}\frac{K_{P_2}}{T_{P_2}} \\ 0 & 0 & -\frac{1}{T_{T_1}} & 0 & 0 & \frac{1}{T_{T_1}} & 0 & 0 & 0 & 0 & 0 & 0 & 0 \\ 0 & 0 & 0 & -\frac{1}{T_{T_2}} & 0 & 0 & \frac{1}{T_{T_2}} & 0 & 0 & 0 & 0 & 0 & 0 \\ 0 & 0 & 0 & 0 & -\frac{1}{T_{T_3}} & 0 & 0 & \frac{1}{T_{T_3}} & 0 & 0 & 0 & 0 & 0 \\ 0 & 0 & 0 & 0 & 0 & -\frac{1}{T_{T_4}} & 0 & 0 & \frac{1}{T_{T_4}} & 0 & 0 & 0 & 0 \\ -\frac{1}{2\pi R_1 T_{G_1}} & 0 & 0 & 0 & 0 & 0 & -\frac{1}{T_{G_1}} & 0 & 0 & 0 & 0 & 0 & 0 \\ -\frac{1}{2\pi R_2 T_{G_2}} & 0 & 0 & 0 & 0 & 0 & 0 & -\frac{1}{T_{G_2}} & 0 & 0 & 0 & 0 & 0 \\ 0 & -\frac{1}{2\pi R_3 T_{G_3}} & 0 & 0 & 0 & 0 & 0 & 0 & -\frac{1}{T_{G_3}} & 0 & 0 & 0 & 0 \\ 0 & -\frac{1}{2\pi R_4 T_{G_4}} & 0 & 0 & 0 & 0 & 0 & 0 & 0 & -\frac{1}{T_{G_4}} & 0 & 0 & 0 \\ \frac{B_1}{2\pi} & 0 & 0 & 0 & 0 & 0 & 0 & 0 & 0 & 0 & 0 & 0 & 1 \\ 0 & \frac{B_2}{2\pi} & 0 & 0 & 0 & 0 & 0 & 0 & 0 & 0 & 0 & 0 & -1 \\ \frac{T_{12}}{2\pi} & -\frac{T_{12}}{2\pi} & 0 & 0 & 0 & 0 & 0 & 0 & 0 & 0 & 0 & 0 & 0 \end{bmatrix}$$

4.2 PID Control Scheme

PID is a simple and straightforward control technique that is widely employed in a variety of applications. The PID controller can be written as,

$$G_{\mathrm{PID}}(s) = K_P + K_I/s + K_D s \tag{10}$$

where K_P, K_I, and K_D = PID parameters for effective performance of PID, its parameters should be obtained optimally, therefore, in this work BBBC algorithm is used to decide parameters by minimizing fitness function, given in (11).

$$F = \frac{1}{m}\sum_{i=1}^{m}[(\mathrm{ACE}_i)^2] = \frac{1}{m}\sum_{i=1}^{m}[(B_i\Delta f_i + \Delta P_{\mathrm{tie}_{i\text{-error}}})^2] \tag{11}$$

Followings are the steps of BBBC algorithm:

Step 1 The population is generated for controller,

$$x_{ij}^{(k)} = x_{i(\min)}^{(k)} + \mathrm{rand.}\left(x_{i(\max)}^{(k)} - x_{i(\min)}^{(k)}\right) \tag{12}$$

Step 2 The fitness function (11) is tested for each population.
Step 3 Center of mass is computed at this step, as given in (13).

$$X_{\text{com}} = \frac{\sum_{j=1}^{p} \frac{x_{ij}^{(k)}}{F_j}}{\sum_{j=1}^{p} \frac{1}{F_j}} \tag{13}$$

Step 4 A new population around X_{com} is generated at this step.

$$x_{ij\,(\text{new})}^{k} = X_{\text{com}} + \frac{r.\alpha\left(x_{i(\max)}^{k} - x_{i(\min)}^{k}\right)}{K} \tag{14}$$

Step 5 Next parameters have been generated at this step.

$$x_{ij\,(\text{next})}^{k} = \min\left\{F\left(x_{ij\,(\text{previous})}^{k}\right), F\left(x_{ij\,(\text{new})}^{k}\right)\right\} \tag{15}$$

Step 6 It is the step when the algorithm calculates the error and stops if it is less than the specified value.

A 30-population, 3-variables, and 10-parameter limiting size have been used as BBBC parameters to obtain optimal PID control scheme. The optimal parameters of the designed parameters are KP = −11.04, KI = −2.558, and KD = −11.878.

4.3 *Fuzzy Logic Cascaded PID Control Approach*

Fuzzy logic controller is centered on fuzzy logic and offers a procedure which converts the expert knowledge-based linguistic strategy into an automatic control strategy. Figure 2 shows the various components of fuzzy control. In this chapter, seven functions of triangular type membership are taken for each input, which is shown in Fig. 3. In Fig. 4, the surface diagram of both the input and output is given.

Fig. 2 Fuzzy scheme

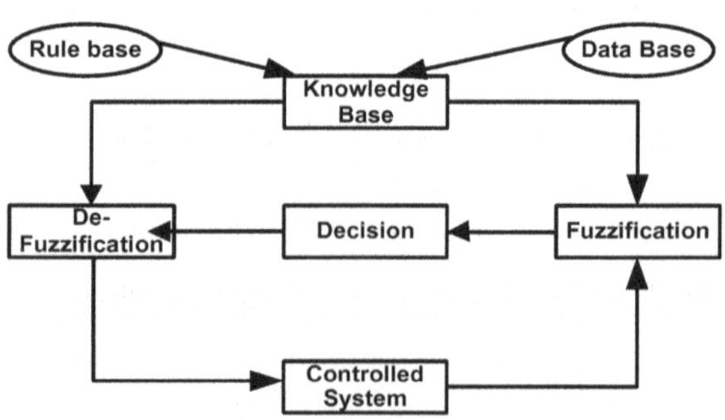

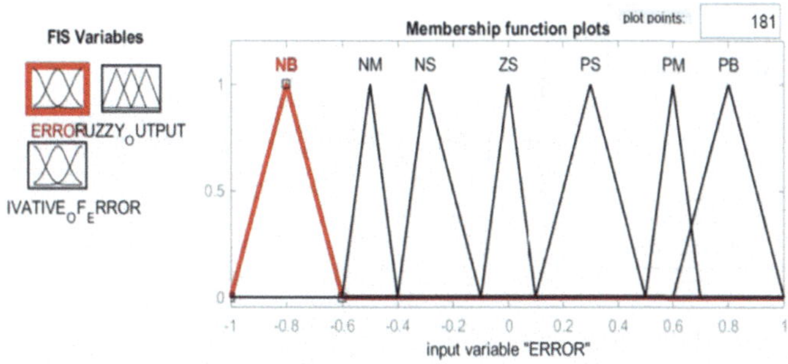

Fig. 3 Membership function for load frequency control

Fig. 4 Surface diagram of input/output

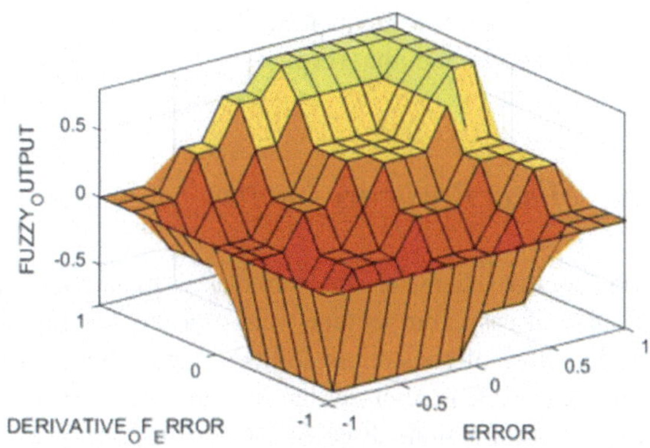

A total of 49 fuzzy rules were adopted for the input–output relationship based on the seven membership function, where NB is (−ve large), NM is (−ve medium), NS is (−ve small), Z is (Zero), PS is (+ small), PM is (+ medium), PB is (+ Big), as given in Table 1.

Table 1 Rule base for LFC

Area control error	Derivative of area control error						
	NB	NM	NS	Z	PS	PM	PB
NB	NB	NB	NM	NB	ZS	NS	NM
NM	NB	NB	NB	NM	PM	ZS	NS
NS	NB	NB	NM	NS	PM	PB	ZS
Z	NB	NM	NS	ZS	PS	PM	PB
PS	NM	NS	ZS	PB	PS	PS	PM
PM	NS	ZS	PS	PM	PS	PS	PS
PB	ZS	PM	PM	PS	PS	PS	PS

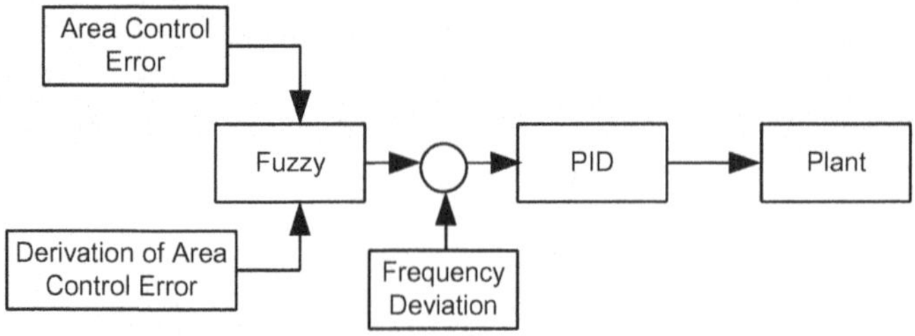

Fig. 5 Block diagram of Fuzzy cascaded PID

In this study, ACE and derivation of area control error (dACE) were used as inputs to a fuzzy cascaded PID control method, and the output of this has been compared with the deviation in frequency. The error works as an input to the PID control to obtain the optimal output (Fig. 5).

4.4 Model Predictive Control

MPC control scheme has been used as a satisfactory control application in industries because of easy computation, real-time implementations, incorporation of delays, and handling of constraints abilities. Figure 6 shows the structure of MPC used in this study. MPC predicts system's future output on the basis of past-present I/O and future control actions. Free and forced responses are two main components of the forecast and are required for the overall forecast. Further, the optimizer calculates the best future control on minimizing a cost function given in (16) subjected to constraints given in (17), means MPC minimizes the error in prediction with minimum control.

The objective function that is optimized by MPC is given as:

Fig. 6 Simple structure of MPC scheme

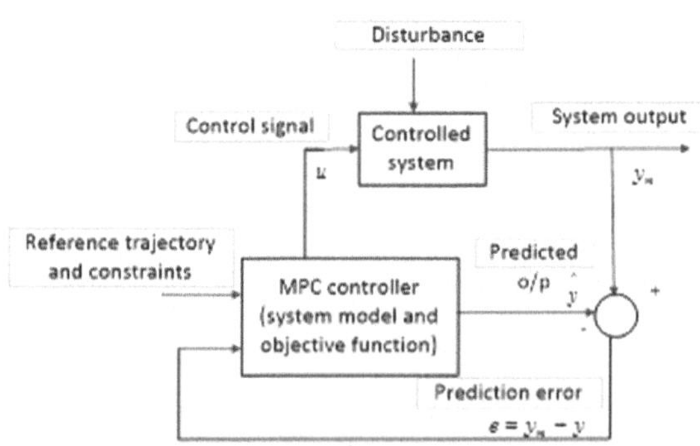

$$J(L_1, L_2, N_u) = \sum_{j=L_1}^{l_2} \alpha(j)[y(k+j|k) - w(k+j)]^2 + \sum_{j=1}^{N_u} \beta(j)[k+j| - 1]^2 \tag{16}$$

where $L_1 - L_2$ = lower/upper prediction horizons, N_u = control horizon, $\alpha(j) - \beta(j)$ = weighting factors, $w(k+j)$ = trajectory over the future horizon, $\Delta u(k+j) = 0$ for $l \geq N_u$ = relation used to give the number of future control. The control and output constraints are given in (17).

$$\begin{aligned} u_{\min} &\leq \Delta u(k) \leq u_{\max} \\ y_{\min} &\leq \Delta y(k) \leq y_{\max} \end{aligned} \tag{17}$$

5 Results Demonstration and Discussion

Based on Fig. 1, the above approaches were designed for a two-area LFC scheme. Both the governor-turbine units and the regions are identical. In each location, two Gencos and two Discos are considered. The parameters for the two-area LFC scheme were derived from [15, 16].

5.1 Case 1: Load Demand Increases in Area-1

The load in only area-1 has been considered in this scenario, with a step perturbation of 0.2 pu. The DPM depicts the numerous bilateral agreements between Gencos and Discos.

$$\text{DPM} = \begin{bmatrix} 0.5 & 0.5 & 0 & 0 \\ 0.5 & 0.5 & 0 & 0 \\ 0 & 0 & 0 & 0 \\ 0 & 0 & 0 & 0 \end{bmatrix}$$

To keep up with the changing loads, Gencos modify their generation as follows:

$$(\Delta P_{G_i})_{b_i} = \sum_j \text{cpf}_{ij} \Delta P_{L_j} \quad i = 1, 2, 3, 4 \tag{18}$$

Based on (18), total generation change in area-1 and area-2 is,

$$(\Delta P_{G_1}) = 0.5 \times 0.1 + 0.5 \times 0.1 + 0 \times 0.1 + 0 \times 0.1 = 0.1$$

$$\left(\Delta P_{G_2}\right) = 0.5 \times 0.1 + 0.5 \times 0.1 + 0 \times 0.1 + 0 \times 0.1 = 0.1$$
$$\left(\Delta P_{G_3}\right) = 0 \text{ pu}, \left(\Delta P_{G_4}\right) = 0 \text{ pu}.$$

The total tie-line power can be determined using (19)

$$\Delta P_{\text{tie}1-2,\text{schd}} = \sum_{i=1}^{2}\sum_{j=3}^{4} \text{cpf}_{ij}\Delta P_{L_j} - \sum_{i=3}^{4}\sum_{j=1}^{2} \text{cpf}_{ij}\Delta P_{L_j} = 0 \text{ pu} \tag{19}$$

As shown in Fig. 7, the frequency of both areas on a load disturbance diverges from their rated values. LFC initiates a generation change to restore frequency to its scheduled value. With a PID controller, it is clear that the frequency disparities between area-1 and area 2 experience fewer oscillations. The MPC control technique, on the other hand, quickly eliminates both areas of deviation. Figures 8, 9, and 10 show the change in tie-line flow as well as the change in generation. It can be seen that the power change of Gencos and variation in tie-line flow immediately returned to their desired values.

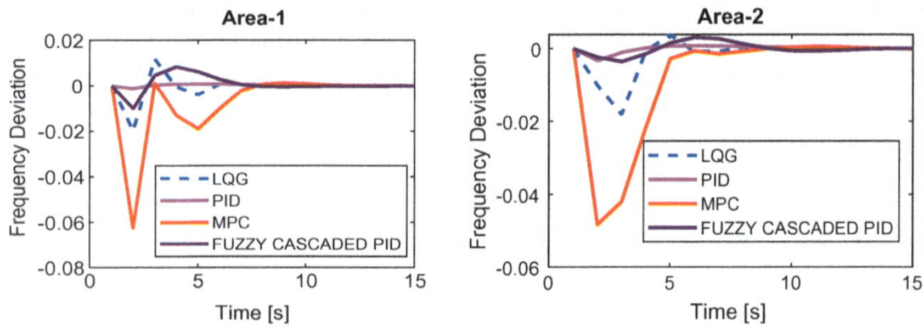

Fig. 7 Frequency deviation

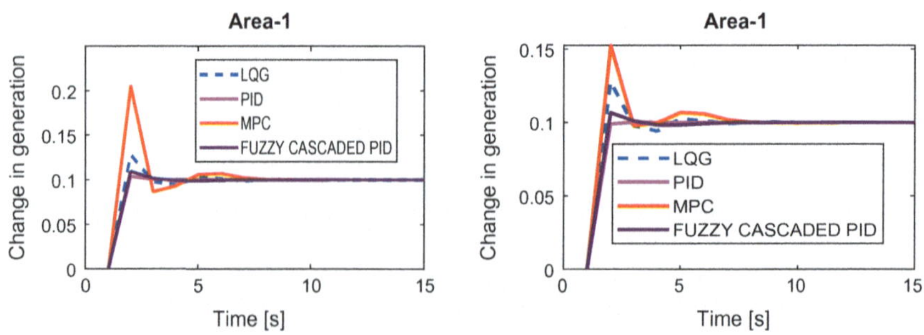

Fig. 8 Generation change

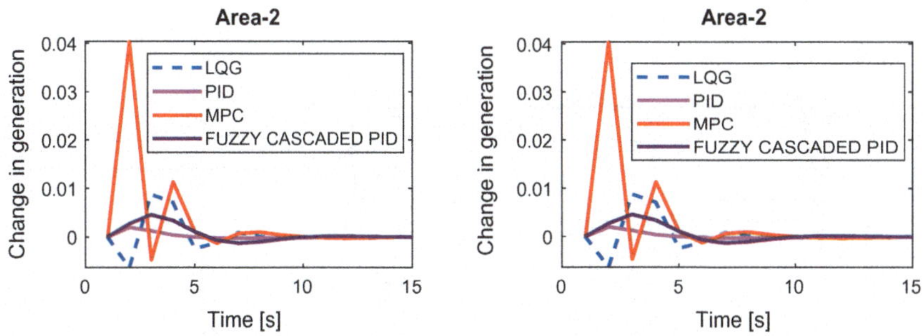

Fig. 9 Generation change

Fig. 10 Tie-line exchange

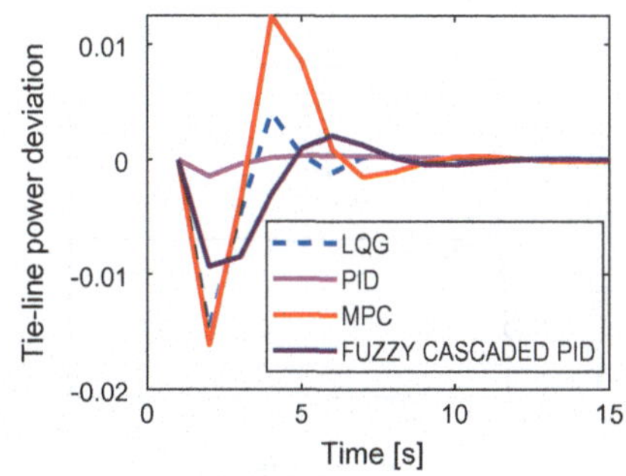

5.2 *Case 2: Load Demand Increases in Both Areas*

In this scenario, a 0.2 pu step load has been considered. The DPM depicts the numerous bilateral agreements between Gencos and Discos.

$$\text{DPM} = \begin{bmatrix} 0.5 & 0.25 & 0 & 0.3 \\ 0.2 & 0.25 & 0 & 0 \\ 0 & 0.25 & 1 & 0.7 \\ 0.3 & 0.25 & 0 & 0 \end{bmatrix}$$

Based on (18), this change in load will be taken by Gencos of both areas, respectively.

$$\begin{aligned}
\Delta P_{G(\text{area-1})} &= 0.5(0.1) + 0.25(0.1) + 0 + 0.3(0.1) = 0.105 \\
\Delta P_{G(\text{area-1})} &= 0.045 \\
\Delta P_{G(\text{area-2})} &= 0 \times 0.1 + 0.25 \times 0.1 + 1 \times 0.1 + 0.7 \times 0.1 = 0.195 \\
\Delta P_{G(\text{area-2})} &= 0.3 \times 0.1 + 0.25 \times 0.1 + 0 \times 0.1 + 0 \times 0.1 = 0.055
\end{aligned}$$

Similarly using (19), the tie-line power flow can be determined.

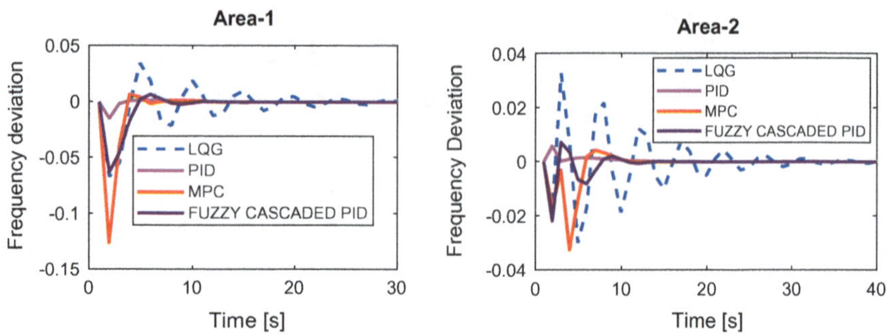

Fig. 11 Frequency deviation

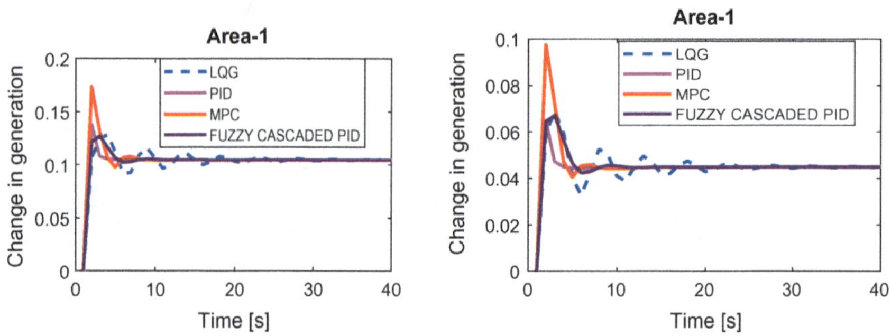

Fig. 12 Generation change

$$\Delta P_{\text{tie}1-2,\text{schd}} = \sum_{i=1}^{2}\sum_{j=3}^{4} \text{cpf}_{ij}\Delta P_{L_j} - \sum_{i=3}^{4}\sum_{j=1}^{2} \text{cpf}_{ij}\Delta P_{L_j} = -0.05 \text{ pu}$$

On a load perturbation, the frequencies of both areas deviate and settled down at desired values, as seen in Fig. 11. LFC initiates a generation change to restore frequency to its scheduled value, as seen in Figs. 12 and 13. As seen in Fig. 14, the tie-line power deviation settles at steady state. The frequency disparities between area-1 and area-2 are easily eliminated when utilizing the MPC controller.

5.3 Case 3: Violation of Load Demand in Area-1

In this case, Disco1 of area-1 violates the contract by drawing excess power of 0.1 pu excess power. For power transactions, the identical DPM that was used in case 2 was used. As a result, Gencos in Area1 will oversupply in order to meet this increased demand, with Genco 1 (G_1) and Genco 2 (G_2) in area 1 providing 75 and 25%, respectively, of the additional demand. A demand of 0.3 pu in area-1 and 0.2 pu in

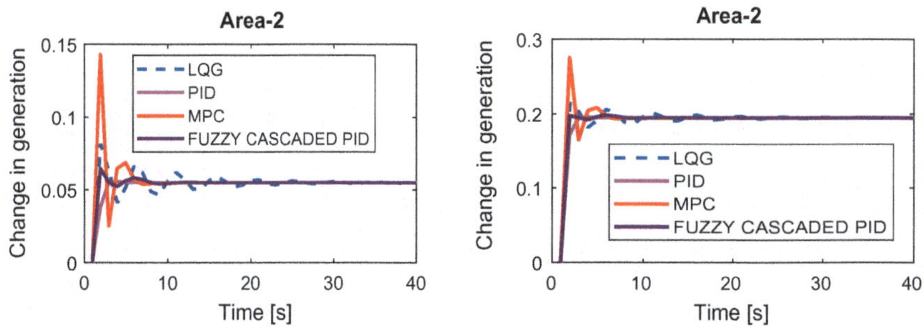

Fig. 13 Generation change

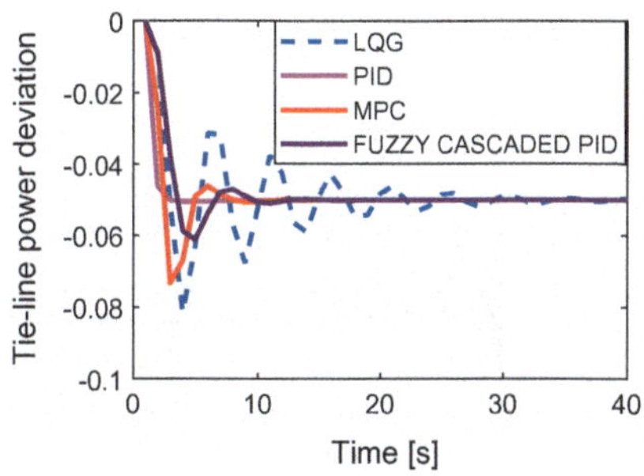

Fig. 14 Tie-line exchange

area-2 was used to replicate this scenario. Figure 15 depicts the frequency deviations of the comparable locations. The Gencos of area-1 are the only ones who follow the excess load of 0.1 pu. G_1 and G_2 of area-1 enhance their generation to 0.18 and 0.07 pu, as compared to case 2. Area-2's Gencos similarly changed their powers and landed on the same values as in example 2. Figures 16 and 17 depict the increase of Gencos' power. Figure 18 shows the tie-line flow exchange.

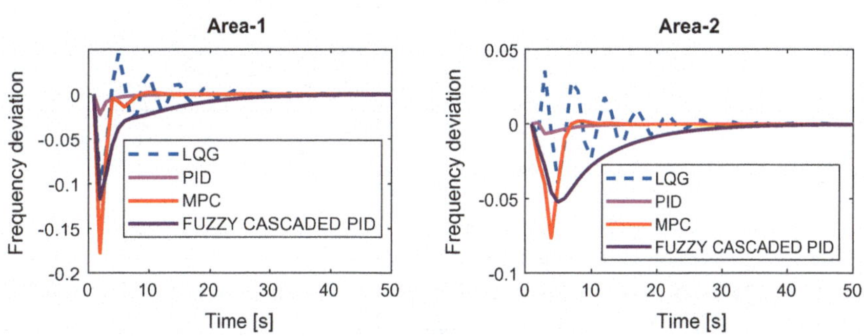

Fig. 15 Frequency deviation

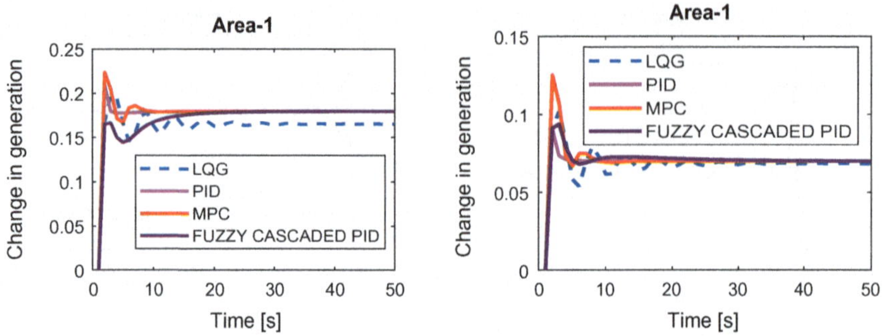

Fig. 16 Change in generation

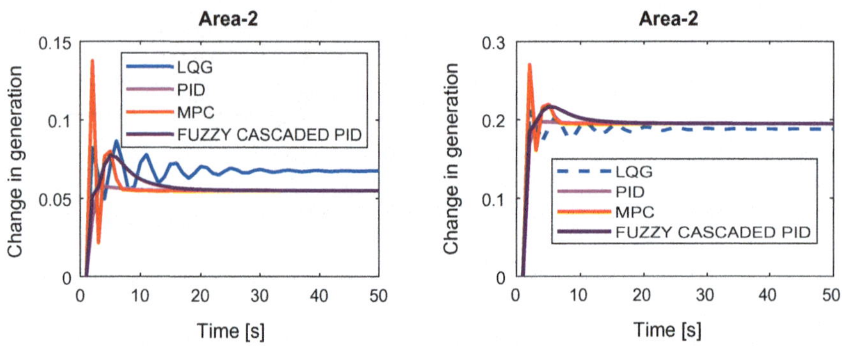

Fig. 17 Change in generation

Fig. 18 Tie-line exchange

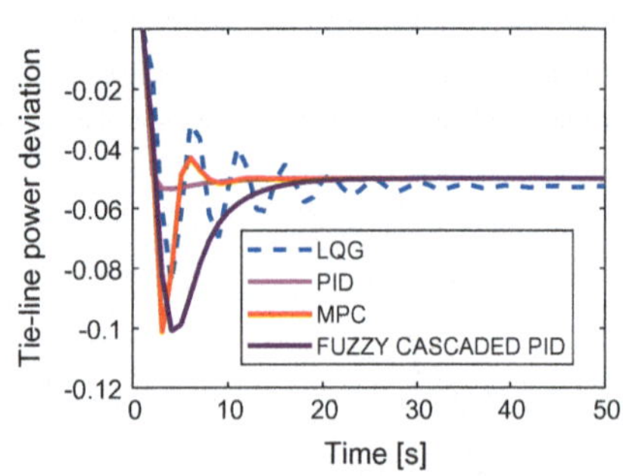

Table 2 gives analysis overshoot, undershoot, and settling time of frequency and tie-line exchange. It is evident that apart from some instances of peaky shoots, MPC control approach is performing better than other control approaches and settling down the deviation and generation at their desired values in smaller settling time than others.

Table 2 Comparative analysis of all control schemes (normal case)

Controller	Signal	Overshoot			Undershoot			Settling time		
		Case 1	Case 2	Case 3	Case 1	Case 2	Case 3	Case 1	Case 2	Case 3
LQG	Δf_1	0.012	0.0343	0.046	−0.02	−0.067	−0.098	11	40	45
	Δf_2	0.0034	0.0328	0.038	−0.0181	−0.0298	−0.0398	14	40	45
	ΔP_{tie}	0.004	0.0	0.0	−0.0148	−0.0812	−0.085	13	40	50
PID	Δf_1	0.0	0.0	0.0	−0.001	−0.015	−0.022	13	24	25
	Δf_2	0.0	0.006	0.0	−0.0033	0.0	−0.0065	15	20	25
	ΔP_{tie}	0.0	0.0	0.0	−0.0015	−0.052	−0.052	15	19	25
MPC	Δf_1	0.0	0.00668	0.0022	−0.063	−0.125	−0.1772	13	18	18
	Δf_2	0.0	0.0044	0.0021	−0.0485	−0.033	−0.0798	14	18	18
	ΔP_{tie}	0.0125	0.0	0.0	−0.0165	−0.073	−0.1012	14	15	20
Fuzzy cascaded PID	Δf_1	0.0084	0.00665	0.0	−0.01	−0.063	−0.115	14	20	30
	Δf_2	0.0031	0.0072	0.0	−0.0035	−0.0218	−0.053	15	20	30
	ΔP_{tie}	0.0021	0.0	0.0	−0.0095	−0.06	−0.101	15	18	30

6 Parameters Variations

There is a possibility that the parameter of power system may change due to abnormal conditions. Therefore, the designed control scheme must cope up with this scenario to handle this kind of occurring. If such condition occurs on system and the existing control scheme is not able to sustain in this, there is no meaning of that control scheme, means a system that requires a robust control scheme which can handle these variations too. Therefore, to check the robustness of the designed controlled schemes governor time-constant in area-1/2 (T_{g_1} and T_{g_2}), and power system gain (K_{P_1} and K_{P_2}) have been altered as shown in Table 3. All three situations were simulated again after the parameters in the system were changed to check the performance of all the controllers for resilience. Figures 19, 20, 21, 22, 23, and 24 depict frequency deviations and tie-line power variances. With conventional controllers, the damping is slow, but with MPC in AGC, it has improved. It is also clear that the MPC has the highest level of robustness when compared to other systems.

6.1 Case 1: Load Demand Increases in Area-1 Only

The same case was considered in Sect. 5.1 has been simulated to check the robustness of designed control approaches against parameter variations. The responses

Table 3 The variation in the parameters

Parameters	Variation
Governor/time-constant (area-1)	+ 30%
Governor/time-constant (area-2)	– 25%
Power system gain (area-1)	+ 15%
Power system gain (area-2)	– 15%

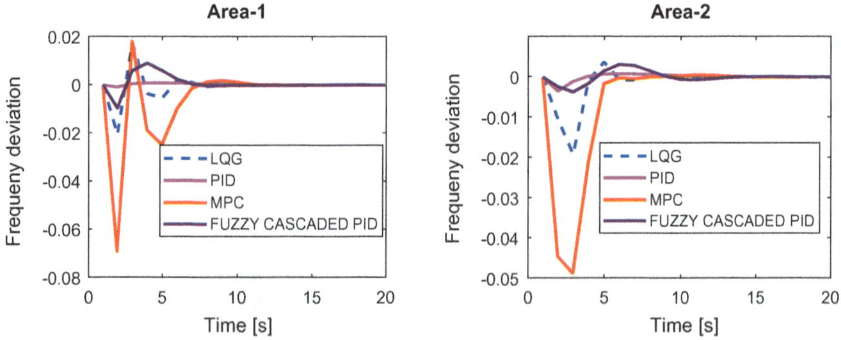

Fig. 19 Frequency deviation

Fig. 20 Tie-line exchange

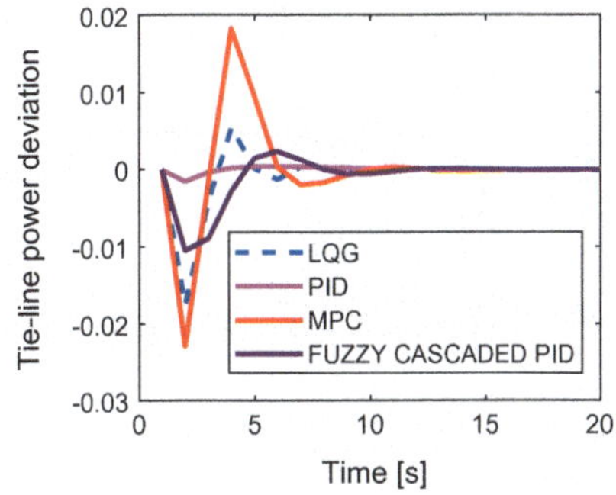

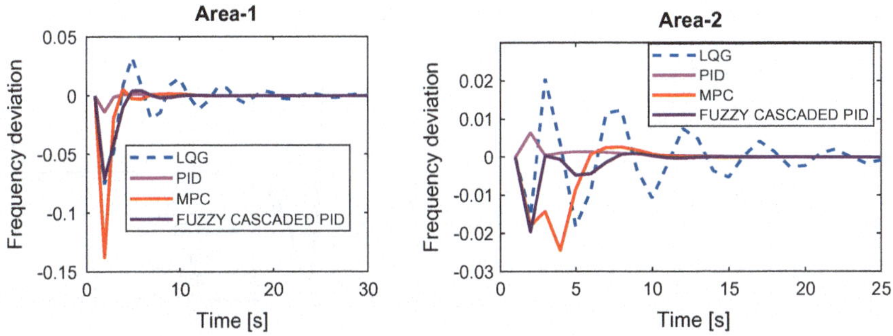

Fig. 21 Frequency deviation

Fig. 22 Tie-line exchange

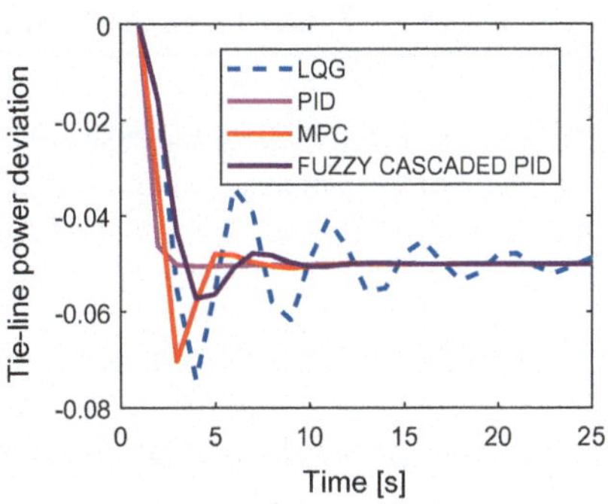

of frequency deviation and tie-line deviation have been shown in Figs. 19 and 20, respectively. Further, the analysis of different time-domain parameters has also been carried out and given in Table 4 as well. It is evident from the result obtained in Figure as well as tables that all control approaches except LQG are working satisfactorily against parameters variation too, however, MPC seems to be working better than other in overall analysis.

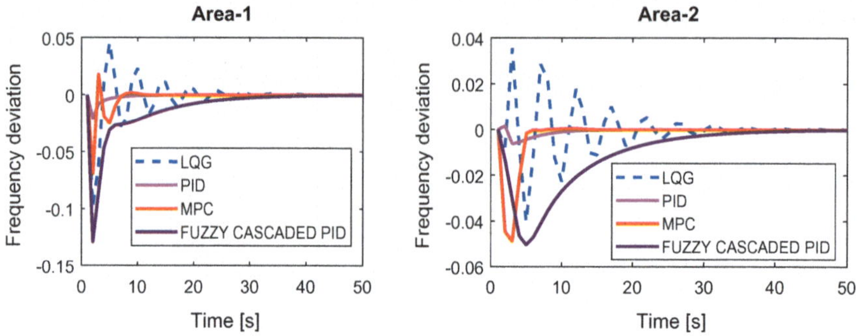

Fig. 23 Frequency deviation

Fig. 24 Tie-line exchange

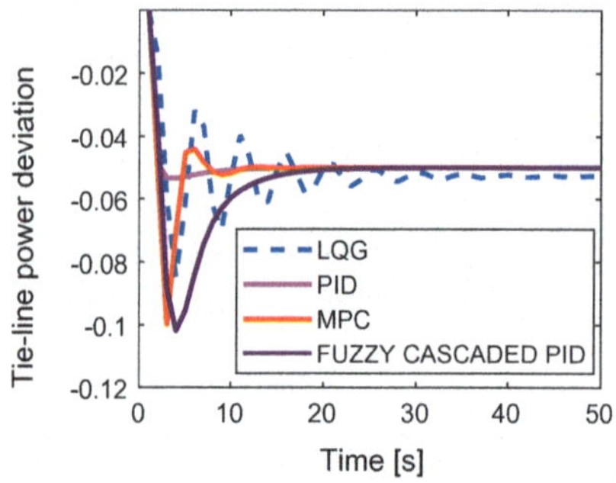

6.2 Case 2: Load Demand Increases in Both Areas

In this section analysis, for the case given in Sect. 5.2 has been carried out to check the designed control approaches against parameter variation. It is evident from Figs. 21 and 22 that MPC control approach is performing better than other approaches and settles down the deviations in much smaller time than other approaches.

6.3 Case 3: Violation of Load Demand in Area-1

In this section analysis, for the case given in Sect. 5.3 has been carried out to check designed control approaches against parameter variation. It is evident from Figs. 23 and 24 that MPC control approach is performing better than other approaches and settles down the deviations in much smaller time than other approaches. The same has been verified in Table 4.

The various results obtained in all cases, i.e., case 1, case 2, and case 3 with and without parameter variation show that in all cases the deviation in frequencies and tie-line settle down at the steady state at the specified values. It is evident that the

Table 4 Comparative analysis of all control schemes (parameter variation case)

Controller	Signal	Overshoot			Undershoot			Settling time		
		Case 1	Case 2	Case 3	Case 1	Case 2	Case 3	Case 1	Case 2	Case 3
LQG	Δf_1	0.01765	0.03	0.05	−0.0209	−0.08	−0.1	13	40	45
	Δf_2	0.004	0.02	0.035	−0.0198	−0.016	−0.04	17	40	45
	ΔP_{tie}	0.005	0.0	0.0	−0.017	−0.075	−0.09	16	40	50
PID	Δf_1	0.0	0.0	0.0	0.001	−0.015	−0.02	14	25	27
	Δf_2	0.0	0.005	0.0	−0.0035	0.0	0.052	20	23	30
	ΔP_{tie}	0.0	0.0	0.0	−0.0014	−0.051	−0.052	20	22	27
MPC	Δf_1	0.0182	0.006	0.02	−0.0692	−0.135	−0.07	14	22	25
	Δf_2	0.0	0.001	0.0	−0.0486	−0.024	−0.05	17	20	25
	ΔP_{tie}	0.0168	0.0	0.0	−0.023	−0.07	−0.1	18	20	20
Fuzzy cascaded PID	Δf_1	0.009	0.0063	0.0	−0.0095	−0.07	−0.13	15	25	40
	Δf_2	0.003	0.0	0.0	−0.0037	−0.023	−0.05	20	25	35
	ΔP_{tie}	0.0	0.0	0.0	−0.010	−0.058	−0.12	20	22	35

responses with MPC scheme are approaching to the desired values at faster rate but with bigger magnitude of undershoot than other approaches.

7 Conclusion

The study of frequency deviation in two-area thermal deregulated power system with different (four) control schemes has been carried out in this chapter. A variety of load change has been considered. At first a step load change in area-1 has been considered while case 2 and case 3 dealt with the step load change in both areas and a contract violation scene in area-1, respectively. For the above-mentioned cases, the authors devised LQG, PID, Fuzzy PID, and MPC control strategies. The LQG scheme was designed using a state space model of a proposed model of a two-area power system. The objective function for designing PID control has been chosen as the minimizing of area control error. The BBBC approach was used to find the best controller parameters. ACE and dACE have been considered as an input to design Fuzzy control scheme. MPC control approach has also been designed utilizing prediction and control model of two area power system. In time-domain simulations, the output of each controller is analyzed from the dynamic behavior of frequency, generation, tie-line power, ACE. Comparison between the results obtained using all approaches is shown in Tables 3 and 4, respectively. It is evident from the comparison that apart from few instances of increased magnitude in oscillations, and MPC-based control scheme provides better performance in terms of (overshoot/undershoot/settling time) and can be effective choice as an AGC/LFC control scheme.

References

1. M. Farahani, S. Ganjefar, M. Alizadeh, PID controller adjustment using chaotic optimisation algorithm for multi-area load frequency control. IET-Control Theory App. **6**, 1984–1992 (2012). https://doi.org/10.1049/iet-cta.2011.0405
2. R.D. Christie, A. Bose, Load frequency control issues in power system operations after deregulation. IEEE Trans. Power Syst. **11**, 1191–1200 (1996). https://doi.org/10.1109/59.535590
3. N. Kumar, B. Tyagi, V. Kumar, Multi-area deregulated automatic generation control scheme of power system using imperialist competitive algorithm based robust controller. IETE J. Res. T&F **64**, 528–537 (2017). https://doi.org/10.1080/03772063.2017.1362965
4. S.K. Pandey, S.R. Mohanty, N. Kishor, A literature survey on load-frequency control for conventional and distribution generation power systems. Renew. Sust. Energ. Rev. **25**, 318–334 (2013). https://doi.org/10.1016/j.rser.2013.04.029
5. H. Gozde, M.C. Taplamacioglu, Automatic generation control application with craziness based particle swarm optimization in a thermal power system. Int. J. Electr. Power Energy Syst. **33**, 8–16 (2011). https://doi.org/10.3182/20140313-3-IN3024.00025
6. S.K. Aditya, D. Das, Design of load frequency controllers using genetic algorithm for two area interconnected hydropower system. Electr. Power Compon. Syst. **31**, 81–94 (2003). https://doi.org/10.1080/15325000390112071

7. Z.W. Geem, J.H. Kim, G.V. Loganathan, A new heuristic optimization algorithm: harmony search. SIMULATION **76**, 60–68 (2001). https://doi.org/10.1177/003754970107600201
8. B. Tyagi, S.C. Srivastava, A LQG based load frequency controller in a competitive electricity environment. Int. J. Emerg. Electr. Power Syst. **2**, 1–13 (2005). https://doi.org/10.2202/1553-779X.1044
9. W. Tan, Tuning of PID load frequency controller for power systems. Energ. Convers. Manag. **50**, 1465–1472 (2009). https://doi.org/10.1016/j.enconman.2009.02.024
10. N. Kumar, B. Tyagi, V. Kumar, Multi area AGC scheme using imperialist competition algorithm in restructured power system. Appl. Soft Comput. **48**, 160–168 (2016). https://doi.org/10.1016/j.asoc.2016.07.005
11. N. Kumar, B. Tyagi, V. Kumar, Deregulated multiarea AGC scheme using BBBC-FOPID controller. Arab. J. Sci. Eng. **42**, 2641–2649 (2016). https://doi.org/10.1007/S13369-016-2293-1
12. L.A. Zadeh, Fuzzy sets. Inf. Control **8**, 338–353 (1965)
13. E. Mamdani, S. Assilian, An experiment in linguistic synthesis with a fuzzy logic controller. Int. J. Man-Mach. Stud. **7**, 1–13 (1975). https://doi.org/10.1016/S0020-7373(75)80002-2
14. M. Ma, X. Liu, C. Zhang, LFC for multi-area interconnected power system concerning wind turbines based on DMPC. IET Gen. Transm. Distrib. **11**, 2689–2696 (2017). https://doi.org/10.1049/iet-gtd.2016.1985
15. A.S. Mir, N. Senroy, Adaptive model predictive control scheme for application of SMES for load frequency control. IEEE Trans. Power Syst. **6**, 526–533 (2018). https://doi.org/10.1109/TPWRS.2017.2720751
16. A.K. Yadav et al., *Soft Computing in Condition Monitoring and Diagnostics of Electrical and Mechanical Systems*, 1E, Advances in Intelligent Systems and Computing, vol. 1096 (Springer Nature, Berlin, 2020), 496 p. ISBN 978-981-15-1532-3. https://doi.org/10.1007/978-981-15-1532-3
17. A. Iqbal et al., *Meta Heuristic and Evolutionary Computation: Algorithms and Applications*, 1E, Studies in Computational Intelligence, vol. 916 (Springer Nature, Berlin, 2020), 849 p. ISBN 978-981-15-7571-6. https://doi.org/10.1007/978-981-15-7571-6
18. J.A. Alzubi, *AI and Machine Learning Paradigms for Health Monitoring System: Intelligent Data Analytics*, 1E, Studies in Big Data, vol. 86 (Springer Nature, Berlin, 2020), 513 p. ISBN: 978-981-33-4412-9. https://doi.org/10.1007/978-981-33-4412-9
19. N. Fatema et al., *Intelligent Data-Analytics for Condition Monitoring: Smart Grid Applications*, 1E (Academic Press, 2021). ISBN: 978-0-323-85510-5. https://doi.org/10.1016/C2020-0-02173-0
20. S. Srivastava et al., *Applications of Artificial Intelligence Techniques in Engineering*, vol. 2, 1E, Advances in Intelligent Systems and Computing, vol. 697 (Springer Nature, Berlin, 2018), 647 p. ISBN 978-981-13-1822-1. https://doi.org/10.1007/978-981-13-1822-1
21. T. Mahto et al., Fractional order control and simulation of wind-biomass isolated hybrid power system using particle swarm optimization, in *Applications of Artificial Intelligence Techniques in Engineering*, Advances in Intelligent Systems and Computing, vol. 698 (2018), pp. 277–287. https://doi.org/10.1007/978-981-13-1819-1_28
22. Vigya et al., Renewable generation based hybrid power system control using fractional order-fuzzy controller. Energy Rep. **7C**, 641–653 (2021). https://doi.org/10.1016/j.egyr.2021.01.022
23. T. Mahto et al., Load frequency control of a solar-diesel based isolated hybrid power system by fractional order control using particle swarm optimization. J. Intell. Fuzzy Syst. **35**(5), 5055–5061 (2018). https://doi.org/10.3233/JIFS-169789
24. Gopal et al., Digital transformation through advances in artificial intelligence and machine learning. J. Intell. Fuzzy Syst. (2021). https://doi.org/10.3233/JIFS-189787

Mathematical Approach-Based Power System Analysis: A Review of Short-Term Hydro Scheduling

Prahlad Mundotiya, Parul Mathuria, and Harpal Tiwari

Abstract The short-term hydro scheduling (STH'S) issue is intended to decide optimum electrical generation schedule for a single unit hydro or an assimilated power plant of cascaded watersheds over the period of time duration from one day to single week. Conventionally, the consolidated hydropower plant definition is typically used in the framework of the STH'S challenge. Hydropower plant units in the plant are grouped together as a single unit. So, numerous hydro companies are now involved in both the electricity and capacity sectors. It underlines the need for a detailed measurement of the energy transfer and the usable power of every unit. Formulation of the STH'S issue on single units will reliably represent the physical and operational characteristics. A comprehensive description of mathematical programming strategies for modelling and solving unit-based STH'S challenges in the context of power system are described in this chapter. This chapter gives a detailed comparative analysis and explanation of each particular topic in the framework of STH'S. This outline is intended to be the initial step for discovering additional computationally fixable and efficient approaches to resolve these problems of the unit-based STH'S issue.

Keywords Short-term hydro scheduling · Electrical energy · Reserves · Hydropower plant · Multi-objective programming

1 Introduction

Nowadays, the renewable energy incorporation in the electrical power system is playing a momentous role in the operation of economic dispatch. Researchers and policymakers are concentrating to rise the part of sustainable energy sources, which is known as renewable energy sources in the electrical power system for replacement of thermal power generation across the globe. Some treaties and agreements have also been done by the eminent leaders of the world in the last decade to reduce carbon

P. Mundotiya (✉) · P. Mathuria · H. Tiwari
Malaviya National Institute of Technology, Jaipur, India

H. Malik et al. (eds.), *Intelligent Data Analytics for Power and Energy Systems*, Lecture Notes in Electrical Engineering 802,
https://doi.org/10.1007/978-981-16-6081-8_5

emission in the environment from the energy industry to control global warming. For this objective, hydro scheduling is very important to coordinate the hydropower plant with thermal generation for peak saving and economic dispatch along with the reduction in carbon emission [1]. Hydro resources have a "storage" nature, allowing the device to react quickly. Their technology has progressed to the point that it can be used in manufacturing processes all over the globe. Furthermore, as a result of the liberalization of electricity power markets and the incorporation of new electrical power firms [2], some effective work is needed to covenant by the competitive behaviour of these electricity power markets in reference to confirm the system's robustness, maximize revenues, and reduce losses or costs [3].

The evolution of electrical power generation has a significant impact on economic, social and living standards of people of any country. Its history dates in last of eighteenth century when the effect of induction was discovered. The initial approach to electrical power production was focussed on limited generation and utilization. As a result, electricity is often in surplus and other times in shortage. However, in the early nineteenth century, the need for energy uses raised with reference increased the population across the globe and industry's reliance on electrical power. The incorporation of electrical energy sources and the optimal economic generation scheduling began as a result. Optimal economic dispatch is a key panorama in the electrical energy or power grid for optimum generator schedule that meets the objective target of cost minimization while still meeting power grid and constraint of grid network's [4, 5].

Electrical power generation schedule is used to enhance power production for the individually unit of generation so that the total price of electrical power production systems is as low as possible while still meeting system constraints. It is done by lowering those objective functions while maintaining acceptable device efficiency in terms of power limits and thermal generator output. Economic costs, device security or other objectives can be represented by objective functions also known as cost and emission functions. Renewable energy sources like solar, wind, tidal and biofuel are now playing an increasingly important role in electrical power generation in most of the countries across the globe. Because of its controllable, storable and flexible properties, hydropower plants play a significant role in ensuring the system's security. Often hydro plants are held for the reserve to accommodate variations in the power grid [6, 7].

The brief out line of this article are as:

- The overview in this article focusses on individual hydro unit modelling tools and the accompanying solution methodologies provided in chapters since 2000.
- Heuristic methods are not explored in this article because the focus is on accurate procedures.
- Uncertainties in natural inflow forecasts, market prices and RE power generation are not taken into account.
- The full description of computationally efficient deterministic models will serve as a strong technical reference for incorporating the stochastic aspect of the STHS issue as well as for optimization in a hydrothermal-RE linked power system.

2 Types of Hydroelectric Power Plant

Hydroelectric power plants with huge reservoirs (HR's), run-of-river hydroelectric power plants (RRH's) and small hydroelectric power plants (SH's) are the three major types of hydroelectric plants that are identified in STH'Ss.

2.1 *Hydroelectric Power Plant with Huge Reservoir (HR's)*

Water can be stored in hydroelectric plants with huge reservoirs (HR's), and the amount of obtainable water can be utilized to generate electrical power when it is required. HRs transform the potential energy of dam water in to electrical power by running a water turbine and a generator, accounting for a significant portion of hydropower. The height difference between the reservoir's top and the level of water discharge, known as the reservoir's head, determines the amount of electricity produced.

2.2 *Run-of-River Hydroelectric Power Plant (RRHs)*

Reservoirs are also present in RRHs, but they are tiny in comparison to those found in other hydro plants. Water cannot be stored in an RRH, so the flow of water either produces electric power or spilled water, subject to the generation plant capability. In most RRH, water is collected at the water inlet structure and transported to the powerhouse via a buried penstock. If the water is not collected at the intake in this scheme, the river will continue to flow normally. The dam is necessary to maintain that the penstock receives sufficient water. Water is transported from a higher altitude to a lower altitude, at which a water turbine is mounted through the penstock. The heads of the dam of a run-of-river hydroelectric power plant are height difference, which is important for the potential energy that the hydro plant generator converts into electric power.

2.3 *Small Hydroelectric Power Plant (SHs)*

Electricity output varies among SHs. Plants with power outputs of less than 1 MW are also available. The most commonly installed capacities vary between 1 and 30 MW. These hydroelectric plants are close to RRHs due to the lack of reservoir space. SHs, on the other hand, are typically extracted from RRHs by the need for special equipment to satisfy the criteria of high power performance, environmental restrictions and consistency all at the same time. SHs have no detrimental impact on the environment. The power generation of SHs power plants is vastly close to the naturally drift of the waterway, making them disposed to periodic fluctuations, which is a downside of this method that is greater than that of RRHs [8, 9] (Fig. 1).

Hydrothermal scheduling is widely used to optimize the cost of power generation. This scheduling technique plans the generation as per the availability of water for hydro generation. The available water is diverted for irrigation, drinking, etc., and this reduces the possibilities of flood and soil erosion too. However, there is a considerable number of problems too [10, 11].

Conventionally, short-term hydro scheduling (STH'S) makes use of optimum generation schedules as per the availability of the resources to make the process more economical. STH'S is commonly used to help spot bidding on the day ahead market up to include a closing shipping schedule after the clearance procedure of electricity market [12].

Generally, the hydro generation model can be classified into, unit or plant-based [13]. The majority of STH'S problems in the literature [14–16] are plant-based, specifically in the short-term estimating of the hydro-thermal electrical power systems [15] or the hydrothermal-RE interconnected power system [17, 18]. To address this problem, a novel approach using aggregated plant model is suggested by this article. Here, these turbines of hydro generator entities are accumulated as a single unit in a hydropower plant. One of the considerable benefits of this idea is that it greatly decreases the possible problem size of the STH'S.

STH'S relates with two concerns that are defined as (i) the problem of unit commitment (UC), which shows on or off position of the generating entities for every time period; and (ii) the problem of unit-load-dispatch (ULD) which is used to decide the concerning dispatch of the generating units [19, 20]. Both the issues together are known as a hydro unit commitment (HUC) [21–23].

STH'S may be considered as a sub-problem when incorporated into thermal/wind power generators [24], whereas it could be the major issue in the case of the hydropower plant.

An STH'S consist of general set constraints and problem objectives. There are kinds of literature addressing some of the issues [24, 25]. The modelling techniques suggested, however, are classified according to their aims and limitations. This approach can give a detailed differentiation and understanding of the problem in the formulation of STH'S.

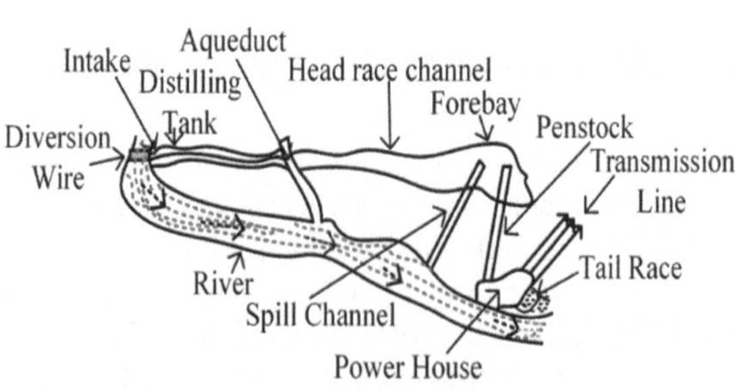

Fig. 1 Schematic of run-of-river small hydro power plant

It should be noted that an alternative of this accumulation has been developed in recent times [26, 27], which does not work as plant-based aggregation; instead, the calculation of input water and output power is done at the plant level, and all the units to be aggregated are assumed to have similar characteristics [12, 14]. Operational units in the model are not dealt with individually but are combined together. However, the calculations describe the individual features so they are considered unit-based STH'S problems. In addition, while chapters [28, 29] are modelled on the level of the plant, it is presumed that every plant works with or equal to a single generating unit.

The STH'S problem objectives are mainly with the electrical power system features and operative requisites. The system operator gives the target of generation on a per hour basis for every hydro plant in a centralized system like in Brazil [30, 31] while not producing any related cost of electricity. The STH'S problem aims at finding the best economical scheduling in the hydro plant units to balance the demand by satisfied all the constraints of the system. The optimized use of water and the reduction in start-ups/shutdowns of units is expected in the economical scheduling.

In decentralized power grids like Spain [32], Scandinavia [7] and Canada [33, 34], distribution of electricity takes place after a market clearing process. Production is carried out by various producers after the demands, and an offer from one of the participating players is being contemplated. The market-clearing price (MCP) is observed from the analysis of the supply and demand relation. The hydropower producers optimize their gains by dealing with electrical power in the competition of the power market while assuming they cannot affect the market rates with their market power.

Considering the need for precise energy calculation, the usable each units of electrical power is calculated by involving in both the power and energy markets. It demands a precise and comprehensive presentation of hydropower generation. Here, effect on the electrical power generated by every unit is shown in terms of hydraulic losses, head variance, efficiency curves and prohibited operational zones. The simulated results should show each unit's dispatch along with the available capacity.

Hence, it is proposed to have a unit-based optimization and modelling, rather than a conventional aggregated plant-based one, to accurately simulate the properties of the unit. With advanced developments in computer hardware, software and technology, sophisticated mathematical simulations may be completed in a reasonable period of time.

3 Mathematical Formulation of STH'S

Based on the mathematical formulation in the available literature, STH'S problem can be categorized as follows:

3.1 Revenue Maximization

The optimization of the revenue is frequently utilized in the ever-growing competition in the market of electricity power [29–31]. This can be obtained by selling the electricity in the market.

$$\text{Max } fn_1 = \sum_{t \in T} Mp_t.\Delta T.P_t^{\text{sl}} \tag{1}$$

where Mp_t is forecasted market price of electricity in period t (€/MWh), ΔT is length of each time period (hours, ht) and P_t^{sl} is power sold to market in period t (MW).

3.2 Operational Cost Minimization

Production cost in the case of hydropower may be considered negligible if the worth of the water stocked is not considered [35]. In the case of STH'S, the cost for start-up and shutdown makes the most significant addition to the overall cost of production [36]. Moreover, start-up and shutdown increases the cost of maintenance and decreases the time served by a machine. Hence, it is advised to reduce the incidents of start-up and shutdowns in the case of STH'S [33, 35, 37–40]. The start-up expenses of the generating unit can be estimated by considering past record of the maintenance and repairing cost corresponding to the incidents of start-ups [33], or their calculation can also depend on the generating unit's nominal power output [38]. In few scenarios, the amount of water released to start up one generating unit (m^3) is used to state the expenses in the start-up cost in place of monetary units [24].

$$\text{Min } fn_2 = \sum_{t \in T} \sum_{h \in H} \sum_{i \in I_h} SU_{i,h} \cdot \mu_{i,h,t} \tag{2}$$

where $SU_{i,h}$ is start-up cost of unit I in plant h (€) and $\mu_{i,h,t}$ is start-up decision of unit i in plant h in period t (1 if it is started up in period t, 0 otherwise). T, H, I_h are set of time period, hydro plant and hydro turbine generator unit in plant.

3.3 Minimize the Used Energy or Spilled

Since the market of electrical power is controlled centrally, the obligation of the load is already decided and nil or a certain quantity of power is available for sale to the spot market. For this situation, the prime concern is to reduce the value of water used by the spilled and turbines [38]. Alternatively, it can be said that the aim is to optimize the potential energy for future income, which provides another form.

$$\text{Min } fn_{3a} = \sum_{r \in R} W_{\text{mf}_{r,t}}.E_h.(V_{\text{in}_{r,0}} - v_{r,t}) \tag{3a}$$

$$\text{Max } fn_{3b} = \sum_{r \in R} W_{\text{mf}_{r,t}}.E_h.v_{r,t} \tag{3b}$$

where W_{mf} is marginal value of reservoir r at the end of scheduling horizon t (€/MWh), E_h is energy conversion factor for plant h (MWh/m^3), $V_{\text{in}_{r,0}}$ is initial water storage of reservoir r (m^3) and $v_{r,t}$ is water volume of reservoir r at the end of period t (m^3).

3.4 Other Problems

While optimizing the operation of a solitary hydropower plant having more than one units, the main aim is to optimize the generated electrical power for the flow considered [18, 21, 26], to decrease the total release of the units for a prescribed load [23, 30, 31], or to increase the energy conversion efficiency of the complete power plant [20]. If the loss of power is the main criteria, then the aim is to reduce the losses in the process of generating hydropower due to elevation of tailrace, penstock (head) loss and generator efficiency variations [17]. The main concern is to optimize the weighted technical efficiency where the main factor considered is the electrical power generated by every plant per period.

The various aims be fulfilled may be carefully selected based on their feasibility. These aims and objectives can be selected by taking into account the features of the electrical power market and the operative situations. For more than one participating criteria in a design, it becomes a multi-parameter optimization challenge. For these problems, a single solution cannot fulfil all the criteria at the same time. Alternatively, collections of Pareto optimal explanations are achieved. The solution of multi-objective optimization challenge states to mainly two types: (1) How and when to discover an optimal characteristic collection of Pareto results that can estimate the whole front of Pareto and (2) How to choose most suitable outcome of all Pareto's optimal solutions [41].

The technique often used for multi-objective optimization of the basic STH'S issue is by using weighting coefficients which transform the multi-objective issue into a single objective optimization issue or through considering some of the criteria as constraints. An STH'S model introduced for multiple power grids' peak reduction [38] can be considered as a specific multi-objective issue. The objective function of a single power grid could compromise with others, as well as fuzzy logic optimization approach is implemented to determine Pareto's feasible solution. In the Ref. [39], the authors used flood-management in STH'S and tested a tri-multi-objective optimization framework. A multi-objective analytical expression is founded on the choice of the resolution maker and the approach of decomposition is advanced.

4 Different Techniques for Solving STH'S Problem

Mostly the objective assignment constituents for the STH'S in a contentious electrical grid incorporate the earnings obtained from selling electrical power, the expenses in a start-up and the worth of electrical energy reserves at the end of the study. These expenses due to penalty regarding violation of several restrictions may be optionally incorporated in the objective function.

On the above enormous discussion, our prime objective is now too focussed on finding the simultaneous solution of the problem such as minimization of the operational cost, revenue maximization, along the minimum used energy. Therefore, our main concern is the economic dispatch of hydrothermal generation for the fulfilment of the power demand (Fig. 2).

In electrical power system operation and control, economic load dispatch (ELD) [40] is a prominent concern in the optimization issues. ELD has to allocate the demanded load in a very inexpensive way between the generators committed while fulfilling the physical and operational limitations in a single area. It is a general approach to divide the generators on the basis of the regions of generation and interconnect those using tie-lines. Economic dispatch in many areas simultaneously may be considered as an addition to economic dispatch where the level of generating power and the power to interchange is optimized to keep the total fuel cost to the minimum.

Transmission constraints are generally ignored while optimizing the economic dispatch problem. However, there are reports on the transmission capacity considering the constraints. Economic dispatch problems are usually solved by taking into account export and import constraints in different areas (Table 1).

This research offers a systematic framework of multi-area power generation planning and a context for multi-area issues [42]. Application of linear programming to a transmission restricted cost study of output has been suggested [43]. A decomposition method for solving multi-region generation scheduling with tie-line constraints using the familiar system is also reported [44]. Newton–Raphson's procedure has also used to solve the economic load dispatch issue of multi-area system for short-range marginal cost [45]. The evolutionary programming is used to resolve multi area economic load dispatch including tie-line constraints [46]. Some researchers have been investigating the efficiency of many evolutionary algorithms for multi-area-economic load dispatch (MAELD) issues [47].

In view of the short-term hydropower UC issue, with a timeframe of usually one as well few more days, such that the inflow of the reservoir is believed to be defined as well as a probabilistic solution could be used. In comparison to the cases dealing with huge reservoirs characteristic of Brazilian electrical power system, the

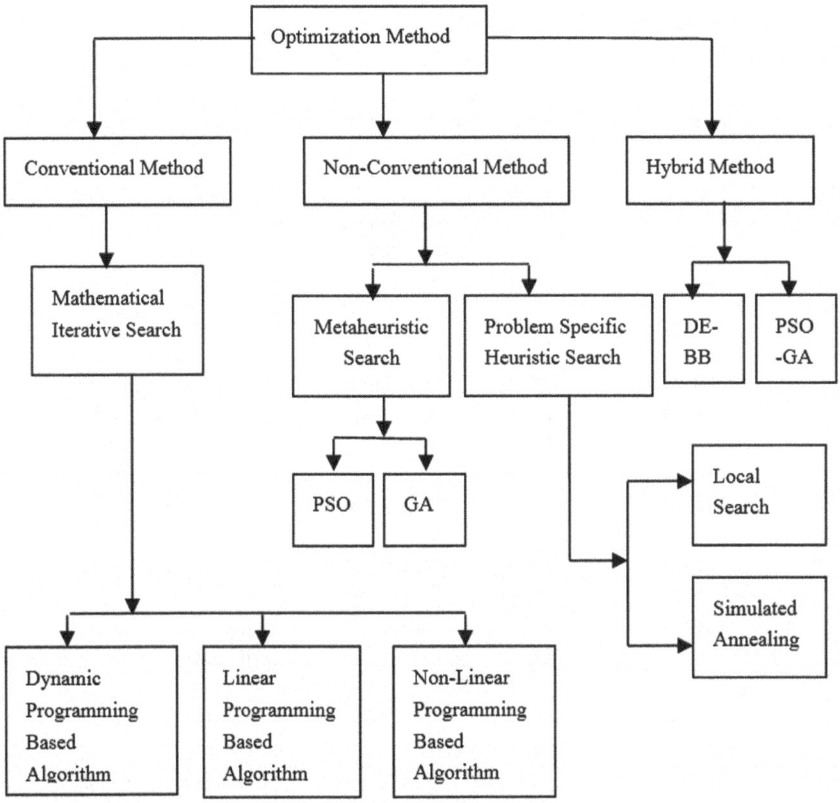

Fig. 2 Different optimization methods and their subsections

water head effects on the power station outcome may be ignored without considering pump storage [41, 47–52]. It also implies coordinated scheduling of several reservoirs situated in each water basin, carried out by an entity. Furthermore, the operational environment should not be cost based, such that, the purpose should not be to increase the producer's profits. Therefore, the primary goal is to optimize the productive utilization of water supplies. For this reason, it also presumes that the energy target in MWh established in the midterm analyses for every power station, in addition to the predicted energy requirements, should be met [41, 47, 48].

Few researchers defined different methodologies for formulation of UC issue of hydropower plant [41, 49–51]. It provides an alternate solution approach that eliminates the use of heuristic approaches or simple dynamic programming without any issue of decomposition. This technique may turn out to be a computational complexity ineffective technique when several generators are included. In the above references, the suggested solution is designed to minimize the loss-of-performance (LoP) associated with hydropower generation such as (i) the elevation of tailrace, (ii) the penstocks head loss, and (iii) variation in efficiency of turbine-generator. The issue

Table 1 Model and solution techniques for STH'S problems

References	Model formulation	STH'S problem	Solution method	Type of hydro plant	Name of country
[28]	LP	ULD	CPLEX Solver	Cascade	Canada
[13]	NLP	ULD	DP	One plant	Brazil
[14]	NLP	ULD	DP, Decomposition algorithm	One plant	Canada
[15]	NLP	ULD	PSO and DP	One plant	China
[27]	NLP	ULD + UC	DP	One plant	Spain
[16]	NLP	ULD	Hill-Climbing	Unit-plant	Brazil
[2]	MILP	ULD + UC	CPLEX-Solver	Pump storage with Cascade system	Switzerland and New Zealand
[24]	MILP	ULD + UC	CPLEX Solver	Cascade plant	Spain
[25]	MILP	ULD + UC	CPLEX Solver	Cascade plant	Spain
[31]	MILP	ULD + UC	CPLEX Solver	Unit-plant with pump storage	N/A
[29]	MILP	ULD + UC	CPLEX Solver	Cascade plant	Canada
[32]	MILP	ULD + UC	CPLEX Solver	Cascade plant	China
[17]	MILP	ULD + UC	LINGO, DP	One plant	China
[34]	(i) NLP and MILP (ii) MINLP	ULD + UC	Pre-processing DP, IPOPT, Xpress Solvers	Cascade plant	Canada
[20]	MILP	ULD + UC	LINGO	One plant	China
[21]	MILP	ULD + UC	GUROBI Solver	Cascade plant	Brazil
[24]	MILP decomposed to MILP and LP	ULD + UC	LR (CPLEX-Solver)	Cascade plant	Canada
[30]	MILP	ULD + UC	CPLEX	Cascade plant	Norway
[35]	MINLP	ULD + UC	RPG Method	Unit plant	Brazil
[38]	MINLP, MILP and NLP	ULD + UC	LR (CPLEX Solver, SQP)	Cascade	Brazil

(continued)

Table 1 (continued)

References	Model formulation	STH'S problem	Solution method	Type of hydro plant	Name of country
[36]	MINLP	ULD + UC	CPLEX Solver	Cascade plant	Spain
[37]	MINLP	ULD + UC	LR, IAL	Cascade plant	Brazil
[22]	MINLP	ULD + UC	(1) Spatial B&B (2) CPLEX, BARON Solver	Cascade plant	Brazil
[26]	MINLP	ULD + UC	N/A	One plant	Brazil
[18]	MINLP	ULD + UC	LR (CPLEX Solver, IAL)	Cascade plant	Brazil
[19]	MINLP	ULD + UC	DICOPT Solver	Cascade plant	Brazil

associated with this process would be that the LoP function of a given hydropower plant is not a constant and analytical process. Rather, it assumes the form of a set of mappings, one for each generator unit combinations in the process. In order to add to it, this chapter suggests a novel technique established on the integration of the different LoP-mappings into the specific target. This is done by describing a regular surface using the greatest suitable approaches that roughly include all independent LoP-curves. In this contrast, the many generating units in the process shall be regarded as an internal optimization parameter. This objective function is including the on and off expenses of electrical power generating the system [50–53]. The above outcomes in such a mixed-integer non-linear programming (MINLP) issue that is decided by suitable computing methods. Ultimately, it generalizes the framing of the issue to include the actual scenario of the power generation plant consisting of different sets of similar generators.

5 Conclusions

This chapter is compared with the widespread articles published on the STH'S issue of power system observed at the consolidated unit level. Although, considerably, fewer articles of literature discuss the optimum scheduling of independent hydro unit generators. However, it is essential to analyse the expression of each and every unit individually in modelling precisely to achieve efficient and feasible outcomes. In addition, the presence of hydro players in both the electricity generation and capacity market can be considered. Along with, these sectors also have a capacity the rapid adaptation to the evolving markets which underscore the need and significance of unit wise design. Therefore, the chapter provides a comprehensive analysis of the various methodological strategies for modelling and solving the unit-based STH'S issue. Although, it provides significant consideration for the potential goals, limitations and different modelling approaches have been suggested in the available literature. Therefore, the most common approaches with mathematical and analytical programming are summarized in this chapter.

References

1. A. Wood, B. Wollenberg, *Power Generation, Operation and Control* (Wiley, New York, 1996)
2. D.P. Kothari, J.S. Dhillon, *Power System Optimization* (Prentice Hall of India Private Limited, New Delhi, 2004)
3. C.E. Davist, R. Pronovost, The stochastic dynamic programming procedures for long term reservoir management. IEEE Trans. Power Apparatus Syst. **PAS-91**(6) (1972)
4. A. Turgeon, Optimal operation of multi reservoir power systems with stochastic inflows. Water Resour. Res. **16**(2) (1980)
5. P.S. Nigam, *Handbook of Hydroelectric Engineering*, 2nd edn. (Nem Chand & Bros, Roorkee 1985)
6. K. Kong, H.I. Skjelbred, O.B. Fosso, An overview on formulations and optimization methods for the unit-based short-term hydro scheduling problem. Electr. Power Syst. Res. **178**, 106027 (2020)
7. A. Iqbal et al., *Renewable Power for Sustainable Growth*, 1E, Lecture Notes in Electrical Engineering, vol. 723 (Springer Nature, Berlin), 805 p. ISBN: 978-981-33-4080-0. https://doi.org/10.1007/978-981-33-4080-0
8. G.W. Chang, M. Aganagic, J.G. Waight, J. Medina, T. Burton, S. Reeves, M. Christoforidis, Experiences with mixed integer linear programming based approaches on short-term hydro scheduling. IEEE Trans. Power Syst. **16**, 743–749 (2001). https://doi.org/10.1109/59.962421
9. A.L. Diniz, M.E.P. Maceira, A four-dimensional model of hydro generation for the short-term hydrothermal dispatch problem considering head and spillage effects. IEEE Trans. Power Syst. **23**, 1298–1308 (2008). https://doi.org/10.1109/tpwrs.2008.922253
10. J.P.S. Catalao, S. Mariano, V.M.F. Mendes, L. Ferreira, Scheduling of head-sensitive cascaded hydro systems: a nonlinear approach. IEEE Trans. Power Syst. **24**, 337–346 (2009). https://doi.org/10.1109/tpwrs.2008.2005708
11. J.P.S. Catalao, H.M.I. Pousinho, V.M.F. Mendes, Mixed-integer nonlinear approach for the optimal scheduling of a head-dependent hydro chain. Electr. Power Syst. Res. **80**, 935–942 (2010). https://doi.org/10.1016/j.epsr.2009.12.015
12. J.P.S. Catalao, H.M.I. Pousinho, V.M.F. Mendes, Hydro energy systems management in Portugal: profit-based evaluation of a mixed-integer nonlinear approach. Energy **36**, 500–507 (2011). https://doi.org/10.1016/j.energy.2010.10.014
13. G.J. Osorio, J.C.O. Matias, J.P.S. Catalao, A review of short-term hydro scheduling tools, in *2013 48th International Universities Power Engineering Conference (UPEC)*, Dublin, Ireland (2013). https://doi.org/10.1109/UPEC.2013.6714906
14. A. Hamann, G. Hug, S. Rosinski, Real-time optimization of the Mid-Columbia hydropower system. IEEE Trans. Power Syst. **32**, 157–165 (2017). https://doi.org/10.1109/tpwrs.2016.2550490
15. M. Nazari-Heris, B. Mohammadi-Ivatloo, G.B. Gharehpetian, Short-term scheduling of hydro-based power plants considering application of heuristic algorithms: a comprehensive review. Renew. Sustain. Energy Rev. **74**, 116–129 (2017). https://doi.org/10.1016/j.rser.2017.02.043

16. S. Das, A. Bhattacharya, A.K. Chakraborty, Fixed head short-term hydrothermal scheduling in presence of solar and wind power. Energy Strategy Rev. **22**, 47–60 (2018). https://doi.org/10.1016/j.esr.2018.08.001
17. A. Arce, T. Ohishi, S. Soares, Optimal dispatch of generating units of the Itaipu hydroelectric plant. IEEE Trans. Power Syst. **17**, 154–158 (2002). https://doi.org/10.1109/59.982207
18. M. Breton, S. Hachem, A. Hammadia, Accounting for losses in the optimization of production of hydroplants. IEEE Trans. Energy Convers. **19**, 346–351 (2004). https://doi.org/10.1109/tec.2004.827043
19. C.T. Cheng, S.L. Liao, Z.T. Tang, M.Y. Zhao, Comparison of particle swarm optimization and dynamic programming for large scale hydro unit load dispatch. Energy Convers. Manage. **50**, 3007–3014 (2009). https://doi.org/10.1016/j.enconman.2009.07.020
20. E.C. Bortoni, G.S. Bastos, T.M. Abreu, B. Kawkabani, Online optimal power distribution between units of a hydro power plant. Renew. Energy **75**, 30–36 (2015). https://doi.org/10.1016/j.renene.2014.09.009
21. E.C. Finardi, E.L. da Silva, Unit commitment of single hydroelectric plant. Electr. Power Syst. Res. **75**, 116–123 (2005). https://doi.org/10.1016/j.epsr.2005.01.008
22. A. Borghetti, C. D'Ambrosio, A. Lodi, S. Martello, An MILP approach for short-term hydro scheduling and unit commitment with head-dependent reservoir. IEEE Trans. Power Syst. **23**, 1115–1124 (2008). https://doi.org/10.1109/tpwrs.2008.926704
23. X. Li, T.J. Li, J.H. Wei, G.Q. Wang, W.W.G. Yeh, Hydro unit commitment via mixed integer linear programming: a case study of the Three Gorges Project, China. IEEE Trans. Power Syst. **29**, 1232–1241 (2014). https://doi.org/10.1109/tpwrs.2013.2288933
24. R.M. Lima, M.G. Marcovecchio, A.Q. Novais, I.E. Grossmann, On the computational studies of deterministic global optimization of head dependent short-term hydro scheduling. IEEE Trans. Power Syst. **28**, 4336–4347 (2013). https://doi.org/10.1109/tpwrs.2013.2274559
25. I.A. Farhat, M.E. El-Hawary, Optimization methods applied for solving the short term hydrothermal coordination problem. Electr. Power Syst. Res. **79**, 1308–1320 (2009). https://doi.org/10.1016/j.epsr.2009.04.001
26. R. Taktak, C. D'Ambrosio, An overview on mathematical programming approaches for the deterministic unit commitment problem in hydro valleys. Energy Syst. **8**, 57–79 (2017). https://doi.org/10.1007/s12667-015-0189-x
27. S. Seguin, P. Cote, C. Audet, Self-scheduling short-term unit commitment and loading problem. IEEE Trans. Power Syst. **31**, 133–142 (2016). https://doi.org/10.1109/tpwrs.2014.2383911
28. A. Marchand, M. Gendreau, M. Blais, G. Emiel, Fast near-optimal heuristic for the short-term hydro-generation planning problem. IEEE Trans. Power Syst. **33**, 227–235 (2018). https://doi.org/10.1109/tpwrs.2017.2696438
29. A.J. Conejo, J.M. Arroyo, J. Contreras, F.A. Villamor, Self-scheduling of a hydro producer in a pool-based electricity market. IEEE Trans. Power Syst. **17**, 1265–1272 (2002). https://doi.org/10.1109/tpwrs.2002.804951
30. J. Garcia-Gonzalez, E. Parrilla, A. Mateo, Risk-averse profit-based optimal scheduling of a hydro-chain in the day-ahead electricity market. Eur. J. Oper. Res. **181**, 1354–1369 (2007). https://doi.org/10.1016/j.ejor.2005.11.047
31. E.C. Finardi, M.R. Scuzziato, Hydro unit commitment and loading problem for day-ahead operation planning problem. Int. J. Electr. Power Energy Syst. **44**, 7–16 (2013). https://doi.org/10.1016/j.ijepes.2012.07.023
32. M.M. Cordova, E.C. Finardi, F.A.C. Ribas, V.L. de Matos, M.R. Scuzziato, Performance evaluation and energy production optimization in the real-time operation of hydropower plants. Electr. Power Syst. Res. **116**, 201–207 (2014). https://doi.org/10.1016/j.epsr.2014.06.012
33. J.I. Perez-Diaz, J.R. Wilhelmi, L.A. Arevalo, Optimal short-term operation schedule of a hydropower plant in a competitive electricity market. Energy Convers. Manage. **51**, 2955–2966 (2010). https://doi.org/10.1016/j.enconman.2010.06.038
34. Z.K. Shawwash, T.K. Siu, S.O.D. Russell, B.C. The, Hydro short term hydro scheduling optimization model. IEEE Trans. Power Syst. **15**, 1125–1131 (2000). https://doi.org/10.1109/59.871743

35. D. De Ladurantaye, M. Gendreau, J.Y. Potvin, Optimizing profits from hydroelectricity production. Comput. Oper. Res. **36**, 499–529 (2009). https://doi.org/10.1016/j.cor.2007.10.012
36. H.I. Skjelbred, J. Kong, O.B. Fosso, Dynamic incorporation of nonlinearity into MILP formulation for short-term hydro scheduling. Int. J. Electr. Power Energy Syst. (2019)
37. B. Tong, Q.Z. Zhai, X.H. Guan, An MILP based formulation for short-term hydro generation scheduling with analysis of the linearization effects on solution feasibility. IEEE Trans. Power Syst. **28**, 3588–3599 (2013). https://doi.org/10.1109/tpwrs.2013.2274286
38. O. Nilsson, D. Sjelvgren, Hydro unit start-up costs and their impact on the short term scheduling strategies of Swedish power producers. IEEE Trans. Power Syst. **12**, 38–43 (1997). https://doi.org/10.1109/59.574921
39. X.Y. Wu, C.T. Cheng, J.J. Shen, B. Luo, S.L. Liao, G. Li, A multi-objective short term hydropower scheduling model for peak shaving. Int. J. Electr. Power Energy Syst. **68**, 278–293 (2015). https://doi.org/10.1016/j.ijepes.2014.12.004
40. Y.T. Qi, J.G. Luo, Y.X. Wei, Q.G. Miao, Maximizing hydropower generation in flood control operation using preference based multi-objective evolutionary algorithm, in *International Conference on Sustainable Development on Energy and Environment Protection (SDEEP)*, Yichang, P.R. China (2017). https://doi.org/10.1088/1755-1315/86/1/012037
41. S. Srivastava et al., *Applications of Artificial Intelligence Techniques in Engineering*, vol. 1, 1E, Advances in Intelligent Systems and Computing, vol. 698 (Springer Nature, Berlin, 2018), 643 p. ISBN 978-981-13-1819-1. https://doi.org/10.1007/978-981-13-1819-1
42. A.K. Yadav et al., *Soft Computing in Condition Monitoring and Diagnostics of Electrical and Mechanical Systems*, 1E, Advances in Intelligent Systems and Computing, vol. 1096 (Springer Nature, Berlin, 2020), 496 p. ISBN 978-981-15-1532-3. https://doi.org/10.1007/978-981-15-1532-3
43. S. Srivastava et al., *Applications of Artificial Intelligence Techniques in Engineering*, vol. 2, 1E, Advances in Intelligent Systems and Computing, vol. 697 (Springer Nature, Berlin, 2018), 647 p. ISBN 978-981-13-1822-1. https://doi.org/10.1007/978-981-13-1822-1
44. A.L. Desell, E.C. McClelland, K. Tammar, P.R. Van Horne, Transmission constrained production cost analysis in power system planning. IEEE Trans. Power Apparatus Syst. **103**(8), 2192–2198 (1984)
45. N. Fatema et al., *Intelligent Data-Analytics for Condition Monitoring: Smart Grid Applications*, 1E (Academic Press, 2021). ISBN: 978-0-323-85510-5. https://doi.org/10.1016/C2020-0-02173-0
46. J. Wernerus, L. Soder, Area price based multi-area economic dispatch with tie line losses and constraints, in *IEEE/KTH Stockholm Power Tech Conference*, Sweden (1995), pp. 710–715
47. A. Iqbal et al., *Meta Heuristic and Evolutionary Computation: Algorithms and Applications*, 1E, Studies in Computational Intelligence, vol. 916 (Springer Nature, Berlin, 2020), 849 p. ISBN 978-981-15-7571-6. https://doi.org/10.1007/978-981-15-7571-6
48. A. Borghetti, C. d'Ambrosio, A. Lodi, S. Martello, An MILP approach for short-term hydro scheduling and unit commitment with head-dependent reservoir. IEEE Trans. Power Syst. **23**(August (3)), 1115–1124 (2008)
49. J.P.S. Catalão, H.M.I. Pousinho, V.M.F. Mendes, Mixed-integer nonlinear approach for the optimal scheduling of a head-dependent hydro chain. Electr. Power Syst. Res. **80**(8), 935–942 (2010)
50. E.C. Finardi, E.L. Da Silva, Solving the hydro unit commitment problem via dual decomposition and sequential quadratic programming. IEEE Trans. Power Syst. **21**(May (2)), 835–844 (2006)
51. S. Soares, C.T. Salmazo, Minimum loss pre dispatch model for hydro electric power systems. IEEE Trans. Power Syst. **12**(August (3)), 1220–1228 (1997)
52. A. Arce, T. Ohishi, S. Soares, Optimal dispatch of generating units of the Itaipu hydro electric plant. IEEE Trans. Power Syst. **17**(February (1)), 154–158 (2002)
53. S. Soares, T. Ohishi, M. Cicogna, A. Arce, Dynamic dispatch of hydro generating units, in *IEEE Bologna Power Tech Conference Proceedings*, June, vol. 2 (2003), 6 p

Novel Approach for Power System Stability Analysis and Improvement: A Case Study Based on UPFC Application

Nikhil Kumar Sinha, Ramesh Devarapalli, and Pulivarthi Nageswara Rao

Abstract The exponential rise in the electrical power demand has led to various types of instability due to overburdening of the existing interconnected power systems. Voltage instability is one such phenomenon that can result in major blackouts. The primary cause of this problem is the collapsing of the transmission system due to the frequent and abrupt variations in the load, which gives rise to deviation in the normal operating conditions. In this chapter, the main focus is on detecting the most sensitive node (bus) in a modified IEEE-14 bus system and investigating a better and effective implementation of UPFC to improve the voltage profile and, thus, the voltage stability. For analyzing this condition, PSAT, a MATLAB-based Simulink and Simulation toolbox, which utilizes L-index method for voltage stability analysis and sensitive nodes determination, are used. The above setup is tested and compared for load variation in the step size of 5% from original power flow (OPF) to 40% load change with and without UPFC, and suitable conclusions are drawn from the results obtained so far.

Keywords L-Index · Sensitive node · Reactive power · Voltage stability · PSAT

Nomenclature

ACMV	Air-conditioning and mechanical ventilation
ANN	Artificial neural network
CO_2	Carbon dioxide
CRAE	Correlation attribute evaluation
DE	Differential evolution
FNN	Feedforward neural network

N. K. Sinha · R. Devarapalli (✉)
B.I.T. Sindri, Dhanbad, Jharkhand 828123, India
e-mail: ramesh.ee@bitsindri.ac.in

P. N. Rao
GITAM (Deemed to be University), Visakhapatnam, Andhra Pradesh 530045, India

H. Malik et al. (eds.), *Intelligent Data Analytics for Power and Energy Systems*, Lecture Notes in Electrical Engineering 802,
https://doi.org/10.1007/978-981-16-6081-8_6

GA	Genetic algorithm
GRAE	Gain ratio attribute evaluation
HR	Humidity ratio
HVAC	Heating, ventilation, and air-conditioning
IGAE	Information gain attribute evaluation
ORAE	One R attribute evaluation
PIR	Passive infrared
ppm	Part per million
PSO	Particle swarm optimization
RFAE	Relief F attribute evaluation
RH	Relative humidity
SA	Simulated annealing
StdDev	Standard deviation
SUAE	Symmetrical uncertainty attribute evaluation
Temp.	Temperature

1 Introduction

Due to the tremendous increment in electricity demand, almost all the existing power systems are operating in close proximity to their respective threshold limits. Thus, it becomes imperative to consider the usage of faster controlling and protective devices like PSS, FACTS, and relays. FACTS are a comparatively modern concept in the power system. Electronic devices work in coordination with conventional PSS devices and improve the overall system stability by reducing the transient period and attaining the steady state faster than PSS alone [1]. Various modifications have been made in designing and implementing these controllers for their versatility in different operating conditions [2]. For the analysis purpose of any power system, the first and foremost step is the load flow study (LFS) [3]. In this regard, a term called 'sensitivity' mark those nodes that generate a comparatively low voltage than the remaining nodes in the power system. A sensitive node in a power system depicts that node that is most severely affected by the load fluctuations. Nowadays, due to exponential and continuous increment in the demand of electricity, it is not feasible to reconstruct all the power system after dismantling the existing ones. Thus, to meet this demand, the existing power systems are interconnected via transmission lines. This, in turn, makes the transmission line, and thus power system, complex in terms of structure, maintenance, and control. This has to be done in order to make sure that during the transfer of electricity from one system to another, failure or collapse of transmission lines does not occur. This also means that optimal loading of the power system becomes quite imperative to increase the security and reliability of the transmission network [4]. Along with this, various kinds of indices are also suggested to prevent voltage collapse and possible blackouts [5].

Several blackouts have occurred since the first establishment of power system in 1889. These power failures covering the large spatial area have a multi-dimensional impact on the economy of the region as well because of heavy monetary losses faced by the industries affected by these blackouts. One such incident occurred in India on July 29th, 2012, and continued till July 31st, 2012 [6]. After reading the report of this blackout, the following conclusions can be drawn: Certain factors are responsible for the occurrence of this blackout. First and foremost is the weak inter-regional power transmission between northern region (NR) and western region (WR). This is due to several lines were taken out for maintenance work leading to the scheduled outage of the line. In addition to this, some lines were disturbed due to the occurrence of faults leading to the forced outage. The second factor was the over-tension in the 400 kV Bina–Gwalior–Agra link. This led to false operating of the under-voltage relay. This was mainly due to the lack of coordination between state load dispatch centers (SLDCs) and regional load dispatch centers (RLDCs). The next factor that further worsened the situation was re-routing the power supply via the eastern region (ER). This led to an increment in the load angle of the WR due to increased line reactance, which further led to the phenomenon of power swings leading to false operation of under-voltage relays. This facilitates the voltage collapse and further load shedding and abrupt disbalance between load and supply, and further tripping of under-frequency relays and over-frequency relays in NR and ER regions, respectively. It can be concluded from the above scenario that such blackout could be avoided if the weakest link, i.e., the node which is most sensitive to load fluctuations, was known beforehand and some suitable protection scheme had been applied. In this chapter, the main focus is on the identification of the most sensitive node in an IEEE-14 bus system by altering the test system data by 5, 10, 15%, and so on. Then the obtained result is compared with the original power flow results of the IEEE-14 bus system [7] and effective usage of UPFC, a kind of FACTS devices, for the improvement of the voltage stability due to its flexible nature and faster control characteristics [8, 9].

In this chapter, some of the notable works done are as follows:

- Identification of weakest bus in IEEE-14 bus power system using L-index.
- Contribution of UPFC between the tie of weakest buses in voltage stability improvement of the whole power system.
- In normal and over loaded conditions, improvement of voltage profile with UPFC connected at different locations has also been studied.

Section 2 explains the identification of weak bus L-index, Sect. 3 gives a brief description on UPFC, Sect. 4 illustrates the results, and the analysis of the obtained result is done. Section 5 provides adequate concluding remarks.

2 L-Index: Weak Bus Identification

Various kinds of indices are being used for estimating the voltage stability of a power system. Along with this, the distance of the operating condition from the critical operating value (safety margin) is also being studied via these indices. Voltage stability indices (VSIs) can be broadly categorized into three types, i.e., Bus VSI, Line VSI, and Overall VSI. One such index belonging to the first type, i.e., Bus VSI, was proposed by the authors of [10].

In a power system of n-bus system if all the load connected buses are brought at the head (i.e., $\alpha_L = \{1, 2, 3, \ldots, N_{L-1}, N_L\}$) and the remaining generator connected buses are brought at the end (i.e., $\alpha_G = \{N_{L+1}, N_{L+2}, \ldots, N_{n-1}, N_n\}$), then the following equation can be derived at [11]

$$\begin{bmatrix} V^L \\ I^G \end{bmatrix} = H\begin{bmatrix} I^L \\ V^G \end{bmatrix} = \begin{bmatrix} Z^{LL} & F^{LG} \\ K^{GL} & Y^{GG} \end{bmatrix}\begin{bmatrix} I^L \\ V^G \end{bmatrix}$$

where

V^L	Voltage at load bus
I^G	Current at generator bus
I^L	Current at load bus
V^G	Voltage at generator bus
H	Hybrid parameter matrix
Z^{LL}	Impedance at load buses
Y^{GG}	Admittance at generator buses
F^{LG}	Voltage gain
K^{GL}	Current gain.

With the help of above equation, L-index of any bus $j\epsilon$ can be found by the following equation

$$L_j = |l_j| = \left|1 - \frac{\sum_{i\epsilon\alpha_G} F_{ji} V_i}{V_j}\right| = \left|\frac{S_j^+}{Y_{jj}^{+*} V_j^2}\right|$$

where

$$S_j^+ = \left(\sum_{k\epsilon\alpha_L} \frac{Z_{jk}^* S_k}{Z_{jj}^* V_k}\right) V_j$$

with S_k and V_k as the complex power and voltage of the buses, respectively.

The value of L-index is given as $L = \max_{j\epsilon\alpha_L}(L_j)$.

The value of L-index obtained through the above equation lies in the range of 0 and 1. A bus with 0 as its L-index indicates no-load condition, whereas a bus with its L-index approaching 1, depicts very high vulnerability to disturbances and very high probability of voltage collapse. The lesser the L-index of a bus, the higher its stability margin [12]. It can be concluded that the bus with highest value of L-index is the most critical bus in the power system.

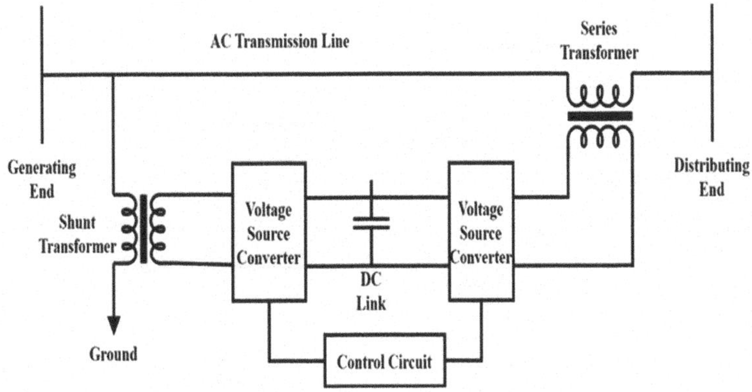

Fig. 1 A modified IEEE-14 bus test system

3 Unified Power Flow Controller (UPFC)

In any power system, series and shunt compensators controls the power flow and voltage profile of a transmission line, respectively. UPFC is a unique kind of controller that alone can fulfill the function of both series and shunt compensators. In other words, UPFC can efficiently control all the parameters of a transmission line. This is so because of its series and shunt voltage converters connected via a common DC link. The schematic diagram of UPFC is given in Fig. 1. UPFC can correct voltage profile by injection of reactive current at AC coupling, thus acting as shunt compensator. Similarly, by injection of voltage between the buses, UPFC acts as a series compensator [13].

4 Result Demonstration and Analysis

In the proposed chapter, an experiment with a modified IEEE-14 bus system is done. For power system analysis, PSAT, a MATLAB-based Simulink and Simulation tool [14] is used. The under-considered system of modified IEEE-14 bus system is designed as a Simulink model, and the standard test data have been used as shown in Fig. 2.

As mentioned earlier, the first and foremost thing during the analysis of any power system, is the load flow studies (LFS). So, the LFS is done with the help of Newton–Raphson method (NRM). This method has been chosen because of its quadratic convergent nature, which means the faster achievement of results for a complex system. The data obtained so far have been presented in the tabulated form in Table 1.

After obtaining the original power flow (OPF) data via LFS, the next thing to be considered is the deviation in the OPF value with incremental change in the load.

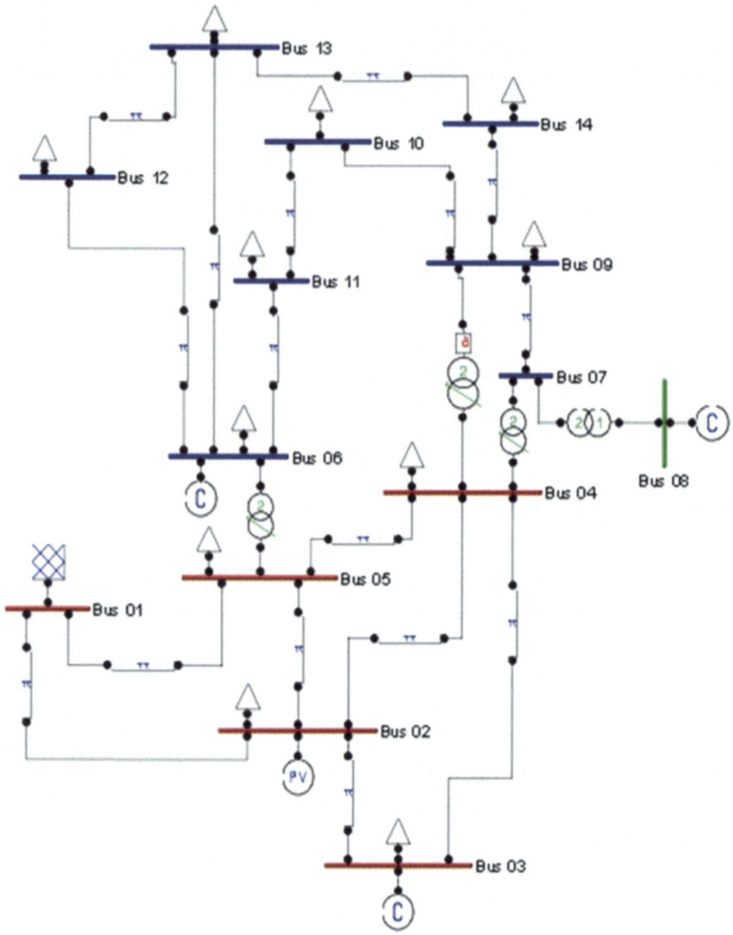

Fig. 2 A modified IEEE-14 bus test system

Hence, the load has been changed in the step size of 5% till 40% is reached, i.e., 5% (case 1), 10% (case 2), 15% (case 3), 20% (case 4), 25% (case 5), 30% (case 6), 35% (case 7), and 40% (case 8), and at each increment, LFS has been done. The obtained data is given in Table 2 and plotted in Fig. 3. The average value of voltage deviation and its difference from OPF are also obtained and given in Table 2.

In the next case, UPFC is inserted between Bus 9 and Bus 14 and all the steps performed above are repeated and data is tabulated in Table 3 and plotted in Fig. 4.

In this case, UPFC is connected between bus 5 and bus 6 and similar to the above two steps, data is obtained and tabulated in Table 4 and plotted in Fig. 5.

Table 1 Power flow result of IEEE-14 bus model

Bus No.	Voltage (p.u.)	Phase (rad)	P. Gen (MW)	Q. Gen (MW)	P. Load (MW)	Q. Load (MW)
1	1.060	0	3.5205	0.2789	0	0
2	1.045	−0.135	0.4000	0.9513	0.3038	0.1778
3	1.010	−0.331	0	0.5979	1.3188	0.2660
4	0.9972	−0.263	0	0	0.6692	0.0560
5	1.0024	−0.227	0	0	0.1064	0.02240
6	1.070	−0.379	0	0.4426	0.1568	0.1050
7	1.0347	−0.354	0	0	0	0
8	1.090	−0.354	0	0.3424	0	0
9	1.0111	−0.401	0	0	0.4130	0.2324
10	1.0105	−0.404	0	0	0.1260	0.0812
11	1.0346	−0.395	0	0	0.0490	0.0252
12	1.0461	−0.401	0	0	0.0854	0.0224
13	1.0362	−0.403	0	0	0.1890	0.0812
14	0.99568	−0.428	0	0	0.2086	0.0700

Table 2 Power flow result showing voltage at various percentage change in load without UPFC for determination of sensitive node

	OPF	Case 1	Case 2	Case 3	Case 4	Case 5	Case 6	Case 7	Case 8	AVG	DIFF
1	1.060	1.060	1.060	1.060	1.060	1.060	1.060	1.060	1.060	1.060	0
2	1.045	1.045	1.045	1.045	1.045	1.045	1.045	1.045	1.045	1.045	0
3	1.010	1.010	1.010	1.010	1.010	1.010	1.010	1.010	1.010	1.010	0
4	0.998	0.995	0.992	0.989	0.986	0.983	0.979	0.976	0.973	0.985	0.0133
5	1.002	0.997	0.997	0.994	0.991	0.988	0.985	0.982	0.979	0.989	0.0131
6	1.070	1.070	1.070	1.070	1.070	1.070	1.070	1.070	1.070	1.070	0
7	1.035	1.032	1.029	1.027	1.024	1.021	1.018	1.015	1.013	1.026	0.0127
8	1.090	1.090	1.090	1.090	1.090	1.090	1.090	1.090	1.090	1.090	0
9	1.011	1.007	1.003	0.999	0.996	0.992	0.987	0.983	0.979	0.994	0.0175
10	1.010	1.007	1.003	0.999	0.996	0.992	0.988	0.983	0.979	0.993	0.0170
11	1.035	1.032	1.030	1.028	1.026	1.023	1.021	1.019	1.017	1.025	0.0098
12	1.046	1.045	1.043	1.042	1.041	1.039	1.038	1.036	1.035	1.040	0.0059
13	1.036	1.034	1.032	1.030	1.028	1.026	1.024	1.022	1.021	1.027	0.0086
14	0.996	0.991	0.987	0.982	0.978	0.973	0.968	0.963	0.961	0.975	0.0202

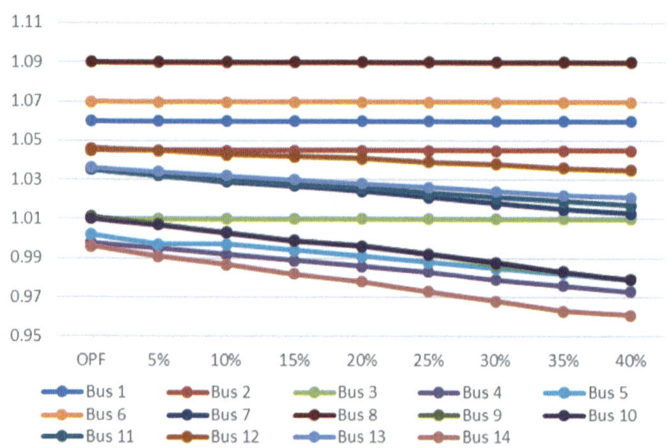

Fig. 3 Power flow result showing voltage at various percentage change in load without UPFC

In this case, UPFC is connected between bus 4 and bus 7 and similar to above two steps, data is obtained and tabulated in Table 5 and plotted in Fig. 6.

Table 3 Power flow result at various percentage change in load with UPFC connected between bus 9 and bus 14

	OPF	Case 1	Case 2	Case 3	Case 4	Case 5	Case 6	Case 7	Case 8	AVG	DIFF
1	1.060	1.060	1.060	1.060	1.060	1.060	1.060	1.060	1.060	1.060	0
2	1.045	1.045	1.045	1.045	1.045	1.045	1.045	1.045	1.045	1.045	0
3	1.010	1.010	1.010	1.010	1.010	1.010	1.010	1.010	1.010	1.010	0
4	0.997	0.995	0.992	0.989	0.986	0.983	0.979	0.976	0.973	0.984	0.0134
5	1.002	0.999	0.997	0.994	0.991	0.988	0.985	0.982	0.978	0.989	0.0120
6	1.070	1.070	1.070	1.070	1.070	1.070	1.070	1.070	1.070	1.070	0
7	1.034	1.031	1.028	1.026	1.023	1.019	1.017	1.014	1.011	1.021	0.0124
8	1.090	1.090	1.090	1.090	1.090	1.090	1.090	1.090	1.090	1.090	0
9	1.009	1.005	1.001	0.997	0.993	0.989	0.985	0.981	0.976	0.991	0.0170
10	1.009	1.005	1.001	0.997	0.993	0.989	0.985	0.981	0.977	0.991	0.0170
11	1.034	1.031	1.029	1.027	1.025	1.022	1.020	1.018	1.015	1.023	0.0100
12	1.047	1.046	1.045	1.043	1.042	1.041	1.039	1.038	1.037	1.041	0.0057
13	1.038	1.036	1.035	1.033	1.031	1.029	1.027	1.026	1.024	1.030	0.0081
14	1.005	1.001	0.997	0.993	0.989	0.985	0.980	0.976	0.972	0.986	0.0180

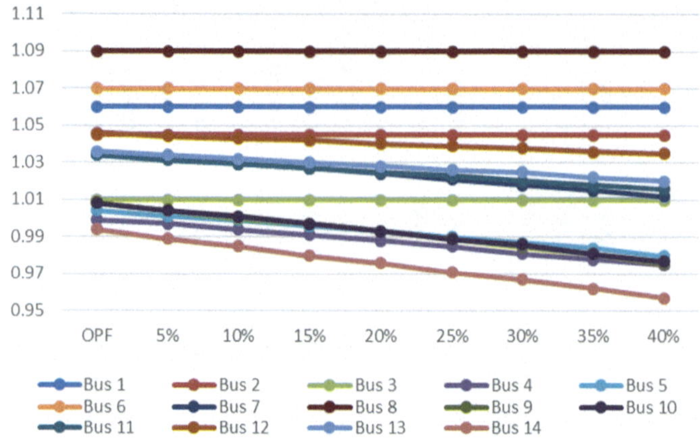

Fig. 4 Power flow result at various percentage change in load with UPFC connected between bus 9 and bus 14

Table 4 Power flow result at various percentage change in load with UPFC connected between bus 5 and bus 6

	OPF	Case 1	Case 2	Case 3	Case 4	Case 5	Case 6	Case 7	Case 8	AVG	DIFF
1	1.060	1.060	1.060	1.060	1.060	1.060	1.060	1.060	1.060	1.060	0
2	1.045	1.045	1.045	1.045	1.045	1.045	1.045	1.045	1.045	1.045	0
3	1.010	1.010	1.010	1.010	1.010	1.010	1.010	1.010	1.010	1.010	0
4	1.005	1.005	1.002	0.999	0.997	0.994	0.991	0.987	0.984	0.995	0.0105
5	1.015	1.016	1.013	1.010	1.007	1.005	1.002	0.999	0.996	1.005	0.0092
6	1.070	1.070	1.070	1.070	1.070	1.070	1.070	1.070	1.070	1.070	0
7	1.038	1.036	1.034	1.031	1.028	1.026	1.023	1.019	1.017	1.027	0.0116
8	1.090	1.090	1.090	1.090	1.090	1.090	1.090	1.090	1.090	1.090	0
9	1.015	1.010	1.006	1.003	0.998	0.995	0.996	0.986	0.982	0.996	0.0185
10	1.014	1.009	1.005	1.002	0.998	0.994	0.989	0.986	0.982	0.996	0.0181
11	1.036	1.033	1.031	1.029	1.026	1.024	1.022	1.019	1.017	1.025	0.0101
12	1.046	1.045	1.044	1.042	1.041	1.039	1.038	1.037	1.036	1.040	0.0060
13	1.037	1.034	1.032	1.030	1.029	1.027	1.025	1.023	1.021	1.028	0.0091
14	0.998	0.993	0.988	0.984	0.979	0.975	0.970	0.965	0.960	0.977	0.0212

In this last case, UPFC is connected between bus 4 and bus 9 and similar to above two steps, data is obtained and tabulated in Table 6 and plotted in Fig. 7.

In the above four cases, UPFC has been connected between buses 9 and 14, 5 and 6, 4 and 7, 4 and 9 as provided in Tables 3, 4, 5 and 6 and graph 2, 3, 4 and 5, respectively. After obtaining the power flow result at each percentage change in load, the voltage profile at each change is tabulated and average of each change is

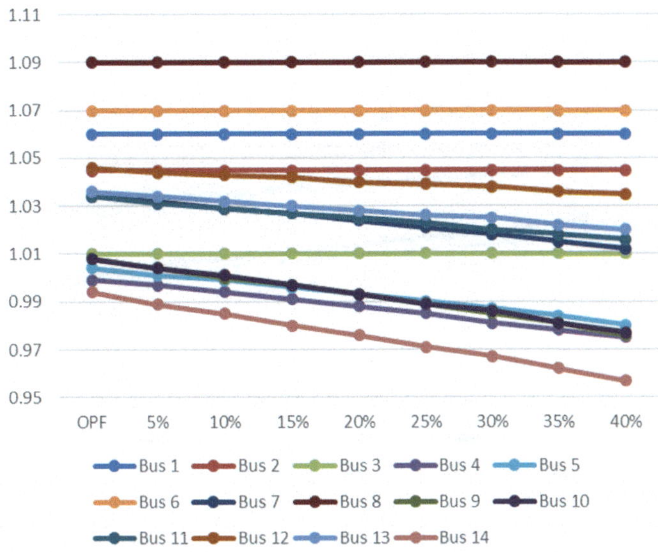

Fig. 5 Power flow result at various percentage change in load with UPFC connected between bus 5 and bus 6

Table 5 Power flow result at various percentage change in load with UPFC connected between bus 4 and bus 7

	OPF	Case 1	Case 2	Case 3	Case 4	Case 5	Case 6	Case 7	Case 8	AVG	DIFF
1	1.0600	1.0600	1.0600	1.0600	1.0600	1.0600	1.0600	1.0600	1.0600	1.0600	0
2	1.0450	1.0450	1.0450	1.0450	1.0450	1.0450	1.0450	1.0450	1.0450	1.0450	0
3	1.0100	1.0100	1.0100	1.0100	1.0100	1.0100	1.0100	1.0100	1.0100	1.0100	0
4	1.0013	0.9985	0.9957	0.9928	0.9898	0.9867	0.9835	0.9802	0.9768	0.9880	0.0132
5	1.0047	1.0021	0.9994	0.9966	0.9937	0.9907	0.9876	0.9844	0.9810	0.9919	0.0127
6	1.0700	1.0700	1.0700	1.0700	1.0700	1.0700	1.0700	1.0700	1.0700	1.0700	0
7	1.0258	1.023	1.021	1.018	1.015	1.012	1.009	1.007	1.003	1.024	0.0014
8	1.0900	1.0900	1.0900	1.0900	1.0900	1.0900	1.0900	1.0900	1.0900	1.0900	0
9	1.0062	1.002	0.9990	0.995	0.991	0.987	0.983	0.979	0.974	0.988	0.0174
10	1.0065	1.003	0.999	0.996	0.992	0.988	0.984	0.979	0.976	0.989	0.0169
11	1.0327	1.031	1.028	1.026	1.024	1.022	1.019	1.017	1.015	1.023	0.0098
12	1.0457	1.044	1.043	1.042	1.040	1.039	1.038	1.036	1.035	1.039	0.0060
13	1.0355	1.034	1.032	1.029	1.028	1.026	1.024	1.022	1.019	1.027	0.0085
14	0.9926	0.988	0.984	0.979	0.975	0.970	0.965	0.960	0.956	0.972	0.0203

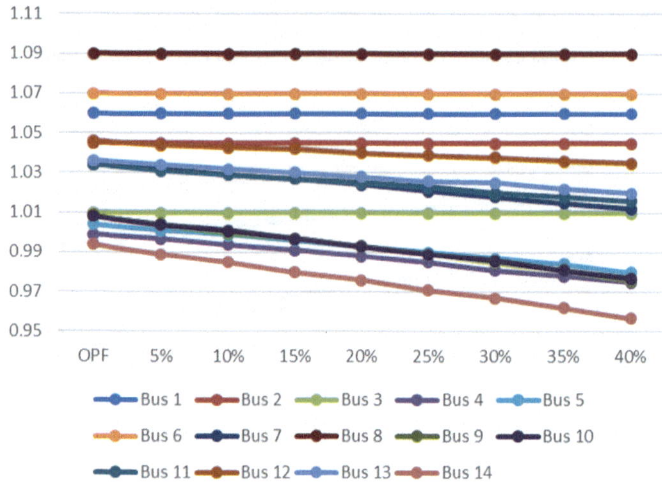

Fig. 6 Power flow result at various percentage change in load with UPFC connected between bus 4 and bus 7

Table 6 Power flow result at various percentage change in load with UPFC connected between bus 4 and bus 9

	OPF	Case 1	Case 2	Case 3	Case 4	Case 5	Case 6	Case 7	Case 8	AVG	DIFF
1	1.060	1.060	1.060	1.060	1.060	1.060	1.060	1.060	1.060	1.060	0
2	1.045	1.045	1.045	1.045	1.045	1.045	1.045	1.045	1.045	1.045	0
3	1.010	1.010	1.010	1.010	1.010	1.010	1.010	1.010	1.010	1.010	0
4	0.999	0.997	0.994	0.991	0.988	0.985	0.981	0.978	0.975	0.986	0.0134
5	1.004	1.001	0.999	0.996	0.993	0.990	0.987	0.984	0.980	0.991	0.0128
6	1.070	1.070	1.070	1.070	1.070	1.070	1.070	1.070	1.070	1.070	0
7	1.034	1.032	1.029	1.027	1.024	1.021	1.018	1.015	1.012	1.022	0.0119
8	1.090	1.090	1.090	1.090	1.090	1.090	1.090	1.090	1.090	1.090	0
9	1.008	1.004	1.000	0.997	0.993	0.989	0.985	0.981	0.976	0.991	0.0170
10	1.008	1.004	1.001	0.997	0.993	0.989	0.986	0.981	0.977	0.991	0.0160
11	1.034	1.031	1.029	1.027	1.025	1.023	1.020	1.018	1.016	1.024	0.0090
12	1.046	1.044	1.043	1.042	1.040	1.039	1.038	1.036	1.035	1.039	0.0059
13	1.036	1.034	1.032	1.030	1.028	1.026	1.025	1.022	1.020	1.027	0.0086
14	0.994	0.989	0.985	0.980	0.976	0.971	0.967	0.962	0.957	0.973	0.0202

estimated. And then we have subtracted the average power flow value to the original power flow result. As stated earlier, the bus having highest voltage difference is considered as the most sensitive load. For the ease of comparison, a graph showing the differences in the values of voltage is plotted as Fig. 8.

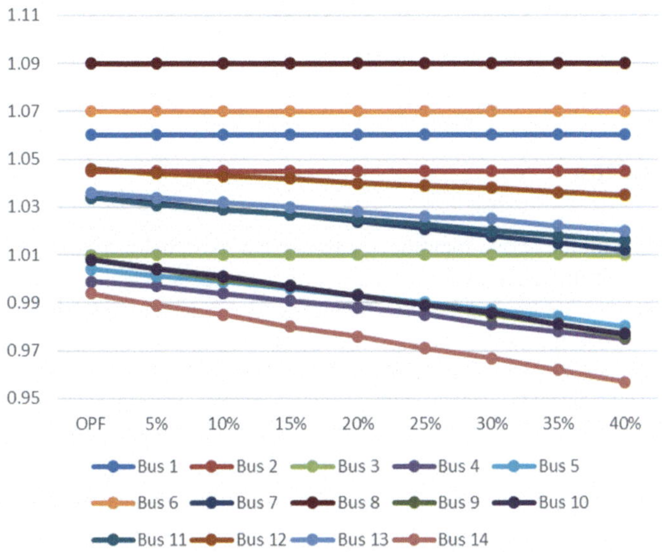

Fig. 7 Power flow result at various percentage change in load with UPFC connected between bus 4 and bus 9

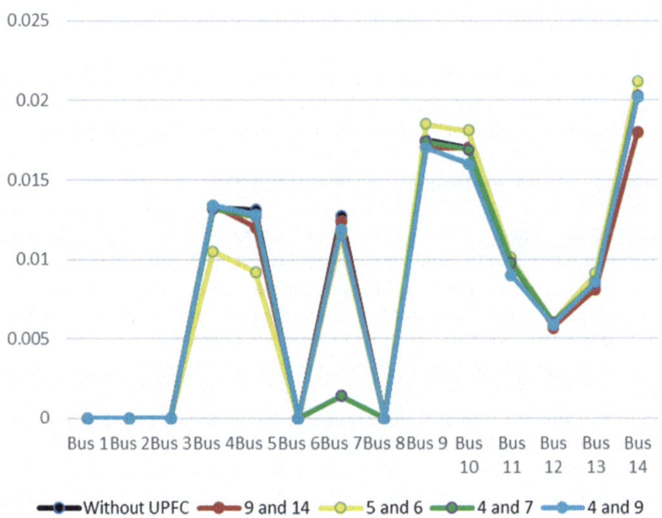

Fig. 8 Comparison of effectiveness of UPFC

The graph shows that for lower-order buses (like bus 6, 7 and 8) UPFC connected between bus 4 and 7 is effective. But its effectiveness become gloomy for order buses (like bus 14). Also, the voltage profile of bus 9 and 14 are pretty much higher than rest of the buses. So, the performance of UPFC connected between bus 9 and 14, seems to serve its purpose most efficiently.

5 Conclusion

After studying all the above-mentioned cases, it can be concluded that buses having higher deviation (difference) from their original power flow value under an incremental change in the load, are most vulnerable to voltage collapse. Hence, special attention is needed for them in order to reduce the possibility of voltage collapse and avoid the severeness of blackouts. Further, the implementation of FACTS devices helps in the reduction of such possibility if correctly coordinated. In case of modified IEEE-14 bus system, bus 9 and 14 are more vulnerable to the phenomenon of voltage collapse due to their higher L-index value. Hence, the UPFC should be connected between these buses for overall effective improvement of voltage profile.

References

1. N.G. Hingorani, L. Gyugyi, *Understanding FACTS: Concepts and Technology of Flexible AC Transmission Systems*
2. D.V.N. Ananth, L.V. Sureshkumar, M. Boddepalli, Modelling and design of static compensator and UPFC based FACTS devices for power system oscillations damping and voltage compensation, in *Intelligent Computing in Control and Communication*, Singapore (2021), pp. 357–371. https://doi.org/10.1007/978-981-15-8439-8_29
3. J.J. Grainger, *Power System Analysis* (McGraw-Hill, 1999)
4. J. Vishnu et al., A strategy for optimal loading pattern of a typical power system—a case study, in *2014 International Conference on Advances in Green Energy (ICAGE)*, Thiruvananthapuram, India, Dec 2014, pp. 112–117. https://doi.org/10.1109/ICAGE.2014.7050152
5. C. Reis, A. Andrade, F.P.M. Barbosa, Methods for preventing voltage collapse, in *2009 International Conference on Electrical and Electronics Engineering—ELECO 2009*, Nov 2009, pp. I-368–I-372. https://doi.org/10.1109/ELECO.2009.5355333
6. M. Parihar, M.K. Bhaskar, Review of power system blackout. Int. J. Res. Innov. Appl. Sci. **3**, 8–12 (2018)
7. P. Mishra, H.N. Udupa, P. Ghune, Calculation of sensitive node for IEEE–14 bus system when subjected to various changes in load (2013)
8. B. Singh, N.K. Sharma, A.N. Tiwari, K.S. Verma, D. Singh, Enhancement of voltage stability by coordinated control of multiple FACTS controllers in multi-machine power system environments, in *International Conference on Sustainable Energy and Intelligent Systems (SEISCON 2011)*, Chennai, India (2011), pp. 18–25. https://doi.org/10.1049/cp.2011.0328
9. P.K. Iyambo, R. Tzoneva, Transient stability analysis of the IEEE 14-bus electric power system, in *AFRICON 2007*, Windhoek, South Africa, Sept 2007, pp. 1–9. https://doi.org/10.1109/AFRCON.2007.4401510
10. P. Kessel, H. Glavitsch, Estimating the voltage stability of a power system. IEEE Trans. Power Deliv. **PWRD-1**(3), 346–352 (1986)
11. J. Hongjie, Y. Xiaodan, Y. Yixin, An improved voltage stability index and its application. Int. J. Electr. Power Energy Syst. **27**(8), 567–574 (2005). https://doi.org/10.1016/j.ijepes.2005.08.012

12. A. Oukennou, A. Sandali, Assessment and analysis of voltage stability indices in electrical network using PSAT software, in *2016 Eighteenth International Middle East Power Systems Conference (MEPCON)*, Cairo, Egypt, Dec 2016, pp. 705–710. https://doi.org/10.1109/MEPCON.2016.7836970
13. S. Parvathy, K.C.S. Thampatty, T.N.P. Nambiar, Design and implementation of partial feedback linearization controller for unified power flow controller. Electr. Power Syst. Res. **187**, 106438 (2020). https://doi.org/10.1016/j.epsr.2020.106438
14. F. Milano, An open source power system analysis toolbox. IEEE Trans. Power Syst. **20**(3), 1199–1206 (2005). https://doi.org/10.1109/TPWRS.2005.851911

Intelligent Modelling and Analysis of *P–Q* Control Technique for SPV Plant Supplying Power to Grid

Nasir Ul Islam Wani, Farhad Ilahi Bakhsh, Pallavi Choudekars, and Ruchira

Abstract In remote areas, losses due to transmission of electrical energy from main grid are very high. In such cases, local generators such as solar photovoltaic (SPV) generators, wind generators, and diesel generators are used to meet up local demand. Individual operation of these generators is highly unreliable, and hence, individual generator is linked to the grid. Here, the operation of SPV plant feeding power to the grid using *P–Q* control technique has been analyzed. For analysis, the system is modelled in MATLAB/Simulink. The capacity of SPV plant is kept 100 kW_p, and its output is given to the maximum power point tracker (MPPT). The MPPT tracks the maximal output power of SPV plant using perturb and observe (P&O)-based algorithm and supplies it to the grid. In the control algorithm, the grid side voltage is transformed from *abc* to $\alpha\beta$ by means of parks transformation, and then, on implementing phase-locked loop (PLL) the required V_d and V_q values are obtained. In similar manner, the inverter currents are decoupled along *d–q* axis. The reference value for the *d–q* frame is generated from the differences of $P_{\text{reference}}$ (reference active power) and i_{actual} (actual active power) using PI controller. The compared output is used to generate the triggering pulses for the inverter, so that the inverter supplies required active power to the grid. The obtained outcomes demonstrate that the SPV plant effectively controls the active power fed to the grid by implementation of proposed *P–Q* control algorithm.

Keywords Solar photovoltaic (SPV) plant · Phase-locked loop (PLL) · Active power control · *P–Q* control algorithm · Grid connected

N. U. I. Wani (✉) · P. Choudekars · Ruchira
Electrical Engineering Department, Amity University, Noida, Uttar Pradesh 201313, India

P. Choudekars
e-mail: pachoudekar@amity.edu

F. I. Bakhsh
Electrical Engineering Department, National Institute of Technology Srinagar, Hazratbal, Srinagar, J&K 190006, India
e-mail: farhad@nitsri.ac.in

H. Malik et al. (eds.), *Intelligent Data Analytics for Power and Energy Systems*, Lecture Notes in Electrical Engineering 802,
https://doi.org/10.1007/978-981-16-6081-8_7

Nomenclature

V_D, V_Q	Direct and quadrature components of voltage
I_d, I_q	Direct and quadrature components of current
V_{op}	Voltage output at boost converter
V_{inp}	Voltage input from SPV plant
D	Duty ratio
I_{op}	Current output at boost converter
I	$I_{Maximum} - I_{minimum}$
f_c	$\frac{f_{sw}}{10}$
f_{sw}	Switching frequency
α	Firing angle
I_{max}	Maximum current of inductor
L	Filter inductance
C	Filter capacitance
V_{inv}	Inverter output voltage
P&Q	Active and reactive power

1 Introduction

In remote areas, losses due to transmission of electrical energy from main grid are very high. When the consumers are stationed far remote from the main grid, then the transmission of energy proves to be costly and many times breakdown of the grid occurs due to overburdening of transmission lines which effects the resoluteness of power systems. Thus, ample experimentation is being effectuated for extended supply of power to the load. In such cases, local generators such as solar photovoltaic (SPV) generators can be used to meet up the local demand [1–7]. These local generating stations, i.e., SPV, can be operated with the main grid and thus forms a microgrid. The microgrids are developing largely due to their environmental protection and reliable operation. Moreover they can provide network support during large demands and aiding restoration after faults. Microgrid can be of two types AC or DC microgrid. Since most of electrical loads are being operated with AC power supply, and thus, AC microgrid is formed. A microgrid can be operated in various modes as in grid linked mode where a principal grid remains connected to it and in islanded mode where the principal grid is isolated from the system. In grid connected mode, microgrid supplies and draws power according to generation and load demand. Further, it has also be seen that the changeover between the different modes of operation needs proper design of controllers that may be useful in transition of modes. The voltage and frequency regulation is maintained by the principal grid.

The leading flaw of solar photovoltaic (SPV) plant is ambiguity in accessibility of sun energy due to the earth's phenomenon of rotation, various circumstances such as fleeting clouds and weather conditions. Consequently, to counter play such circumstance, MPPT algorithms are contrived. The diversified MPPT algorithms are deliberated in [8–10], illustrating the benefits and pitfalls of respective procedures in terms of sensed specifications, accurateness, computation and complicacy. Perturb and observe and incremental conductance are tactics where minimal calculation are recommended and provide minimum complexity at the stake of downgraded accuracy.

In [11] and [12], the inverters control is provided by employing droop control schemes in both modes of operation. The main superiority of droop control scheme is that a decentralized control can be achieved and can be used for the system using multiple DG's. The coordinated operation of the inverter modules [13] shows the viability of power allocation between parallel connected modules. In [14], an autonomous control strategy for controlling the active power has been formulated so as to achieve management of the power in a decentralized manner.

The control schemes are necessary to assure a convenient progression from constant *P–Q* state to constant *f–V* state and appropriate application in grid linked and islanded mode [15]. The proper control algorithms are required in the progression from the grid linked state to the islanded state. The most popular control algorithm are discussed [16, 17]. In the collaborated *V–f* and *P–Q* state, SPV plants with MPPT shows the operation of SPV for tracking maximum power [18–22]. Jiang and Yu [19] shows the methodology for controlling the parallel inverters in order to achieve an eminent regulation of voltage in the microgrid. An advanced active power and voltage control strategy for inverters and droop control method for allocating power between the parallel inverter has also been demonstrated. The intelligent based approach has been incorporated in [23–27].

In this chapter, the operation of SPV plant feeding power to the grid using *P–Q* control technique has been analyzed. In this research a centralized control strategy is developed for a microgrid in a grid connected mode. This control strategy ensures maximum power tracking with various irradiance and temperature. For the operation of the inverter at changeable frequency and measurement of frequency of output line currents, a phase-locked loop (PLL) is required. As soon as inverter is linked to the power grid, the grid manages voltage and inverter caters to the power control.

This chapter has been divided into six sections. Section 2 presents the system description. Section 3 gives the proposed control strategy for both the DC–DC converter and DC–AC inverter. Section 4 implements the simulation model of 100 kW_p SPV plant connected to grid. Section 5 shows the results obtained using proposed control technique. At last, Sect. 6 summarizes the chapter.

2 System Description

Figure 1 represents block diagram of grid connected SPV plant. The SPV plant is connected to the DC–DC converter and then a DC–AC inverter. The switching of DC–AC converter will result in a large number of harmonics and the tuned LC filter will reduce the harmonics. The converters are guided by dint of *P–Q* control

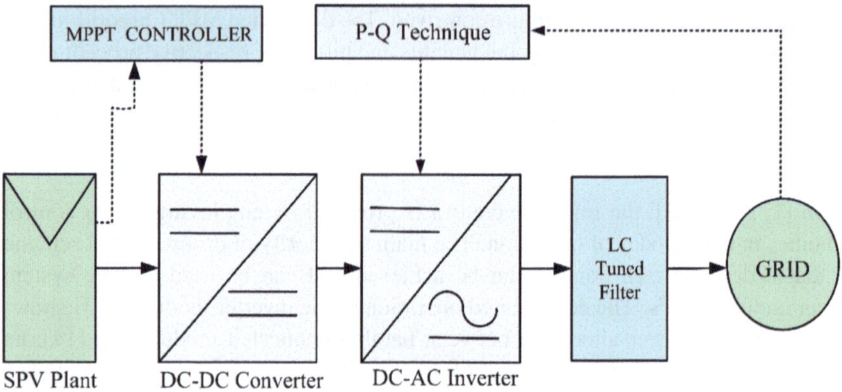

Fig. 1 Block diagram of a SPV plant connected to the grid

technique which controls the duty ratio of DC–DC converter and triggering switches for AC–DC inverter.

2.1 SPV Plant

The SPV plant is a system used to supply electrical power by converting the solar radiations directly into the electrical power. It implements a series and parallel combination of PV arrays. The 100 kW_p SPV plant uses 305 Sun power module (SPR-305E-WHT-D). The plant comprises of 66 parallel strings of5series connected modules per string. The SPV plant has two inputs, i.e., irradiation in Watt/metre^2 (W/m^2) and temperature in degree Celsius (°C). The signal builder block brings forth the irradiance and temperature profiles which is coupled to SPV plant inputs.

Figure 2 shows the input provided to the SPV. During $t = 0$ s to $t = 0.5$ s, irradiance is kept constant at 1000 W/m^2 then during $t = 0.5$ s to $t = 1$ s, it decreases linearly from 1000 to 600 W/m^2. Along $t = 1$ s to $t = 4$ s, irradiance is kept constant at 600 W/m^2. At $i = 4$ s, irradiance is increased from 600 W/m^2 to 800 W/m^{2}and during $t = 4$ s to $t = 8$ s, irradiance remains constant at 800 W/m^2. At $t = 8$ s, irradiance is increased from 800 to 1000 W/m^{2}and it remains constant at 1000 W/m^{2}from $t = 8$ s to $t =$ 10 s. The input temperature is varied between the 25 and 45 °C. During $t = 0$ s to $t =$ 2 s, the temperature is kept at of 25 °C. From $t = 2$ s to $t = 3$ s, temperature escalates linearly from 25 to 45 °C. During $t = 3$ s to $t = 4$ s temperature is maintained constant at 45 °C. From $t = 4$ s to $t = 5$ s, temperature decreases linearly from 45 to 25 °C. During $t = 5$ s to $t = 10$ s, it is kept constant at 25 °C.

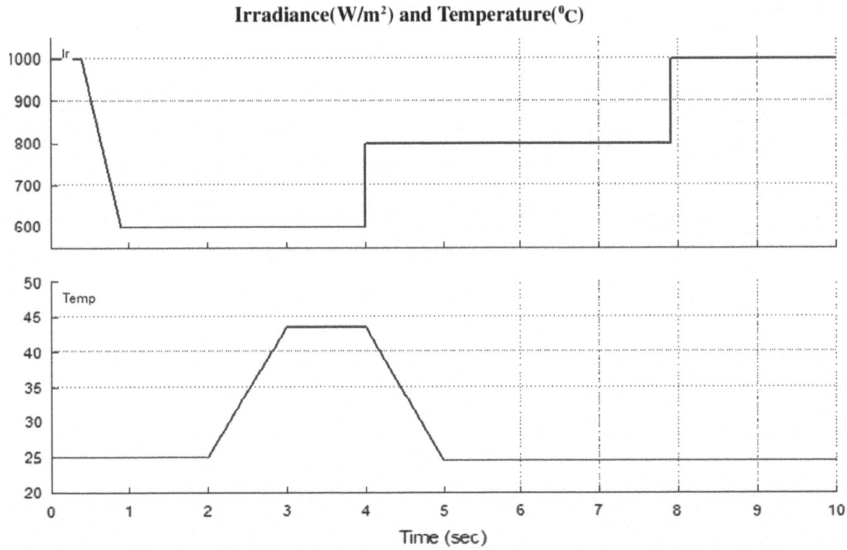

Fig. 2 Variation of input irradiance and temperature given to the SPV plant

2.2 DC–DC Converter

A DC–DC converter is employed to generate required DC voltage from varying DC supply (coming from SPV plant). Here, a boost converter is used. The DC–DC boost converter is a step-up converter with the output voltage (average value) exorbitant than input voltage. The output voltage, inductance (L) and capacitance (C) are calculated as [20]:

$$V_{\text{op}} = \frac{V_{\text{inp}}}{1 - D} \tag{1}$$

$$L = \frac{V_{\text{inp}}\left(V_{\text{op}} - V_{\text{inp}}\right)}{f_{\text{sw}} \times \Delta I \times V_{\text{op}}} \tag{2}$$

$$C = \frac{I_{\text{op}}\left(V_{\text{op}} - V_{\text{inp}}\right)}{f_{\text{sw}} \times \Delta I \times V_{\text{op}}} \tag{3}$$

In SPV, for each output value of voltage and current, there is a spot at where the maximal power can be notched from it. Thus, operation of SPV is done at the point of maximum power, and this is called as maximal power tracing. This is done by using MPPT algorithm by regulating the duty ratio (D) of DC–DC converter [21]. The V_{PV} and I_{PV} are given to the MPPT, and the required duty ratio is effectuated for regulating the DC–DC converter.

2.3 DC–AC Inverter

The DC–AC inverter transubstantiates the DC output of boost converter into required C supply. In this work, three-phase PWM converter is used which is modelled by means of 3-arms bridge using IGBT switches. The output voltage of inverter is given by Chen and Wang [18]:

$$V_{\text{inv}} = \frac{V_{\text{op}} \times \pi}{3\sqrt{6}\cos\alpha} \tag{4}$$

2.4 Utility Grid and Filter

The utility grid is having a three-phase, 415 V_{rms} (L–L), 50 Hz supply. The SPV plant supplies power to grid through inverter and filter. The switching of the inverter leads to a large number of harmonics. These harmonies are eliminated by an LC fitter. The corresponding values of L and C are calculated as [18]:

$$L \le \frac{0.03V_{\text{inv}}}{2 \times f_{\text{c}} \times I_{\max}} \tag{5}$$

$$C \le \frac{1}{4\pi f_{\text{c}}^2 L} \tag{6}$$

3 Converter and Inverter Controller

For the control of DC–DC converter and DC–AC inverter, a proper control strategy is required. The separate control algorithm are applied to guide duty ratio of DC–DC converter and switching of inverter switches. The two controllers used are:

3.1 Control of DC–DC Converter

The DC–DC converter controller is utilized for implementing the maximum power point tracking. The operation of SPV plant at maximum power point P is done by controlling the duty ratio (D). The typical duty ratio D is calculated as:

$$D = D_{\text{old}} - \Delta D \quad \forall \Delta p < 0 \text{ and } \Delta V < 0 \tag{7}$$

$$D = D_{\text{old}} + \Delta D \quad \forall \Delta p < 0 \text{ and } \Delta V < 0 \tag{8}$$

$$D = D_{\text{old}} - \Delta D \quad \forall \Delta p > 0 \text{ and } \Delta V > 0 \tag{9}$$

$$D = D_{\text{old}} + \Delta D \quad \forall \Delta p < 0 \text{ and } \Delta V > 0 \tag{10}$$

3.2 Control of DC–AC Inverter

In controlling of DC–AC inverter, the grid voltage is transformed from *abc* to $\alpha\beta$ by means of Parks transformation, and then, on implementing PLL, the required V_{D} and V_{Q} values are obtained. In similar manner, the inverter currents are decoupled to d–q axis. The inverter operates in current control mode, and the PLL is used for matching the frequency. The transformed currents are used for regulation of $P_{\text{calculated}}$ and $Q_{\text{calculated}}$ of the inverter. The $P_{\text{calculated}}$ and $Q_{\text{calculated}}$ power are given by equations as [13]:

$$P_{\text{calculated}} = 1.5 \times (V_{\text{D}} \times I_{\text{d}} + V_{\text{Q}} \times I_{\text{q}}) \tag{11}$$

$$Q_{\text{calculated}} = 1.5 \times (V_{\text{Q}} \times I_{\text{d}} - V_{\text{D}} \times I_{\text{q}}) \tag{12}$$

Figure 3 shows the overall regulation algorithm for the DC–AC inverter. First, the grid voltage is transformed into V_{D} and V_{Q} components, and the inverter currents are transformed to I_{d} and I_{q} components. Using Eqs. (11) and (12), the active and reactive power are obtained. The obtained P and Q are compared with reference P_{ref} and Q_{ref}, respectively. The errors of P and Q are given to a PI controller as shown. Similarly, the currents are compared and added to the voltages to initiate the required PWM signal for controlling the firing of the inverter.

4 Simulation Model of SPV Plant Connected to Grid

The complete simulation model of a SPV plant linked to the grid is displayed in Fig. 4. SPV plant is concatenated to a DC–DC converter for enforcement of MPPT, so the maximal power of the SPV plant is attained during the varying irradiation and temperature. The DC–DC converter is modelled using boost (average) block whose output is given to DC–AC inverter. The grid is modelled using a three-phase source block.

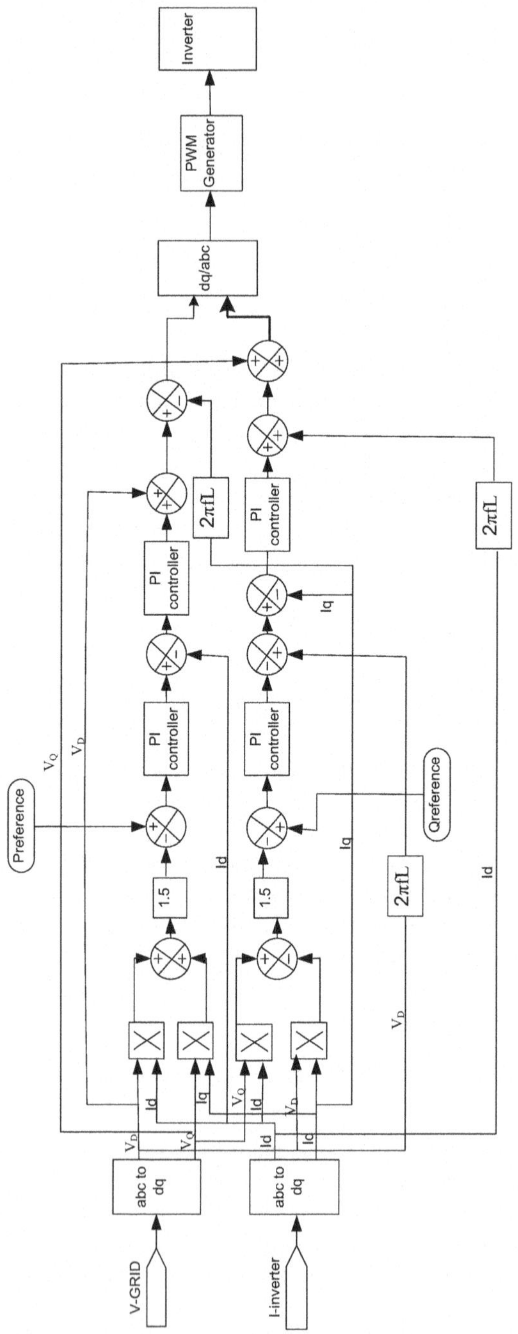

Fig. 3 Control algorithm for the DC–AC inverter

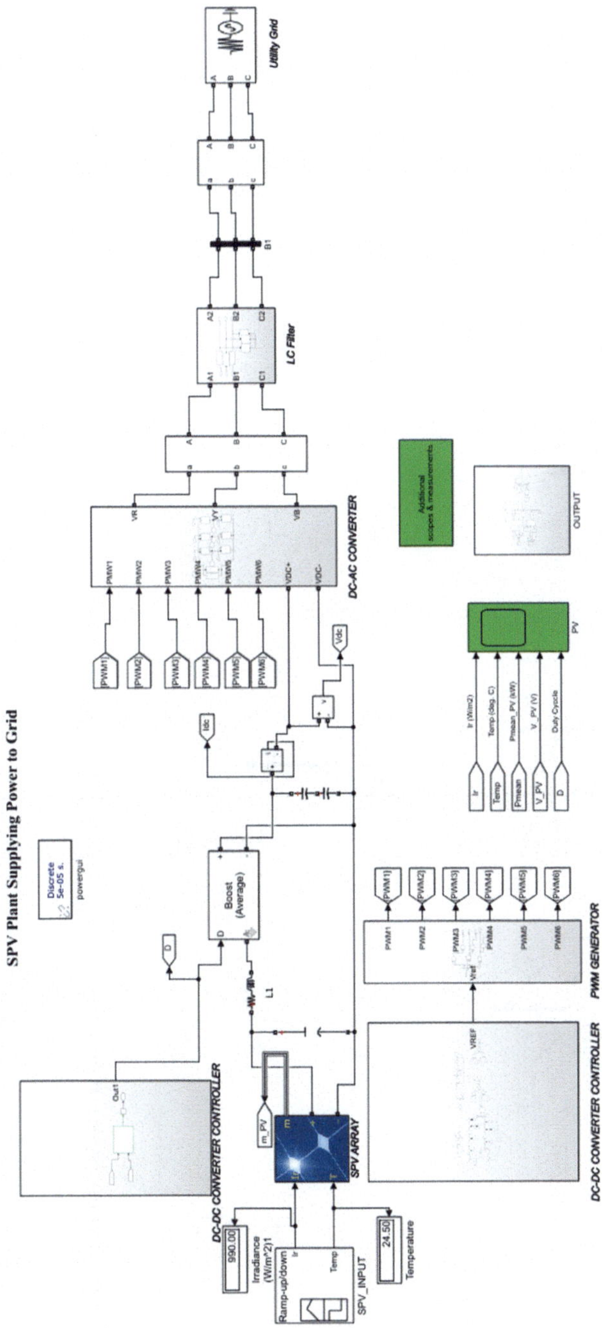

Fig. 4 Simulation model of SPV plant supplying power to the grid

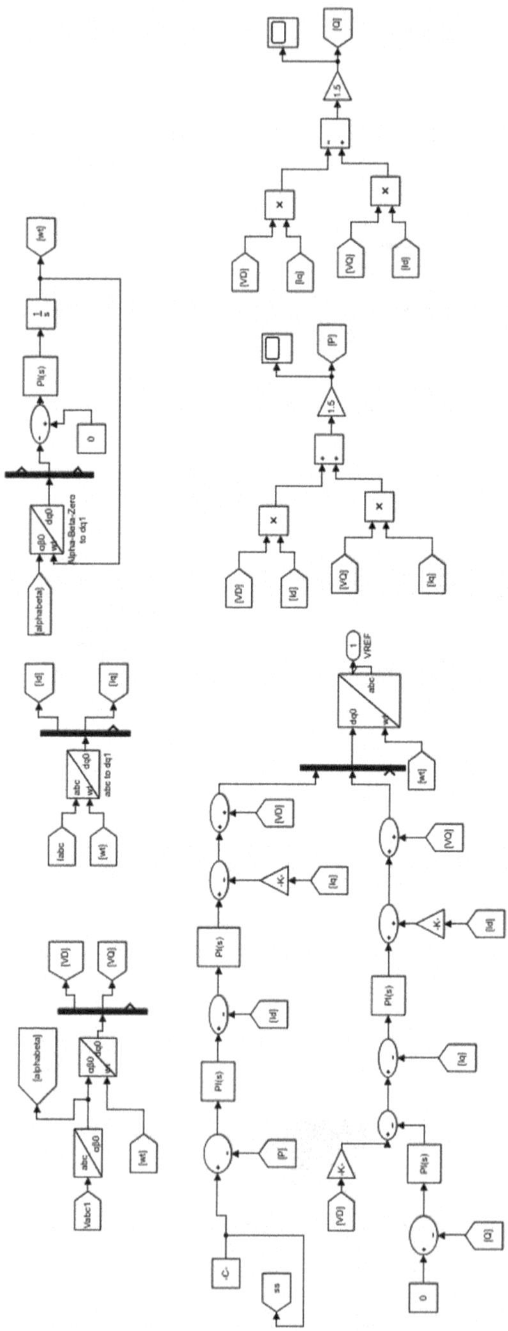

Fig. 5 Controller for the DC–AC inverter

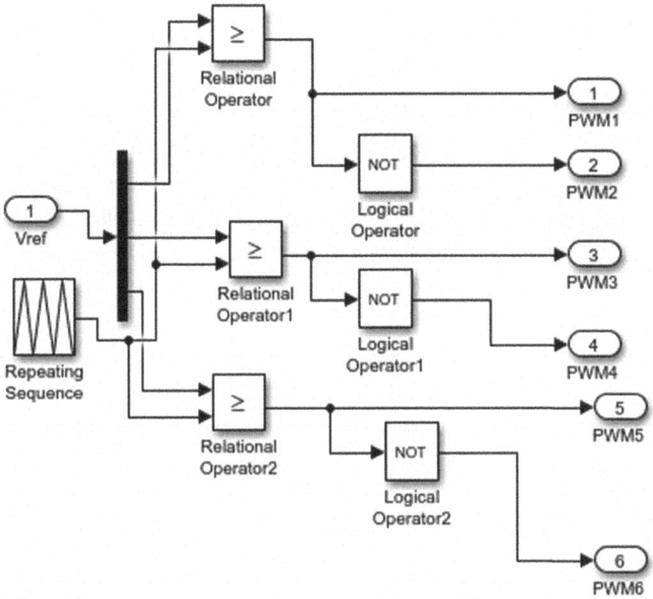

Fig. 6 PWM generator block subsystem

The DC–AC inverter controller shown in Fig. 5. The grid voltages (V_{abc1}) and inverter current (I_{abc}) are transformed to *d*–*q* coordinates using *abc* to $\alpha\beta_0$ transformation and from $\alpha\beta_0$ to *dq* transformation. The PLL is a closed-loop system, to mark phase and frequency of V_{abc1}. In grid-linked approach, the inverter operates in current control mode (CCM) where reference frequency and voltage are generated using a PLL. The active power (*P*) and reactive power (*Q*) are measured by different blocks as displayed in Fig. 5. The output is again transformed to *abc* domain to generate the reference voltage (V_{ref}).

Figure 6 presents the PWM generator block. The repeating sequence block which provides the repeating carrier signal and V_{ref} generated from Fig. 4 are given inputs to the relational operator. The alternate output of relational operator are given to logic NOT operator. And hence, the PWM signals for the inverter are generated (Table 1).

5 Results

This section discusses the overall results of the developed simulated model of SPV plant supplying power to grid. During the entire simulation time, it can be seen that change in irradiance and temperature will affect the output power of the SPV plant.

From Fig. 2, it is clear that starting with $t = 0$ s to $t = 0.5$ s, when the irradiance is at 1000 W/m^2 and temperature is also constant at 25 °C, the output voltage of SPV plant is 272 V, and current output of SPV plant is 368 A, and the power output of SPV plant is nearly100 kW (Fig. 7). During $t = 0.5$ s to $t = 1$ s, the irradiance

Table 1 Simulation parameters

Parameters	Value
Rated frequency	50 Hz
Rated voltage of SPV plant	275 V
Rated power of SPV plant	100 kW
Grid	3-phase, 415 V_{rms} (L–L), 50 Hz
L-filter	6.9 mH/phase
C-filter	4.9 mF/phase
Switching frequency of converter	50×10^3 Hz

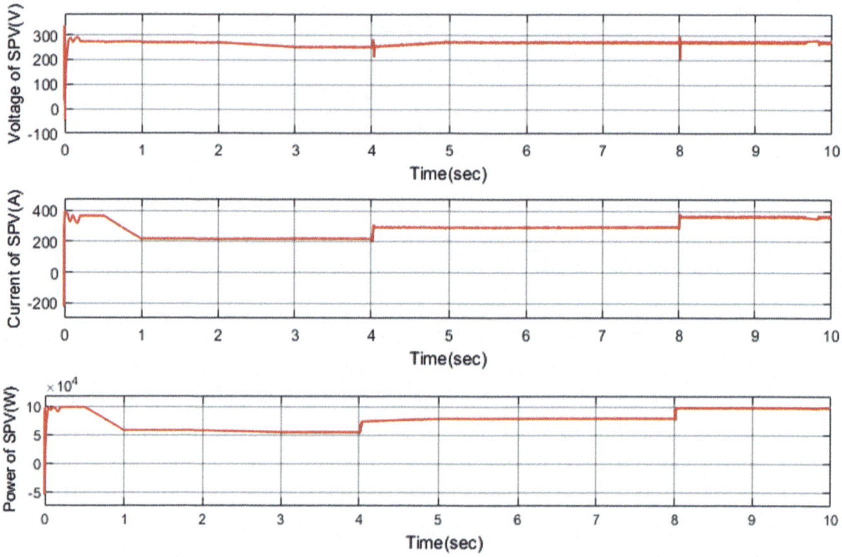

Fig. 7 Voltage, current, and power outputs of SPV plant

is linearly subdued from 1000 to 600 W/m^2, and temperature is consistent at 25 °C (Fig. 2). The voltage output of SPV plant reduces to 270 V from 272 V, and current output of SPV plant linearly decreases from 368 to 220.2 A. Thus, the power output SPV plant also reduces linearly from 100 to 59.9 kW (Fig. 7). From $t = 2$ s to $t = 3$ s, the irradiance is unaltered at 600 W/m^2 and temperature is linearly escalated from 25 to 45 °C (Fig. 2), the voltage output of SPV plant linearly decreases from 270 to 256.1 V, and current output of SPV plant remains constant at 220.2 A. Hence, the power output of SPV plant also decreases from 59.9 to 56.39 kW (Fig. 7). During $t = 3$ s to $t = 4$ s, the irradiance is consistent at 600 W/m^2, temperature is consistent at 45 °C (Fig. 2), the voltage output of SPV plant remains constant at 256.1 V, and current output of SPV plant remains constant at 220.2 A. Hence, the power output of SPV plant remains constant at 56.39 kW (Fig. 7).

During $t = 4$ s to $t = 5$ s, the irradiance is instantaneously elevated from 600 to 800 W/m^2, temperature is linearly abated from 45 to 25 °C (Fig. 2), the voltage output of SPV plant linearly increases from 256.1 to 269.1 V, and current output of SPV plant instantaneously increases to 300 A from 220.2 A. Hence, the power output of SPV plant increases from 76.84 to 80.07 kW (Fig. 7). During $t = 5$ s to $t = 8$ s, the irradiance is kept at 800 W/m^2 and temperature is kept constant 25 °C (Fig. 2). The voltage output and current output of SPV plant remains constant at 269.1 V and 300 A, respectively. Therefore, the power output SPV plant also remains constant at 80.07 kW (Fig. 7). At $t = 8$ s, the irradiance is instantaneously escalated to 1000 W/m^2 and temperature is maintained constant at 25 °C (Fig. 2). The voltage output of SPV plant increases instantly and becomes 272 V from 269.1 V, and current output of SPV plant increases from 300 to 368 A. Hence, the SPV plant output power increases to 100.09 kW from 80.07 kW (Fig. 7). From $t = 8$ s to $t = 10$ s, the irradiance and temperature are unaltered at 1000 W/m^2 and 25 °C, respectively (Fig. 2). Hence, the output voltage, output current, and output power of SPV plant remains constant at 272 V, 368 A, and 100.09 kW, respectively (Fig. 7). The current and power curve of DC–DC converter are presented in Fig. 8.

The results of Fig. 9 demonstrates the variation of the grid voltage, grid current, and active power (P_{actual}) with change in the reference active power ($P_{\text{reference}}$). During $t = 0$ s to $t = 2$ s, the reference active power is kept at −40 kW (negative sign shows that the power is fed to grid). At this reference value, the grid voltage becomes 178.19 V_{rms} and the rms value of grid current becomes 268.7 A. The actual active power (P_{actual}) fed to grid follows the reference value and becomes equal to −40 kW at around $t = 0.2$ s. At $t = 2$ s, $P_{\text{reference}}$ is increased from −40 to −60 kW. Thus, the grid voltage increases from 178.19 to 185.26 V_{rms} and the grid current becomes 288.49 A from 268.7 A. The P_{actual} of the grid follows the reference value and becomes equal to it at around $t = 2.05$ s. From $t = 2$ s to $t = 4$ s, the reference active power is maintained constant, and thus, grid voltage, grid current, and power fed to grid also remain constant. At $t = 4$ s, $P_{\text{reference}}$ is changed from −60 to −70 kW and maintained constant till $t = 6$ s. With the change in the value of reference power,

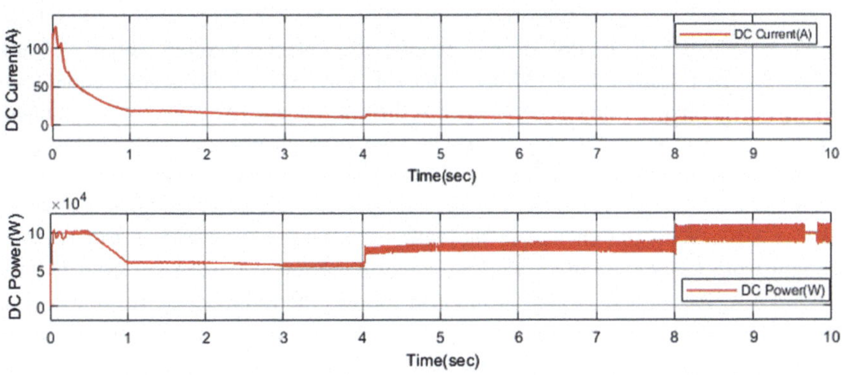

Fig. 8 Variation of current and power outputs at the DC–DC converter

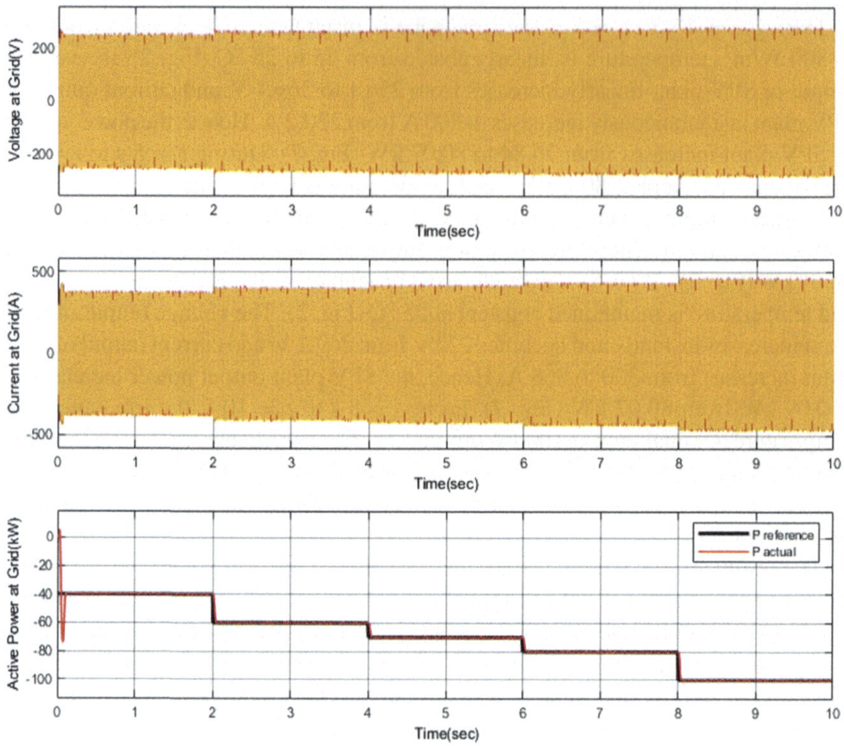

Fig. 9 Variation of the grid voltage, grid current, and active power with the change in reference active power

the grid voltage changes from 185.26 to 190.91 V_{rms} and the grid current changes from 288.49 to 300.52 A. Here again, P_{actual} of the grid follows the reference active power and becomes equal to it at around $t = 4.05$ s. From $t = 4$ s to $t = 6$ s, the grid voltage, grid current, and power fed to grid remains constant.

At $t = 6$ s, $P_{reference}$ is increased from −70 to −80 kW. Hence, the grid voltage increases from 190.91 to 194.45 V_{rms} and the grid current becomes 311.12 A from 300.52 A. The P_{actual} of the grid follows the reference value and becomes equal to it (around $t = 6.05$ s). From $t = 6$ s to $t = 8$ s, the reference active power is maintained constant at −80 kW, and therefore, grid voltage, grid current, and power fed to grid also remains constant at 194.45 V_{rms}, 311.12 A, and −80 kW, respectively.

At $t = 8$ s, $P_{reference}$ is changed from −80 to −100 kW and maintained constant till $t = 10$ s. With the change in the value of reference power, the grid voltage changes from 194.45 to 199.41 V_{rms} and the grid current also increases from 311.12 to 325.26 A. Again here, P_{actual} of the grid follows the reference active power and becomes equal to it at around $t = 8.05$ s. From $t = 8$ s to $t = 10$ s, the grid voltage, grid current, and power fed to grid remains constant.

6 Conclusion

In this chapter, the active power management scheme is implemented on a SPV plant supplying power to a grid. For its implementation, first SPV plant, grid, P–Q control strategy is drafted in MATLAB/Simulink. Then, the developed model is applied for the control of active power supplied by SPV plant to the grid. The control strategy converts the three-phase currents and voltages of the grid into d–q components by first converting *abc* components into $\alpha\beta 0$ components (stationary frame) and then from $\alpha\beta 0$ components to d–q components. This conversion makes the control very easy. The obtained outcomes of the developed simulation model predicts that using the proposed control scheme one can regulate the active power fed to grid under varying inputs (irradiation and temperature) of SPV plant.

References

1. M. Data et al., A frequency-control approach by photovoltaic generator in a PV–diesel hybrid power system. IEEE Trans. Energy Convers. **26**(2), 559–571 (2011). https://doi.org/10.1109/pecon.2008.4762600
2. S.B. Kjjaer et al., A review of single-phase grid-connected inverters for photovoltaic modules. IEEE Trans. Ind. App. **41**(5), 1292–1306 (2005). https://doi.org/10.1109/tia.2005.853371
3. Q. Lii, P. Woolfs, A review of the single phase photovoltaic module integrated converter topologies with three different DC link con. IEEE Trans. Power Elecron. **23**(3), 1320–1333 (2008). https://doi.org/10.1109/tpel.2008.920883
4. S. Mehta, F.I. Bakhsh, N. Sharma, Efficient energy management and grid interconnection of solar PV system. Int. J. Eng. Tech. **7**, 1273–1277 (2018). https://doi.org/10.14419/ijet.v7i3.12.17869
5. A. Kumar, M. Andleeb, F.I. Bakhsh, Design and analysis of solar PV rooftop in Motihari. J. Phys. **1817**(1), 1–14 (2021). https://doi.org/10.1088/1742-6596/1817/1/012019
6. S. Ahmad et al., A novel DC-to-AC controlled multi-level inverter, in *IMPACT*, AMU, India, 23–25 Nov 2013, pp. 315–319. https://doi.org/10.1109/MSPCT.2013.6782144
7. N. Sharma et al., Efficiency enhancement of a solar power plant using maximum power point tracking techniques, in *CCTES-18*, Integral University, Lucknow, India, 14–15 Sept 2018, pp. 106–111. https://doi.org/10.1109/CCTES.2018.8674146
8. M.A.G. de Briito et al., Evaluation of the main MPPT techniques for photovoltaic applications. IEEE Trans. Industr. Electron. **60**(3), 1156–1167 (2013). https://doi.org/10.1109/tie.2012.2198036
9. L. Giil-Anntonio, M.B. Saaldivar-Maarquez, O. Poortillo-Roodriguez, Maximum power point tracking techniques in photovoltaic systems: a brief review, in *13th International Conference on Power Electronics*, June 2016, pp. 317–322. https://doi.org/10.1109/ciep.2016.7530777
10. M. Nisha, M. Andleeb, F.I. Bakhsh, Effect of partial shading on a PV array and its maximum power point tracking using particle swarm optimization. J. Phys. **1817**(1), 1–14 (2021). https://doi.org/10.1088/1742-6596/1817/1/012025
11. J.M. Guerrero et al., Control strategy for flexible microgrid based on parallel line-interactive UPS systems. IEEE Trans. Industr. Electron. **56**(3), 726–736 (2009). https://doi.org/10.1109/tie.2008.2009274
12. K. Jaehong et al., Mode adaptive droop control with virtual output impedances for an inverter-based flexible ac microgrid. IEEE Trans. Power Electron. **26**(3), 689–701 (2011). https://doi.org/10.1109/tpel.2010.2091685

13. S. Haider et al., A dual control strategy for power sharing improvement in islanded mode of AC microgrid. Prot. Control Modern Power Syst. **3**, 10 (2018). https://doi.org/10.1186/s41601-018-0084-2
14. D. Wu et al., Autonomous active power control for islanded AC microgrids with photovoltaic generation and energy storage system. IEEE Trans. Energy Convers. **29**(4), 882–892 (2014). https://doi.org/10.1109/tec.2014.2358612
15. S. Adhikaari, F. Li, Coordinated V-f and P-Q control of solar photovoltaic generators with MPPT and battery storage in micro grids. IEEE Trans. Smart Grid **5**(3), 1270–1281 (2014). https://doi.org/10.1109/tpel.2010.2073488
16. K. Jong-Yuul et al., Cooperative control strategy of energy storage system and micro sources for stabilizing the microgrid during islanded operation. IEEE Trans. Power Electron. **25**(12), 3037–3048 (2010). https://doi.org/10.1109/tpel.2010.2073488
17. H. Kannchev et al., Energy management and operational planning of a microgrid with a PV-based active generator for smart grid applications. IEEE Trans. Industr. Electrons. **58**(10), 4583–4592 (2011). https://doi.org/10.1109/tie.2011.2119451
18. M.R. Chen, Z. Wang, Optimal P-Q control of grid-connected inverters in a microgrid based on adaptive population extremal optimization. *Energies* **11** (2107). https://doi.org/10.3390/en11082107
19. Z. Jiang, X. Yu, Active power—voltage control scheme for islanding operation of inverter-interfaced microgrids, in *2009 IEEE Power & Energy Society General Meeting*, Calgary, AB, Canada (2009), pp. 1–7. https://doi.org/10.1109/pes.2009.5275754
20. N. Saxena et al., Implementation of a grid-integrated PV-battery system for residential and electrical vehicle applications. IEEE Trans. Industr. Electron. **65**(8), 6592–6601 (2018). https://doi.org/10.1109/tie.2017.2739712
21. N. Hamouda et al., Predictive control of a grid connected PV system incorporating active power filter functionalities, in *2019 ICSRESA*, Tebessa, Algeria (2019), pp. 1–6. https://doi.org/10.1109/icsresa49121.2019.9182655
22. A. Iqbal et al., *Meta Heuristic and Evolutionary Computation: Algorithms and Applications*, 1E, Studies in Computational Intelligence, vol. 916 (Springer Nature, Berlin, 2020), 849 p. ISBN 978-981-15-7571-6. https://doi.org/10.1007/978-981-15-7571-6
23. J.A. Alzubi et al., *AI and Machine Learning Paradigms for Health Monitoring System: Intelligent Data Analytics*, 1E, Studies in Big Data, vol. 86 (Springer Nature, Berlin, 2020), 513 p. ISBN: 978-981-33-4412-9. https://doi.org/10.1007/978-981-33-4412-9
24. A. Iqbal et al., *Renewable Power for Sustainable Growth*, 1E, Lecture Notes in Electrical Engineering, vol. 723 (Springer Nature, Berlin), 805 p. ISBN: 978-981-33-4080-0. https://doi.org/10.1007/978-981-33-4080-0
25. N. Fatema et al., *Intelligent Data-Analytics for Condition Monitoring: Smart Grid Applications*, 1E (Academic Press, 2021). ISBN: 978-0-323-85510-5. https://doi.org/10.1016/C2020-0-02173-0
26. M.A. Alotaibib et al., Power quality disturbance analysis using data-driven EMD-SVM hybrid approach. J. Intell. Fuzzy Syst. (2021). https://doi.org/10.3233/JIFS-189739
27. P. Kaushal et al., A hybrid intelligent model for power quality disturbance classification, in *Applications of Artificial Intelligence Techniques in Engineering*. Advances in Intelligent Systems and Computing, vol. 697 (2018), pp. 55–63. https://doi.org/10.1007/978-981-13-1822-1_6

Intelligent Approach for Fuel-Constrained Economic Emission Dispatch Analysis Using Multi-objective Differential Evolution

C. Jena, Abhishek Rajan, Sarita Samal, Babita Panda, Arjyadhara Pradhan, and Lipika Nanda

Abstract Decades before, researchers have only given emphasis on optimizing the cost of energy production from thermal generating units. In view of the massive global warming and environmental pollution issues, government regulations are being imposed on the utilities which restricts the generating companies to emit toxic gases like oxides of sulfur and nitrogen. Hence, it becomes utmost important to optimize environmental emissions together with the generating cost. Single-objective optimization optimizes only one objective at a time. Therefore, to comply with the government regulation, multi-objective optimization needs to be utilized to optimize two objectives at a same time. This work aims to solve a fuel-constrained economic emission dispatch (FCEED) problems with some standard load constraints for thermal generating units using multi-objective optimization. FCEED is highly nonlinear non-convex constrained optimization problem which tries to find the optimum generating schedule for committed generators such that the operating cost and environmental emissions are minimum for certain loading condition. Differential evolution (DE) is designed as the multi-objective optimization technique to solve the FCEED optimization problem. After comparison, the result shows that the consumption of fuel could be restricted by varying the output power of different generating units, for which the power system will be operated within its contractual constraints and fuel limitations. It is analyzed that to get the same power demand from the objective functions (level of emission and fuel cost), one can be increased and the other can be decreased. However, the penalty levied for failing to maintain the fuel contract could be compensated. Numerical findings have been examined for the various test systems and comparing the results to those obtained from the strength pareto the evolutionary algorithm 2.

Keywords Fuel constrained · Multi-objective differential evolution · Economic emission dispatch

---

C. Jena · S. Samal · B. Panda · A. Pradhan · L. Nanda
School of Electrical Engineering, KIIT University, Bhubaneswar, India

A. Rajan (✉)
National Institute of Technology Sikkim, Ravangla, Sikkim, India

H. Malik et al. (eds.), *Intelligent Data Analytics for Power and Energy Systems*, Lecture Notes in Electrical Engineering 802,
https://doi.org/10.1007/978-981-16-6081-8_8

Nomenclature

F_{im}	Fuel delivered to thermal unit i in interval m
$F_i^{\min}$, $F_i^{\max}$	Lower and upper fuel delivery limits of thermal unit i
F_{D_m}	Fuel delivered in interval m
F_c	Total cost of fuel
F_e	Total emission of fuel
P_{im}	O/P power of thermal unit i in interval m
$P_i^{\min}$, $P_i^{\max}$	Lower and upper generation limits of thermal unit i
P_{D_m}	Load demand in interval m
t_m	Duration of subinterval m.
V_{im}	Fuel storage of thermal unit i in interval m
$V_i^{\min}$, $V_i^{\max}$	Lower and upper fuel storage limits of thermal unit i
V_i^0	Initial fuel storage of thermal unit i
a_i, b_i, c_i, d_i, e_i	Cost coefficients of ith thermal unit
$\alpha_i, \beta_i, \gamma_i, \sigma_i, \theta_i$	Emission coefficients of ith thermal unit
η_i, δ_i, μ_i	Fuel consumption coefficients of thermal unit i

1 Introduction

In economic emission, dispatch generations can be scheduled by optimizing both the cost of fuel and level of emission simultaneously, satisfying different operating constraints. Due to sudden concern of fuel shortages, a new dispatch problem is introduced with some power utilities which is more significant than later one. Increase of budget of fuel in the total operating cost along with environmental effect and fuel-constrained economic emission dispatch issue has emerged as a bigger issue.

A significant number of articles on the scheduling of fuel for thermal units are published [1–5]. The fuel-constrained economic dispatch issue is a multi-objective function, in which each objective function consists of one or more than one variable from one phase only. Some constraints consist of variables taken from a single step, while others cover two or more than two steps. The problem of the fuel-constrained economic emission dispatch can be overcome by splitting the overall time into discrete-time.

Several researchers are interested in creating a multi-objective evolutionary search algorithm in recent years. The pioneering multi-objective techniques are used to overcome the economic emission dispatch issue, and these are non-dominating sorting genetic algorithm II (NSGA II) [6], multi-objective evolutionary algorithm (MOEA) [7], multi-objective particle swarm optimization [8], and fuzzy clustering-based particle swarm optimization (FCPSO) [9], etc. The techniques are population-based approaches and several pareto-optimal solutions that may be achieved in one single round.

The method of normalization was used in [10] for resolving the issue of combined economic and emission dispatch. In [11], hybrid DE-HS algorithm (differential evolution with harmony search) has been used to address the multi-objective environmental/economic dispatch issues. In [12], a dynamic population size artificial bee colony algorithm is discussed for solving the issue of economic and emission dispatch. In [13], HPSO-GSA (hybrid particle swarm optimization and gravitational search algorithm) with GSA and PSO attributes has been used to address the economic emission load dispatch issue by taking different functional limitations into consideration.

The associated new member of the evolutionary algorithm family, differential evolution (DE) [14–16], was introduced as a new method for numerical optimization by Storn and Price in Berkeley during 1994–1996. In [17] differential evolution (DE) algorithm was studied for solving economic load dispatch (ELD) problems in power systems. This is a population-based approach and is usually referred to as a stochastic search optimizer; however, it is very basic yet efficient. DE's key benefit is its ability to solve optimization problems that involve minimization processes with nonlinear and multi-modal objective functions.

Due to simplicity and ability to converge quickly high-quality solution, DE is used successfully to solve several power systems issues like economic dispatch [18, 19], combined heat and power economic dispatch [20], short-term cascaded hydroelectric system scheduling [21], multi-objective VAR management [22], short-term optimal hydrothermal scheduling [23], etc. The present chapter has the following contribution:

- Proposes a nonlinear multi-objective differential evolution (MODE) approach [24, 25] to address the fuel-constrained economic emission dispatch (FCEED) issue of the thermal generation unit.
- Applying the proposed approach to a test system is validated.
- Comparisons between the results is achieved by the intelligent approach and those achieved by strength pareto evolutionary algorithm 2 (SPEA 2). In [26] presents differential evolution with Gaussian mutation (DEGM) to solve economic dispatch problem of thermal generating units with non-smooth/non-convex cost functions due to valve-point loading, taking into account transmission losses and nonlinear generator constraints such as prohibited operating zones.

2 Problem Formulation

The scheduling of generation is carried out among the committed thermal generating unit in order to simultaneously optimize both the emission level and cost of fuel to meet the standard fuel constraint as well as load constraints. The complete scheduling cycle is divided into M time intervals with constant load demand and N number of sub-intervals of thermal generating units.

The function of fuel cost for each fired generating unit fossil fuel can be expressed in accordance with the valve point effect [27].

$$F_c = \sum_{m=1}^{M} \sum_{i=1}^{N} t_m \left[a_i + b_i P_{im} + c_i P_{im}^2 + \left| d_i \sin\left\{ e_i \left(P_i^{\min} - P_{im} \right) \right\} \right| \right] \tag{1}$$

Primary sources of fossil-based generating stations are nitrogen oxides (NO_x) which emission is considered to be the selected index [28] from an environmental point of view. So, the total emission level for each unit can be expressed as

$$F_e = \sum_{m=1}^{M} \sum_{i=1}^{N} t_m \left[\alpha_i + \beta_i P_{im} + \gamma_i P_{im}^2 + \sigma_i \exp(\theta_i P_{im}) \right] \tag{2}$$

At each interval, power balance constraints are the active power generation balancing demand within its limits. Similarly, fuel delivery to each unit balances the fuel supplied.

$$\sum_{i=1}^{N} P_{im} - P_{Dm} = 0, \quad m \in M \tag{3}$$

$$\sum_{i=1}^{N} F_{im} - F_{Dm} = 0, \quad m \in M \tag{4}$$

Again, fuel storage constraints at each unit are the fuel volume at the starting of every interval plus the supply of fuel to the unit minus the burned fuel on the unit that supplies the remaining fuel at the starting of the following interval.

$$V_{im} = V_{i(m-1)} + F_{im} - t_m \left(\eta_i + \delta_i P_{im} + \mu_i P_{im}^2 \right), \quad i \in N, \quad m \in M \tag{5}$$

3 Principle of Multi-objective Optimization

Optimizing multiple objective functions is non-consistent and in general, competing and opposing, rather than one, since with respect to all objective functions, no better solution is considered than any others. These optimum solutions are referred to as solutions of pareto-optimal. Multi-objective optimization issue comprising generally of several objectives as well as many equality-and-inequality-related constraints expressed as with decision vector (x) and objective function f_i:

$$\text{Minimize } f_i(x) \quad i = 1, \ldots, N_{\text{obj}} \tag{6}$$

$$\text{Subjected to } \begin{cases} g_k(x) = 0 \; k = 1, \ldots, K \\ h_l(x) \le 0 \; l = 1, \ldots, L \end{cases} \tag{7}$$

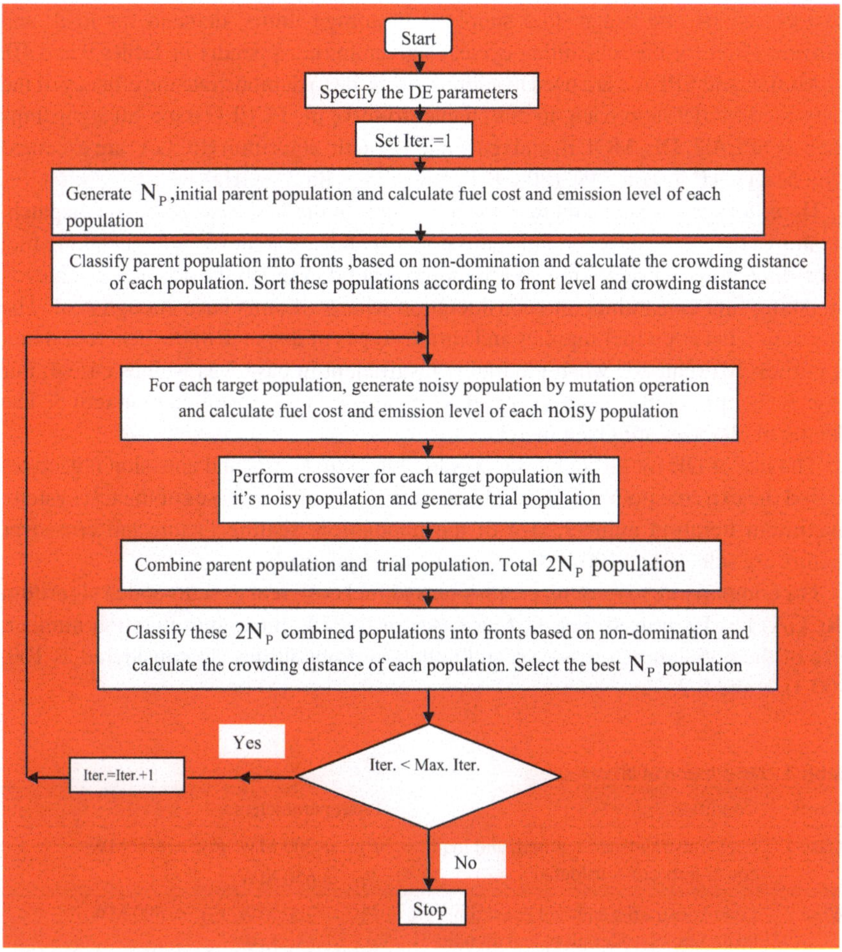

Fig. 1 Flowchart of multi-objective differential evolution

4 Optimization Technique

Flowchart (Fig. 1).

5 Simulation Results

The suggested approach was applied to a 5-coal burning unit test device that has been remained on line for 3 weeks. The generator data includes cost coefficients, coal emissions and usage, fuel supply and storage limits, demand for load, and delivered fuel for the scheduling cycle, as given in the appendix in Tables 9 and 10.

MODE and SPEA 2 are used to address the issue to demonstrate the efficacy of the suggested MODE approach. In this chapter to solve the FCEED issue, all algorithms such as SPEA 2, DE, MODE, and real coded genetic algorithm (RCGA) are executed with MATLAB 7.0 on a PC (80 GB Pentium-IV with 3.0 GHz).

Here, four cases are taken into consideration. In the first case, emission dispatch, economic dispatch, and economic emission dispatch are achieved without taking fuel constraints into consideration. While in the second case, all the above are achieved by taking fuel constraints into consideration where all units have enough coal. The interaction between fuel supplies and various types of dispatch of the generation unit is purposely organized when the fuel is insufficient in case 3 as well as case 4. For case 3, fuel in unit 4 is insufficient, and for case 4, fuel is insufficient in unit 5. The details of all cases are given in Table 1.

The use of DE for all four cases would reduce fuel costs and emission objectives to find the extreme points of the trade-off surface. In all four cases of the test system, maximum iteration number, size of the population, scaling factor, and crossover factors are selected as 100, 100, 0.9, and 1.0, respectively.

For comparison, to minimize emission and fuel cost, real genetic-coded algorithm (RCGA) has been used. For all four cases in RCGA, maximum iteration number, size of the population, crossover, and mutation probabilities are considered as 100, 100, 0.9, and 0.2.

Table 1 Description of all cases

Cases	Condition	Input parameters
Case 1	Fuel constraints are ignored, and load constraints are considered	$P_{D_1} = 700$ MW, $P_{D_2} = 800$ MW, $P_{D_3} = 650$ MW
Case 2	Load constraints and fuel constraints are taken into consideration, and all units have sufficient coal	$P_{D_1} = 700$ MW, $P_{D_2} = 800$ MW, $P_{D_3} = 650$ MW, $F_{D_1} = F_{D_2} = F_{D_3} = 7000$ ton, $V_1^0 =$ 2000 ton, $V_2^0 = 5000$ ton, $V_3^0 = 5000$ ton, $V_4^0 = 8000$ ton, $V_5^0 = 8000$ ton
Case 3	Load constraints and fuel constraints are taken into consideration, and there is shortage of fuel at unit 4	$P_{D_1} = 700$ MW, $P_{D_2} = 800$, $P_{D_3} = 650$, $F_{D_1} = F_{D_2} = F_{D_3} = 7000$ ton, $V_1^0 =$ 2000 ton, $V_2^0 = 5000$ ton, $V_3^0 = 5000$ ton, $V_4^0 = 500$ ton, $V_5^0 = 8000$ ton
Case 4	Load constraints and fuel constraints are considered, and there is shortage of fuel at unit 5	$P_{D_1} = 700$ MW, $P_{D_2} = 800$ MW, $P_{D_3} = 650$ MW, $F_{D_1} = F_{D_2} = F_{D_3} = 7000$ ton, $V_1^0 =$ 2000 ton, $V_2^0 = 5000$ ton, $V_3^0 = 5000$ ton, $V_4^0 = 8000$ ton, $V_5^0 = 500$ ton

For all four situations, MODE is implemented to simultaneously optimize the costs and the emission objectives. For all four cases, maximum iteration number, size of the population, scaling factor, and crossover in the suggested MODE approach are considered as 50, 20, 0.75, and 1.0.

SPEA 2 is used to address the issue by comparing it. For all four cases in the test system in SPEA 2, maximum iteration number, size of the population, crossover, and mutation probabilities are considered as 50, 20, 0.9, and 0.2.

Table 2 summarizes dispatch and emission dispatch solutions obtained using DE and RCGA for case 1. Table 2 demonstrates dispatch solutions obtained from economic emission dispatch using MODE and SPEA2 for case 1. Figure 2 demonstrates the convergence cost achieved by DE and RCGA from case 1. Figure 3 indicates the convergence of emission achieved by case 1 with DE and RCGA. The distribution of 20 non-dominated MODE and SPEA 2 solutions achieved in the last generation for case 1 is indicated in Fig. 4.

Table 3 summarizes dispatch solutions derived from economic dispatch and dispatch of emissions using DE and RCGA in case 2. Dispatch solutions obtained from economic emission dispatch using MODE and SPEA2 are depicted in Table 4 in case 2. Figure 5 illustrates cost convergence derived using DE and RCGA from case 2. Figure 6 indicates the convergence of emission achieved in case 2 with DE and RCGA. The distribution of 20 non-dominated MODE and SPEA 2 solutions achieved in the last generation for case 2 is indicated in Fig. 7.

The fuel costs and emission level achieved by emission dispatch, economic dispatch as well as economic emission dispatch are similar to one another as described in Tables 2, 3, and 4.

Table 5 displays dispatch solutions derived from emission dispatch and economic dispatch with DE and RCGA for case 3. Table 6 outlines dispatch solutions achieved from economic emission dispatch using MODE and SPEA2 for case 3. Figure 8 shows the convergence of costs obtained using DE and RCGA from case 3. Figure 9 indicates the convergence of emissions obtained using DE and RCGA for case 3. The distribution for case 3 of 20 non-dominated solutions achieved from the last generation of suggested SPEA 2 and MODE as illustrated in Fig. 10.

Tables 2, 3, 4, 5, and 6 indicate that cost of fuel is increased and the level of emission is reduced in case 3 than those obtained from case 1 and case 2 for economic dispatch. The cost of fuel is decreased, and the level of emission is increased for case 3 than those obtained from case 1 and case 2 for emission dispatch. It is also seen that for economic emission dispatch, the fuel cost is decreased and the emission level is increased for case 3 than those obtained from case 1 and case 2.

Table 7 shows dispatch solutions derived by emission dispatch and economic dispatch with DE and RCGA for case 4. Dispatch solutions obtained from economic emission dispatch using MODE and SPEA2 for case 4 are described in Table 8. Figure 11 representations of cost convergence achieved with DE and RCGA in case 4. Figure 12 indicates the convergence of emission obtained using RCGA and DE for case 4. The distribution for case 4 of 20 non-dominated solutions achieved from the last generation of suggested SPEA 2 and MODE as illustrated in Fig. 13.

Table 2 Dispatch solutions for case 1 without considering fuel constraints

Interval	Economic dispatch				Emission dispatch				Economic emission dispatch		Economic emission dispatch	
	(DE)		(RCGA)		(DE)		(RCGA)		(MODE)		(SPEA2)	
	Generation (MW)	Objective	Generation (MW)	Objective	Generation (MW)	Objective	Generation (MW)	Objective	Generation (MW)	Objective	Generation (MW)	Objective
1	$P_1 =$ 50.3045	Cost = 1,057,633 $	$P_1 =$ 48.7070	Cost = 1,057,760 $	$P_1 =$ 75.0000	Cost = 1,127,536 $	$P_1 =$ 74.9805	Cost = 1,127,095 $	$P_1 =$ 70.3819	Cost = 1,085,076 $	$P_1 =$ 62.7705	Cost = 1,086,926 $
	$P_2 =$ 125.000		$P_2 =$ 124.9947		$P_2 =$ 107.5394		$P_2 =$ 108.2674		$P_2 =$ 121.5701		$P_2 =$ 110.6314	
	$P_3 =$ 175.000	Emission = 854,052 lb	$P_3 =$ 174.9226	Emission = 847,189 lb	$P_3 =$ 148.3539	Emission = 561,219 lb	$P_3 =$ 147.1719	Emission = 561,342 lb	$P_3 =$ 151.2118	Emission = 637,173 lb	$P_3 =$ 158.9422	Emission = 624,523 lb
	$P_4 =$ 49.6955		$P_4 =$ 52.6383		$P_4 =$ 208.4844		$P_4 =$ 207.8208		$P_4 =$ 145.6957		$P_4 =$ 139.9314	
	$P_5 =$ 300.000		$P_5 =$ 298.7373		$P_5 =$ 160.6224		$P_5 =$ 161.7594		$P_5 =$ 211.1404		$P_5 =$ 227.7245	
2	$P_1 =$ 75.0000		$P_1 =$ 74.8560		$P_1 =$ 75.0000		$P_1 =$ 74.9881		$P_1 =$ 69.6273		$P_1 =$ 74.8493	
	$P_2 =$ 125.0000		$P_2 =$ 124.9843		$P_2 =$ 120.3204		$P_2 =$ 119.3675		$P_2 =$ 117.1540		$P_2 =$ 121.2990	
	$P_3 =$ 175.0000		$P_3 =$ 174.9952		$P_3 =$ 162.6283		$P_3 =$ 165.1740		$P_3 =$ 160.8912		$P_3 =$ 174.0485	
	$P_4 =$ 125.0000		$P_4 =$ 125.2769		$P_4 =$ 234.2381		$P_4 =$ 232.9771		$P_4 =$ 170.3515		$P_4 =$ 174.1388	
	$P_5 =$ 300.0000		$P_5 =$ 299.8876		$P_5 =$ 207.8132		$P_5 =$ 207.4933		$P_5 =$ 281.9759		$P_5 =$ 255.6644	
3	$P_1 =$ 31.8215		$P_1 =$ 34.3568		$P_1 =$ 75.0000		$P_1 =$ 74.9018		$P_1 =$ 63.7779		$P_1 =$ 53.9416	

(continued)

Table 2 (continued)

Interval	Economic dispatch				Emission dispatch				Economic emission dispatch		Economic emission dispatch	
	(DE)		(RCGA)		(DE)		(RCGA)		(MODE)		(SPEA2)	
	Generation (MW)	Objective	Generation (MW)	Objective	Generation (MW)	Objective	Generation (MW)	Objective	Generation (MW)	Objective	Generation (MW)	Objective
	$P_2 =$ 120.0214		$P_2 =$ 123.2650		$P_2 =$ 100.0806		$P_2 =$ 100.3692		$P_2 =$ 110.764		$P_2 =$ 119.1979	
	$P_3 =$ 175.0000		$P_3 =$ 173.1910		$P_3 =$ 140.2288		$P_3 =$ 139.9243		$P_3 =$ 142.2972		$P_3 =$ 153.1367	
	$P_4 =$ 40.0000		$P_4 =$ 40. 2155		$P_4 =$ 193.4620		$P_4 =$ 193.7269		$P_4 =$ 118.8558		$P_4 =$ 139.2248	
	$P_5 =$ 283.1571		$P_5 =$ 278.9726		$P_5 =$ 141.2286		$P_5 =$ 141.0778		$P_5 =$ 214.3050		$P_5 =$ 184.4991	

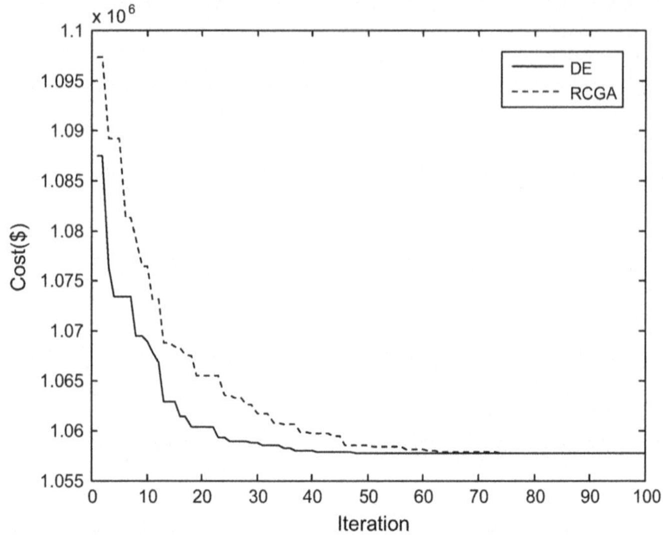

Fig. 2 Cost convergence for case 1

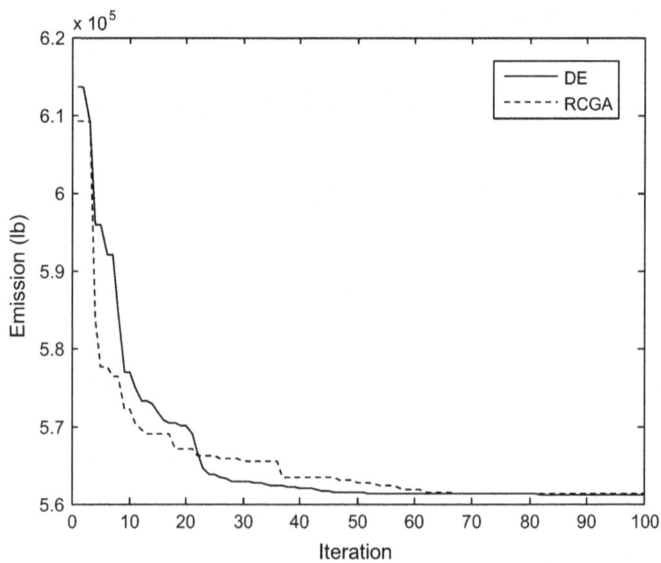

Fig. 3 Emission convergence for case 1

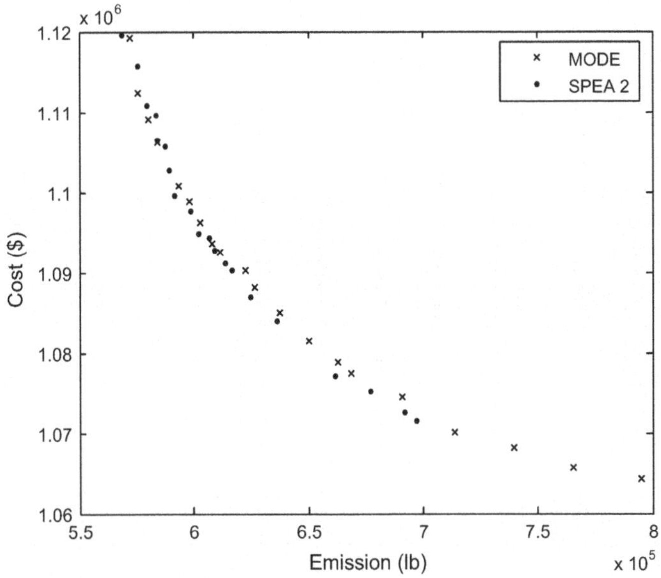

Fig. 4 Pareto-optimal front of the last generation for case 1

Tables 2, 3, 4, 7, and 8 indicate that cost of fuel is increased and the level of emission is reduced in case 4 than those obtained from case 1 and for economic dispatch from case 2. For emission dispatch, the fuel cost is decreased slightly, while the emission level is increased slightly for case 4 than those obtained from case 1 and case 2. It is also observed that the cost of fuel is increased and the level of emission is decreased in case 4 than those obtained from case 1 and case 2 for economic emission dispatch. However, the penalty otherwise levied for not upholding the fuel contract can be paid.

Optimum emission dispatch, optimum economic dispatch as well as optimum economic emission dispatch for both Case 3 and Case 4 are not reached. However, this may compensate for the penalty that can be levied when the fuel contract is not maintained.

Table 3 Fuel-constrained economic dispatch and emission dispatch solutions for case 2 with initial fuel storage (tons) $V_1^0 = 2000$, $V_2^0 = 5000$, $V_3^0 = 5000$, $V_4^0 = 8000$, $V_5^0 = 8000$

Interval	Economic dispatch						Emission dispatch					
	DE			RCGA			DE			RCGA		
	Generation (MW)	Fuel delivered (tons)	Objective	Generation (MW)	Fuel delivered (tons)	Objective	Generation (MW)	Fuel delivered (tons)	Objective	Generation (MW)	Fuel delivered (tons)	Objective
1	$P_1 =$ 50.3298	$F_1 =$ 803.3	Cost = 1,057,633 $	$P_1 =$ 50.3862	$F_1 =$ 778.1	Cost = 1,057,804 $	$P_1 =$ 75.0000	$F_1 =$ 971.1	Cost = 1,133,665 $	$P_1 =$ 74.8411	$F_1 =$ 700.3	Cost = 1,127,496 $
	$P_2 =$ 125.0000	$F_2 =$ 400.4		$P_2 =$ 124.4792	$F_2 =$ 932.4		$P_2 =$ 107.5394	$F_2 =$ 183.5		$P_2 =$ 106.5009	$F_2 =$ 899.9	
	$P_3 =$ 175.0000	$F_3 =$ 1925.5	Emission = 854,047 lb	$P_3 =$ 174.4406	$F_3 =$ 1534.5	Emission = 850,138 lb	$P_3 =$ 148.3558	$F_3 =$ 1935.3	Emission = 561,219 lb	$P_3 =$ 145.1616	$F_3 =$ 1447.0	Emission = 561,511 lb
	$P_4 =$ 49.6707	$F_4 =$ 1190.2		$P_4 =$ 50.9235	$F_4 =$ 1329.0		$P_4 =$ 208.4572	$F_4 =$ 2775.5		$P_4 =$ 210.4224	$F_4 =$ 2950.7	
	$P_5 =$ 299.9995	$F_5 =$ 2680.6		$P_5 =$ 299.7705	$F_5 =$ 2426.0		$P_5 =$ 160.6476	$F_5 =$ 1134.5		$P_5 =$ 163.0741	$F_5 =$ 1002.1	
2	$P_1 =$ 74.9996	$F_1 =$ 170.0		$P_1 =$ 74.8391	$F_1 =$ 569.4		$P_1 =$ 75.0000	$F_1 =$ 149.6		$P_1 =$ 74.9405	$F_1 =$ 998.2	
	$P_2 =$ 125.0000	$F_2 =$ 20.1		$P_2 =$ 124.9490	$F_2 =$ 918.2		$P_2 =$ 120.3054	$F_2 =$ 996.2		$P_2 =$ 121.4427	$F_2 =$ 415.1	
	$P_3 =$ 175.0000	$F_3 =$ 2000.0		$P_3 =$ 174.8974	$F_3 =$ 1755.2		$P_3 =$ 162.5958	$F_3 =$ 2000.0		$P_3 =$ 163.5206	$F_3 =$ 1266.0	
	$P_4 =$ 125.0005	$F_4 =$ 3000.0		$P_4 =$ 125.5740	$F_4 =$ 2558.6		$P_4 =$ 234.2259	$F_4 =$ 2043.9		$P_4 =$ 235.2615	$F_4 =$ 2488.1	

(continued)

Table 3 (continued)

Interval	Economic dispatch						Emission dispatch					
	DE			RCGA			DE			RCGA		
	Generation (MW)	Fuel delivered (tons)	Objective	Generation (MW)	Fuel delivered (tons)	Objective	Generation (MW)	Fuel delivered (tons)	Objective	Generation (MW)	Fuel delivered (tons)	Objective
	$P_5 =$ 299.9999	$F_5 =$ 1810.0		$P_5 =$ 299.7405	$F_5 =$ 1198.7		$P_5 =$ 207.8730	$F_5 =$ 1810.2		$P_5 =$ 204.8347	$F_5 =$ 1832.6	
3	$P_1 =$ 31.8285	$F_1 =$ 191.6		$P_1 =$ 31.4473	$F_1 =$ 411.8		$P_1 =$ 75.0000	$F_1 =$ 996.7		$P_1 =$ 74.7648	$F_1 =$ 945.6	
	$P_2 =$ 120.0186	$F_2 =$ 646.3		$P_2 =$ 122.4461	$F_2 =$ 872.2		$P_2 =$ 100.0379	$F_2 =$ 96.3		$P_2 =$ 102.7998	$F_2 =$ 361.9	
	$P_3 =$ 174.9998	$F_3 =$ 1978.8		$P_3 =$ 174.7436	$F_3 =$ 1781.6		$P_3 =$ 140.2360	$F_3 =$ 796.8		$P_3 =$ 139.8154	$F_3 =$ 1714.0	
	$P_4 =$ 40.0000	$F_4 =$ 1961.0		$P_4 =$ 41.0864	$F_4 =$ 1318.9		$P_4 =$ 193.4661	$F_4 =$ 2166.8		$P_4 =$ 190.4028	$F_4 =$ 1878.9	
	$P_5 =$ 283.1530	$F_5 =$ 2222.4		$P_5 =$ 280.2767	$F_5 =$ 2615.5		$P_5 =$ 141.2599	$F_5 =$ 2943.4		$P_5 =$ 142.2173	$F_5 =$ 2099.6	

Table 4 Fuel-constrained economic emission dispatch solutions for case 2 with initial fuel storage (tons) $V_1^0 = 2000$, $V_2^0 = 5000$, $V_3^0 = 5000$, $V_4^0 = 8000$, $V_5^0 = 8000$

Interval	Economic emission dispatch (MODE)			Economic emission dispatch (SPEA2)		
	Generation (MW)	Fuel delivered (tons)	Objective	Generation (MW)	Fuel delivered (tons)	Objective
1	$P_1 = 65.0870$	$F_1 = 374.9$	Cost = 1,085,760 $	$P_1 = 64.6209$	$F_1 = 976.7$	Cost = 1,086,487 $
	$P_2 = 118.3058$	$F_2 = 821.3$		$P_2 = 115.2044$	$F_2 = 914.4$	
	$P_3 = 161.9117$	$F_3 = 2000.0$	Emission = 628,291 lb	$P_3 = 159.0814$	$F_3 = 1621.1$	Emission = 627,198
	$P_4 = 137.7627$	$F_4 = 1045.6$		$P_4 = 160.1219$	$F_4 = 2952.3$	
	$P_5 = 216.9328$	$F_5 = 2758.1$		$P_5 = 200.9714$	$F_5 = 535.5$	
2	$P_1 = 72.4646$	$F_1 = 63.2$		$P_1 = 74.5789$	$F_1 = 891.6$	
	$P_2 = 122.3103$	$F_2 = 603.9$		$P_2 = 120.9823$	$F_2 = 602.7$	
	$P_3 = 175.0000$	$F_3 = 1674.1$		$P_3 = 170.2571$	$F_3 = 1257.0$	
	$P_4 = 186.1858$	$F_4 = 2067.3$		$P_4 = 162.5534$	$F_4 = 2708.3$	
	$P_5 = 244.0392$	$F_5 = 2591.4$		$P_5 = 271.6283$	$F_5 = 1540.4$	
3	$P_1 = 61.1643$	$F_1 = 963.3$		$P_1 = 62.7666$	$F_1 = 936.4$	
	$P_2 = 123.2043$	$F_2 = 1000.0$		$P_2 = 105.5265$	$F_2 = 762.7$	
	$P_3 = 138.9994$	$F_3 = 1867.7$		$P_3 = 169.4441$	$F_3 = 956.4$	
	$P_4 = 114.3206$	$F_4 = 2016.5$		$P_4 = 122.2385$	$F_4 = 2175.0$	
	$P_5 = 212.3115$	$F_5 = 1152.6$		$P_5 = 190.0243$	$F_5 = 2169.5$	

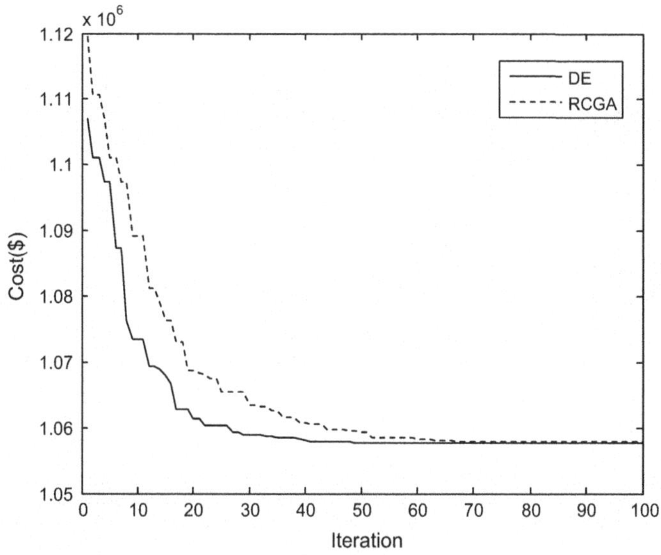

Fig. 5 Cost convergence for case 2

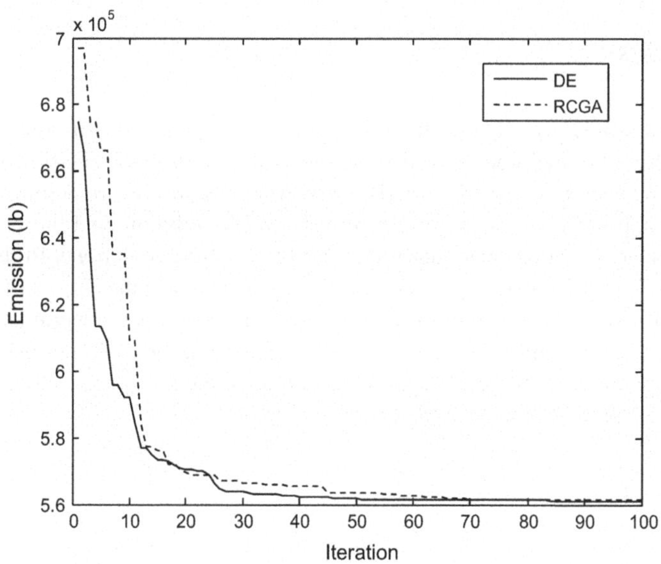

Fig. 6 Emission convergence for case 2

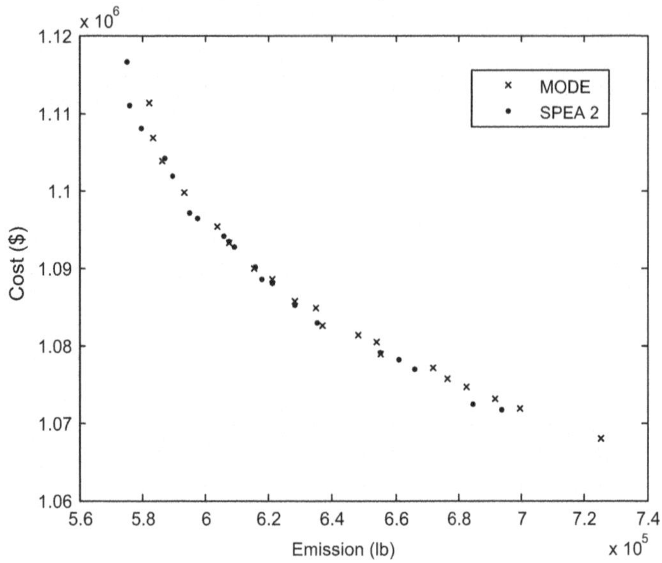

Fig. 7 Pareto-optimal front of the last generation for case 2

6 Conclusion

The present article investigates the utility of the multi-objective differential evolution to address the fuel-constrained economic emission dispatch (FCEED) issue of the thermal generation unit. Comparisons between the results are achieved by the intelligent approach, i.e., multi-objective differential evolution, and those achieved by strength pareto evolutionary algorithm 2 (SPEA 2) have been done. Both costs of fuel and the level of emission are optimized simultaneously in the case of FCEED, in view of sufficient fuel consumption controlling to meet supplier constraints. It is not always possible to achieve optimum economic emission dispatch, but usually, this is well below the penalty for breaching constraints of the fuel system. Better results are obtained using MODE than SPEA2.

Table 5 Fuel-constrained economic dispatch and emission dispatch solutions for case 3 with initial fuel storage (tons) $V_1^0 = 2000$, $V_2^0 = 5000$, $V_3^0 = 5000$, $V_4^0 = 500$, $V_5^0 = 8000$

Interval	Economic dispatch						Emission dispatch					
	DE			RCGA			DE			RCGA		
	Generation (MW)	Fuel delivered (tons)	Objective	Generation (MW)	Fuel delivered (tons)	Objective	Generation (MW)	Fuel delivered (tons)	Objective	Generation (MW)	Fuel delivered (tons)	Objective
1	$P_1 =$ 39.8274	$F_1 =$ 86.5	Cost = 1,059,205 $	$P_1 =$ 46.2967	$F_1 =$ 490.8	Cost = 1,058,819 $	$P_1 =$ 75.0000	$F_1 =$ 1000.0	Cost = 1,093,126 $	$P_1 =$ 74.8447	$F_1 =$ 546.6	Cost = 1,092,863 $
	$P_2 =$ 118.4045	$F_2 =$ 160.6		$P_2 =$ 124.1988	$F_2 =$ 944.8		$P_2 =$ 113.6015	$F_2 = 2.3$		$P_2 =$ 117.2953	$F_2 =$ 962.3	
	$P_3 =$ 165.2075	$F_3 =$ 1570.2	Emission = 837,673 lb	$P_3 =$ 174.8694	$F_3 =$ 1586.1	Emission = 832,642 lb	$P_3 =$ 155.6017	$F_3 =$ 1990.5	Emission = 614,091 lb	$P_3 =$ 170.0582	$F_3 =$ 1924.2	Emission = 613,710 lb
	$P_4 =$ 78.0233	$F_4 =$ 2372.5		$P_4 =$ 57.8008	$F_4 =$ 1088.1		$P_4 =$ 173.5134	$F_4 =$ 2999.5		$P_4 =$ 140.4073	$F_4 =$ 2998.2	
	$P_5 =$ 298.5374	$F_5 =$ 2810.2		$P_5 =$ 296.8343	$F_5 =$ 2890.1		$P_5 =$ 182.2834	$F_5 =$ 1007.7		$P_5 =$ 197.3945	$F_5 =$ 568.7	
2	$P_1 =$ 75.0000	$F_1 =$ 981.9		$P_1 =$ 74.0376	$F_1 =$ 750.3		$P_1 =$ 74.9805	$F_1 =$ 633.0		$P_1 =$ 74.9631	$F_1 =$ 884.7	
	$P_2 =$ 125.0000	$F_2 =$ 217.0.9		$P_2 =$ 124.4470	$F_2 =$ 553.6		$P_2 =$ 124.9818 $F_2 =$ 949.5		$P_2 =$ 124.9673	$F_2 =$ 996.7		
	$P_3 =$ 175.0000	$F_3 =$ 1998.9		$P_3 =$ 174.6488	$F_3 =$ 1865.4		$P_3 =$ 170.4118	$F_3 =$ 2000.0		$P_3 =$ 170. 9546	$F_3 =$ 1721.6	
	$P_4 =$ 125.0001	$F_4 =$ 2059.5		$P_4 =$ 128.0763	$F_4 =$ 2596.9		$P_4 =$ 148.1378	$F_4 =$ 2997.7		$P_4 =$ 180.4024	$F_4 =$ 2998.6	

(continued)

Table 5 (continued)

Interval	Economic dispatch						Emission dispatch					
	DE			RCGA			DE			RCGA		
	Generation (MW)	Fuel delivered (tons)	Objective	Generation (MW)	Fuel delivered (tons)	Objective	Generation (MW)	Fuel delivered (tons)	Objective	Generation (MW)	Fuel delivered (tons)	Objective
	$P_5 =$ 300.0000	$F_5 =$ 1741.8		$P_5 =$ 298.7903	$F_5 =$ 1233.8		$P_5 =$ 281.4881	$F_5 =$ 419.8		$P_5 =$ 248.7126	$F_5 =$ 398.4	
3	$P_1 =$ 31.9035	$F_1 =$ 813.9		$P_1 =$ 38.6909	$F_1 =$ 669.2		$P_1 =$ 74.9942	$F_1 =$ 535.3		$P_1 =$ 74.6438	$F_1 =$ 660.2	
	$P_2 =$ 119.9734	$F_2 =$ 216.8		$P_2 =$ 106.0992	$F_2 =$ 732.6		$P_2 =$ 108.2428	$F_2 =$ 999.6		$P_2 =$ 107.8920	$F_2 =$ 992.6	
	$P_3 =$ 174.9968	$F_3 =$ 1998.1		$P_3 =$ 174. 5749	$F_3 =$ 1168.9		$P_3 =$ 149.5267	$F_3 =$ 1850.7		$P_3 =$ 151.6888	$F_3 =$ 1990.5	
	$P_4 =$ 40.0005	$F_4 =$ 1516.3		$P_4 =$ 51.0046	$F_4 =$ 2612.0		$P_4 =$ 147.9900	$F_4 =$ 2997.2		$P_4 =$ 147.4801	$F_4 =$ 2994.7	
	$P_5 =$ 283.1257	$F_5 =$ 2454.9		$P_5 =$ 279.6304	$F_5 =$ 1817.2		$P_5 =$ 169.2462	$F_5 =$ 617.2		$P_5 =$ 168.2953	$F_5 =$ 362.1	

Table 6 Fuel-constrained economic emission dispatch solutions for case 3 with initial fuel storage (tons) $V_1^0 = 2000$, $V_2^0 = 5000$, $V_3^0 = 5000$, $V_4^0 = 500$, $V_5^0 = 8000$

Interval	Economic emission dispatch (MODE)			Economic emission dispatch (SPEA2)		
	Generation (MW)	Fuel delivered (tons)	Objective	Generation (MW)	Fuel delivered (tons)	Objective
1	$P_1 = 59.4445$	$F_1 = 549.7$	Cost = 1,073,303 $	$P_1 = 73.0084$	$F_1 = 844.6$	Cost = 1,074,008 $
	$P_2 = 125.0000$	$F_2 = 351.9$		$P_2 = 122.6514$	$F_2 = 801.8$	
	$P_3 = 174.8987$	$F_3 = 638.0$	Emission = 684,693 lb	$P_3 = 168.2436$	$F_3 = 1702.3$	Emission = 673,024 lb
	$P_4 = 103.5303$	$F_4 = 2626.5$		$P_4 = 118.2999$	$F_4 = 1972.3$	
	$P_5 = 237.1265$	$F_5 = 2833.9$		$P_5 = 217.7967$	$F_5 = 1679.1$	
2	$P_1 = 69.9770$	$F_1 = 68.1$		$P_1 = 74.7721$	$F_1 = 432.3$	
	$P_2 = 119.0024$	$F_2 = 344.3$		$P_2 = 124.6438$	$F_2 = 848.4$	
	$P_3 = 175.0000$	$F_3 = 1307.3$		$P_3 = 174.6266$	$F_3 = 1980.4$	
	$P_4 = 153.0358$	$F_4 = 2697.4$		$P_4 = 146.2138$	$F_4 = 2976.6$	
	$P_5 = 282.9848$	$F_5 = 2583.0$		$P_5 = 279.7437$	$F_5 = 762.5$	
3	$P_1 = 59.0161$	$F_1 = 859.9$		$P_1 = 72.7733$	$F_1 = 491.8$	
	$P_2 = 116.8183$	$F_2 = 133.0$		$P_2 = 118.0090$	$F_2 = 795.5$	
	$P_3 = 143.8090$	$F_3 = 1223.1$		$P_3 = 162.5505$	$F_3 = 1888.8$	
	$P_4 = 109.3848$	$F_4 = 2712.6$		$P_4 = 80.7817$	$F_4 = 2696.7$	
	$P_5 = 220.9717$	$F_5 = 2071.3$		$P_5 = 215.8855$	$F_5 = 1127.3$	

Fig. 8 Cost convergence for case 3

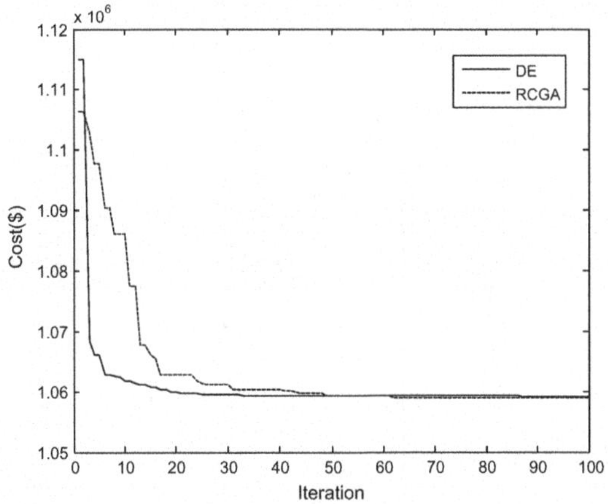

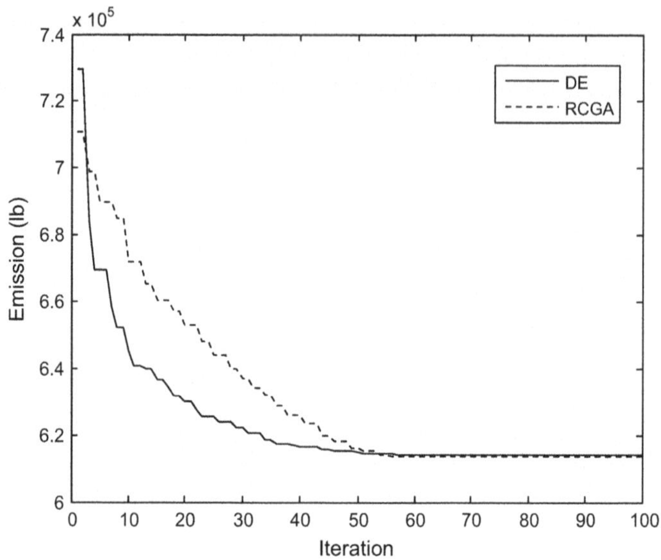

Fig. 9 Emission convergence for case 3

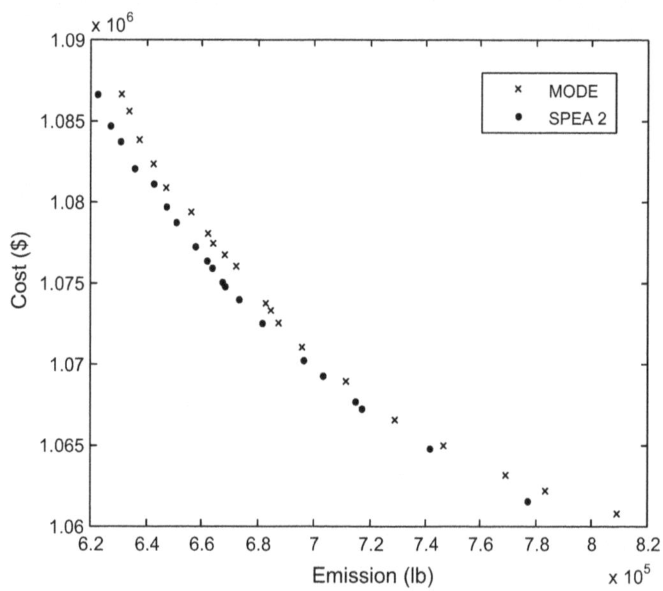

Fig. 10 Pareto-optimal front of the last generation for case 3

Table 7 Fuel-constrained economic dispatch and emission dispatch solutions for case 4 with initial fuel storage (tons) $V_1^0 = 2000$, $V_2^0 = 5000$, $V_3^0 = 5000$, $V_4^0 = 8000$, $V_5^0 = 500$

Interval	Economic dispatch						Emission dispatch					
	DE			RCGA			DE			RCGA		
	Generation (MW)	Fuel delivered (tons)	Objective	Generation (MW)	Fuel delivered (tons)	Objective	Generation (MW)	Fuel delivered (tons)	Objective	Generation (MW)	Fuel delivered (tons)	Objective
1	$P_1 =$ 64.6327	$F_1 =$ 959.6	Cost = 1,072,011 $	$P_1 =$ 74.5460	$F_1 =$ 670.7	Cost = 1,072,842 $	$P_1 =$ 73.6410	$F_1 =$ 370.03	Cost = 1,126,446 $	$P_1 =$ 74.7600	$F_1 =$ 781.4	Cost = 1,131,378 $
	$P_2 =$ 121.5749	$F_2 =$ 897.1		$P_2 =$ 124.3769	$F_2 =$ 967.1		$P_2 =$ 112.3754	$F_2 =$ 321.7		$P_2 =$ 112.7311	$F_2 =$ 684.0	
	$P_3 =$ 174.5428	$F_3 =$ 1994.6	Emission = 694,273 lb	$P_3 =$ 173.9435	$F_3 =$ 1546.7	Emission = 686,620 lb	$P_3 =$ 150.7188	$F_3 =$ 1395.2	Emission = 562,921 lb	$P_3 =$ 143.8396	$F_3 =$ 1065.3	Emission = 563,327 lb
	$P_4 =$ 101.2611	$F_4 =$ 151.3		$P_4 =$ 119.3743	$F_4 =$ 824.1		$P_4 =$ 201.5249	$F_4 =$ 2188.4		$P_4 =$ 211.8778	$F_4 =$ 2387.1	
	$P_5 =$ 237.9886	$F_5 =$ 2997.5		$P_5 =$ 207.7592	$F_5 =$ 2991.4		$P_5 =$ 161.7399	$F_5 =$ 2724.7		$P_5 =$ 156.7915	$F_5 =$ 2082.2	
2	$P_1 =$ 74.8386	$F_1 =$ 982.8		$P_1 =$ 71.3752	$F_1 =$ 952.5		$P_1 =$ 74.9254	$F_1 =$ 994.9		$P_1 =$ 74.4660	$F_1 =$ 778.6	
	$P_2 =$ 124.9129	$F_2 =$ 17.5		$P_2 =$ 124.9989	$F_2 =$ 810.3		$P_2 =$ 120.1896	$F_2 = 8.2$		$P_2 =$ 112.4141	$F_2 =$ 908.6	
	$P_3 =$ 174.7765	$F_3 = 0.1$		$P_3 =$ 173.8995	$F_3 =$ 1918.0		$P_3 =$ 161.1406	$F_3 =$ 51.7		$P_3 =$ 161.8784	$F_3 =$ 1501.6	
	$P_4 =$ 168.3754	$F_4 =$ 2999.9		$P_4 =$ 146.6616	$F_4 =$ 328.2		$P_4 =$ 236.1880	$F_4 =$ 2991.8		$P_4 =$ 243.5720	$F_4 =$ 900.2	

(continued)

Table 7 (continued)

Interval	Economic dispatch						Emission dispatch					
	DE			RCGA			DE			RCGA		
	Generation (MW)	Fuel delivered (tons)	Objective	Generation (MW)	Fuel delivered (tons)	Objective	Generation (MW)	Fuel delivered (tons)	Objective	Generation (MW)	Fuel delivered (tons)	Objective
	$P_5 =$ 257.0966	$F_5 =$ 2999.7		$P_5 =$ 283.0648	$F_5 =$ 2990.9		$P_5 =$ 207.5564	$F_5 =$ 2953.3		$P_5 =$ 207.6695	$F_5 =$ 2911.0	
3	$P_1 =$ 57.4434	$F_1 =$ 499.9		$P_1 =$ 72.1086	$F_1 =$ 765.7		$P_1 =$ 72.2428	$F_1 =$ 288.2		$P_1 =$ 74.3946	$F_1 =$ 457.4	
	$P_2 =$ 124.7904	$F_2 =$ 935.6		$P_2 =$ 115.2357	$F_2 =$ 893.1		$P_2 =$ 98.6504	$F_2 =$ 701.4		$P_2 =$ 98.6743	$F_2 =$ 866.6	
	$P_3 =$ 175.0000	$F_3 =$ 1818.7		$P_3 =$ 172.5911	$F_3 =$ 1962.1		$P_3 =$ 131.3219	$F_3 =$ 1227.1		$P_3 =$ 145.1968	$F_3 =$ 969.6	
	$P_4 =$ 63.5782	$F_4 =$ 756.5		$P_4 =$ 67.4497	$F_4 =$ 381.2		$P_4 =$ 196.2013	$F_4 =$ 2994.2		$P_4 =$ 196.9188	$F_4 =$ 2877.4	
	$P_5 =$ 229.1879	$F_5 =$ 2989.2		$P_5 =$ 222.6150	$F_5 =$ 2997.8		$P_5 =$ 151.5836	$F_5 =$ 1789.1		$P_5 =$ 134.8154	$F_5 =$ 1829.0	

Table 8 Fuel-constrained economic emission dispatch solutions for case 4 with initial fuel storage (tons) $V_1^0 = 2000$, $V_2^0 = 5000$, $V_3^0 = 5000$, $V_4^0 = 8000$, $V_5^0 = 500$

Interval	Economic emission dispatch (MODE)			Economic emission dispatch (SPEA2)		
	Generation (MW)	Fuel delivered (tons)	Objective	Generation (MW)	Fuel delivered (tons)	Objective
1	$P_1 =$ 68.8702	$F_1 =$ 625.8	Cost = 1,096,650 \$	$P_1 =$ 56.4094	$F_1 =$ 302.1	Cost = 1,096,509 \$
	$P_2 =$ 114.7729	$F_2 =$ 236.3		$P_2 =$ 116.0625	$F_2 =$ 330.4	
	$P_3 =$ 166.3807	$F_3 =$ 738.3	Emission = 596,646 lb	$P_3 =$ 159.7154	$F_3 =$ 1606.5	Emission = 618,125 lb
	$P_4 =$ 143.4241	$F_4 =$ 2497.8		$P_4 =$ 160.5276	$F_4 =$ 1826.7	
	$P_5 =$ 206.5521	$F_5 =$ 2901.9		$P_5 =$ 207.2852	$F_5 =$ 2934.3	
2	$P_1 =$ 74.2855	$F_1 =$ 659.7		$P_1 =$ 73.4619	$F_1 =$ 999.1	
	$P_2 =$ 121.4536	$F_2 =$ 928.1		$P_2 =$ 110.6655	$F_2 =$ 377.9	
	$P_3 =$ 170.0481	$F_3 =$ 1477.8		$P_3 =$ 174.4181	$F_3 =$ 717.7	
	$P_4 =$ 208.1990	$F_4 =$ 1294.0		$P_4 =$ 219.8668	$F_4 =$ 2663.9	
	$P_5 =$ 226.0138	$F_5 =$ 2640.5		$P_5 =$ 221.5877	$F_5 =$ 2241.5	
3	$P_1 =$ 62.6368	$F_1 =$ 952.5		$P_1 =$ 59.0782	$F_1 =$ 808.5	
	$P_2 =$ 104.5039	$F_2 =$ 193.1		$P_2 =$ 84.7793	$F_2 =$ 556.8	
	$P_3 =$ 158.7797	$F_3 =$ 1916.9		$P_3 =$ 170.6290	$F_3 =$ 1323.7	
	$P_4 =$ 144.5260	$F_4 =$ 1164.5		$P_4 =$ 114.8903	$F_4 =$ 1459.2	
	$P_5 =$ 179.5516	$F_5 =$ 2773.1		$P_5 =$ 220.6232	$F_5 =$ 2851.9	

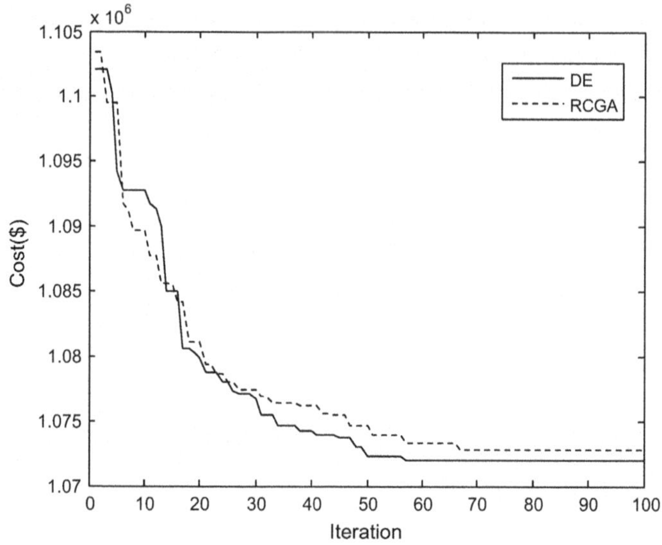

Fig. 11 Cost convergence for case 4

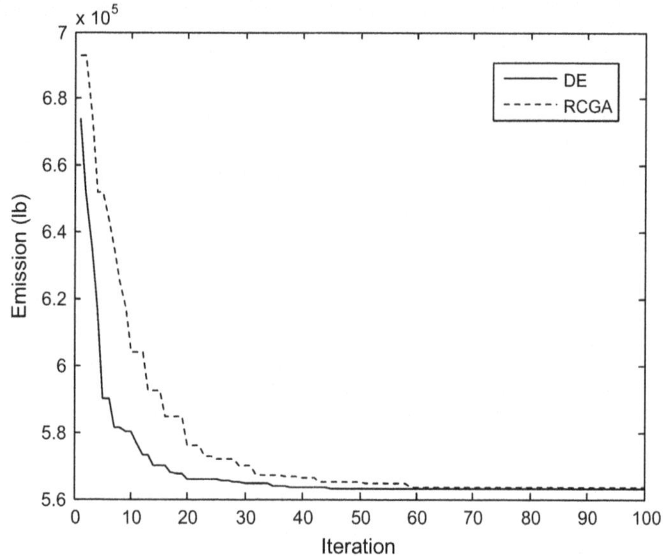

Fig. 12 Emission convergence for case 4

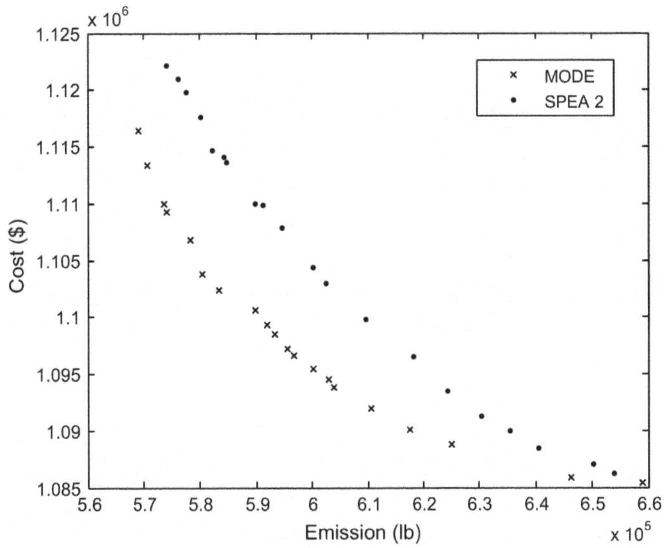

Fig. 13 Pareto-optimal front of the last generation for case 4

Appendix

See Tables 9 and 10.

Table 9 Generator characteristics

Unit	$P_i^{\min}$ (MW)	$P_i^{\max}$ (MW)	a_i ($/h)	b_i ($/MWh)	c_i $/(MW)^2 h	d_i ($/h)	e_i (rad/MW)	α_i (lb/h)	β_i (lb/MWh)	γ_i (lb/(MW)^2 h)
1	20	75	25	2.0	0.0080	10	0.012	80	−0.805	0.0180
2	20	125	60	1.8	0.0030	20	0.010	50	−0.555	0.0150
3	30	175	100	2.1	0.0012	30	0.009	70	−0.955	0.0115
4	40	250	120	2.2	0.0040	40	0.008	45	−0.600	0.0080
5	50	300	40	1.8	0.0015	50	0.007	30	−0.555	0.0120

Unit	σ_i (lb/h)	θ_i (1/MW)	η_i (ton/h)	δ_i (ton/MWh)	μ_i (ton/(MW)^2 h)	$F_i^{\min}$ (ton)	$F_i^{\max}$ (ton)	$V_i^{\min}$ (ton)	$V_i^{\max}$ (ton)
1	0.008	0.0735	0.83612	0.066889	0.00026756	0	1000	0	10,000
2	0.009	0.0655	2.00669	0.060200	0.00010033	0	1000	0	10,000
3	0.010	0.0504	3.34448	0.070230	0.00004013	0	2000	0	20,000
4	0.015	0.0340	4.01338	0.073578	0.00013378	0	3000	0	30,000
5	0.017	0.0285	1.33779	0.060200	0.00005017	0	3000	0	30,000

Table 10 Load demand and fuel delivered during scheduling period

Interval	Duration (h)	Load demand P_D (MW)	Fuel delivered F_D (ton)
1	168	700	7000
2	168	800	7000
3	168	650	7000

References

1. F.J. Trefny, K.Y. Lee, Economic fuel dispatch. IEEE Trans. Power Apparatus Syst. PAS-100 **7**, 3468–3477 (1981)
2. S. Vemuri, A.B.R. Kumar, D.F. Hackett, J.T. Eisenhauer, R. Lugtu, Fuel resource scheduling. Part-I—Overview of an energy management problem. IEEE Trans. Power Apparatus Syst. PAS-103 **7**, 1542–1548 (1984)
3. A.B.R. Kumar, S. Vemuri, Fuel resource scheduling. Part-II—Constrained economic dispatch. IEEE Trans. Power Apparatus Syst. PAS-103 **7**, 1549–1555 (1984)
4. A.B.R. Kumar, S. Vemuri, L.A. Gibbs, D.F. Hackett, J.T. Eisenhauer, Fuel resource scheduling. Part-III—The short-term problem. IEEE Trans. Power Apparatus Syst. PAS-103 **7**, 1556–1561 (1984)
5. A.J. Wood, B.F. Wollenberg, *Power Generation, Operation and Control* (Wiley, New York, 1996)
6. T.F. Robert, A.H. King, C.S. Harry, Rughooputh, K. Deb, Evolutionary multiobjective environmental/economic dispatch: stochastic versus deterministic approaches. KanGAL, Rep. 2004019 (2004), pp. 1–15
7. M.A. Abido, Multiobjective evolutionary algorithms for power dispatch problem. IEEE Trans. Evol. Comput. **10**(3), 315–329 (2006)
8. L. Wang, C. Singh, Stochastic economic emission load dispatch through a modified particle swarm optimization algorithm. Electr. Power Syst. Res. **78**, 1466–1476 (2008)
9. S. Agrawal, B.K. Panigrahi, M.K. Tiwari, Multiobjective particle swarm algorithm with fuzzy clustering for electrical power dispatch. IEEE Trans. Evol. Comput. **12**(5), 529–541 (2008)
10. Y.-C. Liang, J.R.C. Juarez, A normalization method for solving the combined economic and emission dispatch problem with meta-heuristic algorithms. Electr. Power Energy Syst. (54), 163–186 (2014)
11. S. Sayah, A. Hamouda, A. Bekrar, Efficient hybrid optimization approach for emission constrained economic dispatch with nonsmooth cost curves. Electr. Power Energy Syst. **56**, 127–139 (2014)
12. D. Aydin, S. Özyön, C. Yasar, T. Liao, Artificial bee colony algorithm with dynamic population size to combined economic and emission dispatch problem. Electr. Power Energy Syst. **54**, 144–153 (2014)
13. S. Jiang, Z. Ji, Y. Shen, A novel hybrid particle swarm optimization and gravitational search algorithm for solving economic emission load dispatch problems with various practical constraints. Electr. Power Energy Syst. **55**, 628–644 (2014)
14. R. Storn, K.V. Price, Differential evolution—a simple and efficient heuristic for global optimization over continuous spaces. J. Global Optim. **11**(4), 341–359 (1997)
15. K.V. Price, R. Storn, J. Lampinen, *Differential Evolution: A Practical Approach to Global Optimization* (Springer, Berlin, 2005)
16. J. Ronkkonen, S. Kukkonen, K.V. Price, Real-parameter optimization with differential evolution, in *Proceeding IEEE Congress Evolutionary Computation*, Edinburgh, Scotland, Sept 2005, pp. 506–513
17. N. Nomana, H. Iba, Differential evolution for economic load dispatch problems. Electr. Power Syst. Res. **78**(3), 1322–1331 (2008)

18. A. Srinivasa Reddy, K. Vaisakh, Shuffled differential evolution for large scale economic dispatch. Electr. Power Syst. Res. **96**, 237–245 (2013)
19. M. Basu, Combined heat and power economic dispatch by using differential evolution. Electr. Power Compon. Syst. **38**(8), 996–1004 (2010)
20. Y. He, Q. Xu, S. Yang, A. Han, L. Yang, A novel chaotic differential evolution algorithm for short-term cascaded hydroelectric system scheduling. Electr. Power Energy Syst. **55**, 628–644 (2014)
21. H. Singh, L. Srivastava, Modified differential evolution algorithm for multi-objective VAR management. Electr. Power Energy Syst. **55**, 731–740 (2014)
22. J. Zhang, S. Lin, W. Qiu, A modified chaotic differential evolution algorithm for short-term optimal hydrothermal scheduling. Electr. Power Energy Syst. **65**, 159–168 (2015)
23. B.V. Babu, B. Anbarasu, Multi-objective differential evolution (MODE): an evolutionary algorithm for multi-objective optimization problems (MOOPs), in *Proceedings of the Third International Conference on Computational Intelligence, Robotics, and Autonomous Systems (CIRAS-2005)*, Singapore, 13–16 Dec 2005
24. B.V. Babu, *Process Plant Simulation* (Oxford University Press, New Delhi, 2006)
25. E. Zitzler, M. Laumanns, L. Thiele, SPEA2: Improving the Strength Pareto Evolutionary Algorithm. Swiss Federal Institute of Technology (ETH), Zurich, Switzerland. Technical Report TIK-Report 103, May 2001
26. M. Basu, C. Jena, C.K. Panigrahi et al., Differential Evolution with Gaussian Mutation for Combined Heat and Power Economic Dispatch, Soft computing, Vol. 20, Issue-2, (Springer, 2016), pp 681-688
27. M.R. Gent, J.W. Lamont, Minimum emission dispatch. IEEE Trans. PAS (90), 2650–2660 (1971)
28. A.K. Yadav et al., *Soft Computing in Condition Monitoring and Diagnostics of Electrical and Mechanical Systems*, 1E, Advances in Intelligent Systems and Computing, vol. 1096 (Springer Nature, Berlin, 2020), 496 p. ISBN: 978-981-15-1532-3. https://doi.org/10.1007/978-981-15-1532-3
29. D.C. Walters, G.B. Sheble, Genetic algorithm solution of economic dispatch with valve point loading. IEEE Trans. Power Syst. **8**(3), 1325–1332 (1993)

Intelligent Adaptive Critic Scheme Implementation and Investigation Using Policy Iteration Technique: A Case Study of Multi-area Automatic Generation Control (AGC) Problem

Nagendra Kumar, Akhilesh Singh, B. P. Joshi, and B. K. Singh

Abstract An adaptive critic system based on the policy iteration (PI) technique has been developed in this study. The proposed control technique has been effectively applied to the problem of multi-area automated generation control (AGC). A linear quadratic regulator (LQR)-based optimal control technique demands a thorough understanding of the system, policy iteration (PI) control method, and includes policy evaluation and policy improvement phases in order to deliver the best control solution without knowing the entire system's information (i.e., matrix A). For policy updates, just matrix B is required. As a result, given partially unknown dynamics, the policy iteration, optimum control strategy becomes model-free and useful. The designed control scheme has been checked on isolated and interconnected power system. A comparison study had also been presented between a policy iteration (PI) optimal control scheme and a LQR optimal control scheme, which demonstrated that the PI optimal control scheme can effectively dampen deviations with better transient performance and faster settling time than the LQR-based optimal control scheme and improve system responses effectively. The compatibility of policy iteration control with North American Electric Reliability Council (NERC) criteria has also been evaluated. The policy iteration method appears to meet both the CPS1 and CPS2 control criteria.

Keywords AGC · PI · Power system · Control scheme

N. Kumar (✉)
Department of Electrical and Electronics Engineering, G. L. Bajaj Institute of Technology & Management Greater Noida, Greater Noida 201306, India

A. Singh
Department of Electrical Engineering, Seemant Institute of Technology Pithoragarh, Pithoragarh, Uttarakhand 262501, India

B. P. Joshi
Department of Applied Science, Seemant Institute of Technology Pithoragarh, Pithoragarh, Uttarakhand 262501, India

B. K. Singh
Department of Electrical Engineering, Bipin Tripathi Kumaun Institute of Technology Almora, Dwarahat, Uttarakhand 262501, India

H. Malik et al. (eds.), *Intelligent Data Analytics for Power and Energy Systems*,
Lecture Notes in Electrical Engineering 802,
https://doi.org/10.1007/978-981-16-6081-8_9

1 Introduction

Load frequency control (LFC) has long been regarded as an important service in the power system, serving two purposes: maintaining frequency and avoiding tie-line variations. Matching generated power to demand is an essential control problem known as LFC. Load disruptions vary the frequency and scheduled power exchange, which is detrimental to the power system's success. As a result, load disruptions alter the system's normal operating point. Since power system has wide and complex structure, these deviations incurred by disturbances may lead to system black out if not properly handled. For secure and successful operation, the system must cope with load and disturbances. This is where automatic generation control (AGC) plays a role to keep system frequency and tie-line power within pre-specified values by regulating the real power of generators. The concept of AGC scheme has been reported in [1]. Nowadays, power system is being restructured. LFC problem in restructured environment is given in [2, 3]. Controller plays a vital role in AGC scheme; therefore, controller design using various approaches has attracted researchers so far. A classical approach to design a controller is given in [4], but this approach suffers with overshoots and transient frequency deviation. To deal with complex and vast structure of power system, various control optimization schemes such as imperialist competition algorithm (ICA), harmony search (HS), big bang–big crunch (BBBC), and others are also being used to obtain optimal controller parameters for AGC scheme in power system [5–8]. The other and important parameter of a good control scheme is its optimality. Therefore, an optimal control scheme for two-area thermal power system is reported in [9]. Optimal control scheme requires complete information of the system states which seems impractical; therefore, lot of work has been reported regarding suboptimal control approaches which remove practical limitation in implementation of full-order state feedback [10]. Since load on a power system is of variable type i.e. it changes regularly, therefore, for a good control scheme it is desirable to have a track of operating point which emerged adaptive and self-tuning control schemes [11]. Further, to overcome the deficiencies of adaptive and optimal control schemes, online optimal adaptive control schemes like approximate dynamic programming (ADP), reinforcement learning (RL), and adaptive critic design (ACD) are reported in the literature [12–16]. Policy iteration (PI) technique is a class of RL approach to provide adaptive optimal control solution [17, 18]. Several AI approaches can find their application in LFC as well as other fields of engineering [19–22]. This chapter is structured as follows: Introduction regarding literature review, relevance of chapter is given in Sect. 1. Motivation, novelty, and graphical abstract are given in Sect. 2. Mathematical modeling of single-area and two-area power system is presented in Sect. 3. Policy iteration-based control scheme is given in Sect. 4. NERC's control performance standards (CPS) are described in Sect. 5. Sections 6 and 7 deal with the comparative results; based on the results, the conclusion is presented in Sect. 8.

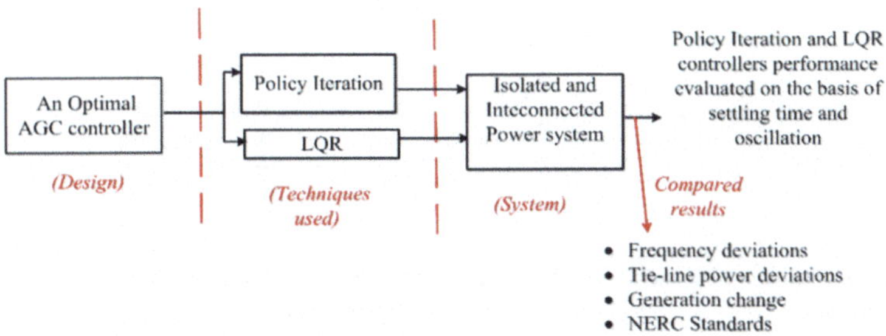

Fig. 1 Graphical representation of the work

2 Motivation, Novelty, Highlights, and Graphical Abstract

Many strategies for creating a robust, optimal AGC controller are available in the literature. It is critical that the control technique used be capable of dealing with uncertainties and nonlinearities in the system. As a result, continuous attempts are conducted to identify new optimal control design strategies that can provide improved performance in any setting. The adaptive critics approach is a step in this direction. According to recent literature, the controller created using this method is capable of handling parameter uncertainty, is robust, and effectively eliminates steady-state errors. It also improves stability. The authors decided to study adaptive critic approach in AGC systems because of its versatility and efficiency in other fields. It was found that PI approach ensured AGC's stability and dynamic performance.

Novelty: The use of policy iteration (PI)-based adaptive critic scheme for optimal control design for AGC system for single-area and two-area power systems is provided in this chapter. This study contributes by offering a complete performance analysis of adaptive optimum control employing a policy iteration scheme. For both models of the AGC system, modeling, simulation results, and analysis are presented.

The main highlights of the presented work are:

- A controller for isolated and two-area power system is designed using policy iteration (PI) approach.
- The LQR approach has also been used.
- A comparison of policy iteration (PI) optimal controller and control scheme based on LQR is carried out.
- Results show the validity of the policy iteration (PI) optimal control approach over LQR (Fig. 1).

3 Modeling of Power System

The block diagram of a conventional ith area can be represented as shown in Fig. 2.

State space characterization of the considered systems can be written as

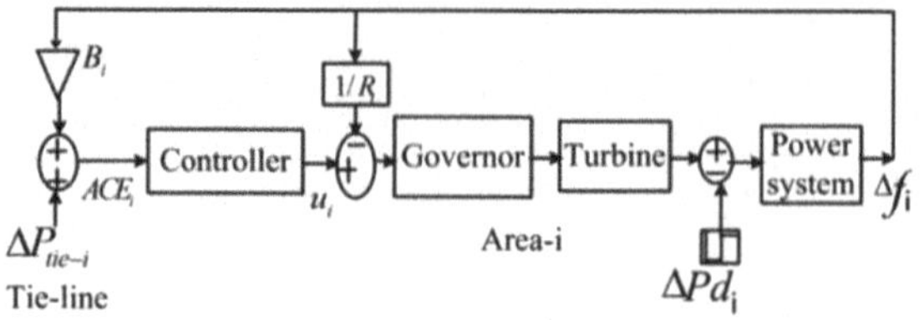

Fig. 2 Conventional LFC scheme for area-i where Δf = deviation/frequency, $\Delta P_{\text{tie-}i}$ = deviation/tie-line, ΔPd_i = change/load, R_i = governor regulation, B_i = frequency bias, and u_i = control signal. The weighted sum of tie-line and frequency deviations forms area control error (ACE_i).

Fig. 3 AGC scheme of an isolated power system

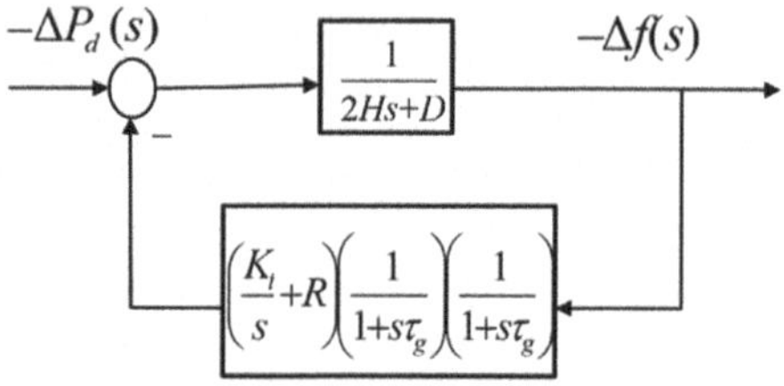

$$\dot{x}(t) = Ax(t) + Bu(t) + Fd(t) \tag{1}$$

The state, control, and disturbance vectors are x, u, and d, respectively. A linearized model of the power system [1] has been studied for the modeling. Figure 3 shows the AGC equivalent block diagram for a single-area system.

The transfer function representation of Fig. 3 can be written as

$$\frac{\Delta f(s)}{-\Delta P_d(s)} = \frac{s\left(1 + s\tau_g\right)(1 + s\tau_T)}{s(2Hs + D)\left(1 + s\tau_g\right)(1 + s\tau_T) + K_i + s/R} \tag{2}$$

For implementing policy iteration (PI) algorithm, the transfer function model given in (2) has been changed to state space model, which is given in the result section. For the modeling of two-area AGC scheme, the considered state variable, control, and disturbance vector are given as [1]

$$x = \left[x_1\ x_2\ x_3\ x_4\ x_5\ x_6\ x_7\ x_8\ x_9\right]^{\mathrm{T}}, u = \left[u_1\ u_2\right]^{\mathrm{T}}, d = \left[d_1\ d_2\right]^{\mathrm{T}}$$

where $x_1 = \Delta f_1$ (frequency deviation in area-1), $x_2 = \Delta Pt_1$ (Change in Turbine power area-1), $x_3 = \Delta Pg_1$ (Change in governor power in area-1), $x_4 = \Delta f_2$ (frequency deviation in area-2), $x_5 = \Delta Pt_2$ (Change in Turbine power in area-2), $x_6 = \Delta Pg_2$ (Change in Governor power in area-2), $x_7 = \Delta P_{\text{tie}(1,2)}$ (Change in Tie-line flow between area-1 and area-2), $x_8 = \int \text{ACE}_1 \text{d}t$ (Integration of ACE_1), $x_9 = \int \text{ACE}_2 \text{d}t$ (Integration of ACE_2), $d_1 = \Delta Pd_1$ Disturbance in area-1, $d_2 = \Delta Pd_2$ Disturbance in area-2.

By using these states, disturbance and control variables matrices have been formed:

$$A = \begin{bmatrix} -\frac{1}{Tp_1} & \frac{Kp_1}{Tp_1} & 0 & 0 & 0 & 0 & -\frac{Kp_1}{Tp_1} & 0 & 0 \\ 0 & -\frac{1}{Tt_1} & \frac{1}{Tt_1} & 0 & 0 & 0 & 0 & 0 & 0 \\ -\frac{1}{R_1 Tg_1} & 0 & -\frac{1}{Tg_1} & 0 & 0 & 0 & 0 & 0 & 0 \\ 0 & 0 & 0 & -\frac{1}{Tp_2} & \frac{Kp_2}{Tp_2} & 0 & \frac{Kp_2}{Tp_2} & 0 & 0 \\ 0 & 0 & 0 & 0 & -\frac{1}{Tt_2} & \frac{1}{Tt_2} & 0 & 0 & 0 \\ 0 & 0 & 0 & -\frac{1}{R_2 Tg_2} & 0 & -\frac{1}{Tg_2} & 0 & 0 & 0 \\ 2\pi T^2 & 0 & 0 & -2\pi T^2 & 0 & 0 & 0 & 0 & 0 \\ B_1 & 0 & 0 & 0 & 0 & 0 & 1 & 0 & 0 \\ 0 & 0 & 0 & B_2 & 0 & 0 & -1 & 0 & 0 \end{bmatrix}$$

$$F = \begin{bmatrix} -\frac{Kp_1}{Tp_1} & 0 & 0 & 0 & 0 & 0 & 0 & 0 & 0 \\ 0 & 0 & 0 & -\frac{Kp_2}{Tp_2} & 0 & 0 & 0 & 0 & 0 \end{bmatrix}^{\tau}$$

$$B = \begin{bmatrix} 0 & 0 & \frac{1}{Tg_1} & 0 & 0 & 0 & 0 & 0 & 0 \\ 0 & 0 & 0 & 0 & 0 & \frac{1}{Tg_2} & 0 & 0 & 0 \end{bmatrix}^{\tau}$$

4 Policy Iteration-Based Adaptive Control Scheme

Optimal control problem can be solved in two ways; one is dynamic programming (DP), and the other is Pontryagin's minimum principle [14–18]. Due to the computationally complex solution of Bellman's dynamic programming, approximate dynamic programming (ADP) approach came into existence. Further, adaptive critic design (ACD) came into existence where critic and actor are used to approximate the value related to cost and the control policy. Policy iteration (PI) algorithm evaluates the cost of a given control policy, updates it, and is repeated until the policy improvement step converges optimally [17, 18]. The steps to design control scheme using policy iteration (PI) approach are given as follows:

4.1 Optimal Control Problem

An LTI system can be represented as

$$\dot{x}(t) = Ax(t) + Bu(t) \tag{3}$$

For this LTI system, an optimum control problem and the cost function to be minimized are presented in (4) and (5), respectively.

$$u^*(t) = \underset{t0 \le i \le \infty}{\arg\min} V(t_0, x(t_0), u(t)) \tag{4}$$

$$V(x(t_0), t_0) = \int_{t_0}^{\infty} \left(x^T(\tau) Q x(\tau) + u^\tau(\tau) R u(\tau)\right) \mathrm{d}\tau \tag{5}$$

Bellman's optimality principle is utilized to give the solution to the problem given in (4).

$$u(t) = -Kx(t) \quad \text{with } K = R^{-1} B^T P \tag{6}$$

where P is obtained on solving Riccati equation.

$$A^T P + PA - PBR^{-1} B^T P + Q = 0 \tag{7}$$

Solution of (7) gives stable controller determined in (6).

4.2 Policy Iteration (PI) Technique

Equations (6) and (7) are represented by actor and critic networks in the policy iteration (PI) technique and are used to update the cost and control policy. If (1) is stabilizable and K is the gain, the cost can be stated as follows:

$$V(x(t)) = \int_{t}^{\infty} x^\tau(\tau)\left(Q + K^T R K\right) x(\tau) d\tau = x^T(t) P x(t) \tag{8}$$

where P represents the solution of Lyapunov equation given in (9)

$$(A - BK)^T P + P(A - BK) = -\left(K^T R K + Q\right) \tag{9}$$

$V(x(t))$ serves as a Lyapunov function for (3). Finally, the quadratic cost function (8) can be expressed as (10)

$$V(x(t)) = \int_{t}^{t+T} x^T(\tau)\left(Q + K^T R K\right) x(\tau) d\tau + V(x(t+T)) \tag{10}$$

Based on (10), the policy iteration technique can be determined as

1. **Policy evaluation**

$$x_t^T P_i x_i = \int_t^{t+T} x_i^T (Q + K_i^T R K_i) x_\tau \mathrm{d}\tau + x_{t+T}^T P_i x_{i+T} \tag{11}$$

2. **Policy improvement**

$$K_{i+1} = R^{-1} B^T P_i \tag{12}$$

It is seen that matrix A is not required in (11)–(12) and only matrix B is necessary to update policy (gain) K.

4.3 Policy Iteration-Based Online Adaptive Control Scheme

It has been proved that system matrix A does not needed in policy iteration approach. States $x(t)$ and $x(t + T)$ are observed online in order to obtain system dynamics information. The PI algorithm, which occurs in (12), requires B matrix. The left-hand side of (11) can be written as (13) to determine critic parameters.

$$x^T(t) P_i x(t) = \overline{p}_i^T \overline{x}(t) \tag{13}$$

where $\overline{x}(t) = \{x_i(t) x_j(t)\}_{i=1,n;\, j=1,n}$ is polynomial basis vector and $\overline{p} = v(P)$ with $v(.)$ is a matrix function. Therefore, using (13), (11) can be rewritten as

$$\overline{p}_i^T (\overline{x}(t) - \overline{x}(t + T)) = \int_t^{t+T} x^T(\tau)\big(Q + K_i^T R K_i\big) x(\tau) \mathrm{d}\tau \tag{14}$$

where $\overline{p}_i$ = vector of unknown and $(\overline{x}(t) - \overline{x}(t + T)$ =regression vector. Right-hand side of (14) is used as target function $d(\overline{x}(t), K_i)$.

$$d(\overline{x}(t), K_i) = \int_t^{t+T} x^T(\tau)\big(Q + K_i^T R K_i\big) x(\tau) \mathrm{d}\tau \tag{15}$$

Using right-hand side of (14) at $N \geq n(n + 1)/2$ points $\overline{x}^i$ in the state space, solution is obtained as

$$\overline{p}_i = \big(X X^T\big)^{-1} X Y \tag{16}$$

where $X = \big[\, \overline{x}_\Delta^1 \; \overline{x}_\Delta^2 \; \ldots \; \overline{x}_\Delta^N \,\big]$

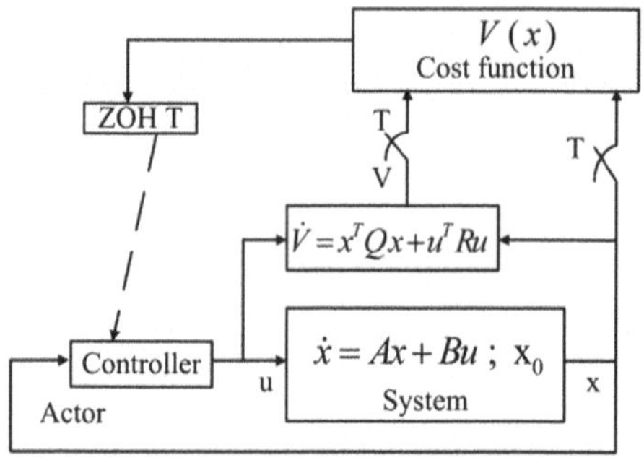

Fig. 4 Actor–critic structure-based adaptive control scheme

$$\overline{x}^i_\Delta = \overline{x}^i(t) - \overline{x}^i(t+T)$$
$$Y = \left[d\left(\overline{x}^1, K_i\right) \;\; d\left(\overline{x}^2, K_i\right) \ldots d\left(\overline{x}^N, K_i\right)\right]^T \tag{17}$$

The policy iteration (PI) requires only knowledge of the observed target $d(\overline{x}(t), K_i)$ as well as measurements at time, t and $t+T$, which makes policy iteration algorithm partially model free. An adaptive optimal control scheme is shown in Fig. 4.

The above policy iteration (PI) approach is used to create a control system.

5 NERC Standard

The goal of the secondary controller is to keep frequency within the acceptable range if the load is changing continually and unpredictably. CPS1 and CPS2 (control performance standards) were proposed by NERC in 1997, and the AGC scheme must meet these control requirements [23].

The CPS1 can be determined as

$$\text{CPS1} = \left(2 - \frac{\text{CF}_{12\text{-month}}}{\varepsilon_1^2}\right) * 100\% \tag{18}$$

where CF is the compliance factor and can be determined as

$$\text{CF}_{12\text{-month}} = \text{avg}_{12\text{-month}}\left[\left(\frac{\text{ACE}}{-10B}\right) * \Delta f\right] \tag{19}$$

where B = frequency bias setting (MW/0.1 Hz), ε_1 =constant, Δf = average of frequency deviation in clock-minute, and $\left(\frac{\text{ACE}}{-10B}\right)$ =clock-minute average of ACE. As per the NERC standard, any AGC scheme must have the 100% compliance of CPS1.

For CPS2 compliance, the average of ACE for each of the six ten-minute periods during the hour must be under a limit (L10). CPS2 compliance of 90% or greater is necessary for the AGC scheme. The CPS2 can be calculated as

$$\text{CPS2} = \left[1 - \frac{\text{violations}}{(\text{total periods} - \text{unavailable periods})}\right] * 100\% \tag{20}$$

Violations clock-ten minutes = 0 if

$$\text{ACE}_{\text{av}} = \left|\frac{\sum \text{ACE}}{n_{\text{ samples in 10 minutes}}}\right| \leq L_{10} \tag{21}$$

Violations clock-ten minutes = 1

$$\text{ACE}_{\text{av}} = \left|\frac{\sum \text{ACE}}{n_{\text{ samples in 10 minutes}}}\right| > L_{10} \tag{22}$$

where L_{10} is defined as

$$L_{10} = 1.65 * \varepsilon_{10} * \sqrt{(-10 * B_i)(-10 * B_s)} \tag{23}$$

In (23), $B_i = B_s$ represents the frequency bias. Violations are the number of periods in which average ACE exceeds L_{10} value in 10 min. The total period is taken as 1 day and no unavailable period.

6 Results Demonstration and Discussions

The proposed control scheme is implemented for AGC scheme for isolated power system and interconnected two-area power systems. Different load perturbation cases have been considered to evaluate the performance of the designed schemes.

6.1 AGC Scheme for Isolated Power System

Policy iteration (PI)-based optimal controller performance has been checked on isolated power system. The state, control, and disturbance vectors for considered isolated power system are $x = \left[x_1\ x_2\ x_3\ x_4\right]^T$, u and d, where state vector 'x' is a (4×1) vector and considered state variables are

$$x_1 = \Delta f \text{ (Frequency deviation)}, x_2 = \Delta P_t \text{ (Change in turbine power)}$$
$$x_3 = \Delta Pg \text{ (Change in governor power) } x_4 = \Delta e \text{ (Error)}$$

Table 1 Isolated power system parameters

$\tau_g = 0.2$ s Governor time constant	$D = 0.8$ Frequency-dependent load
$H = 5$ s Inertia constant	$\tau_t = 0.5$ s Turbine time constant
$R = 0.05$ Speed regulation	$B = 20.8$ Frequency bias constant

The various parameters of isolated power system are given in Table 1.

On placing the value of various parameters given in Table 1 and converting transfer function model, the state space model has been developed. Therefore, A (plant matrix), B (control matrix), and C (output matrix) for isolated power system can be represented as

$$A = \begin{bmatrix} -7.08 & -10.56 & -20.8 & -7 \\ 1 & 0 & 0 & 0 \\ 0 & 1 & 0 & 0 \\ 0 & 0 & 1 & 0 \end{bmatrix}, B = \begin{bmatrix} 1 \ 0 \ 0 \ 0 \end{bmatrix}', \ C = \begin{bmatrix} 0.1 \ 0.7 \ 1 \ 0 \end{bmatrix} \quad (24)$$

For implementing policy iteration (PI) algorithm for the system defined in (24), the following assumptions have been considered: Initially, the states, cost, and critic parameters are given as $x_0 = \begin{bmatrix} 0 \ 0.1 \ 0 \ 0 \ 0 \end{bmatrix}$; $P = [0000; 0000; 0000; 0000]$, sample time $T = 0.05$ s, $Q = R = I$ (weighting factors), and simulation length = 100 samples

The positive definite solution and controller gains obtained using LQR and PI techniques are obtained as given below

$$P_{lqr} = \begin{bmatrix} 0.1297 & 0.4270 & 0.2043 & 0.0711 \\ 0.4270 & 4.2445 & 4.1005 & 1.4206 \\ 0.2043 & 4.1005 & 9.7058 & 3.7699 \\ 0.0711 & 1.4206 & 3.7699 & 2.9227 \end{bmatrix}, P = \begin{bmatrix} 0.1023 & 0.3914 & 0.1122 & 0.0343 \\ 0.3914 & 4.1983 & 3.9812 & 1.3735 \\ 0.1122 & 1.3735 & 9.3967 & 3.6469 \\ 0.0343 & 4.6983 & 7.2937 & 1.4379 \end{bmatrix}$$

$$K_{lqr} = [0.1297 \quad 0.4270 \quad 0.2043 \quad 0.0711], K = [0.1288 \quad 0.4259 \quad 0.2014 \quad 0.0705]$$

where P_{lqr} is the solution of Riccati given in (7) and P is the adaptive critic matrix obtained using (11), (12), and (17). K_{lqr} is a vector of actor parameters of LQR design by (6) and (7), and K is the vector of actor parameters of adaptive scheme using policy iteration (PI) in (12). Simulation results for system given in (24) are shown in Figs. 5–9. The system trajectories (states and control) converge toward zero in Figs. 5 and 6. Figure 7 depicts closed-loop poles. Figure 8 depicts the convergence of matrix P's critic parameters toward ideal values. The updating of P parameters with iteration is depicted in Fig. 9, where * indicates an update and * indicates no change. Figure 10 depicts the isolated power system's frequency deviation response to a 0.2 pu load disturbance.

Performance of policy iteration (PI) and LQR schemes has also been checked for parameter uncertainty case by changing the system matrix parameter $A(1, 3) = -18$

Fig. 5 System states

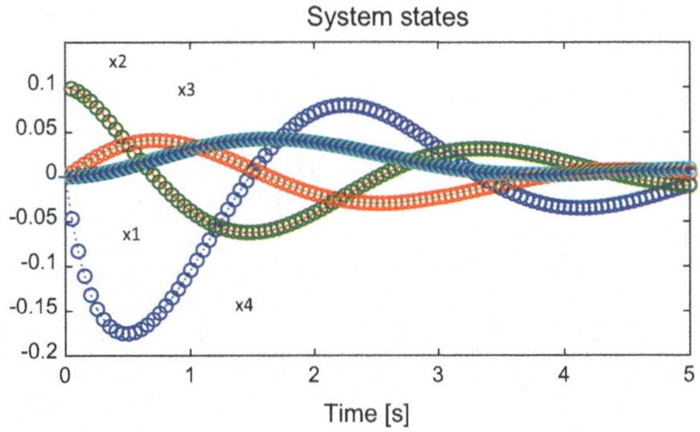

Fig. 6 Control signal

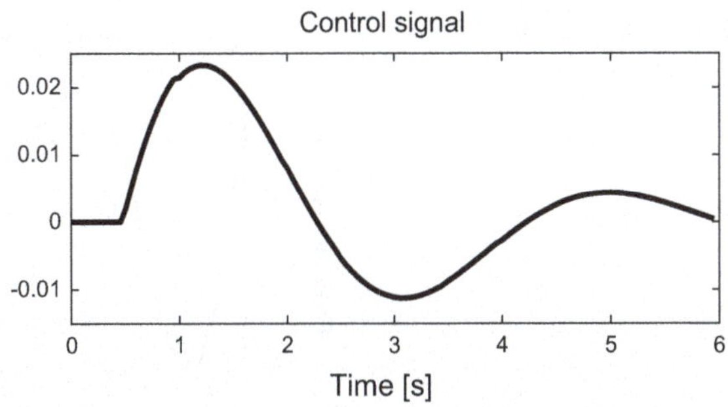

Fig. 7 System poles

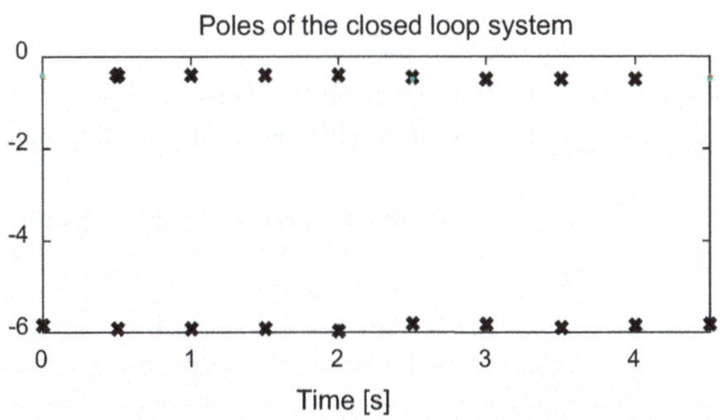

Fig. 8 Critic parameters

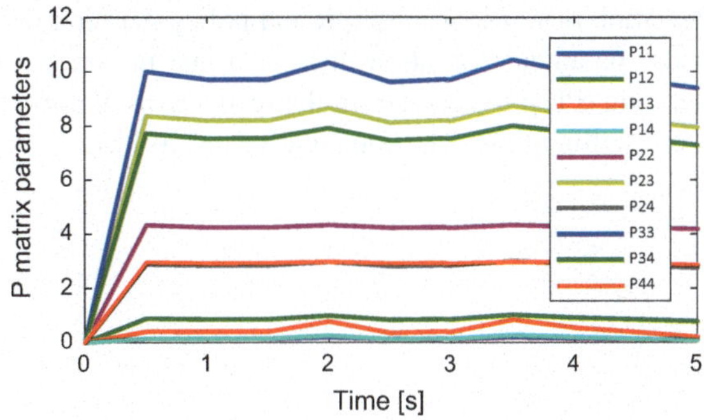

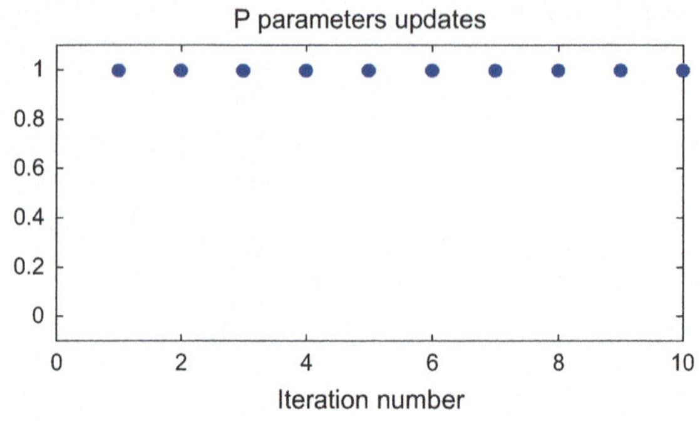

Fig. 9 Updation of critic parameters

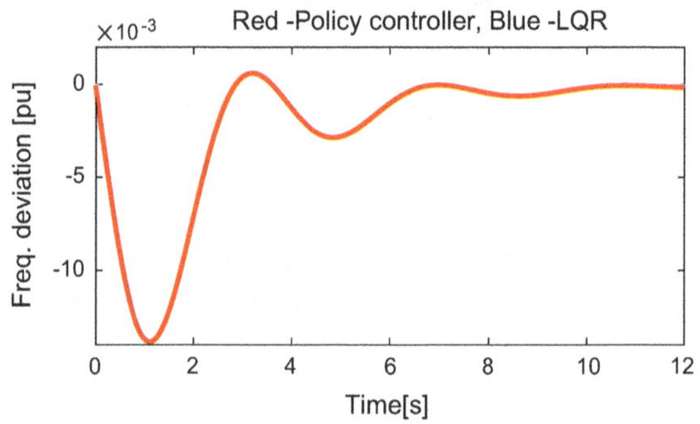

Fig. 10 Deviation in frequency

at sample $k = 41$. For this considered case, unique positive definite solution and actor gains using policy iteration (PI) and LQR are obtained as

$$P_{lqr} = \begin{bmatrix} 2.257 & 18.034 & 42.115 & 0.0711 \\ 18.034 & 150.124 & 352.551 & 16.468 \\ 42.115 & 352.551 & 863.159 & 128.270 \\ 0.0711 & 16.468 & 128.270 & 296.519 \end{bmatrix}, P = \begin{bmatrix} 0.1023 & 0.3914 & 0.1122 & 0.0343 \\ 0.3914 & 4.1983 & 3.9812 & 1.3735 \\ 0.1122 & 1.3735 & 9.3967 & 3.6469 \\ 0.0343 & 4.6983 & 7.2937 & 1.4379 \end{bmatrix}$$

$$K_{lqr} = [2.2578 \quad 18.0340 \quad 42.1151 \quad 0.0711], K = [0.1288 \quad 0.4259 \quad 0.2014 \quad 0.0705]$$

System poles at sample $k = 41$ are shown in Fig. 11. Figure 12 shows the frequency deviation response using LQR and policy iteration (PI) obtained optimal controller under parameter variation. It is seen that in parameter variation condition, policy iteration (PI) provides the similar response as of normal condition, while LQR takes more settling time. The comparative results have been given in Table 2.

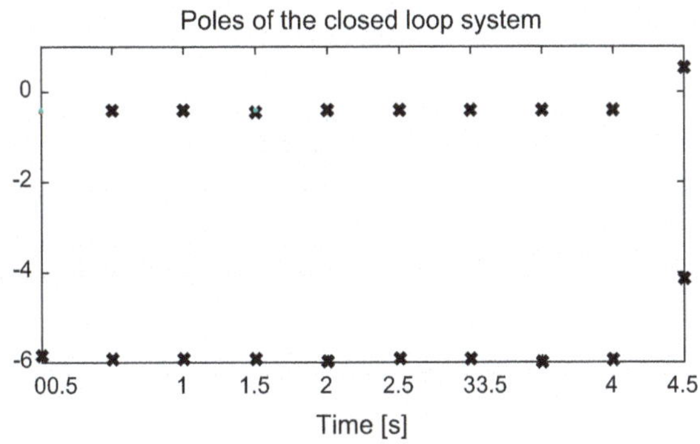

Fig. 11 Poles of closed-loop system

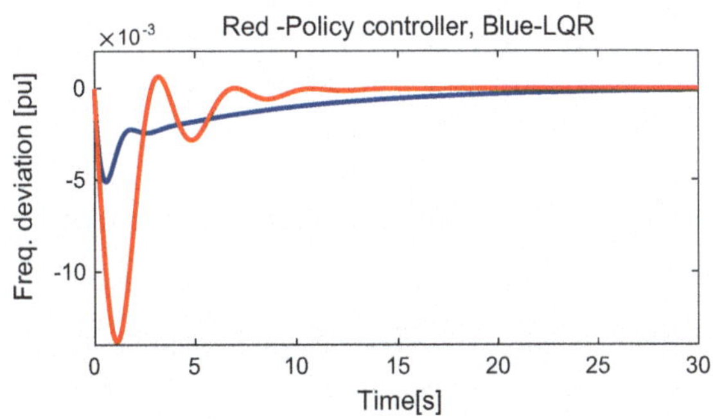

Fig. 12 Frequency deviation response under parameter variation at $k = 41$

Table 2 Comparative study of LQR and policy control schemes (isolated power system)

Controller	Settling time (sec)		Overshoots		Undershoot	
	Frequency		Frequency		Frequency	
	w/o change in parameter	With change	w/o change in parameter	With change	w/o change in parameter	With change
LQR	12	30	$6.4 * 10^{-4}$	0.0	−0.0137	0.0
Policy	12	15	$6.4 * 10^{-4}$	$0.5 * 10^{-3}$	−0.0137	−0.0137

6.2 AGC Scheme for Interconnected Two-Area Power System

The performance of policy iteration control scheme has also been evaluated for interconnected two-area power system. Table 3 lists the specifications for a two-area power system. Using parameters of two-area system, the system matrix A and input matrix B can be written as

Table 3 Two-area power system parameters

$Tg = 0.08$ s governor time constant	$Tp = 20$ s power system time constant
$K_P = 120$ s power system gain constant	$Tt = 0.4$ s turbine time constant
$R = 2.4$ speed regulation	$B = 0.425$ frequency bias constant
$T^0 = 0.707$ synchronizing constant	

$$A = \begin{bmatrix} -0.05 & 5 & 0 & 0 & 0 & 0 & -5 & 0 & 0 \\ 0 & -2 & 2 & 0 & 0 & 0 & 0 & 0 & 0 \\ -0.83 & 0 & -2.5 & 0 & 0 & 0 & 0 & 0 & 0 \\ 0 & 0 & 0 & -0.05 & 5 & 0 & 5 & 0 & 0 \\ 0 & 0 & 0 & 0 & -2 & 2 & 0 & 0 & 0 \\ 0 & 0 & 0 & -0.83 & 0 & -2.5 & 0 & 0 & 0 \\ 0.05 & 0 & 0 & -0.05 & 0 & 0 & 0 & 0 & 0 \\ 0.425 & 0 & 0 & 0 & 0 & 0 & 1 & 0 & 0 \\ 0 & 0 & 0 & 0.425 & 0 & 0 & -1 & 0 & 0 \end{bmatrix} \quad B = \begin{bmatrix} 0 & 0 & 2.5 & 0 & 0 & 0 & 0 & 0 & 0 \\ 0 & 0 & 0 & 0 & 0 & 2.5 & 0 & 0 & 0 \end{bmatrix}^{\mathrm{T}} \tag{25}$$

The following assumptions have been used to implement the policy iteration (PI) control scheme: sample time $T = 0.06$ s, $Q = R = I$, simulation length in samples = 120 and initial condition for states and cost function, and critic parameters are given as follows:

$$x_0 = [0.01\ 0.01\ 0.01\ 0.01\ 0.01\ 0.01\ 0.05\ 0.05\ 0\ 0];$$
$$P = [000000000;\ 000000000;\ \ldots\ldots.000000000]$$

Using this information, the positive definite solution and controller gains are obtained as

$$Pl_{qr} = \begin{bmatrix} 1.0362 & 1.3337 & 0.4004 & -0.0249 & -0.0222 & -0.0008 & 0.0341 & 0.9020 & -0.0051 \\ 1.3337 & 2.4063 & 0.8682 & -0.0222 & -0.0348 & -0.0076 & -1.0182 & 1.1697 & -0.0032 \\ 0.4004 & 0.8682 & 0.5358 & -0.0008 & -0.0076 & -0.0026 & -0.6210 & 0.4000 & -0.0000 \\ -0.0249 & -0.0222 & -0.0008 & 1.0362 & 1.3337 & 0.4004 & -0.0341 & -0.0051 & 0.9020 \\ -0.0222 & -0.0348 & -0.0076 & 1.3337 & 2.4063 & 0.8682 & 1.0182 & -0.0032 & 1.1697 \\ -0.0008 & -0.0076 & -0.0026 & 0.4004 & 0.8682 & 0.5358 & 0.6210 & -0.0000 & 0.4000 \\ 0.0341 & -1.0182 & -0.6210 & -0.0341 & 1.0182 & 0.6210 & 36.9848 & 1.1256 & -1.1256 \\ 0.9020 & 1.1697 & 0.4000 & -0.0051 & -0.0032 & -0.0000 & 1.1256 & 3.1100 & 0.1272 \\ -0.0051 & -0.0032 & -0.0000 & 0.9020 & 1.1697 & 0.4000 & -1.1256 & 0.1272 & 3.1100 \end{bmatrix}$$

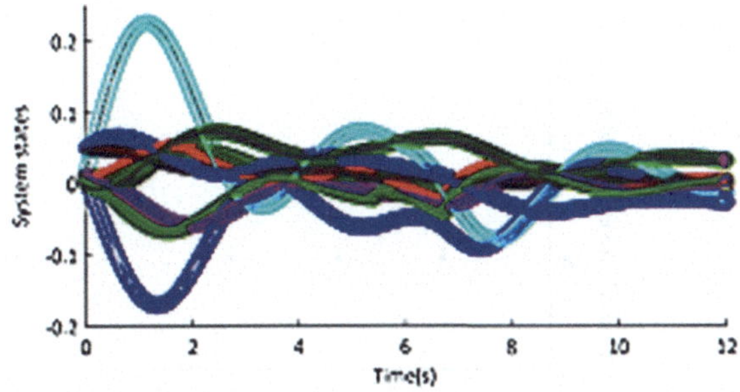

Fig. 13 System states

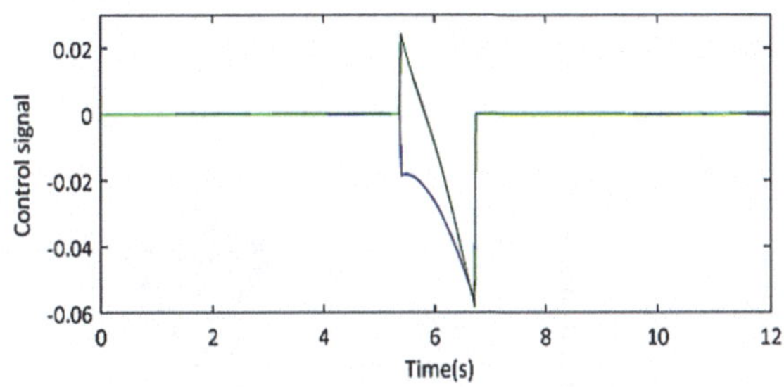

Fig. 14 Control signal

$$P = \begin{bmatrix} -0.0014 & 0.0256 & 0.0021 & 0.0046 & 0.0217 & -0.0023 & -0.0177 & 0.0097 & -0.0043 \\ 0.0256 & -0.0508 & 0.0406 & 0.0130 & -0.0060 & -0.0028 & 0.0662 & 0.0017 & -0.0031 \\ 0.0021 & -0.0031 & 0.0931 & -0.0040 & 0.0908 & 0.0220 & 0.0137 & 0.0247 & -0.0043 \\ 0.0046 & 0.0247 & -0.0043 & 0.0100 & 0.0105 & 0.0052 & 0.0057 & -0.0029 & 0.0016 \\ 0.0217 & 0.0057 & -0.0058 & 0.0016 & 0.0144 & 0.0478 & 0.0183 & -0.0005 & -0.0025 \\ -0.0023 & 0.0478 & 0.0365 & -0.0005 & -0.0025 & -0.0051 & 0.0417 & -0.0084 & 0.0109 \\ -0.0177 & -0.0025 & -0.0051 & 0.0417 & -0.0084 & 0.0109 & -0.0520 & -0.0017 & 0.0007 \\ 0.0097 & 0.0417 & -0.0168 & 0.0109 & -0.0260 & -0.0017 & 0.0007 & -0.0008 & -0.0014 \\ -0.0043 & -0.0084 & 0.0217 & -0.0260 & -0.0017 & 0.0007 & -0.0004 & -0.0014 & 0.0011 \end{bmatrix}$$

$$K_{lqr} = \begin{bmatrix} 1.0009 & 2.1706 & 1.3394 & -0.0020 & -0.0190 & -0.0065 & -1.5526 & 1.0000 & -0.0000 \\ -0.0020 & -0.0190 & -0.0065 & 1.0009 & 2.1706 & 1.3394 & 1.5526 & -0.0000 & 1.0000 \end{bmatrix}$$

$$K = \begin{bmatrix} 0.0052 & -0.0078 & 0.2328 & -0.0100 & 0.2270 & 0.0551 & 0.0342 & 0.0618 & -0.0108 \\ -0.0058 & 0.1194 & 0.0913 & -0.0012 & -0.0063 & -0.0128 & 0.1042 & -0.0210 & 0.0272 \end{bmatrix}$$

Figures 13–17 illustrate the simulation results for the system described in (25). The responses in Figs. 13 and 14 illustrate that the system trajectories (states and control) converge to zero. Figure 15 depicts closed-loop poles. Figure 16 depicts the convergence of matrix P's critic parameters toward ideal values. The updating of P parameters with iteration is depicted in Fig. 17.

6.2.1 Load Occurs in Area-1 Only (Without Parameter Variation)

Let us consider a load demand of 0.1 pu in area-1. This load demand affects the frequency and tie-line power exchange. Figure 18 shows the frequency deviation

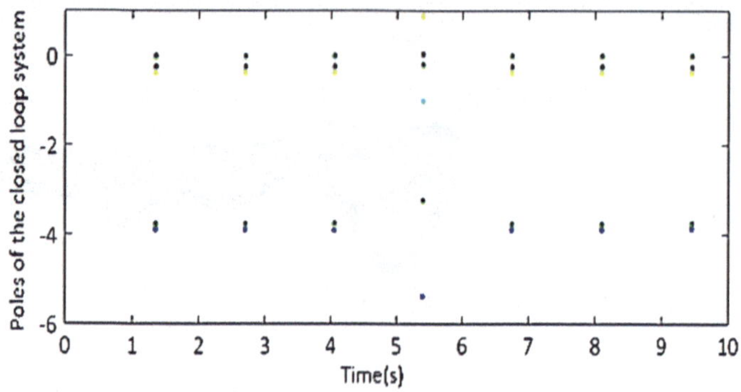

Fig. 15 Closed-loop poles

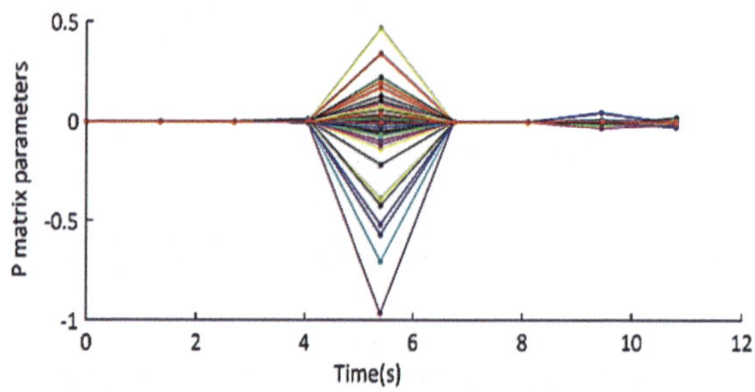

Fig. 16 Critic parameters

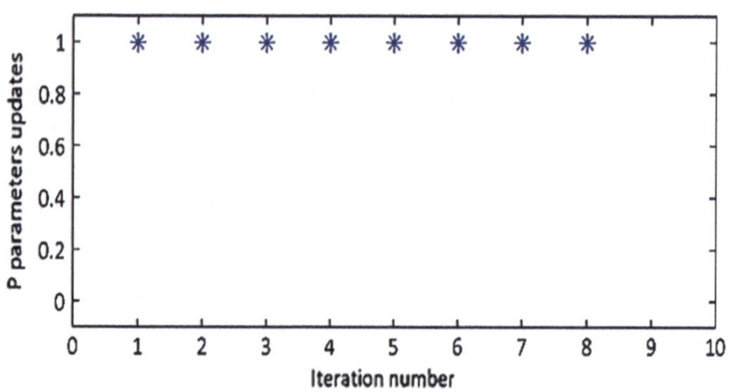

Fig. 17 Critic parameters updation

response using LQR and policy iteration (PI) optimal adaptive controller. Figure 19 shows that area-1 generation settles to the required load demand at steady state. Figure 20 shows the tie-line power deviation causes by load demand.

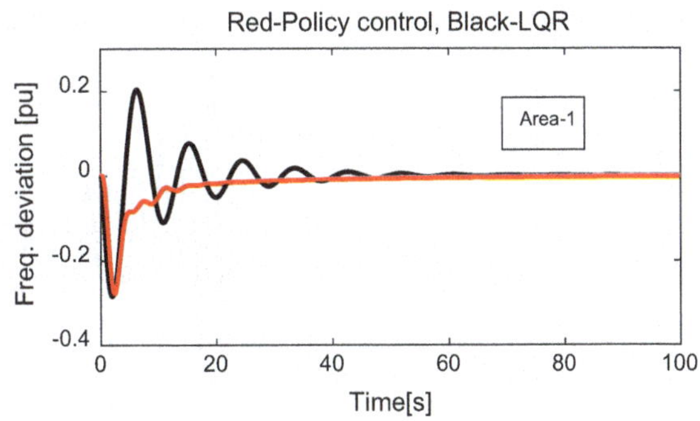

Fig. 18 Frequency deviation

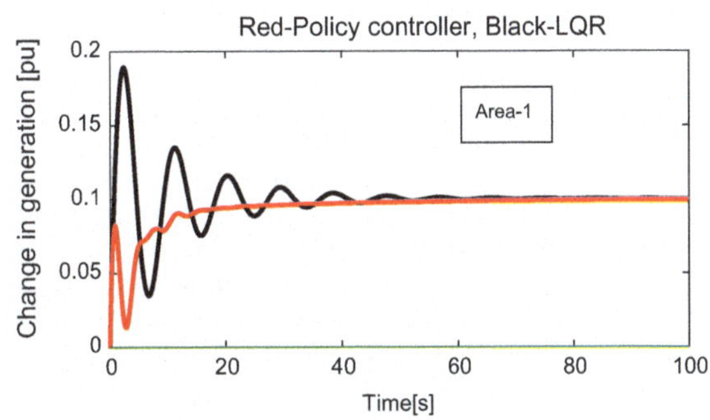

Fig. 19 Change in generation

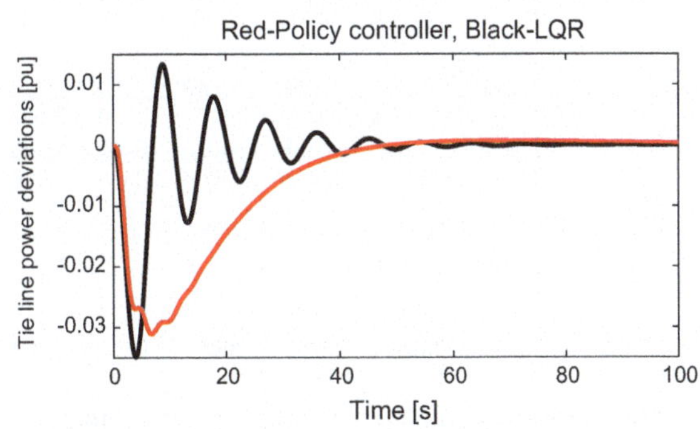

Fig. 20 Tie-line power deviation

6.2.2 Load Occurs in Area-1 Only (with Parameter Variation)

Let us consider the same load demand of 0.1 pu in area-1 as given above. For this case, system parameter has been changed by changing system matrix parameter $A(1, 2) = 6$ at sample $k = 21$. For this case, positive definite solution and actor parameters determined using LQR and policy iteration (PI) technique are given as

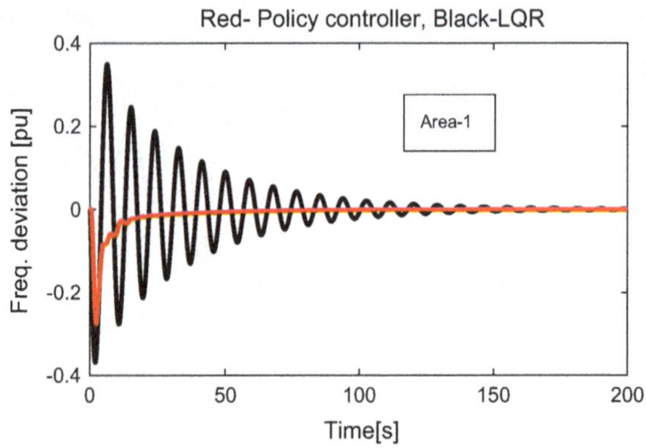

Fig. 21 Frequency deviation

$$P_{lqr} = \begin{bmatrix} 0.9374 & 1.3818 & 0.3942 & 0.0251 & -0.0198 & 0.0004 & 0.2067 & 0.8218 & -0.0055 \\ 1.3818 & 2.8505 & 0.9943 & -0.0309 & -0.0412 & -0.0077 & -0.7334 & 1.2224 & -0.0053 \\ 0.3942 & 0.9943 & 0.5779 & -0.0043 & -0.0108 & -0.0031 & -0.4985 & 0.4000 & -0.0005 \\ -0.0251 & -0.0309 & -0.0043 & 1.0351 & 1.3323 & 0.3999 & -0.0076 & -0.0059 & 0.9016 \\ -0.0198 & -0.0412 & -0.0108 & 1.3323 & 2.4044 & 0.8676 & 1.0490 & -0.0025 & 1.1694 \\ 0.0004 & -0.0077 & -0.0031 & 0.3999 & 0.8676 & 0.5355 & 0.6305 & 0.0005 & 0.4000 \\ 0.2067 & -0.7334 & -0.4985 & -0.0076 & 1.0490 & 0.6305 & 35.3545 & 1.3827 & -1.2079 \\ 0.8218 & 1.2224 & 0.4000 & -0.0059 & -0.0025 & 0.0005 & 1.3827 & 3.0339 & 0.1401 \\ -0.0055 & -0.0053 & -0.0005 & 0.9016 & 1.1094 & 0.4000 & -1.2079 & 0.1401 & 3.0976 \end{bmatrix}$$

$$P = \begin{bmatrix} -0.0014 & 0.0256 & 0.0021 & 0.0046 & 0.0217 & -0.0023 & -0.0177 & 0.0097 & -0.0043 \\ 0.0256 & -0.0508 & 0.0406 & 0.0130 & -0.0060 & -0.0028 & 0.0662 & 0.0017 & -0.0031 \\ 0.0021 & -0.0031 & 0.0931 & -0.0040 & 0.0908 & 0.0220 & 0.0137 & 0.0247 & -0.0043 \\ 0.0046 & 0.0247 & -0.0043 & 0.0100 & 0.0105 & 0.0052 & 0.0057 & -0.0029 & 0.0016 \\ 0.0217 & 0.0057 & -0.0058 & 0.0016 & 0.0144 & 0.0478 & 0.0183 & -0.0005 & -0.0025 \\ -0.0023 & 0.0478 & 0.0365 & -0.0005 & -0.0025 & -0.0051 & 0.0417 & -0.0084 & 0.0109 \\ -0.0177 & -0.0025 & -0.0051 & 0.0417 & -0.0084 & 0.0109 & -0.0520 & -0.0017 & 0.0007 \\ 0.0097 & 0.0417 & -0.0168 & 0.0109 & -0.0260 & -0.0017 & 0.0007 & -0.0008 & -0.0014 \\ -0.0043 & -0.0084 & 0.0217 & -0.0260 & -0.0017 & 0.0007 & -0.0004 & -0.0014 & 0.0011 \end{bmatrix}$$

$$K_{lqr} = \begin{bmatrix} 0.9854 & 2.4858 & 1.4448 & -0.0108 & -0.0269 & -0.0077 & -1.2462 & 1.0000 & -0.0012 \\ 0.0009 & -0.0194 & -0.0077 & 0.9998 & 2.1690 & 1.3389 & 1.5762 & 0.0012 & 1.0000 \end{bmatrix}$$

$$K = \begin{bmatrix} 0.0052 & -0.0078 & 0.2328 & -0.0100 & 0.2270 & 0.0551 & 0.0342 & 0.0618 & -0.0108 \\ -0.0058 & 0.1194 & 0.0913 & -0.0012 & -0.0063 & -0.0128 & 0.1042 & -0.0210 & 0.0272 \end{bmatrix}$$

The various parameter results are demonstrated in Figs. 21, 22, and 23, respectively.

The obtained results show that at parameter variation robustness of policy iteration (PI) optimal adaptive controller is not affected while as LQR performance detroit.

6.2.3 Different Loads Occur in Area-1 and Area-2

Let us consider that 0.1 pu load demand increases in area-1 and 0.1 pu load demand decreases in area-2. Figure 24 a, b shows the frequency deviation of each area. Figure 25 a, b shows the change in generation of each area, respectively. It is evident

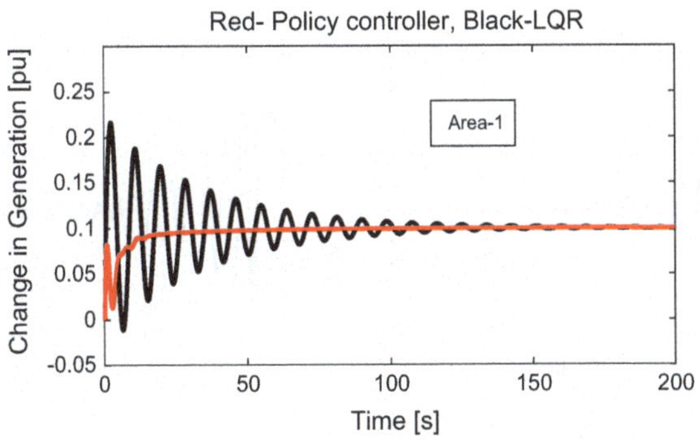

Fig. 22 Generation change

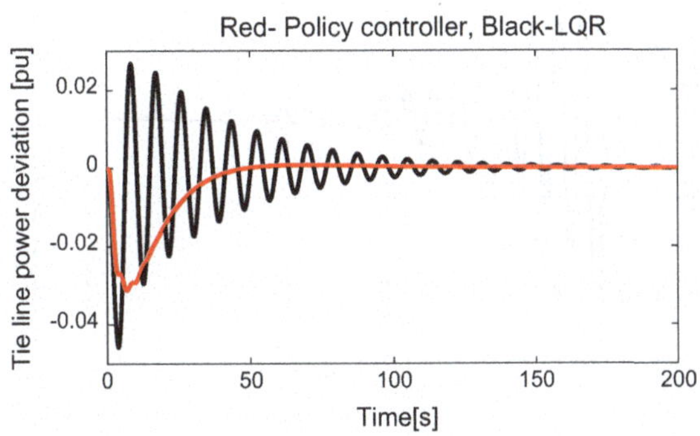

Fig. 23 Tie-line exchange

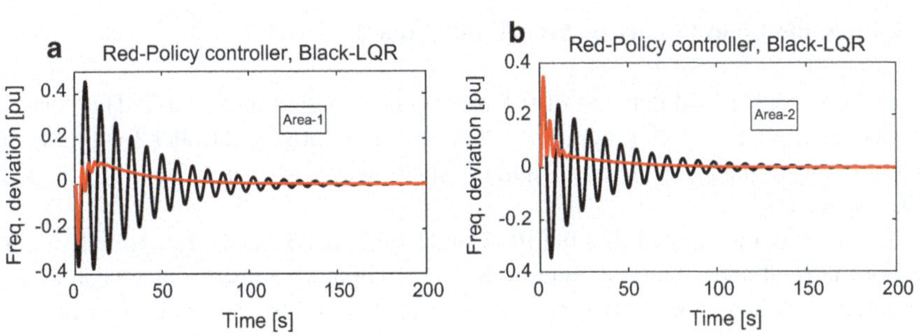

Fig. 24 **a** Frequency deviation (area-1). **b** Frequency deviation (area-2)

that generators reach to their required generation much faster by having policy iteration (PI) as an optimal adaptive controller than LQR. Figure 26 shows the tie-line which settles down to zero at steady state.

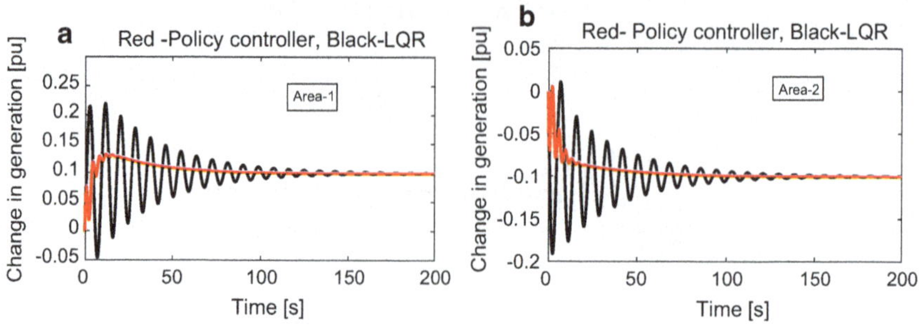

Fig. 25 **a** Generation change (area-1). **b** Generation change (area-2)

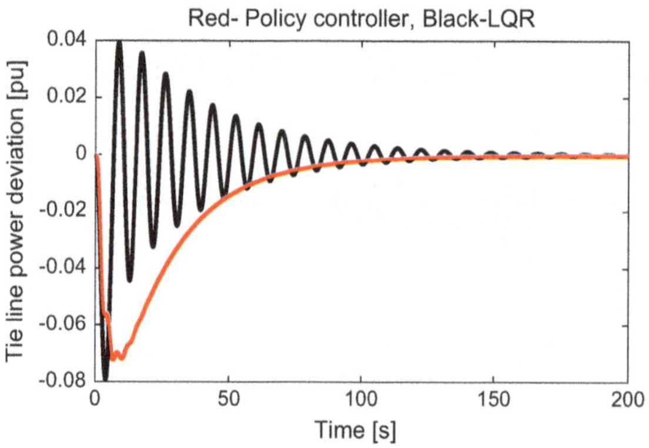

Fig. 26 Tie-line exchange

6.2.4 Same Load Occurs in Area-1 and Area-2

Let us consider a load demand of 0.1 pu occurs in area-1 and area-2. The performance of LQR and policy iteration (PI) optimal adaptive controller for this load demand on frequency, generation, and tie-line power is shown in Figs. 27, 28, and 29, respectively.

It can be demonstrated that the PI optimal scheme converges to optimal values in both normal and abnormal conditions. The comparative results have been given in Tables 4–5, which show that policy iteration (PI) optimal controller gives better performance than LQR.

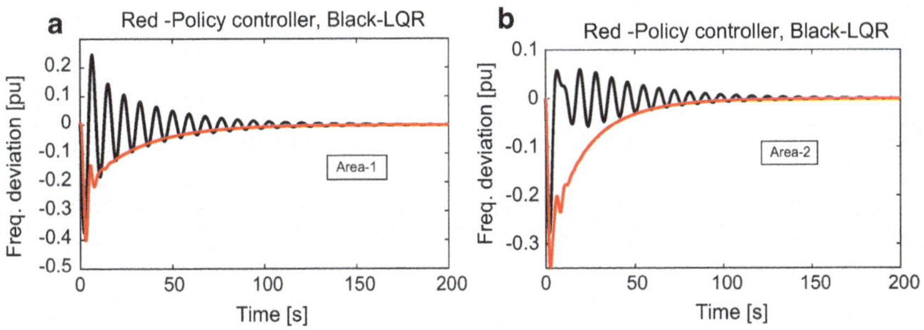

Fig. 27 **a** Frequency deviation (area-1). **b** frequency deviation (area-2)

Fig. 28 Generation change (area-1 and area-2)

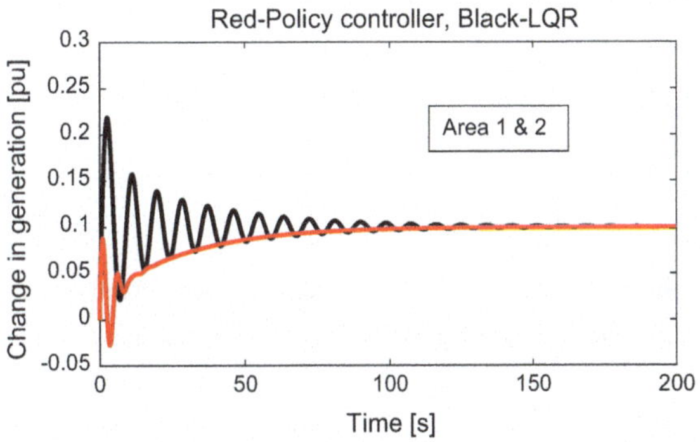

Fig. 29 Tie-line exchange

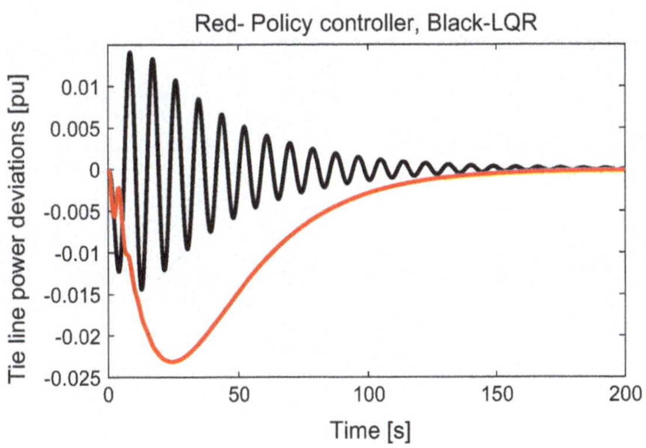

7 Compliance with NERC Standard

Compliance of the NERC standards for policy iteration (PI)-based and LQR-based AGC schemes has been established on two-area power system. The profile of load variation has been taken as given in Fig. 30

Table 4 Comparative study of LQR and policy control schemes (two-area power system)

Controller	Settling time (sec)		Overshoots		Undershoot	
	Frequency		Frequency		Frequency	
Load occurs in both areas	Area-1/Δf_1	Area-2/Δf_2	Area-1/Δf_1	Area-2/Δf_2	Area-1/Δf_1	Area-2/Δf_2
At same load LQR	190	185	0.25	0.05	−0.38	−0.28
At same load Policy	120	125	0.0	0.0	−0.40	−0.35
At different load LQR	190	195	0.48	0.30	−0.38	−0.34
At different load LQR	120	125	0.10	0.35	−0.26	0.0

Table 5 Comparative study of LQR and policy control schemes (two-area power system)

Controller	Settling time (sec)		Overshoots		Undershoot	
	Frequency		Frequency		Frequency	
Load occurs in area-1 only	w/o change in parameter	With change	w/o change in parameter	With change	w/o change in parameter	With change
LQR	80	170	0.20	0.35	−0.285	−0.368
Policy	60	100	0.0	0.0	−0.275	−0.278

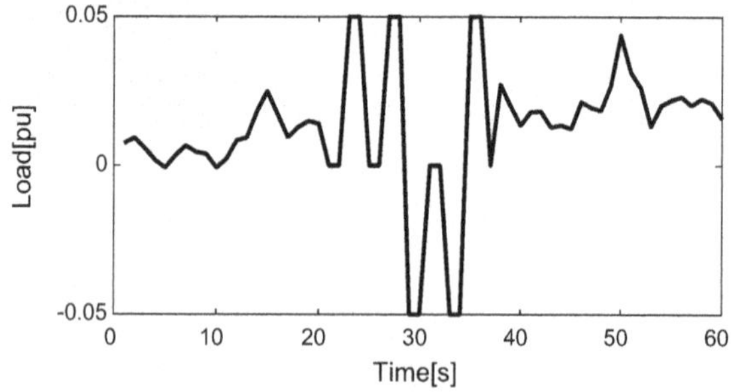

Fig. 30 Random load variations

CPS1, as described in (19), and CPS2, as defined in (20), were calculated using the computed CF value. Typical values of $\varepsilon_{10} = 0.0025$ Hz and $\varepsilon_1 = 0.0131$ Hz were taken from [23]. This gives $L_{10} = 0.01753$. Value of $\mathrm{ACE_{av}}$ has been determined for load changes for each 10 min. Figure 31 shows variation in $\mathrm{ACE_{av}}$. On comparing L_{10} with $\mathrm{ACE_{av}}$, it is seen that PI scheme has no violation, while LQR has one violation.

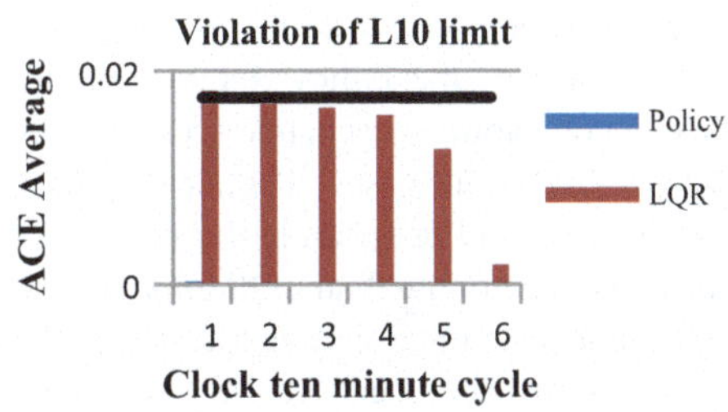

Fig. 31 Violation of L_{10} limit

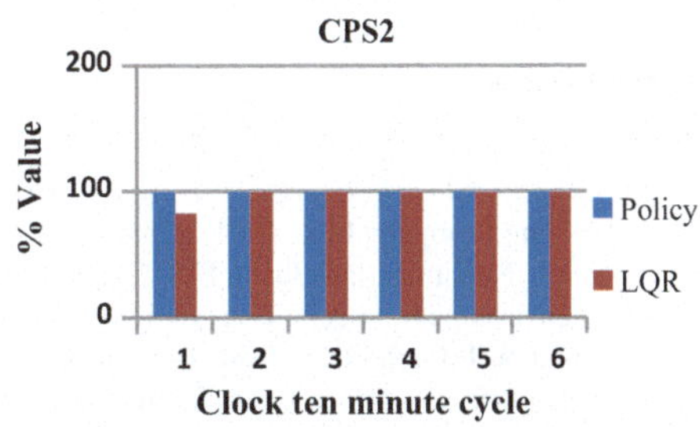

Fig. 32 Comparison of CPS2

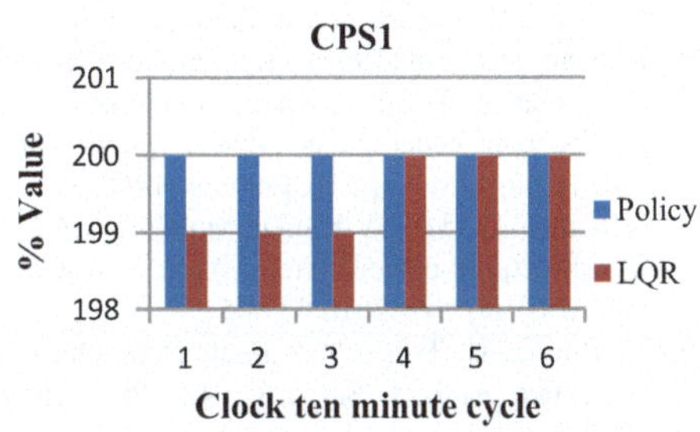

Fig. 33 Comparison of CPS1

The CPS2 values for the LQR and PI schemes are shown in Fig. 32. Figure 33 depicts CPS1.

8 Conclusion

This chapter discusses frequency-related issues in a power system. An investigation of AGC problem has been carried out with policy iteration (PI) optimal control scheme and LQR control scheme in isolated power system and two-area power system. Policy iteration (PI) technique uses policy evaluation and policy improvement steps to evaluate and update the cost and control policy and provides optimal control without using the complete states. Thus, adaptive optimal control scheme based on PI technique is partially a model-free approach. Both the control schemes have been compared in terms of the dynamic response of the system. Further, the performance of the policy iteration-based optimal control scheme has also been checked on parametric change scenario. It is observed that frequency error at steady state is zero in all cases. The simulation shows that the designed control scheme is successful and provides better results, i.e., smaller settling time, smaller oscillations, faster damping than LQR approach. The simulation results justify the effective performance of the policy iteration-based optimal control in both the studied case, and it can be concluded that the same approach can be utilized in AGC scheme.

References

1. O.I. Elgerd, *Electric Energy Systems Theory: An Introduction*, 2nd edn. (Tata McGraw-Hill Publishing Company Ltd., 2004) (Chapter 9)
2. A. Singh, N. Kumar, B.P. Joshi, B.K. Singh, Load frequency control with time delay in restructured environment, Special issue on intelligent tools and techniques for signals, machines and automation. J. Intel. Fuzzy Syst. 1–7, 26 July 2018. https://doi.org/10.3233/JIFS-169778
3. R.D. Christie, A. Bose, Load frequency control issues in power system operations after deregulation. IEEE Trans. Power Syst. **11**(3), 1191–1200 (1996). https://doi.org/10.1109/59.535590
4. N. Cohn, Considerations in the regulation of interconnected area. IEEE Trans. Power Syst. **PAS-86**, 1527–38 (1957). https://doi.org/10.1109/TPAS.1967.291937
5. N. Kumar, B. Tyagi, V. Kumar, Optimization of PID parameters using BBBC for multi-area AGC scheme in deregulated power system. Turkish J. Electr. Eng. Comput. Sci. **24**, 4105–4116 (2015). https://doi.org/10.3906/elk-1408-36
6. N. Kumar, B. Tyagi, V. Kumar, Multi area AGC scheme using imperialist competition algorithm in restructured power system. Appl. Soft. Comput. Elsevier **48**, 160–168 (2016). https://doi.org/10.1016/j.asoc.2016.07.005
7. N. Kumar, B. Tyagi, V. Kumar, Deregulated multiarea AGC scheme using BBBC-FOPID controller. Arab. J. Sci. Eng. **42**, 2641–2649 (2016). https://doi.org/10.1007/s13369-016-2293-1
8. N. Kumar, B. Tyagi, V. Kumar, Multi-area deregulated automatic generation control scheme of power system using imperialist competitive algorithm based robust controller. IETE J. Res. **64**, 528–537 (2017). https://doi.org/10.1080/03772063.2017.1362965
9. C. Fosha, O.I. Elgerd, The megawatt-frequency control problem: a new approach via optimal control theory. IEEE Trans. Power Appar. Syst. **89**(4), 563–577 (1970). https://doi.org/10.1109/TPAS.1970.292603
10. Ibraheem, P. Kumar, N. Hassan, Nizamuddin, Sub-optimal automatic generation control of interconnected power system using output vector feedback control strategy. Electric Power Componen. Syst. **40**, 977–94 (2012). https://doi.org/10.1080/15325008.2012.675802
11. R.R. Shoults, J.A.J. Ibarra, Multi-area adaptive LFC developed for a comprehensive AGC simulator. IEEE Trans. Power Appar. Syst. **8**(2), 541–547 (1993). https://doi.org/10.1109/59.260829
12. F.Y. Wang, H. Zhang, D. Liu, Adaptive dynamic programming: an introduction. IEEE Comput. Intell. Mag. **4**(2), 39–47 (2009). https://doi.org/10.1109/MCI.2009.932261
13. A. Singh, N. Kumar, B.P. Joshi, K.S. Vaisla, AGC using adaptive optimal control approach in restructured power system. Special issue on intelligent tools and techniques for signals, machines and automation. J. Intel. Fuzzy Syst. 1–10 (2018). https://doi.org/10.3233/JIFS-169779

14. N. Kumar, B. Tyagi, and V. Kumar, Deregulated AGC scheme using dynamic programming controller, in *IEEE Conference on Power System (NPSC)*, Guwahati, Assam, India, pp. 1–6, 18–20 Dec. 2014. https://doi.org/10.1109/NPSC.2014.7103791
15. N. Kumar, A. Varshney, K. Sharma, B. Prasad, "DHP based controller design for interconnected AGC scheme, in *2nd IEEE International Conference on Power, Energy, Environment and Intelligent Control (PEEIC-2019)*, Greater Noida, India, 18–19 October 2019. DOI: https://doi.org/10.1109/PEEIC47157.2019.8976721
16. N. Kumar, A. Varshney, K. Sharma, B. Prasad, Dual heuristic programming based controller for deregulated AGC scheme, in *2nd IEEE International Conference on Power, Energy, Environment and Intelligent Control (PEEIC-2019)*, Greater Noida, India, 18–19 October 2019. https://doi.org/10.1109/PEEIC47157.2019.8976728
17. D. Vrabie, F. Lewis, Neural network approach to continuous-time direct adaptive optimal control for partially unknown nonlinear systems. Neural Netw. **22**(3), 237–246 (2009). https://doi.org/10.1016/j.neunet.2009.03.008
18. L.B. Prasad, B. Tyagi, H.O. Gupta, Adaptive critic design using policy iteration technique for LTI systems: a comprehensive performance analysis. J. Control Autom. Electr. Syst. **26**, 134–148 (2015). https://doi.org/10.1007/s40313-015-0167-5
19. A.K. Yadav et al., *Soft Computing in Condition Monitoring and Diagnostics of Electrical and Mechanical Systems* (Springer Nature, Part of the Advances in Intelligent Systems and Computing, 2020), vol. 1096, p. 496. ISBN 978-981-15-1532-3. https://doi.org/10.1007/978-981-15-1532-3
20. A. Iqbal et al., *Renewable Power for Sustainable Growth* (Springer Nature, Part of the Lecture Notes in Electrical Engineering), vol. 723, 805 p. ISBN: 978-981-33-4080-0. https://doi.org/10.1007/978-981-33-4080-0
21. N. Fatema et al., *Intelligent Data-Analytics for Condition Monitoring: Smart Grid Applications* (Academic Press, 2021). ISBN: 978-0-323-85510-5. https://doi.org/10.1016/C2020-0-02173-0
22. S. Srivastava et al., *Applications of Artificial Intelligence Techniques in Engineering—volume 2* (Springer Nature, Part of the Advances in Intelligent Systems and Computing, 2018), vol. 697, 647 p. ISBN 978-981-13-1822-1. https://doi.org/10.1007/978-981-13-1822-1
23. T. Sasaki, Enomoto, Dynamic analysis of generation control performance standard. IEEE Trans. Power Syst. **17**(3), 806–811 (2002). https://doi.org/10.1109/TPWRS.2002.800946

Intelligent Method for Installation and Investigation of VSC-HVDC Converter Using Metaheuristic Algorithm: A Case Study of Unified Optimal Power Flow Problem

Sunilkumar Agrawal and Prasanta Kundu

Abstract The modern power system is facing significant structural change in its transmission side. Renewable energy sources are available at remote locations and diverse, resulting in the increment of a sizeable asynchronous area in the network. The bulk power transfer from the remote location to consumption end with controllable power-sharing between multiple asynchronous grids is only possible through high-voltage direct current (HVDC) networks. The voltage source converter technology is developing and increasing its share in the HVDC transmission network. This significant change must require detailed planning and expansion strategy before being installed. This study analyzes the new methodology for installing a VSC-HVDC converter based on the power system performance indices. The unified optimal power flow model is developed for a hybrid network and solved using the PSO algorithm to satisfy the desired objective functions. The analysis has been performed on modifying the 5-bus and IEEE-30 bus test system. Each AC line of the network is replaced by an HVDC line independently for point-to-point (P2P) transmission configuration. The obtained results have proven the effectiveness of the suggested methodology by improving the value of performance indices and identified the impact of a particular location of the converter on power system performance.

Keywords VSC-HVDC · HVDC optimal power flow · Power system optimization · PSO algorithm · Performance indices

1 Introduction

Optimal power flow problems are related to an economic analysis of power systems while satisfying operational constraints. It is a nonlinear, non-convex, and large-scale static optimization problem with continuous and discrete control variables.

S. Agrawal (✉) · P. Kundu
Department of Electrical Engineering, SVNIT-SURAT, Surat, Gujarat, India

P. Kundu
e-mail: pk@eed.svnit.ac.in

H. Malik et al. (eds.), *Intelligent Data Analytics for Power and Energy Systems*,
Lecture Notes in Electrical Engineering 802,
https://doi.org/10.1007/978-981-16-6081-8_10

OPF problems are initially solved with conventional methods such as gradient base, linear programming, and quadratic programming. The newton-based algorithm was developed for solving OPF problems with equality constraints only, and inequality constraints are formulated as penalty terms in quadratic form. The most effective numerical interior-point method that solves OPF problem using Lagrange multiplier. In this method, the Lagrangian augmented function is formed and solved using an iterative approach. In the above all conventional method, the selection of initial point is crucial; wrong selection of initial point may lead to converging the algorithm for local optimal point. Various optimization techniques have been developed to overcome the mathematical limitation of traditional methods. Many works of literature have proven the effectiveness of optimization techniques in the power system for OPF problems.

HVDC transmission technologies integrate the AC grid with the DC grid. HVDC system has reported various technical and economic benefits than high-voltage alternating current (HVAC) system [1, 2]. HVDC converter technology is categorized into two main groups: (1) line-commutated converter-based HVDC (LCC-HVDC); (2) voltage source converter-based HVDC (VSC-HVDC). The use of VSC technology for transmission networks is started in 1997. VSC technology is equipped with insulated gate bipolar transistor (IGBT) that can be turned on/off multiple times for each frequency cycle. Because of advancements in switching methods, VSC technology finds the preferable solution for multi-terminal HVDC networks. VSC structure's advantages and the number of projects installed worldwide are reported in [3, 4]. However, the VSC system has more power losses during switching and conduction operations than the CSC system due to high switching operations. VSC technology is commercially provided by ABB (HVDC Light® and Siemens HVDC Plus®).

With the advancement in power electronics devices, VSC topology becomes more advanced, and the rating of VSC-HVDC has also increased. Many projects are installing with VSC-HVDC converter, especially for offshore wind farm integration. The development and increasing share of VSC-HVDC inspire the researcher to develop various power flow algorithms and optimal power flow models for the new hybrid transmission grid. HVDC power flow can be solved either by sequential method or simultaneous method. The simultaneous method solves the power flow problem in a single run, while the sequential methods require to call multiple independent loops associated with separate AC and DC networks independently.

1.1 Related Work

The classical formulation and solution methods of the OPF require that the target function must be convex. A few literary works have been reviewed to understand the mathematical model of the OPF problem and identify the correlation between conventional and heuristic methods with their advantages and disadvantages. The Newton–Raphson-based OPF problem formulation is given in [5] and widely adopted by researchers. An improved interior-point (IP) method is suggested in [6]. In this method, tolerance limits and initial solutions are modified for better optimize results. Both classical methods need great care of a selection of initial points. An improper initial point may not converge the program, or else it may converge to local minima. The application of metaheuristic algorithm for solving OPF problems for AC transmission networks has reported better optimal results for multiple objective functions than numerical methods. They solve the OPF problem for single as well as multi-objective problems efficiently. The evolutionary programming-based algorithm is suggested in [7] for solving OPF problem. Different case studies have been performed to analyze the advantage of evolutionary programming over conventional methods. The results are promising to compare to the conventional method, but it also possesses the disadvantage of proper selection of initial conditions for a variable. The genetic algorithm is applied in [8], which has shown significant advantages in solving an optimization problem. The algorithm creates mutation and cross-over between them to optimize the objective function. In this algorithm, OPF results are much faster than the evolutionary programming method. The PSO algorithm is introduced in [9] which was applied by [10] for solving OPF problem in AC power system. In [10], various techno-economic objective functions are optimized and the advantages of PSO algorithm are described by comparing results with different algorithms. Successively other heuristic-based methods like teacher learner-based optimization (TLBO) [11], Tabu search algorithm [12], and gray wolf optimizer [13] for solving AC OPF problems. The in-depth analysis of different metaheuristic methods in various power system problems is discussed in [14–18].

The sequential power flow models for the VSC-HVDC network are suggested in [19, 20], and few authors also have developed simultaneous power flow models [21, 22]. They both have their respective merits and demerits. The sequential model developed in [19] is applied to solve OPF problems using MATADC software [23]. The first VSC-MTDC base OPF model is suggested in [24] with an objective function as the total active dispatch cost. AC and DC grid constraints are included for analysis, but converter voltages, converter angles, modulation index, and MVA rating of converters are excluded from the list of constraints. The AC-DC OPF is formulated as a mixed-integer linear optimization problem, and the formulation is coded in the GAMS Platform. An extended OPF model incorporating a detailed model of VSC- MTDC is proposed and defined as M-OPF in [25]. A cost–benefit analysis that calculates the benefits to cost ratio is applied to determine the preferred VSC-MTDC alternative for installing into an existing AC system. The OPF model proposed in [26] has been implemented with a Newton-based algorithm. The augmented Lagrangian function includes a nodal power mismatch equation, power exchanged between converters, the voltage drop in DC link, transformation equation for AC-DC side voltage generated by PWM control, and MVA rating of converters. VSC is modeled as synchronous reactance with source, the effect of transformer is not included. Two generator model is suggested in [27], where the AC side of VSC is modeled as the AC generator, and the DC side is modeled as a DC generator. The OPF problem is solved for two different objectives: minimizing total cost and minimizing total generation. The problem is solved with an inbuilt MATLAB fmincon solver using the IP method. The extended security-constrained analysis is done for different contingencies in [28]. The OPF models suggested in [24–28] have mainly applied a numerical approach to solve an optimization problem. They also not considered any performance-based objective function in their analysis. Limited efforts have been applied for solving the OPF problem for a hybrid AC-DC network using the metaheuristics algorithm. The model suggested in [29, 30] considered the CSC-HVDC network, and equations are solved by applying a sequential approach. The VSC-HVDC network OPF models with metaheuristic methods are solved in [31, 32]. They also applied a sequential approach for solving the power flow equations. The selection criteria are not clearly given for the replacement of a particular AC line with an HVDC line.

Many works of literature have explained the importance of power system performance indices to maintain the stability and security of the power system. The optimal value of power system performance indices guarantees the power system is operated within its desirable limits. Voltage stability indicator is introduced in [33], and authors have evaluated composite power system reliability indices incorporating the voltage stability margin criteria. This indicator helps to find out the voltage collapse condition based on the load margin analysis. The index is defined as L-index, used in many case studies for optimal placement of FACTs controller. Linear programming (LP)-based OPF algorithm for corrective FACTS control is explained in [34]. The corrective action design for relief in overloading and voltage violation situations by minimizing the average loadability index. The average loadability index is introduced and used as a deciding factor for the optimal placement of UPFC device.

The reach of the metaheuristic approach for solving different optimization problems like minimization of generator operational cost, active power loss minimization, voltage deviation minimization, stability index improvement, optimal reactive power dispatch, and the optimal location of FACTS device for conventional AC grid is enormous. Comparatively very few application is available for OPF problem for hybrid AC-DC network. HVDC grid is a next-generation technology that would go through a profound change in technical, economic, and operational aspects. Proper economic and operational planning is required before its application. This gap is minimized here with novel methodology, and contribution is briefed as follows:

- The application of the unified OPF model developed for a hybrid AC-DC network with a VSC-HVDC converter is extended for performance indices-based OPF problems.
- The unified OPF model is solved to minimize the voltage stability index and average loadability index using the PSO algorithm.
- A novel methodology is developed for optimal placement of VSC-HVDC converter by comparing their impact on power system performance indices by replacing each AC line independently with an HVDC line.

2 Unified OPF Modeling for VSC-HVDC

2.1 VSC-HVDC Power Flow Equations

VSC-HVDC network unified power flow equations are developed by modifying the static load flow equations of AC power system. These equations are modified for AC buses connected through a point of common coupling (PCC), as shown in the schematic diagram of the converter station in Fig. 1. The unified power flow equation mainly constitutes for equality constraints of OPF problem. The active and reactive power injection equations of any lth AC bus are formulated as per Eqs. (1)–(2)

$$P_{G_l} - P_{D_l} - V_l \sum_{m=1}^{N_{ac}} V_m \cdot (G_{lm}\cos(\delta_l - \delta_m) + B_{lm}\sin(\delta_l - \delta_m)) = 0 \tag{1}$$

$$Q_{G_l} - Q_{D_l} - V_l \sum_{m=1}^{N_{ac}} V_m \cdot (G_{lm}\sin(\delta_l - \delta_m) - B_{lm}\cos(\delta_l - \delta_m)) = 0 \tag{2}$$

Equations (1) and (2) are modified for hybrid AC–DC networks by inserting the power transfers to the PCC terminal from the AC grid. The modified equation can be formulated as follows:

$$P_{G_l} - P_{D_l} - V_l \sum_{m=1}^{N_{ac}} V_m \cdot (G_{lm}\cos(\delta_l - \delta_m) + B_{lm}\sin(\delta_l - \delta_m)) - P_{sl} = 0 \tag{3}$$

$$Q_{G_l} - Q_{D_l} - V_l \sum_{m=1}^{N_{ac}} V_m \cdot (G_{lm}\sin(\delta_l - \delta_m) - B_{lm}\cos(\delta_l - \delta_m)) - Q_{sl} = 0 \tag{4}$$

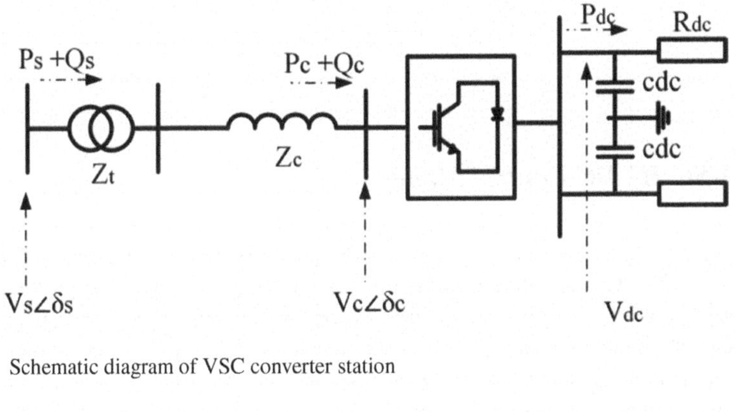

Fig. 1 Schematic diagram of VSC converter station

In Eqs. (3)–(4), P_s and Q_s represent the active and reactive power injecting to the PCC terminal from AC grid. P_s and Q_s values are considered zero for AC grid at which the converter is not connected. For the remaining buses, they are calculated based on Eqs. (5) and (6), respectively

$$P_s = V_s^2 \cdot G_{tc} - V_s \cdot V_c \cdot [G_{tc} \cos(\delta_s - \delta_c) + B_{tc} \sin(\delta_s - \delta_c)] \tag{5}$$

$$Q_s = -V_s^2 \cdot B_{tc} - V_s \cdot V_c \cdot [G_{tc} \sin(\delta_s - \delta_c) - B_{tc} \cos(\delta_s - \delta_c)] \tag{6}$$

The exchange of active power on AC and DC sides of the converter plays an essential role in defining exact unified modeling. The power exchange contains three terms: (1) power received at the AC side of the converter terminal, (2) power injected to DC bus, (3) converter power losses. Many pieces of literature have neglected the losses at converter terminals during power flow studies. Here, these losses are also considered in the form of the average loss model [19]. The power balance at the converter terminal can be represented by Eq. (7).

$$P_c - P_{\text{dc}} - P_{\text{closs}} = 0 \tag{7}$$

$\mathbf{P}_c$, $\mathbf{P}_{\text{dc}}$, and $\mathbf{P}_{\text{closs}}$ are vectors of active power received at AC side of converter terminal; power injected at DC terminal; and converter power loss. The length of mentioned vectors depends on the number of converter stations connected during operation. The value of each member of these vectors is calculated as per the following Eqs. (8–10), respectively,

$$P_c = -V_c^2 \cdot G_{tc} + V_s \cdot V_c \cdot [G_{tc} \cos(\delta_s - \delta_c) - B_{tc} \sin(\delta_s - \delta_c)] \tag{8}$$

$$P_{\text{dc}}^k = V_{\text{dc}}^k I_{\text{dc}}^k \tag{9}$$

$$P_{\text{closs}} = a_i + b_i |I_c| + c_i |I_c|^2 \tag{10}$$

2.2 VSC-HVDC Control Modes

Each converter can control four variables independently: (1) Active power, (2) Reactive power, (3) DC side voltage, and (4) AC side terminal Voltage. Depending upon the type of converter, various groups are formed for each converter. These groups are classified into four categories: (1) V_{dc}–Q control mode, (2) V_{dc}–V_{c} control mode, (3) P_s–Q_s control mode, (4) P_s–V_c control mode. In the multi-terminal configuration, one converter must be operated in DC voltage control mode, define as a slack DC bus that maintains the DC power balance in a grid.

Equations (11) to (14) are formulated for each converter based on respective control modes and included as equality constraints.

$$P_s^{sp} - P_s = 0 \tag{11}$$

$$Q_s^{sp} - Q_s = 0 \tag{12}$$

$$V_c^{sp} - V_c = 0 \tag{13}$$

$$(V_{dc}^{sp}) - V_{dc} = 0 \tag{14}$$

3 Problem Formulation

The global OPF problem is designed to minimize a particular objective function value by satisfying all operational constraints that can be expressed as follows:

$$\begin{aligned} &\text{minimize } f(x, u) \\ &\text{such that } g'(x, u) = 0 \\ &h'(x, u) \leq 0 \\ &f(x, u) = \{f_1(x, u), f_2(x, u), \ldots., f_n(x, u)\} \end{aligned} \tag{15}$$

3.1 Objective Function

$f(x, u)$ in Eq. (15) can be a single objective or multi-objective function. The two different case studies are carried out independently for the respective objective function.

3.1.1 Average Line Loadability Index

The average line loadability of any line indicates the security margin of that transmission line against overloading. The definition of average loadability is given in [30], which is used for the optimal installation of UPFC devices. The minimization of the average loadability index indicated the reduction in the overall congestion transmission network. By diverting the power through a lightly loaded line, overload on other transmission lines can be reduced, reducing the average loadability of all transmission lines. HVDC lines can carry higher power than AC lines; they can also reschedule power-sharing between AC and DC lines. There have been proven that HVDC lines are more economical for longer transmission.

The average laudability index of any transmission line can be calculated based on Eq. (16)

$$S_{\text{avg}} = \frac{1}{nl} \sum_{j=1}^{nl} \frac{S_j}{S_j^{\max}} \tag{16}$$

S_{avg}, S_j, and $S_j^{\max}$ are the average loadability index of all lines, actual, and maximum apparent power flow in the *j*th line. The total number of transmission lines is represented by *nl*. Apparent power on transmission include both sending end to receiving end (S_{ft}) and receiving end to sending end (S_{tf}) apparent powers, so apparent power flows through line calculated as per Eq. (17),

$$S_j = \frac{1}{2}\left(S_{ft} + S_{tf}\right) \tag{17}$$

The congestion or overloading of a transmission line is reduced by reducing the average loadability index value. The objective function considered for analysis is formulated as below,

$$f_1 = \min(S_{\text{avg}}) \tag{18}$$

3.1.2 Voltage Stability Index

The voltage stability index mainly decides the proximity of load bus close to the stability margin. The index indicates the amount of loading is acceptable before collapsing load buses. Power system operators use the index value as a proximity indication, either online or offline, to check the real-time operation of power systems. They also design and plan the power system operation by examining the index value for given generation and load data. The voltage stability indicator for load bus can be formulated as per Eq. (19)

$$L_j = \left|1 + \frac{V_{oi}}{V_i}\right| \tag{19}$$

The detailed derivation for the above index is explained in [33]. It concludes that the index value is ideally near to zero and less than unity for an overall system stability condition.

$$f_2 = \min(\max(L_i)), \quad i = 1, 2, 3..., N_L \tag{20}$$

3.2 Constraints

Equality constraints mentioned in Eq. (15) are defined from the power flow equations. g' (x, u), equality constraints set that constitutes modified power flow equations derived for each AC bus in Eqs. (3)–(4). It also constitutes the power balance equations at each converter terminal, converter control mode equations, and power balance equations of DC grid. The physical significance of inequality constraints is the limitation imposed on the operational state and controlling power system devices for stability and security assurance. h' (x, u) is a set of inequality terms that can be represented in general form as per Eqs. (21) and (22).

$$x^{\text{low}} \leq x \leq x^{\text{up}} \tag{21}$$

$$u^{\text{low}} \leq u \leq u^{\text{up}} \tag{22}$$

x and u are state and control variables of hybrid AC-DC systems, respectively. They can be represented by Eqs. (23) and (24).

$$\begin{aligned} x &= \left[x^{\text{AC}}, x^{\text{DC}}\right] \\ x^{\text{AC}} &= \left[P_{G1}, v_{L1}, v_{L2}, \ldots, v_{L,Nl}, Q_{G1}, Q_{G2}, \ldots G_{GNg}\right] \\ x^{\text{DC}} &= \left[v_{c1}, v_{c2}, \ldots v_{c,Nc}, v_{\text{dc1}}, v_{\text{dc2}}, v_{\text{dc3}}, \ldots, v_{\text{dc}}, N_{\text{dc}}\right] \end{aligned} \tag{23}$$

$$\begin{aligned} u &= \left[u^{\text{AC}}, u^{\text{DC}}\right] \\ u^{\text{AC}} &= \left[P_{G2}, P_{G3}, \ldots, P_{G,Ng}, v_{G1}, v_{G2}, \ldots, v_{G,Ng}, Q_{c'1}, Q_{c'2}, \ldots Q_{c',Nqc}\right] \\ u^{\text{DC}} &= \left[p_{c1}, p_{c2}, \ldots p_{c,Nc}, q_{c1}, q_{c2}, \ldots q_{c,Nc}, v_{\text{dc}}^{\text{slack}}, v_{\text{dc2}}, v_{\text{dc3}}, \ldots, v_{\text{dc},Ndc}\right] \end{aligned} \tag{24}$$

Transmission line limits and economic constraints are for AC side, DC transmission line limits and converter ratings for DC side are additional inequality constraints considered for OPF problem.

$$S_{Lij} \leq S_{Lij}^{\text{up}} \tag{25}$$

$$\sum_{i=1}^{N_g} c(P_{gi}) \leq \left(\sum_{i=1}^{N_g} c(P_{gi})\right)_{\text{base}} \tag{26}$$

$$p_{dcij} \leq p_{dcij}^{\text{up}} \tag{27}$$

$$|S_c| \leq S_c^{\text{rated}} \tag{28}$$

The violation of inequality constraints is calculated as penalty cost and added with the primary objective function with the respective penalty factor. The augmented function is developed for each state variable limit violation. The control variables are self-constrained and found within permissible limits by the algorithm.

$$\overline{f} = f + \lambda_l \sum_{l=1}^{N_{\text{eq}}} (g_l'(\overline{x}, \overline{u}))^2 + \lambda_l \sum_{l=1}^{N_{\text{ieq}}} (h_l'(\overline{x}, \overline{u}))^2 \tag{29}$$

3.3 PSO Algorithm

Particle swarm optimization (PSO) algorithm is a nature-inspired metaheuristic technique introduced in [9]. This algorithm is developed from a flock of birds and a school of fish movement. Initially, random particles are generated for the number of control variables, and each particle finds its own personal best value of an objective function. Each particle's optimal value is defined as its personal best, and the lowest value among all particles is defined as a global best value. Each particle moves in search space during successive iteration to match its personal best value with the global best position. The position and velocity of each particle are updated based on its personal and global best value obtained in the previous iteration. Particles improve their position and velocity based on Eqs. (30) and (31), respectively

$$X_j^{k+1} = X_j^k + V_j^{k+1} \tag{30}$$

$$\begin{aligned} V_j^{k+1} &= \omega V_j^k + c_1 \cdot \text{rand}_1() \times (P_{B,j}^k - X_j^k) \\ &+ c_2 \cdot \text{rand}_2() \times (G_B^k - X_j^k) \end{aligned} \tag{31}$$

j and k are the numbers of population and iterations, respectively, for defined OPF problem, c_1 and c_2 are acceleration coefficients. ω is the inertia weight that controls the movement of particles. It does not allow particles to move away from search space. The inertia weight parameter is updated by Eq. (32) after each iteration.

$$\omega(k) = \omega_{\max} - \left(\frac{\omega_{\max} - \omega_{\min}}{iter_{\max}} \right) \times k \tag{32}$$

The flowchart for the unified OPF problem with PSO algorithm is shown in Fig. 2.

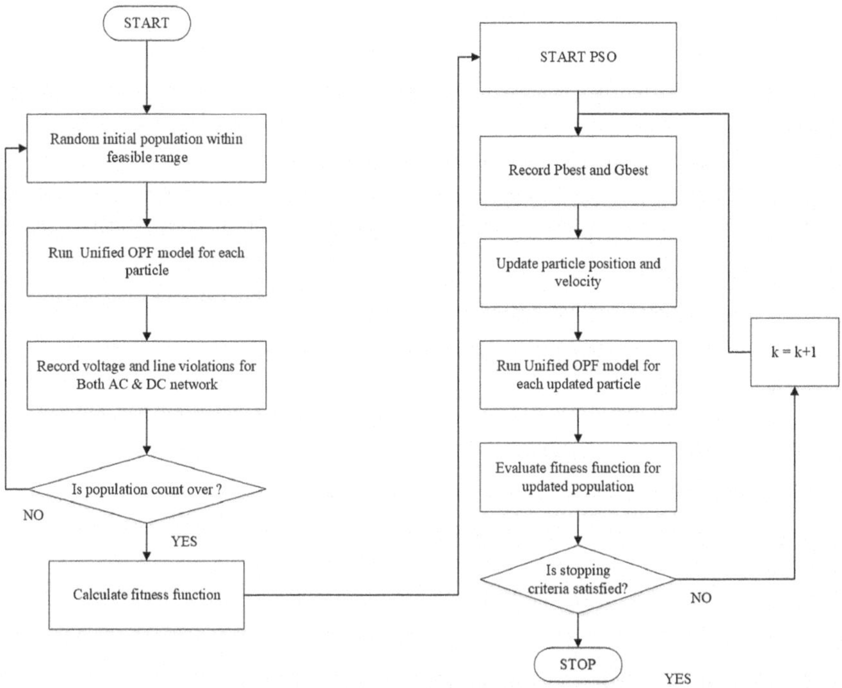

Fig. 2 Flowchart of unified OPF problem with PSO algorithm

4 Case Studies and Result Demonstration

In this analysis, a novel methodology has been developed to identify the optimal location of the converter station. This methodology has considered different power system performance indices as a deciding parameter. As discussed in the earlier section, the power system operator has a vital role in maintaining secure operations with minimum optimal cost. These performance indices decide the level of stability and secure state of a power system. As explained earlier, there is no standard test system available for testing and comparison in a hybrid HVDC system. Different configurations have been selected and tested for power flow and optimal power flow analysis in various literature. In this analysis, each line of AC system is replaced by a two-terminal HVDC line, and all the performance indices are optimized and analyzed. The primary objective functions are selected as follows: minimization of average loading on the transmission line and minimization of voltage stability index (L_{max}).

Case I: Minimization of average loading on transmission line.

Case II: Minimization of the voltage stability index.

4.1 Modified 5-bus Test System

The 5-bus test system network given in [31] is used and modified for this analysis. The data used for the analysis are given in Appendix. Upper and lower limits of control and state variables for hybrid AC-DC system considered as per Table 1.

The 5-bus system has 7 transmission lines, and each transmission line is replaced by HVDC line, and independent analysis is performed to optimize the objective function. This independent analysis is categorized as different test system configurations. In this chapter, the obtained result of the top four most optimal test configurations is only represented for comparison purposes. The detailed classification of different test configuration for both case studies is given as follows:

Test System I: Base system where no HVDC included.

Test System II: Line 2–5 is replaced with HVDC.

Test System III: Line 2–4 is replaced with HVDC.

Test System IV: Line 2–3 is replaced with HVDC.

Test System V: Line 1–3 is replaced with HVDC.

4.1.1 Test System I: Base System Where No HVDC Included

A basic OPF analysis has been done, and the optimal cost is obtained for a base case without an HVDC system. The obtained cost is considered maximum allowed economic constraints for other test systems while optimizing other performance

Table 1 Upper and lower limits of hybrid system variables

AC system variables	Low	High
North—P_{G1}	0.1	2.0
South—P_{G2}	0.1	2.0
North—Q_{G1}	−3.0	3.0
North—Q_{G2}	−3.0	3.0
North—V_{G1}	0.90	1.10
South—V_{G2}	0.90	1.10
Lake—V_{D3}	0.90	1.10
Main—V_{D4}	0.90	1.10
Elm—V_{D5}	0.90	1.10
DC system variables	Low	High
Converter 1—P_c	−50	50
Converter 2—Q_c	−25	25
Converter 1—V_{DC}	0.9	1.1
Converter 2—V_c	0.9	1.1

Table 2 Optimal control variables and state variable (test system I)

Variables and objective functions	Case 0 Cost optimization	Case I Minimization of average line loadability index	Case II Minimization of voltage stability index
P_{G1}	79.63068	64.8976	79.93095
P_{G2}	88.45025	102.9681	88.1378
V_{G1}	1.1	1.099998	1.1
V_{G2}	1.099999	1.093984	1.094848
V_{D3}	1.082685	1.078247	1.078792
V_{D4}	1.083098	1.07828	1.078908
V_{D5}	1.077796	1.071981	1.072811
Cost ($/h)	748.0626	750.0711	748.1332
Average line loadability index	0.3039	0.283174	0.3071
Voltage stability index (L_{max})	0.0653	0.0653	0.064663
Voltage deviation (p.u.)	0.2305	0.2285	0.2436

indices. The generation cost obtained in case 0 is 748.062 $/h. All cases are individually analyzed for the same base system. The optimal value of the average line loadability index is 0.283174, and for voltage, stability index is 0.064663. The optimal setting of control and state variables obtained during the test system are represented in Table 2. An additional case 0 is related to the minimization of total generator cost for the base system.

4.1.2 Test Systems II, III, IV, and V: Modified 5-Bus with HVDC

An HVDC line is installed in this analysis by replacing each AC line independently, and different test configurations are analyzed. The converters are connected in P2P configuration. In test configuration II, the converter connected to the south bus, i.e., second terminal, is operated in PQ control mode, and the converter connected at the fifth terminal is operated as a slack DC control mode. The results of two case studies for the respective objective function are given in Table 3. Table 3 represents the optimal settings of control variables and state variables of hybrid AC-DC network and the optimal values of objective functions for test configuration II.

The OPF analysis is carried out for other test configurations by replacing the individual AC line with HVDC line. The obtained results of OPF problem with other test configurations are represented in Tables 4, 5, and 6, respectively. The values of both indices are desired to be as small as possible. The average loadability index values obtained with HVDC test configurations are found much lesser than the base case. The converters connected to generator buses 1 or 2 are operated in PQ control

Table 3 Optimal settings of control variables and state variable for test configuration II (2–5)

Variables and objective functions	Case I Minimization of average line loadability index	Case II Minimization of voltage stability index
P_{G1}	80.252	10.00356
P_{G2}	79.74757	159.4644
V_{G1}	1.074493	1.046175
V_{G2}	1.065767	0.9937
V_{D3}	1.044314	0.996667
V_{D4}	1.043063	0.996309
V_{D5}	1.02	1.02
P_{c1}	50	49.99947
Q_{c1}	0.9053	9.168667
V_{c1}	1.020121	1.039751
V_{c2}	1.066322	0.985251
V_{dc1}	1	1
V_{dc2}	1.000555	1.000555
Cost ($/h)	745.6284	798.3069
Average line loadability index	0.2809	0.3789
Voltage stability index (Lmax)	0.1355	0.0629
Voltage deviation (p.u.)	0.1074	0.0270

modes. It is found that they act as a rectifier for particular test configurations. The converters are connected to either third, fourth, or fifth buses are operated in slack DC control mode. All test configurations result from tables clearly show that hybrid AC-DC system variables are operated within their permissible limits and satisfy the power system's operational constraints.

4.1.3 Comparative Analysis

The comparative analysis with all the test configurations is represented in Table 7 for two individual objective functions. While comparing base system (without HVDC system) optimal results with different HVDC test configurations, it can be seen that both performance indices are improved with the HVDC system. The minimum value of these performance indexes indicates that the system is more stable and secure against abnormal conditions.

The average loadability index for test configuration V (1–3) is 0.2652 implies that the AC lines are averagely loaded 26.52% from their maximum rating. The values obtained for both objective functions with test configuration V are the least

Table 4 Optimal settings of control variables and state variable for test configuration III (2–4)

Variables and objective functions	Case I Minimization of average line loadability index	Case II Minimization of voltage stability index
P_{G1}	78.89967	9.976503
P_{G2}	88.47565	159.9231
V_{G1}	1.04716	1.020324
V_{G2}	1.037656	1.065767
V_{D3}	1.020123	1.010301
V_{D4}	1.02	1.02
V_{D5}	1.013362	0.962064
P_{c1}	50	50
Q_{c1}	0.001504	7.261027
V_{c1}	1.018642	1.072591
V_{c2}	1.038291	0.957461
V_{dc1}	1	1
V_{dc2}	1.000555	1.000555
Cost ($/h)	745.2885	800.3584
Average line loadability index	0.2667	0.4355
Voltage stability index ($L_{\max}$)	0.1132	0.0324
Voltage deviation (p.u.)	0.0535	0.0682

among all test configurations. The average loadability index and voltage stability index values in test configuration V are 8.35 and 44.70% more optimal than the average of other topmost optimal HVDC configurations. The observation suggests that the test configuration V is the most optimal configuration for a 5-bus test system for selected objective functions.

4.2 *Modified IEEE 30-Bus Test System*

The algorithm's efficacy is tested with the IEEE-30 bus test system that many researchers use for various OPF problem studies. The bus and line data for IEEE-30 bus taken from MATPOWER® [35] and the generator cost coefficients and transmission line limits are considered as given in [36]. The maxima and minima limits of all control variables for the hybrid AC-DC system are shown in Table 8. All values are given in p.u.

Table 5 Optimal settings of control variables and state variable for test configuration IV (2–3)

Variables and objective functions	Case I Minimization of average line loadability index	Case II Minimization of voltage stability index
P_{G1}	80.62245	30.80007
P_{G2}	86.34924	138.8781
V_{G1}	1.042419	1.033616
V_{G2}	1.036355	0.981192
V_{D3}	1.02	1.02
V_{D4}	1.01963	1.008268
V_{D5}	1.012259	0.970068
P_{c1}	49.99998	49.99935
Q_{c1}	0.10039	−13.4597
V_{c1}	1.022196	1.082207
V_{c2}	1.042947	1.047278
V_{dc1}	1	1
V_{dc2}	1.000555	1.000555
Cost ($/h)	743.5284	777.8488
Average line loadability index	0.2652	0.4441
Voltage stability index (L_{max})	0.0985	0.0240
Voltage deviation (p.u.)	0.0519	0.0582

The following different test configurations are analyzed and represented with results for the modified IEEE-30 bus.

Test System I: Base System where no HVDC included.

Test System II: Line 1–3 is replaced with HVDC.

Test System III: Line 2–4 is replaced with HVDC.

Test System IV: Line 2–5 is replaced with HVDC.

Test system V: Line 2–6 is replaced with HVDC.

4.2.1 Test System I: Base System Where No HVDC Included

In this test configuration, an OPF analysis is performed for a base AC system. There is no HVDC line connected. The additional case is also analyzed to obtain the optimal generation cost of power plants.

The optimal value for generation cost is obtained as 799.0718 $/h. The values of other objective functions with their optimal control settings are represented in Table

Table 6 Optimal settings of control variables and state variable for test configuration V (1–3)

Variables and objective functions	Case I Minimization of average line loadability index	Case II Minimization of voltage stability index
P_{G1}	78.69165	10.00056
P_{G2}	88.77165	160.5145
V_{G1}	1.044063	1.017417
V_{G2}	1.03415	0.960712
V_{D3}	1.02	1.02
V_{D4}	1.019214	1.00395
V_{D5}	1.010686	0.954189
P_{c1}	49.99999	44.59545
Q_{c1}	0.007884	−1.28862
V_{c1}	1.021008	1.084111
V_{c2}	1.034789	0.962709
V_{dc1}	1	1
V_{dc2}	1.000555	1.000495
Cost ($/h)	745.6664	803.2107
Average line loadability index	0.3205	0.4805
Voltage stability index ($L_{\max}$)	0.1170	0.0349
Voltage deviation (p.u.)	0.0499	0.0698

Table 7 Comparative analysis with all the test systems

Performance indices	Test system I Base case (without HVDC)	Test system II (2–5)	Test system III (2–4)	Test system IV (2–3)	Test system V (1–3)
Average line loadability index	0.283174	0.2809	0.2667	0.3205	0.2652
Voltage stability index	0.064663	0.0629	0.0324	0.0349	0.0240

9. The cost is considered as an economic constraint for the remaining two objective function case studies.

Table 8 Maxima and minima of variables for hybrid AC-DC system

	Maximum	Minimum		Maximum	Minimum
AC system variables					
P_{g1}	2.00	0.50	$T_{12\,(6\text{–}10)}$	1.1	0.9
P_{g2}	0.80	0.20	$T_{15\,(4\text{–}12)}$	1.1	0.9
P_{g5}	0.50	0.15	$T_{36\,(28\text{–}27)}$	1.1	0.9
P_{g8}	0.35	0.10	$Q_{C'10}$	0.05	0
P_{g11}	0.30	0.10	$Q_{C'12}$	0.05	0
P_{g13}	0.40	0.12	$Q_{C'15}$	0.05	0
V_{g1}	1.1	0.9	$Q_{C'17}$	0.05	0
V_{g2}	1.1	0.9	$Q_{C'20}$	0.05	0
V_{g5}	1.1	0.9	$Q_{C'21}$	0.05	0
V_{g8}	1.1	0.9	$Q_{C'23}$	0.05	0
V_{g11}	1.1	0.9	$Q_{C'24}$	0.05	0
V_{g13}	1.1	0.9	$Q_{C'29}$	0.05	0
$T_{11\,(6\text{–}9)}$	1.1	0.9			
HVDC system variables					
V_{dc1}	1.1	0.9	P_{S2}	1.00	−1.00
Q_{S1}	0.25	−0.25	Q_{S2}	0.25	−0.25

4.2.2 Test Systems II, III, IV, and V: Modified IEEE 30-Bus with HVDC

Analysis for different HVDC test configurations is carried out as done for the 5-bus test system. The top four most optimal test configuration results are represented here for both case studies. The applicability of the proposed methodology and algorithm is also tested here for different control modes of the converter. In this analysis, DC slack bus controls reactive power injected to the ac side instead of voltage control. Converters connected to generator buses are operated in PQ control mode only. Test configuration II is designed by replacing the line between first bus and third bus with HVDC line. The results obtained for minimization of average line loadability index with top for most optimal HVDC test configurations are shown in Table 10. The values indicated in bold letters specify the particular objective function selected for analysis. The average line loadability index values are different for the different test configurations, and significant improvements are observed compared with the base system.

Similarly, results obtained for minimization of voltage stability index with top four most optimal test configurations of HVDC systems are represented in Table 11. The voltage stability indicator values are improved with the inclusion of HVDC lines. Tables 10 and 11 verify that the control variables and state variables are operated within their permissible limits. Hence, the hybrid AC-DC transmission network is satisfying the operational constraints of a power system. The values of active and

Table 9 Optimal control variables and state variable for base system (IEEE 30 bus)

Control variables and objective functions	Case 0 Cost optimization	Case I Minimization of average loadability index	Case II Minimization of voltage stability index
P_{g1}	177.057537	114.728579	177.537996
P_{g2}	48.69661	57.51307	48.69406
P_{g5}	21.30453	39.75301	21.09115
P_{g8}	21.08168	34.50077	19.89096
P_{g11}	11.88447	29.99876	12.63861
P_{g13}	12	12.00203	12.26988
V_{g1}	1.1	1.080041	1.099702
V_{g2}	1.087849	1.06621	1.083322
V_{g5}	1.061656	1.031083	1.060017
V_{g8}	1.069396	1.039361	1.069612
V_{g11}	1.1	1.050947	1.093783
V_{g13}	1.1	1.048822	1.099444
$T_{11\,(6–9)}$	1.044673	1.014958	0.960828
$T_{12\,(6–10)}$	0.9	0.930897	0.949453
$T_{15\,(4–12)}$	0.986288	0.981489	0.965915
$T_{36\,(28–27)}$	0.965675	0.989758	0.961366
$Q_{C'10}$	5	3.216249	4.970797
$Q_{C'12}$	4.999984	4.999534	4.987773
$Q_{C'15}$	5	1.712886	5
$Q_{C'17}$	5	4.979505	4.984046
$Q_{C'20}$	4.999992	4.248759	4.993012
$Q_{C'21}$	4.999999	4.998413	4.996354
$Q_{C'23}$	3.819419	1.441237	5
$Q_{C'24}$	4.999997	4.999733	4.986767
$Q_{C'29}$	2.740102	2.522037	4.992797
Cost ($/h)	799.0718	850.0289	799.3055
Average line loadability index	0.3398	0.2849	0.3496
Voltage stability index (L_{max})	0.1164	0.1304	0.1137
Voltage deviation (p.u.)	1.8575	0.7797	2.0449

Table 10 Optimal settings of control variable for different HVDC test configurations under case I

Control variables and objective functions	Test configuration II Line (1–3)	Test configuration III Line (2–4)	Test configuration IV Line (2–5)	Test configuration V Line (2–6)
P_{g1}	118.3232	117.1518	99.07416	115.9616
P_{g2}	50.93734	54.99417	79.76759	55.1456
P_{g5}	42.21613	40.26259	27.8153	42.22163
P_{g8}	34.37675	34.24263	33.82744	33.67589
P_{g11}	29.5311	29.84499	29.97065	28.18031
P_{g13}	12.19492	12.15164	16.10459	12.30929
V_{g1}	1.028723	1.049986	1.07112	1.056028
V_{g2}	1.02999	1.040232	1.055325	1.040682
V_{g5}	1.008142	0.996026	1.06932	1.022442
V_{g8}	1.017948	1.006772	1.046042	1.037549
V_{g11}	1.024643	1.007453	1.040373	1.032565
V_{g13}	1.008832	1.006494	1.045384	1.033385
$T_{11\,(6–9)}$	0.97897	0.93067	0.954883	0.957417
$T_{12\,(6–10)}$	1.01732	1.024155	1.007507	1.032668
$T_{15\,(4–12)}$	1.010194	1.015363	1.014039	1.014089
$T_{36\,(28–27)}$	0.977372	0.990109	0.99381	0.98373
$Q_{C'10}$	3.944859	4.157461	4.409696	4.304674
$Q_{C'12}$	0.510866	2.43046	0.943804	1.822555
$Q_{C'15}$	3.245583	2.007317	1.344191	2.433296
$Q_{C'17}$	2.108037	2.59488	4.969615	2.909228
$Q_{C'20}$	1.706782	0.476429	2.280415	1.040188
$Q_{C'21}$	5	2.933357	4.94119	1.277057
$Q_{C'23}$	2.247447	1.970666	2.595416	1.144202
$Q_{C'24}$	4.825009	4.87038	4.663003	4.9923
$Q_{C'29}$	1.832357	3.713344	2.583517	1.661478
V_{dc1}	0.991545	1.035471	1.026078	1.098832
P_{S2}	95.75711	23.67381	93.63801	−78.1269
Q_{S1}	−9.91106	−4.85998	11.03366	17.36227
Q_{S2}	−10.9966	9.245409	−5.88594	−11.7935
Cost ($/h)	849.5761	849.5337	848.7199	849.7257
Average line loadability index	0.2814	0.2845	0.2661	0.2648
Voltage stability index (L_{max})	0.1956	0.3100	0.2331	0.2373
Voltage deviation (p.u.)	0.2611	0.2394	0.7860	0.5444

Table 11 Optimal setting of control variable for case II with different HVDC test configurations

Control variables and objective functions	Test configuration II Line (1–3)	Test configuration III Line (2–4)	Test configuration IV Line (2–5)	Test configuration V Line (2–6)
P_{g1}	87.24294	55.5951	56.42975	55.96809
P_{g2}	79.35614	80	79.30267	78.37855
P_{g5}	62.99273	49.7993	49.69833	49.75771
P_{g8}	37.12007	34.99401	34.95246	34.98917
P_{g11}	34.96358	29.92119	29.94393	29.84499
P_{g13}	30	39.91242	39.8647	39.90803
V_{g1}	1.09975	1.099722	1.099636	1.099177
V_{g2}	1.098661	1.04712	1.053188	1.051029
V_{g5}	1.092774	0.97872	0.980204	0.986296
V_{g8}	1.068373	0.996028	1.003783	1.014811
V_{g11}	1.050204	1.067365	0.994024	1.054894
V_{g13}	0.988081	1.005626	1.048306	1.02155
$T_{11\,(6\text{–}9)}$	1.010299	0.955747	0.946077	0.911022
$T_{12\,(6\text{–}10)}$	0.914996	0.952867	1.059063	1.015183
$T_{15\,(4\text{–}12)}$	1.085437	1.058817	1.074225	1.1
$T_{36\,(28\text{–}27)}$	1.080428	0.91132	0.913996	0.924361
$Q_{C'10}$	0.939289	1.051939	4.62991	4.104513
$Q_{C'12}$	2.416124	1.187054	4.611933	2.713878
$Q_{C'15}$	3.680139	0.86444	2.81905	4.082075
$Q_{C'17}$	0.874492	2.915959	3.521378	2.374747
$Q_{C'20}$	0.663836	4.821057	4.557373	4.246134
$Q_{C'21}$	4.696756	2.783218	4.861828	0.433594
$Q_{C'23}$	4.948682	5	2.957939	3.429219
$Q_{C'24}$	4.11618	1.448308	2.081884	4.436435
$Q_{C'29}$	0.938045	4.823452	4.947271	4.964026
V_{dc1}	1.057072	0.97806	0.992329	1.022453
P_{S2}	98.81929	99.74586	99.96398	−99.9379
Q_{S1}	−10.0451	1.380043	22.62782	17.04122
Q_{S2}	−11.9516	9.244108	17.70635	1.575732
Cost ($/h)	882.5521	975.2297	973.0609	968.1167
Average line loadability index	0.3751	0.3891	0.3923	0.3693
Voltage stability index (L_{max})	0.0812	0.0983	0.0978	0.0662
Voltage deviation (p.u.)	1.4477	0.3584	0.8656	1.0630

Table 12 Comparative analysis with all the test systems

Performance indices	Test configuration I base case (without HVDC)	Test configuration II Line (1–3)	Test configuration III Line (2–4)	Test configuration IV Line (2–5)	Test configuration V Line (2–6)
Average line loadability index	0.2849	0.2814	0.2845	0.2661	0.2648
Voltage stability index	0.1137	0.0812	0.0983	0.0978	0.0662

reactive powers are given in terms of MW and MVAR, respectively. The bus voltages are indicated in their p.u. form. Detail comparison between HVDC test configuration and base system is given in Table 12.

4.2.3 Comparative Analysis

The comparison table clears that the location of the converter station affects the power system performance. The most optimized result for both indices is obtained for test system V (line 2–6 replaced with HVDC). The other HVDC test configurations have optimized both indices as compared with a base case but less optimal than Test system V. The average line loading index with the most optimal test system V is improved to 7.05% of base case and 4.52% of the average of other topmost optimal HVDC configuration. The overloading of transmission lines is also reduced in the same proportionate manner. The voltage stability index improved to 41.77% of the base case and 71.67% of the average of other topmost optimal HVDC configurations.

5 Conclusion

In this analysis, a novel methodology is developed to found the optimal location of the converter station. The unified OPF analysis is performed to reduce the average line loadability and voltage stability index by replacing each AC line with an HVDC line while satisfying operational and economic constraints. The unified OPF model equations are solved with the PSO algorithm that has converged and executed adequately for the suggested mathematical model. The suggested OPF model is converged for different control modes and for different test configurations. The impact of optimal location of converter station on proper AC bus could enhance the stability and security of transmission system. The results obtained for all optimization cases have shown that the proper replacement of HVDC lines can improve the economic and technical performance of the overall system.

Appendix

The details of 5-bus test system [37] are represented as below (Fig. 3 and Tables 13 and 14).

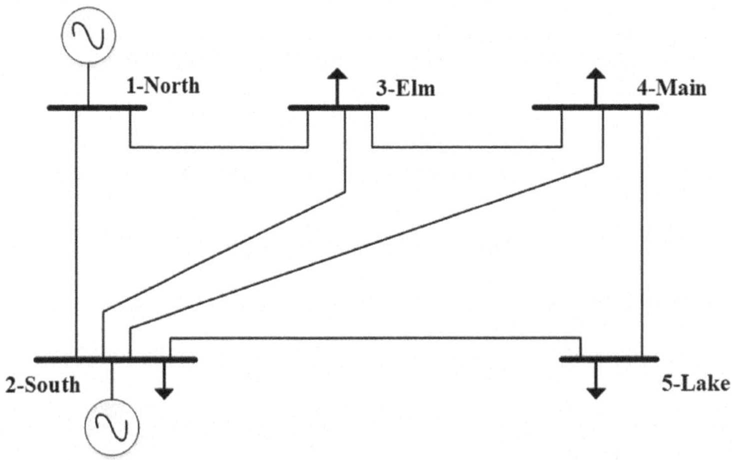

Fig. 3 Single line diagram of 5-bus test system

Table 13 5-bus system bus data

Number	Type	Real demand (MW)	Reactive demand (MVAR)	Q_{min} (MVAR)	Q_{max} (MVAR)
1—North	1	0	0	−300	300
2—South	2	20	10	−300	300
3—Elm	3	45	15	0	0
4—Main	3	40	5	0	0
5—Lake	3	60	10	0	0

Table 14 5-bus system line data

Line from—to bus	Z (p.u.)	B (p.u.)
1–2	0.02 + j * 0.06	0.06
1–3	0.08 + j * 0.24	0.05
2–3	0.06 + j * 0.25	0.04
2–4	0.06 + j * 0.18	0.04
2–5	0.04 + j * 0.12	0.03
3–4	0.01 + j * 0.03	0.02
4–5	0.08 + j * 0.24	0.05

References

1. D. Krug, S. Bernet, S. Dieckerhoff, Comparison of state-of-the-art voltage source converter topologies for medium voltage applications, in *38th IAS Annual Meeting on Conference Record of the Industry Applications Conference*, 2003, vol. 1, pp. 168–175. IEEE
2. K.R. Padiyar, *FACTS Controllers in Power Transmission and Distribution* (New Age International, 2007)
3. N. Flourentzou, V.G. Agelidis, G.D. Demetriades, VSC-based HVDC power transmission systems: an overview. IEEE Trans. Power Electron. **24**(3), 592–602 (2009)
4. X. Zheng, C. Hairong, Review and applications of VSC HVDC. High Voltage Eng. **33**(1), 1–10 (2007)
5. D.I. Sun, B. Ashley, B. Brewer, A. Hughes, W.F. Tinney, Optimal power flow by Newton approach. IEEE Trans. Power Apparatus Syst. **10**, 2864–2880 (1984)
6. X. Yan, V.H. Quintana, Improving an interior-point-based OPF by dynamic adjustments of step sizes and tolerances. IEEE Trans. Power Syst. **14**(2), 709–717 (1999)
7. J. Yuryevich, K.P. Wong, Evolutionary programming based optimal power flow algorithm. IEEE Trans. Power Syst. **14**(4), 1245–1250 (1999)
8. S. Paranjothi, K. Anburaja, Optimal power flow using refined genetic algorithm. Electr. Power Compon. Syst. **30**(10), 1055–1063 (2002)
9. J. Kennedy, The particle swarm: social adaptation of knowledge, in *Proceedings of 1997 IEEE International Conference on Evolutionary Computation (ICEC'97)*, 1997, pp. 303–308. IEEE
10. A. Iqbal et al., *Meta Heuristic and Evolutionary Computation: Algorithms and Applications* (Springer Nature, Part of the Studies in Computational Intelligence, 2020), vol. 916, p. 849. ISBN 978-981-15-7571-6
11. A. Iqbal et al., *Renewable Power for Sustainable Growth* (Springer Nature, Part of the Lecture Notes in Electrical Engineering), vol. 723, p. 805. ISBN: 978-981-33-4080-0
12. N. Fatema et al., *Intelligent Data-Analytics for Condition Monitoring: Smart Grid Applications* (Academic Press, Cambridge, 2021). ISBN: 978-0-323-85510-5
13. S. Srivastava et al., *Applications of Artificial Intelligence Techniques in Engineering—volume 1* (Springer Nature, Part of the Advances in Intelligent Systems and Computing, 2018), vol. 698, 643 p. ISBN 978-981-13-1819-1
14. S. Srivastava et al., *Applications of Artificial Intelligence Techniques in Engineering—Volume 2* (Springer Nature, Part of the Advances in Intelligent Systems and Computing, 2018), vol. 697, 647 p. ISBN 978-981-13-1822-1
15. M.A. Abido, Optimal power flow using particle swarm optimization. Int. J. Electr. Power Energy Syst. **24**(7), 563–571 (2002)
16. H. Bouchekara, M. Abido, M. Boucherma, Optimal power flow using teaching-learning-based optimization technique. Electr. Power Syst. Res. **114**, 49–59 (2014)
17. M. Abido, Optimal power flow using tabu search algorithm. Electr. Power Componen. Syst. **30**(5), 469–483 (2002)
18. A.A. El-Fergany, H.M. Hasanien, Single and multi-objective optimal power flow using grey wolf optimizer and differential evolution algorithms. Electr. Power Compon. Syst. **43**(13), 1548–1559 (2015)
19. J. Beerten, S. Cole, R. Belmans, A sequential AC/DC power flow algorithm for networks containing multi-terminal VSC HVDC systems, in *IEEE PES General Meeting* (IEEE, 2010), pp. 1–7
20. J. Beerten, S. Cole, R. Belmans, Generalized steady-state VSC MTDC model for sequential AC/DC power flow algorithms. IEEE Trans. Power Syst. **27**(2), 821–829 (2012)
21. S. Khan, S. Bhowmick, A novel power-flow model of multi-terminal VSC-HVDC systems. Electr. Power Syst. Res. **133**, 219–227 (2016)
22. X.-P. Zhang, Multiterminal voltage-sourced converter-based HVDC models for power flow analysis. IEEE Trans. Power Syst. **19**(4), 1877–1884 (2004)

23. J. Beerten, R. Belmans, Development of an open source power flow software for high voltage direct current grids and hybrid AC/DC systems: MATACDC. IET Gener. Transm. Distrib. **9**(10), 966–974 (2015)
24. M. Baradar, M.R. Hesamzadeh, M. Ghandhari, Modelling of multi-terminal HVDC systems in optimal power flow formulation, in *2012 IEEE Electrical Power and Energy Conference* (IEEE, 2012), pp. 170–175
25. W. Feng, L.B. Tjernberg, A. Mannikoff, A. Bergman, An extended OPF incorporating multi-terminal VSC-HVDC and its application on transmission loss evaluation, in *PowerTech (POWERTECH), 2013 IEEE Grenoble* (IEEE, 2013), pp. 1–6
26. A. Pizano-Martinez, C.R. Fuerte-Esquivel, H. Ambriz-Perez, E. Acha, Modeling of VSC-based HVDC systems for a Newton-Raphson OPF algorithm. IEEE Trans. Power Syst. **22**(4), 1794–1803 (2007)
27. R. Wiget, G. Andersson, Optimal power flow for combined AC and multi-terminal HVDC grids based on VSC converters, in *2012 IEEE Power and Energy Society General Meeting* (IEEE, 2012), pp. 1–8
28. V. Saplamidis, R. Wiget, G. Andersson, Security constrained optimal power flow for mixed AC and multi-terminal HVDC grids, in *PowerTech, 2015 IEEE Eindhoven* (IEEE, 2015), pp. 1–6
29. U. Kılıç, K. Ayan, Optimizing power flow of AC–DC power systems using artificial bee colony algorithm. Int. J. Electr. Power Energy Syst. **53**, 592–602 (2013)
30. U. Kılıç, K. Ayan, Optimal power flow solution of two-terminal HVDC systems using genetic algorithm. Electr. Eng. **96**(1), 65–77 (2014)
31. E.E. Elattar, A.M. Shaheen, A.M. Elsayed, R.A. El-Sehiemy, Optimal power flow with emerged technologies of voltage source converter stations in meshed power systems. IEEE Access **8**, 166963–166979 (2020)
32. Y. Li, Y. Li, G. Li, D. Zhao, C. Chen, Two-stage multi-objective OPF for AC/DC grids with VSC-HVDC: incorporating decisions analysis into optimization process. J. Energy **147**, 286–296 (2018)
33. G.M. Huang, N.-K. Nair, Voltage stability constrained load curtailment procedure to evaluate power system reliability measures, in *2002 IEEE Power Engineering Society Winter Meeting. Conference Proceedings (Cat. No. 02CH37309)* (IEEE, 2002), vol. 2, pp. 761–765
34. W. Shao, V. Vittal, LP-based OPF for corrective FACTS control to relieve overloads and voltage violations. IEEE Trans. Power Syst. **21**(4), 1832–1839 (2006)
35. R.D. Zimmerman, C.E. Murillo-Sánchez, R.J. Thomas, MATPOWER: steady-state operations, planning, and analysis tools for power systems research and education. IEEE Trans. Power Syst. **26**(1), 12–19 (2010)
36. O. Alsac, B. Stott, Optimal load flow with steady-state security. IEEE Trans. Power Apparatus Syst. **3**, 745–751 (1974)
37. G.W. Stagg, A.H. El-Abiad, *Computer Methods in Power System Analysis* (McGraw-Hill, 1968)
38. K. Visakha, D. Thukaram, L. Jenkins, Application of UPFC for system security improvement under normal and network contingencies. Electr. Power Syst. Res. **70**(1), 46–55 (2004)

Implementation and Analysis of TID Controller for Power Apparatus Applications Using Flower Pollination Algorithm

Tirumalasetty Chiranjeevi, Naladi Ram Babu, Ramesh Devarapalli, and Kiran Babu Vakkapatla

Abstract In this chapter, tilt integral derivative (TID) controller is implemented based on flower pollination algorithm (FPA) for speed control of a DC motor, which is one of the power apparatus application. TID parameters K_T, K_I, K_D, and n are tuned by using nature-inspired metaheuristic optimization algorithm (MOA) called as FPA. Desired response of DC motor is obtained by using tuned parameters of TID. Thereafter comparative analysis has been done with other controllers like PID and fractional-order PID (FOPID) in terms of different performance indices (PIs) like integral of square error (ISE), integral of time-weighted absolute error (ITAE), integral of absolute error (IAE), and integral of time-weighted square error (ITSE) and different time domain specifications. After that, the performance of TID based on FPA has been compared with other MOAs like genetic algorithm (GA) and firefly algorithm (FA).

Keywords TID · DC motor · PID · FPA · FOPID

1 Introduction

Fractional-order systems are modeled using fractional-order differential equations and give more accurate results compared with classical systems [1]. Fractional

T. Chiranjeevi
EED, REC Sonbhadra, Churk, Uttar Pradesh, India

N. R. Babu (✉)
Chalapathi Institute of Engineering and Technology, Guntur, Andhra Pradesh, India

R. Devarapalli
EED, BIT Sindri, Dhanbad, Jharkhand, India
e-mail: ramesh.2015dr1103@ee.ism.ac.in

K. B. Vakkapatla
EED, Lingayas Institute of Management and Technology, Madalavarigudem, Andhra Pradesh, India

H. Malik et al. (eds.), *Intelligent Data Analytics for Power and Energy Systems*,
Lecture Notes in Electrical Engineering 802,
https://doi.org/10.1007/978-981-16-6081-8_11

calculus (FC) is one of the most developing fields of mathematics, and it is a generalization of classical calculus. FC deals with the study of fractional derivatives (FDs) and integrals. In olden days, FC is not attracted most of the scientists and engineers because this subject is not taught in colleges and some of the FDs are suitable for some applications only not for all. These days most of the scientists and engineers are much aware of this subject because of memory and hereditary properties of FDs, and we can find various applications of FDs in almost all areas of engineering, science, and technology [1]. The present work is also one of the applications of FC.

Speed control of a power apparatus application, i.e., DC motor, is very essential because these motors are widely used in residential and industrial purposes. Different MOAs-based PID for speed control of a DC motor is reported in the literature [2]. Regarding FOPID, based on various MOAs, very few works are reported in the literature [2–9]. In this respect, various MOAs like earthquake algorithm, genetic algorithm, cuckoo search optimization algorithm, firefly algorithm, gray wolf optimization algorithm, and bee colony algorithm-based PID or FOPID are proposed for DC motor speed control [2–9].

DC motor speed control is obtained by using various MOAs-based PID or FOPID have been considered in above-discussed works. Quick disturbance rejection is achieved by using TID as compared to PID and FOPID. Therefore, there is a scope for the implementation of TID using various MOAs for speed control of a DC motor. In the present work, we implement TID for speed control of a DC motor based on FPA.

The main contributions of this chapter are

- In this chapter, we implement TID based on FPA for speed control of a DC motor.
- Optimization of TID parameters K_T, K_I, $K_{D,}$ and n is carried out by using FPA.
- We obtain desired response of motor by using optimized parameters of TID.
- Later on comparative analysis is done with FPA-based PID and FOPID in the sense of various PIs and time domain specifications.
- After that, the performance of TID based on FPA is compared with the performance of TID based on other MOAs like GA and FA.

The rest of the chapter is coordinated as follows. Basic fundamentals of PID, FOPID, and TID controllers and different PIs are given by Sect. 2. Modeling of DC motor is described in Sect. 3. Basic idea on FPA is presented in Sect. 4. The results and comments are presented in Sect. 5. Conclusions of the work are given in Sect. 6.

2 Main Heading Basics of PID, FOPID, and TID Controllers and Various PIs

Despite the fact that numerous approaches have been developed, PID is the best controller for industrial applications. However, PID has several disadvantages, such as high peak overshoot and a higher sensitivity to controller parameters.

In this respect, FOPID is proposed which is a generalization of PID [1]. In FOPID, the order of integrator is λ and order of differentiator is μ, both λ and μ are non-integer values. In FOPID, λ and μ are equal to one then we get PID. FOPID gives more accurate results as compared to PID.

However, TID combines the benefits of both PID and FOPID and it is a type of FOPID. Quick disturbance rejection is achieved by using TID as compared to PID and FOPID. TID is designed similarly to PID, with the distinction that K_P in PID is substituted with $K_T\ (1/S)^n$ TID. The PID, FOPID, and TID transfer functions and block diagrams are shown below (Fig. 1).

$$T/F|_{\mathrm{PID}} = K_{\mathrm{P}} + \frac{K_{\mathrm{I}}}{S} + K_{\mathrm{D}}S \tag{1}$$

$$T/F|_{\mathrm{FOPID}} = K_{\mathrm{P}} + \frac{K_{\mathrm{I}}}{S^{\lambda}} + K_{\mathrm{D}}S^{\mu} \tag{2}$$

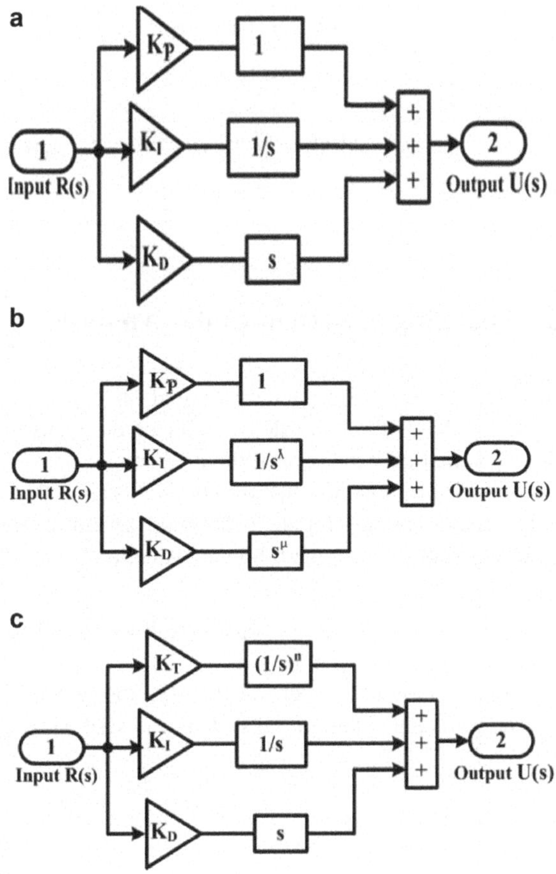

Fig. 1 Block diagram representations of **a** PID, **b** FOPID, **c** TID

$$T/F|_{\text{TID}} = K_{\text{T}}\left(\frac{1}{S}\right)^{n} + \frac{K_I}{S} + K_{\text{D}}S \tag{3}$$

We may consider various PIs like ISE, ITAE, IAE, and ITSE for measuring controller performance in control system. ISE is preferred in order to suppress the large errors. In case of small errors, IAE is preferred. ITAE and ITSE are preferred in case of errors that remain for a long time. These PIs are described as follows

$$\text{ISE} = \int_0^\infty [\text{Error}]^2 \text{d}t = \int_0^\infty [z_{\text{s}}(t) - z(t)]^2 \text{d}t \tag{4}$$

$$\text{ITSE} = \int_0^\infty t[\text{Error}]^2 \text{d}t = \int_0^\infty t[z_{\text{s}}(t) - z(t)]^2 \text{d}t \tag{5}$$

$$\text{IAE} = \int_0^\infty |\text{Error}| \text{d}t = \int_0^\infty |z_{\text{s}}(t) - z(t)| \text{d}t \tag{6}$$

$$\text{ITAE} = \int_0^\infty t|\text{Error}| \text{d}t = \int_0^\infty t|z_{\text{s}}(t) - z(t)| \text{d}t \tag{7}$$

3 Modeling of System Under Analysis

Armature controlled separately excited DC Motor is the system under analysis. Figure 2 shows the equivalent circuit of the analyzing system. The following is the mathematical modeling of the analyzing system. This is the electromechanical system. While modeling the system, we assume that electrical system consists of only armature circuit. Figure 3 shows the armature equivalent circuit.

Using KVL, we get the following equation

$$\vartheta_{a=}\mathcal{R}_a\iota_a + \mathcal{L}_a\frac{d\iota_a}{dt} + \mathbf{e}_b \tag{8}$$

Figure 4 shows the mechanical part of the system's equivalent circuit.

The mechanical part of the system is described by the equation

$$\tau = \Im\frac{d\wp}{dt} + \mathcal{B}\wp \tag{9}$$

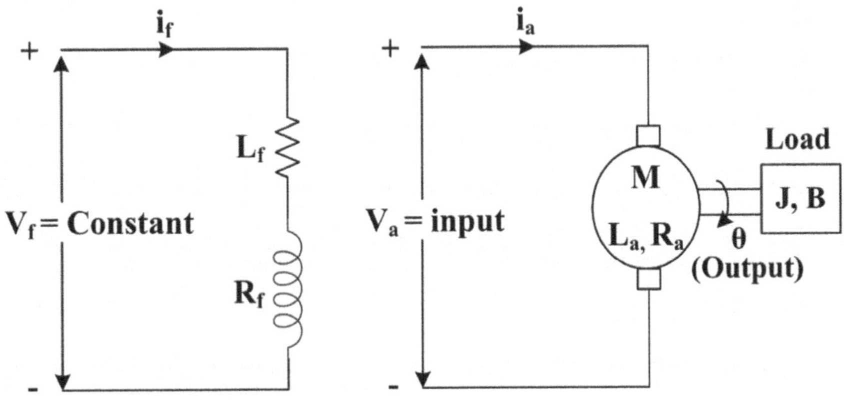

Fig. 2 DC motor equivalent circuit

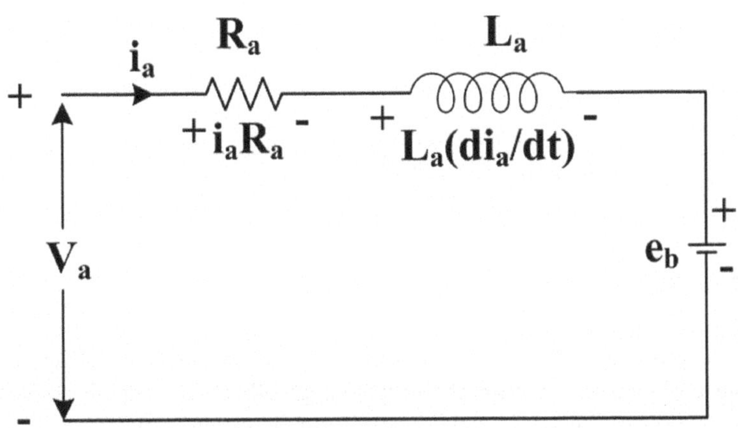

Fig. 3 Armature equivalent circuit

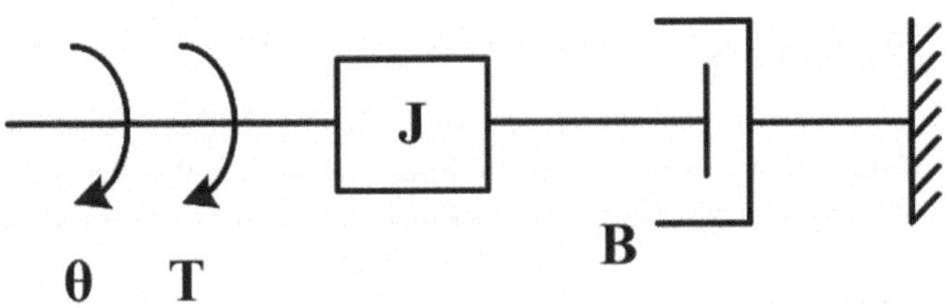

Fig. 4 Mechanical part of the system's equivalent circuit

where $\wp = \frac{d\theta}{dt}$.

The torque and back emf equations are provided by

$$\tau = \rho \imath_a \tag{10}$$

$$e_b = \varphi \wp \tag{11}$$

The transfer function of the analyzing system is obtained over simplifying Eqs. (8)–(9) as [2]

$$\frac{\wp(s)}{\vartheta_a(s)} = \frac{\rho}{(\mathcal{R}_a + s\mathcal{L}_a)(\Im s + \mathcal{B}) + \varphi\rho} \tag{12}$$

Consider the analyzing system specifications as $\rho = 1$, $\varphi = 1$, $\mathcal{R}_a.2\Omega$, $\mathcal{L}_a = 0.5H$, $\mathcal{B} = 0.3$, $\Im = 0.2$.

After substituting the above specifications, Eq. (12) becomes

$$\frac{\wp(s)}{\vartheta_a(s)} = \frac{1}{0.1s^2 + 0.55s + 1.6} \tag{13}$$

4 FPA Basics

Yang et al. proposed FPA [10]. It is a nature-propelled metaheuristic technique that emulates the conduct of fertilization of blooming/pollination plants. Blooming can be done by insects, birds, etc., and is of two forms as shown in Fig. 5.

4.1 Cross-blooming

This kind of blooming can be caused by birds and insects that can travel longer distances with various velocities. This led to global pollination with Levy flight behaviors. The most extreme number of plants about 90% in world can be submitted to this pollination type.

Step 1. Initialize the parameters with switching probability p ∈[0 , 1]
Step 2. Generate initial population of flowers randomly
Step 3. Evaluate initial population and find the current best solution gbest
Step 4. while loop (for stopping criteria)
Step 5. For each flower
Step 6. if rand( ) < p
Global pollination based on Levy step: $x_i^{t+1} = x_i^t + L(x_i^t - g_{best})$
else
Step 7. Selection of two random solutions x_j^t and x_k^t
Step 8. Local pollination
end if
Step 9. Evaluate new solutions
Step 10. Update solutions with better new ones
end for
Step 11. Keep the current best solution
end while

Fig. 6 FPA pseudo-code [10]

4.2 *Self-blooming*

In this, flowers do not require external sources. Around 10% of plants have these blooming properties. Wind is a major source, and distance covered by the pollinators is very less.

4.3 *FPA Rules*

Rule 1: Global blooming will be achieved by cross-blooming by considering Levy flight behaviors.

Rule 2: Local blooming will be achieved by self-blooming.

Rule 3: Closeness among two blossoms is relative to its reproduction likelihood.

Rule 4: Global of local blooming can be overseen by choosing switch likelihood that lies among (0, 1) (Fig. 6).

4.3.1 Mathematical Form of FPA

Global optimum can be achieved by rule 1, 3 with $\varkappa$ as solution matrix with t as iteration and is given by (14)

$$\varkappa_i^{t+1} = \varkappa_i^t + L\gamma\left(\varkappa_i^t - gbest\right) \tag{14}$$

where L is Levy index and γ is Levy scale factor.

The equations for local optima will be given by rule 2, 3 and are given by (15)

$$\varkappa_i^{t+1} = \varkappa_i^t + \varepsilon\left(\varkappa_j^t - \varkappa_k^t\right) \quad (15)$$

where random numbers $(\varkappa_j^t, \varkappa_k^t)$ and $0 \leq \varepsilon \leq 1$.

In this chapter, various PIs are considered to optimize PID, FOPID, and proposed TID settings with limitations set given by Eqs. (16) and (17). Figure 7 shows the flowchart of FPA.

$$0 \leq K_P \leq 1; 0 \leq K_I \leq 1; 0 \leq K_D \leq 1; 0 \leq K_T \leq 1 \quad (16)$$

$$0 \leq n \leq 7; 0 < \lambda < 2; 0 < \mu < 2 \quad (17)$$

5 Simulation Analysis

See Fig. 8.

5.1 *Subheading Controller Comparison Among PID, FOPID, and the Proposed TID*

The system in Fig. 8a is considered and is provided with different controllers like PID, FOPID, and the proposed TID. Investigations are performed by considering ISE as PI. FPA is considered for optimization. Optimum values of the corresponding controllers are listed in Table 1, and its corresponding system responses are plotted in Fig. 9. Careful investigations of Fig. 9 reveal that the responses with the proposed TID outperform over PID and FOPID. Hence, the rest of the studies are carried by considering TID.

5.2 *Comparison of Various PIs*

The study system in Fig. 8 is considered and is provided with TID and FPA in all the cases. Various PIs are considered one at a time. Optimum values of the proposed TID considering various PIs are given in Table 2. Figure 10 shows the responses with TID. From Fig. 10, it is observed that the responses with ISE as PI enhances system dynamics and converges faster than ITAE, IAE, and ITSE. Hence, the rest of the studies are investigated with ISE as PI.

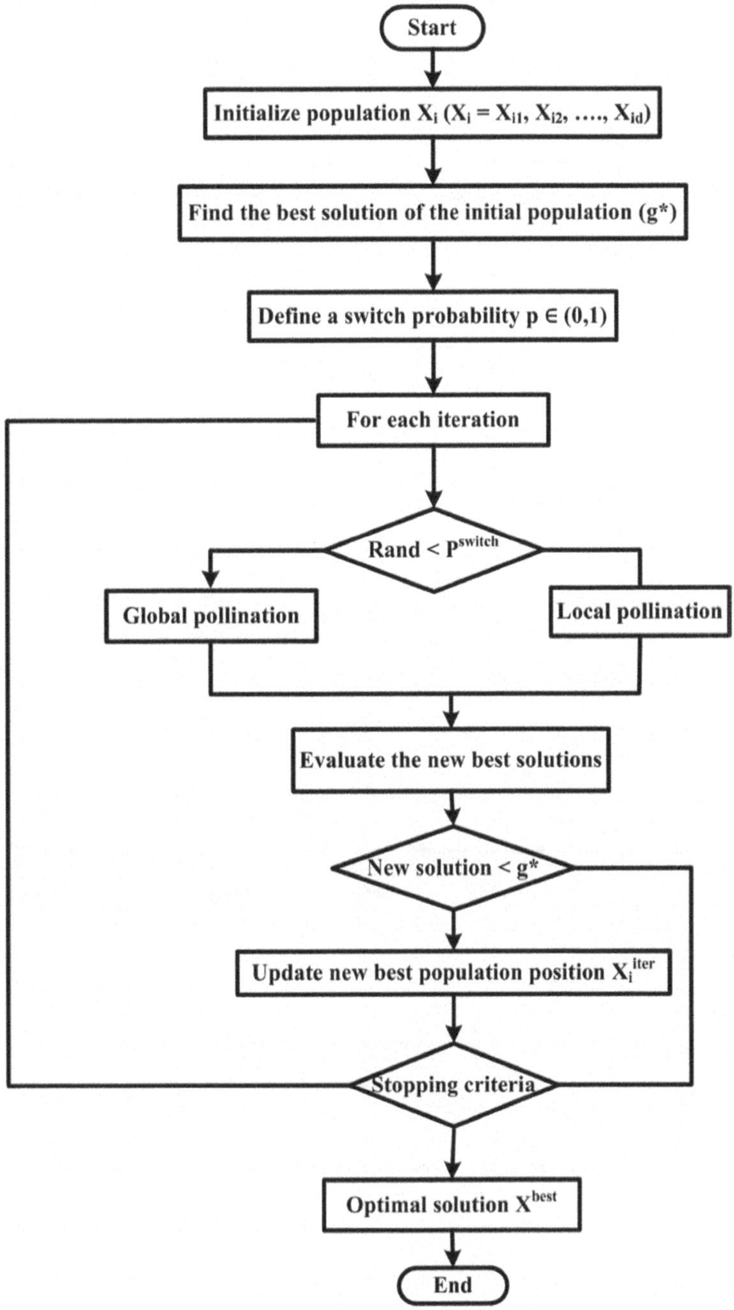

Fig. 7 FPA flowchart [10]

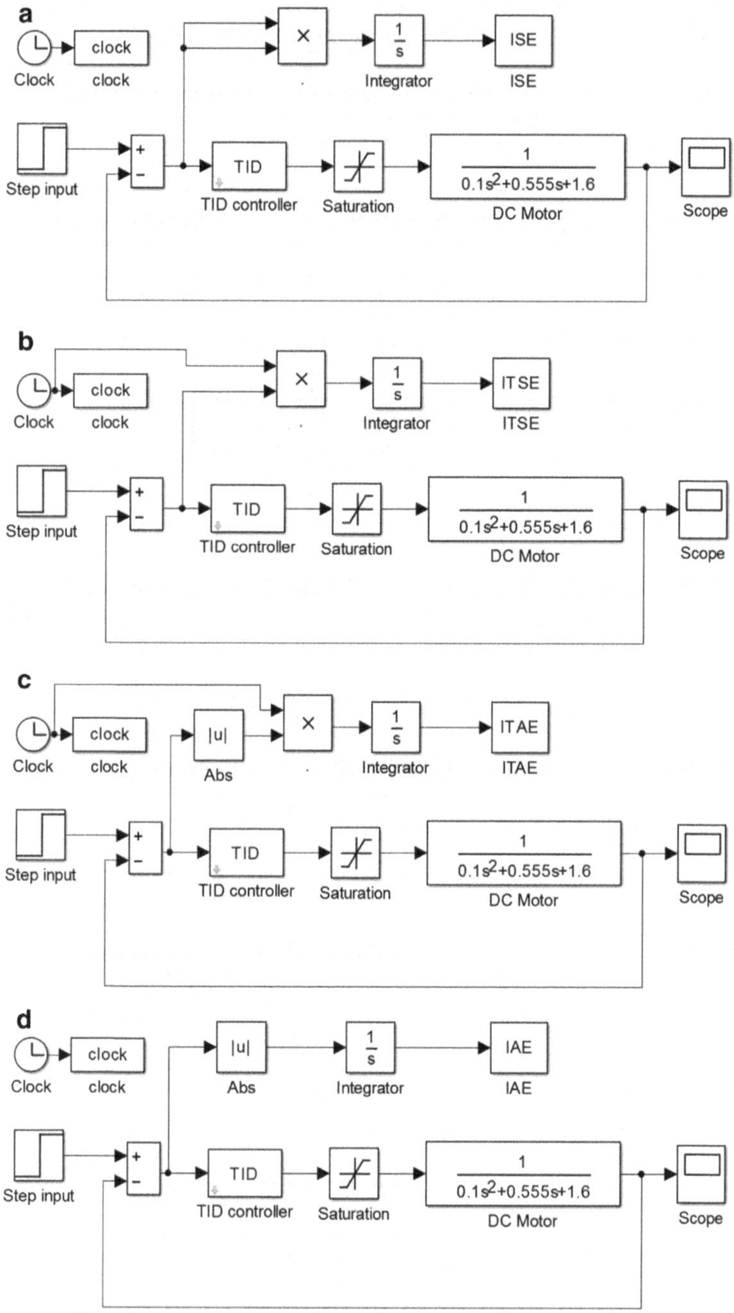

Fig. 8 Simulink diagrams of DC motor speed control with **a** ISE, **b** ITSE, **c** ITAE, **d** IAE

Table 1 Optimum values of PID, FOPID, and TID considering ISE

PID	FOPID	TID
$K_p = 0.6323$	$K_p = 0.5888$	$K_T = 0.0686$
$K_i = 0.0975$	$K_i = 0.5690$	$n = 1.3554$
$K_d = 0.2784$	$\lambda = 0.3211$	$K_i = 0.1030$
	$K_d = 0.765$	$K_d = 0.5622$
	$\mu = 0.94372$	

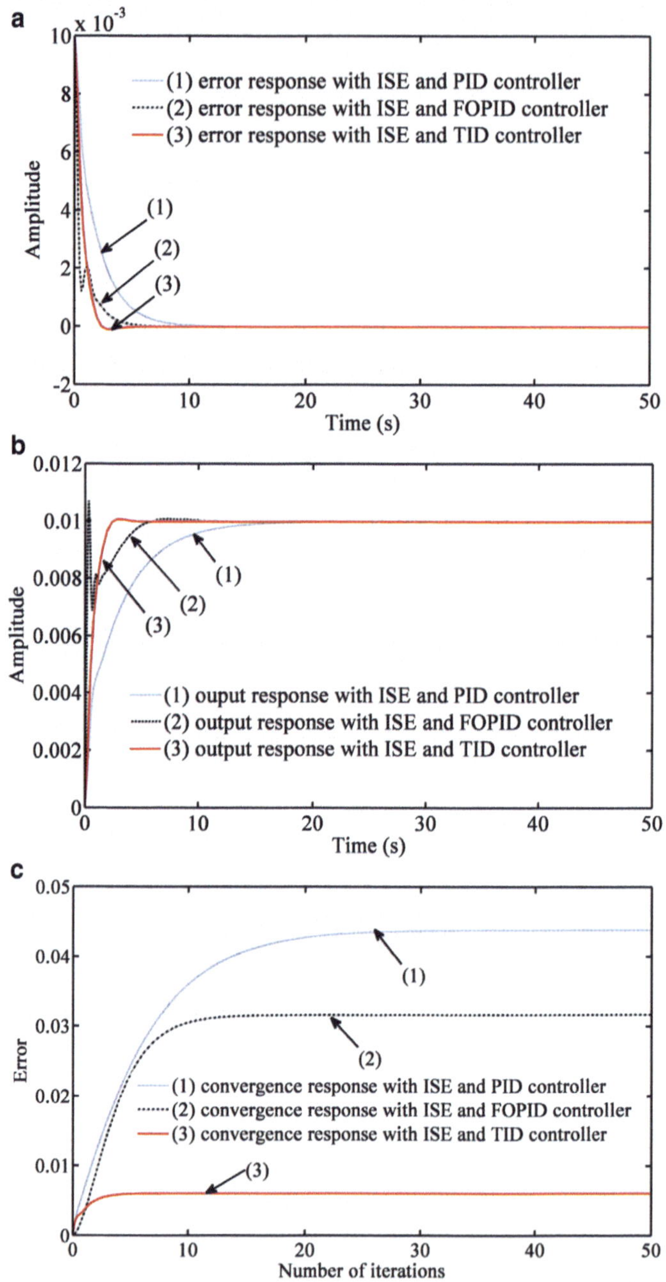

Fig. 9 DC motor response with PID, FOPID, and TID based on FPA, **a** error, **b** output, **c** convergence analysis

Table 2 FPA optimized TID values considering ISE, IAE, ITSE, and ITAE

ISE	$K_T = 0.0686$	$n = 1.3554$	$K_i = 0.1030$	$K_d = 0.5622$
ITSE	$K_T = 0.4094$	$n = 5.7746$	$K_i = 0.6041$	$K_d = 0.2200$
IAE	$K_T = 0.0451$	$n = 3.5019$	$K_i = 0.6819$	$K_d = 0.6145$
ITAE	$K_T = 0.3168$	$n = 1.6672$	$K_i = 0.7553$	$K_d = 0.7653$

5.3 Performance Comparison with Various MOAs Like FPA, GA, and FA

Figure 8a is considered and is provided with ISE and TID for the speed control of DC Motor. Various MOAs like FPA, GA, and FA are considered for the optimization of TID parameters. The obtained optimum values are listed in Table 3, and its corresponding system responses are shown in Fig. 11. Careful observations of Fig. 11 witness that the responses with FPA exhibit better responses over FA and GA.

6 Conclusion

Presented study conclusion. In this chapter, we have designed TID based on FPA and implemented for speed control of one of the power apparatus application called as DC motor. TID parameters are optimized using FPA and then by considering optimized parameters simulation responses are obtained using MATLAB Simulink. In simulation analysis, three case studies, i.e., PID, FOPID and the proposed TID comparison based on FPA, various PIs comparison using FPA-based TID and performance comparison with various MOAs like FPA, GA, and FA are discussed. From the simulation responses, we can conclude the following remarks.

- The responses with the proposed TID outperform over PID and FOPID.
- The responses with ISE as PI enhance system dynamics and converges faster than ITAE, IAE, and ITSE.
- The responses with FPA exhibit better responses over FA and GA.

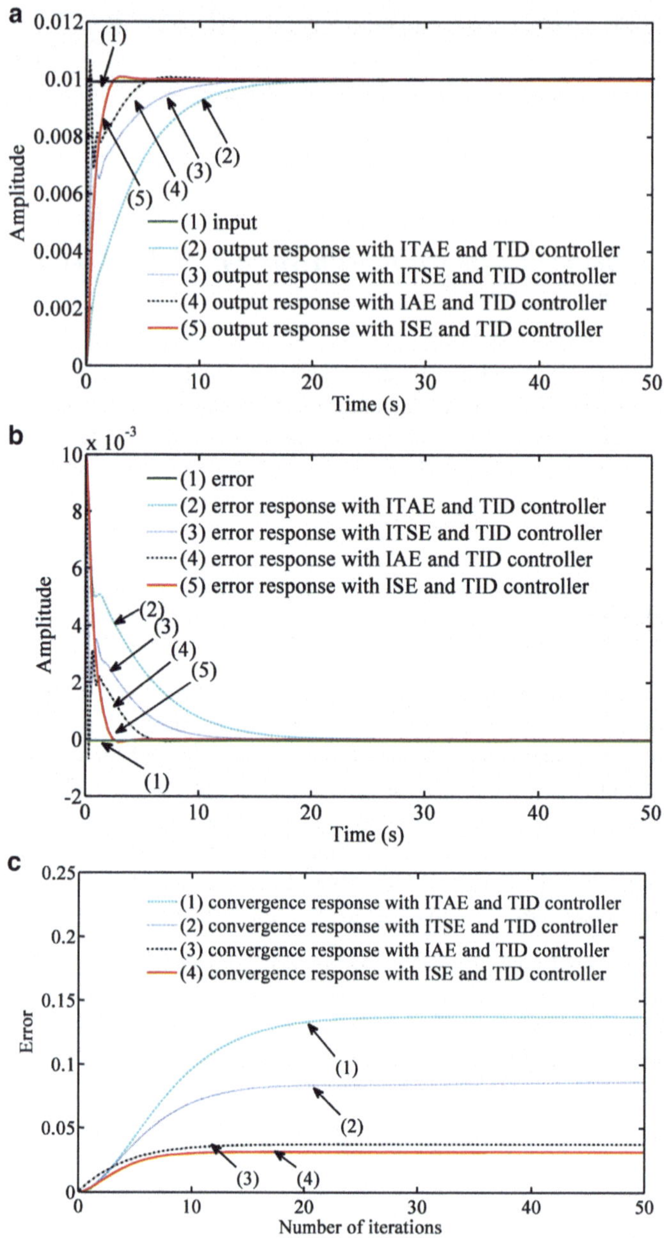

Fig. 10 DC motor response with TID based on FPA by considering various PIs, **a** output, **b** error, **c** convergence analysis

Table 3 Optimized TID values considering FA, GA, and the proposed FPA

GA	$K_T = 0.5822$	$n = 2.7916$	$K_i = 0.0299$	$K_d = 0.3791$
FA	$K_T = 0.5615$	$n = 1.4278$	$K_i = 0.4964$	$K_d = 0.5376$
FPA	$K_T = 0.0686$	$n = 1.3554$	$K_i = 0.1030$	$K_d = 0.5622$

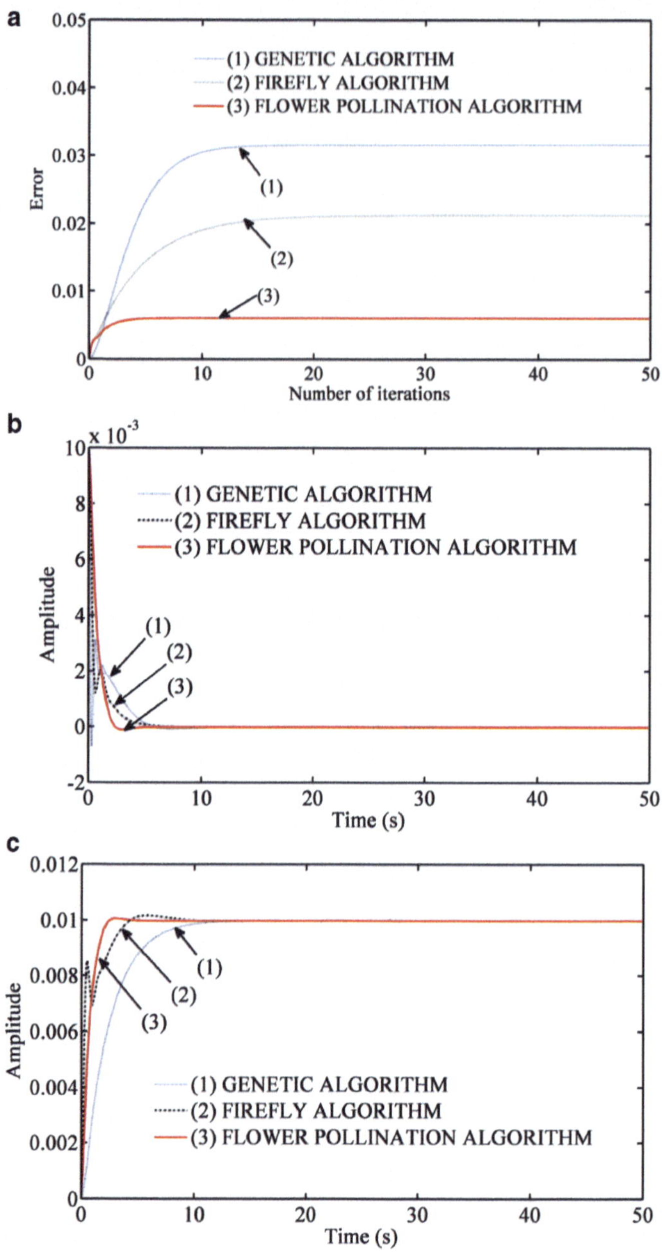

Fig. 11 DC motor response with TID based on FPA, FA, and GA, **a** convergence analysis, **b** error, **c** output

References

1. I. Podlubny, *Fractional Differential Equations* (Academic Press, New York, 1999)
2. T. Chiranjeevi, N. Ram Babu, S.K. Pandey et al., Maiden application of flower pollination algorithm based tilt integral derivative controller with filter for control of electric machines. Mater. Today Proc. (2021). https://doi.org/10.1016/j.matpr.2021.05.049
3. A.K. Yadav et al., *Soft Computing in Condition Monitoring and Diagnostics of Electrical and Mechanical Systems* (Springer Nature, Part of the Advances in Intelligent Systems and Computing, 2020), vol. 1096, p. 496. ISBN 978-981-15-1532-3. https://doi.org/10.1007/978-981-15-1532-3
4. A. Iqbal et al., *Meta Heuristic and Evolutionary Computation: Algorithms and Applications* (Springer Nature, Part of the Studies in Computational Intelligence, 2020), vol. 916, p. 849. ISBN 978-981-15-7571-6. https://doi.org/10.1007/978-981-15-7571-6
5. J.A. Alzubi, *AI and Machine Learning Paradigms for Health Monitoring System: Intelligent Data Analytics* (Springer Nature, Part of the Studies in Big Data, 2020), vol. 86, p. 513. ISBN: 978-981-33-4412-9. https://doi.org/10.1007/978-981-33-4412-9
6. A. Iqbal et al., *Renewable Power for Sustainable Growth* (Springer Nature, Part of the Lecture Notes in Electrical Engineering, 2021), vol. 723, 805 p). ISBN: 978-981-33-4080-0. https://doi.org/10.1007/978-981-33-4080-0
7. N. Fatema et al., *Intelligent Data-Analytics for Condition Monitoring: Smart Grid Applications* (Academic Press, 2021). ISBN: 978-0-323-85510-5. https://doi.org/10.1016/C2020-0-02173-0
8. S.Srivastava et al., Applications of Artificial Intelligence Techniques in Engineering- volume 1, 1E, Springer Nature, Part of the Advances in Intelligent Systems and Computing, vol. 698, 643 pages, 2018, ISBN 978–981–13–1819–1. Doi: https://doi.org/10.1007/978-981-13-1819-1.
9. S. Srivastava et al., *Applications of Artificial Intelligence Techniques in Engineering—Volume 2* (Springer Nature, Part of the Advances in Intelligent Systems and Computing, 2018), vol. 697, 647 p. ISBN 978-981-13-1822-1. https://doi.org/10.1007/978-981-13-1822-1
10. M. Abdel-Basset, L.A. Shawky, Flower pollination algorithm: a comprehensive review. Artif. Intell. Rev. **52**, 2533–2557 (2019)

Intelligent Technique for Eccentricity Fault Diagnosis of Power Apparatus Using Signal Processing Method

Khadim Moin Siddiqui, Rafik Ahmad, and Farhad Ilahi Bakhsh

Abstract In the present time, an efficient health monitoring system can be made by using advanced intelligent technique for the diagnosis of induction motor eccentricity fault in the early stages. If the fault is diagnosed in the early stages, then one can save the industry for millions of dollars. The prime intention of the researchers for expanding a non-intrusive health monitoring system of induction motor health detection in relatively low cost and also ought to be powerful for detection of developing online faults in the early stages. In the induction motor, the unbalanced magnetic pull creates airgap eccentricity fault, it becomes harsh if not taken place timely observation, and it will also lead to large revenue losses in the industries. This issue has been addresses in this research chapter, and an attempt is made to give an efficient health monitoring technique by using advanced intelligent technique. To achieve better results, hybrid technique has been used to extract relevant information of the fault from the raw signal in the developing stage. The EMD and wavelet algorithm has been used jointly for efficient health monitoring purpose for inverter-fed induction motor machine. Two techniques have been used for fault diagnosis purpose one which is FFT technique and other is hybrid technique. It has been observed that the hybrid technique has given encouraging results over FFT technique.

Keywords Eccentricity fault diagnosis · Empirical decomposition · Hybrid technique · Advanced intelligent technique · Signal processing · Induction motor

K. M. Siddiqui (✉)
Electrical Engineering Department, SRMGPC, Lucknow, Uttar Pradesh, India

R. Ahmad
Electrical Engineering Department, BBDITM, Lucknow, Uttar Pradesh, India

F. I. Bakhsh
Electrical Engineering Department, National Institute of Technology Srinagar, Hazratbal, Srinagar, J&K 190006, India
e-mail: farhad@nitsri.ac.in

H. Malik et al. (eds.), *Intelligent Data Analytics for Power and Energy Systems*, Lecture Notes in Electrical Engineering 802,
https://doi.org/10.1007/978-981-16-6081-8_12

1 Introduction

Constant assessment of the well-being of the related induction motor throughout their repair life is called health monitoring. The capability to identify faults as they are still rising is called intrusive failure finding [1]. One can provide enough warning of forthcoming failure by implementing an efficient health monitoring. Hence, there is possibility to make plan for future protective safeguarding and repair work for induction machine [2, 3].

In the past, several methods have been used for fault identification purpose and effectively applied to perceive the motor failures at diverse points. The variable motor parameters such as voltage, current, vibrations, noise, and torque have already been used for fault diagnosis purpose. Often, induction machine faults produce one or more analytical symptoms for example excessive heating, increased losses, line currents, torque fluctuations, and prejudiced airgap voltages. Mainly, in the induction motor, damages are occurring because of electrical and mechanical pressures [4–10].

As per IEEE and EPRI reports [11, 12], bearings are most common reason of rotating machine faults, and the eccentricity irregularity constitutes a substantial component of the three-phase rotating electrical machines. In the healthy motor, the airgap eccentricity inherently occurs up to 10%, but it depends upon the construction of the rotating machine.

In the eccentricity fault, the uneven airgap occurs in between stator and rotor. It is the major problem of induction motor. If unequal airgap occurs in the induction motor, then it produces imbalance and due to this electromagnetic forces are produced between stator and rotor. The axis movement varies the electromagnetic forces, and the rotor axis will be shifted away from the stator axis for eccentricity fault. Many substantial effects have also been occurred on the eccentricity due to loading, slotting, and winding arrangements [13, 14].

The electromagnetic force works linking stator and rotor in uneven way, and it drags rotor beyond its original position, and it is called unbalanced magnetic pull (UMP). The more enlargements in UMP further damage the machine due to excessive generated vibrations. The eccentricity generally appears due to incorrectly positioned bearings or worn, manufacturing tolerances, and inaccuracy of installation [4–6, 15–19].

Two types of eccentricities take place in the induction motor, first is static, and other is dynamic. In static eccentricity, the rotor rounds in its own axis, and also the rotor will be in fixed position from the midpoint of the stator. In the dynamic eccentricity, rotor will not be in its original position, but it tranquil rounds around the stator axis [20, 21].

If the eccentricity fault occurs, then the magnetic flux density allocation will be changed in between stator and rotor. Because of this reason, the resultant radial force is generated, and it operates at the least airgap side. The stable drag on the rotor side is generated by the static eccentricity, whereas the rotor rotates with its rotor velocity with the help of rotating force vector. The electromagnetic force in the induction

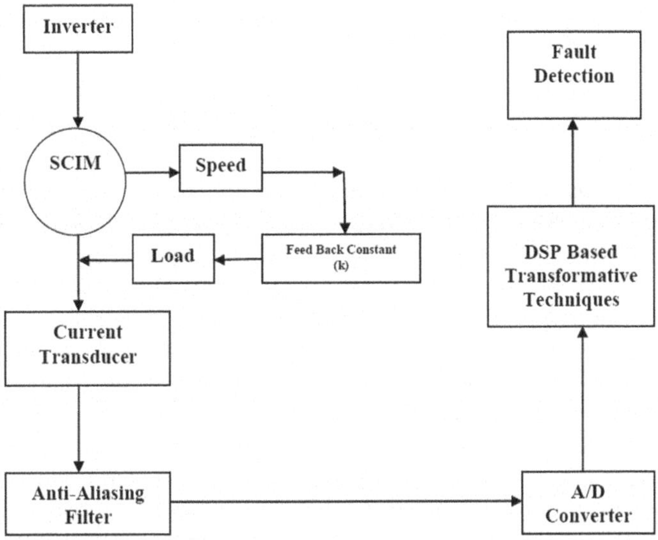

Fig. 1 Proposed inverter-fed induction motor setup

motor can be calculated by two methods, first is analytical method, and second is numerical method. Both methods have their own merits and demerits [22].

The present chapter deals analysis of eccentricity fault of asynchronous motor by advanced intelligent technique. The results have been compared from old widely used intelligent technique and found that the proposed approach may be very useful for diagnosis of eccentricity faults in the early stages. Mainly:

- The eccentricity fault analysis has been done in transient and steady-state conditions for diagnosing faults in early stages.
- Theoretical detailed analysis has been done for wavelet and EMD combined.
- Hybrid technique has been as an advanced intelligent technique.
- The combination of wavelet and EMD has been used for effective eccentricity fault diagnosis purpose
- Eccentricity fault has been diagnosed for inverter-fed induction motor drives in the different conditions.
- The fault has also been diagnosed by FFT algorithm.
- Performance comparison has been done for FFT and hybrid technique of eccentricity fault analysis.

2 Proposed Adjustable Speed Induction Motor Drive Setup

For airgap eccentricity fault diagnosis, the following setup (shown in Fig. 1) is made. The analysis has been done for an induction motor having rating 3 HP, 50 Hz, and 1430 RPM.

From this setup, analysis of eccentricity fault has conducted through advanced signal processing-based intelligent technique. The processes of the proposed fault diagnostic system may be understood by Fig. 1. The fault diagnostic system comprises with PWM inverter, a squirrel cage induction motor, current transducer, anti-aliasing filter, and analog-to-digital converter. The information obtained by the output of analog-to-digital converter is applied to chosen signal processing techniques, and it will provide desired results. From obtained results, one can do fault analysis by non-invasive technique.

In the proposed fault diagnostic system, the PWM inverter is used for supplying the induction motor. The line current is sensed by the closed loop sensor with merits such as fast response and greater accuracy. Further, the bandwidth of the signal is limited through anti-aliasing filter for satisfying sampling theorem. The digital output will be obtained in the output of A/D converter. In the proposed system, two approaches have been used for fault detection purpose. First is FFT technique, and second is hybrid technique. The hybrid technique is the combination of wavelet algorithm and MALLET algorithm. The hybrid technique will remove the strong fundamental component and will provide band limited signals. These obtained signals will be completely free from any kind of noise or aliasing or interference. Hence, one will get best results by using hybrid technique.

3 Applied Fundamental Theory of FFT

The signal interpretation is the most important issue to take into consideration for apposite fault detection purpose of induction motor components failure by signal processing-based intelligent techniques. The stator current signal of induction motor creates dissimilar frequencies of sinusoidal continuous waves by Fourier transform. The different frequencies components play major role in detecting of motor faults. Therefore, by FFT, one may diagnose all induction motor faults after observing the frequencies around the fundamental frequencies. The FFT based methodology may be adaptable to acclimatize to dissimilar operating states. If one does the stator current sampling, then the dissimilar frequencies magnitudes is attained. Hence, the magnitude and frequency waveform can be obtained and used for fault analysis purpose. The original signal can be recovered with the help of FFT. The FFT based intelligent technique was generated by JW Cooley and JW Tuckey in 1965. The older direct Fourier transform (DFT) is used to do direct calculation with $2N$ operations. The N is termed as number of samples. But, the FFT is used to do direct operation with

$2N\text{Log}\,(N)$. Due to this advancement in DFT, the computation time was significantly reduced.

The fundamental mathematical analysis can be understood as follows:

The following is an introduction to Fourier series representation of discrete periodic signals and for discrete periodic signals, the DFT satisfies the condition $x[n] = x[n+N]$, where N is fundamental period having fundamental frequency ω_0.

$$\omega_0 = \frac{2\pi}{N} \tag{1}$$

$\theta_k[n]$ Equation is given as follows:

$$\theta_k[n] = \mathrm{e}^{j\omega_0 kn} = \mathrm{e}^{\frac{j2\pi}{N}kn} \tag{2}$$

where $k = b, b+1, \ldots (N-1)+b$.

b is integer value.

$$\theta_k[n] = \theta_{k-lN}[n]$$

l is integer value.

The $x[n]$ is defined by the following equation:

$$x[n] = \sum_{k=N} a_k \theta_k[n] \tag{3}$$

where a_k can be defined as follows:

$$a_k = \frac{1}{N} \sum_{n=N} x[n]\mathrm{e}^{-j\omega_0 kn} \tag{4}$$

$a_k = a_{k-n}$ and $a_k = a_{-k}$, these conditions will only be satisfied when n is real.

The DFT is defined as:

$$x[k] = \sum_{n=N} x[n]\mathrm{e}^{-j\frac{2\pi}{N}kn} \tag{5}$$

$$x[k] = \sum_{n=N} x[n]\mathrm{e}^{-j\omega_0 kn} \tag{6}$$

The IDFT can be written as:

$$x[n] = \frac{1}{N} \sum_{n=N} X[k]\mathrm{e}^{j\frac{2\pi}{N}kn} \tag{7}$$

$$x[n] = \frac{1}{N} \sum_{n=N} X[k] e^{j\omega_0 kn} \tag{8}$$

4 Results and Discussion

A. *Eccentricity Fault Diagnosis by FFT*

The present section presents the eccentricity fault analysis by the older signal processing based intelligent technique. The FFT based intelligent technique mainly used to diagnose motor faults in the steady-state conditions when the load is not variable. The time domain analysis of the proposed setup is given in [4].

The line current sidebands are monitored through FFT technique. The sensed time domain signal has been applied in the FFT algorithm for extraction of frequency information which was not displayed in the raw signal. In the time domain analysis, one cannot acquire the information at what time which frequency exists.

Therefore; there is a need of FFT for extraction of frequency information. Earlier, this technique was utilized for much motor fault analysis which were operated directly by the mains and able to diagnose faults in the steady-state condition. Hence, it is called steady-state fault diagnosis analysis method. In this work, this intelligent technique has been applied for inverter-fed induction motor drives. It will be interesting to see how FFT will give results for variable speed induction motor because there is a challenge of high switching frequency. The fault analysis has been done for 30, 50 and 70% airgap eccentricity.

The solution of the above question has been explained by the results shown in Figs. 2 and 3. The healthy condition power spectrum is shown in Fig. 2, and one can observe the amplitude corresponding to the frequencies.

If one observes obtained results of Fig. 3 for eccentricity faulty state of the induction motor then can say that the amplitude of the side lobes are decreasing as per the nature of faults. Therefore, one can say that it is a non-intrusive technique of the detection of airgap irregularities; hence, all the results are different from the healthy power spectrum of the motor. Therefore, an effective method has been applied for the diagnosis of airgap irregularities of the induction motor. The main drawback of this method is only used for diagnosis of fault in the steady-state condition and also unable to give at what time incident takes place. In the other words, one can say that this method is unable to give time–frequency information simultaneously and also not able to diagnose eccentricity fault in the transient conditions. This problem has been trounced in the next section by the anticipated investigation.

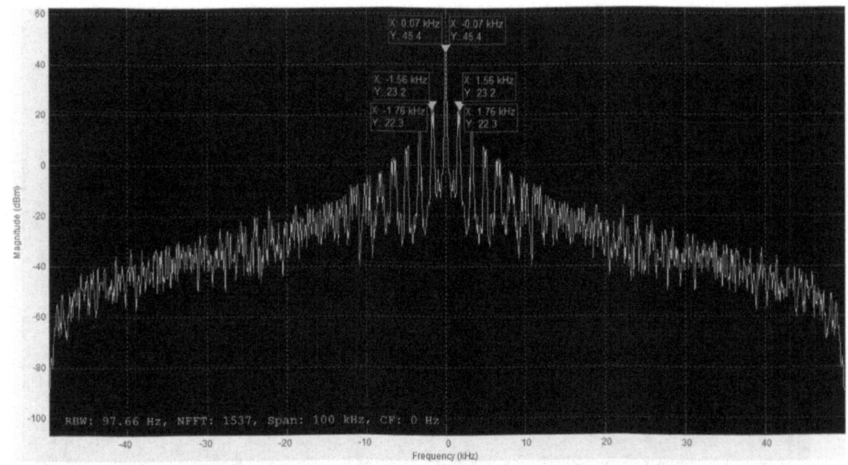

Fig. 2 Power spectrum for healthy state of the motor

5 Applied Theory of Wavelet Algorithm and EMD

A. *Applied Theory of Wavelet Transform*

In the past, fast Fourier transform (FFT)-based signal processing technique is extensively used in the condition monitoring of stator inter-turn fault of induction motor with some limitations. However, with the FFT, one cannot do analysis of abrupt changing signals in the effective way. The FFT based information is not localized in time or space. Due to this main problem, researchers concentrate toward wavelet for time and frequency information simultaneously and also for analyzing signals precisely which are having sudden changing nature. A quickly rotting wave subsists for finite duration having zero mean value and that broaden to infinity is called wavelet. Wavelet comes in diverse dimension and forms. Some illustrious wavelets are shown in Fig. 4.

Since, the wavelets are presented in broad variety as shown in Fig. 4; hence, it shows the usefulness of it. For different applications, one can use corresponding wavelet for better results. In this chapter, Daubechies wavelet has been used in fault analysis and also seen that this wavelet is very suitable for electric machine applications without loosing wavelet coefficients.

The mother wavelet $\Psi(t)$ is given in Eq. (9). The dilations and shifting both are most important features of wavelet. Both features make wavelet very powerful.

$$\psi_{a,b}(t) = \frac{1}{\sqrt{a}}\psi\left(\frac{t-b}{a}\right) \tag{9}$$

In the above equation, a is scaling and b is shifting parameter. The scaling and shifting are presented in details:

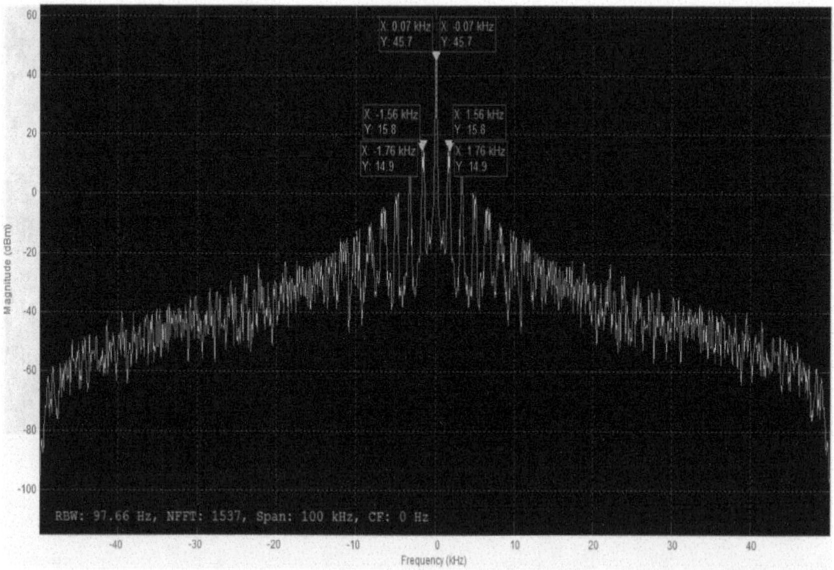

(a)

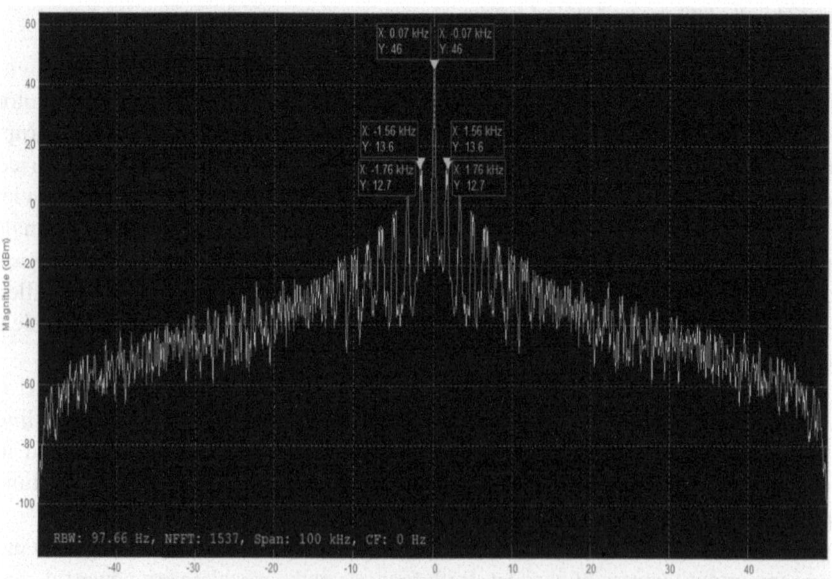

(b)

Fig. 3 Power spectrum for eccentricity fault, **a** 30% airgap eccentricity, **b** 50% airgap eccentricity, **c** 70% airgap eccentricity

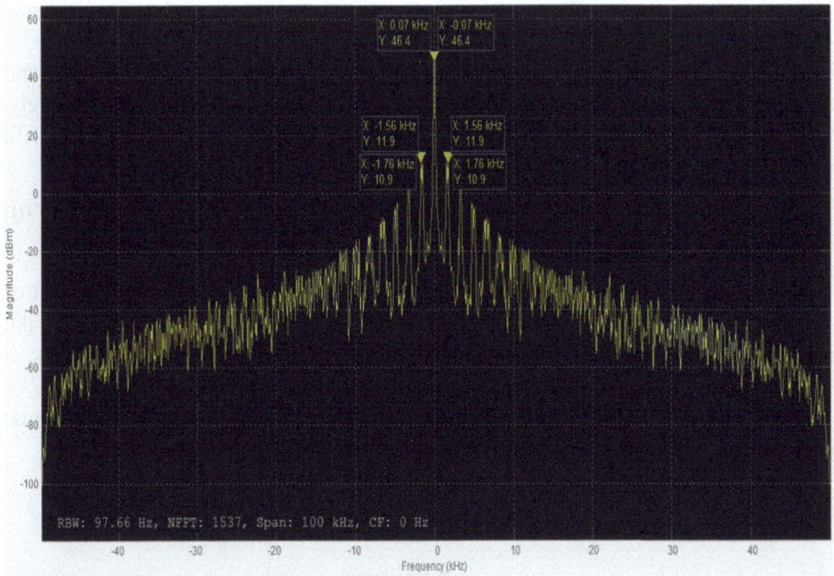

(c)

Fig. 3 (continued)

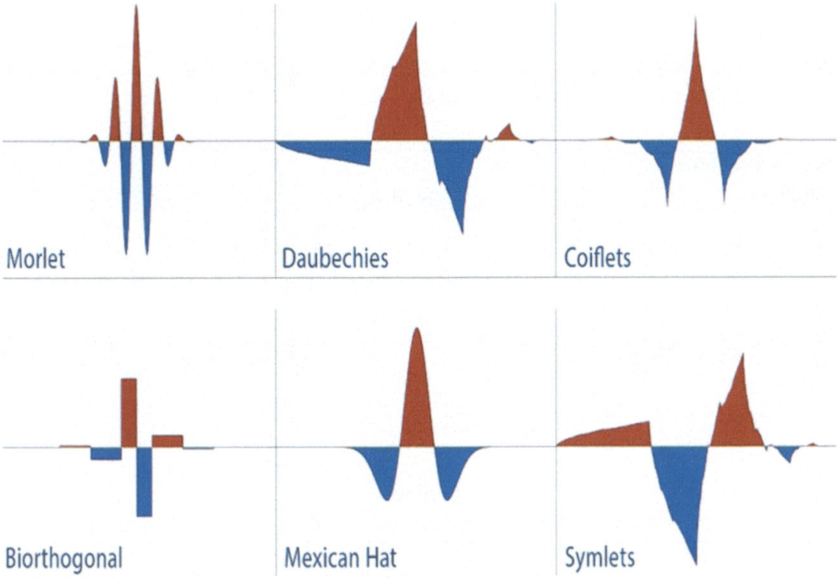

Fig. 4 Wavelets used in different applications

Scaling

The stretching and shrinking the signal in time is called scaling. The stretching and shrinking process for the given signal $\psi(t)$ is presented in Figs. 5 and 6, respectively.

The relationship between mother wavelet $\psi(t)$ and scaling s is given in Eq. (10).

$$\psi\left(\frac{t}{s}\right) \text{ where, } s > 0 \tag{10}$$

The scale factor and frequency have inversely proportionality. This constant of proportionality is termed as the center frequency. The below example shows the scaling of sine wave by 2, and due to this, the original frequency will be reduced half.

In the frequency domain, the wavelet exhibits band pass features, and it can be easily understood by Fig. 7.

Fig. 5 Signal $\psi(t)$

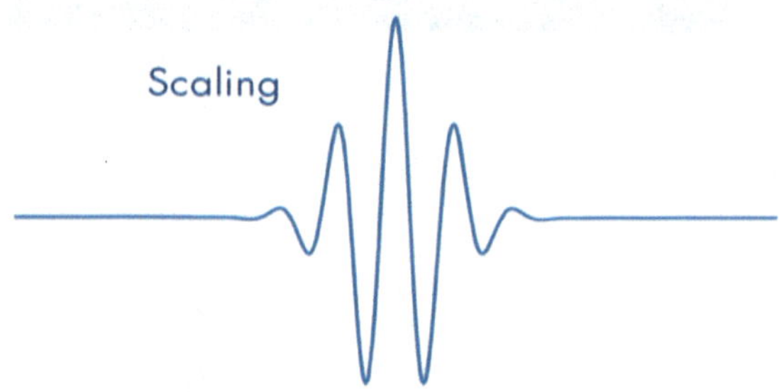

Fig. 6 **a** Stretching **b** Shrinking

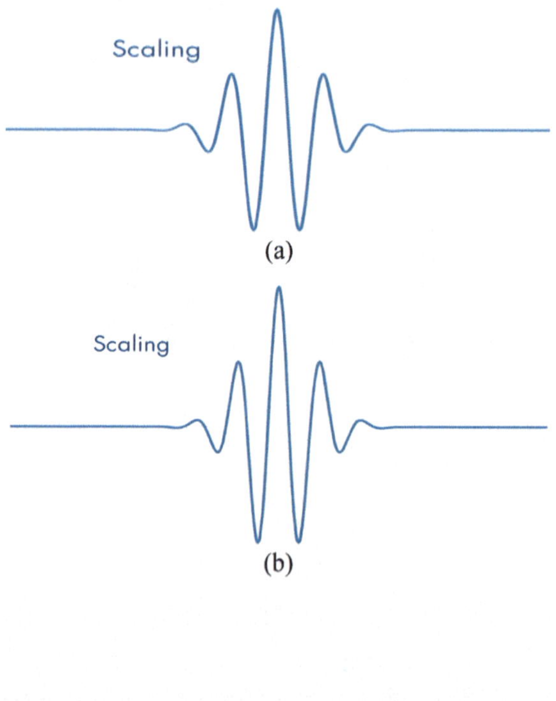

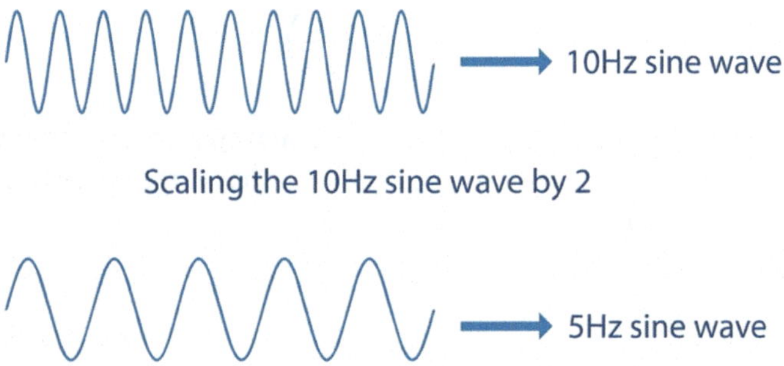

Fig. 7 Scaling the sine wave

The equivalent frequency relation is shown in Fig. 8.

$$F_{\text{eq}} = \left(\frac{C_f}{s\delta t}\right) \tag{11}$$

where C_f is center frequency, s is wavelet scale, and δt is sampling interval.

The relation between wavelet scale and equivalent frequency is shown in the following Table 1. Table 1 shows how equivalent frequency is changing with the wavelet scale correspondingly. When the wavelet scale is set 2 at that time, equivalent frequency will be half and so on. It is very powerful feature of wavelet algorithm.

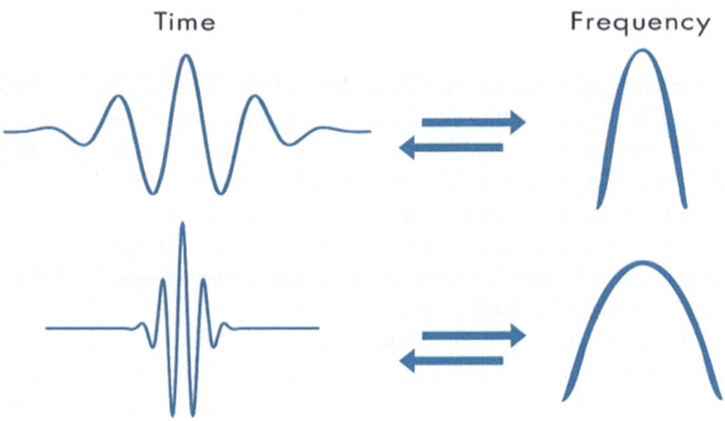

Fig. 8 Band characteristics of wavelet

Table 1 Wavelet scale and equivalent frequency relationship

Wavelet scale	2	4	8	16
Equivalent frequency (F_{EQ})	$\frac{F_{\text{eq}}}{2}$	$\frac{F_{\text{eq}}}{4}$	$\frac{F_{\text{eq}}}{8}$	$\frac{F_{\text{eq}}}{16}$

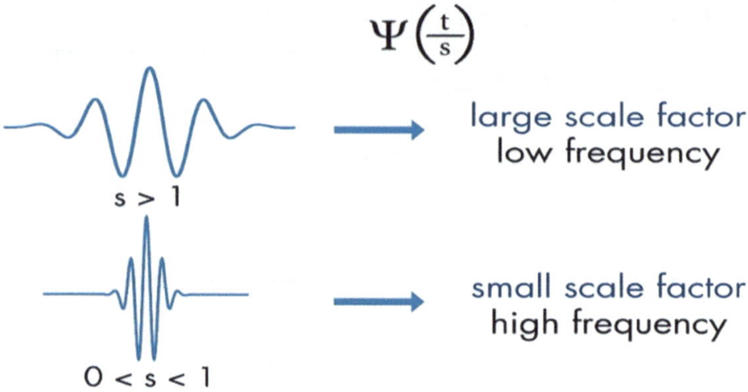

Fig. 9 Scale factor relationship with frequency

Figure 9, presents the relationship between scale factor (low and high frequency band). For low scale factor, the frequency will be high or vice versa. The wavelet is stretched for high scale factor and shrunken for low scale factor.

For capturing slowly changing signal applications, a stretched wavelet is suitable, and it also can be better understood by Fig. 10.

Figure 10 shows the construction of different scales of a given signal by wavelet function. In Fig. 10, the wavelet is moving with time across the finite original signal and capturing transient and any drift changes. Due to this feature, the wavelet gives better results in the fault analysis of AC machines in the transient condition.

Shifting

The feature of delaying or advancing the mother wavelet along the length of the signal is called shifting. During feature extraction from the given signal, the wavelet should be aligned along with the signal for extraction of the features in the effective way. The different shifting properties of a wavelet are shown in Fig. 11. Due to shifting property of the wavelet algorithm, the wavelet is moved and complete path of the original signal for finding out any hidden fault information. Therefore, this method has become extremely useful in the analysis of non-stationary signals. The value of the k can be any integer number such as $k = 0, 1, 2, 3, 4\ldots.$

In DWT, base scale is 2, and it can be varied as 2^j, where $j = 1, 2, 3\ldots$ and the scaling occurs at integer multiples represented by $2^j m$, where, $m = 1, 2, 3, 4\ldots.$

This kind of sampling is very useful, and it removes the coefficients redundancy. Due to this feature, DWT requires less memory. The decomposition process for a given signal "s" by DWT is presented in Fig. 12. The high pass and low pass filters are used to create high pass and low pass subbands for the signal s. The high pass is referred as "D1," and low pass is referred as "A1". In the first stage of decomposition, half number of coefficients will be discarded as per nyquist criteria. Due to this feature, computational performance will be enhanced greatly because of remaining

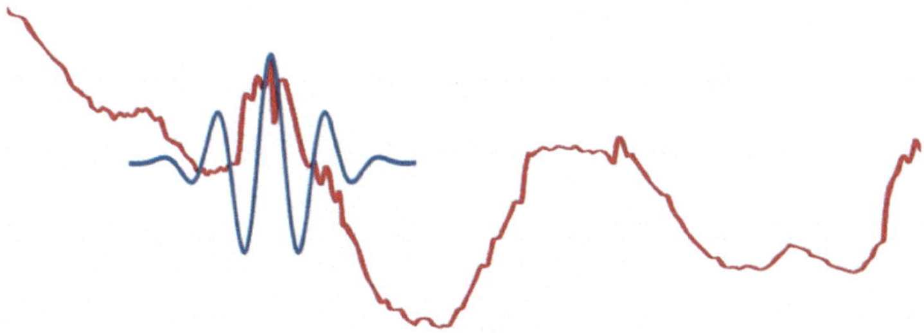
(a)

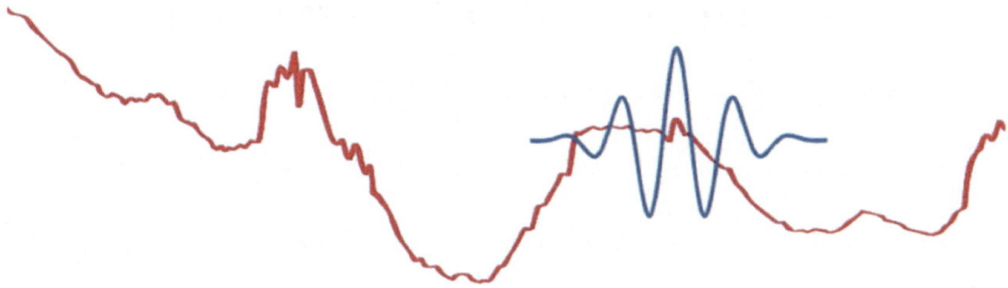
(b)

(c)

Fig. 10 Constructing different scales

fewer coefficients. These filters are also liable to reconstruct original signal. The down sampling process is used for removing aliasing if there is any.

Figure 13 shows the other stages used for signal decomposition. From Fig. 13, we can observe the decomposition process; the original signal is decomposed with the help of DWT for extracting high frequency band information through high pass filter and for extracting low frequency band information through low pass filter. The high pass filter signal is called detailed signal, and low pass filter is called approximation signal. This process will be repeated in the similar way until or unless we do not receive clear information of the fault. The signals are produced with the improved resolution with time and frequency. In the chapter, the above-discussed theory has been applied for the stator inter-turn fault diagnosis.

B. ***Empirical Mode Opposition (EMD)***

The EMD is a data adaptive multiresolution approach. The EMD does decomposition of the signal and provides significant information. By EMD, the analysis of

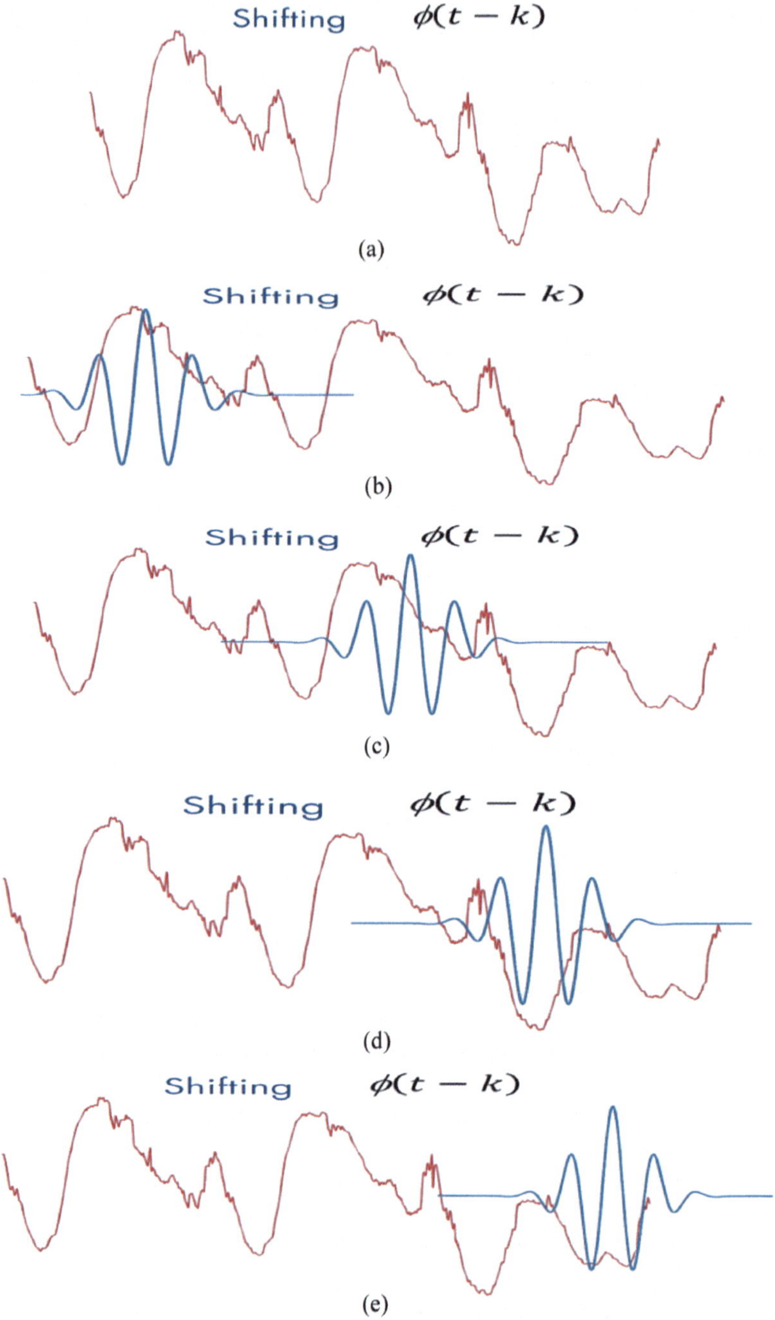

Fig. 11 Shifting properties of wavelet

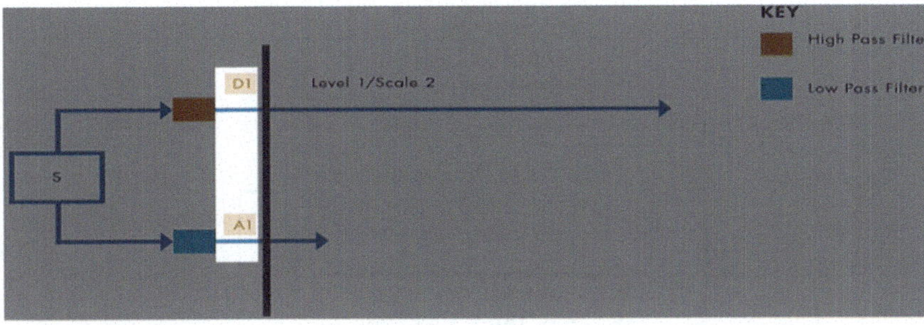

Fig. 12 First stage of decomposition

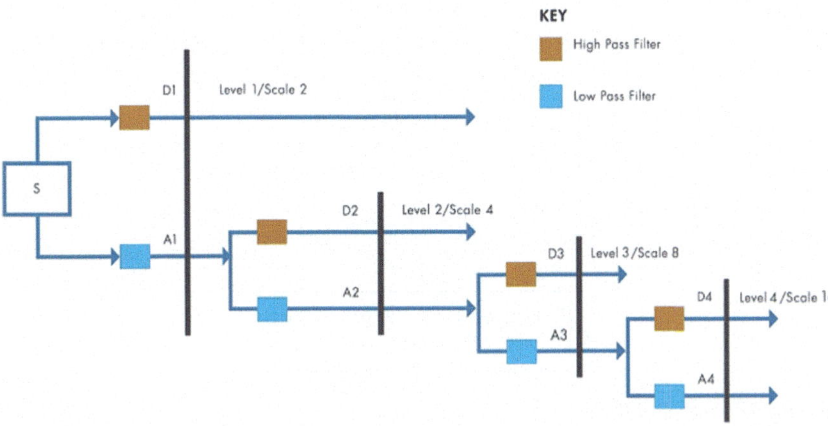

Fig. 13 Other stages of signal decomposition

stationary and non-stationary signals is possible in the effective way. The empirical mode decomposition extracts different components of the signal with different resolutions. The EMD may be very useful in the fault analysis of electric machines. By EMD, one can extract time–frequency information from the time domain signal effectively. In the EMD, the time and scale components of the original signal are same, and it makes analysis easy. The main assumptions of EMD are non-stationary and nonlinear time series and keeps different simple modes of oscillations. The EMD algorithm can be understood from Fig. 14, and complete details are given in [23].

6 Eccentricity Fault Diagnosis by Hybrid Technique

The empirical decomposition technique (EMD) with wavelet algorithm is becoming the most economical, accurate, and attractive technology in induction motor fault monitoring. This MCSA based approach monitors the line current by non-intrusive

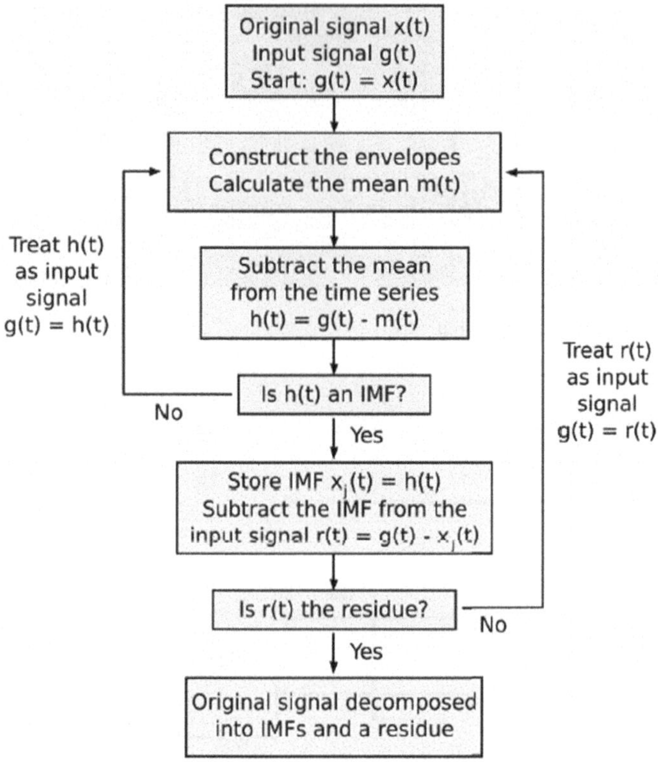

Fig. 14 Flowchart of EMD [23]

technique, but this technique provide us the time and frequency information simultaneously. Therefore, at what time incident takes place may observe through multiresolution analysis efficiently. In this technique, no additional hardware is required to the machine and very suitable for cage-type induction motors in terms of profitable system improvement.

This approach is very suitable because only a separable current transformer is employed to sense the current indicator and also not needed to admittance to the rotating induction machine. The main advantage of this technique in terms of safety of the motor is that the current is computed in the source region with no any trouble to the motor operation. This same technique is used in all stator current monitoring techniques, but the main difference is the applied signal processing technique for extraction of the required information. In this chapter, empirical decomposition technique with wavelet transform has been used for fault diagnosis purpose in the efficient way. In this chapter, hybrid fault diagnosis method has been applied for eccentricity fault diagnosis purpose. The hybrid fault diagnosis technique consists of wavelet transform and empirical mode decomposition. Since, the wavelet transform analysis is commonly used for processing signals, and this transform maps any signal to a set of base functions.

By the mother wavelet's dilation and translation feature, one can obtained base functions and hence can achieve a reasonable decomposition of the signals in different frequency bands at different time points. Though, intricate aliasing may be present in the high frequency portion. Therefore, as a signal processing method, EMD provides better performance for nonlinear and non-stationary signals as compared to conventional method.

Therefore, hybrid technique has been proposed and used in this chapter to achieve better results. The results obtained by hybrid technique for healthy and airgap eccentricity faulty conditions is presented by Fig. 15a–d. Daubechies mother wavelet (db10) has been used for this application due to their advantages over other wavelets, and up to 12th level, empirical decomposition has been done for extraction of relevant information related to fault. The high frequency band stator current signal has been used for differentiating healthy motor from faulty motor. The analysis has been done for 30, 50, and 70% airgap eccentricity faulty conditions. If one compares all results from healthy condition of the motor then can say that all the results are different from each other. The harmonics are increasing in the high frequency band detailed signal as per nature of the fault. For better understanding purpose, detailed coefficients have also been shown along with high frequency band detailed signal in Fig. 15a–d.

The power density has also been estimated for above waveforms achieved by hybrid technique. The change in power density can be understood by Table 2.

Hence, an efficient approach has been applied to diagnose airgap eccentricity fault in the transient conditions with time–frequency information simultaneously, and in future, this proposed technique may be implemented in the industries for induction motor real-time online monitoring purpose.

7 Conclusions

This research chapter presented a novel hybrid technique for detection of airgap eccentricity fault of the squirrel cage induction motor in a variable speed induction motor drive. The hybrid fault detection approach joining EMD and wavelet algorithm is proposed to address the non-stationary and nonlinear fault signals of induction motor's eccentricity fault. The line current signal is first analyzed using wavelet algorithm, and then, EMD is applied to reduce aliasing and interference. The FFT and hybrid techniques have been used for airgap eccentricity fault detection purpose. The results based on hybrid technique for airgap eccentricity fault are able to diagnose in the early stages accurately. The high frequency band detailed signal has given accurate results as per nature of the fault. Hence, one can state that the efficient approach has been proposed for detection of eccentricity fault for adjustable electric drive, and now it can avert the motor before reaching in the disastrous conditions and may save large revenue for industries.

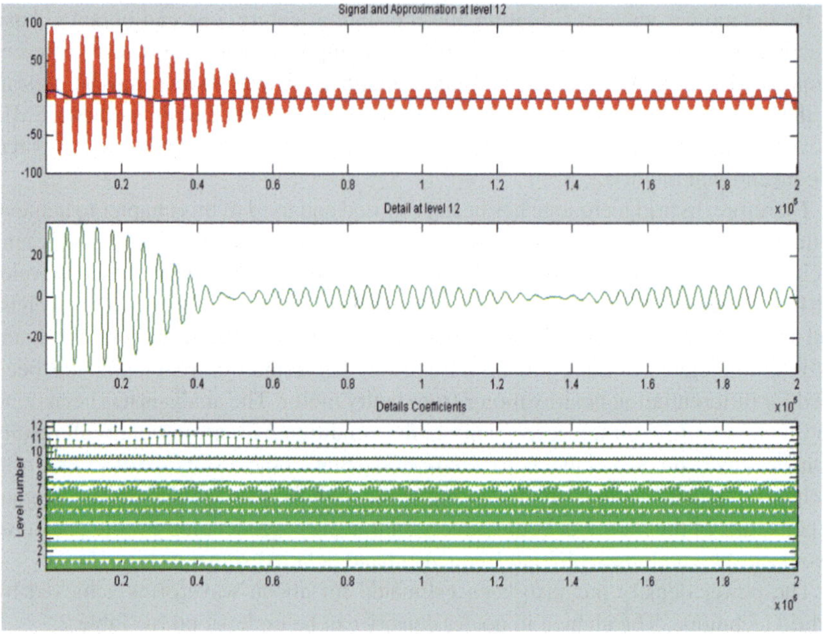

(a)

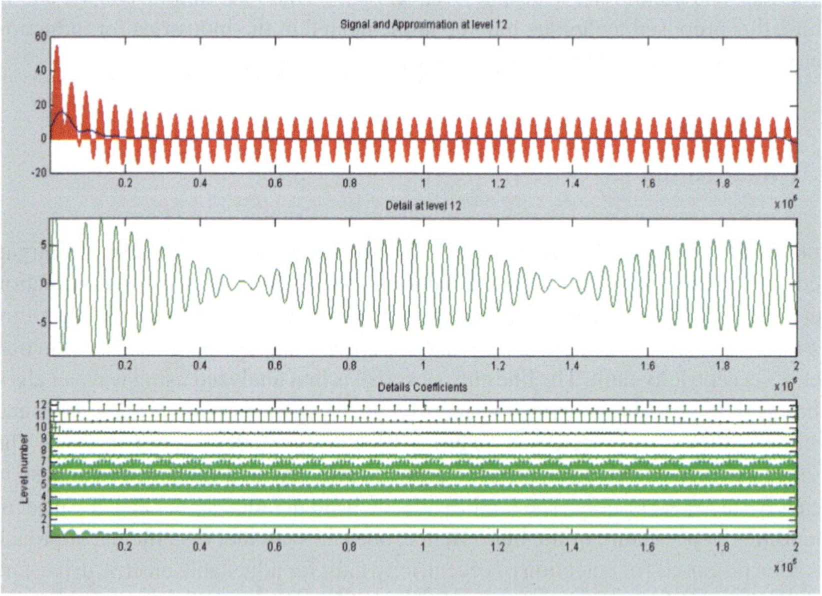

(b)

Fig. 15 High frequency band stator current signal **a** healthy mode, **b** 30% airgap eccentricity mode, **c** 50% airgap eccentricity mode, **d** 70% airgap eccentricity mode

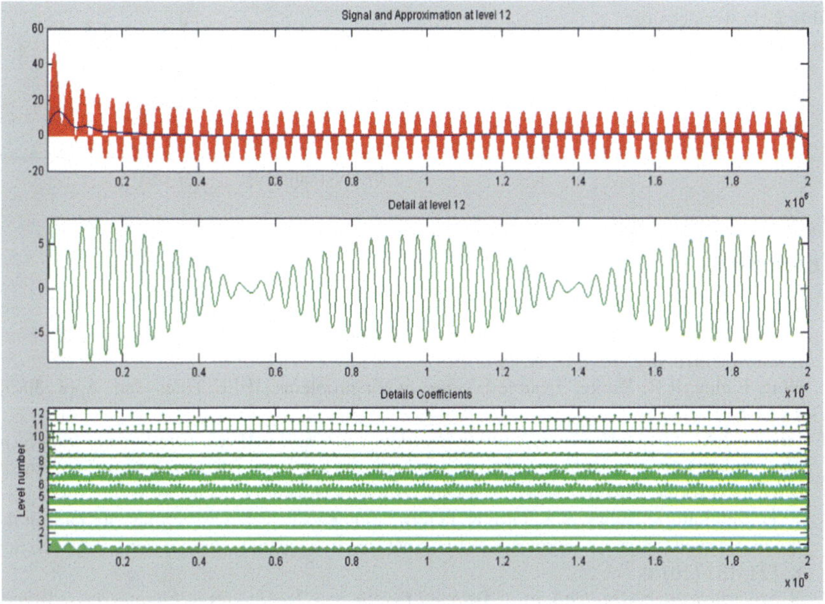

(c)

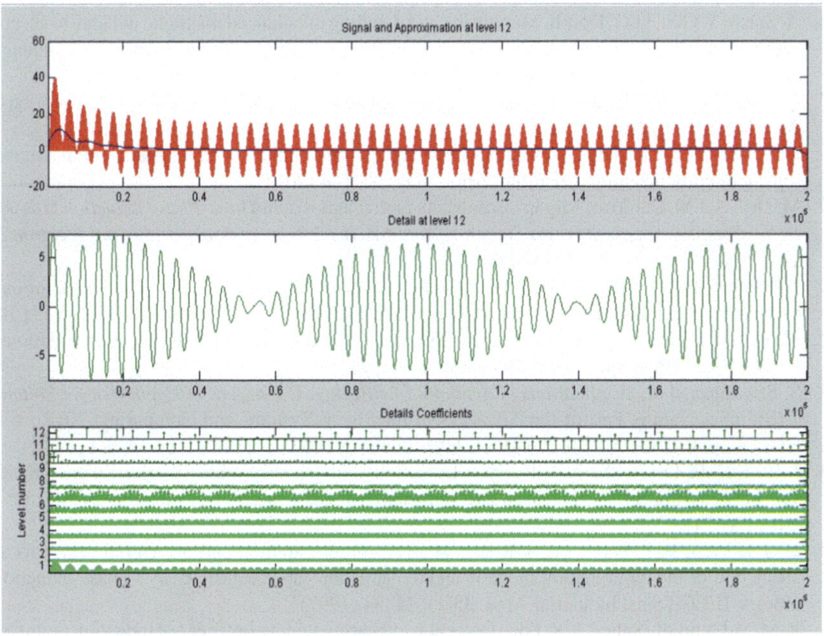

(d)

Fig. 15 (continued)

Table 2 Power density estimation

S. No.	Motor condition	Estimated density
1	Healthy	0.084
2	30% eccentricity	0.109
3	50% eccentricity	0.124
4	70% eccentricity	0.135

References

1. P. Vas, *Parameter Estimation, Condition Monitoring, and Diagnosis of Electrical Machines* (Oxford University Press, USA, 1993)
2. W.R. Finley, R.R. Burke, Troubleshooting motor problems. IEEE Trans. Ind. Appl. **30**(5), 1383–1397 (1994)
3. Alzubi, J.A., *AI and Machine Learning Paradigms for Health Monitoring System: Intelligent Data Analytics* (Springer Nature, Part of the Studies in Big Data, 2020), vol. 86, p. 513. https://doi.org/10.1007/978-981-33-4412-9. ISBN: 978-981-33-4412-9
4. K.M. Siddiqui, K. Sahay, V.K. Giri, N. Gothwal, Diagnosis of airgap eccentricity fault in the inverter driven induction motor drives by transformative techniques. Elsevier's Perspect. Sci. **8**, 127–131 (2016)
5. S. Srivastava et al., *Applications of Artificial Intelligence Techniques in Engineering—Volume 2* (Springer Nature, Part of the Advances in Intelligent Systems and Computing, 2018), vol. 697, 647 p. ISBN 978-981-13-1822-1. https://doi.org/10.1007/978-981-13-1822-1
6. A. Salah, Y. Guo, D.G. Dorell, Monitoring and damping unbalanced magnetic pull due to eccentricity fault in induction machines—a review, in *Proceedings of 20th International Conference on Electrical Machines and Systems* (2017)
7. Q. Yang, D. An, EMD and wavelet transform based fault diagnosis for wind turbine gear box. Adv. Mech. Eng. **2013**, 1–9 (2015)
8. D.G. Dorrell, O. Kayani, Measurement and calculation of unbalanced magnetic pull in wound rotor induction machine. IEEE Ind. Appl. Mag. **50**, 1–4 (2014)
9. M. Drif, A.J.M. Cardoso, Airgap-eccentricity fault diagnosis, in *Three-Phase Induction Motors, by the Complex Apparent Power Signature Analysis, IEEE Transaction on Industrial Electronic*, vol. 15, no. 3, pp. 1404–1410 (2008)
10. N. Fatema et al., *Intelligent Data-Analytics for Condition Monitoring: Smart Grid Applications* (Academic Press, 2021). https://doi.org/10.1016/C2020-0-02173-0. ISBN: 978-0-323-85510-5
11. M. Group, Report of large motor reliability survey of industrial and commercial installations, Part II. IEEE Trans. Ind. Appl. **21**(4), 865–872 (1985)
12. S. Srivastava et al., *Applications of Artificial Intelligence Techniques in Engineering—Volume 1* (Springer Nature, Part of the Advances in Intelligent Systems and Computing, 2018), vol. 698, 643 p. ISBN 978-981-13-1819-1. https://doi.org/10.1007/978-981-13-1819-1
13. J. Hong, S.B. Lee, C. Kral, A. Haumer, Detection of airgap eccentricity for permanent magnet synchronous motors based on the d-axis inductance. IEEE Trans. Power Electron. **27**(5), 2605–2612 (2012)
14. D.G. Dorrell, W.T. Thomson, S. Roach, Analysis of air-gap flux, current, vibration signals as a function of the combination of static and dynamic air-gap eccentricity in 3-phase induction motors. IEEE Trans. Industrial Appl. **33**(1), 24–34 (1997)
15. K.M. Siddiqui, K. Sahay, V.K. Giri, Detection of bearing fault in inverter fed induction motor by transformative techniques, in 12th IEEE International Conference on Proceedings. Electronics, Energy, Environment, Communication, Computer, Control (E^3-C^3), pp. 1–6 (2015)
16. M. Trabelsi, M. Boussak, A. Chaari, High performance single and multiple faults diagnosis in voltage source inverter fed induction motor drives, in *XXth IEEE International Conference, Proceedings on Electrical Machines (ICEM-2012)* (2012), pp. 1717–1723

17. A. Iqbal et al., *Renewable Power for Sustainable Growth* (Springer Nature, Part of the Lecture Notes in Electrical Engineering), vol. 723, 805 p. https://doi.org/10.1007/978-981-33-4080-0. ISBN: 978-981-33-4080-0
18. A. Iqbal et al., *Meta Heuristic and Evolutionary Computation: Algorithms and Applications* (Springer Nature, Part of the Studies in Computational Intelligence, 2020), vol. 916, p. 849. https://doi.org/10.1007/978-981-15-7571-6. ISBN 978-981-15-7571-6
19. A. Sinervo, A. Laiho, A. Arkkio, Low-frequency oscillation in rotor vibration of a two-pole induction machine with extra four pole stator winding. IEEE Ind. Appl. Mag. **47**(9), 2292–2302 (2011)
20. M.E.H. Benbouzid, A review of induction motors signature analysis as a medium for faults detection. IEEE Trans. Ind. Appl. **47**(5), 984–993 (2000)
21. K.M. Siddiqui, K. Sahay, V.K. Giri, Health monitoring and fault diagnosis in induction motor—a review. Int. J. Adv. Res. Electr. Electron. Instrum. Eng. **3**(1), 6549–6565 (2014)
22. A.K. Yadav et al., *Soft Computing in Condition Monitoring and Diagnostics of Electrical and Mechanical Systems* (Springer Nature, Part of the Advances in Intelligent Systems and Computing, 2020), vol. 1096, p. 496. ISBN 978-981-15-1532-3. https://doi.org/10.1007/978-981-15-1532-3
23. A. Zeiler, R. Faltermeier, I.R. Keck, A.M. Tom´e, Member, IEEE, C.G. Puntonet, E.W. Lang, Empirical mode decomposition-an introduction, in *The 20210 International Joint Conference on Neural Networks (IJCNN)*, Spain

Intelligent Methods for Power System Analysis: Advancement in Optimization and Its Application

Ushakiran Huiningsumbam, Ashish Mani, and Anjali Jain

Abstract Optimization problems often entail scenarios wherein the user's purpose is to decrease and/or maximize not a single but multiple, usually conflicting, objective functions simultaneously. Optimization problems often arise in real-world situation concerning engineering and design, chemical operation, and so forth. In this scenario, attaining an optimum for one goal/objective function necessitates certain compromises on one or more of the others and in such cases, we will end up with a collection of equally good solutions rather than just one in such issues. It is usually hard to address the multi-objective optimization problems since its optimal goals are frequently discordant. From a mathematical standpoint, finding an optimal solution that meets all objectives is troublesome and problematic. The chapter begins with a brief history of global optimization. The combined influence of innovative ideas and challenging applications has led to the establishment/formation of multi-objective optimization techniques. The emerging methods are predicated upon evolutionary algorithms that were inspired by nature. This also contributes to the study of numerous optimization challenges and the development/advancement of multiple cutting-edge intelligent algorithms in the effort of power system analysis. Furthermore, we provide an outline of multiple intelligent and advanced optimization methods used in different perspective and its application to electrical engineering among other goals. Finally, a comprehensive assessment of current challenges in the multi-objective optimization problem is provided, as well as ideas or direction for future research. The chapter is intended for readers (students, researchers or any), who wish to acquire the foundation or basic knowledge about the conceptual fundamentals of optimization as well as cutting-edge methodologies of evolutionary multi-objective optimization. The vision is to offer a general framework for research in this vibrant field, as well as to assist the advanced researchers and scholars in identifying untapped research opportunities.

U. Huiningsumbam (✉) · A. Mani · A. Jain
Department of EEE, Amity University, Noida, India

A. Mani
e-mail: amani@amity.edu

H. Malik et al. (eds.), *Intelligent Data Analytics for Power and Energy Systems*,
Lecture Notes in Electrical Engineering 802,
https://doi.org/10.1007/978-981-16-6081-8_13

Keywords Single-objective optimization (SOO) · Intelligent optimization algorithm · Multi-objective optimization (MOO) · Evolutionary algorithm · Pareto-optimality · Decision maker · Swarm optimization

1 Introduction

Almost all real-world challenge entails optimizing several incompatible and competing objectives at the same time. And it occurs that artificial intelligence majorly solves most of the issues experienced by the conventional grid while integrating with RES relatively, thereby providing a reliable, sustainable, and economically feasible power system [1, 2]. While the best solution in single-objective optimization (SOO) is usually evident, it isn't the case in multi-objective optimization (MOO). In lieu of single optimum, there are numerous trade-offs to consider, which are generally alluded as Pareto-optimal solutions. Once all these objective are taken into account, the solutions remain ideal or optimal in the context that no solution outperforms them in the search-space. In today's world, production maximization, environmental preservation, cost-saving, output maximization, resource-energy conservation, and sustainable developmental challenges, among other things, are become increasingly important in a variety of disciplines. As a result, we as humans must organize our manufacturing, production, research, and social pursuit/activities in a far more economically viable and environmentally effective manner. Moreover, these scientific research and engineering improvement difficulties pertain to resource, material, product, and energy utilization optimization concerns. Conversely, as technology and economies around the world become more competitive, optimization issues grow more hard as the scale, fault-detection and diagnosis in intelligent systems become more problematic, nonlinearity, and number of objects and constraints increases [3]. The essential notions of multi-objective optimization (MOO) are clearly outlined and the principles of MOO are discussed in this section. Following that classical ways to estimating the sets of Pareto-optimal solutions, with their potential drawbacks is reviewed. Additionally, evolutionary algorithms are described as a modern optimization technique with few attractive characteristics for this type of problems.

Optimization issues can be continuous or sequential (with discrete or continuous choice variables), bounded or unbounded, single-objective or multi-objective, dynamic or static (on-line or off-line), linear or nonlinear [4]. Two approaches are generally considered to solve optimization problems: To specify the object space, decision space, and the mapping linkage among them, we first analyse the optimization issue and convert it into standard terms/expressions; then we solve those structured data using efficient and appropriate algorithm to acquire the optimal solution or remedies. MOOPs are a common occurrence. Determine the design of an intricate software and/or hardware system, such as those seen in cell phones and automobiles. The systems cost is frequently kept to a minimum, while maximum performance is sought. Other goals, like power dissipation and reliability, may be relevant depending on the application [5]. They could be presented explicitly as independent optimization criteria or as limitations, such as the system's size must not exceed specified dimensions. The majority of real time issues, such as design, control, scheduling, maintenance, optimization, etc., are intrinsically defined according to the achievement of many conflicting objectives. Due to uncontrollable causes, the parameters or variables used to solve these problems are typically imprecise, resulting in more sophisticated problem formulation. The most difficult aspect of solving MOOP, in contrast to SOOP, is there's no mathematically interpretable solution given the sheer complexity of simultaneously satisfying numerous solutions/objectives, which are often in conflict with one another. As a result, drawing up a sets of assumption which depicts ideal/optimal solutions forming the Pareto frontier is required. Each objective has a different value since all points along the frontiers are theoretically equivalent and viable. Despite the fact that the gained solutions are not preferable, decision makers normally only require one final solution depending on the personal preferences for the conflicting and incompatible targets.

In addition, regions around the globe are already focusing on the questions of cost and energy saving and reducing emissions, a significant strategy to achieve sustainable development objectives. As conventional energy sources such as natural gas, oil, coal, etc., have become scarce, and the use of renewable sources like PV cell/solar, tidal, recyclable water, wind, etc., are being developed. The subject of how to enhance resource efficiency, particularly constrained energy efficiency, has become increasingly crucial in process enterprises and our everyday routines. Human production and living activities, on the other hand, have already had great impact, notably detrimental affects to the planet. Protecting the environment has always been and will continue to be of tremendous vitality to sustainable earth development as well as to all its inhabitants. The achievement of energy savings and reduced emissions of industrial and live activity is therefore helpful for sustainable development [6, 7].

This chapter helps to the research of MOO issues and multiple advanced cutting-edge intelligent algorithms in the effort of power system analysis and to its electrical domain. Furthermore, we provide an outline of MOO methods used in different perspective and its application to electrical engineering among other goals. Finally, a comprehensive assessment of current challenges in the MOOP is provided, along with the ideas or direction for future research.

1.1 Multi-objective Optimization (MOO) and Single-Objective Optimization (SOO)

See Tables 1 and 2.

Table 1 SOOP and MOOP similarities [8]

S. No.	Attributes	Similarities
1	Constraints	Equality and/or inequality constraints may or may not exist in the problem
2	Decision variables	Factors that are continuous and/or numeric, with or without constraints
3	Solution strategies	Differential, Linear, nonlinear, and/or integral equations can be used to solve objectives and/or constraints
4	Equation forms	Optimization approaches that are both stochastic and deterministic (metaheuristics) can be applied
5	Ideal/ optimal solution	A distinct (or) multiple optimal solutions can exist for a SOOP (such as local and global optima). Similarly, local and global Pareto optimum fronts might exist in a MOOP (each with many optimum solution.)

Table 2 Moop versus Soop [8]

Attributes	MOOP	SOOP
Amount of objectives	Two or more objectives, either minimization or maximization	There is just one objective to maximise or minimize
Method development	MOOP could be solved by 2 techniques. One way is for MOOP to be converted into a SOOP for solution with a SOO method. Another method is to alter stochastic (metaheuristic) optimization algorithms to deal with solving MOOP	Optimization approaches were initially devised for SOOP with/without constraints
Multi-dimensional spaces (MDS)	2 MDS, 1 for decision variables and another for objectives	1 MDS for decision variables
Number of optimal solutions	Numerous optimal solutions. If all the objectives do not conflict with each other then there's only 1 optimal solution	Mostly 1 optimal solution
Optimal solution computational time	In practice, the processing time is likely to grow in proportion to the number of objective. Deterministic approaches are better when only a few Pareto-optimal solution are sought, while for many Pareto-optimal solutions, stochastic approaches may be faster	When compared to stochastic techniques, deterministic approaches are faster, but stochastic techniques are more likely to find the global solution

(continued)

Table 2 (continued)

Attributes	MOOP	SOOP
Selection of an Implementation Solution	Selecting/identifying one of the non-dominated (optimal) solutions that are highly desirable from the standpoint of objectives in the MOO problem, will require additional preferences and procedures. Choosing one of the non-dominated solutions is itself an optimization problem as there are so many possibilities for normalization, weighting, and ranking	It is simple as one or a few optimal solutions are identified. In far too many cases, the global optimum is preferable
Optimum Solutions Knowledge	Numeric variance and trade-offs objectives in the Pareto optimum front, which might be local or global	generally, limited to 1 optimal solution, (local /global optimal solution)

2 Multi-objective Optimization

2.1 Terminology and Basic Concepts

MOOPs are a common occurrence in both scientific and technical discipline. Optimization issues can be classified into two sub-groups, according to its number of optimized objective functions: SOO and MOO.

An MOO is predicated on the basis of multiple decision-making criteria where its objectives are acknowledged (or) considered as a factor of quantitative formulation (mathematical expressions) of errors or susceptibility of the system performances comprising more than single-optimization problem to be optimized concurrently [9].

The current MOO strategies require optimal-solution to sidesteps the clash between goals/objectives. The use of a specific optimization approach will rely on its benefits, and a metric of maintaining a low-computing cost. Also multi-objective approaches are deemed to be more suited for technical advancement as they seek for global optimization of the system-improving the general performance and additionally controlling and overcoming the conflicts of objectives. As established, many practical optimization problems usually have several conflicting objectives. Generally, there is no method for multi-target optimization that can simultaneously optimize all objective functions. A solution that works well for one function may be ineffective for another or multiple ones. Consequently, finding a solution which fulfills all the MOOP objective functions is quite difficult. There're a number of viable options in this case as given in the expressions below [10].

In general, MOOPs consist of sets of parameters 'n', 'k', and 'm'.

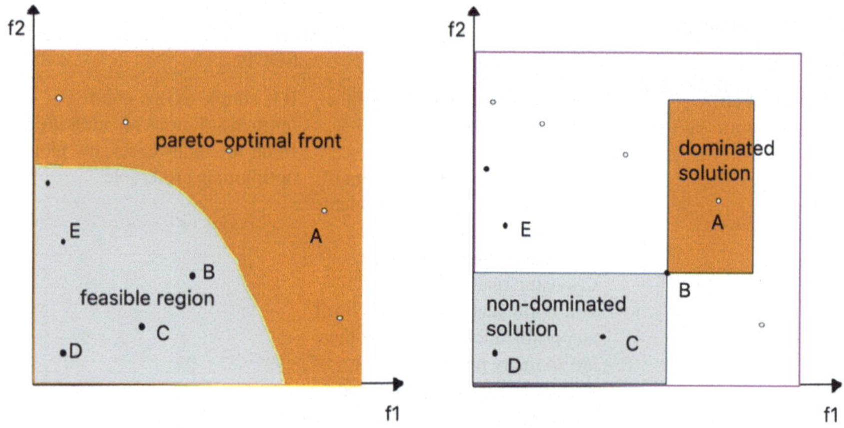

Fig. 1 Relations between Pareto-optimality and its possible solution in objective space [10]

Assuming, 'n' = the decision variable, 'k' = the objective function, 'm' = constraints, where 'k' and 'm' is the function of 'n'.

And the optimization objective is given by the expression:

$$\text{Maximize} \quad y = f(x) = (f_1(x), f_2(x), f_3(x) \ldots f_k(x)) \tag{1}$$

$$\text{Subject to} \quad e(x) = (e_1(x), e_2(x), e_3(x) \ldots e_m(x)) \leq 0 \tag{2}$$

where '$x = (x_1, x_2 \ldots x_n) \in X$' and '$y = (y_1, y_2 \ldots y_k) \in Y$'.

Considering that 'X' = Decision space and 'x' = Decision vector, and 'Y' = Objective space and 'y' = Objective vector.

The set of feasible solution is defined by the constraint '$e(x) \leq 0$.'

Figure 1 demonstrates the relations of the Pareto-optimality in objective space and its probable solution (left and the right graphs, respectively). Here, precise mapping linkage were issued among these 2 spaces. The shaded region illustrated in the left graph represents the feasible solution region which were established by optimization problems constraints. The dominance relations were illustrated by mapping the five points into the objective function space given in the decision-space. The light gray box on the right graph encompasses the area in the objective space dominated by 'B'. The objective vectors whose corresponding decision vector dominates the solution associated with 'B' are contained in the orange rectangle. The solution of 'B' is indifferent to all solutions where its-resulting objective vector is in neither region. Pareto-optimal solutions are given/depicted by the white dots in figure. They are mutually indistinguishable. This shows the major contrast to SOPs: no optimum solution is possible, but instead an optimal collection of trade-offs. Unless preference knowledge is provided, none of them can be classified as superior than the rest (such as ranking of the objectives). MOO methods aim for solution that are distributed uniformly and as similar to Pareto-optimal front as possible. Prior to the discovery of the Pareto-optimal set, decision makers must choose the definitive results from among these optimal alternatives based on real optimization problems or individual preferences. Also the convergence and diversity will be good for such techniques.

- Feasible set is regarded as the set of 'x' which satisfies the constraint '$e(x)$'.

  Where the feasible set is expressed as:

$$X_f = \{x \in X | e(x) \leq 0\} \tag{3}$$

  And the image in 'X_f' is expressed as:

$$Y_f = f(X_f) = \cup_{x \in X_f}\{f(x)\}[0, 1]$$
$$\text{[which is the feasible region in the objective space]}$$

  The feasible set are entirely (fully) sorted as per the objective function 'f'in single-objective optimization for 2 solutions: $\alpha, \beta \in Xf$ where either $f(\alpha) \geq f(\beta)(or)f(\beta) \geq f(\alpha)$. The scope is to discover the solution which maximizes the values of 'f'. When multiple objectives are involved, then, the dynamic shifts, i.e., 'Xf' is rarely entirely ordered, but it is frequently partially order. The approach depicted by point 'B' is far more superior to the solutions depicted by point 'C' where it outperforms at a lesser cost. It would be even better if it simply improved one goal, like in the example of 'C and D', where 'C' outperforms 'D', while having the same cost. It would be ideal if only one goal is improved, as in the instance of 'C and D' where 'C' achieves a better performance than 'D' despite the equal cost.

  For 2 objectives function u and v, $u = v$ iff $\forall i \in \{1, 2, 3, 4, \ldots, k\} : u_i = v_i$

$$u \geq v \text{ iff } \forall i \in \{1, 2, 3, 4, \ldots, k\} : u_i \geq v_i \tag{4}$$

$$u > v \text{ iff} u \geq v \wedge u \neq v$$

  The relations were presented as: '$B > C$', '$C > D$', '$B > D$' also '$B \not> E$' and '$E \not> B$' (thus 'B' and 'E' are not superior to each other). Also, as already stated that 'B' outperforms 'E', the MOOPs relations to its possibilities of its 2 decision vectors $\alpha \& \beta$ is presented: $f(\alpha) \geq f(\beta)$, $f(\beta) \geq f(\alpha)$ (or) $f(\alpha) \not\geq f(\beta)$, $f(\beta) \not\geq f(\alpha)$.

- Pareto dominance is a solution where a specific optimal can't be enhanced in any other objective sans compromising at least one other objective. Let's consider 2 decision vectors α and β when its PD is given in the expression below:

$$\begin{aligned} &\alpha \succ \beta(\alpha \text{dominates} \beta) \text{ iff} f(\alpha) > f(\beta) \\ &\alpha \succeq \beta(\alpha \text{weakly dominates} \beta) \text{iff} f(\alpha) \geq f(\beta) \\ &\alpha \sim \beta(\alpha \text{is indifferent to} \beta) \text{iff} f(\alpha) \not\geq f(\beta) \wedge f(\beta) \not\geq f(\alpha) \end{aligned} \tag{5}$$

The optimality criterion of MOOPs can be established on the notion of Pareto Dominance. From Fig. 1, above it can be seen that 'A' is distinct from 'B', 'C', 'D', and 'E' and that its decision vector 'α' is unaffected by any other decision-vector.

- 'The Pareto-optimal 'set' (or) 'Pareto optimality' is the collection of all Pareto-optimal solutions, where the matching objective vectors make up the Pareto-optimal frontier. Here, a decision vector '$x \in X_f$' is referred to be non-dominated in respect to set $A \subseteq X_f$ iff

$$\nexists \alpha \in A : \alpha \succ x \tag{6}$$

where 'x' is the Pareto-optimal only when 'x' is non-dominated regarding 'X_f'.

- Non-dominated sets and fronts: Here, considering '$A \subseteq X_f$,' then function '$p(A)$' provides non-dominated sets of decision-vectors in 'A', i.e.,

$$p(A) = \{\alpha \in A | \alpha \text{ is non-dominated w.r.t } A\} \tag{7}$$

from figure, it could be explained that the non-dominated set '$p(A)$' is the non-dominated front, and the associated sets of objective vectors '$f(p(A))$' is the non-dominated set. The set '$X_p = p\left(X_f\right)$' is also known as the Pareto-optimal set, whereas the set '$Y_p = f\left(X_p\right)$' is known as the Pareto-optimal front.

2.2 Search and Decision Making

There are two sorts of problem complexity that can be recognized when resolving a MOOPs [11]: search and DM. The first step is the optimization process, which involves sampling the feasible set for Pareto-optimal solutions. Massive and complicated search spaces, often renders searching problematic, precluding the use of precise heuristics like linear programming in SOO [12]. The difficulty of picking an appropriate compromised solution from the Pareto-optimal set is confronted in the second process. To negotiate the usually tough trade-offs among opposing objectives, a human decision maker is indispensable. MOO approaches are further split into three-categories based on how optimizing and the decision-making processes are linked [11, 13].

- *Decision-making before search*: The MOP's objectives are combined to a single goal, which implicitly incorporates the DM's preference information.
- *Search before decision making*: With no preference information, the optimization is conducted. The search procedure yields a set of candidate solutions (preferably Pareto-optimal) from which the DM calls the shot.
- *Making decisions during search*: During the interactive optimization process, the DM can express preferences. Following each optimization phase, the DM presents

variety of trade-offs, depending on which the DM defines more preference criteria and, as a result, drives the search.

2.3 MOO Algorithm

An MOO is based on multiple decision-making criteria where its objectives are acknowledged as a function of quantitative variables of sensibility or error of the process performances pertaining more than single-objective function to be optimized synchronously [9]. The current MOO approaches necessitate optimal solutions to sidesteps the clash between objectives. Additionally, use of a specific optimisation strategy relies upon its benefits and a criterion of maintaining a moderate computational cost. Also multi-objective approaches are deemed to be more suited for power system advancement as they seek for global optimization of the system-improving the general performance and additionally controlling and overcoming the conflicts of objectives. Figure 2 shows various MOO method.

2.3.1 A Priori Methods

The classification chart shown in Fig. 2 shows the various types of a priori methods.

Weighted Sum Method (WSM)

In this technique, by generating a linear combination (or) the weighted sum of the objectives, the fundamental MOOP is altered or modified into SOOP [10].

$$\begin{aligned} \text{Max, } y &= \{w_1 \cdot f_1(x) + w_2 \cdot f_2(x) + \cdots w_k \cdot f_k(x)\} \\ &= f(x), \text{ subjecting to } x \in X_f \end{aligned} \tag{8}$$

Here, w_i = weights and we then proceed by normalizing $\sum w_i = 1$, with no generality loss. If an exact-optimization procedure is employed and all weights are positive, then this strategy results on to produce Pareto-optimal solutions that are easy to demonstrate. Further if performing the above optimization problem for a specific array of possible weight combinations will produce a set of solutions. Considering that for a given weight combination, if a feasible decision vector 'a' maximizes 'f' which is non Pareto optimum. Then, there exist a solution 'b' that leads 'a', i.e., $f_1(b) > f_1(a)$ and $f_i(b) \geq f_i(a)$ for $i = 2, \ldots, k$ with no loss in generality. As a result, $f(b) > f(a)$ contradicting the premise that '$f(a)$' is the maximum.

With non-convex trade-off surfaces, this technique's fundamental drawback is that it can't yield all Pareto-optimal solutions. Based on the embedded system design example, this is depicted in Fig. 3. As in figure shown, considering fixed weights 'w_1' and 'w_2'if we set the value of 'x' to maximize $y = w_1 \cdot f_1(x) + w_2 \cdot f_2(x)$. Then,

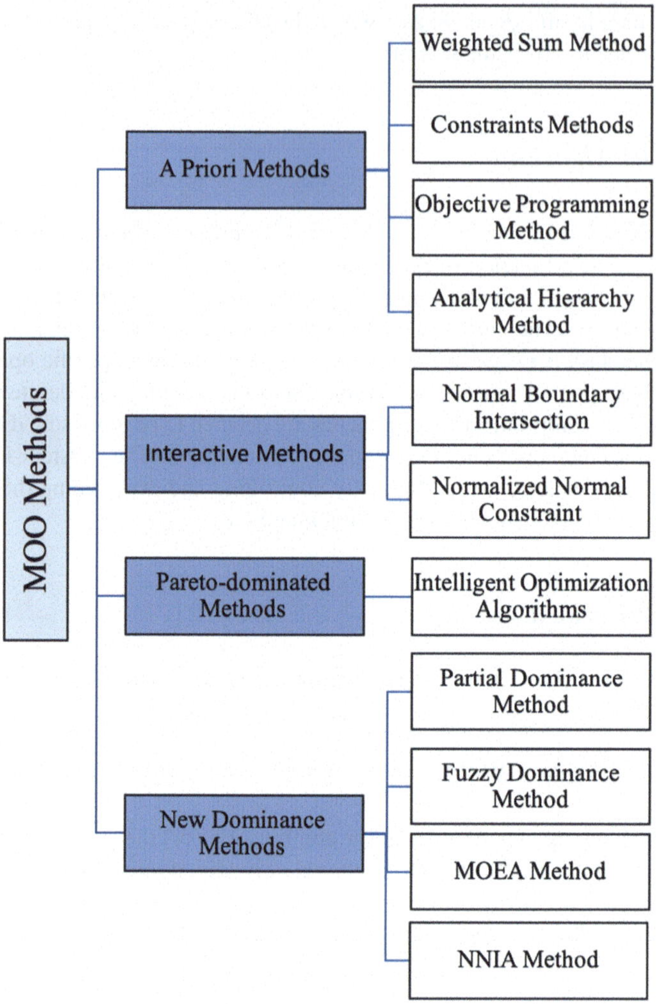

Fig. 2 Optimization methods

the expression can be reformulated as $f_2(x) = -\frac{w_1}{w_2} f_1(x) + \frac{y}{w_2}$, thereby defining a line in slope '$-\frac{w_1}{w_2}$' with intercept '$\frac{y}{w_2}$' in objective space, i.e., the solid line depicted in figure. Shifting this line upward until no feasible objective vector is upward and at most one feasible objective vector (i.e., w and z) is on it, pertains to the optimization process. The points x and y, on the other hand, could never maximize f. When the slope is increased, z obtains a higher 'f' value (upper orange line); when the slope is lowered, w obtains an 'f' value higher than x and z (lower orange line).

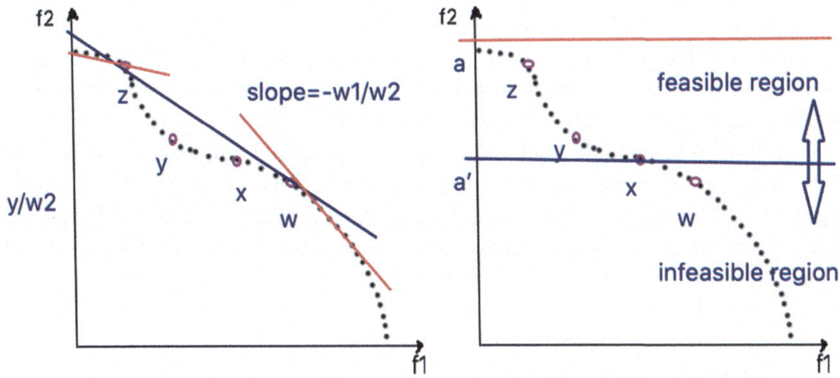

Fig. 3 Graphical representation of the weighting method (left) and the constraint method (right) [10]

Constraint Method

The ε-Constraint approach, which require experiences to determine bounds for each objective function is devised in [14]. It transforms the '$k - 1$' of the 'k' objectives into constraint. On the other hand, 'ε_i' is the parameters that the optimizer alters in order to generate and identify several Pareto-optimal solutions.

$$\begin{aligned} &\text{Maximize, } y = f_h(x) = f(x) \\ &\text{ST}\quad e_1(x) = f_h(x) \geq \varepsilon_i, (1 \leq i \leq k, i \neq h) \\ &x \in X_f \end{aligned} \tag{9}$$

The constraint technique can generate solutions associated with non-convex regions of the trade-off curve, as depicted in Fig. 3 on the right graph. If the lower bounds are not specified correctly, then the resulting new feasible set may be vacant, implying that the accompanying SOP has no solution. An acceptable range of values for the 'ε_i' must be known ahead of time in order to prevent this issue.

Objective Programming Method

In this approach, the DM initially set the optimum value of each objective, i.e., $do = [do_{1,}do_2, \ldots, do_m]^{\mathrm{T}}$, and adding them in to the optimization objectives.

$$\begin{aligned} &\min f(x) \Rightarrow \min \sum_{i=1}^{m} |f_i(x) - do_i| \\ &\text{Subject to}\quad x \in X_f \end{aligned} \tag{10}$$

To efficiently attain the Pareto-optimal solution, the required values must be set within the feasible range. Consequently, setting the desired values is tough. Furthermore, the objective programming method works well for linear programming problems but not so well for nonlinear complicated issues.

Even though these traditional approaches for engaging with MOOPs are convenient, they can only address convex optimization issues by integrating several objectives based on prior experience, limiting their applicability [15]. More intelligent algorithms, as well as new dominance mechanisms, have been suggested and implemented to tackle MOOP.

2.3.2 Interactive Methods (IM)

In this method, there's 2 distinguishing phases of interactive solution processes [16]: learning phase and decision-making phase. DM seeks to comprehend problems structure by exploring alternative feasible solution to determine an area of interest during the learning phase. The substantial variety of desire knowledge imparted by the DM marks this phase. The DM's task in the decision phase is fine-tune the search in the desired space to identify the most preferable option. The increasing expansion in volume and intricacy of MOO's decision-making problems has resulted in a greater reliance on data in the solution processes. Technological developments enable the operation of broader and more complicated systems, as well as the collection and processing of greater volumes of data and the solution of more complicated optimization issues. These method agents' foremost task is to govern the interactive method and deliver created optimization algorithm to the coordinator station. The IM agent is used to monitor the interaction between the coordinator station and the IM. On request from the coordinator agent, it executes 1 iteration of the IM with valued information, received parameter values and possible solutions. Many interactive approaches, as previously noted, turn the original MOOP into SOOP by integrating the given valued information where it's sub-problems is solved by using relevant SOO algorithms to provide candidate solutions representing the DM's preferences [17]. The normal boundary intersection (NBI) and normalized normal constraint (NNC) approaches are two extensively used interactive approaches for handling decision issues in multi-objective. In comparison with the WSM, these are new scalarization approaches that reformulate a MOOP into a parametric SOOP.

NBI Method

This method can produce a relatively close Pareto-optimal frontier spread, rendering it a more appealing strategy to solving non-convex, high-dimensional MOOPs than the WSM. In [18], the researchers presented the NBI method, which took a geometrically intuitive approach to multi-objective issues. The method builds normal lines (quasi) to the plane after first constructing the convex hull of individual minima

(CHIM). Its justification is that the Pareto-optimal point is projected to be the intersection of any juncture on the CHIM's (quasi) normal and the border of the feasible objective space nearest to the origin.

NNC Method

The NNC approach was proposed by Messac et al. [19], which is comparable to the NBI approach but included features from the ε-Constraint technique. The ε-Constraint approach minimizes the most significant objective function 'f_k', while adding inequality constraints of the form $f_i \leq \varepsilon_i$ to the remaining objectives. The NNC approach builds a plane termed the utopia hyper-plane out of all normalized individual minima, and evenly scattered points '$\overline{f_p}$' in this hyper-plane are chosen by adjusting the weights continuously. However, for other goal functions, '$m - 1$' hyper-planes are constructed.

To overcome the disadvantages of the WSM, the NBI and NNC approaches are combined in a stochastic multi firing optimum control in [20]. Along the non-convex Pareto front, the consolidated IM techniques were able to achieve balanced share. It can handle equality/inequality constraints as well as boundary value problems with stringent local and global optimality tolerances.

2.4 Pareto-Dominated Methods

IOA (Intelligent optimization algorithm) is heuristic algorithms that seeks global optimal solutions, unlike classical and analytical mathematical methods, which needs exact numerical calculation or predetermined start search-values (Fig. 4).

2.4.1 Biology Inspired Algorithms

It is influenced by biological processes in the micro and macro domains (like evolution behaviors) or significant development and structural characteristics [21]. There are two sorts of algorithms: evolution-based and swarm-based [22].

Evolution-Based Algorithms

Evolutionary algorithm (EA) is a deterministic search algorithm which replicate natural ecosystems' survival of the fittest process. Evolutionary strategy (ES) [23], membrane computing (MC) [24], differential evolution (DE), genetic algorithm (GA), membrane computing (MC) [24], evolutionary programming (EP) [25], harmony search algorithm (HS), and other algorithms show significant adaptability

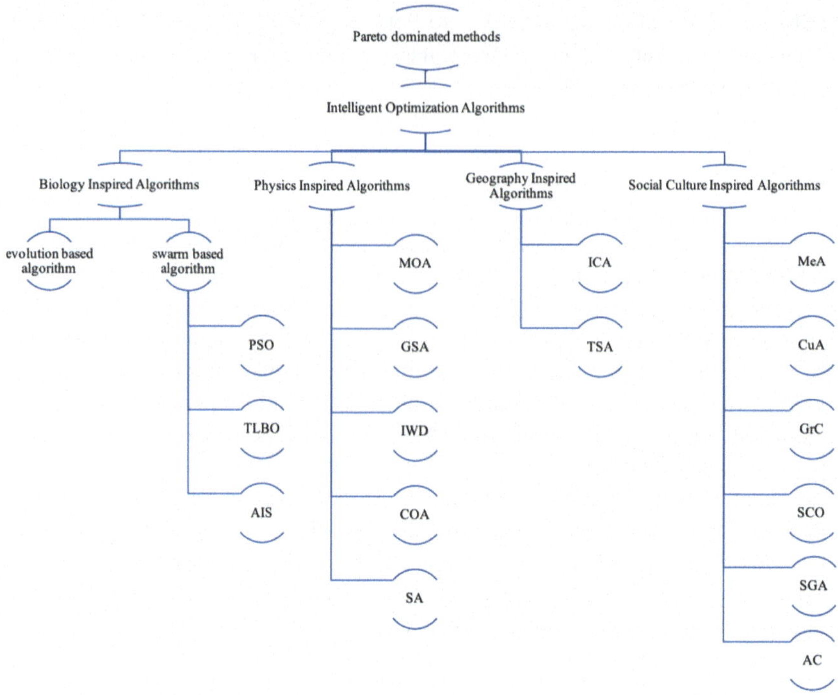

Fig. 4 Classification of Pareto-dominated methods

and self-organization. Furthermore, in [26], the author proposed improved differential evolution (DE) search ability in the initial phase and explore ability in the latter phase by adding a chaos neighborhood searching strategy. In [27], the author coupled DE and PSO to create an ensemble/hybrid method that enhanced performance and increased the speed of searching. The hybrid approach performed particularly well when addressing optimization problems with constraints. Also, the authors in [28] devised a differential harmony search (DHS) algorithm that combines DE with harmony search processes. The DHS strengthens the algorithm's exploring capabilities.

Swarm-Based Algorithms (SBA)

This strategy was primarily influenced by social dynamics and emulate the collective behavior of species like bird flocks, ant colonies, and honey bees, among others. These agents (swarm individuals) work together to find food, which is essential for their existence, as well as to stay safe from other agents. SBA includes cuckoo search algorithm (CS) [29], particle swarm optimization algorithm (PSO), artificial bee colony algorithm (ABC) [30], teaching–learning based optimization algorithm (TLBO), artificial immune system (AIS), ant colony optimization algorithm (ACO) [31], shuffled frog leaping algorithm (SFLA) [32], firefly algorithm (FA) [33], coral reef optimization algorithm (CROA) [34], bacteria foraging optimization algorithm (BFO) [35], pigeon inspired optimization (PIO) [36], etc. The evolution-based algorithms outlined above have been driven by Darwinian evolution, while swarm intelligence is built by mimicking the behaviors of social swarming [37]. Swarm intelligence [38] indicated intelligent activities, which were exhibited beyond central control through straightforward behaviors of persons in the mass. These people were responsible for solving difficulties in food, search, visiting, transport, and transportation. Swarm-based algorithms were then built on the basis of swarm intelligence.

- Particle swarm optimization (PSO): The PSO algorithm, whose conceptual model was an information sharing paradigm, was proposed in [39, 40]. This method has been widely utilized to address various types of optimization issues in a various discipline. To avoid local optima or untimely convergence, numerous variations of the PSO algorithm have been devised to retain or improve the heterogeneity. PSO and NSGA-II were coupled in Li et al. [41], and the experimental results revealed that perhaps the hybrid technique outperformed NSGA-II. The multi-objective PSO (MOPSO) was proposed in [42] that integrated an exogenous population with flexible and dynamic grids.
- Teaching–learning based optimization (TLBO): it was initially introduced in [43]. The TLBO is divided into 2 phase: Teacher and learner phase. In the teacher phase, learners acquired knowledge from the teacher, and in the learner phase, learner learns from one another. The teacher is found to be an excellent alternative out of the complete population so far. In order to achieve global optimal solutions, a revised TLBO approach is presented in [44], which included a mutation process described in [45], which included two mutation processes and two crossover processes to improve the system's local and global search capability. It also performs well in humongous optimization issues with minimal communication complexity [46].
- Artificial immune system (AIS): It is centered on immunology and functions as an adaptive system that solves complex issues by mimicking the functions, principles, and model of immunology [47]. It's been used in domains including image classification, cyber security, machine learning, and optimizations, among others. Evolution of AIS is presented in [48, 49] as multi-objective immune system algorithm
  and non-dominated neighborhood immune algorithm, i.e., MISA AND NNIA, respectively. The NNIA is particularly useful for handling MOOP involving more than 3 objectives.

In comparison with traditional numerical methods, EA is a type of population-based probability search algorithm. These techniques don't require any additional start points or search space gradient details. Thus, they're good for optimization issues which can't be addressed using traditional statistical approaches. Furthermore, evolution-based algorithms feature parallelism and distribution properties that make them suited for handling large-scale (or) high-dimensional optimization problems.

2.4.2 Physics-Inspired Algorithms

This algorithm is also known as Bayesian algorithm usually initiated for solving optimization problems. They mirror matter's behavioral patterns and attributes, or they adhere to the same concept as that of physics' laws. Magnetic optimization algorithm (MOA) [50], chaotic optimization algorithm (COA) [77], gravitational search algorithm (GSA) [51, 52], intelligent water drops algorithm (IWD) [53], simulated annealing (SA) [54, 55], and others are examples of physics-inspired algorithms.

2.4.3 Geography Inspired Algorithms

It is a type of metaheuristic algorithm that generates arbitrary outcomes throughout the global search space. These approaches for optimization are categorized as imperialistic competition algorithm (ICA) [56], tabu search algorithm (TSA) [31, 57, 58], etc., where the ICA is motivated by imperialism, which is stated as the doctrine of expanding authority and denotes a government's duty. An imperialist's power is determined by the number of colonies he or she has. When one imperialist's authority is strengthened, it weakens the authority of other imperialists. And TSA was a local-search-based metaheuristic technique. Through a series of steps, it investigates all viable solutions in the search-space. For evading from local minima, a sequence of steps is barred at each sampling interval.

2.4.4 Social Culture Inspired Algorithms

This strategy, which incorporates cultural evolution theory into optimization algorithms, is motivated by economic, cultural systems, social, etc., among others. Some developments of these algorithms include memetic algorithm (MeA) in [59] proposed by Moscato where it is motivated by the social culture. Cultural algorithm (CuA) where its source of inspiration is the signal, knowledge, etc., of the social culture initially proposed by Reynolds in [60]. Granular computing (GrC) which is proposed by Lin [61] which is inspired by the information cognition. Affective computing (AC) is introduced by Picard [62] where its inspiration is swayed by social and cultural affection. Selfish gene algorithm (SGA) is based on selfish gene in mankind proposed by Corno et al. in [63]. Social cognitive theory (SCO) by Xie [64] is motivated by social cognitive process.

2.5 New Dominance Method

The ε-dominance mechanism is developed as the Pareto-dominated approaches are unable to concurrently tackle the distribution and convergence concerns of optimization algorithms [65]. This method differs from the Pareto-dominance method except that the initial is built on the notion of space-hyper grids, while the latter is not. Following that in [66], the coupled Pareto-adaptive dominance mechanism is suggested. The 'ε'was a vector in the mechanism whose size equaled the quantity of hyper grids. The geometrical shapes of the pareto front was related to the quantities of the parameter in the vector. The numbers were self-adjusted based on information about the dominating power of different areas of the Pareto front. To tackle high-dimensional MOOP, also in [67, 68] the researchers integrated the primary component analysis (PCA) approach, evolutionary algorithms and the correntropy PCA approach. Partial dominance is later explored in [69]. Thereafter, the theory of fuzzy dominance is presented [68] where the weights to every objective function is provided by the fuzzy dominance function and then the weighted objectives were added together. Zhang and Li [70] suggested a multi-objective evolutionary algorithm (MOEA) based on decomposition technique that merged EA with traditional MOO approaches. The MOEA/D technique broke down multi-objective issues that resemble the Pareto front into SOO sub-problem, which were then solved using an EA. The widths along weighted sub-problems vectors were used to create neighborhood relations between sub-problems. Each sub-optimization problem's phase was completed by collaboration with its neighboring sub-problem. Mlakar [71] introduced a Gaussian process-differential evolution multi-objective algorithm (GP-DEMO) which is based on surrogate models. To analyse solutions under uncertainty, the freshly defined association with structuring element was employed instead of Pareto predominance. Because of the faulty surrogate model approximation, error ratio was able to be reduced. The evolutionary algorithm was utilized to find, assess, and employ optimal solutions for updating the surrogate-model.

2.6 Steps Involved in MOOP

The steps involved in the structured approaches for multi-objective optimization (MOO) operations are shown in Fig. 5. The arched arrow on the right indicates the necessity for repetition of its previous-phase in Fig. 5.

3 Related Works in Power System Domain

Multi-objective approaches are deemed more suitable for power system advancement as they seek for global optimization of the system-improving the general performance and additionally controlling and overcoming the conflicts of objectives. In dependent HGPS (i.e., hybrid green power system), the authors in [72] used the fuzzy logic method to compromise economic and environment emission objectives with varying stages of relevance. RES production processes, in contrast to other energy optimization production systems, energy production performances are more targeted, which is expressed primarily by the optimization of RES design. Along with economic

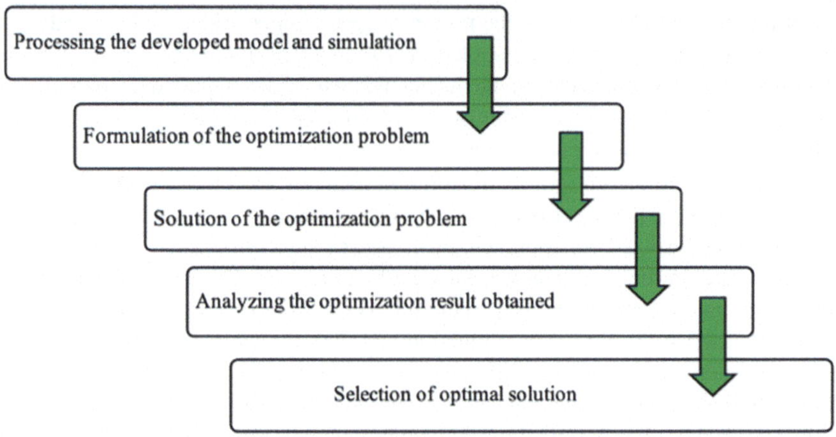

Fig. 5 Procedure for the multi-objective optimization

and environmental goals, energy efficiency is becoming increasingly important in the development of renewable resources. Stand-alone power systems, such as wind farms, smart grids, power grids or microgrids, and biomass production processes, as well as distributed generation networks or hybrid energy systems that incorporate various renewable, are the examples of these processes. In [73], the authors provide a multi-proposal approach to improve the micro-grid power quality in an islanded mode. They employ a current scheme and cascade voltage to control frequency and voltage in inverter based micro-grid (IBMG) system. Here, the hybrid multi-objective symbiotic search algorithm is introduced to optimize specific objective such as PI parameters tuning, voltage overshoot and undershoot, ITAE, settling time and rise time. The resultant outcome shows robustness within the applied controller, where fluctuating sources and harmonics are considered as the disturbances in the power system.

The authors of [74] presented a multi-objective control technique for multi-functional grid-connected inverters to improve their performance. To provide satisfying demands for RES interactions in power grid and for catering auxiliary power quality operations, the devices offer regulated capacity. A cooperative approach is shown with communication support among VSI to handle the power quality requirements on distribution lines. Relationship between the holistic power quality assessment objectives and tiered management for regulating wattage demand in an MG is presented in the chapter. Parameter computations required by regional operators to conduct optimal remuneration are used to manage every DG system, to satisfy the demands of the system operator the control technique also computed THD on the electrical grid based on voltage and current readings at PCC. To satisfy the demands of the system operator the Pareto's front solution is solved by the tertiary control. The control scheme technique was used to solve the energy management difficulty of optimizing a complete power quality evaluation index. To assess the parameters that increase MG power-quality., this efficacy indicator was developed using catastrophe decision theory. The findings revealed that the constrained capacity of grid-connected multi-functional inverters was successfully leveraged to conceal the ideal power quality criteria on MGs. As previously stated, MGs either feed power to the electric grid or operate autonomously in an islanded mode. The transitions of these 2 operation modes caused transient leaps which jeopardize electric grid reliability. The aforementioned urged the researchers in [75] to design a multi-objective control scheme supporting a linear quadratic based: optimal bumpless controllers with 2 degrees of freedom. Here, the MG's 2 operation modes control scheme is suggested. It measures voltages and currents. This proposed strategy enhances the reference and controller tracking error basing on optimum linear quadratic schemes. It usually covers the island mode and grid-connected mode demands such as power sharing, voltage or frequency synchronization with the traditional grid and active or reactive power flow, respectively.

In [76], the authors implemented an adaptive controller between MPC and ANN. It's done in an IBMG to keep the voltage and frequency in check. The Lyapunov criterion is further employed to ensure that the system remains stable. In order to developed a control law in inverter model on the basis of dynamics of the power controller, the inverter model is analysed by the MPC, to avoid uncertainties problem, ANNs is introduced, and the process conditions is usually administered by the controller. So for model validation, cases like sudden load changes, variations in gains, various load types and communication delays were tested by the authors.

Zio et al. [77] proposed a MOO methodology for integrating RESs into power transmission system based on differential development. The framework looked for the best scale and location for various microgrids, taking into consideration of both accessibility of power demands, primary micro sources (or RES), component failures, and other uncertainties, as well as the economic issue. The predicted global cost was used to gauge economic efficacy, whereas the DCVaR (CG) index was used to measure uncertainty (conditional value at risk deviation of global cost). The report also predicted that the optimization framework would be expanded in future, taking into account technical, social, and geographic factors. In [78], the chapter states that the power oscillations are produced by the continual change in generation and demand, which are commonly assessed by difference in steady and dynamic generation in the utility grid, which is given by:

$$v_D(x(t), t) = \pi_{\Delta P} \frac{|P_{\text{EPS}}(t) - P_{\text{EPS}}(t)(t - \Delta t)|}{\Delta t} \tag{11}$$

where ESP = electrical power system, πp = cost/kW squared of power P_{EPS} to PCC. This is usually adopted to keep the voltage and frequency in check and to guarantee stability in MG. Also, reduction on peak power brings economical benefits to the grid where its quantified objectives is given by:

$$v_P(x(t), t) = \pi_P P_{\text{EPS}}^2(t) \tag{12}$$

This strategy seeks an optimal control law to regulate the MG variables also it is the best option to minimize error of the o/p values.

In [79], the authors proposed a method for voltage control. The prime focus of this research is to find ways to help in the reactive power compensation region by strengthening local Volt/VAr controllers. Dynamical system approach is utilized (which is usually non-linear) with non-incremental local iVolt/VAr control and is used to validate equilibrium in casting the local control dynamics as a game where Each node acts as a 'selfish player,' using its local control function as a finest approach for minimizing its very own performance index and using the fixed-point hypothesis as well as a mapping justification directly related to contraction. DVR is defined by the authors in [80] as a mechanism that adjusts voltage sags by infusing three phase voltage in series and synchronism with the voltages at the distribution feeder. The primary goal of DVR is to adjust for voltage sag during fault occurrence. In this chapter, the authors proposed DVR approach with 2 goals, i.e., on voltage sags and voltage THD (total harmonic distortion) where a modified and enhanced version of PSO algorithm were used to measure the proportional integral controller coefficients of iDVR. The precedent urged the author in [81] to establish a suitable objective function to increased its performance, so they optimized by fuzzifying the objectives. In [81], bi-objective algorithm proves to provide more accurate result on DVR control study (Tables 3 and 4).

4 Conclusion and Future Scope

This chapter includes a discussion of MOOP as well as a definition of solutions. MOOP intelligent optimization methods like swarm based and EA methods were suggested and have been refined consistently for tackling the issues with solid execution/performances due to the orthodox, complexity, and mostly nonlinearity of MOOP. We also provide a quick overview of the MOO technique used for RESs integration to the power grid. The relevant performance indices for accessing the efficiency of MOO algorithms are summarized, as are several current test problems containing mathematical functions. In future work, we can explore renewable energy based smart grids by optimizing and controlling multiples objectives to correct the power grid issues without limiting to only power. The focus will be over integration of the various control schemes for effective, reliable and efficient use of renewable sources to fulfill the local-load demands. Moreover, it could also be expanded to designing scalable renewable energy sources with sources such as wind, sun, etc., to reduce energy cost from electricity and diesel from the utility grid. It could also be focused on energy optimization and energy storage of smart grid.

Table 3 Comparison of various multi-objective method on S_G

Method	Limitations	Benefits
Pareto-optimality [82–84]	The amount of weight vectors rises rapidly with the size of the objective space, and prior knowledge of the Pareto front in objective space is required	It provides robust platform for algorithm design and incorporates scalarization methods
ε-constraint [85–87]	The objective function encoding is confined to a few objectives, high-computational cost	It is a straightforward approach that has been used in a variety of engineering domains
Linear combination of weights [88]	When there's scarce information regarding the problem, it is hard to ascertain acceptable weight coefficients	Implementation, formulation and design, usability, and computing efficiency are all simplified
MOEA [89–91]	Each non-dominated solution in the Pareto optimum set projects unknown convergence behavior, poor performance and slow convergence	Environmental selection, directed search, and continual population optimization are all part of the performance index
MOGA [92, 93]	High-computational cost	The algorithm's versatility allows it to locate multiple-members of the Pareto optimum set 1 iteration
Global criterion [95]	The formulation of desirable goals necessitates additional computational work; if the goals are specified in a viable domain, a solution may be non-dominated, limiting their applicability	Since the Pareto ranking technique is not necessary, it is simple and effective
NSGA-II [96, 97]	More Pareto-optimal results are achieved when there are more members in the initial non-dominated set	Elitism preserves Pareto-optimal solutions, additional diversity control is not required

Table 4 Application of MOO methods in RES optimization

Methods	Optimized system	Objectives
EA	optimization of wind farm layout	capital investment, production of energy and operational costs
PSO	stand-alone RESs	life-time carbon-dioxide emissions and reduction of total net present cost (NPC)
MOPSO	hybrid micro-grid system	renewable energy reliability and costs
MOODE	RESs integration into electric power networks	sought for ideal position and size of various DG technologies, considering the uncertainties (or) risks linked to primary RESs, power consumption, economic concerns and component failures
IOA	DG systems	financial, technical objectives

References

1. Y.R. Sood, M. Hasmat, A. Ahmad, S. Srivastava, *Applications of Artificial Intelligence Techniques in Engineering—volume1: Part of the Advances in Intelligent Systems and Computing* (Springer, Berlin, 2018). https://doi.org/10.1007/978-981-13-1819-1
2. Y.R. Sood, M. Hasmat, A. Ahmad, S. Srivastava, *Applications of Artificial Intelligence Techniques in Engineering—volume2: Part of the Advances in Intelligent Systems and Computing* (Springer, Berlin, 2018). https://doi.org/10.1007/978-981-13-1822-1
3. H. Malik, A.K. Yadav, A. Iqbal, *Soft Computing in Condition Monitoring and Diagnostics of Electrical and Mechanical Systems: Part of the Advances in Intelligent Systems and Computing* (Springer, Berlin, 2020). https://doi.org/10.1007/978-981-15-1532-3
4. R.T. Marler, J.S. Arora, Survey of multi-objective optimization methods for engineering. Struct. Multidiscip. Optim. **26**(6) (2004). https://doi.org/10.1007/s00158-003-0368-6
5. Ó. Gonzales-Zurita, J.-M. Clairand, E. Peñalvo-López, G. Escrivá-Escrivá, Review on multi-objective control strategies for distributed generation on inverter-based microgrids. Integr. Manage. Distrib. Energy Resour. Power Syst. Energies **13**, 3483 (2020). https://doi.org/10.3390/en13133483
6. H. Malik, S. Bayhan, K. Abdellah, A. Riyaz, A. Iqbal, *Renewable Power for Sustainable Growth: Part of the Lecture Notes in Electrical Engineering* (Springer, Berlin, 2020). https://doi.org/10.1007/978-981-33-4080-0
7. H. Malik, A. Iqbal, N. Fatema, *Intelligent Data-Analytics for Condition Monitoring: Smart Grid Applications* (Elsevier, Amsterdam, 2021). https://doi.org/10.1016/C2020-0-02173-0
8. G.P. Rangaiah, A.F. Hoadley, Z. Feng, Multi-objective optimization applications in chemical process engineering: tutorial and review. Processes **8**, 508 (2020). https://doi.org/10.3390/pr8050508
9. Y. Cui, Y. Han, Q. Zhu, Z. Geng, Multi-objective optimization methods and application in energy saving. Energy **125**, 681 (2017) https://doi.org/10.1016/j.energy.2017.02.174
10. E. Zitzler, Evolutionary algorithms for multiobjective optimization: methods and applications (1999). https://doi.org/10.3929/ethz-a-003856832
11. J. Horn, Evolutionary Computation Applications. F1. 9. Multicriterion Decision Making, in *Handbook of Evolutionary Computation* (1997)
12. R.E. Steuer, Multiple criteria optimization. Theor Comput Appl (1986). https://doi.org/10.1002/oca.4660100109
13. C.L. Hwang, A.S. Md. Masud, *Methods for Multiple Objective Decision Making* (Springer, Berlin, 1979). https://doi.org/10.1007/978-3-642-45511-7_3

14. L. Haimes, Y. Yacov, *On a-Bicriterion-Formulation-of the-Problems of-Integrated System Identification and System Optimization. (IEEE 1971)*. https://doi.org/10.1109/TSMC.1971.4308298
15. F. Wang, L. Zhang, Q. Tang, X. He, Y. Rao, Fast construction method of Pareto non-dominated solution for multi-objective decision-making problem (2016)
16. B. Jürgen, S. Roman, K. Miettinen, K. Deb, *Multiobjective Optimization: Interactive and Evolutionary Approaches* (Springer, Berlin, 2008)
17. K. Miettinen, *Nonlinear Multiobjective Optimization* (Springer, Berlin, 2012)
18. I. Das, J.E. Dennis, Normal-boundary intersection: a new method for generating the Pareto surface in nonlinear multicriteria optimization problems. SIAM J. Optim. (1998) https://doi.org/10.1137/S1052623496307510
19. I.-Y. Amir, A. Messac, A. Mattson Christopher, The normalized normal constraint method for generating the Pareto frontier. Struct. Multidiscip. Optim. (2003)
20. J.F. Van Impe., P.M.M. Van Erdeghem, F. Logist, Efficient deterministic multiple objective optimal control of (bio) chemical processes. Chem. Eng. Sci. (2009)
21. S. Olariu, A.Y. Zomaya, Biology-derived algorithms in engineering optimization, in *Handbook of Bioinspired Algorithms and Applications* (2005)
22. F. Neri, C. Cotta, Memetic algorithms and memetic computing optimization: a literature review. Swarm Evolut. Comput. (2012)
23. A.A. Nikolaus, D.V. Arnold, Hansen, Evolution strategies, in *Springer Handbook of Computational Intelligence* (2015)
24. H.-P. Schwefel, B. Thomas, An overview of evolutionary algorithms for parameter optimization. Evolut. Comput. (1993)
25. J.W. Michael, A.J. Fogel, O. Lawrence, Intelligent decision making through simulation of evolution. Behav. Sci. (1966)
26. D. Jia, G. Zheng, M.K. Khan. An effective memetic differential evolution algorithm based on chaotic local search. Inform. Sci. **181**(15), 3175–3187 (2011)
27. Y. Wang, et al., Hybridizing particle swarm optimization with differential evolution for constrained numerical and engineering optimization. Appl. Soft. Comput. (2010)
28. L. Li, L. Wang, An effective differential harmony search algorithm for the solving non-convex economic load dispatch problems. Int. J. Electr. Power Energy Syst. (2013)
29. S. Deb, X.-S. Yang, Engineering optimisation by cuckoo search. Int. J. Math. Model. Numer. Optim. (2010)
30. K. Dervis, B. Basturk, On the performance of artificial bee colony (ABC) algorithm. Appl. Soft Comput. (2008)
31. A. Iqbal, et al., *Metaheuristic and Evolutionary Computation: Algorithms and Applications* (Springer Nature, Part of the Studies in Computational Intelligence, 2020). https://doi.org/10.1007/978-981-15-7571-6
32. T.-H. Huynh, *A Modified Shuffled Frog Leaping Algorithm for Optimal Tuning of Multivariable PID Controllers* (IEEE, 2008)
33. X.-S. Yang, Firefly Algorithms for multimodal optimization, in *International Symposium on Stochastic Algorithms* (Springer, Berlin, 2009)
34. S. Salcedo-Sanz, J. Del Ser, I. Landa-Torres, S. Gil-López, J.A. Portilla-Figueras, The coral reefs optimization algorithm: a novel metaheuristic for efficiently solving optimization problems. Sci. World J. (2014)
35. K.M. Passino, Bacterial foraging optimization. Int. J. Swarm Intel. Res. (IJSIR) **1**(no. 1), 1–16 (2010)
36. H. Duan, P. Qiao, Pigeon-inspired optimization: a new swarm intelligence optimizer for air robot path planning. Int. J. Intel. Comput. Cybern. **7**, 24–37 (2014)
37. E. Bonabeau, M. Dorigo, G. Theraulaz, *Swarm Intelligence: From Natural to Artificial Systems* (Oxford University Press, 1999)
38. M.M. Millonas, *Swarms, Phase Transitions, and Collective Intelligence* (1993)
39. E. Russel, J. Kennedy, *Particle Swarm Optimization.* ICNN'95 (IEEE, 1995)
40. E. Russell, J. Kennedy, A new optimizer using particle swarm theory, in *MHS'95* (IEEE 1995)

41. X. Li, A non-dominated sorting particle swarm optimizer for multi-objective optimization, in *Genetic and Evolutionary Computation Conference* (Springer, Berlin, 2003)
42. M. Salazar Lechuga, G.T. Pulido, C.A. Coello, *Handling Multiple Objectives with Particle Swarm Optimization* (IEEE 2004)
43. D.P. Vakharia, R.R. Venkata, V.J. Savsani, Teaching–learning-based optimization: a novel method for constrained mechanical design optimization problems. Comput. Aided Des. (2011)
44. H. Hosseinpour, M. Jabbari, T. Niknam, Multi-objective fuzzy adaptive PSO for placement of AVRs considering DGs, in *Power Engineering and Automation Conference '11* (IEEE, 2011)
45. T. Niknam, A. Kavousi Fard, A. Baziar, Multi-objective stochastic distribution feeder reconfiguration problem considering hydrogen and thermal energy production by fuel cell power plants. Energy (2012)
46. J.A. Martín García, A.J. Gil Mena, Optimal distributed generation location and size using a modified teaching–learning based optimization algorithm. Int. J. Electr. Power Energy Syst. (2013)
47. D. Dasgupta, *An Overview of Artificial Immune Systems and Their Applications* (1993)
48. N.C. Cortés, C.A. Coello, Solving multiobjective optimization problems using an artificial immune system, in *Genetic Programming and Evolvable Machines* (2005)
49. M. Gong, L. Jiao, D. Haifeng, L. Bo, Multi Objective immune algorithm with nondominated neighbor-based selection. Evol. Comput. **16**(2), 225–255 (2008)
50. M. Tayarani, R. Akbarzadeh, H. Mohammad, Magnetic optimization algorithms a new synthesis, in *IEEE World Congress on CI'08* (IEEE, 2008)
51. E. Rashedi, H. Nezamabadi-Pour, S. Saeid, GSA: a gravitational search algorithm. Inform. Sci. (2009)
52. R. Esmat, H. Nezamabadi-Pour, S. Saeid, Filter modeling using gravitational search algorithm. Eng. Appl. Artif. Intel. (2011)
53. S.-H. Hamed, Optimization with the nature-inspired intelligent water drops algorithm. Evolut. Comput. (2009)
54. S. Kirkpatrick, C. Daniel Gelatt, M.P. Vecchi, Optimization by simulated annealing. Science 220 (1983)
55. S.S. Sanghamitra, U.M. Bandyopadhyay, K. Deb, A simulated annealing-based multiobjective optimization algorithm: AMOSA. IEEE Trans. Evolut. Comput. (2008)
56. E. Atashpaz-Gargari, C. Lucas, Imperialist competitive algorithm: an algorithm for optimization inspired by imperialistic competition, in *IEEE Congress on Evolutionary Computation'07* (IEEE 2007)
57. F. Glover, Heuristics for integer programming using surrogate constraints. Decis. Sci. (1977)
58. F. Glover, T. Eric, *A User's Guide to Tabu Search.* Ann. Oper. Res. (1993)
59. M. Pablo, On evolution, search, optimization, genetic algorithms and martial arts: Towards memetic algorithms, in *Caltech Concurrent Computation Program* (1989)
60. R.G. Reynolds, An introduction to cultural algorithms, in *Proceedings of the Third Annual Conference on Evolutionary Programming* (World Scientific, NJ, 1994)
61. T.Y. Lin, Granular computing on binary relations II: rough set representations and belief functions. Rough Sets Knowl. Discover. **1** (1998)
62. R. Picard, *Affective Computing* (Massachusetts Institute of Technology, Cambridge, 1997)
63. F. Corno, S.R. Matteo, S. Giovanni, The selfish gene algorithm: a new evolutionary optimization strategy, in *Proceedings of ACM Symposium on Applied Computing'98* (1998)
64. Z.-L.Y. Xie, W.-J. Zhang, X. Feng, Social cognitive optimization for nonlinear programming problems, in *Proceedings of International Conference on MLC'02* (IEEE, 2002)
65. E. Zitzler, M. Laumanns, L. Thiele, K. Deb, L. Marco, Combining convergence and diversity in evolutionary multiobjective optimization. Evol. Comput. (2002)
66. A.G. Hernndez-Daz, L.V. Santana-Quintero, C.A. Coello, J. Molina, Pareto-adaptive ϵ-dominance. Evol. Comput. (2007)
67. K. Deb, D.K. Saxena, On *Finding Pareto-Optimal Solutions Through Dimensionality Reduction for Certain Large-Dimensional Multi-Objective Optimization Problems.* Kangal Report (2005)

68. D.K. Saxena, K. Deb, Non-linear dimensionality reduction procedures for certain large-dimensional multi-objective optimization problems: employing correntropy and a novel maximum variance unfolding, in *International Conference on EMCO'07* (Springer, 2007)
69. D. Brockhoff, E. Zitzler, Are all objectives necessary? On dimensionality reduction in evolutionary multiobjective optimization, in *Parallel Problem Solving from Nature-PPSN IX* (Springer, 2006)
70. Q. Zhang, H. Li, MOEA/D: a multiobjective evolutionary algorithm based on decomposition. IEEE Trans. Evol. Comput. (2007)
71. M. Mlakar, D. Petelin, T. Tušar, B. Filipič, GP-DEMO: differential evolution for multiobjective optimization based on Gaussian process models. Eur. J. Oper. Res. (2015)
72. H. Gharavi, M.M. Ardehali, S. Ghanbari-Tichi, Imperial competitive algorithm optimization of fuzzy multi-objective design of a hybrid green power system with considerations for economics, reliability, and environmental emissions. Renew. Energy **78**, 427–437 (2015)
73. Y. Teekaraman, R. Kuppusamy, S. Nikolovski, Solution for voltage and frequency regulation in standalone microgrid using hybrid multiobjective symbiotic organism search algorithm. Energies **12**, 2812 (2019)
74. Z. Zeng, H. Li, S. Tang, Y. Huan, R. Zhao, Multi-objective control of multi-functional grid-connected inverter for renewable energy integration and power quality service. IET Power Electron. (2016)
75. Y. Wu, J.M. Guerrero, J.C. Vasquez, Y. Wu, Bumpless optimal control over multi-objective microgrids with mode-dependent controllers. Energies 12, 3619 (2019)
76. D.O. Amoateng, M.A. Hosani, M.S. Elmoursi, K. Turitsyn, J.L. Kirtley, Adaptive voltage and frequency control of islanded multi-microgrids. IEEE Trans. Power Syst. (2017)
77. E. Zio, R. Mena, M. Hennebel, Y.-F. Li, A multi-objective optimization framework for risk-controlled integration of renewable generation into electric power systems. Energy (2016)
78. A. Sopinka, L. Pitt. British Columbia electricity supply gap strategy: a redefinition of self-sufficiency. Electricity J. **26**, 81–88 (2013)
79. X. Zhou, J. Tian, L. Chen, E. Dall'Anese, Local voltage control in distribution networks: Aigame-theoretic perspective, in *NAPS'16* (IEEE, 2016)
80. A.K. Ramasamy, R. Mukerjee, V.K. Ramachandaramuthy, R. K. Iyer, Dynamic voltage restorer for voltage sag compensation. J. Power Qual. Res. Group (2005)
81. M.R. Khalghani, M. Ali Shamsi-nejad, M. Hassan Khooban. Dynamic voltage restorer control using bi-objective optimisation to improve power quality indices. IET Sci. Measure. Technol. (2014)
82. M. Ross, C. Abbey, F. Bouffard, G. Jos: Multiobjective optimization dispatch for microgrids with a high penetration of renewable generation. IEEE Trans. Sustain. Energy **6**, 1306–1314 (2015)
83. H.R. Baghaee, M. Mirsalim, G.B. Gharehpetian, H.A. Talebi, Reliability/cost-based multi-objective Pareto optimal design of stand-alone wind/PV/FC generation microgrid system. Energy **115**, 1022–1041 (2016)
84. M. Sedighizadeh, A. Rezazadeh, R.V. Doyran, S.M. Mahdi Alavi, Optimal allocation of passive filters and inverter based DGs joint with optimal feeder reconfiguration to improve power quality in a harmonic polluted microgrid. Renew. Energy Focus (2020)
85. E.J. Agnoletto, D.S. De Castro, R.V. Neves, R.Q. Machado, V.A. Oliveira, *An Optimal Energy Management Technique Using the Constraint Method for Grid-Tied and Stand-Alone Battery-Based Microgrids* (IEEE, 2019)
86. A. Hamidi, D. Nazarpour, S. Golshannavaz, Multiobjective scheduling of microgrids to harvest higher photovoltaic energy. IEEE Trans. Industr. Inf. **14**(1), 47–57 (2017)
87. Z. Garroussi, R. Ellaia, J.-Y. Lucas, A metaheuristic for a bi-objective demand-side optimization for cooperative smart homes. Elect. Eng. (2020)
88. H.P. Geering, *Optimal Control with Engineering Applications* (Springer, 2007)
89. N.C. Ekneligoda, A.M. Dissanayake, Multiobjective optimization of droop-controlled distributed generators in DC microgrids. IEEE Trans. Indus. Inform. (2019)

90. D.I. Brandao, W.M. Ferreira, A.M.S. Alonso, E. Tedeschi, F.P. Marafão, Optimal multiobjective control of low-voltage AC microgrids: power flow regulation and compensation of reactive power and unbalance. IEEE Trans. Smart Grid (2019)
91. W.M. Ferreira, I.R. Meneghini, D.I. Brandao, F.G. Guimarães, Preference cone based multi-objective evolutionary algorithm to optimal management of distributed energy resources in microgrids. Appl. Energy **274**, 115326 (2020)
92. M. Manas, Optimization of distributed generation based hybrid renewable energy system for a DC Micro-grid using particle swarm optimization. Distrib. Gener. Altern. Energy J. **33** (2018)
93. P.P. Vergara, R. Torquato, L.C.P. Da Silva, Towards a real-time energy management system for a microgrid using a multi-objective genetic algorithm, in *Power & Energy Society General Meeting'15* (IEEE, 2015)
94. V.V.S.N. Murty, A. Kumar, Multi-objective energy management in microgrids with hybrid energy sources and battery energy storage systems. Protect. Control Modern Power Syst. (2020)
95. J. Lai, C. Hu, G. Li, X. Lu, H. Zhou, Networked-based distributed cooperative voltage control for power electronics interfaced microgrids, in *ICIEA'16* (IEEE, 2016)
96. B. Ye, X. Shi, X. Wang, H. Wu, Optimisation configuration of hybrid AC/DC microgrid containing electric vehicles based on the NSGA-II algorithm. J. Eng. (2019)
97. F. Zhao, J. Yuan, N. Wang, Dynamic economic dispatch model of microgrid containing energy storage components based on a variant of NSGA-II algorithm. Energies **12**, 871 (2019)

Investigation and Analysis of Harmonics in Different Control Techniques of PMSM Drives: An Application of Power System Health Monitoring

Ritu Tak and Satyendra Singh

Abstract Monitoring of the power system is one of most important accepts in the power grid. Power system monitoring includes the contingency analysis, security concern, optimal operation and the control of the wide area of the power system. Permanent magnet synchronous motor (PMSM) drive scheme is the best choice for the high power application and as a control drive in the power system monitoring. In this chapter, the simulation and modulation of the PMSM drive system have been done and investing or analyzing the presence of the harmonic in torque of the PMSM drive by the different control techniques. Harmonic distortion reduces the performance of the drive, and indirectly, it decreases the reliability of the power system. Basically, torque control method utilizing space and hysteresis vector pulse width modulation in PMSM is compared on the basis of harmonic distortion. The result of analyses shows torque control using a space vector pulse width modulation method is preferred in PMSM for high power applications or for the controlling drives in power system monitoring.

Keywords Torque · Harmonic distortion · Hysteresis current controller · Space vector pulse width modulation · Carrier frequency

1 Introduction

PMSM drives are mostly used as the controller drives in high power application such as in electric vehicles, for motion control system, in rail vehicle drive system because it produces a lot of electricity and is quite efficient as compared to induction motors. In many applications, PMSM drives replace the induction motor drives due to various unique qualities of PMSM [1–3]. In power system monitoring, PMSM drives is preferred as a control drive due to its superior dynamic performance. The existing challenge in the concern of reliability in power system depends on the controlling system of the contemporary area of the power system. In power system monitoring

R. Tak (✉) · S. Singh
School of Electrical Engineering, Bhartiya Skill Development University, Jaipur, India

H. Malik et al. (eds.), *Intelligent Data Analytics for Power and Energy Systems*, Lecture Notes in Electrical Engineering 802,
https://doi.org/10.1007/978-981-16-6081-8_14

including oscillatory stability monitoring analysis, and control techniques, in this paper analysis the presence of harmonics in different control strategy of PMSM drive [4, 5]. The performance of the controlling drives has an impact on the power system's reliability. The effectiveness of the drive depends upon the technique of controlling torque. The reference current trajectory is specified in the torque control arrangement and the current controller forces the load current to follow it [6, 7]. A comparative analysis of the PMSM with the various control strategies has been done and presented in the chapters [8–10] from the simulation results which show that the optimal torque per ampere control approach outperforms to the opponent. Soft computing and advance intelligent systems and computing technique are used for diagnostics of PMSM drive [11] for improving their performance. The two well-known schemes for current control in the PMSM are hysteresis and PWM technique. In the work [12], the torque control system for a surface-mounted PMSM was simulated and investigated. In articles [13, 14], researchers look into and create a mechanism for tracking torque in PMSM drives using the hysteresis current control approach. For PMSM drives, a three-level hysteresis current control method is presented for decreasing harmonic content [15]. The behavior of the PMSM control drives torque using field weakening scheme is analyzed in chapter [16]. PMSM control drives are most popular in controlling the mechanical or hydraulic system; a new scheme is proposed in chapter [17, 18] for reducing the harmonic level. Artificial intelligence techniques like neural network or fuzzy logic system is used in PMSM controller drives for improving the reliability or reducing the harmonics in the power system [19]; if the reliability of the controlling drives PMSM increases, then indirectly the reliability of the power monitoring system enhances. The scheme of torque and speed control of PMSM drive is analyzed by increasing the speed of the drive about 20–30% of nominal rpm at the condition of field weakened; the experiment result shows in [20, 21].

These below points highlighted the work done in this chapter:

- The design of the model for PMSM control drive has considered.
- The harmonic content has examined when modeling the space vector pulse width modulation (SVPWM) control system of the PMSM drive.
- The hysteresis control system of the PMSM drive has been modeled, and the harmonic content has been examined.
- At a diverse set of working situations such as varying the carrier frequency in SVPWM or vary the current window in the hysteresis control technique has been investigated for the performance of the PMSM control drive.
- The pulsation or the ripple content in the torque and current waveforms by using these two techniques have been analyzed and compared.

Organization of the chapter: Sect. 1 describes the literature review or the importance of control drive in power system monitoring. In Sect. 2, the proposed approach is described by using the flow chart; it explains stepwise the work done in this chapter. The super dynamic model or the mathematical presentation of the physical model of PMSM controller drive is discussed in Sects. 3 and 4. The two most popular control techniques of PMSM drive are explained in Sect. 5. In comparative way, the harmonic analysis of SVPWM control scheme and hysteresis control scheme are shown in Sect. 6. In Sect. 7, from the result of analyzes, chapter is concluded.

2 Proposed Approach

The flow chart explains the holistic work done in the chapter in the step-by-step manner.

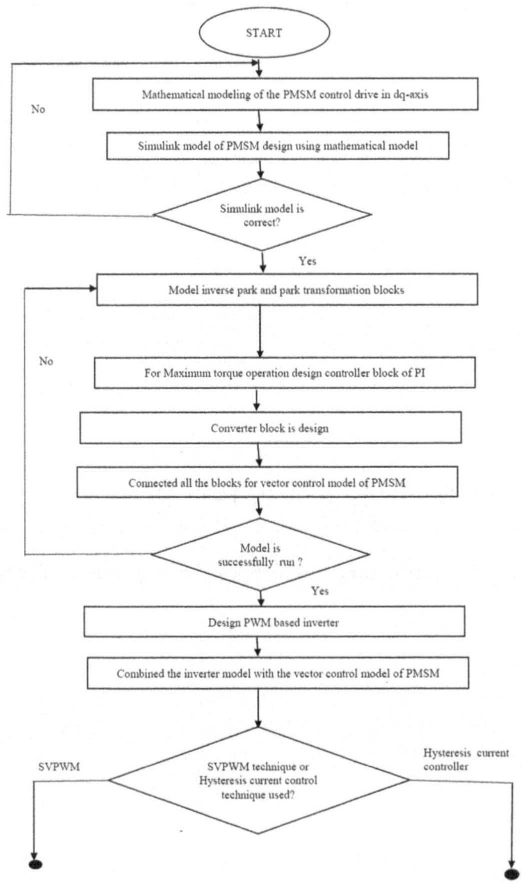

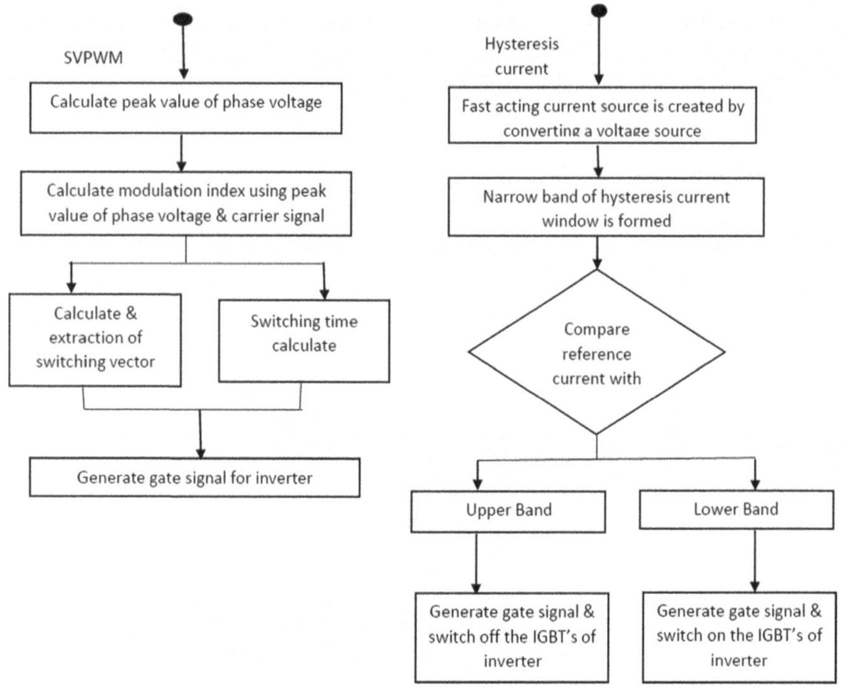

3 Mathematical Derivation of PMSM Drive

The stator voltages and stator flux linkages of the *d*-*q* axis of PMSM drive can be written as [22]:

$$V_{qs} = R_q i_{qs} + \frac{\mathrm{d}}{\mathrm{d}t}\lambda_{qs} \tag{1}$$

$$V_{ds} = R_d i_{ds} + \frac{\mathrm{d}}{\mathrm{d}t}\lambda_{ds} \tag{2}$$

$$\lambda_{qs} = L_{qq} i_{qs} + L_{qd} i_{ds} + \lambda_{af} \sin\theta_{ir} \tag{3}$$

$$\lambda_{ds} = L_{dq} i_{qs} + L_{qd} i_{ds} + \lambda_{af} \cos\theta_{ir} \tag{4}$$

In balanced condition, the resistance of *q*-axis is equal to resistance of *d*-axis. Let it be supposed R_{qd}; the *q*-axis and *d*-axis inductances are also in the surface-mounted PMSM drive; θ_{ir} is the instantaneous rotor situation.

An equation for the voltage of stator in terms of the rotor position is:

$$\begin{bmatrix} V_{qs} \\ V_{ds} \end{bmatrix} = R_s \begin{bmatrix} i_{qs} \\ i_{ds} \end{bmatrix} + \begin{bmatrix} \frac{1}{2}(L_q + L_d) & 0 \\ 0 & \frac{1}{2}(L_q + L_d) \end{bmatrix} \frac{\mathrm{d}}{\mathrm{d}t} \begin{bmatrix} i_{qs} \\ i_{ds} \end{bmatrix} + \lambda_{af}\omega_r \begin{bmatrix} \cos\theta_{ir} \\ -\sin\theta_{ir} \end{bmatrix} \tag{5}$$

The rotor orientation casings are now connected to the stationary orientation edges as follows:

$$i_q d_s = [T^r] i_q^r d_s \tag{6}$$

$$v_q d_s = [T^r] v_q^r d_s \tag{7}$$

$$T^r = \begin{bmatrix} \cos\theta_{ir} & \sin\theta_{ir} \\ -\sin\theta_{ir} & \cos\theta_{ir} \end{bmatrix} \tag{8}$$

$$\begin{bmatrix} v_{qs}^r \\ v_{ds}^r \end{bmatrix} = \begin{bmatrix} R_s + L_q p & \omega_r L_d \\ -\omega_r L_q & R_s + L_d p \end{bmatrix} \begin{bmatrix} i_{qs}^r \\ i_{ds}^r \end{bmatrix} + \begin{bmatrix} \omega_r \lambda_{af} \\ 0 \end{bmatrix} \tag{9}$$

$$\begin{bmatrix} i_{qs}^r \\ i_{ds}^r \\ i_0 \end{bmatrix} = \frac{2}{3} \begin{bmatrix} \cos\theta_{ir} & \cos\left(\theta_{ir} - \frac{2\pi}{3}\right) & \cos\left(\theta_{ir} + \frac{2\pi}{3}\right) \\ \sin\theta_{ir} & \sin\left(\theta_{ir} - \frac{2\pi}{3}\right) & \sin\left(\theta_{ir} + \frac{2\pi}{3}\right) \\ \frac{1}{2} & \frac{1}{2} & \frac{1}{2} \end{bmatrix} \begin{bmatrix} i_{as} \\ i_{bs} \\ i_{cs} \end{bmatrix} \tag{10}$$

4 Scheme of Controlling Torque in PMSM Drive

The stator current that produces flux can be estimated using the current locations and the torque perspective as [23]

$$\begin{bmatrix} i_T^* \\ i_f^* \end{bmatrix} = i_{\text{sref}} \begin{bmatrix} \sin\delta \\ \cos\delta \end{bmatrix} \tag{11}$$

By transformation of *d*-*q* axis to *abc* axis the references of the phase current are obtained as

$$\begin{bmatrix} i_{as}^* \\ i_{bs}^* \\ i_{cs}^* \end{bmatrix} = \begin{bmatrix} \cos\theta_{ir} & \sin\theta_{ir} \\ \cos\left(\theta_{ir} - \frac{2\pi}{3}\right) & \sin\left(\theta_{ir} - \frac{2\pi}{3}\right) \\ \cos\left(\theta_{ir} + \frac{2\pi}{3}\right) & \sin\left(\theta_{ir} + \frac{2\pi}{3}\right) \end{bmatrix} \begin{bmatrix} i_T^* \\ i_f^* \end{bmatrix} = i_{\text{sref}} \begin{bmatrix} \sin(\theta_{ir} + \delta_{\text{ref}}) \\ \sin\left(\theta_{ir} + \delta_{\text{ref}} - \frac{2\pi}{3}\right) \\ \sin\left(\theta_{ir} + \delta_{\text{ref}} + \frac{2\pi}{3}\right) \end{bmatrix} \tag{12}$$

The proportion integral controller is used to control and make the errors in the phase currents to be zero or yield the reference phase voltage. The reference voltage generated by the PI controller creates the gating indications for the inverter through

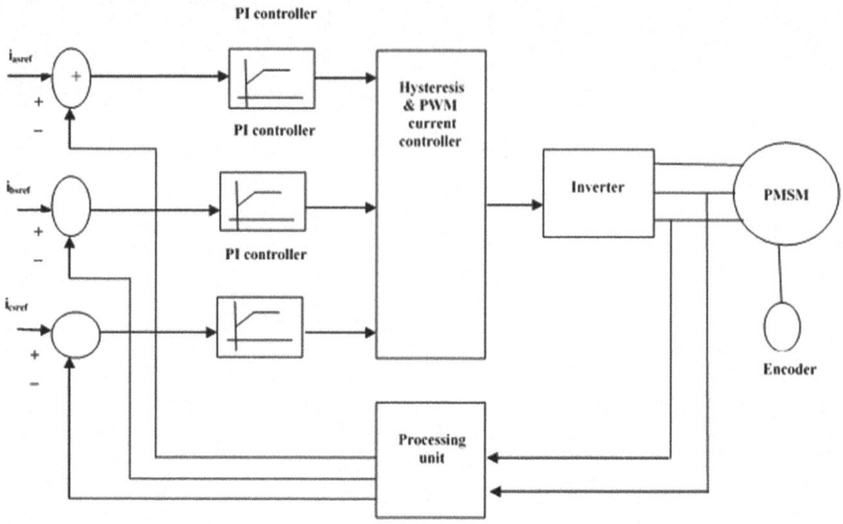

Fig. 1 Torque controlling scheme of PMSM drive

the current controller. The two popular technique hysteresis and SVPWM current controllers are discussed in this chapter.

The schematic flow chart of basic torque design is shown in Fig. 1.

5 Implementation of Controlling Schemes of Torque in PMSM Drive

5.1 Hysteresis Current Controller

A fast-acting current source is produced in the hysteresis controller by translating a voltage source. A small band of excursion is generated from the intended value to regulate the actual current with a hysteresis controller. By summing the stator phase current inside a limited envelope, a hysteresis window is determined and subtracting the preset deviation of current Δi from the stator phase current reference.

As a result, the hysteresis phase current controller's switching logic is as follows:

$$i_{as} - i_{as}^{*} \geq \Delta i \quad \text{set } v_{ao} = \frac{V_{dc}}{2} \tag{13}$$

$$i_{as} - i_{as}^{*} \leq -\Delta i \quad \text{set } v_{ao} = \frac{-V_{dc}}{2}$$

Other stages' implementation can be done in the same way. The voltages on the d- and q-axes, as well as the stator current, may be calculated using the phase and line voltages. The d- and q-axis stator currents, as well as torque and mutual flux connections from the rotor reference frames, are used to calculate stator phase currents.

In the hysteresis controller, only the instantaneous peak current and windowed current are dependent on the magnitude of the hysteresis. The hysteresis current controller performance increases with decreasing current windows as the analysis shows.

5.2 *SVPWM Current Controller*

Harmonics and variations in the fundamental component can be controlled by varying the duration of the useful input voltage. This is achieved by altering the pulse width of the inverter's gate signal, which is known as PWM. There have been several PWM schemes to employ in motor drives. In general, all PWM systems seek to enhance fundamental harmonics while selectively eliminating a few lesser harmonics. The voltage signal generated by the current controller serves as the reference signal for the PWM controller, which interacts with the carrier signal to create gate pulses [24].

The logic of switching for SVPWM to generate the gating signal of inverter as

$$v_{\text{am}} = \frac{V_{\text{dc}}}{2}\frac{v_{\text{pref}}}{v_{\text{pc}}} \tag{14}$$

where

v_{pref} is a phase peak value.
v_{pc} is the carrier signal peak value.

The modulation index of PWM is define as

$$m = \frac{v_{\text{pref}}}{v_{\text{pc}}} \tag{15}$$

By changing the carrier signal, peak value changes modulation index and that changes the amplitude of the fundamental quantity.

6 Analysis of Harmonics in Different Control Techniques of PMSM Drives

The parameter of the PMSM controller drive for analyzing is set to as specified in Table 1.

Table 1 Parameters of PMSM drive

Parameters	Representation	Significance
Rated speed	ω_r	300 rpm
Poles	p	4
q-axis inductance	L_q	0.006 H
Stator resistance	R_s	1.67 Ω
Friction coefficient	B	0.01 Nms
d-axis inductance	L_d	0.0043 H
Inertia moment	J	0.007 kg/m^2
Rotor flux linkage	λaf	0.1400 Wb

The torque-controlled schemes are analyzed for improving the performance of the PMSM drive. Moreover, it enhanced the reliability of the power monitoring arrangement.

6.1 *Torque Control Scheme with Hysteresis Current Controller*

The response of PMSM is observed with the hysteresis current controller for change in hysteresis current window. The variation of torque, speed, current and flux linkage are analyzed for different values of hysteresis current window. The THD levels of the phase current in PMSM compare with different values of the hysteresis current window.

Figure 2 shows the reaction of the PMSM drive with a constant velocity of 314 rad/s is preferred as base velocity at the hysteresis band of 0.6 p.u. The parameters are represented in normalized units on the graph (p.u.).

The THD in phase *a* current with the hysteresis band of 0.6 p.u. is shown in Fig. 3 is 20.22%.

The response of PMSM drive with the hysteresis band 2.4 p.u. and its THD spectrum is revealed in Figs. 4 and 5 correspondingly. Similarly, with the hysteresis band 3.6 p.u. the response of PMSM and its THD spectrum is shown in Figs. 6 and 7.

Table 2 gives the comparative analyzes of the THD level in phase current of the PMSM drive by different values of the hysteresis current window.

The torque variation with different values of the hysteresis current window is shown in Fig. 8. The simulation result shows that the hysteresis current window; if narrow, then the pulsation in torque is less, and it follows the reference torque.

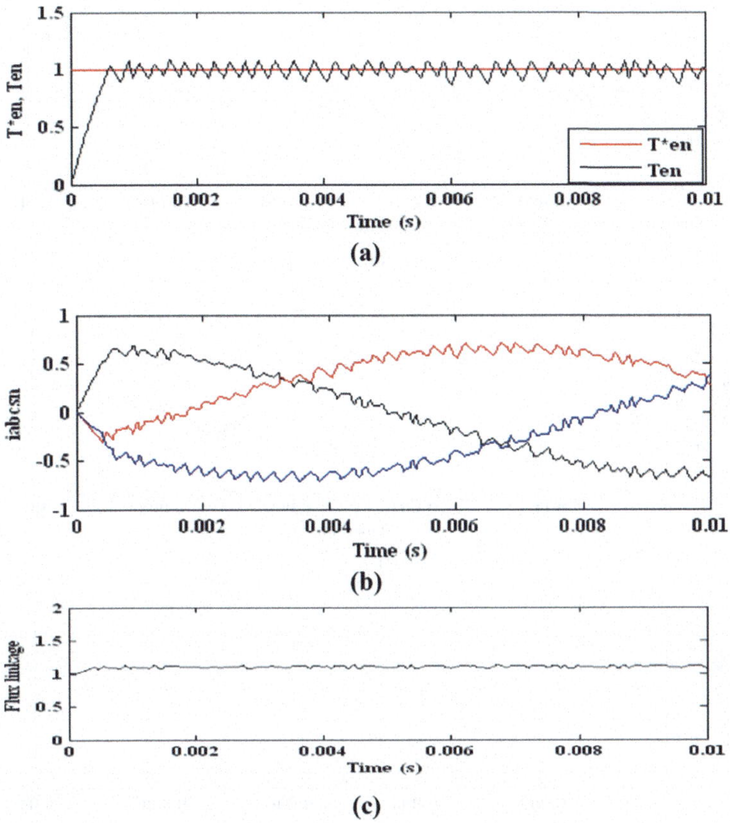

Fig. 2 At hysteresis band of 0.6 p.u. **a** Response for torque, **b** response for phase current, **c** response for flux

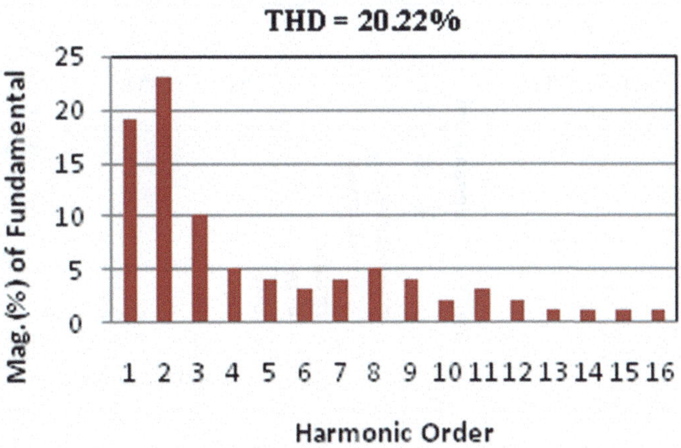

Fig. 3 THD spectrum with hysteresis band 0.6 p.u.

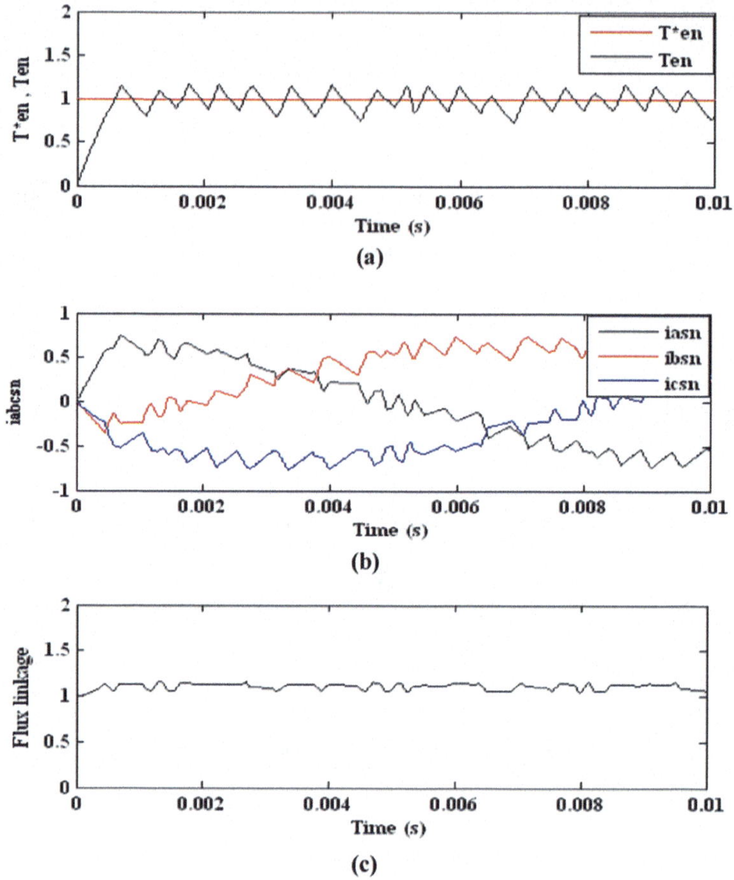

Fig. 4 At hysteresis band of 2.4 p.u. **a** Response for torque, **b** response for phase current, **c** response for flux

Fig. 5 THD spectrum with hysteresis band 2.4 p.u.

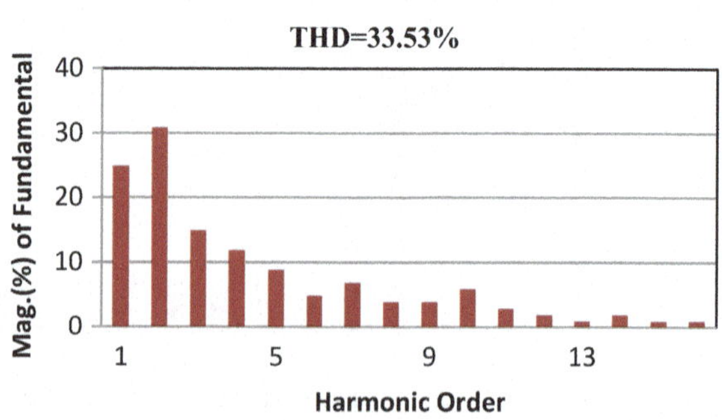

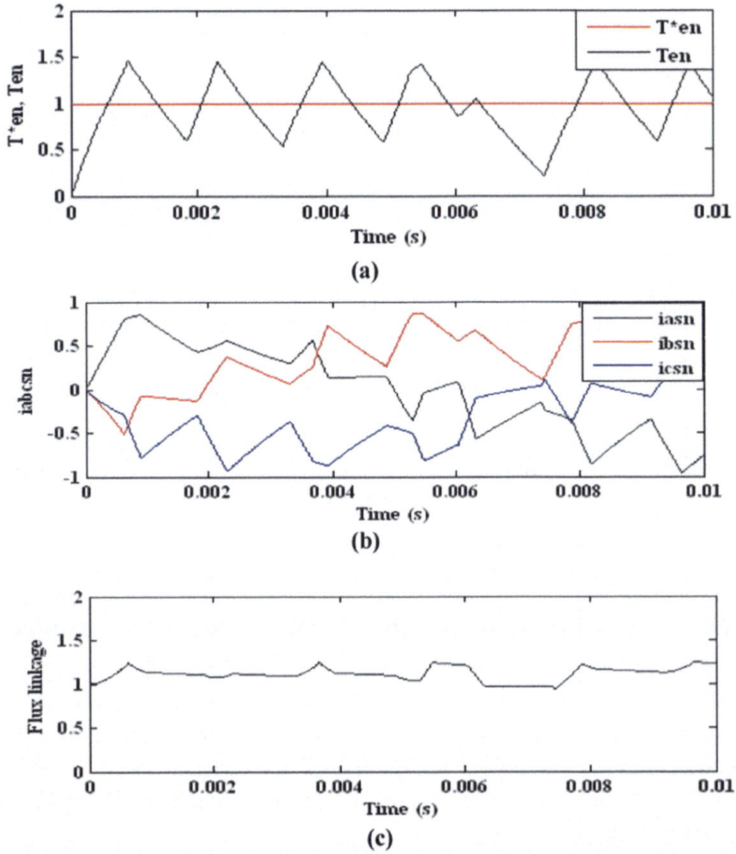

Fig. 6 At hysteresis band of 3.6 p.u. **a** Response for torque, **b** response for phase current, **c** response for flux

Fig. 7 THD spectrum with hysteresis band 3.6 p.u.

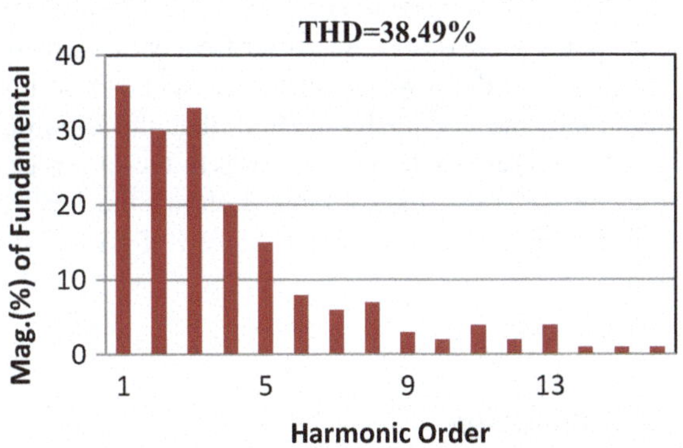

Table 2 Analysis of THD in the current of the PMSM drive with hysteresis current controller

S. No.	Control of hysteresis current	
	Hysteresis band (p.u.)	THD in current %
1	$\Delta i = 0.6$	20.22
2	$\Delta i = 2.4$	33.53
3	$\Delta i = 3.6$	38.49

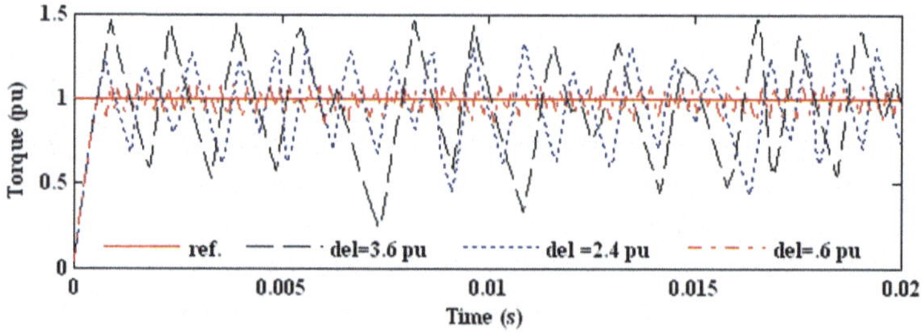

Fig. 8 Torque variation at different hysteresis bands

6.2 *Torque Control Scheme with SVPWM Current Controller*

SVPWM is used to regulate the torque in PMSM drives instead of a hysteresis current controller. The torque, speed, current and flux connections are all affected by changes in the carrier frequency of SVPWM. 314 rad/s is the constant speed of the PMSM drive. Figure 9 depicts the torque, speed and current responses in SVPWM with an 8 kHz carrier frequency.

As shown in Fig. 10, the third harmonic content is very high in phase current of PMSM drive at 8 kHz carrier frequency. The carrier frequency of the PMSM drive is increased to 15 and 25 kHz for analysis. Figures 11 and 13 depict the drive's reaction and from THD bar graph, i.e., Figs. 12 and 14 at these two carrier frequency shows that ripples decreases as the carrier frequency increases.

In Fig. 15, the response of torque is shown at three different values of carrier frequency, and the analyses show that the dissimilarity in torque between the actual and reference torque decreases as the carrier frequency increases in SVPWM controlled PMSM drives. Table 3 shows the level of the THD in PMSM drive considerably reduces up to the 11.32% if SVPWM technique is used in PMSM drive.

7 Conclusion

In this chapter, analysis of harmonic content in the different control strategy is done and proposed the method to reduce the content of harmonics in the control strategy

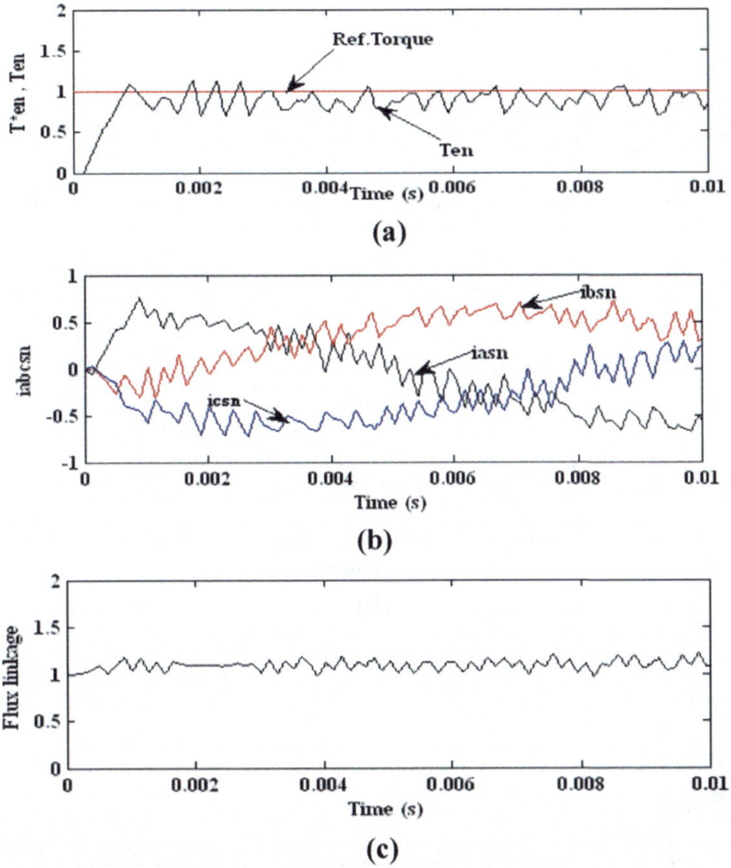

Fig. 9 With an 8 kHz carrier frequency **a** Response for torque, **b** response for phase current, **c** response for flux

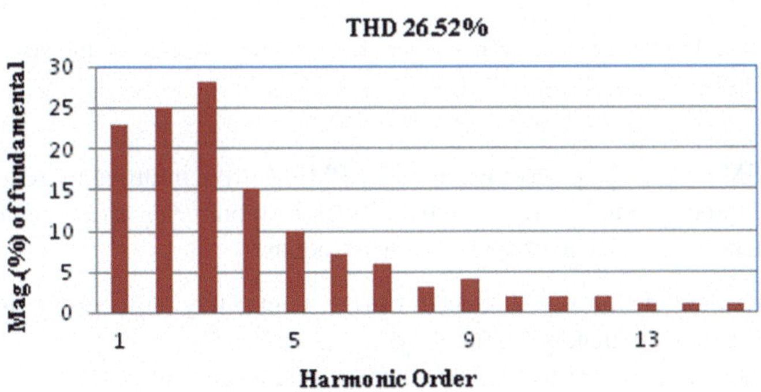

Fig. 10 THD spectrum with 8 kHz carrier frequency

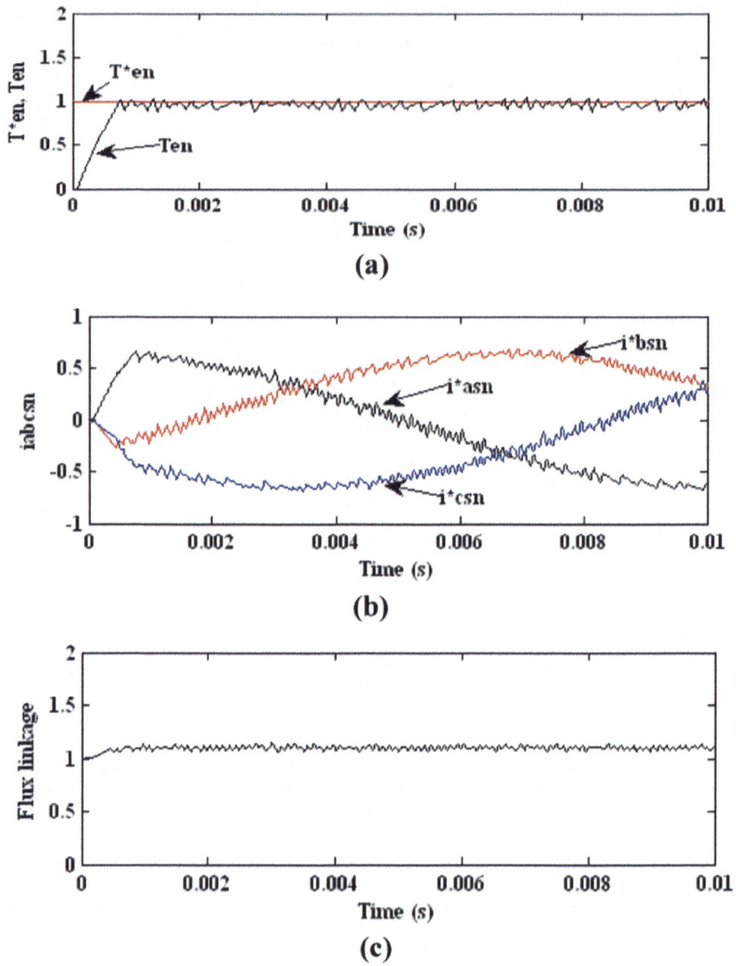

Fig. 11 With an 15 kHz carrier frequency **a** Response for torque, **b** response for phase current, **c** response for flux

of the PMSM drive. The performance of the PMSM drive improve by reducing the harmonic content and indirectly the reliability of the monitoring arrangements of the power system increase. The chapter is concluding as:

- The harmonic distortion produced by the controlling devices in the power monitoring system makes it unhealthy.
- A comparative analysis of SVPWM and hysteresis current control technique of PMSM control drive is presented in this chapter.
- THD factor calculated at different points by varying the hysteresis band and conclude it if increases the band of hysteresis the pulsation and THD factor increases in flux and torque.

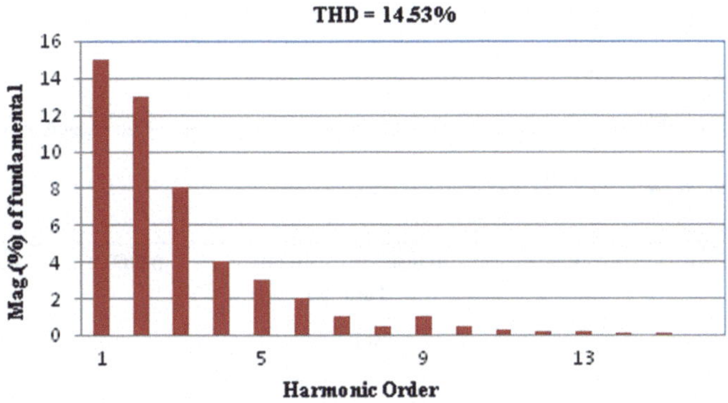

Fig. 12 THD spectrum with 25 kHz carrier frequency

- By varying the carrier frequency of SVPWM, THD factor is calculated at different levels then find it by increasing carrier frequency the pulsation and ripple in torque decreases.
- By improving the controlling system of the PMSM drive, the presentation of the drive is enhanced.
- For the application of the PMSM drive in power monitoring system, it must be high performance drive with minimum ripple factor, and also the torque pulsation is negligible.

The analyses result concluded that PMSM drive by SVPWM controller at higher carrier frequency is preferred for the application of the power monitoring system as this control drives gives the high performance with reduce ripples or torque pulsation.

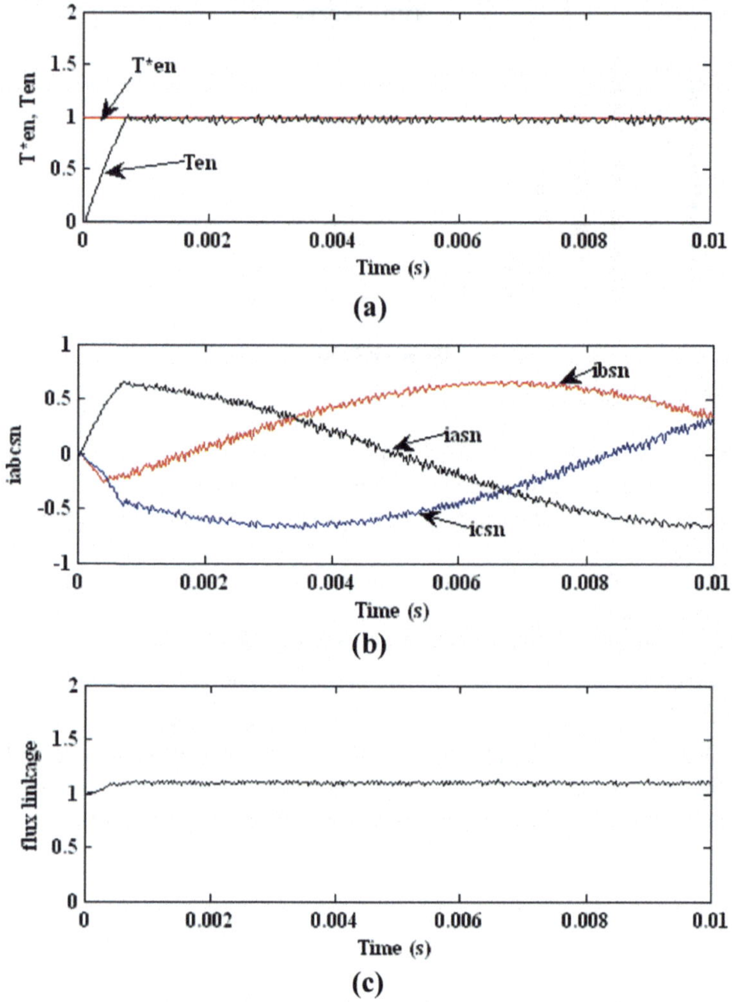

Fig. 13 With an 25 kHz carrier frequency **a** Response for torque, **b** response for phase current, **c** response for flux

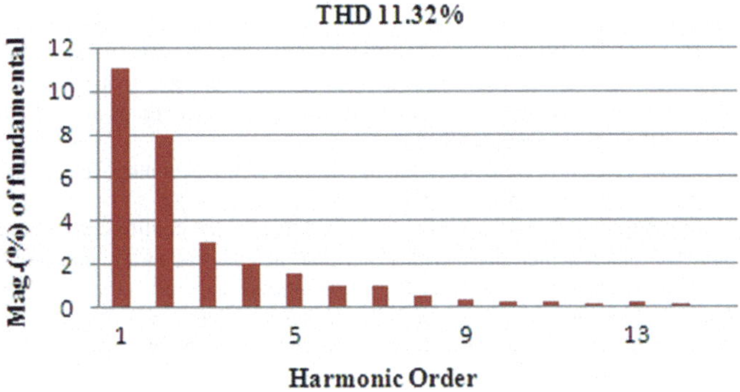

Fig. 14 THD spectrum with carrier frequency of 25 kHz

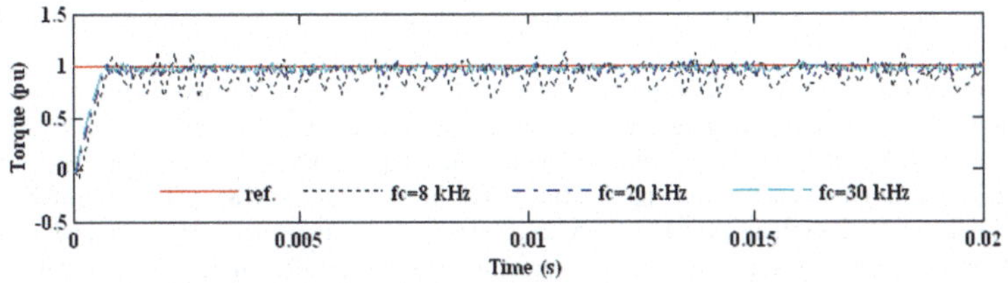

Fig. 15 Torque variation at diverse carrier frequency

Table 3 THD analysis in an SVPWM-controlled PMSM drive

S. No.	Control technique of SVPWM current	
	Carrier frequency (kHz)	THD in current (%)
1	8	26.52
2	15	14.53
3	25	11.32

References

1. G. Pellegrino, A. Vagati, B. Boazzo, P. Guglielmi, Comparison of induction and PM synchronous motor drives for EV application including design examples. IEEE Trans. Ind. Appl. **48**(6), 2322–2332 (2012). https://doi.org/10.1109/TIA.2012.2227092
2. A. Balashanmugham, M. Maheswaran, Permanent-Magnet synchronous machine drives. Mach. Electr. Mob. Solut. (2019). https://doi.org/10.5772/intechopen.88597
3. S.S. Rauth, B. Samanta, Comparative analysis of IM/BLDC/PMSM drives for electric vehicle traction applications using ANN-Based FOC, in *IEEE 17th India Council International Conference (INDICON)*, pp. 1–8 2020. https://doi.org/10.1109/INDICON49873.2020.9342237
4. A. Iqbal et al., *Renewable Power for Sustainable Growth* (Springer Nature, Part of the Lecture Notes in Electrical Engineering), vol. 723, p. 805. ISBN 978-981-33-4080-0. https://doi.org/10.1007/978-981-33-4080-0

5. N. Fatema et al., *Intelligent Data-Analytics for Condition Monitoring: Smart Grid Applications* (Academic Press, 2021). ISBN 978-0-323-85510-5. https://doi.org/10.1016/C2020-0-02173-0
6. I.M. Alsofyani, K. Lee, Predictive torque control based on discrete space vector modulation of PMSM without flux error-sign and voltage-vector lookup table. Electronics **9**, 1542 (2020). https://doi.org/10.3390/electronics9091542
7. P. Karamanakos, T. Geyer, Guidelines for the design of finite control set model predictive controllers. IEEE Trans. Power Electron. **35**(7), 7434–7450 (2020)
8. L. Tang, L. Zhong, M.F. Rahman, Y. Hu, A novel direct torque controlled interior permanent magnet synchronous machine drive with low ripple in flux and torque and fixed switching frequency. IEEE Trans. Power Electron. **19**, 346–354 (2004)
9. C. Ogbuka, M.N. Caiethna, M. Agu, A high performance hysteresis current control of a permanent magnet synchronous motor drive. Turkish J. Electr. Eng. Comput. Sci. (2017). https://doi.org/10.3906/ELK-1505-160
10. J.A. Güemes, A.M. Iraolagoitia, J.I. Del Hoyo, P. Fernandez, Torque analysis in permanent-magnet synchronous motors: a comparative study. IEEE Trans. Energy Convers **26**(1) (2011)
11. A.K. Yadav et al., *Soft Computing in Condition Monitoring and Diagnostics of Electrical and Mechanical Systems* (Springer Nature, Part of the Advances in Intelligent Systems and Computing, 2020), vol. 1096, p. 496, ISBN 978-981-15-1532-3. https://doi.org/10.1007/978-981-15-1532-3
12. R. Pindoriya, A.K. Mishra, B.S. Rajpurohit, R. Kumar, Performance analysis of control strategies of permanent magnet synchronous motor, in *IEEE Conference TENCON* (2016)
13. R. Tak, S.Y. Kumar, B.S. Rajpurohit, Modeling the torque control scheme of surface mounted permanent magnet synchronous motor. Int. J. Mech. Eng. Technol. **8**(5), 1137–1146 (2017)
14. O. Cosmas, N. Cajethan, A. Marcel, A fast hysteresis current-controlled permanent magnet synchronous motor drive based on field orientation. J. Electr. Eng. **67**(2), 69–77 (2016)
15. J.M. Lazi, Z. Ibrahim, M. Sulaiman, Mean and differential torque control using hysteresis current controller for dual PMSM drives. J. Theor. Appl. Inf. Technol. **33**(1) (2011)
16. H. Lin, T.A. Lipo, B. Kwon, S.R. Cheon, Three level hysteresis current control for a three-phase permanent magnet synchronous motor drive, in *Power Electronics and Motion Control 7th International IEEE Conference in China*, 2012
17. A. Ebrahimi, M. Maier, N. Parspour, Analysis of torque behaviour of Permanent magnet synchronous motor in field weakening operation, in *Power and Energy Conference at USA* (IEEE, 2013)
18. G. Pei, L. Li, X. Gao, J. Liu, R. Kennel, Predictive current trajectory control for PMSM at voltage limit. IEEE Access **8**, 1670–1679 (2020)
19. S. Srivastava et al., *Applications of Artificial Intelligence Techniques in Engineering—Volume 1* (Springer Nature, Part of the Advances in Intelligent Systems and Computing, 2018), vol. 698, p. 643. ISBN 978-981-13-1819-1. https://doi.org/10.1007/978-981-13-1819-1
20. Y. Miyama, M. Hazeyama, S. Hanioka, N. Watanabe, A. Daikoku, M. Inoue, PWM carrier harmonic iron loss reduction technique of permanent-magnet motors for electric vehicles. IEEE Trans. Indust. Appl. **52**(4) (2016)
21. R. Tak, S.Y. Kumar, B.S. Rajpurohit, Speed control schemes of four quadrant operating PMSM drive for electric vehicles applications. Int. J. Appl. Eng. Res. **12**(15), 5156–5162 (2017). ISSN 0973-4562
22. Tomaoz Rudnicki, Robert Czerwinski, Dariuoz Polok and Andrzej Sikora, "Performance analysis of a PMSM drive with torque and speed control" , Mixed Design of Integrated circuits and systems, 22nd IEEE International conference, 2015.
23. R. Tak, S.Y. Kumar, B.S. Rajpurohit, Dynamic modeling of surface mounted permanent synchronous motor for servo motor application. Int. J. Sci. Res. Eng. Technol. **6**(8) (2017)
24. R. Tak, S. Shukla, B.S. Rajpurohit, Performance analysis permanent magnet synchronous motor drives with pulse width modulation control technique, in *Algorithms for Intelligent Systems* (Springer). ISSN: 2524-7565

Design and Analysis of Artificial Intelligence Method for MPPT Control: An Application of Grid Connected PV System

Devansh Bhatnagar and Dinanath Prasad

Abstract This chapter emphasizes on design and analysis of intelligent method of MPPT control for integrating multilevel inverter with grid. There are limitations to the efficiency of PV system on account of varying inputs in the surrounding. After research and analysis, the course of action is to execute intelligent methods for maximum power point tracking from the PV array. In this chapter, we proposed fuzzy logic controller for extracting maximum power from designed PV array. It will give required duty as gating signal to boost converter switch and accordingly voltage at maximum power will be generated. In next stage, i.e. DC to AC conversion stage, three-level cascaded multilevel inverter is employed with SPWM technique for controlling switching gate pulse. Afterwards, LCL filter is employed for integrating inverter with single-phase grid. It subsequently enhances the input grid current profile. It also prevents resonance problem between grid and inverter side that leads to overvoltage conditions in grid. Whole work in terms of design and simulation is accomplished through iMATLAB/SIMULINK environment.

Keywords Fuzzy logic control · Partially shaded condition · MPPT · Boost converter · Three-level cascaded inverter · THD · LCL filter design · Grid integration

Nomenclature

AC Alternating current

D. Bhatnagar · D. Prasad (✉)
Ajay Kumar Garg Engineering College, Ghaziabad, India
e-mail: Prasaddinanath@akgec.ac.in

D. Bhatnagar
e-mail: Devansh1721099@akgec.ac.in

D. Prasad
Delhi Technological University, Delhi, India

H. Malik et al. (eds.), *Intelligent Data Analytics for Power and Energy Systems*, Lecture Notes in Electrical Engineering 802,
https://doi.org/10.1007/978-981-16-6081-8_15

Tmp	Ambient Temperature
C	Celsius
CASD	Cascaded
C_{mpp}	Current at maximum power point
DBC	DC to DC Boost Converter
DC	Direct Current
FISY	Fuzzy Inference System
FLC	Fuzzy logic control
FL	Fuzzy Logic
MPPT	Maximum power point tracking
MLI	Multilevel inverter
MPP	Maximum Power Point
MF	Membership functions
MPP	Maximum power point
VOC	Open circuit voltage
PV	Photovoltaic system
P&O	Perturb & observe
RES	Renewable energy system
SIM	SIMULINK
SPV	Solar Photovoltaic array
I_{rrd}	Solar Irradiations
SPM	Sinusoidal pulse width modulation
ISC	Short circuit current
THD	Total Harmonic Distortion
V_{mpp}	Voltage at maximum power point
W/m^2	Watt per square meter
W_p	Watt peak

1 Introduction

National mission and UN's mission aim towards development of self-sustainable strategy based on clean, carbon-free renewable energy such as solar for electrical energy generation owing to increase in consumption realized the need for integration of RES with the grid to meet World's energy requirement without depending on fossil fuels. Hence, integration of RES is future of energy sector, which can be seen as one Sun one World one Grid policy.

The main challenges arouse during deriving of power from solar cells is efficiency, quality, cost and adaptability [1]. To avoid these shortcomings, the much efficient and intelligent techniques have been discovered and deployed in MPPT controllers in a recent decade [1–3]. Response of such intelligent methods is observed as much accurate and optimized for drawing out maximum power from solar PV under varying conditions. Although, MPPT controller based on P&O method is cost effective and able to obtain maximum available power output from designed PV array. However, it is very inefficient and slow in terms of its dynamic response, performance under steady-state condition, accuracy and adaptability to sudden change.

This chapter will be emphasizing on design and comparative analysis of intelligent fuzzy logic control-based approach utilized in MPPT controllers [4–8]. The DC–DC power converters are employed to enhance the performance of RES. Based on the duty signal received by FL controller, DBC will extract the voltage at maximum power by matching source and load impedance [9, 10]. Hence, FL-based algorithm is used in MPPT applied to PV array connected with boost power converter is simulated in MATLAB [11–17]. Now, as we know that more than 90% of loads are inductive and non-linear in nature, power quality issues like current harmonics [THD] and sag voltage are prevalent. Here, we developed three-level CASD H-bridge voltage source inverter with output voltage level as Vdc, 0, − Vdc. Hence, we will get improved harmonic profile of input current to grid, lower electromagnetic interference as the levels of inverter are increased [18–20].

To get stabilized output in terms of harmonic profile [THD] and voltage regulation against the variations in supply and load side, advanced control approaches are utilized to control the gate pulse of switches of inverter employed [21, 22]. Owing to high switching frequency and application in high power situations, IGBT are employed in H-Bridges of three-level inverter with SPM unipolar technique as control scheme for switches [23–31].

Now for successful grid integration, LCL filter are implemented between inverter and grid. This act as decoupler between grid and inverter side inductance to avoid resonance problems due to over voltages [32–36]. LCL also allows reduction of high-frequency current harmonics to grid thereby enabling high sinusoidal current to grid. In this chapter, we designed 5 kWp SPV with DBC employing FL-based MPPT controller feeding CASD three-level inverter, and subsequently, LCL filter will allow successful grid integration of single phase 230 V 50 Hz supply.

- We have executed MPP tracking using fuzzy control logic for integrating multilevel inverter with grid-connected PV system which is novelty of this work.
- This work discussed comparative analysis of different MPPT techniques and concluded efficient, accurate and rapid ways to incorporate system in terms of dynamic variations.
- Performance of FL-based MPPT is calculated and observed in terms of its tracking speed stability, overshoot and effectiveness in extracting maximum power from array.
- THD analysis of grid current is shown subsequently percentage harmonics is brought within limits as per IEEE standards.

1.1 Proposed Methodology

We have proposed the design and analysis of various MPPT controller techniques operations at MPP for better and optimized performance of PV array. Further to that, we will develop CASD multilevel inverter using SPM control strategy with P&O (perturbation and observation technique) and FLC algorithm for maintaining operating point of array at MPP, thereby reducing THD percentage in input current and providing the grid with high sinusoidal current. The proposed simulation and analysis of our work is carried out with a 5 kW peak capacity in SIM.

2 System Modelling Formulation

The chapter provides complete design, modelling and analysis work for realizing intelligent method for maximum power point tracking of solar PV system and integration single-phase grid with multilevel inverter. For that, first step is design of Solar PV array of capacity 5 kW peak capable of delivering 230 V with consistency. Figure 1b shows complete system layout of solar grid connected system.

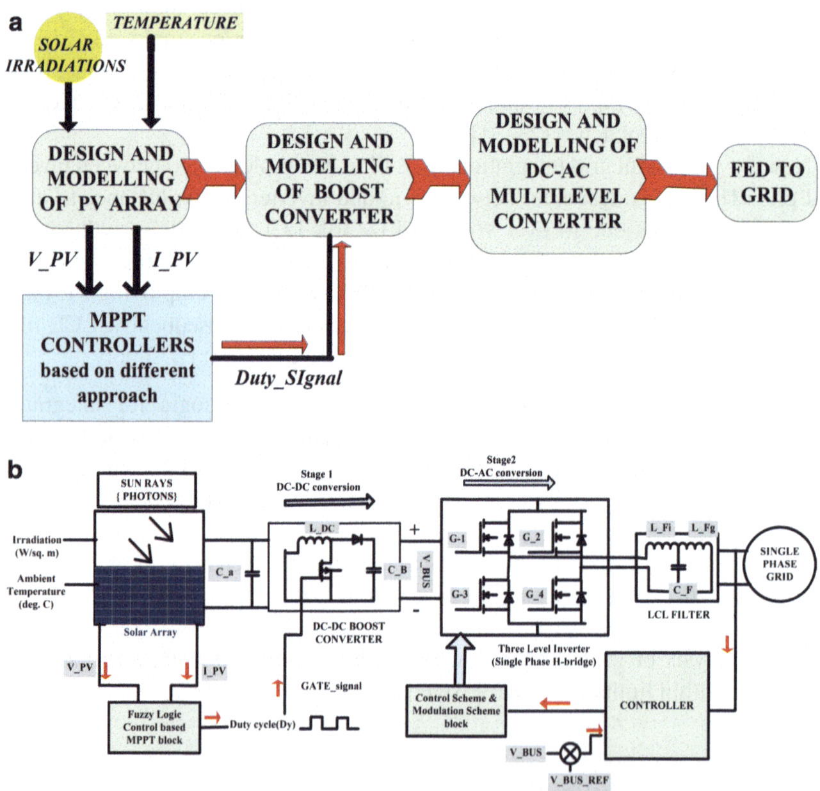

Fig. 1 **a** Graphical flowchart of holistic work done in this chapter, **b** Proposed system layout of fuzzy logic-based solar grid connected system

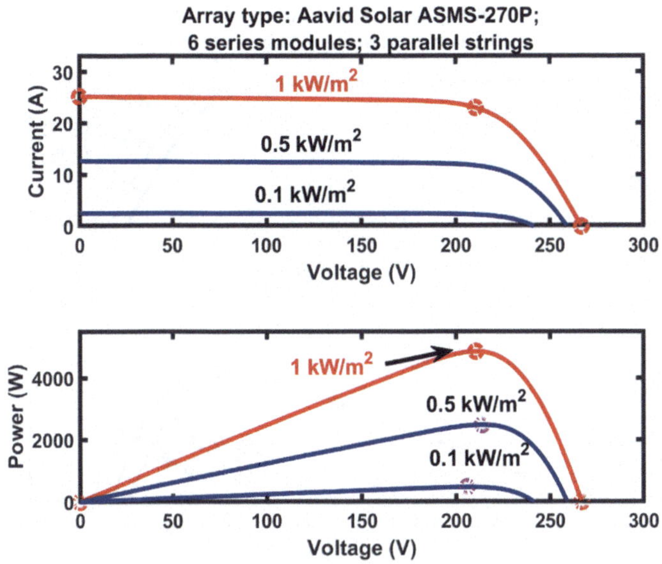

Fig. 2 SPV characteristics at fixed 25 °C and varying solar irradiations

2.1 PV Array Modelling

Considering AVID-SOLAR-ASMS-270P PV module for simulation purpose consisting of 72 cells combined in series to form a module. For a given module, VOC V_{oc} is 44.5 V; ISC I_{sc} is 8.42 A; Vmpp V_{mpp} is 35.1 V; Cmpp I_{mpp} is 7.69 A and Peak Power of a module $P_{peak} = 269.91\ W_{peak}$.

Hence, number of modules that are present in series in a string is 6 and number of such parallel strings is considered as 3, Thus for a given PV array, Vmpp will be written as: V_{mpp} is (6 * 35.1) = 210.6 V; Cmpp will be written as: I_{mpp} is (3 * 7.69) = 23.07 A; Peak Power $P_{peak} = V_{mpp} * I_{mpp} = 4858.542\ W_{peak}$.

In order to study and make analysis of MPPT control based on different techniques, preliminary that should be known is study of PV characteristics. The varying conditions effect on characteristic of SPV must be known like characteristics curve of SPV with varying irradiations as shown in Fig. 2 and that of varying temperatures as shown in Fig. 3, respectively.

2.2 Design of Boost Converter

They consist of an inductor L_{DC}. A MOSFET and a diode and capacitor at input side C_α and a capacitor at output side C_β. There are few assumptions that need to verify before moving onto design:

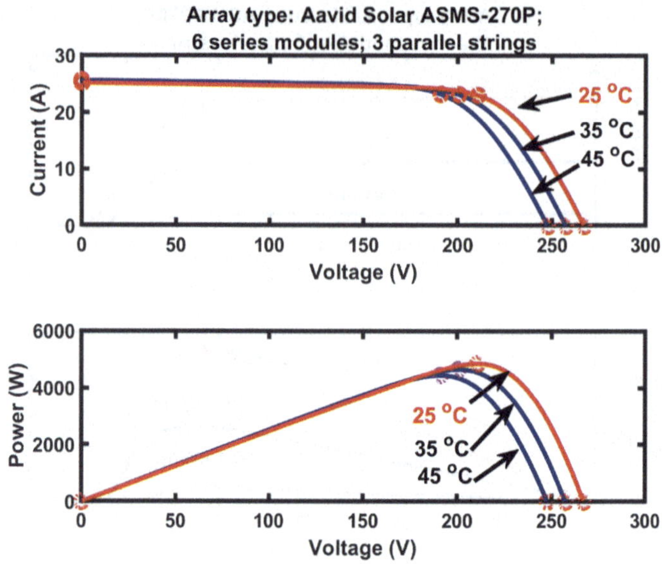

Fig. 3 SPV characteristics at fixed 1000 W/m^2 and varying temperatures

- Switching equipment should be ideal.
- Equivalent series resistance is considered.
- Inductor current to operate in continuous mode.
- All passive elements used should be linear, time invariant and frequency independent.

There are two purposes of using DBC, first to step up the DC level of solar output to operate at maximum power point and second for having steady voltage flow at output of boost terminals in order to give constant supply to grid connected H-bridge MLI. The average output voltage at boost terminals is controlled by duty cycle (D_y) given as gate signal of switch employed (here MOSFET). It may be defined as the ON duration of switch to switching time period taken by switch to on and off. This duty is defined by MPPT controller. Switching is of pulse width modulation type at particular switching frequency pulse of on and off durations.

Here, we have considered switching frequency of MOSFET, $F_{SW} = 10$ kHz. DC voltage fed to single-phase voltage source inverter given by:

$$V_{DC} = \frac{2\sqrt{2} * Vs}{\sqrt{3} * m} \tag{1}$$

where Vs is r.m.s value of input voltage, i.e. 200 V, m is the modulation index of inverter generally $m = 0.95$ putting these values we get, $V_{DC} = 343.78$ V.

Duty cycle of DBC is given by:

$$D_y = 1 - \frac{V_{\text{in}}}{V_{\text{out}}} \tag{2}$$

where ΔI_L is the ripple content in inductor current and is given consideration as 5% of DBC inductor current, i.e.

$$\Delta I_L = 6\% \text{ of } \frac{P_{\text{mpp}}}{V_{\text{mpp}}} \approx 1.393\,\text{A} \tag{3}$$

Calculations of DBC inductor value

$$L_{\text{DC}} = \frac{V_{\text{DC}} * D_y(1 - D_y)}{2 * F_{\text{SW}} * \Delta I_L} \tag{4}$$

Substituting values discussed above in formula (4), we get L_{DC} as 2.48 mH. However, after optimizing model, value used in simulation is $L_{\text{DC}} = 25$ mH. Value of capacitance at boost terminals given by:

$$C_\beta = \frac{I_{\text{dc}}}{2 * \omega * V_{\text{DC(ripple)}}} \tag{5}$$

Whose value comes out to be around 6 μF. However, after optimizing around, 6000 μF is the best suited value as per requirement of chapter. Where I_{dc} is the DC current along DBC input terminals; ω is angular frequency (2 * 3.14 * 50) and $V_{\text{DC(ripple)}}$ is ripple content in output DC voltage, i.e. 2% of $V_{\text{DC}} \approx 6$ V. Value of capacitance C_α is taken as 100 μF for steady in DC input. Equivalent resistance used in series with the capacitor is kept as 0.0001 Ω.

2.3 Calculation for Selection of Filter

LCL filters have better dynamic response and damping efficiency as compared *L* or LC filters. LCL filter is selected with the aim to reduce the high-frequency (or switching frequency) current harmonics into the grid. It also helps in better decoupling between the impedance of inverter and grid. However, to have good response, small values of *L* is selected due to which resonance problems arises leading to unstable system, and C_{F} mainly filters high-frequency harmonics and absorbs reactive power Hence, the values of LCL is optimized by varying parameters.

For selection of AC Inductor at inverter side denoted by:

$$L_{Fi} = \frac{\sqrt{3} * m * V_{\text{DC}}}{12 * h * F_{\text{SW}} * \Delta I_L} \tag{6}$$

where h is the factor of overloading taken as 1.2 and switching frequency of single-phase three-level inverter F_{SW} is 10 kHz; after putting all values in (6), we get $L_{Fi} = 2.81$ mH. Value of inductance for Grid side inductor given by $L_{Fg} = h * L_{Fi} =$ 1.2 times inverter side inductance. Thus, $L_{Fi} = 2$ mH and $L_{Fg} = 3$ mH is used in the simulation after optimization of filter.

Value filter capacitance is given by following equation:

$$C_{F} = \frac{0.05}{\omega_{G} * Z_{base}} \tag{7}$$

where Z_{base} is the $Z_{BASE} = V_{L-L}/P_{active}$; here, V_{L-L} is line to line grid voltage, P_{active} is the inverter rated power (active) and ω_{G} is the operating grid frequency;

Resonant frequency is given by:

$$\omega_{reso} = \sqrt{(L_{Fi} + L_{Fg}) / (L_{Fi} * L_{Fg} * C_{F})} \tag{8}$$

Calculated as 2.1 kHz and $F_{reso} = F_{SW}/10$; Further to that, parallel damping can be done with C_{F} in order to reduce resonance problem value calculated by $R_{D} = 1/(3 * \omega_{reso} * C_{F})$ value is 500 Ω.C_{F} is calculated as 5 µF. However, 2.2 µF is used in the simulation for best performance of LCL filter.

3 Artificial Intelligence Method Formulation Using Fuzzy Logic

Fuzzy logic is an intelligent method of reasoning which gives very precise output whose accuracy depends upon rules input to fuzzy system. One of the most advantageous part of fuzzy is its convergence speed, and its results with accuracy from intermediate range of possibilities between 0 and 1. Coming to application of FL technique for drawing maximum available power from SPV as output. As we know that, DC output of the SPV varies continuously during the day and night due to varying physical conditions like irradiance and ambient temperature. Hence, a DBC is connected after SPV for taking required output from it with help of MPPT algorithm. This helps in achieving the voltage at MPP. Thus, FL-based MPPT easy the situation by providing with accurate duty to power converter switch. This boosts/buck the DC voltage levels by matching equivalent impedance as seen from PV terminals with the load (impedance matching to apply maximum power transfer theorem).

In Fig. 4, outline schematic of FL-based MPPT controller is shown along with other required functional blocks. The FL control involves four process fuzzification. Inference engine with help of rule base fuzzy set is created and then defuzzification to convert fuzzy values into crisp output. Human intelligence and comprehensions are needed to accomplish and develop MF. One must be aware of the issues and

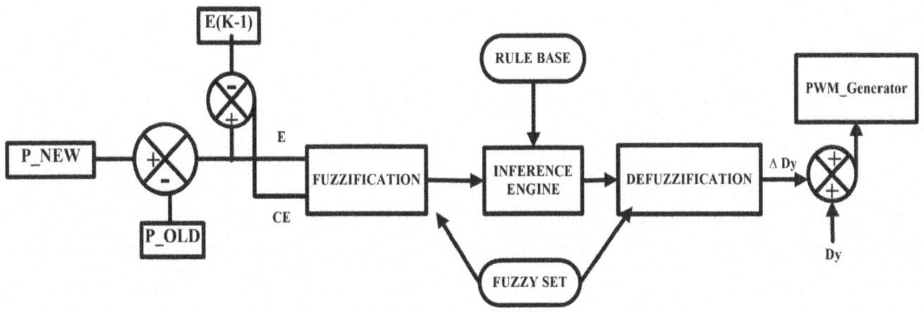

Fig. 4 Outline schematic of FLC-Based MPP controller

expertise needed in order to make linguistic variable that is rule that leads to fuzzy logic [3–5].

3.1 *Fuzzification Process*

Initially, power output from PV array is stored in a memory as previous or old power $P(k-1)$ and power in real time that is instantaneous power output from PV array shown as $P(k)$, difference between two will be the differential change in power represented by $\Delta P(k)$ as follows:

$$\sim \Delta P_{\mathrm{PV}}(k) = P_{\mathrm{PV}}(k) - P_{\mathrm{PV}}(k-1) \tag{9}$$

$$\sim \Delta V_{\mathrm{PV}}(k) = V_{\mathrm{PV}}(k) - V_{\mathrm{PV}}(k-1) \tag{10}$$

Similarly, $\Delta V_{\mathrm{PV}}(k)$ stands for differential change in voltage from reference and instantaneous one.

As per the schematic, there comes out e two for FLC, first input is slope of *P*–*V* curve of SPV which is given by following expression:

$$E_{\mathrm{PV}} = \frac{\Delta P_{\mathrm{PV}}(k)}{\Delta V_{\mathrm{PV}}(k)} = \frac{P_{\mathrm{PV}}(k) - P_{\mathrm{PV}}(k-1)}{V_{\mathrm{PV}}(k) - V_{\mathrm{PV}}(k-1)} \tag{11}$$

Often termed as error or incremental change in power that is to be obtained from SPV. And second input is change in error $\mathrm{CE}_{\mathrm{PV}}$ whose expression is defined below as:

$$\mathrm{CE}_{\mathrm{PV}} = E_{\mathrm{PV}}(k) - E_{\mathrm{PV}}(k-1) \tag{12}$$

Eventually, the MPP will be accomplished when both $\Delta P_{\mathrm{PV}}(k)$ and $\Delta V_{\mathrm{PV}}(k)$ are zero. Fuzzification process bifurcate the input as well as output variables into five (5)

Table 1 Fuzzy rules designed for MPPT

E_{PV}/CE_{PV} ΔD_y	NB_f	NS_f	Z_f	PS_f	PB_f
NB_f	PS_f	PS_f	Z_f	NS_f	NB_f
NS_f	PS_f	PS_f	Z_f	NS_f	NS_f
Z_f	PS_f	Z_f	Z_f	Z_f	PS_f
PS_f	PS_f	PS_f	Z_f	NS_f	NS_f
PB_f	PS_f	PS_f	Z_f	NS_f	NB_f

different fuzzy linguistic sets or rules for fuzzy logic. Here, taken as NB_f {termed as Negative Big}, NS_f {as Negative Small}, Z_f{Zero}, PS_f {as Positive Small}, PB_f {as Positive Big} [6, 7, 18]. Following are fuzzy subsets for inputs error and change in error:

$$E_{PV} = \{NB_f, NS_f, Z_f, PS_f, PB_f\} \tag{13}$$

$$CE_{PV} = \{NB_f, NS_f, Z_f, PS_f, PB_f\} \tag{14}$$

Domain range for both inputs variables is set to be [− 8, 8].

Fuzzy set for output can be written as follows:

$$\Delta D_y = \{NB_f, NS_f, Z_f, PS_f, PB_f\} \tag{15}$$

Domain range for output variable is to be [−3, 3].

3.2 *Fuzzy Rule Set*

In this work, Fuzzification is performed using "Mamdani's Method". Fuzzy rule base defined precisely based on the knowledge of inputs and output in order to generate variation in duty. Henceforth, there are 25 rules defined in FISY to track MPP [5–7] (Table 1).

3.3 *Inference Engine*

Fuzzy inference engine is the main formulating method which takes all the logical decision and formulations based on fuzzy rules setting and range, thence it converts the fuzzy rule base with the help of knowledge base, or we can say inputs into the fuzzy linguistic variables or output according to the various interpretations. The working of fuzzy rule base is shown below properly with help of four cases (Fig. 5).

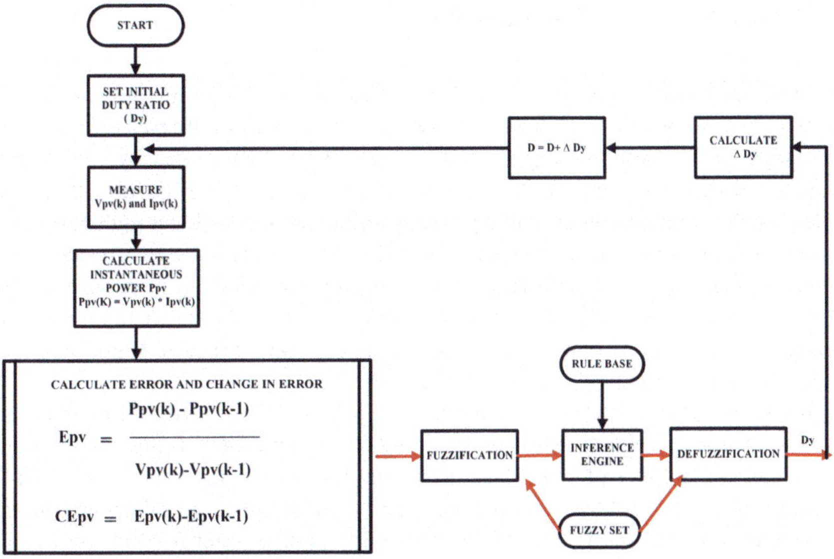

Fig. 5 Flowchart of FL-based MPPT algorithm

Case1: $\sim \frac{dP_{PV}(k)}{dV_{PV}(k)} > \frac{dP_{PV}(k-1)}{dV_{PV}(k-1)}$ and $D_y(k) > D_y(k-1)$ then this implies change in slope of PV curve is positive, and to reach MPP, change in output should be negative.

Case2: $\sim \frac{dP_{PV}(k)}{dV_{PV}(k)} < \frac{dP_{PV}(k-1)}{dV_{PV}(k-1)}$ and $D_y(k) > D_y(k-1)$ then this implies change in slope of PV curve is negative, and to reach MPP, change in output should continue to negative.

Case3: $\frac{dP_{PV}(k)}{dV_{PV}(k)} < \frac{dP_{PV}(k-1)}{dV_{PV}(k-1)}$ and $D_y(k) < D_y(k-1)$ then this implies change in slope of PV curve is negative, and to reach MPP, change in output should be positive.

3.4 *Defuzzification Process*

Defuzzification is the crucial operation of transformation of aggregated output fuzzy set into single crisp value. It makes this possible by "Center of Gravity" or centroid method [7]. It commutes centroid from the range of aggregated fuzzy output values and gives highly related set-in form of a crisp value. Final value of duty cycle using centre of gravity is shown as below:

$$D = \frac{\sum_{j=1}^{n} \mu(D_j) - D_j}{\sum_{j=1}^{n} \mu(D_j)} \tag{16}$$

4 Control Scheme Formulation

The proposed system layout clearly depicts that we are designing and simulation of two-stage solar grid connected systems integrated with multilevel inverters. In first stage, DBC is used to step up voltage level of DC SPV output by controlling duty signal of boost converter's switch which is decided by MPPT algorithms in order to deliver maximum available power from SPV to load side through matching source side and load side impedance. Since H-Bridge needs to be fed constant and steady voltage supply, therefore, need to ensure proper selection of parameters of DBC. During second stage, DC–AC power conversion is performed. Output inverter voltage has three levels now power is fed to single-phase 50 Hz 230 V grid through a LCL filter. Parameters of LCL filter discussed already in Sect. 1.

Total number of voltage levels in output of MLI is "k" defined by formula $k = 2v + 1$, where v is the count of isolated DC fed at input of H-bridge of inverter, or we can say number of H-Bridges used. One H-bridge comprises of four IGBT switches and output voltage level is controlled by gating pulse and switching pattern of inverter topology used. The output voltage levels of inverter used in work is + Vdc, 0, − Vdc. For construction of H-Bridges, switches with high switching speed like IGBT/MOSFET are utilised and also they are used in high-power applications. Given below is the SIM of three-level CASD (means one after another) H-Bridge inverter fed with DC bus output in Fig. 6.

The simulation performed in this work is employing three-level CASD H-bridge MLI controlled by SPM UNIPOLAR injection scheme. These configuration uses only one reference signal to generate SPM signals as gate controlling pulses. This modulation scheme is simulated with close loop control in SIM (Fig. 7).

In PWM control scheme, the IGBT are turned ON and OFF several times and output voltage is controlled by varying the pulse width. The reference or modulation

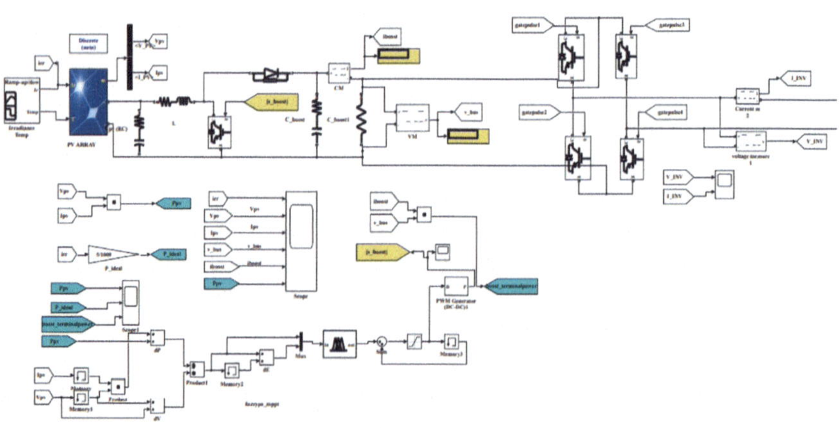

Fig. 6 SIM model of three level CASD H-Bridge inverter

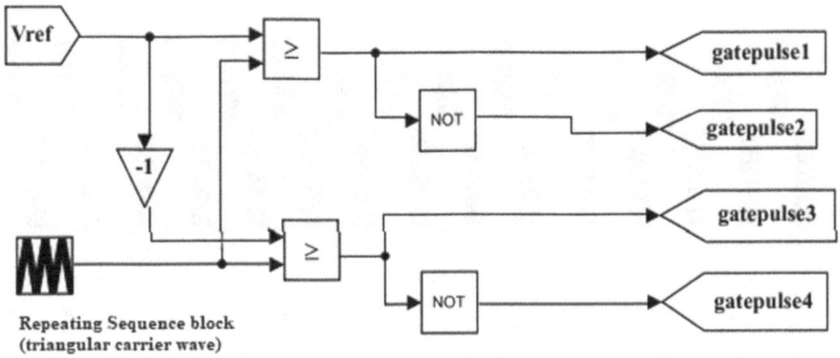

Fig. 7 SPM control scheme for generating gate pulses for inverter

waveform has peak amplitude and fundamental frequency, and it is centred somewhere in between the carrier wave. The general working on which sinusoidal pulse width modulation control is based on is the comparison of a bus voltage (i.e. is boost or reference voltage) as a output from PI controller (compared with voltage we want at grid) with a carrier wave, generally a triangular carrier wave (sample time same as initial one that is $10e-6$) and switching frequency of switch taken as 10 kHz. The reference wave is consistently compared in contrast with carrier wave with help of relational operator [21, 22].

If the reference wave that is at the fundamental frequency (frequency which we require at output of inverter) is higher than the carrier signal (at higher frequency), thereupon the switch in agreement to that carrier wave is turned on implying the pulse will be recorded as high, and on other hand if reference wave is lower than the carrier signal, thereupon the switch in agreement to that carrier will be turned off and pulse will be 0 (Figs. 8 and 9).

5 Result Demonstration and Discussion

The FL-based MPPT for integrating solar grid connected system with MLI three-level H-Bridge inverter injecting active currents into single-phase 230 V, 50 Hz grid via LCL filter is been simulated, and results are discussed as below: simulation and modelling of FLC in SIM is shown in Fig. 10. Further to that, rule base view of FLC using "MAMDANI's method" as inference engine which evaluates and performs reasoning of different rules defined in knowledge of Rule Base. In Fig. 11, Rule Base view of FLC is depicted.

Surface view of FLC is shown in Fig. 12. Showing two inputs variables, i.e. change in error CE and error E, and an output, i.e. differential change in duty dD. Final duty generated D_y + dD is sent to DBC module to optimal the voltage for MPPT.

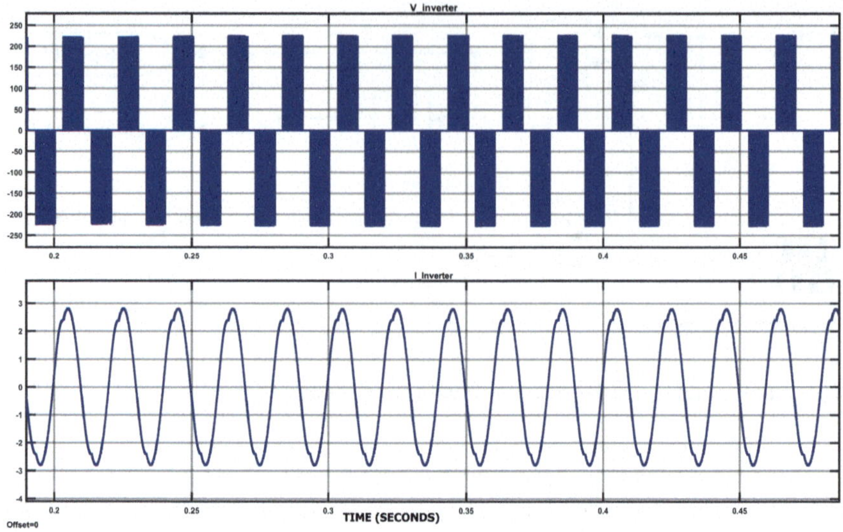

Fig. 8 Inverter output waveform of phase voltage (three levels) and current waveform

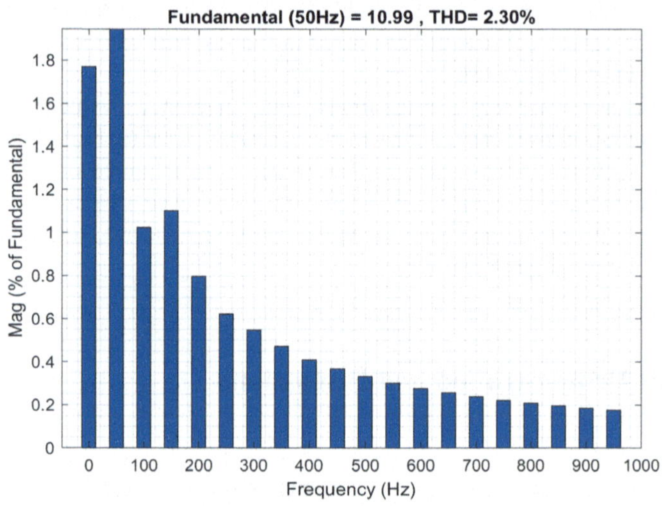

Fig. 9 THD variation through FFT analysis tool in MATLAB

Various MF of FIS variables (CE change in error and *E* error) is depicted in Figs. 13 and 14 and output variable that is change in duty dD is depicted Fig. 15.

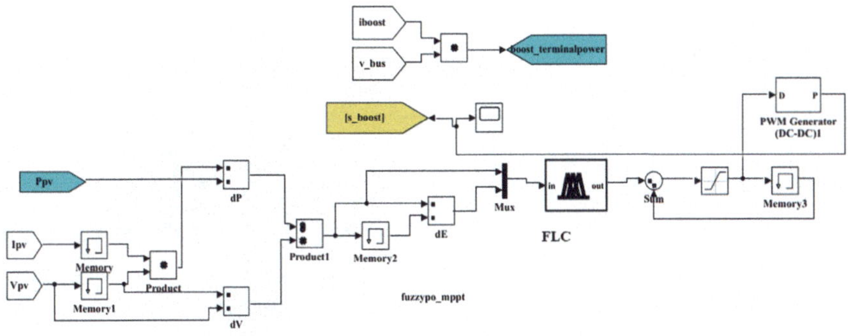

Fig. 10 Model of FL-based MPP controller in SIM

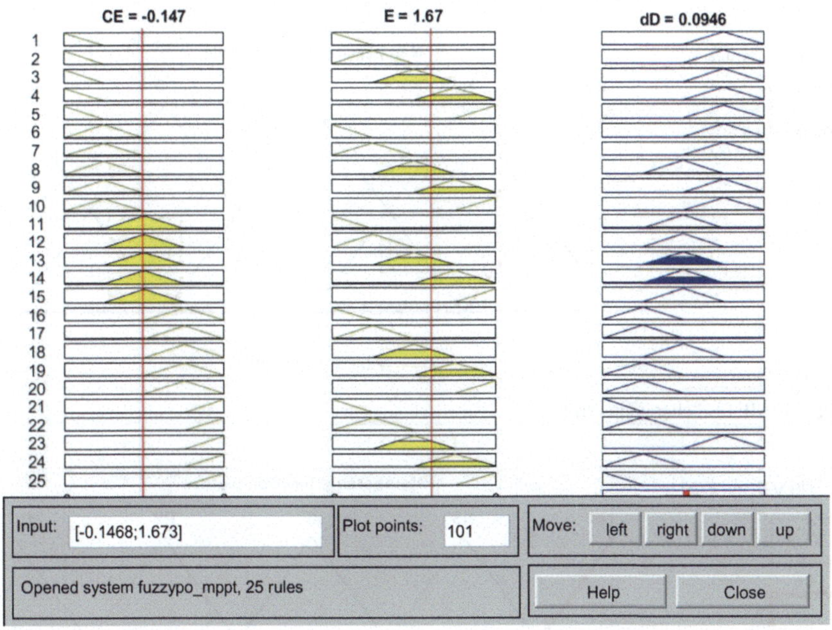

Fig. 11 Rule base view of FLC

5.1 *Dynamic Response of FLC*

In Fig. 16, dynamic response of FLC is shown. It is clearly seen from figure that Vdc reference voltage value is 230 V is perfectly tracked by FLC. Hence, it had been the better algorithm method for good dynamic response as compared P&O method.

It can be noted down from Fig. 16, that fuzzy responses settles down in less than 0.1 s so fast as compared to P&O whose settling time is around 0.44 s in our work.

Fig. 12 Output surface view for fuzzy system of FLC

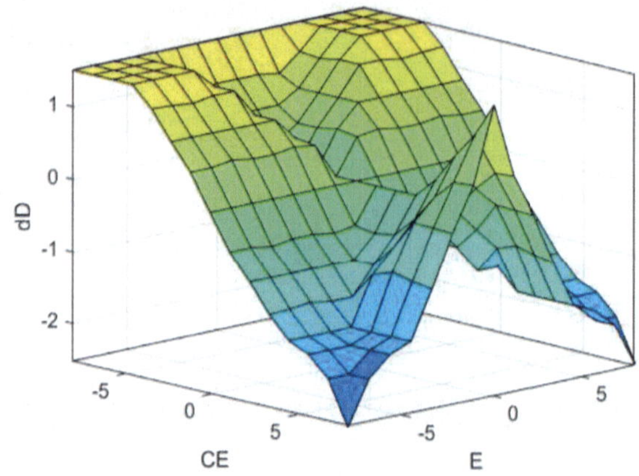

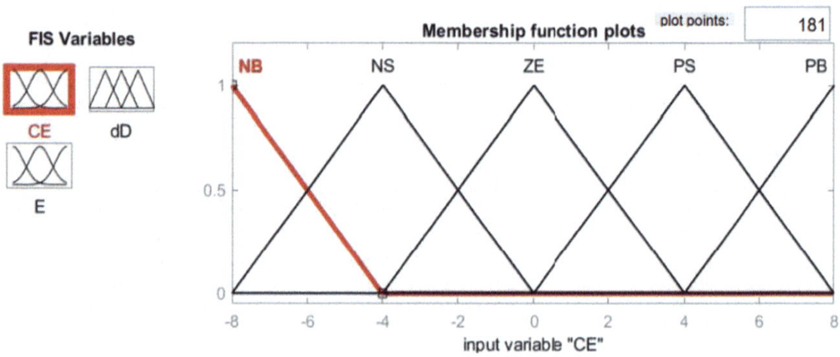

Fig. 13 MF for "change in error" CE

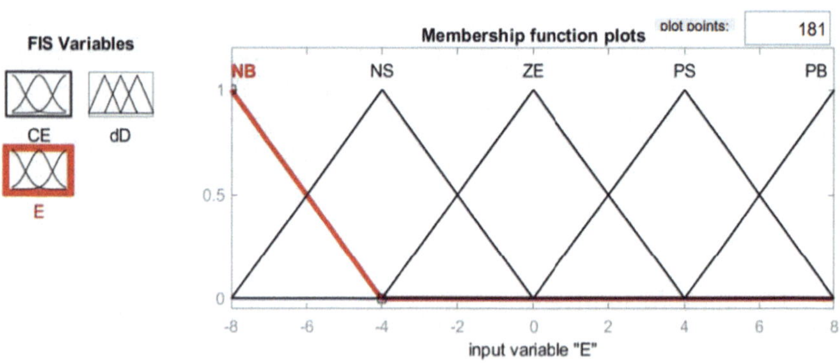

Fig. 14 MF for variable "error" E

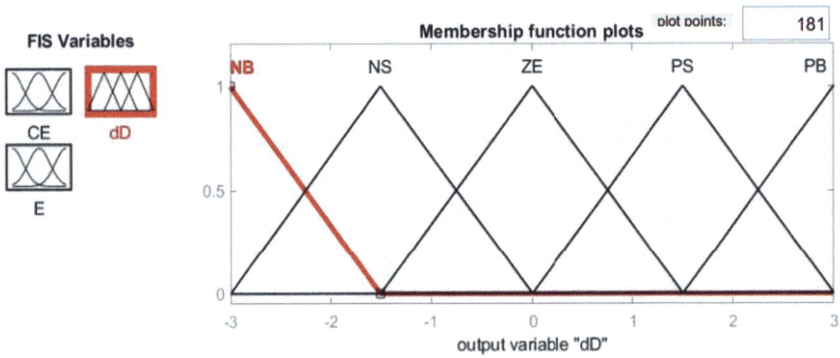

Fig. 15 MF for "change in duty" dD

Fig. 16 Dynamic response of FLC-based MPPT

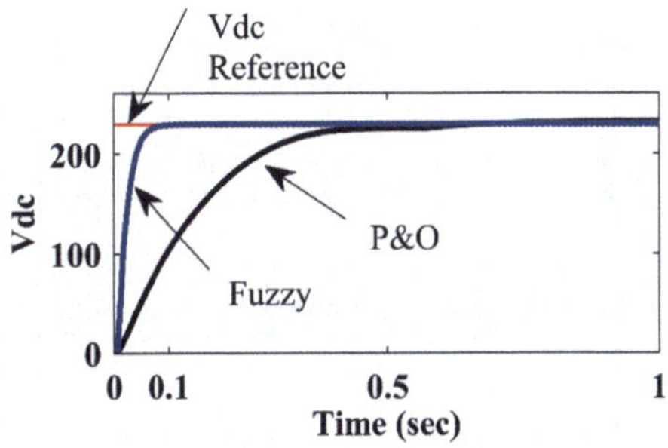

In Fig. 17a, we can see SPV input and output parameters like voltage, current and power output at fixed irradiation 1000 W/m^2 and fixed ambient temperature 25 °C. When irradiation of 1000 W/m^2 is inclined on SPV, it gives voltage approx. 200 V at output terminals and PV current of nearly 24 A. Now BUS voltage at DBC output terminals voltage is around 230 V, boost current is about 20 A and power output is 5000 W_p.

In Fig. 17b, PV performance at varying irradiations can be clearly observed. As the irradiation is dropped to 700 W/m^2 from 0.6 to 1.2 s, there is sudden drop in PV voltage but not up to that extent to which P&O drops. Also, proposed fuzzy controller stabilizes the voltage within ease of time and very minute drop in SPV output curve; hence, we can say FLC is more stable at achieving steady state condition and adaptive to sudden variation in input environmental conditions. Second inference can be concluded that voltage of PV decreases with decrease in insolation and increases with increase in insolation.

In Fig. 17c, it is observed that PV performance is efficient and quick when done with fuzzy as change in ambient temperature of solar cell makes situation difficult for a controller to track MPP as well it takes time to adapt by classical approach, i.e. P&O where as in case of proposed FLC finds accurate MPP that too in good convergence speed, hence, leading to higher generated solar energy. Second inference can be

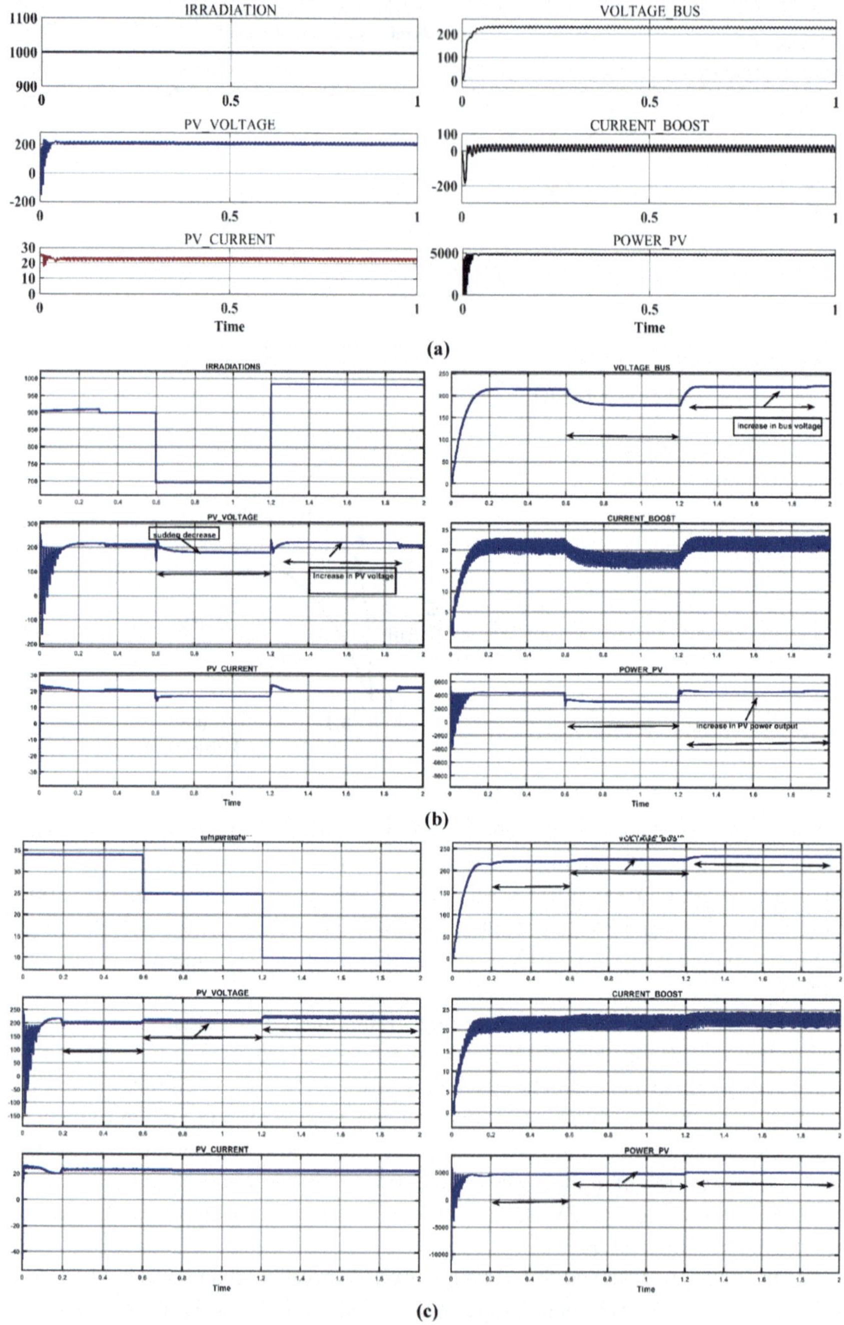

Fig. 17 **a** Shows PV inputs and outputs, DBC output voltage and current and power output waveform at constant 1000 W/m^2 irradiation and constant 25 °C ambient temperature. **b** Shows SPV performance at varying irradiations of 900, 700 and 1000 W/m^2 and fixed temperature that is 25 °C. **c** Shows SPV performance at varying temperature of 35, 25 and 10 °C and fixed irradiations given at 1000 W/m^2

observed as with increase in ambient temperature of solar cell, the value of output and input voltage, and power is reduced and thereby decreasing overall generated power impacting PV performance negatively. However, input current of PV remains unaltered with change in temperatures.

In Fig. 18, we can see PV power output response by fuzzy logic controller has better maximum power tracking capability as from 0.35 to 0.6 s P&O-based MPPT is showing fluctuation from desired value.

From Fig. 19, it is clearly seen that fuzzy response shown by BLUE and P&O by RED. It is verified from above figure that fuzzy response is constant to large extent, and there is a lot of fluctuations in voltage response from P&O.

There is overshoot and consistent fluctuations in bus voltage output between 200 and 225 V from enlarged view, it can be observed feasibly in Fig. 20.

From Fig. 21, it is clearly seen that fuzzy response shown by BLUE and P&O by RED. It is verified from above figure that fuzzy response is constant to large extent, and there is a lot of fluctuations in voltage response from P&O from 0.3 to 0.6 s (Fig. 21).

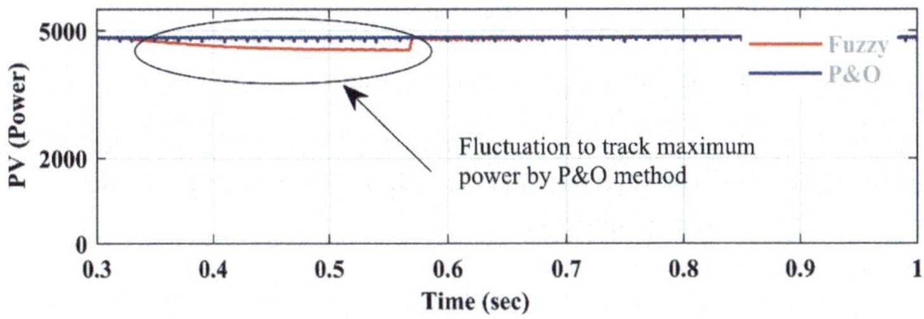

Fig. 18 PV_POWER waveform comparative analysis b/w fuzzy and P&O

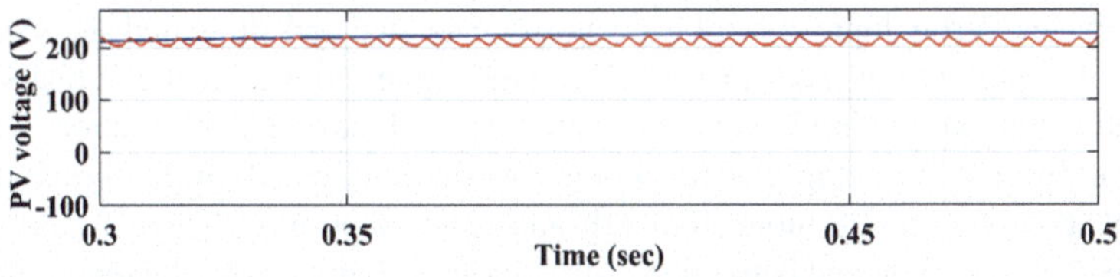

Fig. 19 PV voltage waveform comparative analysis b/w fuzzy and P&O

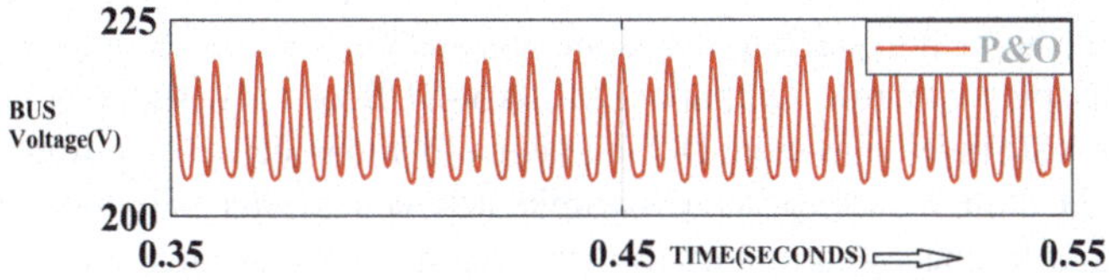

Fig. 20 BUS voltage output showing large fluctuation from enlarged view

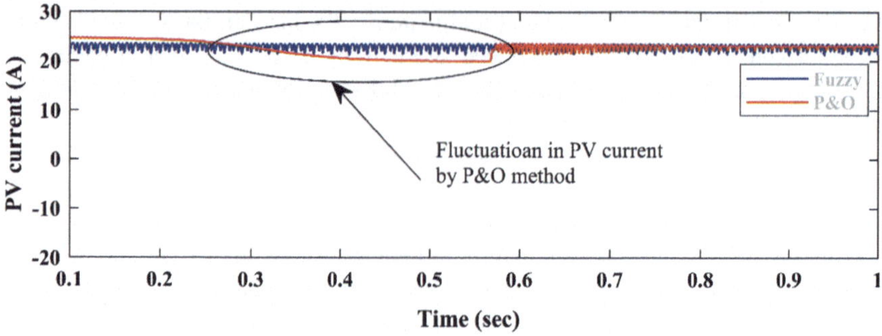

Fig. 21 PV_Voltage waveform comparative analysis b/w fuzzy and P&O

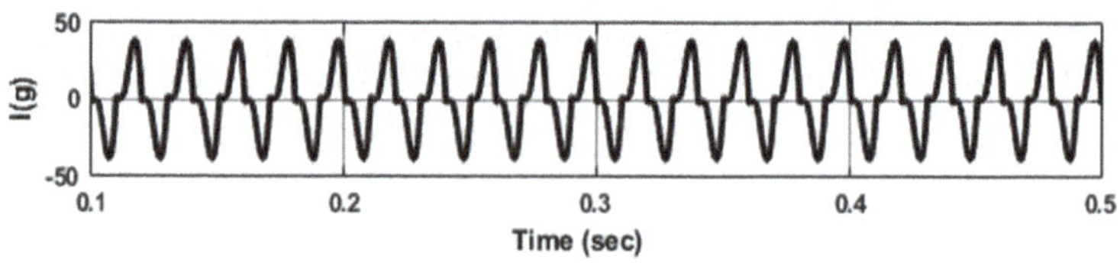

Fig. 22 Grid current waveform injected by three level inverter

Here, we have observed THD reduction below 5%, refer Fig. 9; hence, overall efficiency of power system is improved.

6 Conclusions

In this chapter fuzzy logic control is designed, executed and simulated successfully in SIM. A comparison is made in terms of its convergence time, efficiency, overshoot, cost and complexity to implement. After analysis and thorough observation, we can conclude from above setup that fuzzy logic works intelligently to track maximum available power from SPV more efficiently and rapid adaption to change in conditions in terms of steady state and better dynamic response. Being aware of uncertainties in relation to physical conditions, FLC is successfully seen tracking MPP with accuracy. Thereby, it has increased overall system efficiency. It does not need exact knowhow of SPV characteristics and thereby making easy for solar manufacturers. Along with this, it can be applied in satellite-based models and electric vehicle where precision and accuracy play vital role. Output of DBC is V-BUS fed to H-Bridge of three-level H-Bridge CASD inverter as input DC source in order to inject active power to single-phase grid. LCL filter is accustomed as coupler between inverter and single-phase grid to reduce high-frequency harmonics from inverter output is executed successfully. On comparative analysis of FL and P&O, we found that efficiency, convergence speed, reduced overshoot, speed and performance far better than traditional methods. THD

Table 2 Comparative analysis of different intelligent techniques for MPP tracking

Method	Convergence time	Overshoot	Adaptability to varying conditions	Efficiency (%)	Software implementation
P&O	0.44 s (to reach steady state)	Large	Difficult to adapt and that too slow	97.36	Less complex
FLC	~ 0.1 s (response is fast)	Very less	Adapts quickly to track MPP	98.01	More complex

has been removed well below 5% according to IEEE standards. Complete proposed system is executed successfully in SIM environment as per standards (Table 2).

References

1. H. Malik et al., *Meta Heuristic and Evolutionary Computation: Algorithms and Applications* (Springer Nature, Part of the Studies in Computational Intelligence, 2020), vol. 916, p. 849. ISBN 978-981-15-7571-6. https://doi.org/10.1007/978-981-15-7571-6
2. H. Malik et al., *Soft Computing in Condition Monitoring and Diagnostics of Electrical and Mechanical Systems* (Springer Nature, Part of the Advances in Intelligent Systems and Computing, 2020), vol. 1096, p. 496. ISBN 978-981-15-1532-3. https://doi.org/10.1007/978-981-15-1532-3
3. A. Iqbal et al., *Renewable Power for Sustainable Growth* (Springer Nature, Part of the Lecture Notes in Electrical Engineering), vol. 723, p. 805. ISBN 978-981-33-4080-0. https://doi.org/10.1007/978-981-33-4080-0
4. H. Malik et al., *Intelligent Data-Analytics for Condition Monitoring: Smart Grid Applications* (Academic Press, 2021). ISBN 978-0-323-85510-5. https://doi.org/10.1016/C2020-0-02173-0
5. A.K. Dahiya, D. Malhotra, Fuzzy logic controller based grid integrated PV system with multi level inverter, in *2016 7th India International Conference on Power Electronics (IICPE)*, 2016, pp. 1–6. https://doi.org/10.1109/IICPE.2016.8079469
6. S. Yan, J. Yuan and L. Xu, Fuzzy logic control of MPPT for photovoltaic power system, in *2012 9th International Conference on Fuzzy Systems and Knowledge Discovery*, 2012, pp. 448–451. https://doi.org/10.1109/FSKD.2012.6234208
7. A. Gupta, P. Kumar, R.K. Pachauri, Y.K. Chauhan, Performance analysis of neural network and fuzzy logic based MPPT techniques for solar PV systems, in *2014 6th IEEE Power India International Conference (PIICON)*, 2014, pp. 1–6. https://doi.org/10.1109/POWERI.2014.7117722
8. S. Srivastava et al., *Applications of Artificial Intelligence Techniques in Engineering—Volume 1* (Springer Nature, Part of the Advances in Intelligent Systems and Computing, 2018), vol. 698, p. 643. ISBN 978-981-13-1819-1. https://doi.org/10.1007/978-981-13-1819-1.
9. B.M. Hasaneen, A.A. Elbaset Mohammed, Design and simulation of DC/DC boost converter, in *2008 12th International Middle-East Power System Conference*, Aswan, Egypt, 2008, pp. 335–340.https://doi.org/10.1109/MEPCON.2008.4562340
10. I.A. Ayad, E. Elwarraki, M. Baghdadi, Intelligent perturb and observe based MPPT approach using multilevel DC-DC converter to improve PV production system. J. Electr. Comput. Eng. **2021**, 13 (2021). https://doi.org/10.1155/2021/6673022.10.1155/2021/6673022
11. A.F. Sagonda, K.A. Folly, A.P. Kenneth, Comparison of three techniques for maximum power point tracking of solar PV, in *2018 IEEE International Conference on Fuzzy Systems (FUZZ-IEEE)*, 2018, pp. 1–8. https://doi.org/10.1109/FUZZ-IEEE.2018.8491542

12. K.L. Lian, J.H. Jhang, I.S. Tian, A maximum power point tracking method based on perturb-and-observe combined with Particle Swarm optimization. IEEE J. Photovoltaics **4**(2), 626–633 (2014)
13. S. Motahhir, A. El Ghzizal, S. Sebti et al., Modeling of photovoltaic system with modified incremental conductance algorithm for fast changes of irradiance. Int. J. Photoenergy **2018**, 13 (2018)
14. A.M. Noman, K.E. Addoweesh, H.M. Mashaly, A fuzzy logic control method for MPPT of PV systems, in *IECON 2012—38th Annual Conference on IEEE Industrial Electronics Society*, Montreal, QC, Canada, 2012, pp. 874–880. https://doi.org/10.1109/IECON.2012.6389174
15. A. Gupta, P. Kumar, R.K. Pachauri, Y.K. Chauhan, Performance analysis of neural network and fuzzy logic based MPPT techniques for solar PV systems, in *2014 6th IEEE Power India International Conference (PIICON)*, Delhi, India, 2014, pp. 1–6.https://doi.org/10.1109/POWERI.2014.7117722
16. A. Aryal, A. Hellany, J. Rizk, M. Nagrial, Residential grid connected solar inverter using fuzzy logic DC-DC Converter, in *2016 International Conference on Sustainable Energy Engineering and Application (ICSEEA)*, Jakarta, Indonesia, 2016, pp. 12–16. https://doi.org/10.1109/ICSEEA.2016.7873560
17. O.B.H.B. Kechiche, B. Barkaoui, M. Hamza, H. Sammouda, "Simulation and comparison of P&O and fuzzy logic MPPT techniques at different irradiation conditions, in *2017 International Conference on Green Energy Conversion Systems (GECS)*, Hammamet, Tunisia, 2017, pp. 1–7. https://doi.org/10.1109/GECS.2017.8066266
18. S. Srivastava et al., *Applications of Artificial Intelligence Techniques in Engineering—Volume 2* (Springer Nature, Part of the Advances in Intelligent Systems and Computing, 2018), vol. 697, p. 647. ISBN 978-981-13-1822-1. https://doi.org/10.1007/978-981-13-1822-1
19. H. Malik, *AI and Machine Learning Paradigms for Health Monitoring System: Intelligent Data Analytics* (Springer Nature, Part of the Studies in Big Data, 2020), vol. 86, p. 513. ISBN 978-981-33-4412-9. https://doi.org/10.1007/978-981-33-4412-9
20. N. Prabaharan, A. Rini Ann Jerin, K. Palanisamy, S. Umashankar, Integration of single phase reduced switch multilevel inverter topology for grid connected photovoltaic system. Energy Proc. **138**, 1177–1183 (2017). ISSN 1876-6102. https://doi.org/10.1016/j.egypro.2017.10.231
21. D. Patel, R. Saravanakumar, K.K. Ray, R. Ramesh, Design and implementation of three level CHB inverter with phase shifted SPWM using TMS320F24PQ, in *India International Conference on Power Electronics 2010 (IICPE2010)*, New Delhi, India, 2011, pp. 1–6. https://doi.org/10.1109/IICPE.2011.5728144
22. M.S. Sivagamasundari, P. Mary, Cascaded H-Bridge five level inverter for grid connected PV system using PID controller. JSTS J. Semiconductor Technol. Sci. **16**, 451–462 (2016). https://doi.org/10.5573/JSTS.2016.16.4.451
23. X. Liang, Emerging power quality challenges due to integration of renewable energy sources. IEEE Trans. Indust. Appl. **53**(2), 855–866 (2017)
24. J.M. Kwon, K.H. Nam, B.H. Kwon, Photovoltaic power conditioning system with line connection. IEEE Trans. Ind. Electron. **53**(4), 1048–1054 (2006)
25. R.A. Mastromauro, M. Liserre, T. Kerekes, A. Dell'Aquila, A single-phase voltage-controlled grid-connected photovoltaic system with power quality conditioner functionality. IEEE Trans. Ind. Electron. **56**(11), 4436–4444 (2008)
26. S.B. Kjaer, J.K. Pedersen, F. Blaabjerg, a review of single phase grid-connected inverters for photovoltaic modules. IEEE Trans. Indust. Appl. **41**(5) (2005)
27. S. Wei, B. Wu, S. Rizzo, N. Zargari, Comparison of control schemes for multilevel inverter with faulty cells, in *30th Annual Conference of IEEE Industrial Electronics Society*, 2004, vol. 2. IECON 2004, Busan, Korea (South), 2004, pp. 1817–1822. https://doi.org/10.1109/IECON.2004.1431859
28. A. Şchiop, A control scheme for multilevel inverter, in *2015 International Symposium on Signals, Circuits and Systems (ISSCS)*, Iasi, Romania, 2015, pp. 1–4.https://doi.org/10.1109/ISSCS.2015.7203977

29. C.M. Nirmal Mukundan, P. Jayaprakash, Cascaded H-Bridge multilevel inverter based grid integration of solar power with PQ improvement, in *2018 IEEE International Conference on Power Electronics, Drives and Energy Systems (PEDES)*, Chennai, India, 2018, pp. 1–6. https://doi.org/10.1109/PEDES.2018.8707467
30. C.M. Nirmal Mukundan, P. Jayapraksh, Solar PV fed cascaded H-bridge multilevel inverter and SIMO-SEPIC based MPPT controller for 3-phase grid connected system with power quality improvement, in *2017 National Power Electronics Conference (NPEC)*, 2017, pp. 106–111. https://doi.org/10.1109/NPEC.2017.8310443
31. S. Umashankar, T.S. Sreedevi, V.G. Nithya, D. Vijayakumar, A new 7-level symmetric multilevel inverter with minimum number of switches. Int. Schol. Res. Notices **2013**, 8 (2013). https://doi.org/10.1155/2013/476876
32. K.A. El Wahid Hamza, H. Linda, L. Cherif, LCL filter design with passive damping for photovoltaic grid connected systems, in *IREC2015 The Sixth International Renewable Energy Congress*, Sousse, Tunisia, 2015, pp. 1–4. https://doi.org/10.1109/IREC.2015.7110945
33. W. Yin, Y. Ma, Research on three-phase PV grid-connected inverter based on LCL filter, in *2013 IEEE 8th Conference on Industrial Electronics and Applications (ICIEA)*, Melbourne, VIC, Australia, 2013, pp. 1279–1283. https://doi.org/10.1109/ICIEA.2013.6566564
34. H. Saxena, A. Singh, J.N. Rai, Design and analysis of different PLLs as load compensation techniques in 1-Ø grid-tied PV system. Int. J. Electron. (2019). https://doi.org/10.1080/00207217.2019.1600745
35. M. Liserre, F. Blaabjerg, S. Hansen, Design and control of an LCL-filter-based three-phase active rectifier. IEEE Trans. Indust. Appl. **41**(5), 1281–1291 (2005). https://doi.org/10.1109/TIA.2005.853373
36. E. Fadavi, S. Javadi, A. Hekmati, Maximum power extracting in a solar plant with multilevel inverters and fuzzy method, in *2014 International Congress on Technology, Communication and Knowledge (ICTCK)*, Mashhad, Iran, 2014, pp. 1–6.https://doi.org/10.1109/ICTCK.2014.7033539

Intelligent Tools and Techniques for Data Analytics of SPV Systems: An Experimental Case Study

Ahmad Faiz Minai, T. Usmani, Sami Uz Zaman, and Ahmad Khubaib Minai

Abstract Due to the rapid depletion of fossil fuels, and after the crisis of 1973, renewable energy resources attract researchers for the generation of electricity. Use of geothermal, tidal, wind, solar, OTEC system, etc., has increased since then. Every renewable resource has some advantages and limitations, but the only advantage which presents in all the resources is the unlimited availability and that is why they are called renewable energy resources which can be used again and again. Like advantage, the common disadvantage is reliability of renewable resources, and thus, hybrid systems are used for the sensitive requirements where power availability is necessity. In all, solar PV system is preferred by the consumers because it is available throughout the day considering 300 sunny days out of 365 for India and most of the countries. In installation of solar photovoltaic (SPV) system, the main hurdle is its high cost and site selection. Generally, the cost of the balance of system (BOS) is less than or equal to the cost of solar panels required at any site, so a proper planning is needed for the investment in SPV plant. If this planning is good, the consumer gets payback in a comparatively short time and hence motivates to install the SPV system which results in helping the nation to reduce carbon footprints and promote sustainable development. So in this chapter, two very useful PV system design software, PVsyst and solar PVGIS, are discussed in detail with the case study of already installed rooftop solar PV plant in the city of Lucknow, India. The performance ratio and losses are also calculated with the help of software. Also, module degradation analysis is performed with the help of PVsyst, and at the end, solar PVGIS and PVsyst are compared along with the SCADA system. After reading this chapter, readers can easily understand the planning and methodology involved in the installation of solar PV system at any location, throughout the world.

Keywords PVSYST · Solar PVGIS · SCADA · Intelligent tools · PV module degradation

A. F. Minai (✉) · S. Uz Zaman · A. K. Minai
Department of Electrical Engineering, Integral University, Lucknow 226026, India

T. Usmani
Department of Electronics and Communication Engineering, Integral University, Lucknow 226026, India

H. Malik et al. (eds.), *Intelligent Data Analytics for Power and Energy Systems*, Lecture Notes in Electrical Engineering 802,
https://doi.org/10.1007/978-981-16-6081-8_16

1 Introduction

Solar-powered sector is developing at a gigantic speed with government meaning to accomplish 100 GW objective till 2022. This drive has considered the government to be on sun-based selection as grounds up, rooftop top framework across longitude and latitude of India [1].

Solar photovoltaic system works because of solar radiation, which makes effect on the annual energy relying upon the quality radiation, cloud shading, surrounding temperature, and module innovation. Today there is need of information on climate anticipating/sunlight-based generation estimating as it assists the utility experts with long-term estimation on sun-based energy generation, O&M, with its importance and effect on the grid stability, load adjustment, tending to peak power demands, making of power portfolio inside utility power purchase bin on inexhaustible energy [2, 3].

Energy review of sun-based plant is the key goal viewed for as of now installed facilities when they are viewed their quality generation, as other energy operation based on the sun is too produce data, i.e., radiations, system parameters, and so on, which should be read for viable activity as huge utility power yield impacts power clearing and grid balancing viewpoint for micro- as well as macro-level [4].

Besides there is a need of hybridization of solar-powered system with the innovation like Internet of things (IoT), AI, predictive modeling, estimation, optimization which must be perceived by utility/sun-based plant architects and planners to perform tasks and settling everyday problems [5]. For the estimation purpose, in private organizations and sun-based generation plant, the present situation is utilized by association in making a strategic advantage, wherein a part of market is getting restricted and edges are contracting as time passes just as address long-haul viewpoint of operational productivity, energy review, and financial profit from venture. In the present situation, associations like to locally available experts who are set up in assuming liability at business with less undertaking arrangement costs/learning. This chapter helps reader to identify/estimate the overall system for a specified rating using best possible software and reduce unnecessary efforts which result in the installation of low-cost system, and then its assessment can be done for the payback and PV module degradation analysis [6].

It is anything but help for India being situated in the tropical belt of the globe, in this manner receiving best radiation from the sun changes from 3.5 to 6.8 $kWh/m^2/day$. The same energy potential is around 6000 million GWh of energy each year. Data of this chapter are from one of the SPV plant installed in Lucknow, Uttar Pradesh. However, all the insolation got isn't changed over to energy as the performance of such sun-based PV plants depends on various factors including losses for plan framework and other natural factors. Execution of sun-powered PV plants on globe is filed commonly on two parameters, viz. plant load factor (PLF) and performance ratio (PR). Indeed these days assortment of sun-oriented experts is giving power generation assurance taking base of PR computations. In sun-powered PV industry PLF (individuals also terms it as CUF, capacity utilization factor) isn't a parameter to be quite thought about. The explanation for it is, time factor (i.e., 24 h). Therefore, experts think that it is better to consider PR to investigate the presentation of a sun-based PV plant. Alongside sun-powered insolation, temperature likewise assumes a major part in influencing the productivity of a PV plant. Climate revised report of NREL recommends that completely reliance of PR on temperature brings about a huge seasonal variation in PR, which can go from ± 2 to ± 10% [1–5].

- In this chapter, simulation of 500 kW grid-connected rooftop solar photovoltaic (SPV) system is performed, considering all the necessary design steps using intelligent tools like solar photovoltaic geographical information system (PVGIS) and photovoltaic systems (PVsyst).
- Further, the result of simulation software is compared with the readings obtained through the SCADA (online monitoring) system, installed with the PV plant of same rating in Lucknow.
- PV module degradation analysis is done using PVsyst, and discussion is made at the end of the chapter in results and discussion section, for the better understanding of the readers [7].

2 Methodology

To analyze the data obtained from different intelligent tools and installed system, the following steps are to be taken:

- Understand the impact of the parameters involves in generation and diagnostics analytics for assessment.
- Descriptive modeling and simulation are performed to understand the operation to achieve intelligence on plant data and analyze the output.
- Explore the methods to use intelligent tools for the perfect planning and generation forecasting of the PV system.
- Understand the basics of plant sustainability with predictive modeling to assess solar plant output, solar cell efficiencies and degradation, with the parameter optimization and correction like performance ratio (PR) and capacity utilization factor (CUF) [8, 9].
- Analysis of PR, CUF, and other plant parameters.
- Optimized installation of solar PV plant and comparison in terms of performance ratio (PR) and capacity utilization factor (CUF).
- Check the outcome of installed SPV system and compare it with the simulated results achieved from forecasting software.
- If it is not matched, check the installed SPV system and identify the issues responsible for bad performance.
- Resolve the issues of SPV plant using simulated software parameters.

The proposed approach for the installation and assessment of SPV plant is given in Fig. 1a.

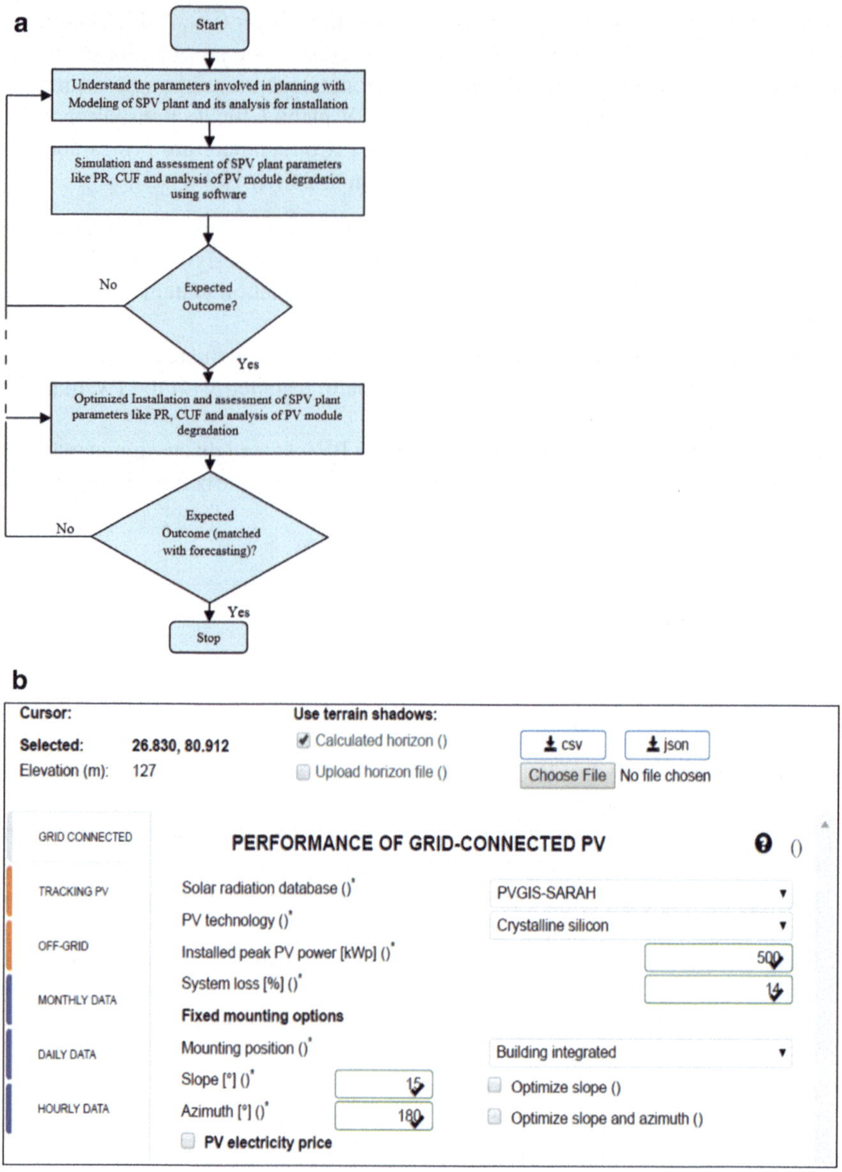

Fig. 1 **a** Proposed approach for the planning/installation and assessment of SPV plant, **b** PVGIS dashboard for input values

3 Design Steps of a 500 kW System Using PVGIS and PVsyst

3.1 Fundamentals and Background of the Intelligent Tool: PVGIS (PV-GIS)

PVGIS is the best free online tool for estimating electricity generation for photovoltaic (PV) stand-alone and grid-interactive system. The annual output power of a solar photovoltaic plant can be simulated with the help of already available parameters for simulation, and maps of solar radiation are also available in PVGIS software regarding forecasting.

In photovoltaic Geological Data Framework, Google Map application made the things simple and utilizable. The region covered by this estimator is practically approx. the whole world. This application estimates the month-to-month and annual potential power generation E [kWh] of a photovoltaic framework with specific panel tilt and direction. On the off chance that you select the menu framework associated, you will get an estimator for fixed sun-oriented modules. With the menu following PV, PVGIS process the power is created by single-axe or double-axes sunlight-based PV trackers. All information and aftereffects of reenactments can be downloaded in Excel (.csv format), pdf or visualize in html documents. Note that PVGIS is just accessible on the web, in Spanish, Italian, German, French, and English.

Note: Now PVGIS covers also America, Asia, and link for PVGIS photovoltaic worldwide simulator is http://re.jrc.ec.europa.eu/pvg_tools/en/tools.html#PVP.

PVGIS for PV distant frameworks anyplace on the earth (America, Asia, Africa, and Europe). By means of the Google Map, it is feasible to figure the sun-powered energy generation for an independent PV framework. This is valuable to get a decent appraisal of the energy power needed to coordinate with the electrical necessities in distant region not connected to the grid [4–6, 10]. In the wake of picking the place on the guide, there is a need to enter the values of solar radiation information base, installed peak PV power, battery limit, discharge cutoff limit [%], consumption of each day, optional hourly utilization record, slope/inclination of modules, azimuth/direction of modules, and solar radiation information base utilizing PVGIS.

Note that PVGIS can incorporate the territory shadows, and it offers two alternatives:

- Feed the value of the determined horizon from PVGIS (it is also available and can be downloaded through the remote horizon data of the site).
- Pass by shadings because of structures or trees are excluded. Browse and use horizon file, if it is available for the predetermined horizon dataset.

When all the necessary input parameters are filled by the user according to the planning of solar PV plant at any location, this software gives the simulation output, as an estimation of electricity generation at a given site. For better understanding of the software, there are three different values of the input parameters (slope angle and azimuth angle) which are taken and shown in Fig. 2a–c, with the obtained value of the annual PV energy generation, annual in-plane irradiation, yearly inconstancy because of different losses are shown and discussed in the result and discussion section with the monthly readings obtained after the simulation of the output electricity generation, monthly irradiation, etc., at a given site.

For a case study, a site in Lucknow, India, is selected and the input parameters of Fig. 2a are considered in this chapter. Two more cases are also considered as shown in Fig. 2b, c, so that it is easy to understand that this software can be used throughout the world, only by changing the site location in the respective field. At a given site, the value of slope angle is 15°, while the azimuth angle is taken 0°. Longitude and latitude are automatically adjusted when the location is selected with the help of map, as shown in Fig. 1b [7].

3.2 Fundamentals and Background of the Intelligent Tool: PVsyst

It is a very systematic and user-friendly tool for the planning/forecasting of a stand-alone or grid-connected solar PV plant. First of all provide all the necessary input parameters like location and specifications of the PV panels, inverters, batteries (in case of off-grid PV system), etc., then after simulation, it gives output in which monthly/yearly electricity generation, irradiation, degradation of PV panels, performance ratio, and different types of losses can be obtained. In this chapter, a 500 kW grid-connected SPV system as shown in Fig. 3 is simulated and compared with other software as well as installed (SCADA) system of the same rating [7].

Following design steps are taken for the simulation using PVsyst software.

3.2.1 Geographical Site Location

First of all latitude and longitude of the site are provided with the tilt/slope angle as input parameters. Then the software automatically gives the different necessary values that can be used for the planning/forecasting of the specific site as shown in Table 1. Horizon line of a given location is also shown in Fig. 4.

3.2.2 Array Sizing

It is a 500 kW rooftop system, so requires large area for the installation of PV panels. So four different roofs of the site are used and array sizing is discussed thoroughly with the help of Fig. 5a–c, with all the possible design considerations as shown in Fig. 6a–d.

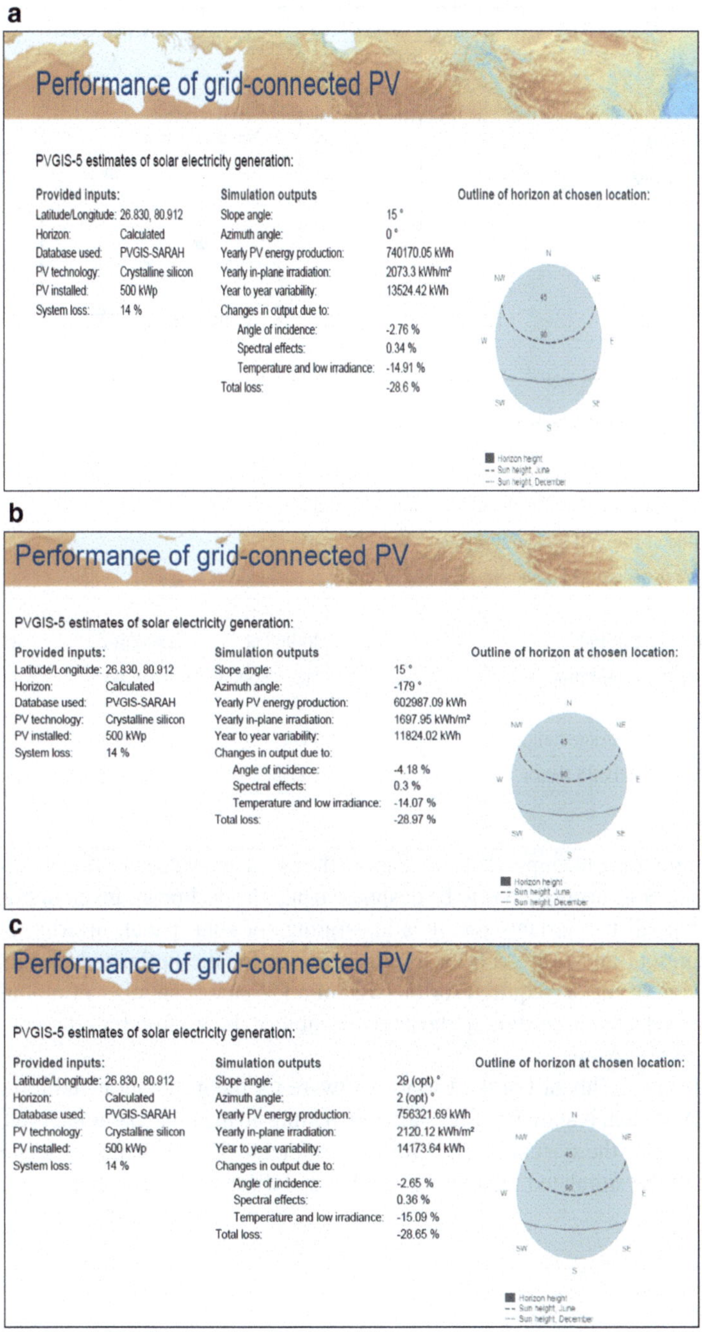

Fig. 2 **a** Estimation of electricity generation using solar for a given site, **b** estimation of electricity generation using solar for a given site, **c** estimation of electricity generation using solar for a given site. *Sun based radiation intuitive guides (http://re.jrc.ec.europa.eu/PVGIS/)

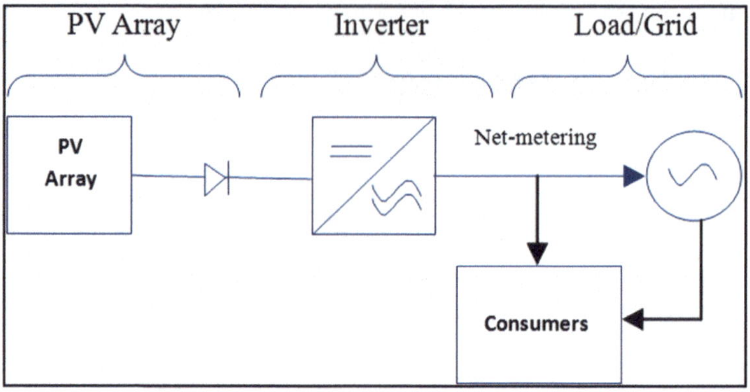

Fig. 3 500 kW grid-interactive SPV system

Table 1 Input parameters for the calculation of horizon line at a given site

Horizon line at Integral University, Lucknow				
Geographical site	Integral University, Lucknow, India			
Situation	Latitude	26.96° N	Longitude	81.00° E
Time defined as	Legal time	Time Zone UT + 5.5	Altitude	116 m
Horizon	Average height	0.0°	Diffuse factor	1.00
	Albedo factor	100%	Albedo fraction	0.00

After providing the input data of location, the estimated value of irradiance, energy is achieved. Now array field is to be designed using the dashboard given in Fig. 6a–d. In this step, all the variants/specifications/ratings of solar panel, inverter, etc., are already present and to be chosen by the designer for the optimized design. After the selection of make (required for the area of a PV panel), number of modules and strings are selected according to the area available at the rooftop for efficient system as given in Fig. 6a–d.

Finally, the dashboard of the PVsyst software for the installation and forecasting of the SPV system is shown in Fig. 7 after providing all the input parameters as shown in Table 2. Now the system is ready for the simulation. For the better understanding, readings of the simulated result are available and discussed in result and discussion section.

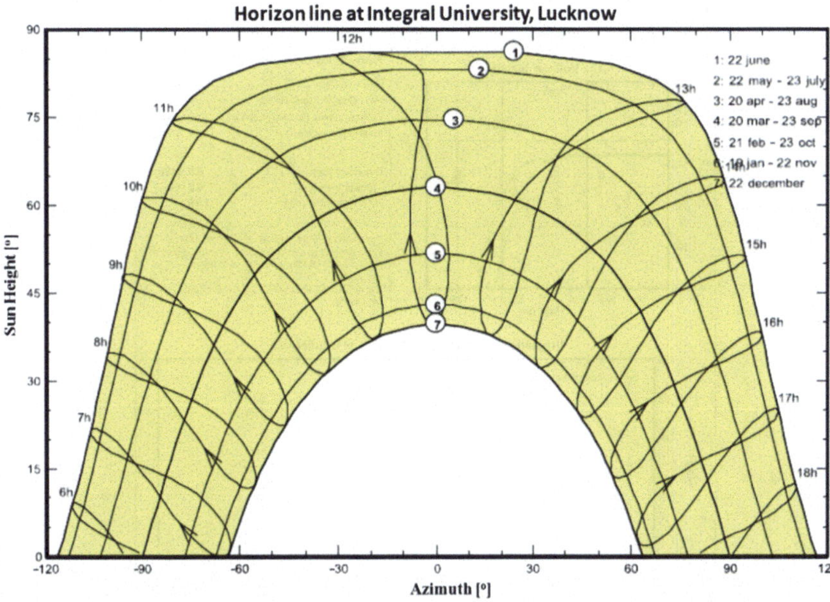

Fig. 4 Horizon line at a given site

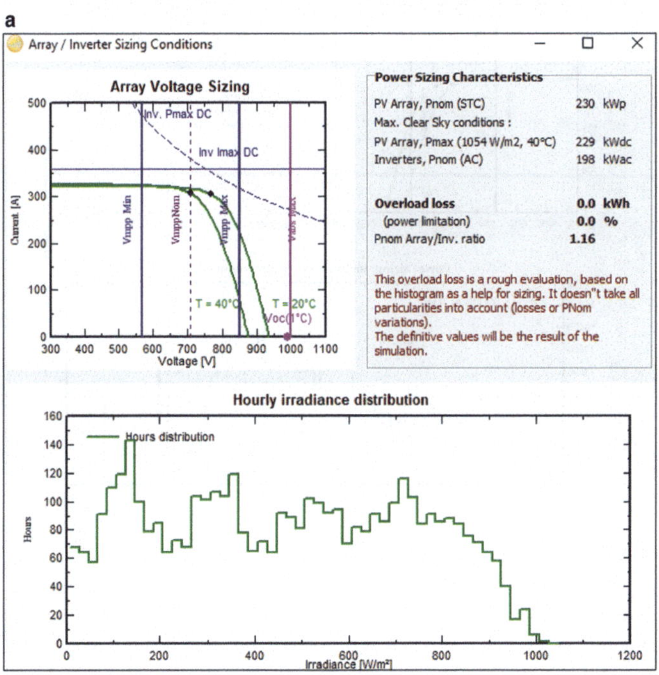

Fig. 5 **a** Hourly irradiance distribution (Irrad. as hours). **b** Irradiations distribution as a function of irradiance (Irrad. As kWh/m^2). **c** Power sizing: inverter output distribution (AC energy as kWh)

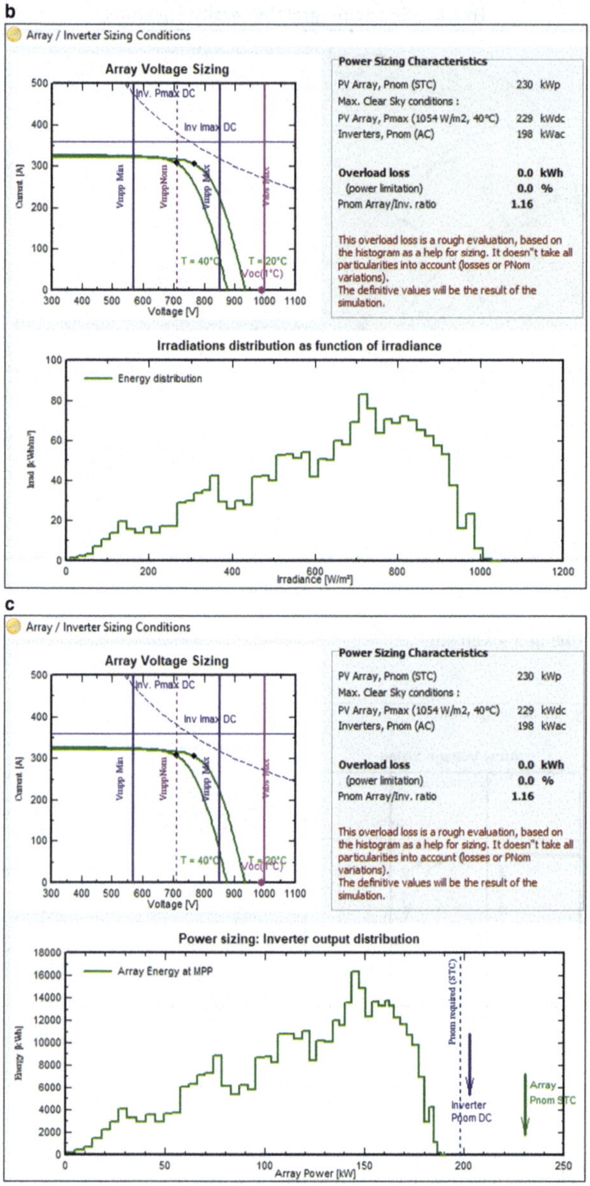

Fig. 5 (continued)

4 Result and Discussion

There are many performance parameters which are used for the analysis and assessment of any solar PV plant like capacity utilization factor (CUF), energy yield (Y_E), plant losses, degradation of panels but the main parameter which is considered everywhere is PR [1]. So the estimation of the PR with the help of predicted electricity generation using software tools PVGIS and PVsyst is performed here, and a comparison is made with the calculated PR of already installed rooftop grid-connected SPV system of the same rating, i.e., 500 kW [11].

$$\mathrm{PR} = \frac{\text{Total Energy (Meter)}}{\text{Total Solar Irradiation x Total Module Area x Module Efficiency}}$$

Table 3 shows the calculation for the performance ratio (PR) of estimated electricity generation using software tools PVGIS and PVsyst. Also, calculation of PR is made for the installed system using SCADA (actual) monitoring system installed with the SPV plant. Figure 8 shows the graphical representation of the PR, and it is seen that the estimated system using PVsyst has the highest PR in comparison with

a

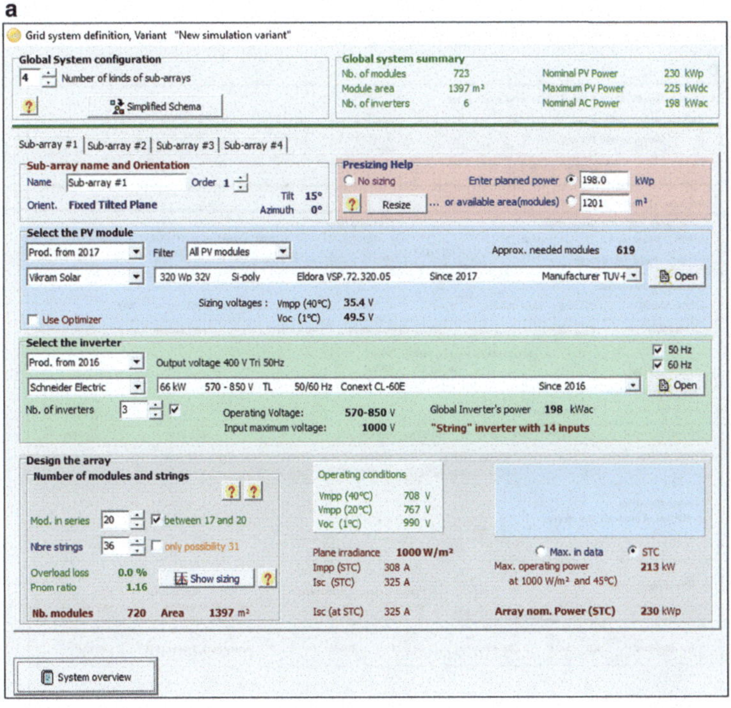

Fig. 6 **a** System parameters for PV array-1. **b** System parameters for PV array-2. **c** System parameters for PV array-3. **d** System parameters for PV array-4

b

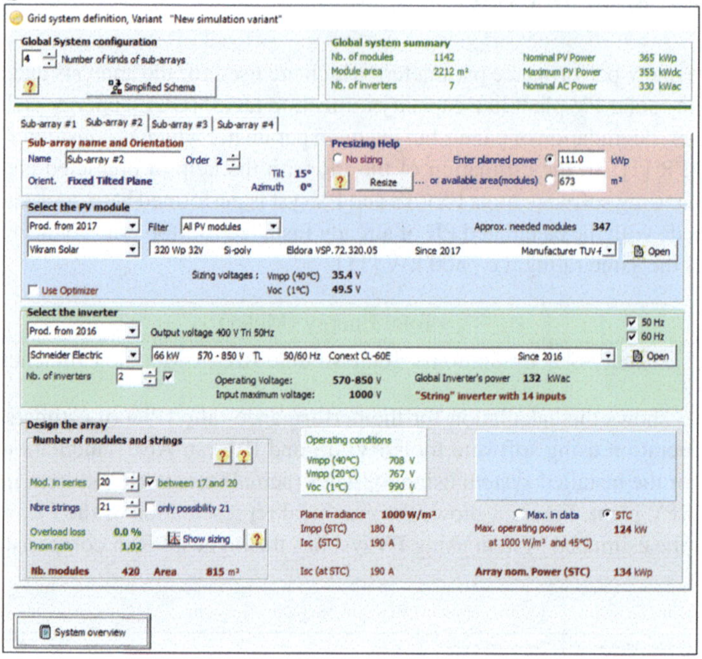

c

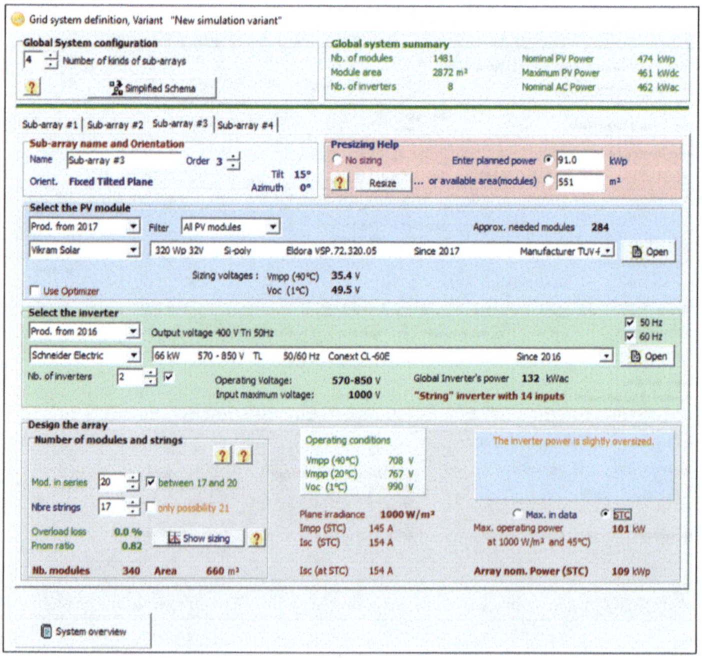

Fig. 6 (continued)

d

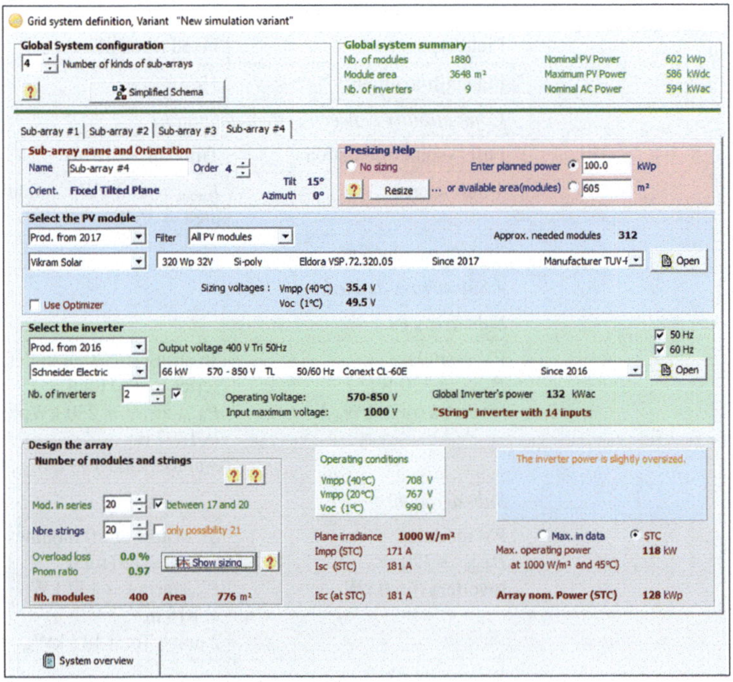

Fig. 6 (continued)

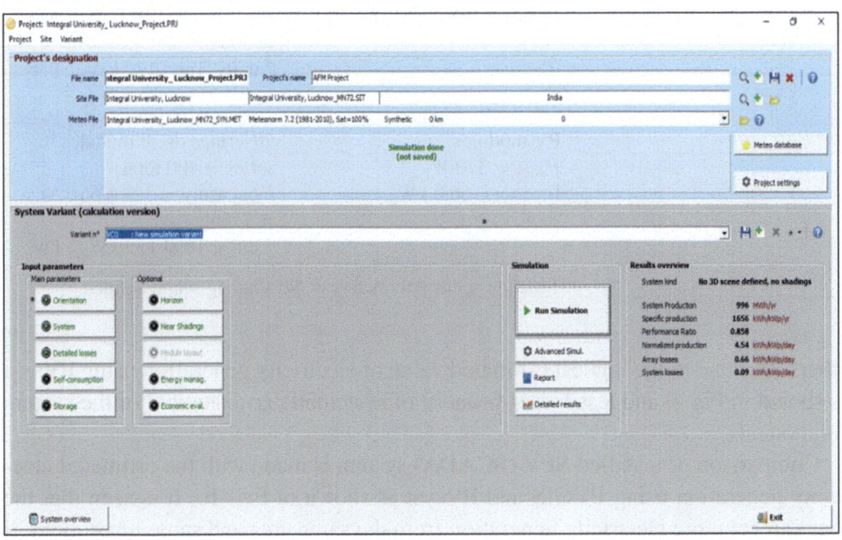

Fig. 7 PVsyst dashboard for simulation

Table 2 System overview of a 500 kW grid-interactive SPV plant

Orientation parameters:	
Field type	Fixed tilted plane
Plane tilt/azimuth	15°/0°
Compatibility between system definitions	
Full system orientation	Tilt/azim = 15°/0°
4 Sub-arrays	P_{nom} = 602 kWp, modules area = 3648 m^2
System parameters	
4 Sub-arrays defined	
Sub–array #1	
PV modules $P_{nom} = 320\ W_p$ Inverters (66.0 kW$_{ac}$)	36 strings of 20 modules in series = 720 total P_{nom} array = 230 kWp, Area = 1397 m^2 3 units, Total 198 kW$_{ac}$, PN
Sub–array #2	
PV modules $P_{nom} = 320\ W_p$ Inverters (66.0 kW$_{ac}$)	21 strings of 20 modules in series = 420 total P_{nom} array = 134 kWp, Area = 815 m^2 2 units, Total 132 kW$_{ac}$, PN
Sub–array #3	
PV modules $P_{nom} = 320\ W_p$ Inverters (66.0 kW$_{ac}$)	17 strings of 20 modules in series = 340 total P_{nom} array = 109 kWp, Area = 660 m^2 2 units, Total 132 kW$_{ac}$, PN
Sub–array #4	
PV modules $P_{nom} = 320\ W_p$ Inverters (66.0 kW$_{ac}$)	20 strings of 20 modules in series = 400 total P_{nom} array = 128 kWp, Area = 776m^2 2 units, Total 132 kW$_{ac}$, PN

*Shading scene parameters = No shading scene defined

others. It is due to the highest estimated value of electricity generation using PVsyst as shown in Fig. 9 and less losses because of systematic arrangement of the system components.

Comparison of installed SPV (SCADA) system is made with the estimated electricity generation using PVGIS and PVsyst as shown in Fig. 10. It is seen that the user can get more electricity generation from the same area and same investment, if proper planning is performed using PVsyst.

Some of the parameters are discussed in next section, which is the unique feature of the PVsyst when it is compared with other forecasting software.

Table 3 Calculation of performance ratio

Parameters	PVGIS	PVsyst	SCADA (actual)
Electricity generated/year	740,170 kWh	996,400 kWh	800,509 kWh
Total irradiation	1866 kWh/m^2	1866 kWh/m^2	1866 kWh/m^2
Total module area	3648 m^2	3648 m^2	3648 m^2
Module efficiency	17%	17%	17%
%PR	$\frac{740,170}{1866\times3648\times0.17}\times100=$ 63.9%	$\frac{996,400}{1866\times3648\times0.17}\times100=$ 86%	$\frac{800,509}{1866\times3648\times0.17}\times100=$ 69%

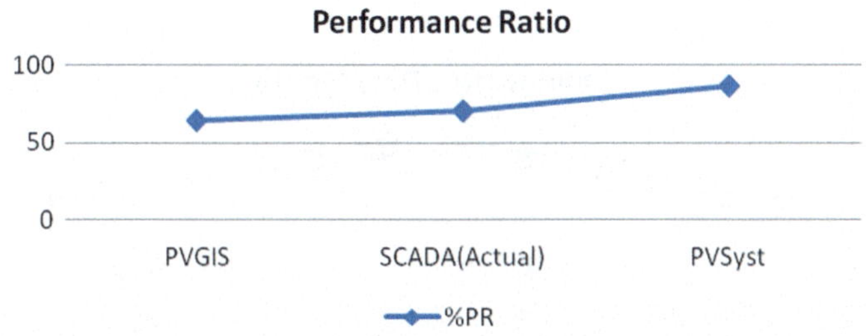

Fig. 8 Performance ratio

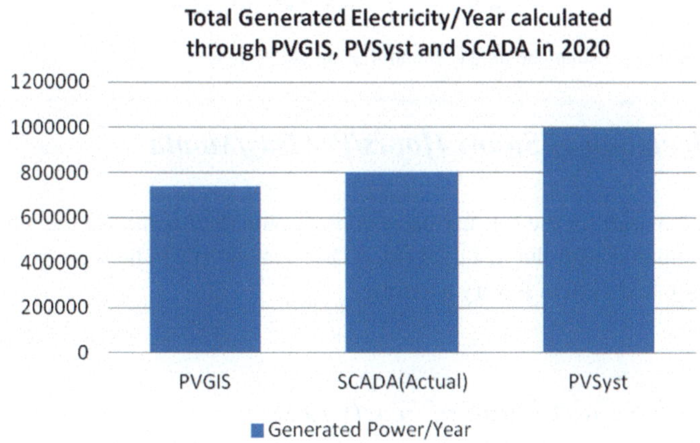

Fig. 9 Total generated electricity/year in 2020

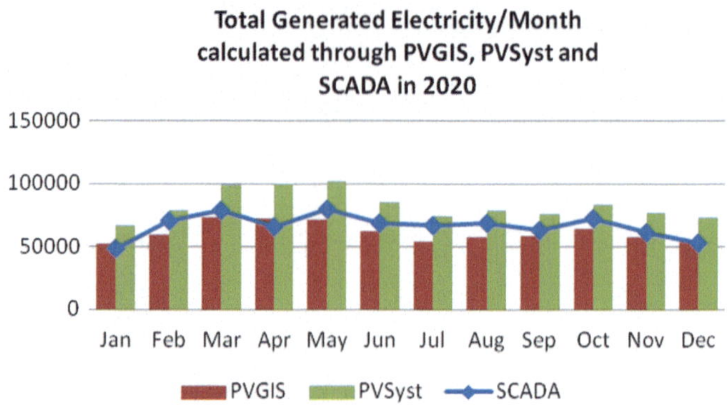

Fig. 10 Total generated electricity/year in 2020

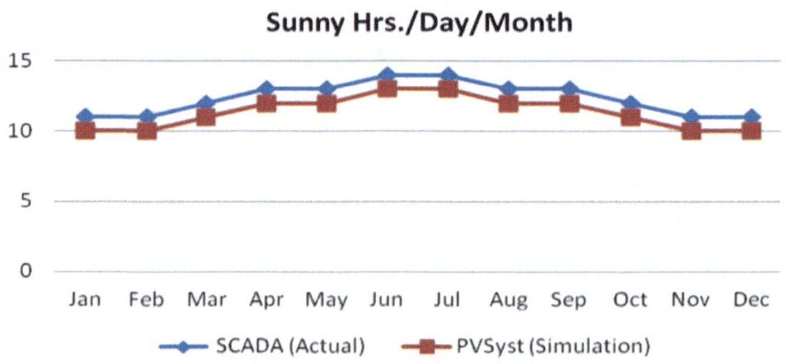

Fig. 11 Estimation of sunny hours/day/month for the whole year

4.1 Calculation of Sunny Hours/Per Day/Month

PVsyst is also used for the estimation of sunny hours/day/month for the whole year, and it is compared with the installed (SCADA) system. It can be seen from Fig. 11, that the estimated system is very accurate.

4.2 Detailed Calculation of Yearly Losses

Forecasting of system losses can be performed for the whole year using PVsyst and shown in Tables 4 and 5.

Table 4 System losses for the whole year using PVsyst

Loss factors (PV array)				
Loss factor (thermal)	U_c (const) *A = Array	29.0 W/m^{2}K	U_v (wind)	0.0 W/m^{2}K/m/s
Wiring loss (ohmic)	*A#1	37 mΩ	% loss	1.5% (STC)
	*A#2	63 mΩ	% loss	1.5% (STC)
	*A#3	78 mΩ	% loss	1.5% (STC)
	*A#4	66 mΩ	% loss	1.5% (STC)
	Global		% loss	1.5% (STC)
Quality loss (module)			% loss	– 0.8%
Loss (module mismatch)			% loss	1.0% at MPP
			% loss	0.10%

IAM (incidence effect): smooth glass (fresnel), $n = 1.526$

0°	30°	50°	60°	70°	75°	80°	85°	90°
1.000	0.998	0.981	0.948	0.862	0.776	0.636	0.403	0.000

Table 5 Estimated losses for the whole year using PVsyst

S. No.	Parameters	Values
1	Global irradiation (horizontal)	1815 kWh/m^2
	Global incident (in coll. plane)	+ 6.4%
	IAM factor (global)	– 3.39%
2	Effective irradiation (collectors)	1866 kWh/m^2 * 3648 m^2 coll
3	Photovoltaic conversion	Efficiency at STC = 16.51%
4	Nominal energy of array (at STC effic.)	1124 MWh
	PV loss (irradiance level)	– 0.48%
	PV loss (temperature)	– 7.94%
	Quality loss in module	+ 0.75%
	Mismatch loss (modules and strings)	– 1.10%
	Wiring loss (ohmic)	– 1.00%
5	Virtual energy of array (at MPP)	1016 MWh
	Loss in inverter (operation)	– 1.91%
	Loss in inverter (over nominal inv. power)	0.00%
	Loss in inverter (max. input current)	0.00%
	Loss in inverter (over nominal inv. volt)	0.00%
	Loss in inverter (power threshold)	0.00%
	Loss in inverter (voltage threshold)	0.00%
	Consumption (night)	0.00%
6	Energy available (inverter output)	996 MWh
7	Energy injected (grid)	996 MWh

4.3 Degradation of Solar Panels/Modules

This is also a good feature of PVsyst that it can be used for the degradation of Solar panels in terms of decreasing efficiency and increasing losses per year, as shown in Fig. 12a, b. So if the efficiency is decreasing more rapidly than these values, installer

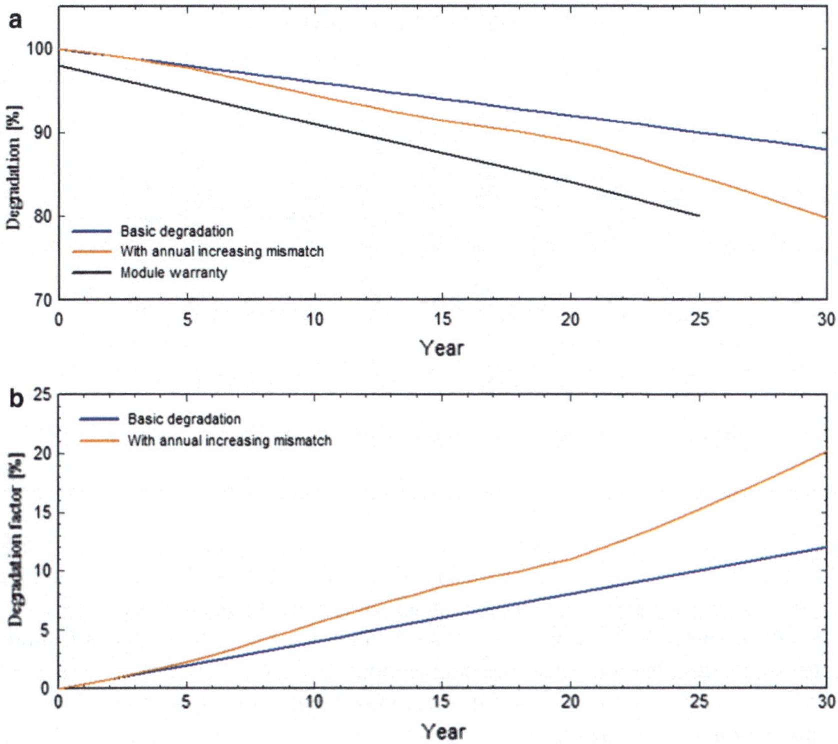

Fig. 12 **a** Decreasing efficiency due to degradation of solar panels. **b** Increasing losses due to the degradation of solar panels

can check the system and fix the issues related to increase of losses. Sometimes due to the shading effect the generation suffers, other reason of the increase of loss is damage of panels, so this analysis helps the installer in increasing the generation as an idea or standard range of losses can be determined with this PVsyst simulation [12–15].

Furthermore, the SCADA system installed with the 500 kW rooftop grid-connected system has all the necessary features for the monitoring of electricity generation, performance ratio (PR), capacity utilization factor (CUF), temperature, insolation, energy yield, etc. System parameters like PR, CUF, and electricity generation are highly dependent on the weather condition of the site. There are possibilities of high chances of the occurrence of the deviation from the estimated values or forecasting for a specific site, but because of the high investment forecasting as well as assessment is necessary to perform time to time. Figure 13 shows the insolation, temperature, and CUF output of the installed system at the site and compared with [16–18].

Where the ratio of generated energy from the plant to the total plant capacity (yearly) is termed as capacity utilization factor (CUF).

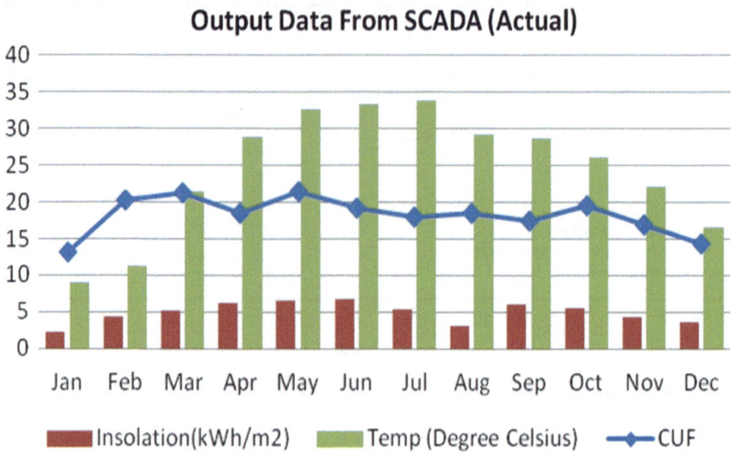

Fig. 13 Insolation, temperature, and CUF output of the installed system

- So finally, it is clearly seen from the Fig. 1b, 2–c for solar PVGIS, that it is simple, user-friendly and has less input parameter's requirement than PVsyst.
- While as shown in Figs. 5a–c, 6a–c and 7, input parameters requirement is more for simulation, but it results in accurate output.
- Also Tables 1 and 2 give clear information of input parameters required for the simulation using PVsyst.
- Losses and accurate module degradation analysis are also a good feature of PVsyst, in which solar PVGIS lags. This is shown in Tables 4, 5 and Fig. 12a, b.
- So when accuracy is required in any system, PVsyst is better than PVGIS, as former have more options of assessment parameters in comparison with PVGIS.
- Major advantage of PVGIS is, it is available free of cost and for PVsyst, and user has to pay according to requirement. This is also a reason that PVsyst has more features than PVGIS.

5 Conclusion

It is shown from figures and tables that the simulation software like solar PVGIS, PVsyst are good enough for the planning and forecasting of SPV plant. When the values are compared with the actual SCADA values of installed SPV plant of the same rating, it is clear that PVsyst software has some more technical features than PVGIS. While accuracy of the forecasting software depends upon the accuracy of input parameters. Since PVGIS is open-access software and it is the major advantage of this tool, which can be used anywhere as discussed earlier in the background section, most of the input parameters are pre-decided or linked with the other weather forecasting websites, which sometimes consider ideal system, and there may be chances of errors in forecasting, so it may be good for non-commercial sites. On the other hand, PVsyst is a paid and accurate software for commercial sites in which there is an option to feed more practical data in comparison with PVGIS. So simulation with PVsyst software requires expertise and that is why it is more complex than PVGIS in some aspects. PVsyst can be used for the forecasting of Electricity generation, degradation of PV panels/modules, losses, performance ratio, number of sunny hours/per day/month, etc., for commercial or non-commercial purposes as discussed in this chapter.

Acknowledgements The authors would like to acknowledge Integral University, Lucknow, for providing the MCN "IU/R&D/2021-MCN0001164." The statements made herein are solely the responsibility of the authors.

References

1. A.K. Sarraf, S. Agarwal, D.K. Sharma, Performance of 1 MW photovoltaic system in Rajasthan: a case study, in *2016 IEEE 7th Power India International Conference (PIICON)*, 2016
2. A.M. Diez-Pascual, R.P. Capilla, P.G. Díaz, Design of simulation tools for teaching in photovoltaic energy engineering. MDPI Proc. **2**, 1364 (2018).https://doi.org/10.3390/proceedings2211364
3. S.M. Bhagavathy, G. Pillai, PV Microgrid design for rural electrification. MDPI Des. **2**, 33 (2018).https://doi.org/10.3390/designs2030033
4. S. Manju, N. Sagar, Progressing towards the development of sustainable energy: a critical review on the current status, applications, developmental barriers and prospects of solar photovoltaic systems in India. Renew. Sustain. Energy Rev. **70**, 298–313 (2017)
5. S. Ahsan, K. Javed, A.S. Rana, M. Zeeshan, Design and cost analysis of 1 kW photovoltaic system based on actual performance in Indian scenario. Perspect. Sci. **8**, 642–644 (2016)
6. A.F. Minai, T. Usmani, M.A. Mallick, Optimum sizing and estimation of a 30 kWp hybrid solar photovoltaic system with multilevel inverter. Int. J. Res. Sci. Innov. (IJRSI), 31–36 (2015). ISSN 912321-2705
7. A.F. Minai, T. Usmani, A. Iqbal, *Performance Evaluation of a 500 kWp Rooftop Grid-Interactive SPV System at Integral University, Lucknow: A Feasible Study Under Adverse Weather Condition. Studies in Big Data*, vol 86 (Springer, Singapore, 2021). https://doi.org/10.1007/978-981-33-4412-9_24
8. D. Dey, B. Subudhi, Design, simulation and economic evaluation of 90kWgrid connected Photovoltaic system. Energy Rep. **6**, 1778–1787 (2020)
9. M. Naseem, M.A. Husain, A.F. Minai et al., Assessment of meta-heuristic and classical methods for GMPPT of PV system. Trans. Electr. Electron. Mater. (2021). https://doi.org/10.1007/s42341-021-00306-3
10. B. Belmahdi, A. El Bouardi, Solar potential assessment using PVsyst software in the Northern Zone of Morocco. 13th Int. Conf. Interdisc. Eng. (INTER-ENG 2019) Proc. Manuf. **46**, 738–745 (2020)
11. A.K. Yadav et al., Soft computing in condition monitoring and diagnostics of electrical and mechanical systems, 1E. Part of the Advances in Intelligent Systems and Computing, vol. 1096 (Springer Nature, 2020), p. 496. ISBN 978-981-15-1532-3. https://doi.org/10.1007/978-981-15-1532-3
12. A. Iqbal et al., Meta heuristic and evolutionary computation: algorithms and applications, 1E. Part of the Studies in Computational Intelligence, vol. 916 (Springer Nature, 2020), p. 849. ISBN 978-981-15-7571-6. https://doi.org/10.1007/978-981-15-7571-6
13. J.A. Alzubi, AI and machine learning paradigms for health monitoring system: intelligent data analytics, 1E. Part of the Studies in Big Data, vol. 86 (Springer Nature, 2020), p. 513. ISBN: 978-981-33-4412-9. https://doi.org/10.1007/978-981-33-4412-9.

14. A. Iqbal et al., Renewable power for sustainable growth, 1E. Part of the Lecture Notes in Electrical Engineering, vol. 723 (Springer Nature), p. 805. ISBN: 978-981-33-4080-0. https://doi.org/10.1007/978-981-33-4080-0
15. N. Fatema et al., Intelligent data-analytics for condition monitoring: smart grid applications, 1E (Academic Press, 2021). ISBN 978-0-323-85510-5. https://doi.org/10.1016/C2020-0-02173-0
16. S. Srivastava et al., Applications of artificial intelligence techniques in engineering-volume 1, 1E. Part of the Advances in Intelligent Systems and Computing, vol. 698 (Springer Nature, 2018), p. 643. ISBN 978-981-13-1819-1. https://doi.org/10.1007/978-981-13-1819-1
17. S. Srivastava et al., Applications of artificial intelligence techniques in engineering-volume 2, 1E. Part of the Advances in Intelligent Systems and Computing, vol. 697 (Springer Nature, 2018), p. 647. ISBN 978-981-13-1822-1. https://doi.org/10.1007/978-981-13-1822-1
18. P. Vinoop et al., PSO-NN-based hybrid model for long-term wind speed prediction: a study on 67 cities of India. Applications of Artificial Intelligence Techniques in Engineering, Advances in Intelligent Systems and Computing **697**, 319–327 (2018). https://doi.org/10.1007/978-981-13-1822-1_29

Intelligent Tools and Techniques for Renewable Energy Analytics: A Review on PV Generation Prediction

Amruta Pattnaik and Anuradha Tomar

Abstract A lot of development and transformation is going on in electrical distribution systems in the last decades. The power conversion of non-renewable energy into hybrid renewable energy resources (RERs)-based device and low voltage (LV) distribution system, which is called electrical electrification, are observing increased penetration of green energy from distributed energy resources (DER). Nowadays, green energy or renewable energy becomes a prominent energy source. PV power is one of the promising RES as compared to other renewable energies as it is low cost, eco-friendly, abundant in nature as well as easy to fix. LV distribution networks with high photovoltaic (PV) penetrations are facing operational and control issues. Further, for optimal operation of microgrids ensuring the efficient utilization of available PV power, accurate PV predictions are essential. In recent years, the application of artificial intelligence techniques and data analytics tools have gained an interest in ensuring techno-economical and safe operation of LV distribution grids, microgrids. This chapter aims to present a detailed review of the various intelligent tools and techniques that have been developed for photovoltaic power generation predictions. Based on the prediction time horizon, PV predictions methods are categorized into three categories, i.e., short-term, medium-term, and long-term forecasting. It is observed that accuracy of PV prediction outcome not only depends upon the applied methodology but also depends upon the optimal selection of features and considered dataset.

Keywords Artificial intelligence · Data analytics · Forecast · Intelligent tools · Photovoltaic prediction

A. Pattnaik (✉)
Dr. Akhilesh Das Gupta Institute of Technology and Management, Shastri Park, FC-26, Delhi 110053, India

A. Tomar
Netaji Subhas University of Technology, Delhi 110078, India

H. Malik et al. (eds.), *Intelligent Data Analytics for Power and Energy Systems*, Lecture Notes in Electrical Engineering 802,
https://doi.org/10.1007/978-981-16-6081-8_17

Nomenclature

AML	Automatic machine learning
ANN	Artificial neural network
AR	Autoregressive
BTM	Behind-the-meter
CNN	Current neural network
DELM	Deep extreme learning machine
DIP	Digital image processing
DKASC	Desert Knowledge Australia Solar Centre
DLC	Deep learning communities
ED	Energy disaggregation
EEMDA	Ensemble empirical mode decomposition algorithm
ELM	Extreme learning machine
GA	Genetic algorithm
GRNN	General regression neural network
LSTM	Long short-term memory
MARS	Multivariate adaptive regression splines
MDE	Mixture design of experiments
MLP	Multilayer perceptron
MLPNN	Multilayer perceptron neural network
MLR	Multiple linear regression
MME	Multimodel ensemble
NARX	Nonlinear autoregressive network with exogenous inputs
NDNN	Novel deep neural network
NN	Neural network
NWP	Numerical weather prediction
PCA	Principal component analysis
PV	Photovoltaic
PVPG	Photovoltaic power generation
RBF	Radial basis function
SPG	Solar power generation
SVM	Support vector machine

1 Introduction

There are many numbers of non-conventional energy resources like solar, tide, wind, biomass, natural gas, etc. However, the use of photovoltaic-based non-conventional energy sources becomes quite popular among society. Photovoltaic (PV) solar cell power generation mainly depends upon sunlight. The sunlight intensity always not the same due to various weather conditions [1]. As a result, PV output is also affected. Solar cell output is not only affected by sunlight intensity but also influenced by the heat generated by the solar cell, the velocity of flowing air, insolation, weather condition, and presence of moisture in the air. Due to the various factors, inconsistency has occurred in the output power of PV plant, and it is quite necessary to provide accurate forecasting/predictions for stability and balancing of solar power output from each second/minutes/hours. The accuracy of the forecast depends upon the past and future database of historic and predicted output of power. Different types of forecasting techniques are reported based on forecasting perspective, historic data of solar light intensity and another climatological data pattern, and approaches applied in the forecast. Depending upon the prediction time horizon, accurate PV forecasting may help in effective planning and operation of low voltage (LV) distribution networks with high PV penetration [2], and energy management of grid connected, or isolated PV-based microgrids/systems [3, 4]. Thus, accurate PV forecasting would result in efficient utilization of available PV capacity and would reduce the operational cost [5], which further results as an increased return on investment [6]. There are two different types of prediction methodologies reported in the literature, i.e., (i) physical and (ii) statistical process. For physical process, the prediction techniques require physical data, the position of the PV panel, reversal efficiency, as well as weather data. Such a physical process methods provide accurate data whereas indicate some intelligent tools and techniques to analyze and define a set of prediction rules. Those statistical techniques are the artificial neural network, machine learning, fuzzy logic, etc. The statistical process has the advantages of short-time basis solar output prediction [7–10]. Many research works are mainly focused on the forecast horizon technique for PV power generation. The prediction perspective or forecast horizon is defined as the predicting the PV output in advance for a duration of time. The aim and correctness of a solar plant predicting models depend on the prediction perspective. According to the many pieces of literature, a prediction/forecasting model should be designed based on prediction time horizon, which is divided into three categories of forecasting techniques, i.e., short, medium (long–short), and long-span-based forecasting or prediction control techniques as presented in Fig. 1 [11–13]. The prediction of PV output for a week is defined as short-term forecasting,

Fig. 1 Types of forecasting for PV power generation

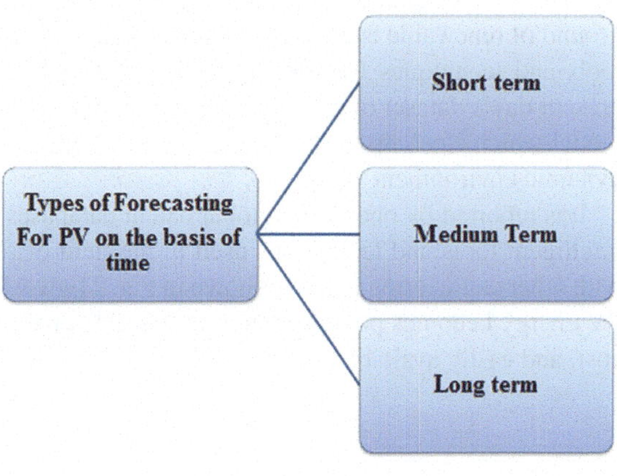

and it is helpful to increase the unit commitment, arranging, and transmitting of energy. Medium-term forecasting technology predicted the PV output in advance of 7–30 days, while long-term forecasting is able to predict the PV output for a month to a year ahead. Such type of long-time span forecasting has the advantages of design a solar plant intake, power plant operation, and control as well as market commitments [14].

This chapter aims to provide a summary of the various PV power prediction methodologies that have been developed by various researchers in order to accurately predict PV power output using different types of intelligent tools and techniques. The whole chapter has been divided into three sections. In Sects. 1 and 2, the overview and latest work in intelligent tools and techniques used in different fields for PV energy forecasting has been discussed. In Sect. 3, different types of tools and techniques used for PV forecasting are illustrated. Sections 4 and 5 are focused on the methodology and key findings of the research gap along with the discussion of this research, respectively. Section 6 concludes the main findings of the review with future scope.

2 Intelligent Data Analytics for RER

Due to climate crisis, the climate of earth is adversely affected. So, the use of green energy or renewable energy is quite increased in a large manner. There are different types of non-conventional energy sources, i.e., solar, wind, biomass, hydel, natural gas, fuel cell, tidal, and many more. The renewable energy has their own advantages over others, but still renewable energies have some limitations. As the sources are abundant in nature, but due to lack of technologies and knowledge, it is unable to get 100% efficiency at the output side. Several studies have been carried out to improve the efficiency of renewable energies. Different types of factors are also encountered to improve the performance of renewable energy resources such as environmental factors, design factors, fabrication factors, raw materials, inducements for assets as well as different types intelligent tools, new practical methods, and many more. However, the designing, implementing, and marketing are a significant process in the ground of renewable energy and challenging too. Furthermore, design an intelligent tools and techniques for forecasting of renewable energy output on the basis of present day's dataset of climate, calculation of output by the help of computational model which can help to save energy, increase the profit as well as will manage the dynamic environment issues too.

It is reported by one of the global patent databases that the maximum number of intelligent tools and techniques used in the field of renewable energy is compared with other engineering fields as shown in Fig. 2 [www.lens.org]. In renewable energy, PV energy becomes popular among renewable energies as it is easy to handle, low cost, and easily available.

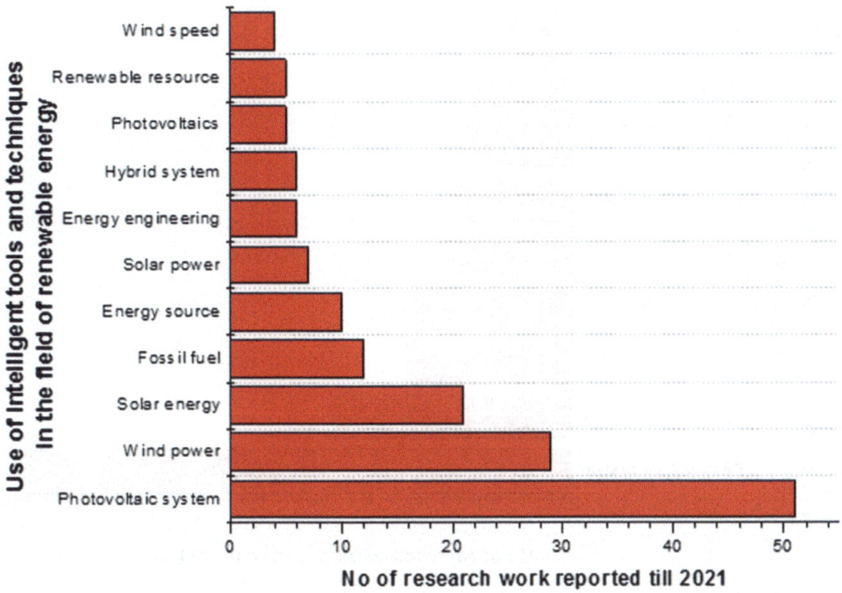

Fig. 2 Use of intelligent tools and techniques in the field of renewable energy reported by global patent database [15]

2.1 *Intelligent Tool and Techniques for PV Generation*

PV output power is completely dependent upon sunlight. PV power will be maximum if the sunlight intensity becomes 1000 W/m^2. The irregularities of sun light create thermalization and transmission losses in PV module. So, the output of PV is affected. Researchers are exploring many fields in PV energy to enhance the PV efficiency. Some of the important areas of research in the field of PV energy are as shown in Fig. 3 [www.lens.org]. A maximum number of patents are reported in the field of PV system. The second and third most important research areas are energy management and energy storage. Presently different types of intelligent tools and techniques are used to analyze, predict the data for certain area or time, load shedding, optimization, islanding detection, technical viability, power loss calculation for PV technologies.

There are different number of tools that have been reported for PV technologies. In Fig. 4, it is observed that application of artificial neural network in PV technology is used maximum number of times for patent. Presently AI, big data analysis, ML, simulation, cloud computing, fuzzy logic, genetic algorithm, and PSO science are the latest research tools in PV technology. Such tools are useful for grouping, collecting, configuration, and forecasting in many fields. Artificial intelligence has many divisions, i.e., ANN, GA, FL, PSO, ML, and many more as shown in Fig. 4. The University College London has reported many works in the field of AI. Likewise, Chinese Academy of Science and ETH Zurich have come in second and third position, respectively, to publish a large number of work in the field of AI.

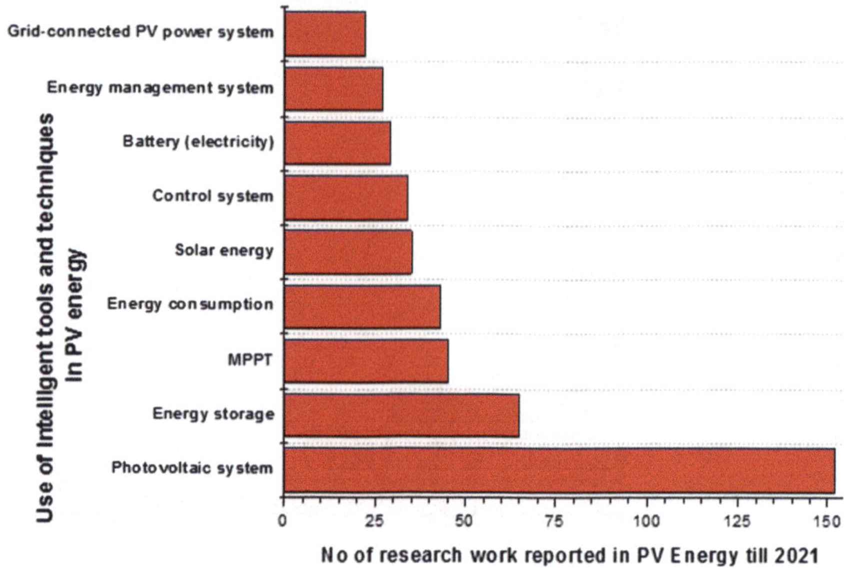

Fig. 3 Bar graph of number of intelligent tools and techniques in the different field of PV and energy as per the global patent database [15]

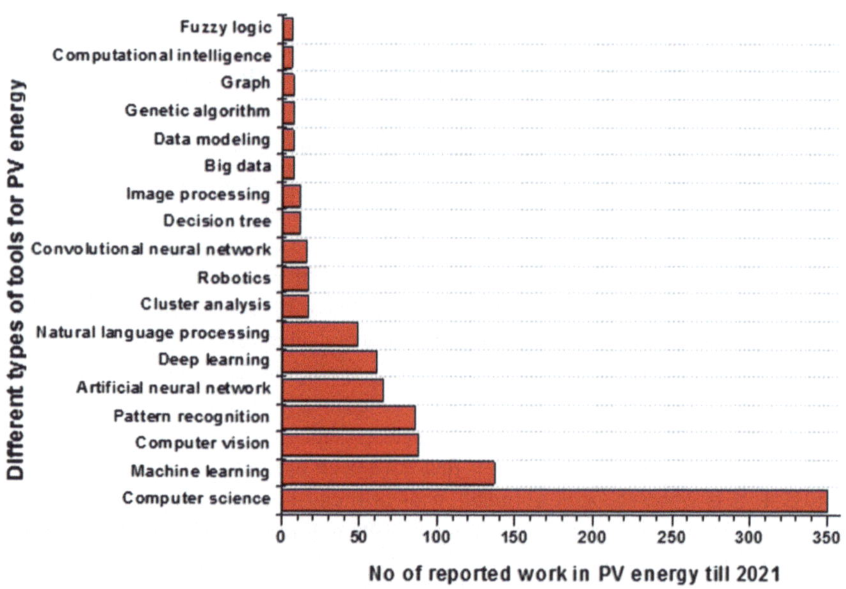

Fig. 4 Bar graph of number of intelligent tools and techniques in the field of PV energy as per global patent database [15]

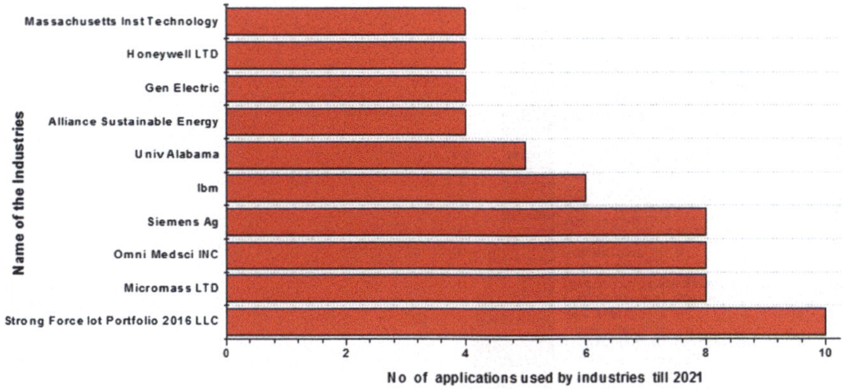

Fig. 5 Bar graph of number of AI applications by industries as per the global patent database [15]

Figure 5 displays the bar graph of topmost industries who are using the AI-based applications. AI-based models are more advantageous over other techniques. World-wide the application of AI has been increased rapidly as it is required less knowledge for internal connection, computational time and labor is less, one solution for many problems. The integration of AI with PV technology becomes a popular design to create intelligent tools and techniques in PV system. By the help of intelligent tools, it can be easily resolved the issues in controlling, supervising, analyzing, monitoring as well as forecasting.

2.2 Intelligent Data Analysis for PV Generation

Next generation of PV industry management process is based upon the principle of PV industry forte with information technology. PV industries have to design new optimizing technology with intelligent tools. It can solve the real-time problems, finding a design pattern; easily adaptable, applicable, predictable, and available are called intelligent data analysis. For the intelligent data analysis, a massive database is required, and such huge database access can be resolved by intelligent data analysis tools and techniques. Thus, PV energy is a promising technology for the use of intelligent data analysis tools. According to global patent and reports [www.len s.org], there are many intelligent data analysis reported till 2021 using a bar chart as shown in Fig. 6 [15].

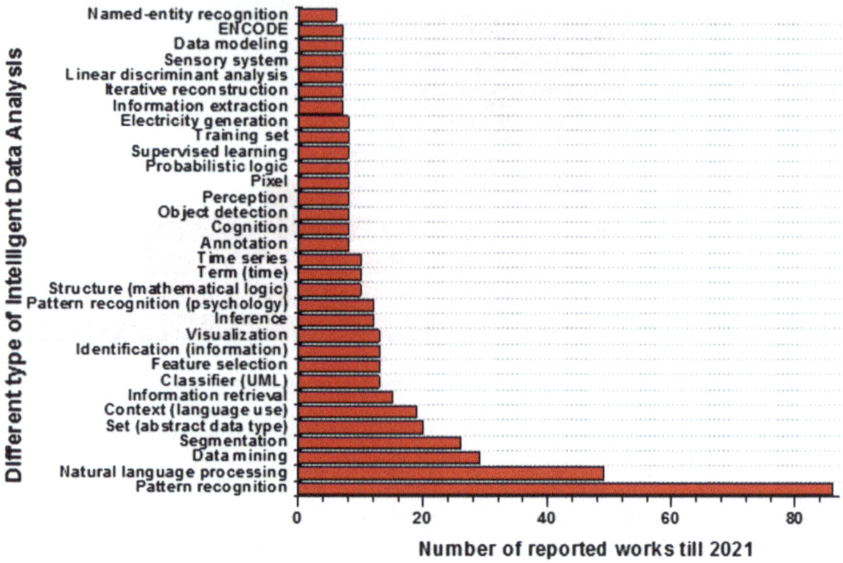

Fig. 6 Use of intelligent data analysis in the different method as per the global patent database [15]

3 Methodology

The intelligent tools for data analysis can identify the pitfall and recover by forecasting of failures which makes the PV power generation maximized. Before applying intelligent data analysis for forecasting, it requires collection of historic clean data and other essential data which will create a useful piece of information for forecasting. Data collection, mining, wrangling, and predicting are the four fundamental analytical steps for PV power generation.

A model of an intelligent data analysis for the improvement PV output is shown in Fig. 7. Data collection is the first step to predict the PV output. As the PV output is influenced by many factors like solar irradiance, area of the PV panel, weather condition, wind velocity, position of panel as well as the raw material used in PV cell too. It needs to collect all the relevant data to PV output by the help of studying, surveying, sampling as well as scheduling. In second stage, it requires data mining. It needs to extract the data from the studying, surveying, sampling as well as scheduling by the help of intelligent tools and techniques. In the third stage of data analysis process, it needs the clean the data. In other words, it removes the bad data and tries to fill the missing values. At the last stage, it needs to predict the output based on input data. In this process, the three different types of methods are reported in the literature to predict the PV output which is achieved by intelligent data analysis tools and techniques. In the below sections, different types of intelligent data analysis tools and techniques are mentioned to generate the PV output.

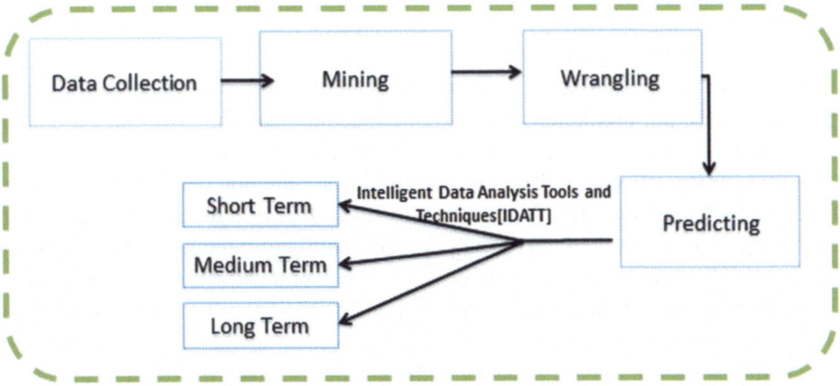

Fig. 7 Model of an intelligent data analysis for PV power generation

4 Related Research Work

Prediction methodologies haven been divided into three classifications, i.e., short-term, medium-term, and long-term prediction, based on the prediction time horizon. Different techniques and tools have been developed by researchers considering the various features for efficient PV predictions considering the short-term, medium-term, and long-term prediction/forecasting time horizon. Those are discussed briefly in following sections.

4.1 Short-Term Forecasting [STF]

The current short-term PV forecasting is important for many applications like congestion management in LV distribution grids with high PV injections, estimation and procurement of flexibility, planning and operation of near real-time operations. Further, it is beneficial for scheduling RER in hybrid energy systems. On the other hand, due to highly intermittent nature of solar PV, short-term PV forecasting becomes a challenging task. Short-term PV prediction is further defined as ultra-short-term and STF based on the prediction time horizon of 0–4 h and 0–7 days, respectively. To increase the utilization of available PV power, it is necessary to control the power generation actions such that to meet the load demand with the help of accurate forecasting of PV power. The short-term PV forecasting involves three-stage procedure such as (i) removal of noise from the signal and dataset processing; (ii) behavior of variable optimization; and (iii) design a training and forecasting method for the network [11]. First stage is related to extraction of useful dataset, filing missing values if any and it is processing as per the adopted prediction methodology. Second step is related to selection of optimum features for prediction algorithm based on feature selection techniques. The third stage consists of diving the dataset into training and testing and tuning the model parameters. In this section, different types of short-term forecasting techniques are discussed.

In the year 2016, Tang et al. reported about the entropy method which was the hybrid of PV output prediction with extreme machine learning [16]. Such a model provided an efficient and reliable method for the long-term as well as a short-term prediction for the PV output. The introduced entropy method consists of sequence of steps which includes data initialization, excites the system over the data after unification, and then predicts the PV output. The results were equaled with radial basis function (RBF) neural network, general regression neural network (GRNN) method, and other methods too. Monteiro et al. [17] approached seven training algorithms for the forecasting of disturbances which were used in artificial neural networks, with nonlinear autoregressive network with exogenous inputs (NARX) architecture during the PV power output estimation [17]. Yadav et al. [18] have discussed the five most influencing forecasting variables in the field of PV power generation which are considered as input of the system [18]. The forecasting variables were used as an input in the five models based upon artificial neural network (ANN) and the other five from multiple linear regression (MLR). Sunlight and wind temperature were considered as input variables for ANN-based prediction of concentrated power by PV unit. One more short-term forecasting model was proposed by Takeda et al. [19] to estimate the PV power generation [19]. A state-space model-based model has been developed which is capable to predict the output power on hourly basis as well as estimation on monthly basis also. Further, hourly basis power model was improved by monthly PV power output capacities. Ensemble Kalman filter was used for the data integration of the state-space model. The proposed method was better as compared to the conventional models which had been used in the year 2017. It was also stated that such a model was suitable for the STF during the analysis of the PV model and its output. Hu et al. [20] stated that a three-stage based ultra-short-term prediction model which was a radial basis function neural network prediction model, and such model was compared with autoregressive model [20]. Such a model provided an accurate forecast for PV during cloudy weather. It had three stages to design the prediction model. In the first stage, the cloud movement was forecasted using the motion vector where the selected cloud was covered by the sun, whereas the second stage informs the dynamic characteristics of the cloud with the help of DIP. At the third stage, radial basis function neural network (NN) was designed. Based on the comparison, 7.4% and 13.6% were noted as the forecasting accurateness of RBF and autoregressive (AR) prototypical, respectively.

Liu et al. [21] applied data-driven ensemble modeling technique with aim to enhance the accurateness of postulation during the PV power generation [21]. Authors also have investigated the impact of inconsistency and uncertainty in the PV power output. Requirement of zero prior separation became the advantage of the model. The proposed technique was based on support vector machine (SVM), multi-layer perceptron (MLP), and multivariate adaptive regression splines (MARS) techniques forming stand-alone procedures to shape the ensemble prototypical. However, the reported model was able to forecast only one-time stamp ahead, i.e., one value at a time (deterministic forecast). The data-driven ensemble model was suitable for large assessment of data as well as random power production due to the minimized risk of the model range. Authors Xu et al. [22] proposed a new rule to get accurate prediction in PV output, i.e., unified photovoltaic forecasting model [22]. Such a model was helpful to minimize the trading cost from the forecasting inaccuracy. The proposed model has designed an integrated system with the combination of the machine learning and power system operation approach where gradient-based optimization along with a feed-forward neural network is applied. It was reported that the whole model was only considered for four hours in advance-based schedule which was happened at the power grid of northwest in China. An ultra-short-term multistep day ahead prediction model was reported by Li et al. [23]. Such model was the combination of boosted colliding bodies optimization, vibrational mode decomposition, and a deep extreme learning machine (DELM). In the investigated model, the forecasting was designed individually according to the weather as well as the data dimensions became less so the computational burden was reduced. The other advantages of the proposed model were reported such as fast speed, high forecasting accuracy, and versatility whereas the only disadvantage was the computational complexity due to the many numbers of complex algorithms. Such a model was suitable to regulate the PV operating ability, planning, and consignment tracing forecasts [23]. Solar irradiation predictors based on both machine learning algorithms and three-dimensional and time-based variables were reported by Rodríguez et al. [24]. Feed-forward neural network spatio-temporal model was one of the suitable models for solar intensity prediction. The model has predicted the solar intensity before ten minutes. They also mentioned that a variation of less than 4% was observed between the original and predicted solar intensity in 82.95% of the inspected days [24].

Hassan et al. [25] suggested a technique using inherently optimized NARX for ultra-short-time-based estimation of output of solar plant [25]. Exogenous models were established and verified with several considered meteorological constraints at five different regions in Algeria and Australia under desert weather. It was observed that the forecasting accuracy of PV output enhanced up to 19% under the consideration of several days as well as confined time. Authors also reported to increase the performance of the suggested model at cloudy climate by re-evaluating as well as re-enforce in larger prediction possibilities. Another model was also proposed to improve the NARX model by collaborative model based on the past climatic conditions. Yu et al. [26] suggested an integrated PV generation prediction model with the improved gray wolf optimization procedure to expand the convergence ability as well as to resolve the impact of extreme learning machine (ELM) arbitrary factors. The proposed procedure has the advantages of both robustness, stability, and high convergence accurateness. The ensemble empirical mode decomposition algorithm (EEMDA) was used to decompose the PV power under altered climate situations. The intricacy of each constituent based upon the model entropy and the constituents were rebuilt to lessen the resolving cost of predicting models [26].

Rodríguez et al. [27] reported a model which was tested with original measured data of temperature in 2017 at a particular site of Vitoria-Gasteiz, Basque Country which was capable to provide a prediction on temperature for 10 min ahead [27]. Both the nearest location and the space between every PV plant and targeted station were concerning variables. The accrued deviance among real and calculated heat was lesser than 1% in 96.60% of the inspected days from the authentication record. The model has the drawback that the design, as well as the accuracy, was last tested in 2017 only.

In this section, different types of short-term forecasting methods have been studied as mentioned in Table 1. It is found that such type of short-term forecasting model is helpful during the development of the PV system along with suitable energy management. Short-term forecasting technique improves the grid resiliency as well.

4.2 Medium-Term Forecasting [MTF]

MTF is also known as long short-term forecasting. Medium-term forecasting techniques become an important method in the field of PV output forecasting for 0–30 days. Such models are applicable for the maintenance, operation, and development of PV plants, traditional power plants, transformers, and transmission lines. Such type of forecasting supports in the design of power system and maintains the design plan by forecasting the obtainability of the upcoming power generation. A day ahead PV forecasting is useful for voltage control, variability of frequency, and power transmission system planning and lowers the need of standby power. Also, it helps to minimize the price of unbalance energy, actual PV power production as well as day ahead energy market assurance. Pierro et al. [28] reported an evaluation of numerous data-driven prediction models by unlike numerical weather prediction (NWP) contribution to form multimodel ensemble (MME) for a 0–24 h ahead PV output prediction. The outcomes improved the accuracy of prediction and brought the skill score from 42 to 46% [28]. To overcome the stochastic nature of the photovoltaic system, the author Liu et al. [29] suggested a power predicting method using a backpropagation (BP) neutral algorithm located in Ashland in an advance of 24-h. It was noted that that the suggested structure had potential to forecast accurately and efficiency for predicting the PV system power output whereas solar irradiance, maximum, minimum, and average day heat, the average daily heat, and the PV output power per hour were considered as the features [29]. Lee et al. [30] proposed hybrid current neural network (HCNN) and long short-term memory (LSTM) networks for the forecasting of PV power generation. They developed the model for studying the time-series data in the DLC. They considered only the unevenly predictable meteorological conditions statistics from the national weather center. Novel deep neural network (NDNN) was designed to predict for a day PV power based upon time-series data from converters and national meteorological data centers [30]. Wang et al. [31] presented three categories of photovoltaic forecasting models, i.e., CNN, LSTM network, and hybrid model based on a CNN to derived data in Desert Knowledge Australia Solar Centre (DKASC), Alice Springs PV system. It observed that that the overall prediction outcome of the CNN-based hybrid model was comparatively most promising and accurate than the other two forecasting models [31]. Gao et al. [32] investigated the methodology for day ahead output power time-series

Table 1 Different types of short-term forecasting

	Short-term forecasting model	Remarks	References
1	Entropy method	Entropy method combining with ELM method possesses higher accuracy and the calculation is faster	Tang et al. [16]
2	Seven training algorithms in AI	Forecasting of disturbances with NRAX model	Monteiro et al. [17]
3	Five artificial neural network (NN) models and multiple linear regression (MLR) models	Solar light intensity and heat in air are found to be most influencing input variables	Yadav et al. [18]
4	State-space models (SSMs)	The hourly PV power models improved by monthly PV purchase volumes model	Takeda et al [19]
5	Ultra-short-term prediction model established on dynamic characteristics of the sun under the cloud	Power forecast accurateness of RBF model: 7.4% and the power forecast accurateness of AR: 13.6%	Hu et al. [20]
6	An ensemble technique	Ensemble model was efficient as compared to existing stand-alone model	Liu et al. [21]
7	Unified photovoltaic forecasting model	Minimize the trading cost	Xu et al. [22]
8	Combination of boosted colliding bodies optimization, vibrational mode decomposition, and a deep extreme learning machine (DELM)	Fast speed, high forecasting accuracy, versatility	Li et al. [23]
9	Solar irradiation predictors	Predicted the solar intensity before ten minutes	Rodríguez et al. [24]
10	NARX for ultra-short-term forecasting of PV power output	Enriched forecast accurateness of static multilayered perceptron neural networks	Hassan et al. [25]
11	Integrated PV generation prediction model with the improved gray wolf optimization procedure	Robustness, stability, and high convergence accurateness	Yu et al. [26]
12	Tested with original measured data of temperature in 2017 at a particular site of Vitoria-Gasteiz, Basque Country	prediction on temperature for 10 min ahead	Rodríguez et al. [27]

forecasting methods to accurately predict photovoltaic power production [32]. Two discrete cases were considered, i.e., the first one was a predicting technique that depends upon climatological data using LSTM networks considering ideal climatic situations and the other one was the time-series method suggested by presenting next day time series and typical climate type statistics under non-ideal climate situations. The result showed that after comparing both the prediction performances, the proposed method obtained the better prediction accuracy in non-ideal weather conditions. Wang et al. [33] proposed an LSTM-CNN network for PV output forecasting which was called as the combined deep learning model. In the projected amalgam model, the time-based structures of the data were taken out first followed by using the spatial features of the solar power statistics [33]. After analyzing the forecast outcomes of the amalgam model with the LSTM model, the CNN model, as well as the hybrid model of the convolutional-LSTM model, showed a better prediction performance. Ziane et al. [34] investigated the impact of meteorological variable quantity on the act of a 6 mWp grid-integrated PV plant by performance evaluation and power predicting in the desert of Adrar. For the prediction of power production, both random forest method and preprocessing methods were implemented whereas meteorological variables applied as inputs. The result showed a robust relationship between the input variable quantity (i.e., weather parameters specifically irradiation and ambient temperature) and the output power production behavior of the photovoltaic station [34]. Zhao et al. [35] constructed a joint method for the next 24 h of solar power generation (SPG) forecast for multiregion photovoltaic plants by applying automatic machine learning (AML) followed by an improved genetic algorithm (GA). The results found that the optimal combined technique provides highly suitable forecast accuracy than other methods used for comparison [35]. Luo et al. [36] proposed a DL-based framework for accurate photovoltaic power generation (PVPG) forecasting. A physics-constrained LSTM (PC-LSTM) was considered to forecast the hourly day-ahead PVPG, and purposes were to overcome the limitation of recent machine learning algorithms. With the obtained results, it was concluded that the proposed PC-LSTM model holds a stronger forecasting precision as compared to the traditional machine learning approaches and with relatively less computational time [36]. In this section, different types of intelligent techniques and tools are reported for the prediction of medium-term PV forecasting and details are mentioned in Table 2. It is observed that several different types of machine learning models such as multilayer perceptron neural network (MLPNN), radial basis function and RNN, and ANN are established by various authors, however accuracy and precision of a particular prediction methodology depend upon the considered features, datasets, and their correlation with the predicting variable.

4.3 Long-Term Forecasting [LTF]

LTF methods have classified under the time horizon of a month to a year. Such prediction models are appropriate for prediction of long-term power generation,

Table 2 Different types of medium-term forecasting

	Medium-term forecasting model	Remarks	References
1	Data-driven prediction models	0–24 h ahead PV output prediction with improved accuracy	Pierro et al. [28]
2	Backpropagation (BP) neural network model	Forecasting of the power output of a PV system located in Ashland in an advance of 0–24 h-ahead	Liu et al. [29]
3	Hybrid current neural network (HCNN) and long short-term memory (LSTM) networks	Studying the time-series data in the deep learning communities (DLC)	Lee et al. [30]
4	Three categories of photovoltaic forecasting models, i.e., CNN, LSTM network, and hybrid model based on a CNN	CNN-based hybrid model was comparatively most promising and accurate	Wang et al. [31]
5	Methodology for day ahead output power time-series forecasting methods	Better prediction accuracy in non-ideal weather conditions	Gao et al. [32]
6	LSTM-CNN network	The CNN model, as well as the hybrid model of the convolutional-LSTM model, showed a better prediction performance	Wang et al. [33]
7	Investigated the impact of meteorological variable quantity	Feature selection and principal component analysis (PCA)	Ziane et al. [34]
8	Automatic machine learning (AML) followed by an improved genetic algorithm (GA)	Highly suitable forecast accuracy	Zhao et al. [35]
9	DL-based framework for accurate photovoltaic power generation (PVPG) forecasting	A physics-constrained LSTM (PC-LSTM) was considered to forecast which provided a stronger forecasting precision	Luo et al. [36]

which supports in decision making related to investments and planning of electrical transmission, and distribution systems. Long-term forecasting has the benefits of foretelling the PV output for the different seasons too but the prediction model has less accuracy due to different climate situations. Long-term solar PV power generation prediction is being developed by applying the appropriate numerical weather forecasting and deep learning techniques which are included with LSTM and RNN. DL in PV forecasting is increasing its popularity. Several methods are also reported for long-term PV forecasting. Some of them are discussed in below sections.

Yona et al. [37] reported the insolation forecasting at twenty-four hours ahead with the help of meteorological parameter's data, fuzzy theory, and NN. Since the fuzzy-based prototypical provided the insolation forecast data, NN smoothly trained the power yield. The appropriate training data plays a vital role in forecasting and if selected in appropriately, it may result in the unstable neural network [37]. Han et al. [11] introduced a forecasting methodology for wind and solar PV hybrid power yield for a long-time horizon, and it was estimated based on copula function and LSTM network. Furthermore, the proposed models were investigated and obtained results were compared with support vector machine prediction models with the finest condition and the results showed the prediction accurateness of the proposed joint prediction model was superior to the autonomous prediction models [11]. Eom et al. [38] reported about an ensemble model-based feature-selective long-term PV power generation methodology which was based on the machine learning algorithms as well as conventional time-series prediction techniques. Accurate predictions were obtained by numerous prediction sources with considering four variables. Authors have applied MLP, and CNN both methods as ensemble models, however CNN-based ensemble models results in the finest estimation of predictive power. Further, the MLP-based ensemble model confirmed better power prediction than the CNN-based ensemble model during the initial prediction time horizon [38]. Aslam et al. [39] reported about the deep learning and machine learning models which were based upon historical solar irradiation dataset and clear sky global horizontal irradiance. A gated recurrent units-based forecasting model provided better efficiency. The projected method acted as a new technique for data management and the use of deep learning methods for one complete year ahead prediction of solar irradiation [39].

Khodayar et al. [40] explored different deep neural network models such as discriminative deep models, convolutional neural networks, and also deep sparse coding algorithms for behind-the-meter (BTM) energy disaggregation (ED). The comparison found that deep generative modeling concentrated less computational complexities and proposed a promising function that shows significant improvement in PV energy prediction [40]. Moreira et al. [41] introduced a new methodology built on an ANN ensemble for photovoltaic forecasting power for the next 7 days. The practical model design approach was applicable for solar panel time-series datasets and ANN factors. Further, the result indicates a mixture design of experiment (MDE) that was implemented to optimize the weights in order to enhance the forecasting determination [41]. Liu et al. [42] contextualized two approaches in the chapter, i.e., high-quality ontology and gated recurrent neural network to increase the accurateness of the last forecast. Investigational outcomes indicated that the prediction accuracy from both approaches was about 5% greater than that of the long short-term memory model and thus this method has higher implication acceptance and also effectively improves the prediction accurateness of the PV system [42].

Different types of long-term forecasting have been reported in this segment and specifics are mentioned in Table 3. PV power yield has strong correlation with solar irradiance and therefore accurate prediction of irradiance is important for long-term PV forecasting. The conventional method like probability methods are used for forecasting but the forecast accuracy rate is poor. Machine learning algorithms have been introduced. The deep learning method provides better accuracy. Long short-term memory and gated recurrent units are the popular deep learning methods. It is observed that direct climatological data from satellites is required for long-term PV forecasting. Appropriate or correct climate reports will improve the accuracy

Table 3 Different types of long-term forecasting

	Long-term forecasting model	Remarks	References
1	Insolation forecasting at twenty-four hours ahead	Appropriate training data plays a vital role in forecasting	Yona et al. [33]
2	Forecasting methodology for wind and solar PV hybrid power	The prediction accurateness of the proposed joint prediction model was superior to the autonomous prediction models	Han et al. [11]
3	Ensemble model-based feature-selective long-term PV power generation methodology	Accurate forecasts	Eom et al. [38]
4	Deep learning and machine learning models	New technique for data management	Aslam et. al. [39]
5	Different deep neural network models	Deep generative modeling concentrated less computational complexities	Khodayar et al. [40]
6	An ensemble for photovoltaic forecasting power	To optimize the weights in order to enhance the forecasting determination	Moreira et. al. [41]
7	High-quality ontology and gated recurrent neural network to	Effectively improve the prediction accurateness of the PV system	Liu et al. [42]

of long-term forecasting. Different types of algorithms were also proposed to find the correct climate data such as correlation analysis, gray correlation, and principal component analysis. Long-term forecasting is characterized by periodicity, nonlinearity, and volatility, the time-varying parameters including the time power item and periodic item. Gray correlation with the time-varying variables provides better accuracy.

5 Research Gap Findings and Discussions

PV technology plays a vital role in power generation. Due to many factors such as sunlight intensity, air temperature, heat waves, and other metrological factors, the photovoltaic plant is unable to provide maximum production of power. Other auxiliary services in PV plant like operating system, frequency control as well as power control are also affected due to the variation of power at the output side. From the literature survey, some of the key research gap are obtained for maximum PV power generation by the help of intelligence data analysis tools and techniques. Those are as follows:

1. Requires large historical datasets for accurate prediction of the PV output: The main limitation of prediction theory is the data collection. In many conditions, it

may be quite difficult to find the required datasets for PV predictions. As the PV industries are quite long-standing industries and the machines and instruments are not the smart devices to collect the accurate and cleaned data for prediction theory, it is a key research gap in the field of prediction theory.

2. Requirement of neat and historical data of last 5–10 years for long-term forecasting: For the long-term forecasting for PV output, it requires the PV output data of last few decades. It is not possible to get accurate data. So, the long-term forecasting is designed based on last few decades' meteorological data. This another key gap during long-term forecasting.
3. Convert the conventional devices in PV industry into smart devices: It is an important research in the field of economy consideration. The installed devices need to convert into smart device which can provide accurate and clean data. Such device data helps to provide accurate prediction on PV output generation.

In renewable energy, predicting is playing a very important role. Nowadays, market commitment, quality of power, energy saving become other main reasons for forecasting in the grid-integrated renewable energy system. As per above literature, correct and accurate prediction should be small in time which comes under the category of short-term forecasting. Short-term forecasting provides the accurate data or forecast before 15 min of PV power generation as well as it does not need any changes in data too. Another level of short-term forecasting is called very short-term or ultra-short-term forecasting which have been used as input for power management in buildings, input variables as well as sun light intensity prediction. Such type of short- and ultra-short-term forecasting provides good and accurate forecasting data. Mid-term forecasting is based upon the forecasting of PV power before some hours or a day or a week. By the help of mid-term forecasting, energy management as well as saving of energy is possible. The quality of power in transmission and distribution of system and power system operation and control is also improved. However, mid-term forecasting is completely based upon both national meteorological condition data which comes from the direct satellite and the spotless sky. These data will helpful to reduce the inaccuracy level in forecasting. However long-term forecasting has its own advantages like such design is able to forecast by correlating the data between actual real power and past weather data which comes under direct long-term forecasting technique, while, indirect long-term forecasting technology applies the hypothetical method to correlate the data between the other climatic as well as environmental factors which are responsible for power fluctuations in PV plant and the computational time is too long to reach the requirement of PV power forecasting. Earlier, ANN model was used to study the solar intensity prediction on daily basis [43, 44], radial basis function neural network (RBFNN) for grid-integrated PV systems [45], forecasting by AI model [46], etc., were reported as forecasting technique. Present days, different types of new intelligent techniques are used on the basis of time such as CNN [47], LSTM NN [48], machine learning techniques [49], deep learning techniques, and choice-based ANN [50] that are reported as latest technology in forecasting technique. Machine learning is quite popular in every type of forecasting model. In other words, it is also named as artificial intelligence model

[51]. Such models are capable to use the complex nonlinear property as well as crate a correlation between input and output proficiently. There are three subparts of machine learning models such as SVM, decision tree, and ANN. The ANN model is one of the frequent techniques for time series. Present days, different new technologies have been introduced and deep learning is one of them. Probabilistic approach is also taking part as forecasting model to determine the meteorological conditions for both wind and solar power forecasting. Such type of models based upon machine or extreme machine or deep learning is able to increase the accuracy of forecasting as well as to reduce the cost of energy price [51–55].

6 Conclusions and Future Scope

PV power prediction is significant for efficient planning and execution of modern electrical generation, transmission, and distribution networks.

Short-term, medium-term, and long-term PV predictions are important for different PV applications and are equally important. Medium- to long-term PV predictions are more challenging task as it requires accurate prediction of weather parameters and uncertainties involved in such predictions would adversely impact the PV forecasting. Some of the key points are observed from the literature review. Those are as follows:

- Solar light intensity, as well as weather conditions like wind speed, temperature, pressure, moisture, humidity are the factors that are responsible for the variation of PV output power.
- Forecasting models with inappropriate weather parameters could also contributes to the uncertainty of solar output forecasting.
- Difficult to eliminate the errors that are occurred due to erroneous measurements or instrument failure.
- ML and DL methods are effective in PV prediction; however, their performance depends upon the availability of accurate and correct datasets which are quite popular for the forecasting of PV output.
- Maximum number of algorithms for optimization techniques are proposed to reduce the computational burden, however it results as increased network complexity.

It can be concluded that intelligent tools are suitable for PV power generation technology. As it is already discussed in the above sections, the artificial intelligence and its subcategories are a suitable tool which can provide a good platform for forecasting. Intelligent tools and techniques can be used to design a single junction PV cell as well as multijunction PV cell which maximizes the PV efficiency. It can also estimate the concentration of doping into PV cell which become the future scope of this research.

References

1. L. Ferrari, A. Bianchini, G. Galli, G. Ferrara, E.A. Carnevale, Influence of actual component characteristics on the optimal energy mix of a photovoltaic-wind-diesel hybrid system for a remote off-grid application. J. Clean. Prod. **178**, 206–219 (2018)
2. A. Tomar, A.N.M.M. Haque, P. Nguyen, Compensation mechanism for active power curtailment in LV distribution networks, in *2020 IEEE PES Innovative Smart Grid Technologies Europe (ISGT-Europe)* (IEEE, 2020), pp. 759–763
3. A. Tomar, S. Mishra, CMPVI-based MIDO scheme under SSE for optimum energy balance and reduced ROI. IEEE Trans. Sustain. Energy **9**(3), 1318–1327 (2017)
4. A. Tomar, S. Mishra, PV energy benefit estimation formulation for PV water pumping system, in *2017 4th IEEE Uttar Pradesh Section International Conference on Electrical, Computer and Electronics (UPCON)* (IEEE, 2017), pp. 44–48
5. L. Oneto, F. Laureri, M. Robba, F. Delfino, D. Anguita, Data-driven photovoltaic power production now casting and forecasting for polygeneration microgrids. IEEE Syst. J. **12**(3), 2842–2853 (2017)
6. A. Tomar, S. Mishra, C.N. Bhende, Techno-economical analysis for PV based water pumping system under partial shading/mismatching phenomena, in *2016 IEEE 7th Power India International Conference (PIICON)* (IEEE, 2016), pp. 1–6
7. M. Morshedizadeh, M. Kordestani, R. Carriveau, D.S.K. Ting, M. Saif, Application of imputation techniques and adaptive neuro-fuzzy inference system to predict wind turbine power production. Energy **138**, 394–404 (2017)
8. A. Khosravi, S. Nahavandi, Load forecasting using interval type-2 fuzzy logic systems: optimal type reduction. IEEE Trans. Industr. Inf. **10**(2), 1055–1063 (2013)
9. C. Paoli, C. Voyant, M. Muselli, M.L. Nivet, Solar radiation forecasting using ad-hoc time series preprocessing and neural networks, in *International Conference on Intelligent Computing* (Springer, Berlin, Heidelberg, 2009), pp. 898–907
10. C. Tao, D. Shanxu, C. Changsong, Forecasting power output for grid-connected photovoltaic power system without using solar radiation measurement, in *The 2nd International Symposium on Power Electronics for Distributed Generation Systems* (IEEE, 2010), pp. 773–777
11. S. Han, Y.H. Qiao, J. Yan, Y.Q. Liu, L. Li, Z. Wang, Mid-to-long term wind and photovoltaic power generation prediction based on copula function and long short term memory network. Appl. Energy **239**, 181–191 (2019)
12. U.K. Das, K.S. Tey, M. Seyedmahmoudian, S. Mekhilef, M.Y.I. Idris, W. Van Deventer, B. Horan, A. Stojcevski, Forecasting of photovoltaic power generation and model optimization: a review. Renew. Sustain. Energy Rev. **81**, 912–928 (2018)
13. M. Louzazni, H. Mosalam, A. Khouya, K. Amechnoue, A non-linear auto-regressive exogenous method to forecast the photovoltaic power output. Sustain. Energy Technol. Assess. **38**, 100670 (2020)
14. F. Barbieri, S. Rajakaruna, A. Ghosh, Very short-term photovoltaic power forecasting with cloud modeling: a review. Renew. Sustain. Energy Rev. **75**, 242–263 (2017)
15. Online (Accessed on 20-June-2021). www.lens.org
16. P. Tang, D. Chen, Y. Hou, Entropy method combined with extreme learning machine method for the short-term photovoltaic power generation forecasting. Chaos Solitons Fractals **89**, 243–248 (2016)
17. R.V. Monteiro, G.C. Guimarães, F.A. Moura, M.R. Albertini, M.K. Albertini, Estimating photovoltaic power generation: performance analysis of artificial neural networks, support vector machine and Kalman filter. Electr. Power Syst. Res. **143**, 643–656 (2017)
18. A.K. Yadav, S.S. Chandel, Identification of relevant input variables for prediction of 1-minute time-step photovoltaic module power using artificial neural network and multiple linear regression models. Renew. Sustain. Energy Rev. **77**, 955–969 (2017)
19. H. Takeda, Short-term ensemble forecast for purchased photovoltaic generation. Sol. Energy **149**, 176–187 (2017)

20. K. Hu, S. Cao, L. Wang, W. Li, M. Lv, A new ultra-short-term photovoltaic power prediction model based on ground-based cloud images. J. Clean. Prod. **200**, 731–745 (2018)
21. L. Liu, M. Zhan, Y. Bai, A recursive ensemble model for forecasting the power output of photovoltaic systems. Sol. Energy **189**, 291–298 (2019)
22. F.Y. Xu, R.X. Tang, S.B. Xu, Y.L. Fan, Y. Zhou, H.T. Zhang, Neural network-based photovoltaic generation capacity prediction system with benefit-oriented modification. Energy **223**, 119748 (2021)
23. Q. Li, X. Zhang, T. Ma, C. Jiao, H. Wang, W. Hu, A multi-step ahead photovoltaic power prediction model based on similar day, enhanced colliding bodies optimization, variational mode decomposition, and deep extreme learning machine. Energy **224**, 120094 (2021)
24. F. Rodríguez, F. Martín, L. Fontán, A. Galarza, Ensemble of machine learning and spatiotemporal parameters to forecast very short-term solar irradiation to compute photovoltaic generators' output power. Energy **229**, 120647 (2021)
25. M.A. Hassan, N. Bailek, K. Bouchouicha, S.C. Nwokolo, Ultra-short-term exogenous forecasting of photovoltaic power production using genetically optimized non-linear autoregressive recurrent neural networks. Renew. Energy **171**, 191–209 (2021)
26. L. Yu, X. Ma, W. Wu, X. Xiang, Y. Wang, B. Zeng, Application of a novel time-delayed power-driven grey model to forecast photovoltaic power generation in the Asia-Pacific region. Sustain. Energy Technol. Assess. **44**, 100968 (2021)
27. F. Rodríguez, M. Genn, L. Fontán, A. Galarza, Very short-term temperature forecaster using MLP and N-nearest stations for calculating key control parameters in solar photovoltaic generation. Sustain. Energy Technol. Assess. **45**, 101085 (2021)
28. M. Pierro, F. Bucci, M. De Felice, E. Maggioni, D. Moser, A. Perotto, F. Spada, C. Cornaro, Multi-Model Ensemble for day ahead prediction of photovoltaic power generation. Solar Energy **134**, 132–146 (2016)
29. L. Liu, D. Liu, Q. Sun, H. Li, R. Wennersten, Forecasting power output of photovoltaic system using a BP network method. Energy Proc. **142**, 780–786 (2017)
30. W. Lee, K. Kim, J. Park, J. Kim, Y. Kim, Forecasting solar power using long-short term memory and convolutional neural networks. IEEE Access **6**, 73068–73080 (2018). https://doi.org/10.1109/access.2018.2883330
31. K. Wang, X. Qi, H. Liu, A comparison of day-ahead photovoltaic power forecasting models based on deep learning neural network. Appl. Energy **251**, 113315 (2019)
32. M. Gao, J. Li, F. Hong, D. Long, Day-ahead power forecasting in a large-scale photovoltaic plant based on weather classification using LSTM. Energy **187**, 115838 (2019)
33. K. Wang, X. Qi, H. Liu, Photovoltaic power forecasting based LSTM-convolutional network. Energy **189**, 116225 (2019)
34. A. Ziane, A. Necaibia, N. Sahouane, R. Dabou, M. Mostefaoui, A. Bouraiou, S. Khelifi, A. Rouabhia, M. Blal, Photovoltaic output power performance assessment and forecasting: impact of meteorological variables. Solar Energy **220**, 745–757 (2021)
35. W. Zhao, H. Zhang, J. Zheng, Y. Dai, L. Huang, W. Shang, Y. Liang, A point prediction method based automatic machine learning for day-ahead power output of multi-region photovoltaic plants. Energy **223**, 120026 (2021)
36. X. Luo, D. Zhang, X. Zhu, Deep learning based forecasting of photovoltaic power generation by incorporating domain knowledge. Energy **225**, 120240 (2021)
37. A. Yona, T. Senjyu, T. Funabashi, C.-H. Kim, Determination method of insolation prediction with fuzzy and applying neural network for long-term ahead PV power output correction. IEEE Trans. Sustain. Energy **4**(2), 527–533 (2013). https://doi.org/10.1109/tste.2013.2246591
38. H. Eom, Y. Son, S. Choi, Feature-selective ensemble learning-based long-term regional PV generation forecasting. IEEE Access **8**, 54620–54630 (2020)
39. M. Aslam, J.M. Lee, H.S. Kim, S.J. Lee, S. Hong, Deep learning models for long-term solar radiation forecasting considering microgrid installation: a comparative study. Energies **13**(1), 147 (2020)
40. M. Khodayar, M.E. Khodayar, S.M.J. Jalali, Deep learning for pattern recognition of photovoltaic energy generation. Electr. J. **34**(1), 106882 (2021)

41. M.O. Moreira, P.P. Balestrassi, A.P. Paiva, P.F. Ribeiro, B.D. Bonatto, Design of experiments using artificial neural network ensemble for photovoltaic generation forecasting. Renew. Sustain. Energy Rev. **135**, 110450 (2021)
42. H. Liu, Q. Gao, P. Ma, Photovoltaic generation power prediction research based on high quality context ontology and gated recurrent neural network. Sustain. Energy Technol. Assess. **45**, 101191 (2021)
43. A.K. Yadav, H. Malik, S.S. Chandel, ANN based prediction of daily global solar radiation for photovoltaics applications, in *2015 Annual IEEE India Conference (INDICON)* (IEEE, 2015), pp. 1–5
44. A.K. Yadav, H. Malik, Comparison of different artificial neural network techniques in prediction of solar radiation for power generation using different combinations of meterological variables, in *2014 IEEE International Conference on Power Electronics, Drives and Energy Systems (PEDES)* (IEEE, 2014), pp. 1–5
45. A.K. Yadav, V. Sharma, H. Malik, S.S. Chandel, Daily array yield prediction of grid-interactive photovoltaic plant using relief attribute evaluator based radial basis function neural network. Renew. Sustain. Energy Rev. **81**, 2115–2127 (2018)
46. A. Sfetsos, A.H. Coonick, Univariate and multivariate forecasting of hourly solar radiation with artificial intelligence techniques. Sol. Energy **68**(2), 169–178 (2000)
47. B. Du, P.D. Lund, J. Wang, Combining CFD and artificial neural network techniques to predict the thermal performance of all-glass straight evacuated tube solar collector. Energy **220**, 119713 (2021)
48. A.H. Elsheikh, V.P. Katekar, O.L. Muskens, S.S. Deshmukh, M. Abd Elaziz, S.M. Dabour, Utilization of LSTM neural network for water production forecasting of a stepped solar still with a corrugated absorber plate. Process Saf. Environ. Prot. **148**, 273–282 (2021)
49. E.M. El-Said, M. Abd Elaziz, A.H. Elsheikh, Machine learning algorithms for improving the prediction of air injection effect on the thermohydraulic performance of shell and tube heat exchanger. Appl. Thermal Eng. **185**, 116471 (2021)
50. R. Mohanty, P.G. Kale, Influence of wind speed on solar PV plant power production—prediction model using decision-based artificial neural network, in *Advances in Computational Intelligence and Communication Technology* (Springer, Singapore, 2021), pp. 3–16
51. H. Malik, A. Iqbal, A.K. Yadav, Soft computing in condition monitoring and diagnostics of electrical and mechanical systems, in *Novel Methods for Condition Monitoring and Diagnostics* (Springer, 2020), p. 499
52. S.M. Saeed, M.Y. Aman, K.A. Ahmad, A. Batari, A.T.A. Yero, A.U. Chinade, Effect of crumb rubber modifier on the fatigue performance of warm mix asphalt, in *Global Civil Engineering Conference* (Springer, Singapore, 2017), pp. 1367–1376
53. A. Iqbal, *Renewable Power for Sustainable Growth: Proceedings of International Conference on Renewal Power (ICRP 2020)* (Springer Nature, 2021)
54. H. Malik, N. Fatema, A. Iqbal, *Intelligent Data-Analytics for Condition Monitoring: Smart Grid Applications* (Elsevier, 2021)
55. H. Malik, S. Srivastava, Y.R. Sood, A. Ahmad (Eds.), *Applications of Artificial Intelligence Techniques in Engineering: SIGMA 2018, Volume 1*, vol. 698 (Springer, 2018)

Design and Performance Analysis of Grid-Connected Solar PV System Using PV Syst Software

Bilal Alam, Safwan Mustafa, Tausif Akram, Alina Naaz, Mohd Tariq, and Md Atiqur Rahman

Abstract An electricity is an important source of energy in daily life. Nowadays, we are totally depending upon the electrical appliances. That is why the whole world is focusing on renewable energy sources such as solar energy to generate electricity. The efficiency of a photovoltaic system depends on the location and type of PV modules used. The PV system is useful in areas with a good amount of sunlight. It turned out that Aligarh has a better access to solar energy. Aligarh receives an average of the sun's rays used annually at a rate of 7.92 MJ/m^2. In this chapter, we copy the grid connected to a solar photovoltaic system using computer software PVsyst v-7.0.10.17617. The total amount of energy produced by the solar-connected system and the different types of losses that occur in the system are given and analyzed.

Keywords Solar photovoltaic · System simulation · Solar radiation · PV syst

1 Introduction

The present energy situation is not sustainable, as fossil fuel reserves are diminishing. In [1, 2] as of 2020, most experts believe that we have around 40–80 years of non-renewable energy available and this may not be ready to satisfy the increasing demand related to economic development and population increase. While the fossil fuels, we consume represent 28% of the world gas emission released annually. Also, fuel combustion causes pollution, which is dangerous for humans, plants, and animals.

The global solar market is burgeoning, and it is predicted that the globe will have 1.5 trillion watts of installed solar PV capacity by 2030. Approximately 5500 trillion kWh/year of energy occurs in the Indian subcontinent most parts receive 4.5–7.5 kWh/sq./Day. Solar photovoltaic power can be successfully tied to generating great

B. Alam · S. Mustafa · T. Akram · A. Naaz · M. Tariq (✉)
Department of Electrical Engineering, Z.H.C.E.T., Aligarh Muslim University, Aligarh, India
e-mail: tariq.ee@zhcet.ac.in

M. A. Rahman
Department of Electrical Engineering, Government Engineering College, Vaishali, India

H. Malik et al. (eds.), *Intelligent Data Analytics for Power and Energy Systems*, Lecture Notes in Electrical Engineering 802,
https://doi.org/10.1007/978-981-16-6081-8_18

capacity in India. Solar also provides flexibility for power acquisition on a distributed basis and enables faster power additions during shorter lead times. Distributed off-grid use, as well as lower temperatures, will be beneficial to the domestic energy vision and meet other energy and heat and cooling needs in both rural and concrete areas. From a energy security standpoint, solar is the safest of all sources because it is widely available. Theoretically, a small portion of a completely different energy event (if taken successfully) could meet energy needs across the country. The National Solar Energy Center has tested about 748 GW of solar energy by taking up 3% of the surface area of the waste to be covered with solar PV modules. Solar power has taken over the Indian National Action Plan for Climate Change and the National Solar Mission with the integration of vital equipment. The installed solar capacity of the solar system was 35,739 MW as atAugust 31, 2020. There are three main types of solar PV and storage systems: grid-tied, grid/hybrid, and stand-alone/off-grid. In this chapter, the grid connected to the design and analysis of the PV system is eliminated with the help of PV syst software. PV syst software can be a tool that enables the study, measurement, and data analysis of complete PV systems. It works with grid-connected, stand-alone, pump, and DC-grid (public transport) PV systems and includes details of meteor and PV, in addition to standard solar power tools. This software is extremely useful for computer modeling and system analysis because it is very reliable and accurate.

This chapter contains four sections in which Sect. 2 outlines the process for the whole process, Sect. 3 contains the operation and analysis of the grid-connected system, and Sects. 4 and 5 contain the results and discussion of the whole program.

2 Grid-Tied PV System

A PV system, inverters, and other balance system components, such as mounting system, charge controller, cables, power transformers, and necessary batteries, are included in the grid-connected-bound PV system. The choice of components depends on the device requirements, and the requirements of the customer are defined. Figure 1 shows a schematic view of a photovoltaic system connected to the grid.

The DC current generated by GCPVS is converted into an AC current with the help of inverters. Solar array compatibility and optimized sizing are a major concern

Fig. 1 Grid-connected solar PV system

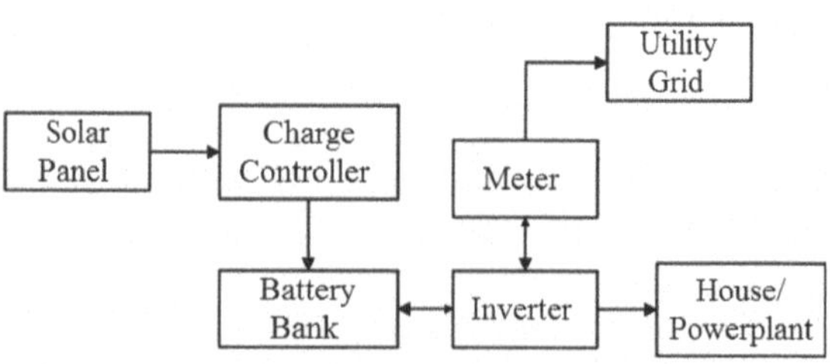

in solar PV system performance system design. The important factor of the inverter must be noted in conjunction with the device design for power and frequency rating. In [3], other connected grid components are also very significant, such as switches, wire, connectors, and the assembly of device mounting.

3 Methodology

In this chapter, we simulate 123 kWp system for location Aligarh whose latitude and longitude is given in Table 1. In place of fixed tilt angle, we take a two-axis variable for improving efficiency and performance.

3.1 PV Syst Software

PV software is generally used for the estimating and simulating the costing, aging, estimation of number of inverters of PV panels. In all over the world, PV syst is used for developing a project. The following steps are needed to obtain a final report.

3.1.1 Irradiation Data

The IR data was derived from the satellite data of Metronorm 7.3. The IR data obtained reflected hourly synthetic value of Aligarh ambient temperature, beam, diffused, and global IR for the whole year as shown in Fig. 2. The statistical equivalent of the deployed power plant GCPVS was measured and simulated using the Metronorn 7.3 satellite data file.

3.1.2 Orientation and Horizon

Figure 3 shows the orientation with fixed tilt for system. The orientation was followed to create the similar simulation in pv syst as in real world.

Table 1 Latitude and longitude of Aligarh

	Longitude	Latitude
Aligarh	27.8974 N	78.0880 E

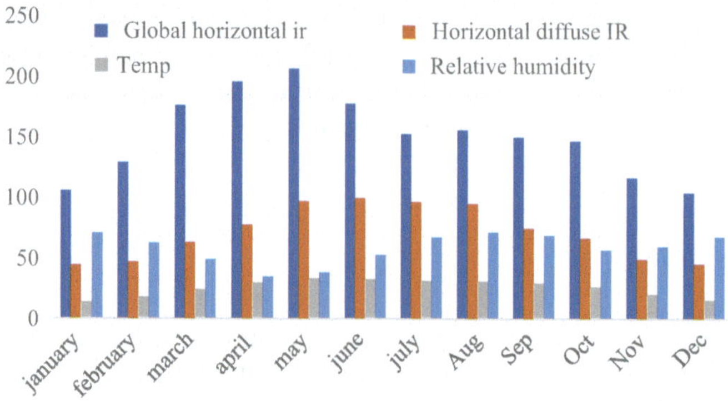

Fig. 2 Synthetic generated meteo data Aligarh

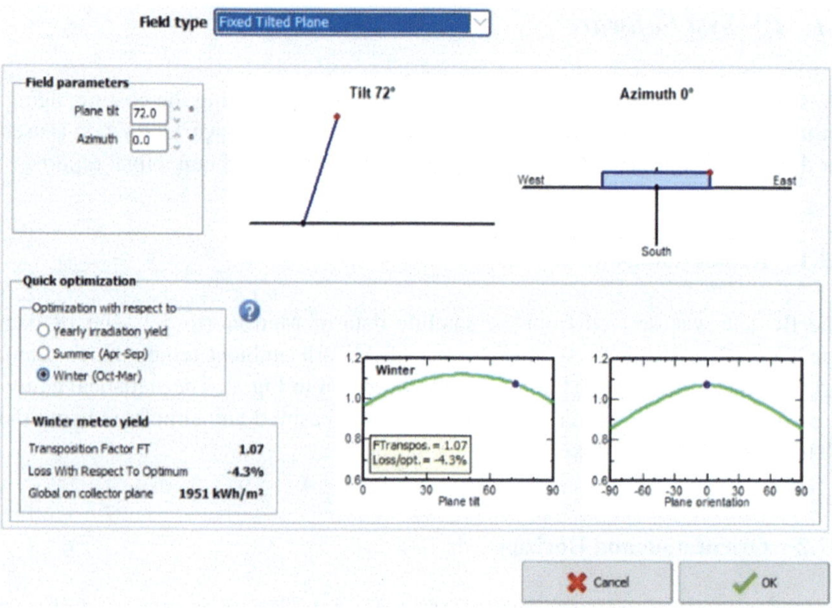

Fig. 3 Orientation with fixed tilt for system

3.2 *Case Study*

In this proposed work, we design a solar PV rooftop grid-connected system for load expenditure of 1828.635 kWh/day which is examined. Table 2 shows the load data and daily energy consumption of the given location.

Total wattage of the whole system = 132.125 kW.

Total daily consumption daily basis = 1828.635 kWh/day.

Table 2 Load data and daily energy consumption

Load	*Q*	*W*	TW	Time	Consumption (Wh/day)
Air purifier	4	20	80	18	1440
Air condition machines (window/ductable split)	58	750	43,500	16	696,000
Air cutter	03	220	660	12	7920
Ceiling fan	223	75	16,725	16	267,600
Pedastal fan	27	55	1485	15	22,275
Ex. fan	40	150	6000	12	72,000
Wall fan	135	30	4050	18	72,900
Cabin fan	31	75	2325	8	18,600
Desert cooler	37	200	7400	12	88,800
Fridge	02	250	500	24	12,000
Hat convector (blower)	04	2000	8000	8	64,000
Electric water heater instant geyser	02	4500	9000	8	72,000
Inverter	05	5000	15,000	2	30,000
Mono block pump 2 HP	01	750	750	2	1500
Stabilizer	12	750	9000	24	216,000
Servo stabilizer	08	450	3600	24	86,400
Water cooler	05	750	3750	24	90,000
Water purifier and aqua guard	02	150	300	24	7200

4 Input Requirement

As each location absorbs varying levels of solar IR radiation, the architecture of the PV system is based on location. This is because of the orientation of precise place (longitude and latitude) with respect to the face of the sun.

4.1 Location

Aligarh is basically located in northern india with latitude 27.8974 N, 78.0880 E, respectively. Aligarh has a plane terrain. This system is designed for 132 kW power for Maulana Azad Library, Aligarh to meet total load of the whole system of the library.

Fig. 4 Tilt and azimuth angle

4.2 Tilt Angle

The angle of incident is known as the angle of slope at which the solar panel is placed or faces the sun. The location of the sun varies with respect to every time of day so mounting angles of the panels often keep changing. The tilt angle is often taken to be equal to the latitude of the location [4]. Tilt angle and latitude angle are taken to optimize the sum of the solar energy on the plates. Based on the optimal value of tilt angle, we have found annually IR yield. Figure 4 shows the tilt angle of the given system.

4.3 Azimuth Angle

The direction of sun decides the azimuth angle; for this proposed work, we take azimuth angle s zero because the panels faces toward the south in northern hemisphere. Figure 4 Describe the azimuth angle.

4.4 Albedo

The calculation of the reflectivity of the surface of the earth is Albedo. It depends on the texture of the earth's position (latitude and longitude). The value fluctuates from 0 to 1. It is higher in snowy regions. We have inferred an albedo value of 0.5 for Aligarh owing to the concrete surface.

4.5 Module and Inverter Specification

For simulating 132 kW of system locating at Aligarh, we use polycrystalline module of Somera VSM.72.350.05 in which each is of 350 w and Generic inverter. Table 3 further explains the specification of inverter and PV panels.

Table 3 Specification of inverter and PV array

A. *PV array characteristics*	
1. Manufacturer 2. Model	Somera VSM.72.350.05
B. *Number of PV modules*	
1. In series 2. In parallel 3. Total Number of module. 4. Unit Nom. Power	18 modules 21 string 378 350 Wp
C. *Inverter*	
1. Manufacture 1. Model	Generic TRIO-TM-50_0-400

Table 4 From MNRE solar rooftop calculator

A. Size of power plant	132.2 kW
B. Worth of the plant	
1. MNRE current benchmark cost 2. Without subsidy 3. With subsidy	36,000 Rs/kW 4,752,000 Rs 4,752,000 Rs
C. Total electricity generation from the plant	
1. Yearly 2. Age (25 years)	182,160 kWh 4554000Kwh
D. Financial savings (tariff @5.5/kWh)	
1. Monthly 2. Annually 3. Life time (25 years)	83,490 Rs 1,001,880 Rs 25,047,000 Rs
E. CO_2 emission	3734 tones
F. The putting in like planting	5975 (teak trees over the life time)

4.6 Economic Analysis

The system is designed for a commercial load of 132 kW in Maulana Azad library at Aligarh. The tilt and the azimuth angle are shown in Table 4. Average solar irradiation (ASI) in UP is 1156.39 W/m^2. 1 kW residential solar plant will create 4.6 kWh of electricity per day on an average over the course of the year (5.5 sunshine hours). Table 4 shows the MNRE solar rooftop (SRT) data.

5 Simulation and Result

PV syst is a software which can provide all the details that needs to design SAPV and GCPVS [5, 6]. It calculates many parameters which was shown in Tables 3 and 4, e.g., size of inverters, losses, size of PV mlodule, MPP, and also economic analysis but this software needs proper tilt angle and some necessary data. Figure 5 shows systematic loss of solar grid-connected PV system. The solar grid-connected PV system system's losses are shown step by step in the loss diagram. There are no. of losses held in array such as wiring loss and battery efficiency loss. PV loss is due to temperature, unused energy. After rectification, the grid input is calculateds.

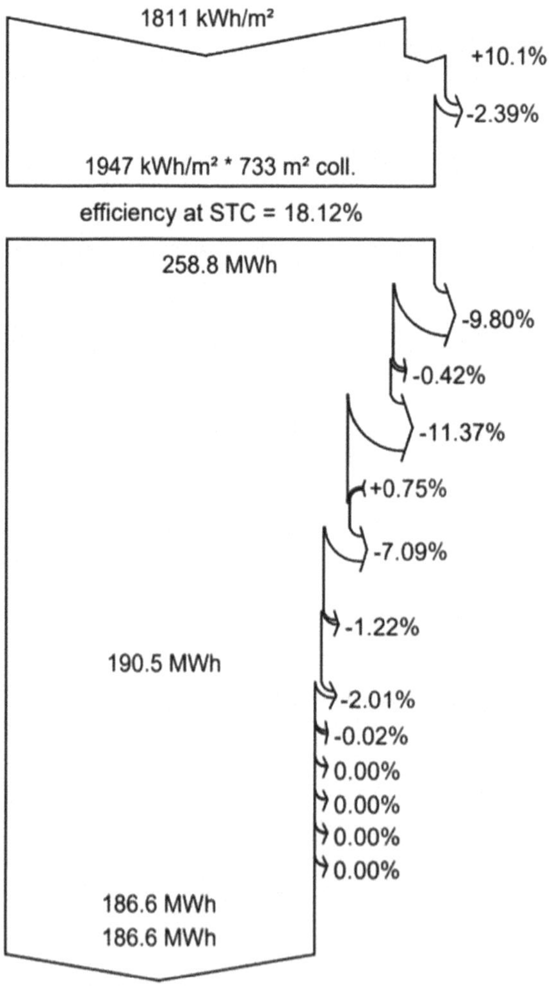

Fig. 5 Loss diagram of the whole year

PR indicates the ratio between output of the inverter and output of the PV array. Figure 6a shows the performance ratio. The required performance ratio is 33.1% which means that 66.9% of power of the total energy generated by the solar power plant is lost. Normalized production is calculated from the simulation study as shown in Fig. 6b. Some of the important factors are shown in the graphical representation such as the unused energy, collection loss, system losses, and battery charging.

Table 5 represents the final and main result for 132 kW solar power plant. This table shows the global horizontal irradiation, energy injected into grid, performance ratio, *T*_amb, etc.

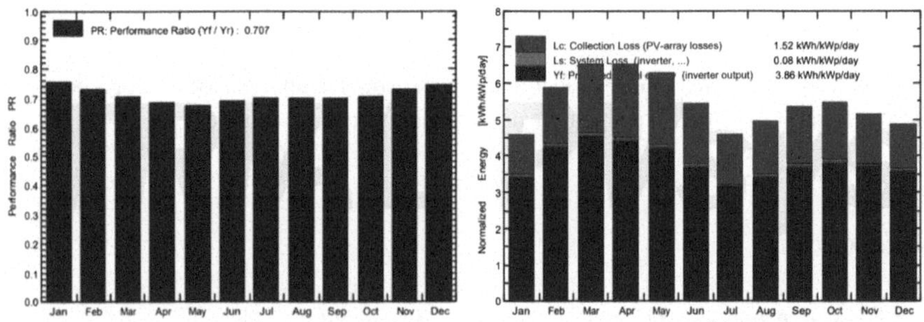

(a) (b)

Fig. 6 **a** Performace ratio and **b** normalized production

Table 5 Balance and main result

	GlobHor (kWh/m^2)	DiffHor (kWh/m^2)	T_Amb (°C)	GlobInc ($kWh/m^{2)}$	GlobEff (kWh/m^2)	E_Array (Wh)	E_Grid (Wh)	PR ratio
January	105.5	48.8	13.73	142.6	139.9	14.52	14.25	0.755
February	129.2	48.0	17.93	165.1	161.9	16.22	15.90	0.728
March	176.7	65.8	24.08	202.6	198.1	19.21	18.82	0.702
April	191.1	82.2	30.09	196.0	191.2	18.09	17.70	0.683
May	206.6	99.0	33.62	195.3	190.0	17.84	17.45	0.675
June	177.5	105.5	32.92	162.7	157.8	15.11	14.80	0.688
July	152.7	101.5	31.90	141.7	137.2	13.37	13.10	0.699
August	156.0	92.9	30.86	153.1	148.7	14.42	14.12	0.697
September	150.1	80.2	29.40	160.3	156.3	15.12	14.81	0.698
October	143.1	71.4	26.40	169.7	166.3	16.18	15.85	0.706
November	116.0	52.0	20.47	154.8	151.7	15.22	14.94	0.729
December	106.3	45.1	15.55	150.4	147.6	15.15	14.87	0.747
Yearly	1810.8	892.5	25.63	1994.3	1946.6	190.45	186.61	0.707

6 Conclusion

In this aricle, we work on design and simulation on Maulana Azad library which is largest library in Asia. Power consumtion of this library is 132 kV per month.

- The ASI in Uttar Pradesh is 1156.39 W/m^2 and 1 kw SRT plant will generate on an average over the year 4.6 kWh of electricity per day.
- The highest energy consumption occurs in May which is 18.35 MWh. And lowest in the month of July which is 13.19 Mwh
- The annual energy supplied to grid is 186.61 Mwh.
- The average annual performance ratio of this solar PV system is 70.7% which is sufficient.

This result indicates that this design and performance analysis of GCS PV system plan will be of profitable to installer due to operational advantage.

References

1. J.L. Sawin, *P40_Renewables 2014 Global Status Report* (2014)
2. N. Kumar, P. Yadav, S.S. Chandel, Comparative analysis of four different solar photovoltaic technologies, in *International Conference on Energy Economics Environment 1st IEEE Uttar Pradesh Section Conference UPCON-ICEEE 2015* (2015). https://doi.org/10.1109/EnergyEconomics.2015.7235077
3. M. Asim, M. Tariq, M.A. Mallick, I. Ashraf, S. Kumari, A.K. Bhoi, in *Critical evaluation of offline MPPT techniques of solar PV for stand-alone applications*, ed. by S. SenGupta, A. Zobaa, K. Sherpa, A. Bhoi. Advances in Smart Grid and Renewable Energy. Lecture Notes in Electrical Engineering. vol. 435 (Springer, Singapore, 2018). https://doi.org/10.1007/978-981-10-4286-7_2
4. M.J. Ahmad, G.N. Tiwari, Optimum tilt angle for solar collectors used in India. Int. J. Ambient Energy **30**(2), 73–78 (2009).https://doi.org/10.1080/01430750.2009.9675788
5. P. Yadav, N. Kumar, S.S. Chandel, Simulation and performance analysis of a 1kWp photovoltaic system using PVsyst, in *4th IEEE Sponsored International Conference on Computation Power, Energy, Information Communication ICCPEIC 2015*, pp. 358–363 (2015). https://doi.org/10.1109/ICCPEIC.2015.7259481
6. V.P. Sharma, A. Singh, J. Sharma, A. Raj, Design and simulation of dependence of manufacturing technology and tilt orientation for 100 kWp grid tied solar PV system at Jaipur, in *2016 International Conference on Recent Advanced Innovation Engigineering ICRAIE 2016* (2016). https://doi.org/10.1109/ICRAIE.2016.7939508

Intelligent Approach-Based Maximum Power Point Tracking for Renewable Energy System: A Review

Kulsoom Fatima, Ahmad Faiz Minai, and Hasmat Malik

Abstract Speedy exaggeration of CO_2 emissions stimulates global warming. In order to slow down global warming, ongoing hydrocarbon deposit systems need to be put back by renewable systems like solar and wind. In addition to above approach, upcoming grid is expected to cater developments like renewable energy production, intelligent energy management systems to satisfy extending requirements. The main motive is to boost efficiency of the renewable energy system; for this, PV systems are always provided with maximum power point tracking algorithm to make sure that PV module operates at its maximum power point. This chapter presents comparison of contrasting maximum power point tracking algorithms centered on artificial intelligence. A detailed comparison is carried out based on the review from available literature and by simulating a model under uniform and partial shading conditions using MATLAB/SIMULINK. The overall objective is to deduce an algorithm that is meaningful, intelligent, optimized, and efficient.

Keywords Maximum power point tracking (MPPT) · Artificial intelligence (AI) · Fuzzy logic control (FLC) · Artificial neural network (ANN) · Genetic algorithm (GA) · Swarm intelligence (SI) · Machine learning (ML)

1 Introduction

Solar power system is an encouraging renewable energy source compared with non-renewable energy resources like fossil fuel and natural gas because of its eco-friendliness, abundant availability, cost-effectiveness, and high efficiency [1]. The two main drawbacks associated with solar power system are: low conversion efficiency (9–17%) [2] and because of the nonlinear characteristic of photovoltaic (PV) cell, power generated by PV module is influenced by atmospheric circumstances.

K. Fatima (✉) · A. F. Minai
Department of Electrical Engineering, Integral University, Lucknow, India

H. Malik
BEARS, University Town, NUS Campus, Singapore

H. Malik et al. (eds.), *Intelligent Data Analytics for Power and Energy Systems*,
Lecture Notes in Electrical Engineering 802,
https://doi.org/10.1007/978-981-16-6081-8_19

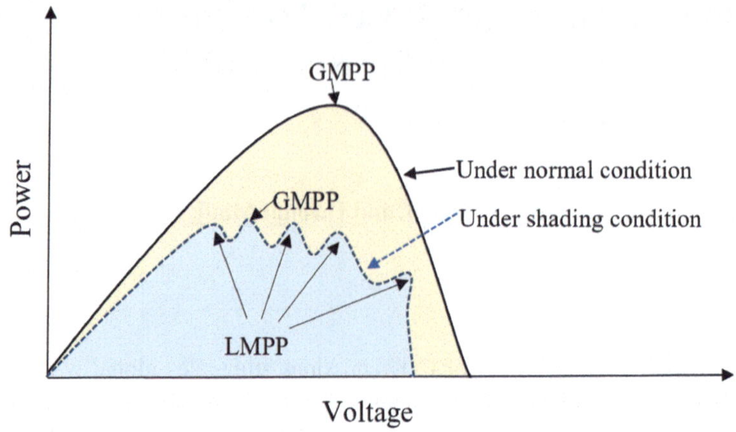

Fig. 1 Power versus voltage curve for a solar power system

The main prospect is to enhance efficiency of system, and therefore, MPPT became research prospect as they enhance efficiency of solar power system and ensure that system always operates at its MPP [2]. Tremendous work has been carried out, and lots of literature are available on tracking maximum power point; this chapter is an attempt for comparison of contrasting AI-centered algorithms.

The conventional hill-climbing MPPT techniques can track efficiently under uniform conditions [3]. However, under PSC, the power output from solar power system generates multiple local peaks including one global MPP (GMPP) that is set forth in Fig. 1, which makes it difficult for hill climbing MPPT technique to search for the real maximum [4].

Conventional hill climbing MPPT includes shortcomings like:

(1) Robust, lack of adaptive and self-learning capabilities.
(2) High steady-state error, oscillation at MPP, and slow transient response.
(3) Inability to find GMPP, trapping at local MPP, and incorrect perturbation direction under PSC or sudden irradiance change due to MPPT failure [5].

To reduce and rectify above-mentioned shortcomings, artificial intelligence (AI)-based MPPT techniques based on evolutionary, heuristic, and metaheuristic techniques are the best possible solution [6]. Apart from implementing MPPTs, solar power system efficiency can be improved by integrating soft-computing weather forecast and adjusting the tilting angle of solar panel to track the sun's direction [7].

AI-based MPPT is also termed as soft-computing MPPT, computational intelligence MPPT, or bio-inspired MPPT. In order to predict and evaluate GMPP throughout the nonlinear P–V curve, AI-based MPPT techniques utilize variable information like solar irradiance, input voltage, and input current of solar power system. Amalgamating AI in MPPT makes the system complex, robust, self-learning, and digitalized that stimulates convergence speed and transient response. Present days, researchers are focusing on AI-based MPPT, and because of this, enough sources are available for comparative literature review, yet there are very limited comparative studies, [8].

This chapter provides an insight of latest development and advancement of AI-based MPPT algorithms suitable for solar power system. AI-based MPPT algorithm is adopted to overcome common drawbacks of conventional MPPT techniques like power fluctuation, inability to operate normally under PSC, and rapid irradiance changes, trapping at one of the local MPPs and oscillation around MPP [9].

2 Proposed Methodology

This chapter provides realization of recent trends and advancement of AI, that are applicable in MPPT so as to boost efficiency of solar power system; initially, utilizations and significance of AI in MPPT for solar power system are reviewed. Recent trends and research areas of AI in MPPT are overviewed; performance evaluation and comparative analysis of contrasting AI algorithm in MPPT techniques are provided. In general, all conventional MPPT techniques exhibit some common disadvantages like power fluctuation, inability to operate normally under PSC, and rapid irradiance changes, trapping at one of the local MPPs and oscillation around MPP; therefore, AI is adopted to overcome these drawbacks. AI-based MPPT techniques are compared and evaluated based on literature undergone and results obtained from MATLAB/SIMULINK. Figure 2 is a flow process representation of proposed methodology used for this review.

3 AI-Based MPPT Techniques

3.1 Fuzzy Logic Controller (FLC)

Control system derived on fuzzy logic capable of converting symmetrical inputs into continuous digital data, i.e., 0 and 1 is a FLC [10]. Certain limitations of HC techniques like oscillation around MPP, greater settling time, and steady-state error were the main prospects for invention of FLC. This technique is simple to outline as information of exact model of MPPT is not required, and this makes it easy to realize and in turn increases the popularity [11]. HC algorithms like P&O and I&C can be integrated with FLC, and it converts HC algorithm into fuzzy rules [12]. Using this approach under varying irradiance, higher power efficiency and load current compared with HC algorithm are obtained [13].

$$\frac{\mathrm{d}P_{\mathrm{PV}}}{\mathrm{d}V_{\mathrm{PV}}} = E_{rr} = \frac{P_{\mathrm{PV}}(k) - P_{\mathrm{PV}}(k-1)}{V_{\mathrm{PV}}(k) - V_{\mathrm{PV}}(k-1)} \tag{1}$$

Fig. 2 Flowchart for proposed methodology

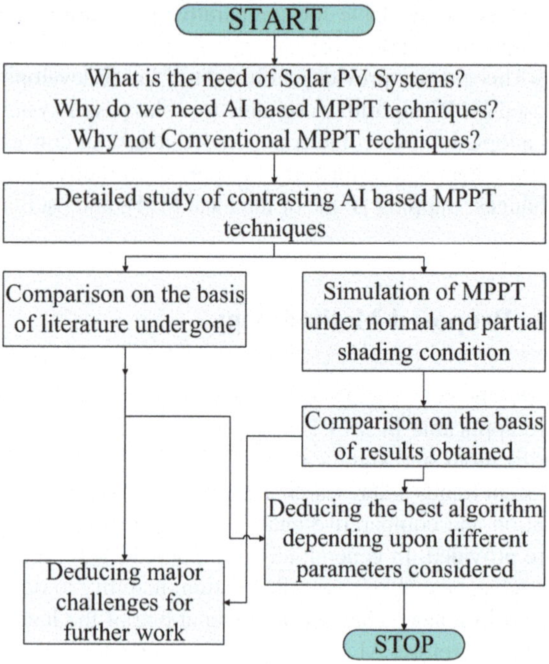

$$\frac{\mathrm{d}P_{\mathrm{PV}}}{\mathrm{d}V_{\mathrm{PV}}} = \Delta E_{rr} = E_{rr}(k) - E_{rr}(k-1) \tag{2}$$

where

E_{rr}	is the number of errors;
ΔE_{rr}	is the rate of change of error;
$P_{\mathrm{PV}}, V_{\mathrm{PV}}$	are output active power, voltage, respectively.

A FLC is applied on any low to medium microcontroller having Arduino Mega and Microchip to control DC–DC converter's duty cycle D based on T and E_e that searches the MPP [14]. Field-programmable gate array (FPGA) makes the FLC reprogrammable making it highly flexible and reconfigurable [15]. FLC is a comparatively easy-going, cost-effective, and traditional MPPT. The basic rules for FLC are as follows,

(1) If $\Delta P > 0$ and $\Delta V > 0$, $\Delta P/\Delta V > 0$, then D is decreased by $-\Delta D$.
(2) If $\Delta P > 0$ and $\Delta V < 0$, $\Delta P/\Delta V < 0$, then D is increased by $+\Delta D$.
(3) If $\Delta P < 0$ and $\Delta V > 0$, $\Delta P/\Delta V < 0$, then D is decreased by $+\Delta D$.
(4) If $\Delta P < 0$ and $\Delta V < 0$, $\Delta P/\Delta V > 0$, then D is increased by $-\Delta D$.
(5) If $\Delta P = 0$, then MPP is achieved.

Here ΔV is variation in voltage, and ΔP is variation in active power.

For every step, considering $E = \Delta P/\Delta V$ along with the sign of ΔP and ΔV, following constraints are deduced,

(1) If $E < 0$, then $D = D + \Delta D$.
(2) If $E > 0$, then $D = D - \Delta D$.
(3) If $E = 0$, then $D = D$.

Another type of FLC is reduced-rule FLC (RR-FLC), which improves the simplicity of FLC by reducing the computational load [16]. Mamdani and Takagi-Sukeno (T-S) design approaches for FLC are also available, among them Mamdani-based FLC is relatively popular [17]. FLC is basically of three steps: fuzzification (input variables are converted into linguistic variables by using various defined membership functions), fuzzy rules (these variables are manipulated based on the rules "if–then" by applying the desired behavior for system), and defuzzification (these variables are converted into numerical variables). The membership functions are significant in affecting the speed and accuracy of FLC [18].

Figure 3, depicts that E_{rr} and ΔE_{rr} are two major FLC input variables and are assigned to several linguistic variable while D of a DC–DC converter is output variable to be manipulated by FLC [19]. The integration of FLC with M5P model tree (Quinlan's M5 algorithm) is investigated to reduce the computation time [20]. Table 1 presents merits and demerits and Table 2 presents recent studies of FLC.

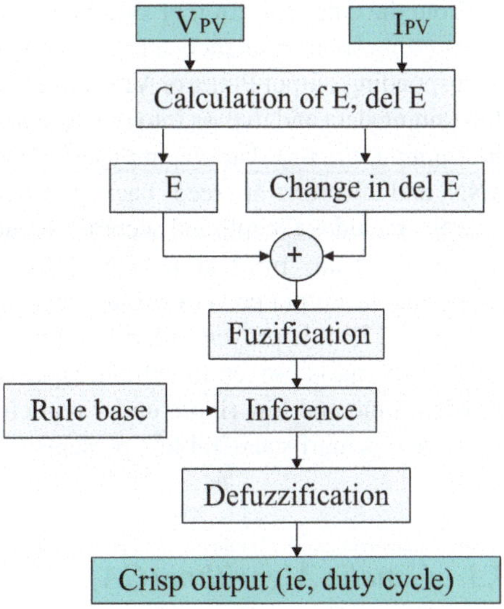

Fig. 3 Flow chart of FLC-based MPPT technique

Table 1 Merits and demerits of FLC-based MPPT technique

Merits	Demerits
High efficiency, small fluctuation in steady state	Difficulty in deriving fuzzy rules, time consuming
Fast tracking speed during sudden change in irradiance	Inability to learn and respond according to environment
Operation with inaccurate input	Complex calculation
Simple design and implementation	Undesirable performance under PSC
Good dynamic performance	Fuzzy rules directly effects system performance

3.2 *Artificial Neural Network (ANN)*

Biological neural networks of human brains gave rise to ANN and neuron as its basic processing element. ANN finds its implementation in all possible fields to resolve complex problems. It is employed to train and test the nonlinear relationship of I-V, P-V characteristics of solar power system. ANN fetches input current, input voltage, irradiance and/or temperature and consistently learns to fit the solar power system for maximum power [23]. FLC can be simulated by using ANN with higher accuracy and simple converter application [24].

From the data collection of simulation or hardware setup, the dataset is acquired by inputting solar irradiance, temperatures, voltage or current to ANN in finding the corresponding output Pmax or Vmax as shown in Fig. 4. These data are converted to the training data and to pass into the designed ANN to teach it how to perform. After the training, the test datasets are used to evaluate the performance of the designed ANN, and the errors are feed- backed to ANN for further adjustment [25].

ANN includes exceptional accuracy in modeling nonlinearity and resolving problems without any prior knowledge [26]. ANN can be utilized in modeling and predicting the output power of solar power system for improving tracking speed and accuracy [27]; it is proven to have better response time and less oscillation around MPP [28]; and is proven to exhibit capability of tracking MPP with the minimum transient time and low ripple under real climatic conditions [29]. Table 3 presents merits and demerits and Table 4 presents recent studies and applications of ANN.

3.3 *Genetic Algorithm (GA)*

GA, another AI-based optimization method based on evolution of chromosomes, is used for distinct optimization problems and is primarily used in MPPT for evaluating the reference voltage of PV panel by modifying population for individual solutions. Voltage-step selection GA algorithm leads to comparatively small oscillations, swift convergence speed, and faster dynamics [32]. Lower population size, simplified mutation processes, and uncomplicated calculation of crossover are some additional advantages with modified GA [33]. GA-based MPPT is competent enough for searching GMPP rather than oscillating around local MPP. GA is short-sighted for optimization of large-scale, highly complex problems in spite of its performance because of simplified algorithmic approach. For optimization of MPPT, GA is initialized by starting the initial parent population as an array:

Table 2 Comparative studies of FLC-based MPPT technique

References	Input parameter I/O sensor	DC–DC converter	MPPT time (s)	Steady-state oscillations (%)	Efficiency (%)	Findings
[21]	V & I	Buck		4.00		FLC is used to control MPPT in a micro grid. The steady-state performance has been improved as compared with conventional P&O method
[14]	V & I	Boost	0.43	1.70	98.5	Single-input T-S FLC is effective in tracking GMPP under PSC compared with conventional P&O algorithm. FLC exhibits less settling time, minimum oscillation
[15]	V & I	Boost	0.3	1.00	98	FPGA-based FLC is flexible as the membership functions and inference rules can be reconfigured by changing very high-speed integrated circuit hardware description language (VHDL)

(continued)

Table 2 (continued)

References	Input parameter I/O sensor	DC–DC converter	MPPT time (s)	Steady-state oscillations (%)	Efficiency (%)	Findings
[17]	V & I	Boost	0.06		99	FLC is efficient in tracking GMPP value with less tracking time, compared with IC and P&O
[22]	V & I	Boost	Less then 0.01		99.37	Improved M5P model (FLC-based MPPT) proves to minimize computation time and lead to energy loss

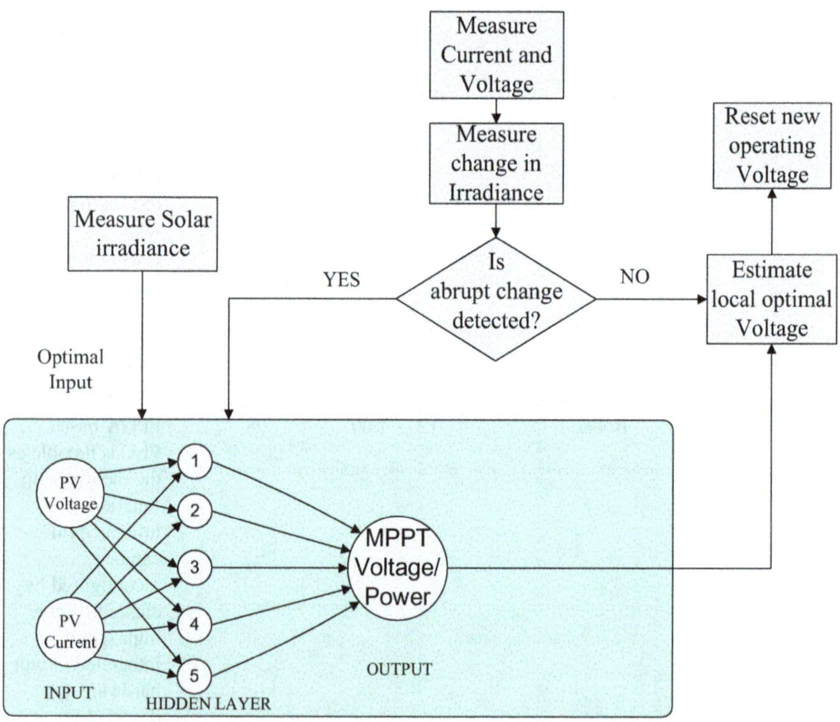

Fig. 4 Block diagram of ANN-based MPPT technique

Table 3 Merits and demerits of ANN-based MPPT technique

Merits	Demerits
Fast response, tracking speed	Massive data set is required
Slight fluctuation in steady state	Complex and time consuming
No need to be reprogrammed	Tracking accuracy is affected by PV panel (system dependable)
	Periodic tuning is required due to environmental change and aging
	Difficult to be trained properly and get training data

$$X^i = \left[\text{parent}^1 \ \text{parent}^2 \ \text{parent}^n \right] \tag{3}$$

where n is population size; and parent^i ($i = 1, 2, \ldots, n$) are voltage values when the algorithm begins optimization; generated output power is objective function $f(Xi)$. Assessment of fitness values for every position is carried out by objective function. They are then employed to develop population gradually and enhance population fitness through generations. Algorithm should be reinitialized particularly for MPPT because of unexpected changes in solar irradiance, load conditions or PSC, (4) and (5) must be satisfied in order to reinitialize GA-based MPPT technique where current measurement is k, next iteration is $k + 1$.

$$|V(k+1) - V(k)| < \Delta V \tag{4}$$

$$\left|\frac{P(k+1) - P(k)}{P(k)}\right| > \Delta P \tag{5}$$

GA process is illustrated in Fig. 5. Firstly, the initial population is encoded in binary. They are decoded into real number, and their fitness values for each chromosome are evaluated. The genetic operations including selection, crossover, and mutation are performed for an optimal solution, specifically in the maximization of power output. Table 5 shows the merits and demerits, and Table 6 illustrates recent research of GA.

3.4 Swarm Intelligence (SI)

3.4.1 Particle Swarm Optimization (PSO)

Particle swarm optimization (PSO) is standardized SI technique that employs heuristic method for resolving MPPT optimization problem. The possible solution is represented by position of a particle and solution space represents duty ratio [36].

Table 4 Recent comparative studies of ANN-based MPPT technique

References	Input parameter I/O sensor	DC–DC converter	MPPT time (s)	Steady-state oscillations (%)	Efficiency (%)	Findings
[30]	V, I, atmospheric parameter	Boost	0.06		Up to 99.68	ANN model is deployed to learn operation variation of a solar power system. PSO is to find optimum initial weights
[24]	V, I	Boost			Above 90	MPPT based on ANN modeled FLC exhibits higher fault tolerance and simpler implementation
[25]	V, I	Buck Boost		0.1		Backpropagation trained neural network can accurately predict MPP of a PV panel. It provides accurate and faster results than P&O-based MPPT
[29]	V, I		0.2–0.4		Above 90	ANN model is based on the input voltage, current, and irradiance to predict GMPP with knowledge learned from training data
[31]	T, Irradiance	Boost	0.03	0.7	Above 90	ANN is trained by using nntool in MATLAB/Simulink. This MPPT controller has less steady-state error, fast response for sudden change in solar temperature and irradiance, compared with IC, P&O

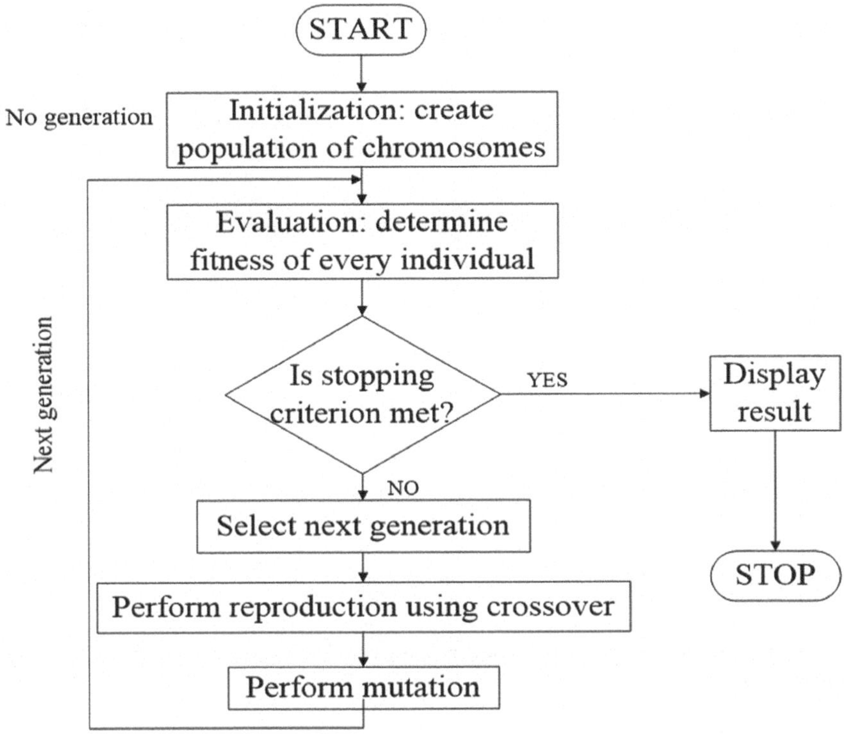

Fig. 5 Flowchart of GA-based MPPT technique

Table 5 Merits and demerits of GA-based MPPT technique

Merits	Demerits
Least computational requirement	Due to series format tracking speed is slow
General and uniform implementation scheme	Depending on initial condition
High stability and rapid response	

PSO is proven to give befitting results after every iteration based on the concept of flocking birds. Population of particles are set forth in PSO and their positions are compared with the best local and global position; these particles are then moved in search space for the best solution [37]. PSO is ready to be integrated with overall distribution, as it rapidly finds the rough region around GMPP [38]. An improved PSO integrates with a nonlinear decreasing inertia weight in improving the search process of the particles [39]. For other modified PSO, the weighting value and learning factor decrease with every iteration. In contrast, the social learning factor is expected to increase. Besides, the weighting value is modified based on the changes in slope and

Table 6 Recent comparative studies of GA-based MPPT technique

References	Input parameter I/O sensor	DC–DC converter	Findings
[34]	T, irradiance		A GA based on large variation radial basis function is used to learn data pattern of temperature and irradiance. The algorithm predicts MPP with high accuracy after dataset training
[32]	V, I	Boost	GA is applied to calculate voltage reference of PV panels to combine MPPT and constant power generation in solar power systems. It imitates the performance of P&O MPPT with small power oscillations and fast dynamics
[35]	V, I		GA is implemented to improve the efficiency especially under PSC which results in an overall reduction of loss energy

power characteristic curve. These modifications enhance tracking speed and stability [40].

3.4.2 Grey Wolf Optimization (GWO)

Grey wolf optimization (GWO) is a modern heuristic optimization technique inspired by the lifestyle of grey wolves. The leader is defined as α, sub-leader as β, lower rank is δ, and the lowest rank is ω. A GWO-based MPPT is dependent on the hunting techniques of the grey wolves, i.e., following the order of α, β, and δ. This algorithm will converge to the prey that is GMPP here.

3.4.3 Firefly Algorithm (FA)

Firefly algorithm (FA) is another type of swarm intelligence technique based on behavior and flashing of fireflies. Concept used is that attractiveness is proportional to brightness of firefly; in this manner, fireflies can unite as optimal solution by attractiveness. Similarly, FA can be utilized to find the optimal MPP.

3.4.4 Modified Cat Swarm Optimization (MCSO)

Modified cat swarm optimization (MCSO) method is system independent, highly efficient in tracking GMPP regardless of the location in search space and tracks GMPP accurately and converges faster [41].

3.4.5 Moth Flame Optimization (MFO)

Moth flame optimization (MFO) is another metaheuristic optimization evolved from convergence of moth toward luminescence [42].

3.4.6 Cuckoo Search (CS)

Cuckoo search (CS) is an emerging swarm intelligence technique based on parasitic reproduction approach that is reproduction strategy of some species of cuckoo birds that lay their eggs in some other bird's nest. The basis of CS is to find the right host nest, which is like looking for food. It is a random process and Lévy flight model is the most common method to model food seeking trajectory of an animal. Mathematically, Lévy flight model represents a random walk, and the step sizes are determined by using Lévy distribution. When using this technique for MPPT, it exhibits faster speed and high tracking accuracy regardless of weather conditions. It is a simpler MPPT technique having three particles and single parameter to be tuned [43]. However, CS method does not guarantees tracking of GMPP and is highly complex to implement [44].

3.4.7 Gravitational Search Algorithm (GSA)

Gravitational search algorithm (GSA) is centered on Newtonian gravity and laws of motion, which states that "particles tend to accelerate towards each other because they attract each other". Standardized steps involved in GSA are: The population size is determined for duty cycle of DC–DC converter ranging between 10 and 90%; solar agents are uniformly positioned between the search space intervals to attain maximum convergence speed; for each agent position, PV output power is calculated, power of MPPT is assumed as mass of the agents; force G acting between agents and net force acting on each agent is computed; acceleration a of each agent is calculated.

Improved GSA has dynamic weight in change factor of gravity constant. The factors of memory and population information are added into updated formula of particle velocity [45].

Other swarm intelligence algorithms like artificial bee colony, bird flocking, animal herding, bacterial growth, microbial intelligence, and crowd or human swarming are inspired by biological behavior for optimization process. Table 7

Table 7 Merits and demerits of SI-based MPPT techniques

Merits	Demerits
Massive dataset is not required	Oscillation because of large random search
Elimination of oscillation around MPP High tracking accuracy and fast convergence	Larger computational burden Requiring huge data
Simple structure, easy implementation, and fast computation ability	Highly complex and time consuming
High ability in searching GMPP regardless where is GMPP	Algorithm parameters need to be carefully set

shows open issues and advantages, and Table 8 illustrates recent studies on swarm intelligence-based MPPT.

3.5 Hybrid MPPT

Hybrid MPPT is a general term used for integration of two or more AI or conventional MPPT techniques. ANN and FLC are suitable to integrate with conventional MPPT methods like P&O and IC. ANN estimates the MPP without any shading conditions or panel temperature, while the HC method improves the result further. A promising hybrid MPPT is "neural network P&O controller" that is an integration of ANN with conventional P&O algorithm; it reduces steady-state oscillation and accelerates tracking speed under sudden irradiance changes or PSC [50]. Other hybrid MPPTs include P&O-ANN and IC-ANN, which integrate with the stacked auto-encoder controller by using deep learning training and building blocks to act as an auto-encoder. It is trained with a greedy layer-wise pattern in extracting maximum power output; after that, backpropagation with supervised learning is used to fine-tune the deep neural network with conventional MPPT (IC and P&O) to reach the maximum power [51].

Another popular hybrid MPPT is adaptive neuro-fuzzy inference system (ANFIS) that sums both ANN and FLC. It has the advantages of both ANN and FLC. ANN is trained to evaluate optimal MPP and used for operating FLC-based MPPT. ANFIS is optimal, flexible, and adaptable to any new configuration for smart power management and solar power system [52]. Neuro-adaptive learning technique is used to model fuzzy procedure in learning all the information about a dataset. It is a process to map all the given dataset from multiple inputs into a single output. ANFIS develops a fuzzy inference system by using input–output datasets. The model determines membership function parameters, which are best fit in allowing FIS to track input and output data [53]. Fuzzy membership function parameters are tuned by utilizing hybrid learning method, including back propagation and least square algorithms [54]. ANFIS-based MPPT is proven to enhance conversion efficiency of solar power

Table 8 Recent studies on swarm intelligence-based MPPT

References	Type of SI	Input parameter I/O sensor	DC–DC converter	MPPT time (s)	Efficiency (%)	Findings
[46]	Pigeon	V, I	Boost	0.1		A pigeon optimization is used to optimize MPPT under PSC. It reduces power oscillation, improves stability, and achieves desirable control results
[36]	PSO	V, I	Boost	1.0		PSO combined with one cycle control is capable of tracking GMPP under varying shading conditions
[37]	PSO	V, I		0.4		PSO is applied for MPPT in obtaining the optimum duty cycle for the Z-source inverter to overcome the shortage of conventional MPPT technique
[39]	PSO	V, I	Buck	1.6	97	PSO is implemented in MPPT to track MPP. PSO has more power and lower power fluctuation compared with FA and P&O

(continued)

Table 8 (continued)

References	Type of SI	Input parameter I/O sensor	DC–DC converter	MPPT time (s)	Efficiency (%)	Findings
[40]	MPSO	V, I	Boost	0.55–1.2		Conventional PSO has been modified to vary weighting value, social learning factor, cognition learning factor based on slope and changes in power
[41]	MCSO	V, I	Boost			The system independent CSO is highly efficient in tracking GMPP regardless of the location in search space. This removes power oscillation around MPP compared with P&O
[42]	MFO	V, I	Boost	0.05	99.91	MFO is applied for optimal exploitation of PV sources under PSC. It exhibits better tracking ability, efficiency compared with IC, FL and PSO
[43]	CS	V, I	Boost	1.8–2.8	99	Deterministic CS is deployed to eliminate random number in voltage calculation of conventional CS method

(continued)

Table 8 (continued)

References	Type of SI	Input parameter I/O sensor	DC–DC converter	MPPT time (s)	Efficiency (%)	Findings
[47]	ACO	V, I	Cuk	0.38		ACO-based MPPT provides optimal power for residential applications from solar panel
[48]	Spider monkey	V, I	Boost	0.20		P&O technique has been improved by PI controller which is tuned by spider monkey algorithm to achieve good response under different atmospheric condition
[49]	AFSA	V, I	Boost	0.04		AFSA can eliminate constraint of multiple local extreme points and catch MPP with high precision
[45]	GSA	T, Irradiance	Boost	0.04		Improved GSA-based MPPT achieves short tracking time and good tracking accuracy in MPPT under various conditions com- pared with GSA and PSO

Table 9 Merits and demerits of hybrid MPPT

Merits	Demerits
Combined advantages of conventional and AI-based MPPT	Relatively complex
Drawbacks of conventional and AI-based MPPT are eliminated	Longer computation time
High accuracy and fast-tracking speed	Costly

system [55]. Fuzzy neural network is also capable of bit error correction in predicting and forecasting weather data for solar power system [56].

ANN is deployable based on hybrid PSO and GSA, along with FLC. For instance, PSO-GSA generates a random initial population first and sends them to ANN for data training [57]. Another hybrid MPPT technique is based on improved open-circuit voltage model-based approach and smart power scanning procedure (checks the voltage values to see whether PSC is prevailing or not) [58]. Apart from ANN, FLC is also versatile to integrate with P&O algorithm as this will lead to added advantages of both techniques together. FLC-based P&O has a variable step size to ensure small oscillation and faster response because big step size ensures fast response but leads to excessive oscillation whereas small step size has slow response and less oscillation [59]. Hybrid IC-PSO algorithm is also available to resolve the inability of conventional MPPT in tracking GMPP under PSC while improving the convergence speed and tracking precision [60].

Another hybrid MPPT is the integration of two powerful machine learning techniques, i.e., coarse-Gaussian support vector machine (CGSVM) and ANN, known as ANN-CGSVM technique. CGSVM is a type of nonlinear support vector machine learning technique categorized as a data mining technique required in generating large and accurate training data for MPPT [61].

A novel combination of the ANFIS controller with HC method estimates the duty ratio with higher accuracy. HC carries on an online fine-tuning of duty cycle, which resolves the problem of conventional MPPT in searching GMPP under PSC since the duty ratio of MPP is estimated offline by the ANFIS technique. Table 9 presents the merits and demerits, and Table 10 presents the recent research of hybrid MPPT.

3.6 *Machine Learning (ML)*

Bayesian ML is a specialized method in unsupervised classification, curve detection, image segmentation, and real-time location-based weather forecasting applicable in MPPT to achieve GMPP [66]. ELM algorithm is utilized to update the weights by different PSO techniques to train a single layer feed-forward network and compare performances with back-propagation forecasting model [67]. Reinforcement learning (RL) method enables autonomous learning by observing the environmental state of

Table 10 Recent studies on hybrid MPPT implementation

References	Type of MPPT	Input parameter I/O sensor	DC–DC converter	MPPT time (s)	Efficiency (%)	Findings
[62]	FLC with PSO,GA	V, I	BOOST	2.00		Parameters of FLC are tuned by using hybrid PSO and GA. It exhibits 2%-8% higher output power with faster response and higher accuracy
[63]	ANFIS-PSO	V, I	Zeta	3.00	98.35	ANFIS-PSO hybrid MPPT is deployed to acquire MPP with zero tracking oscillation
[64]	FLC with P&O	V, I	Boost	1.00	99.6	The designed membership functions of FLC incorporates advantages of P&O and FLC and eliminates their drawbacks
[33]	Modified GA and FA	V, I	Buck	0.089		A fusion algorithm is deployed to integrate three nature-inspired algorithms for MPPT. It simplifies calculation of GA by integration of mutation process of DE and modifies attractive process of FA

(continued)

Table 10 (continued)

References	Type of MPPT	Input parameter I/O sensor	DC–DC converter	MPPT time (s)	Efficiency (%)	Findings
[65]	FLC & variable step P&O	V, I	Boost	0.03		FLC is developed to regulate DC-link while an improved P&O with variable step size is designed to reduce PV power fluctuation
[51]	PO-ANN, IC-ANN	V, I	Cuk	0.4	> 91	Hybrid techniques based on PO-ANN and IC-ANN are utilized in SAEs. It is trained with DL network and building blocks to enables maximum power extraction
[54]	ANFIS	V, I	Buck	0.012	91	ANFIS-based model of IC method and constant voltage method has been proposed for MPPT

the solar power system used to train and adjust perturbation for the maximum output as illustrated in Fig. 6. Table 11 shows the merits and demerits while Table 12 presents the recent studies of ML-based MPPT.

4 Analytical Comparison of AI-Based MPPT Techniques

AI-based MPPT techniques are evaluated in terms of enormous parameters and features like: number of control variables (input sensory parameters), utilized platform (software: MATLAB/Simulink; hardware: arm cortex microcontroller,

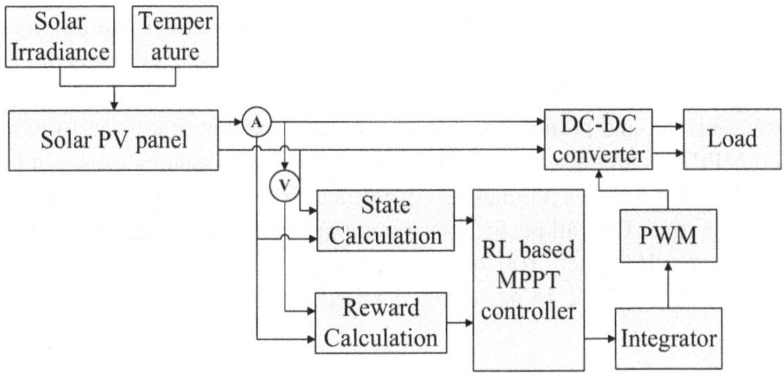

Fig. 6 Flowchart of RL-based MPPT technique

Table 11 Merits and demerits of ML-based MPPT technique

Merits	Demerits
Can consider multiple variables	Highly complex and costly
Weather forecast for MPPT prediction	Large amount of data is required
High accuracy and fast-tracking speed	Longer computation time

Table 12 Recent studies of ML-based MPPT

References	AI-based MPPT	Type of MPPT	Input parameter I/O sensor	DC–DC converter	MPPT time (s)	Efficiency (%)
[66]	ML	BML	V, I	Boost	1.88	98.9
[67]	ML	MBML	V, I			
[68]	Others	DE	V, I	SEPIC	2.00	99
[69]	Others	MFPA	V, I	Boost	0.05	99.1

Arduino, Raspberry Pi, and DSP board-dSpace), solar panel parameters, switching frequency of DC–DC converter (buck, boost, buck-boost, Ćuk or SEPIC), tracking speed or transient time, oscillation, accuracy, and MPPT efficiency. Humidity, shading, cloud, and metrological data are considered as input parameters rather than just current and voltage.

AI-based MPPT techniques are advanced and efficient but are highly complex and costly and require large amount of data. The balance between performance, cost, and/or complexity is critical for the application of MPPT in a particular area. AI-based MPPT techniques are categorized into seven major groups. The family of swarm intelligence is the largest in AI-based MPPT due to fast performance and high accuracy. The hybrid MPPT is relatively versatile as the AI-based MPPT is easily integrated with each other. ML is another popular technique having numerous approaches and techniques to learn from the experience or dataset in order to deliver maximum power output. FLC, ANN, and GA do not have any sub-categories. Bio-inspired algorithms and ML are immensely popular due to sophistication in terms of accuracy, speed, and performance. A graphical representation of performance of AI-based MPPT techniques on the basis of literature undergone is shown in Fig. 7.

All AI-based MPPT techniques are examined in terms of performance and are illustrated in Fig. 8. Overall performance compared with other algorithms are rated between points 0–100 where 100 implies highest performance while 0 implies undesirable performance. Table 13 presents a detailed comparison of AI-based MPPT in

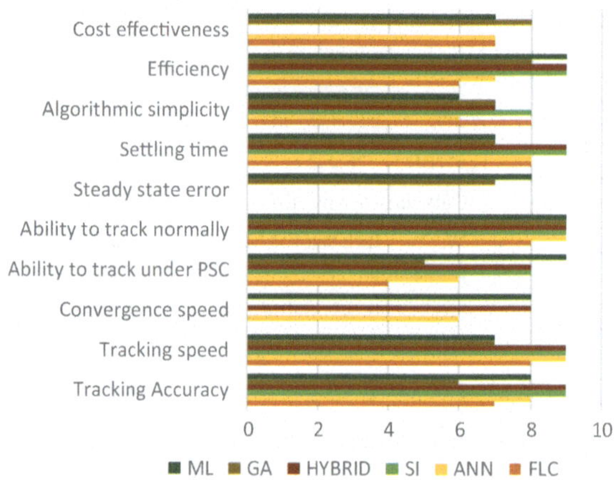

Fig. 7 Performance of AI-based MPPT techniques based on literature undergone

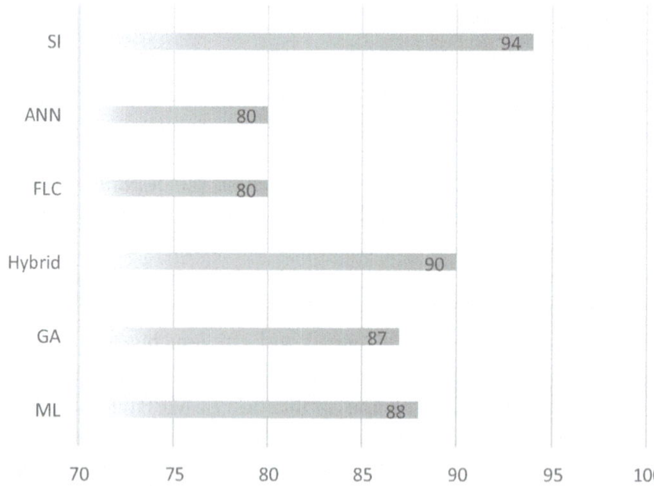

Fig. 8 Overall performance of AI-MPPT techniques

Table 13 Comparison of AI-based MPPT techniques on the basis of literature undergone

Index	FLC	ANN	SI	Hybrid	GA	ML
Tracking accuracy	Moderate	High	High	High	Moderate	High
Tracking speed	Moderate	Fast	Fast	Fast	Moderate	Moderate
Convergence speed	Moderate	Moderate	Fast	Fast	Fast	Fast
Ability to track under PSC	Poor	Poor	High	High	High	High
Ability to track normally	High	High	High	High	High	High
Steady-state oscillation	Small	Small	Almost zero	Small	Moderate	Small
Oscillation around MPP	No	No	No	No	No	No
Settling time	Fast	Fast	Fast	Fast	Fast	Fast
complexity	Moderate	High	Moderate	High	High	High
Input parameter	V, I	Irradiance, T, V, I	V, I (varies)	Varies	V, I (varies)	Varies
Periodic tuning	Yes	Yes	No	No	No	No
Dependency of initial design	High	High	Moderate	Moderate	Moderate	Moderate
System independence	Poor	Poor	High	High	High	High
Efficiency	Poor (PSC)	Poor (PSC)	High	High	High	High
Cost	High	High	Moderate	High	Moderate	High
Time	Moderate	High	Moderate	High	Moderate	High
Complexity	Medium	Medium	Simple	High	High	High
Application	Grid and solar vehicles	Water pump, solar vehicle	Off-grid, on-grid	Off-grid, on-grid	Off-grid, on-grid	Off-grid, on-grid

terms of performance indices like tracking accuracy, tracking speed, convergence speed, ability to track under PSC, etc., based on literature undergone. The results are established based on the literature reviews on existing studies and validated by simulation results on MATLAB/Simulink. It is evident that SI has highest point in average, followed by hybrid, ML, and GA. They are metaheuristic methods capable of adapting the operating environment of solar power system. The balance between algorithm complexity and MPPT performance is achievable by using SI, hybrid, ML, or GA techniques.

It is also perceived that FLC and ANN have relatively poor convergence speed and tracking ability under PSC (continuous periodic tuning process is required for converter switch to track MPP). For ANN, a massive dataset is required to design a proper ANN-based MPPT with high accuracy, difficulty in training, and higher time consumption. For FLC, it is difficult to derive its fuzzy rules accurately and unable to learn actively from the dynamic environment and perform undesirably. In contrast, SI, hybrid, GA, and ML exhibit faster speed and high ability in tracking even under PSC owing to their newer architecture which combines the advantages of conventional HC MPPT and the latest advancement of AI.

5 Simulation Results

5.1 Simulation Setup and Configuration

To validate and compare the performance of AI-based MPPT techniques, an extensive simulation is conducted on MATLAB/Simulink R2020a to study, evaluate, and investigate dynamic behavior of AI-based MPPT under PSC. The PV panel SunPower module (SPR-305E-WHT-D) inputs with varying solar irradiance E_e and temperature T. It is simulated under PSC to emulate the practical environment. A 5 kHz DC–DC boost converter is designed, and its insulated gate bipolar transistor (IGBT) switching devices are controlled by AI-based MPPT controller for most optimized voltage and current outputs.

A DC–AC inverter based on synchronverter topology is deployed to convert optimized solar DC output to AC for supplying three-phase balanced RL. The MPPT controller has been changed from FLC, ANN, SI, hybrid, GA to ML to compare their tracking ability for MPP under PSC, and simulation results are then compared and validated as shown in Table 14.

Nonlinearity of I-V and P-V characteristics of a solar power system is main reason of an AI-based MPPT to search for MPP with different irradiance and temperature. Figure 9a [31] presents the I-V and P-V characteristics, when irradiance varies and temperature remains constant at 25 °C. On the contrary, Fig. 9b [31] presents the I-V and P-V characteristics, when temperature varies and irradiance remains constant at 1000 W/m^2.

Table 14 Comparison of AI-based MPPT techniques on the basis of simulation result

AI-based MPPT	Tracking time (s)	Steady-state oscillation (%)	Affected by PSC
FLC	0.70	3.0	Yes
ANN	0.25	< 1.0	Yes
SI	0.28	1.7	No
Hybrid	0.23	0.1	No
GA	0.80	10.0	No
ML	0.60	1.5	Slightly

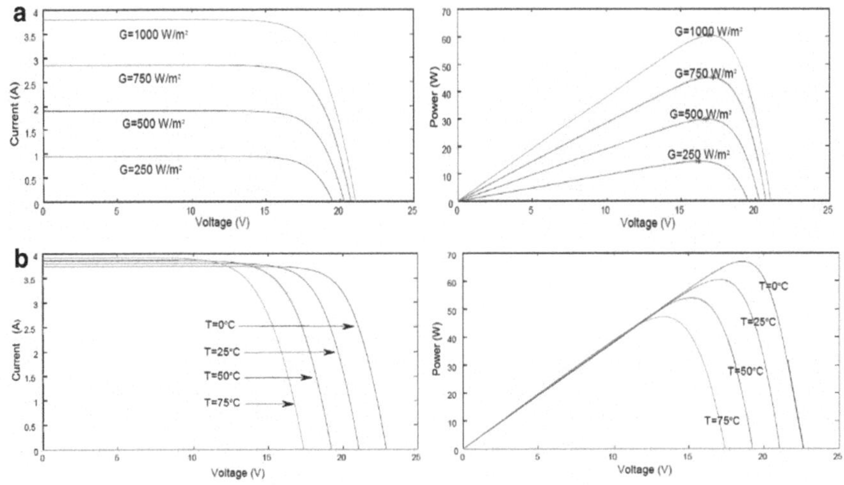

Fig. 9 **a** I-V and P-V curves of solar panel under STC having constant temperature (25 °C) and varying irradiance. **b** I-V and P–V curves of solar panel under STC having constant irradiance (1000 W/m^2) and varying temperature [31]

5.2 PSC Analysis

PSC is analyzed by emulating PSC for inputs of the solar panel. To simulate PSC, current is adjusted to allow multiple peaks in P-V curves. PSC effects on the solar module are accounted for, which enables partial shading on certain cells. The phenomenon is common for the practical environment where partial shading occurs because of dirt, leaf, cloud, tree, and other obstacles which block the sunlight.

SI and hybrid MPPT perform optimally by tracking GMPP, because of population searching ability and combination of different algorithms. GA tracks the local MPP with some steady-state oscillations but ML and ANN are performing well. However, the performance of FLC is relatively unsatisfactory because of its slow transient response and inability to track GMPP; it is trapped at local MPP and results in lower power conversion efficiency.

5.3 MPPT Ability

The tracking ability of AI-based MPPT controller for MPP with constant irradiance is simulated. Figure 10a–f shows MPPT ability of different algorithms. The dotted blue line indicates optimal MPP at approximately 650 W with normal irradiance and temperature, and red line indicates the output power of solar power system. Performance of AI-based MPPT is relatively satisfactory except for FLC.

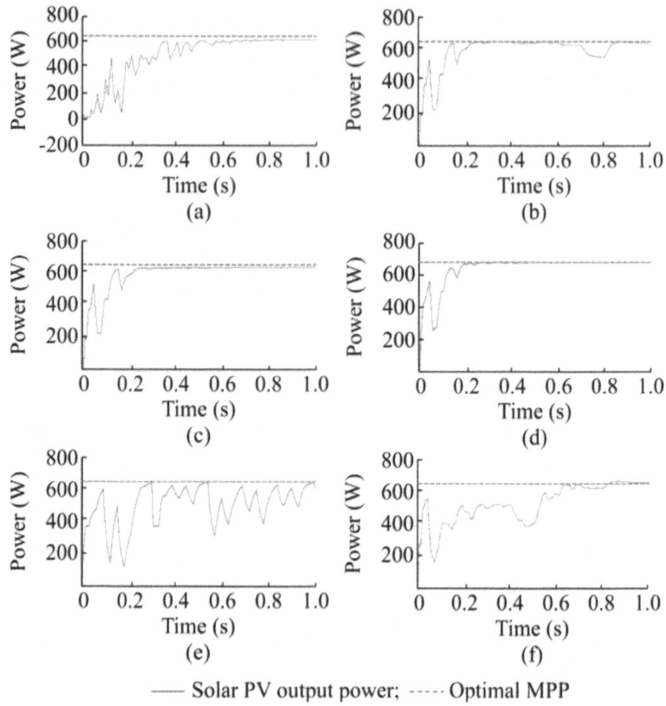

Fig. 10 MPPT ability of **a** FLC, **b** ANN, **c** SI, **d** hybrid, **e** GA, **f** ML

5.4 Comparative Analysis and Validation of Results

Table 14 summarizes and presents the comparison of AI-based MPPT techniques in terms of steady-state oscillation at MPP, ability to resist the adverse effect of PSC or varying irradiance, which occurs at around $t = 0.7$ s. It is observed that FLC is underperforming by exhibiting higher tracking time with high steady-state oscillation and affected by PSC. However, SI and hybrid-based MPPT perform satisfactorily with the minimum tracking time, low steady-state oscillation, and the minimum disturbance under PSC. These analyses are tallied with Table 13, and both simulation results and literature review findings are mutually verified and validated. Nevertheless, different scenarios and applications result in different choices of AI-based MPPT based on the design criteria and requirements.

6 Discussion

The extensive comparison and investigation of different AI-based MPPT have clearly shown that each algorithm has its own merits and demerits. The choice of algorithm solely depends on the design criteria and requirements of the designer. The input parameters like voltage and current, solar irradiance, temperature, humidity, and shading are required to train AI. AI is employed to predict current and voltage considering variable input of MPPT and nonlinear I-V and P-V characteristics. Performance and accuracy of MPPT techniques are influenced by voltage step, i.e., when voltage step is too small, it takes longer span to reach MPP, whereas when voltage step is too large, although it takes shorter time to reach MPP but MPP cannot be reached because of the excessive oscillation around MPP. A sudden change in irradiance is made as an input to observe output of MPPT, in order to check whether the convergence speed is in response to the rapid changes of input.

Another important aspect of AI-based MPPT is to search for GMPP under PSC. The MPPT could fail because of the inability of algorithm to search for GMPP, rather it will be stuck at local MPP and thus cannot produce optimal power output.

Fluctuation in operation point, non-uniform distributed solar insolation, inability of algorithm in identifying GMPP when there are multiple local MPPs are some consequences of oscillation. Oscillation time, in terms of performance parameters, is the period between changes until the output enters into steady state, i.e., no more oscillation. The tracking speed or convergence speed indicates how fast MPPT tracks the real MPP. The criteria for choosing a particular AI-based MPPT are based on complexity of implementation, sensors requirements, ability to detect multiple local maxima, steady-state error, response time, transient time, settling time, overshoot and ripples in output voltage and cost. Conventional or HC methods fail to track GMPP under PSC as they oscillate around MPP during the steady state and require longer time for tracking MPP along with lower efficiency whereas AI-based MPPT techniques do not exhibit the drawbacks of conventional MPPT but require higher cost, complex computation and modeling. Hybrid methods are the best among all algorithms, as it integrates two or more algorithms leading to mutual cancellation of open issues [70–74].

Apart from MPPT, inverter provides an interface between the solar power system and the power grid; therefore, an efficient inverter is required for converting DC power to AC power and as an anti-islanding protection [75]. PID controller is recommended to regulate duty cycle to pulse width modulation from MPPT techniques due to its flexibility, stability, minimum overshoot, fine-tuning characteristics, minimum rise time for output voltage, and performance optimization. AI-based MPPT techniques are applicable for grid-connected (on-grid), standalone (off-grid), and other specialized applications including solar vehicle, solar lamp, water heater, DC motor, and water pump.

7 Recommendations and Future Research Direction

This section aims to recommend AI-based MPPT to be applied in the solar power system and their future research areas. The traditional MPPT techniques are phasing out as the latest AI-based MPPT techniques have better performance and stability.

The development of the AI-based MPPT is dependent on the latest advancement in ML. The main challenges include the ability to search for GMPP and the complexity of the algorithm.

The conventional MPPT techniques such as open current, open voltage, IC, and P&O are recommended for simple and low-cost application which do not require high performance. In order to resolve, optimize, and predict the nonlinearity of the PV cell without staying at local MPP under PSC, the AI-based MPPT techniques are recommended for optimal performance, accuracy, and convergence speed. GA is faster than classical methods, but it tends to stick at local minima. The improved GA requires higher computation resources, and different parameters require tuning. In contrast, DE is fast and accurate without any employment of probability distribution. However, its population can be stagnant in some sub-optimal values. PSO has the highest performance by considering different best positions to update the population, which is also simple to be implemented in hardware and independent from the installed system. However, it tends to converge prematurely and can be trapped at local minima. For the maximum performance, PSO is recommended due to its maturity compared with GA. DE is better than GA in terms of accuracy and computation time while GA is faster than classical methods. GA and DE techniques track GMPP under PSC because of their capabilities of resolving multi-objective problems. For sensitive power fluctuation applications like household appliances, motor, extreme low voltage (ELV), light sources, electro-heat equipment, electrical machine, uninterruptible power source (UPS), computer, and electronic devices, CS and radial movement optimization (RMO) are recommended owing to their faster convergence speed to settle at GMPP with minimal fluctuation.

The parameters of an AI-based MPPT controller are the design complexity, ability to track GMPP, cost-effectiveness, PV panel dependency, prior training requirement, dataset requirement, convergence speed, analog or digital architecture, required sensory information, periodic tuning, stability, and efficiency. While designing AI-based MPPT, balance between complexity and performance of algorithm should be considered, i.e., higher the performance of AI-based MPPT, more complex will be the design algorithm.

8 Conclusion

In this chapter, AI-based MPPT techniques for solar photovoltaic system are reviewed, designed, and compared based on the performance in terms of algorithmic structure, input parameters, cost and complexity, platform, tracking speed, oscillation accuracy, efficiency, and their applications. These techniques are designed in MATLAB/SIMULINK environment under uniform, partial shading condition. As per comparison of literature undergone and simulation results, AI-based MPPT techniques display good convergence speed, small oscillation at steady state, and accurate tracking even under PSC. However, most of the techniques are costly and complex to build and require more datasets compared with conventional MPPT techniques.

Emerging algorithms like hybrid, SI, ML are recommended compared with FLC, ANN, and GA because of adaptive learning capabilities, fully digitalized system, and lesser open issues. ANN and FLC are non-preferable because of periodic tuning requirement, aging architecture, and inability in tracking MPP under PSC. The main challenges still prevailing include balance between complexity and performance of algorithm, i.e., higher the performance of AI-based MPPT, more complex will be the design algorithm and the ability to search for GMPP instead of LMPP. These findings are expected to provide a detailed insight into latest advancement of AI-based MPPT techniques along with their methodologies and characteristics.

Acknowledgements The authors would like to acknowledge Integral University, Lucknow for providing the MCN "IU/R&D/2021-MCN0001165". The statements made herein are solely the responsibility of the authors.

References

1. S. Kanwal, B. Khan, M.Q. Rauf, Infrastructure of sustainable energy development in Pakistan: a review. J. Mod. Power Syst. Clean Energy **8**(2), 206–218 (2020)
2. K.Y. Yap, H. Chua, M.J.K. Bashir et al., Central composite design (CCD) for parameters optimization of maximum power point tracking (MPPT) by response surface methodology (RSM). J. Mech. Continua Math. Sci. **1**(1), 259–270 (2019)
3. A. Ibrahim, S. Obukhov, R. Aboelsaud, Determination of global maximum power point tracking of PV under partial shading using cuckoo search algorithm. Appl. Solar Energy **55**, 367–375 (2020)
4. C. Correa-Betanzo, H. Calleja, C. Aguilar et al., Photovoltaic-based DC micro-grid with partial shading and fault tolerance. J. Mod. Power Syst. Clean Energy **7**(2), 340–349 (2019)
5. L. Zhang, S. Yu, T. Fernando et al., An online maximum power point capturing technique for high-efficiency power generation of solar photovoltaic systems. J. Mod. Power Syst. Clean Energy **7**(2), 357–368 (2019)
6. M. Chen, S. Ma, J. Wu et al., Analysis of MPPT failure and development of an augmented nonlinear controller for MPPT of photovoltaic systems under partial shading conditions. Appl. Sci. **7**(1), 95–116 (2017)
7. S. Srivastava et al., Applications of artificial intelligence techniques in engineering-volume 2, in *Part of the Advances in Intelligent Systems and Computing*, vol. 697 (Springer Nature, 2018), p. 647
8. A. Iqbal et al., Meta heuristic and evolutionary computation: algorithms and applications, in *Part of the Studies in Computational Intelligence*, vol. 916 (Springer Nature, 2020), p. 849
9. O. Abdalla, H. Rezk, E.M. Ahmed, Wind driven optimization algorithm based global MPPT for PV system under non-uniform solar irradiance. Sol. Energy **180**, 429–444 (2019)
10. A. Abbadi, F. Hamidia, A. Morsli et al., MPPT based fuzzy-logic controller for grid connected residential photovoltaic power system, in *International Conference in Artificial Intelligence in Renewable Energetic Systems*, Tipaza, Algeria, Dec 2019, pp. 236–244
11. B.A. Naidu, S.A. Kumar, D.M.S.S. Narayana, Fuzzy intelligent controller for the MPPT of a photovoltaic module in comparison with perturb and observe algorithm. Int. J. Appl. Eng. Res. **13**, 10058–10062 (2018)
12. B. Benlahbib, N. Bouarroudj, S. Mekhilef et al., A fuzzy logic controller based on maximum power point tracking algorithm for partially shaded PV array-experimental validation. Elektronika Ir Elektrotechnika **24**, 38–44 (2018)

13. H. Suryoatmojo, R. Mardiyanto, D.C. Riawan et al., Design of MPPT based fuzzy logic for solar-powered unmanned aerial vehicle application, in *Proceedings of International Conference on Engineering, Applied Sciences, and Technology (ICEAST)*, Phuket, Thailand, Aug 2018, pp. 1–4
14. C.R. Algarín, R.L. Fuentes, A.O. Castro, Implementation of a cost-effective fuzzy MPPT controller on the Arduino board. Int. J. Smart Sens. Intell. Syst. **11**(1), 1–9 (2018)
15. A. Youssefa, M.E. Telbany, A. Zekry, Reconfigurable generic FPGA implementation of fuzzy logic controller for MPPT of PV systems. Renew. Sustain. Energy Rev. **82**, 1313–1319 (2018)
16. E. Kandemir, S. Borekci, N.S. Cetin, Comparative analysis of reduced-rule compressed fuzzy logic control and incremental conductance MPPT methods. J. Electron. Mater. **47**, 4463–4474 (2018)
17. B.K. Naick, K. Chatterjee, T. Chatterjee, Fuzzy logic controller based maximum power point tracking technique for different configurations of partially shaded photovoltaic system. Arch. Electr. Eng. **67**, 307–320 (2018)
18. P. Verma, R. Garg, P. Mahajan, Asymmetrical interval type-2 fuzzy logic control based MPPT tuning for PV system under partial shading condition. ISA Trans. **100**, 251–263 (2020)
19. S. Samal, S.K. Sahu, P.K. Barik, Extraction of maximum power from a solar PV system using fuzzy controller based MPPT technique, in *Proceedings of IEEE International Conference on Technologies for Smart-City Energy Security and Power (ICSESP-2018)*, Bhubaneswar, India, Mar 2018, pp. 1–6
20. M. Ali, A. Talha, E.M. Berkouk, New M5P model tree—based control for doubly fed induction generator in wind energy conversion system. Wind Energy **23**(9), 1831–1845 (2020)
21. J.-C. Kim, J.-C. Kim, J.-S. Ko, Optimization design and test bed of fuzzy control rule base for PV system MPPT in micro grid. Sustainability **12**(9), 3763–3787 (2020)
22. S. Blaifi, S. Moulahoum, R. Benkercha et al., M5P model tree based fast fuzzy maximum power point tracker. Sol. Energy **163**, 405–424 (2018)
23. S. Srivastava et al., Applications of Artificial Intelligence Techniques in Engineering-volume 1, in *Part of the Advances in Intelligent Systems and Computing*, vol. 698 (Springer Nature, 2018), p. 643
24. J.M. Lopez-Guede, J. Ramos-Hernanz, N. Altin et al., Neural modeling of fuzzy controllers for maximum power point tracking in photo-voltaic energy systems. J. Electron. Mater. **47**, 4519–4532 (2018)
25. J.J. Khanam, S.Y. Foo, Modeling of a photovoltaic array in MATLAB Simulink and maximum power point tracking using neural network. Electr. Electron. Technol. Open Access J. **2**(2), 40–46 (2018)
26. P.W. Chtouki, M. Zazi, Comparison of several neural network perturb and observe MPPT methods for photovoltaic applications, in *Proceedings of IEEE International Conference on Industrial Technology (ICIT)*, Lyon, France, Feb 2018, pp. 909–914
27. A.B.M.S. Bouakkaz, O. Boudebbouz, A. Bouraiou et al., ANN based MPPT algorithm design using real operating climatic condition, in *Proceedings of 2nd International Conference on Mathematics and Information Technology (ICMIT)*, Adrar, Algeria, Feb 2020, pp. 159–163
28. C.R. Algarín, S.H. Deimer, D.R. Leal, A low-cost maximum power point tracking system based on neural network inverse model controller. Electron. J. **7**(1), 4–20 (2018)
29. R. Divyasharon, R.N. Banu, D. Devaraj, Artificial neural network based MPPT with CUK converter topology for PV systems under varying climatic conditions, in *Proceedings of IEEE International Conference on Intelligent Techniques in Control, Optimization and Signal Processing (INCOS)*, Tamil Nadu, India, Apr 2019, pp. 1–6
30. S.D. Al-Majidi, M.F. Abbod, H.S. Al-Raweshidy et al., A particle swarm optimization trained feed forward neural network for predicting the maximum power point of a photovoltaic array. Eng. Appl. Artif. Intell. **92**, 103688–103700 (2020)
31. K. Fatima, M.A. Alam, A.F. Minai, Optimization of solar energy using ANN techniques, in *Proceedings of 2nd International Conference on Power Energy, Environment and Intelligent Control (PEEIC)*, Greater Noida, India, Oct 2019, pp. 174–179

32. H.D. Tafti, A. Sangwongwanich, Y. Yang et al., A general algorithm for flexible active power control of photovoltaic systems, in *Proceedings of IEEE Applied Power Electronics Conference and Exposition (APEC)*, San Antonio, USA, Mar 2018, pp. 1115–1121
33. Y. Huang, X. Chen, C. Ye, A hybrid maximum power point tracking approach for photovoltaic systems under partial shading conditions using a modified genetic algorithm and the firefly algorithm. Int. J. Photoenergy, 1–13 (2018)
34. N. Fatema et al., *Intelligent Data-Analytics for Condition Monitoring: Smart Grid Applications* (Academic Press, 2021)
35. A. Harrag, S. Messalti, Adaptive GA-based reconfiguration of photovoltaic array combating partial shading conditions. Neural Comput. Appl. **30**, 1145–1170 (2016)
36. K. Anoop, M. Nandakumar, A novel maximum power point tracking method based on particle swarm optimization combined with one cycle control, in *Proceedings of International Conference on Power, Instrumentation, Control and Computing (PICC)*, Thrissur, India, Jan 2018, pp. 1–6
37. J.A. Alzubi, AI and machine learning paradigms for health monitoring system: intelligent data analytics, in *Part of the Studies in Big Data*, vol. 86 (Springer Nature, 2020), p. 513
38. H. Li, D. Yang, An overall distribution particle swarm optimization MPPT algorithm for photovoltaic system under partial shading. IEEE Trans. Industr. Electron. **66**(1), 265–275 (2018)
39. M. Merchaoui, A. Saklyl, M.F. Mimouni, Improved fast particle swarm optimization based PV MPPT, in *Proceedings of the 9th International Renewable Energy Congress (IREC 2018)*, Hammamet, Tunisia, Apr 2018, pp. 1–6
40. L.-Y. Chang, Y.-N. Chung, K.-H. Chao et al., Smart global maximum power point tracking controller of photovoltaic module arrays. Energies **11**(3), 567–582 (2018)
41. L. Guo, Z. Meng, Y. Sun et al., A modified cat swarm optimization based maximum power point tracking method for photovoltaic system under partially shaded condition. Energy **144**, 501–514 (2018)
42. N. Aouchiche, M. Aitcheikh, M. Becherif et al., AI-based global MPPT for partial shaded grid connected PV plant via MFO approach. Sol. Energy **171**, 593–603 (2018)
43. B.R. Peng, K.C. Ho, Y.H. Liu, A novel and fast MPPT method suitable for both fast changing and partially shaded conditions. IEEE Trans. Industr. Electron. **65**(4), 3240–3251 (2017)
44. J. Ahmed, Z. Salam, An enhanced adaptive P&O MPPT for fast and efficient tracking under varying environmental conditions. IEEE Trans. Sustain. Energy **9**(3), 1487–1496 (2018)
45. A.K.Yadav et al., "Soft Computing in Condition Monitoring and Diagnostics of Electrical and Mechanical Systems", Springer Nature, Part of the Advances in Intelligent Systems and Computing, vol. 1096, pages 496, 2020.
46. A. Tian et al., A novel pigeon-inspired optimization based MPPT technique for PV systems. Processes **8**(3), 356–378 (2020)
47. N. Priyadarshi, V.K. Ramachandaramurthy, S.P.F. Azam, An ant colony optimized MPPT for standalone hybrid PV-wind power system with single CUK converter. Energies **12**(1), 167–189 (2019)
48. T.K. Behera et al., Spider monkey based improve P&O MPPT controller for photovoltaic generation system, in *Proceedings of IEEE International Conference on Technologies for Smart-City Energy Security and Power (ICSESP-2018)*, Bhubaneswar, India, Mar 2018, pp. 1–6
49. M. Mao, Q. Duan, P. Duan et al., Comprehensive improvement of artificial fish swarm algorithm for global MPPT in PV system under partial shading conditions. Trans. Inst. Meas. Control. **40**, 2178–2199 (2017)
50. A.F. Minai et al., Metaheuristics paradigms for renewable energy systems: advances in optimization algorithms, in *Metaheuristic and Evolutionary Computation: Algorithms and Applications. Studies in Computational Intelligence* (Springer, 2020)
51. C. Vimalarani, N. Kamaraj, B.C. Babu, Improved method of maximum power point tracking of photovoltaic (PV) array using hybrid intelligent controller. Optik **168**, 403–415 (2018)
52. S. Nikolovski, H.R. Baghaee, D. Mlakić, ANFIS-based peak power shaving/curtailment in microgrids including PV units and BESSs. Energies **11**(11), 2953–2975 (2018)

53. R. Syahputra, R.O. Wiyagi, I. Soesanti et al., Design of maximum power point tracking based on adaptive neuro-fuzzy system for solar array system. J. Theor. Appl. Inf. Technol. **96**, 4481–4490 (2018)
54. A.A. Aldair, A.A. Obed, A.F. Halihal, Design and implementation of ANFIS-reference model controller based MPPT using FPGA for photovoltaic system. Renew. Sustain. Energy Rev. **82**, 2202–2217 (2018)
55. I. Abadi, C. Imron, R.D. Noriyati et al., Implementation of maximum power point tracking (MPPT) technique on solar tracking system based on adaptive neuro-fuzzy inference system (ANFIS). E3S Web Conf. **43**, 1–8 (2018)
56. A. Ali, A.N. Hasan, Optimization of PV Model using fuzzy-neural network for DC–DC converter systems, in *Proceedings of 9th International Renewable Energy Congress (IREC)*, Hammamet, Tunisia, Mar 2018, pp. 1–6
57. S. Duman, N. Yorukeren, I.H. Altas, A novel MPPT algorithm based on optimized artificial neural network by using FPSOGSA for standalone photovoltaic energy systems. Neural Comput. Appl. **29**, 257–278 (2016)
58. M.E. Başoğlu, B. Çakır, Hybrid global maximum power point tracking approach for photovoltaic power optimisers. IET Renew. Power Gener. **12**(8), 875–882 (2018)
59. J. Macaulay, Z. Zhou, A fuzzy logical-based variable step size P&O MPPT algorithm for photovoltaic system. Energies **11**(6), 1340–1354 (2018)
60. Z. Wang, X. Zhang, B. Hu et al., Control strategy of grid-connected photovoltaic generation system based on GMPPT method. IOP Conf. Ser. Earth Environ. Sci. **121**(4), 1–9 (2018)
61. A.M. Farayola, A.N. Hasan, A. Ali, Efficient photovoltaic MPPT system using coarse Gaussian support vector machine and artificial neural network techniques. Int. J. Innov. Comput. Inf. Control **14**, 323–339 (2018)
62. M. Dehghani, M. Taghipour, G.B. Gharehpetian et al., Optimized fuzzy controller for MPPT of grid-connected PV systems in rapidly changing atmospheric conditions. J. Mod. Power Syst. Clean Energy. https://doi.org/10.35833/MPCE.2019.000086.
63. N. Priyadarshi, S. Padmanaban, J.B. Holm-Nielsen et al., An experimental estimation of hybrid ANFIS-PSO-based MPPT for PV grid integration under fluctuating sun irradiance. IEEE Syst. J. **14**(1), 1218–1229 (2019)
64. S.D. Al-Majidi, M.F. Abbod, H.S. Al-Raweshidy, A novel maximum power point tracking technique based on fuzzy logic for photo-voltaic systems. Int. J. Hydrogen Energy **43**, 14158–14171 (2018)
65. B. Talbi, F. Krim, T. Rekioua et al., A high-performance control scheme for photovoltaic pumping system under sudden irradiance and load changes. Sol. Energy **159**, 353–368 (2018)
66. F. Keyrouz, Enhanced Bayesian based MPPT controller for PV systems. IEEE Power Energy Technol. Syst. J. **5**(1), 11–17 (2018)
67. M.K. Behera, I. Majumder, N. Nayak, Solar photovoltaic power forecasting using optimized modified extreme learning machine technique. Eng. Sci. Technol. **21**, 428–438 (2018)
68. K.S. Tey, S. Mekhilef, M. Seyedmahmoudian et al., Improved differential evolution-based MPPT algorithm using SEPIC for PV systems under partial shading conditions and load variation. IEEE Trans. Industr. Inf. **14**(10), 4322–4333 (2018)
69. T. Pei, X. Hao, Q. Gu, A novel global maximum power point tracking strategy based on modified flower pollination algorithm for photovoltaic systems under non-uniform irradiation and temperature conditions. Energies **11**, 2708–2723 (2018)
70. P. Vinoop et al., PSO-NN-based hybrid model for long-term wind speed prediction: a study on 67 cities of India, in *Applications of Artificial Intelligence Techniques in Engineering, Advances in Intelligent Systems and Computing*, vol. 697, 2018, pp. 319–327.https://doi.org/10.1007/978-981-13-1822-1_29
71. T. Mahto et al., Fractional order control and simulation of wind-biomass isolated hybrid power system using particle swarm optimization, in *Applications of Artificial Intelligence Techniques in Engineering, Advances in Intelligent Systems and Computing*, vol. 698, 2018, pp. 277–287
72. H. Malik, A.K. Yadav, A novel hybrid approach based on relief algorithm and fuzzy reinforcement learning approach for predicting wind speed. Sustain. Energy Technol. Assess. **43** (2020). https://doi.org/10.1016/j.seta.2020.100920

73. H. Malik et al., Multi-step ahead time-series wind speed forecasting for smart-grid application. J. Intell. Fuzzy Syst. (2021). https://doi.org/10.3233/JIFS-189736
74. A. Azeem et al., k-NN and ANN based deterministic and probabilistic wind speed forecasting intelligent approach. J. Intell. Fuzzy Syst. **35**(5), 5021–5031 (2018). https://doi.org/10.3233/JIFS-169786
75. A.F. Minai, T. Usmani, A. Iqbal, M.A. Mallick, Artificial bee colony based solar PV System with Z-source multilevel inverter, in *2020 International Conference on Advances in Computing, Communication and Materials (ICACCM)*, Dehradun, India, 2020, pp. 187–193

A Novel Approach for Estimating and Analyzing the Environmental Parameters: A Case Study for Renewable Energy Prospective

Mohammad Irfan Alam and Amjad Ali Pasha

Abstract On the one end, humans are struggling to reduce carbon footprints, while on the other, energy demand is increasing every day. Humans are looking for renewable energy sources to handle both problems. Static as well as dynamic systems are being built on the principle of self-sufficiency. Moreover, a realistic estimation of environmental parameters is vital for both ground-based applications and space-based applications. A realistic estimate of environmental parameters is crucial in the design and development of aerospace systems and other ground structures and systems. The chapter describes a novel and robust approach for estimating power and energy estimation and other environmental parameters. It starts with describing the environment module in which modeling of operational parameters, viz., temperature, pressure, and ambient wind speed, is explained. The environment module consists of models for atmosphere, wind, and solar radiation. A robust numerical-based method to calculate total solar energy generation from any shape is described. Operating temperature is vital to solar array performance. Therefore, effect of operating temperature on solar efficiency is also discussed at the end.

Keywords Solar energy · Wind model · HWM14

1 Introduction

In general, the operating and performance of any designed systems are affected by the operational parameters. Adverse conditions affect the system performance and even can fail them. Therefore, a realistic estimate is essential. For any aerospace system, accurate estimation of pressure, temperature, density, wind, etc., is crucial for safe and

M. I. Alam (✉)
VIT Bhopal University, Sehore, India
e-mail: mohammad.alam@vitbhopal.ac.in

A. A. Pasha
King Abdul-Aziz University, Jeddah, Saudi Arabia
e-mail: aapasha@kau.edu.sa

H. Malik et al. (eds.), *Intelligent Data Analytics for Power and Energy Systems*,
Lecture Notes in Electrical Engineering 802,
https://doi.org/10.1007/978-981-16-6081-8_20

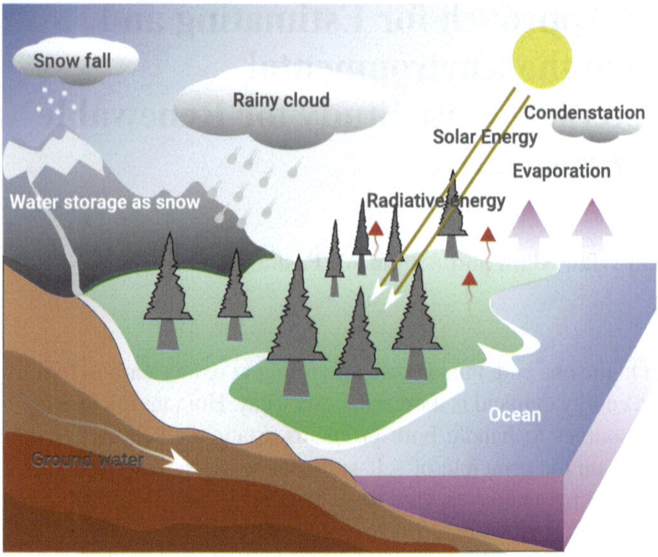

Fig. 1 Standard temperature variation with altitude

optimal performance. Similarly, the design of the ground structure, e.g., skyscrapers, building design, urban planning, airports, windmill bridges without realistic wind consideration, can be catastrophic. The collapse of the narrow Tacoma bridge is a classic example of such failure. Similarly, an ample amount of solar energy on our planet can play a vital role in reducing the environmental impact [1]. It is said to be clean, green, and the future energy source. Solar-based systems are designed and deployed scientifically to maximize their performances [2]. It is easy to say that high intensity of solar radiation is needed to maximize the output. However, it may lead to a significant temperature rise of the solar cell, resulting in efficiency loss. Solar energy is also directly responsible for local weather. Evaporation of water from the earth's surface is caused by solar energy. Precipitation brings the water back to the surface. Figure 1 illustrates the water cycle on our planet. The weather patterns of the earth are largely influenced by it.

Climate and weather can be confusing sometimes. Climate is a temporal and spatial average over a long time. It is the average daily weather for a significantly extended period at any specific location and is much more stable. The weather, on the other hand, is relatively instantaneous that reflects short-term conditions of the atmosphere. It can change seasonally or daily or even hourly and within proximity. Therefore, the weather condition is challenging to predict, especially for the far future. It is essential to keep in mind that actual

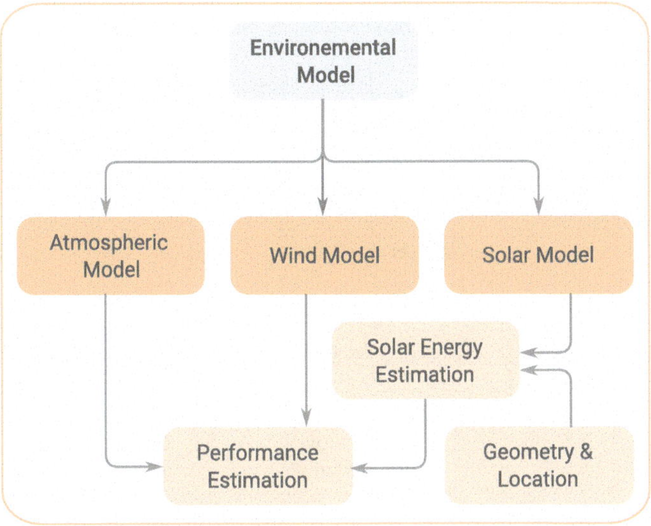

Fig. 2 Organization of the chapters

conditions may vary considerably from the estimate by models presented in the chapter.

This chapter describes a novel method that is capable enough to handle complex geometries. It also provides the atmospheric model to estimate ambient conditions, viz., temperature, pressure, and density. It also describes the wind model for calculating wind speed and the solar model to estimate energy generation. The performance of a solar array might get affected considerably if operating at high temperatures. Therefore, a simple model to calculate the efficiency loss is also discussed toward the end of the chapter. The chapter organization and mutual connections between the models are illustrated in Fig. 2.

2 Methodology

The methodology for estimating various environmental parameters is illustrated in Fig. 3. The first step is to identify the location of interest. This can also be provided as a design requirement or can be interpreted from the problem definition. Then, the corresponding input parameters in a quantitative form are obtained. The input parameters, viz., latitude, longitude, and altitude, are supplied to the environmental model for the given location. The model estimates the temperature, pressure, and density at the given altitude.

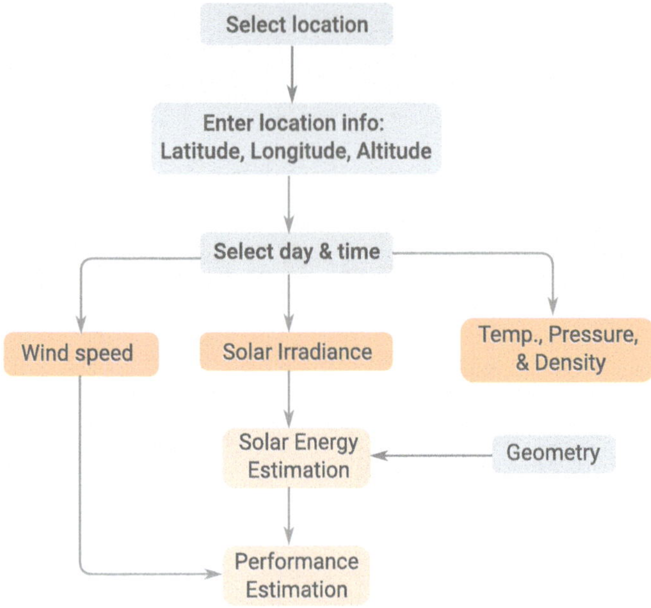

Fig. 3 Process flowchart

For estimating wind and solar irradiance, day number and time are also needed to be defined. For defining day number, January 1 is considered as day one and the last day, i.e., December 31, as day 365. With a set of input parameters, models described in the subsequent sections can obtain the corresponding output parameters, viz., temperature, pressure, density, wind speed, and solar irradiance.

3 Atmospheric Model

The environmental parameters, viz., temperature, pressure, and density at various altitudes can be estimated using the International Standard Atmosphere (ISA) [3]. The model is also known as 'US Standard Atmosphere, 1976,' jointly developed by NOAA [4], NASA, and the USAF. The model assumes standard sea-level conditions. The difference in temperature with the actual one can be taken care through applying ΔISA. Figure 4 illustrates the variation of temperature with altitude along with the lapse rates in different zones.

The equations for temperature calculation are given as:

Temperature:

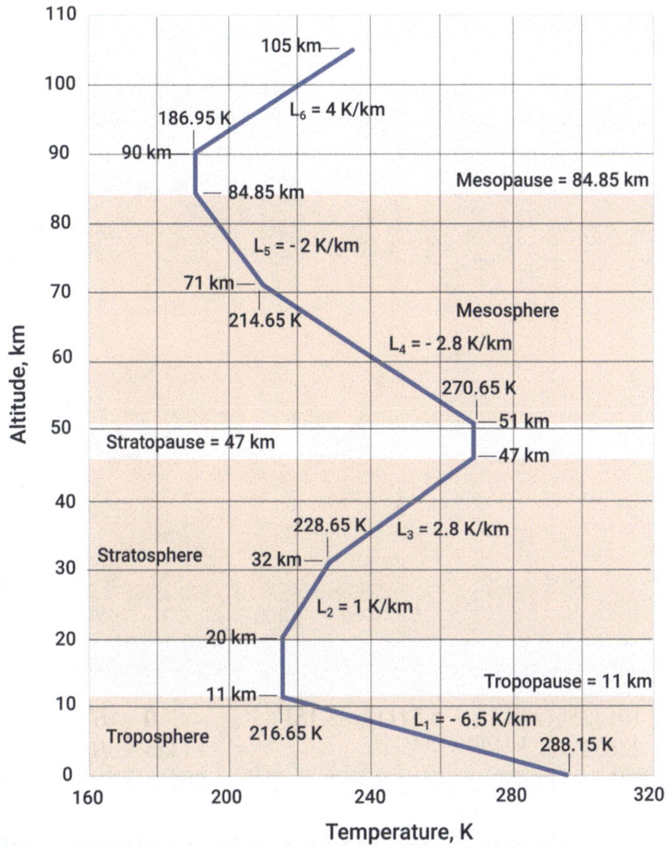

Fig. 4 Standard temperature variation with altitude

$$T_a = \begin{cases} 288.15 - 6.5H & 0 < H \leq 11\,\text{km} \\ 216.65 & 11\,\text{km} < H \leq 20\,\text{km} \\ 216.65 + (H - 20) & 20\,\text{km} < H \leq 32\,\text{km} \end{cases} \tag{1}$$

where H is in km and T_a is in K, which can be further converted in the other units through the following equations:

$$T_a = \begin{cases} T_K - 273.16 & (\text{in}^\circ C) \\ [1.8 \times (T_K - 273.15)] + 32 & (\text{in}^\circ F) \\ [1.8 \times (T_K - 273.16) + 32] + 459.69 & (\text{in}^\circ R) \end{cases} \tag{2}$$

Pressure:
Fig. 5 shows the pressure variation with respect to altitudes.

The respective equations for pressure calculation are as follows:

Fig. 5 Standard temperature variation with altitude

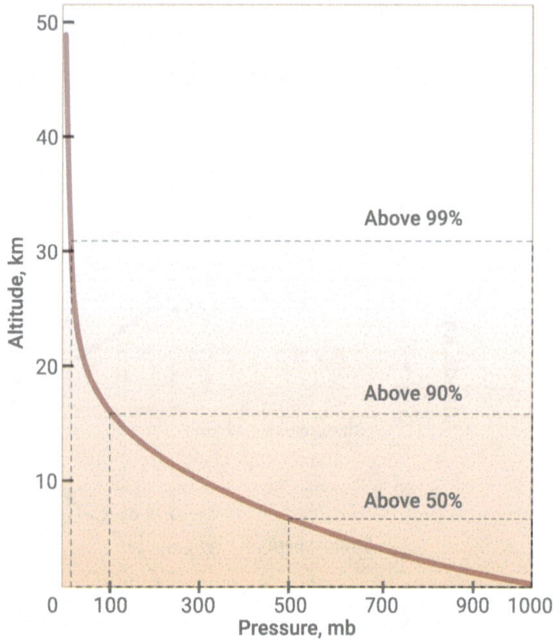

$$P_a = \begin{cases} 101325\,[(288.15 - 6.5H)/288.15]^{5.256} & 0 < H \leq 11\,\text{km} \\ 22632\,\exp^{-0.157688(H-11)} & 11\,\text{km} < H \leq 20\,\text{km} \\ 5474.87\,[1 + 0.004616(H-20)]^{-34.163} & 20\,\text{km} < H \leq 32\,\text{km} \end{cases} \tag{3}$$

Further on the basis of *ideal gas law*, density of air (ρ_{air}) at operating height can be calculated as:

$$\rho_a = P_a/RT_a \tag{4}$$

The air viscosity (μ) can be obtained at given temperature as:

$$\mu = 0.1456 \times 10^{-5} \frac{\sqrt{T}}{1 + \frac{110}{T}} \tag{5}$$

The standard atmosphere model `atmosisa`[1] is available in MATLAB through Aerospace Toolbox.

Listing 20.1 Standard Atmospheric Model in MATLAB

```
% Calculating atmopsheric properties using MATLAB

[T, a, P, rho] = atmosisa(1000)

```

[1] *Standard Atmospheric Model in MATLAB* https://in.mathworks.com/help/aerotbx/ug/atmosisa.html#d123e22745.

```
% Output
T = 281.6500
a = 336.4341
P = 8.9875e+04
rho = 1.1116
```

Similarly, the same atmospheric model is also available in Python as `pyatmos`.[2] The module can be installed using `pip` as '`pip install pyatmos`'. A sample code is presented in the following listing.

Listing 20.2 Standard Atmospheric Model in Python

```
# Calculating atmopsheric properties using Python

from pyatmos import expo
# geometric altitudes by default
rho_gm = expo([0,20,40,60,80])
print(rho_gm) # [kg/m^3]

# output
[1.22500000e+00 7.76098911e-02 3.97200000e-03
    3.20600000e-04
1.90500000e-05]


# geopotential altitudes
rho_gp = expo([0,20,40,60,80],'geopotential')
print(rho_gp)

# output
[1.22500000e+00 7.69385063e-02 3.84131212e-03
    2.97747719e-04
1.59847603e-05]
```

4 Ambient Wind Model

Spatial as well as temporal variations in ambient wind speed are observed. Therefore, the same location, i.e., fixed latitude and longitude, wind speed, and direction, can change significantly over altitude and time. In general, the ambient wind speed is lowest at altitudes around 17–25 km. Though, it may differ with location. Figures 6 and 7 compare the wind condition at the earth's surface and at the altitude of 500 hpa at a given instance.[3]

To estimate ambient wind speed and its direction at any place on earth and above, wind models are used. They often comprise empirical relations obtained by processing wind data. These wind data are collected from the satellites over several years. Aligned with the same idea, a horizontal wind model was developed in 1993 by

---

[2] *Standard Atmospheric Model in Python* https://pypi.org/project/pyatmos/.

[3] https://earth.nullschool.net/.

Fig. 6 Global wind at the surface

Fig. 7 Global wind at 500 hpa

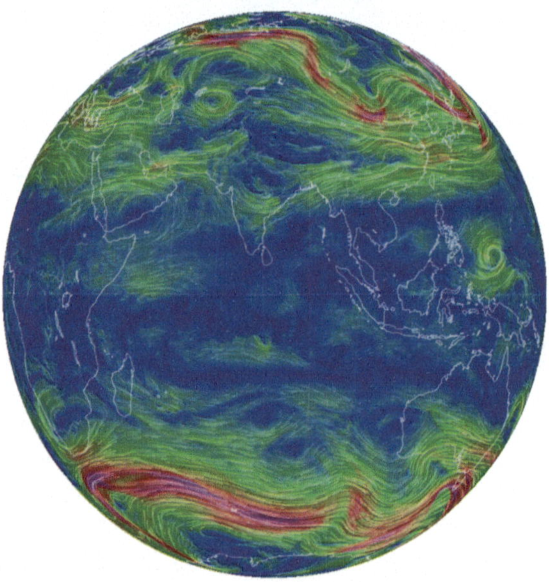

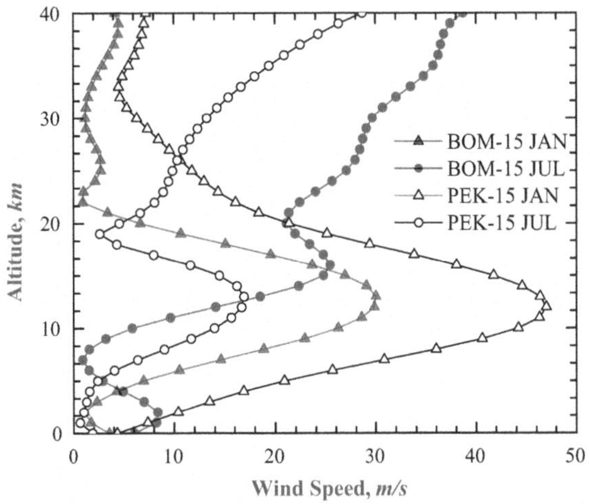

Fig. 8 Variation of wind with altitude

Hedin [5]. Later, in 2007 an updated version `HWM07` and then `HWM14` were introduced by Drob et al. [6, 7]. The present study uses a wind model, viz., HWM2014 developed by Drob et al. [7]. It provides an estimate of the ambient wind speed up to the exosphere (0–s500 km) at any location in the world. There are other studies in the wind model as well as in the open literature [8].

Figure 8 is obtained using the wind model. It shows the wind speed variation at Mumbai (BOM) and Beijing (PEK) with altitudes on two specific days of a year. The plot empowers the statement that wind speed is on the lower side from 17 to 25 km of heights. Therefore, this can be a preferred altitude for the deployment of the stratospheric airships [9, 10]. The wind model `atmoshwm`[4] in MATLAB can be accessed through the Aerospace Toolbox. A sample code to get the wind speed is mentioned in the following listing.

Listing 20.3 Horizontal Wind Model in MATLAB

```
% MATLAB implementation for estimating wind speed
    components

wind_speed = atmoshwm(-45,-85,25000,'day',150,'seconds
    ',39600,'apindex',80,'model','total', 'version', '
    14')

% Output
wind_speed =

3.2874    25.8735
```

Figures 9 and 10 plot the variations of zonal wind (along East–West direction) and meridional wind (along North–South direction) for the entire year and at varying

[4] *Horizontal wind model in MATLAB* https://in.mathworks.com/help/aerotbx/ug/atmoshwm.html.

Fig. 9 Variation of zonal wind at various altitudes over BOM

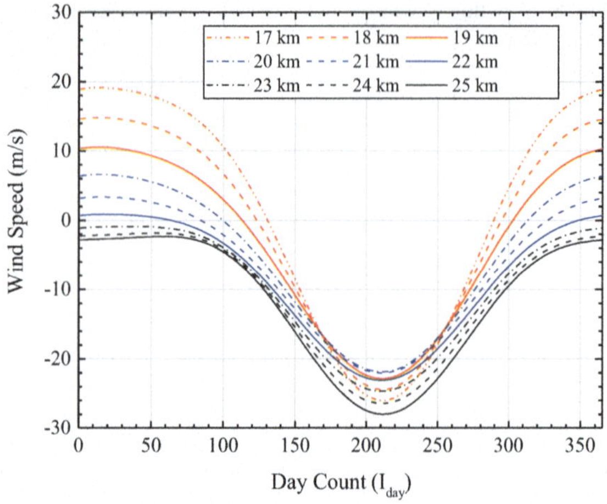

Fig. 10 Variation of meridional wind at various altitudes over BOM

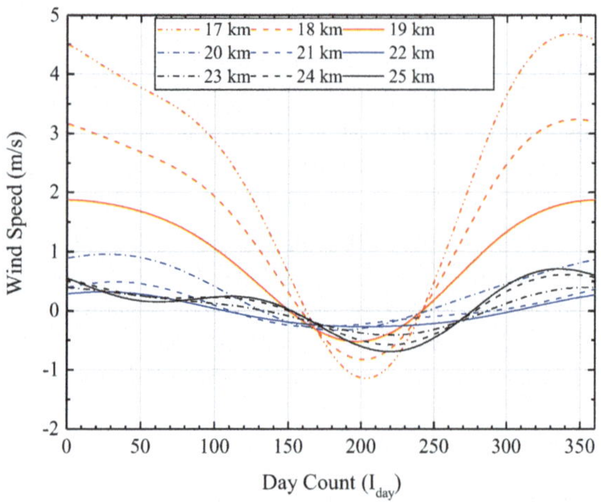

altitudes of BOM, respectively. It is concluded that the zonal wind is significantly higher at stratospheric heights than that of meridional wind.

The source code in Fortran horizontal wind models, viz., `HWM93`, `HWM07`, and `HWM14`, are available.[5] The Fortran source code can be compiled in Python using `f2py` module of `numpy`. Alternatively, a Python wrapper `pyhwm2014`[6] can also be used. It can be installed using `pip` by '`pip install pyhwm2014`'. A sample code in Python is provided in the following listing.

[5] *Horizontal wind model in Python* https://agupubs.onlinelibrary.wiley.com/doi/full/10.1002/2014EA000089.

[6] *Horizontal wind model in Python* https://pypi.org/project/pyhwm2014/.

Listing 20.4 A sample code for using HWM2014 in Python

```
# Python implementation for wind estimation
# Install HWM2014 in Python using 'pip install
   pyhwm2014'

from pyhwm2014 import HWM14, HWM14Plot

# get wind speed at different time
wind_gmt = HWM14(alt=130., ap=[-1, 35], day=323,
     option=3, utlim=[0., 23.45], utstp=.25, verbose=
   False,
     year=1993)

hwm14Gbj = HWM14Plot(profObj=wind_gmt)

# get wind speed at different longittude
wind_long = HWM14( alt=130., ap=[-1, 35], day=323,
     glonlim=[-180., 180.], glonstp=2.,
     option=4, verbose=False, year=1993)

hwm14Gbj = HWM14Plot(profObj=wind_long)
```

The output plot of the above code is shown in Fig. 11.

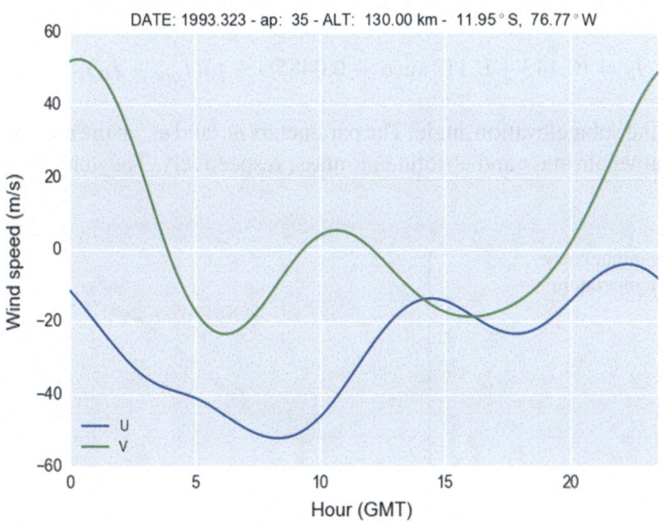

Fig. 11 Wind speed at different instances

5 Solar Model

Solar irradiance is responsible for providing solar energy through photovoltaic panels. Its intensity can vary with location and time. Maximum power point tracking is a popular technique to maximize the power output from any solar panel [11]. Apart from a classical approach and metaheuristic methods [12], artificial intelligence-based techniques [13–15] can also be applied to optimize the solar array layout and positioning. A comprehensive solar irradiance model is described by Ran et al. [16]. A location-specific detailed study of solar irradiance is also carried out [17]. Solar energy causes diurnal temperature variation at any site. Figure 12 illustrates the temperature variation due to solar energy. The energy estimation model described in the chapter uses the solar radiation model developed by Dai et al. [18]. It estimates the total incident solar radiation I, which falls on the solar array surface at a given location and time. It is a sum of direct solar radiation (I_D) and diffused solar radiation (I_d).

$$I = I_D + I_d \tag{6}$$

in which

$$I_D = I_{\text{sun}} \exp\left[-0.103 m_a^{0.571} - 0.081(\omega m_r)^{0.213} - \tau^{0.91} m_r^{0.87}\right] \tag{7}$$

and

$$I_d = (0.143 + 0.113 \sin\alpha - 0.0485\omega + \tau)(I_{sun} - I_D)\sin\alpha \tag{8}$$

where α is the solar elevation angle. The parameters m_r and m_a in the above equations are the relative air mass and absolute air mass, respectively. They can be calculated

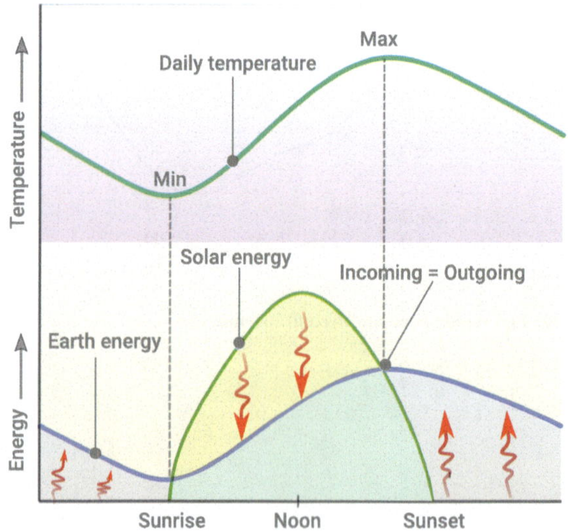

Fig. 12 Day–night energy balance and temperature variation

as:

$$m_r = \left[\sin\alpha + 0.15(3.885 + \alpha)^{-1.253}\right]^{-1} \tag{9}$$

and

$$m_a = m_r(p_a/1013) \tag{10}$$

where p_a is atmospheric pressure in kPa.

In Eq. 7, ω is the vertical water vapor column, which can be calculated as:

$$\omega = \omega_m \exp(-0.44\Delta H) \tag{11}$$

In 8, τ is the *aerosol optical depth* (AOD), which can be calculated as:

$$\tau = \tau_m \exp(-0.691\Delta H) \tag{12}$$

In the above equations, τ_m and ω_m are the values corresponding to the reference location. ΔH is the altitude difference in km from the reference location.

Earth revolves around the sun in an elliptical orbit. Therefore, its actual distance from the sun varies round the year. To estimate the solar radiation at the top of earth's atmosphere (I_{sun}) at any given day, the following equation is used:

$$I_{\text{sun}} = I_n(1.017 + 0.0174\cos\Gamma)^2 \tag{13}$$

I_n in the above equation is the extraterrestrial solar constant averaged for a year. Willson et al. [19] had recommended its value as 1368 W/m^2. The equation also contains Γ known as true anomaly and can be estimated as:

$$\Gamma = \text{MA} + 0.0334\sin(\text{MA}) + 0.000349\sin(2\text{MA}) \tag{14}$$

where $\text{MA} = 2\pi I_{day}/365$ is mean anomaly for a given (I_{day}).

The sun ray's direction ($\hat{\boldsymbol{N}}_{\boldsymbol{I}}$) in Fig. 13 can be represented in East North Up (ENU) frame as:

$$\hat{\boldsymbol{N}}_{\boldsymbol{I}} = \cos\alpha\sin\psi\hat{\boldsymbol{e}} + \cos\alpha\cos\psi\hat{\boldsymbol{n}} + \sin\alpha\hat{\boldsymbol{u}} \tag{15}$$

In the above equation, solar azimuth angle (θ_z) and zenith angle (θ_z) can be represented with reference to Fig. 13 as:

$$\cos\psi = (\cos\theta_z\sin\phi - \sin\delta)/(\sin\theta_z\cos\phi) \tag{16}$$

$$\sin\theta_z = \cos\delta\sin\omega/\sin\psi \tag{17}$$

where ω is the hour angle. It varies from $-\pi/2$ to $+\pi/2$. Figure 14 shows the solar azimuth angle varying to altitude angle during the middle of July, April, October, and January.

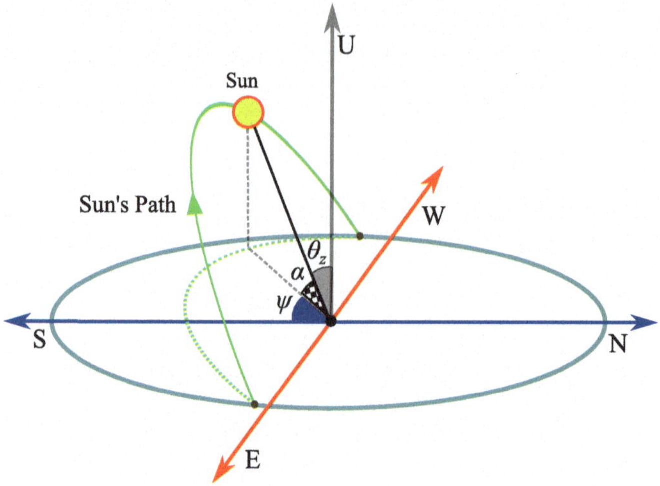

Fig. 13 Various angles related to position of sun

Fig. 14 Solar azimuth for PEK over the season

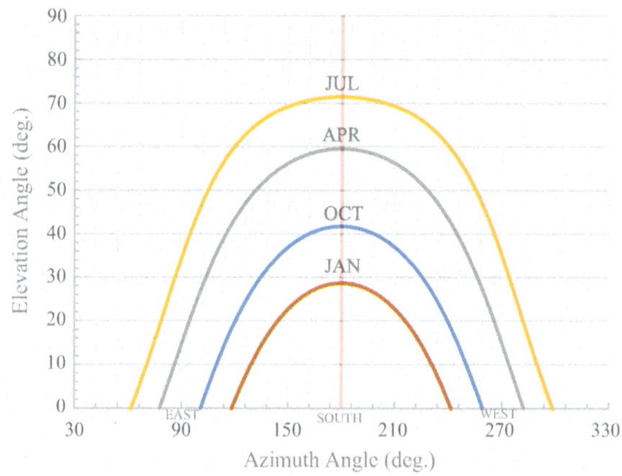

The solar radiation available at BOM and PEK is shown in Fig. 15 for the day of 15 January and 15 July, respectively. It can be noticed that output power can vary considerably at a given location on different deployment dates due to changes in solar radiation. Therefore, it is essential to consider the variation while designing systems intended for a year-round operation. Moreover, a strategic selection of deployment locations of solar-based systems can reduce the overall cost involved significantly.

With altitude, direct solar radiation, as well as diffused solar radiation, varies as shown in Fig. 16. Furthermore, one can conclude that at relatively high altitudes, the diffused radiation becomes only 10% of the direct radiation.

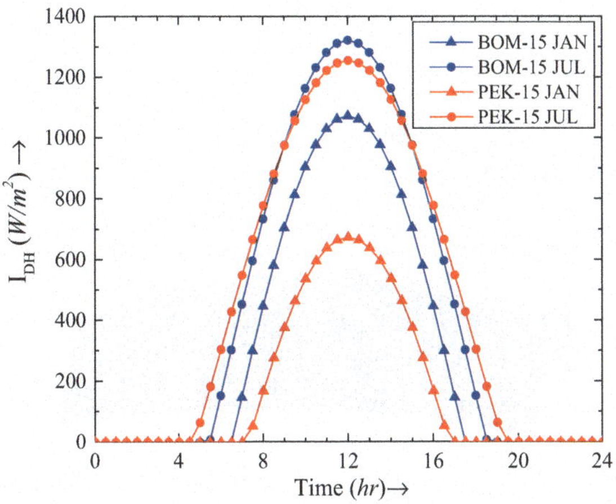

Fig. 15 I_{DH} at BOM and PEK ($h_{alt} = 20$ km)

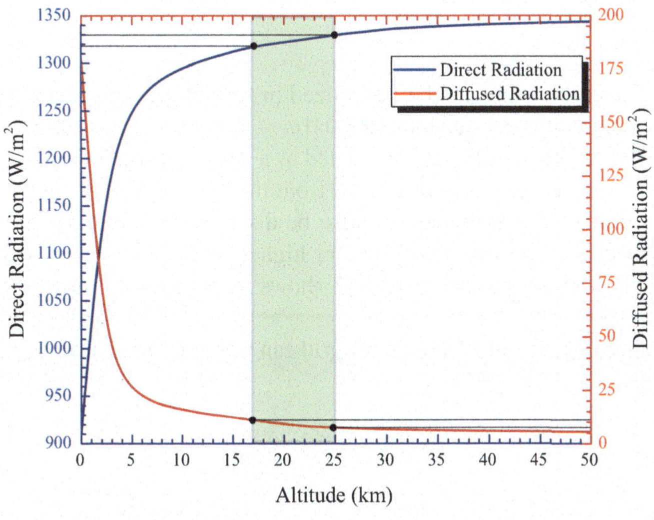

Fig. 16 Variation of solar radiation with altitude

6 Solar Energy Calculation

To calculate solar area (A_{SA}) needed to fulfill the energy demanded by the system, the development of a detailed energy model is crucial. In some previous studies, e.g., Wang et al. [20], the section containing solar array has been approximated as a cylinder for simplicity. The power output from the solar array directly depends on the angle of incidence to the surface normal. Therefore, in the present study, a model based on a numerical approach has been developed to estimate the energy

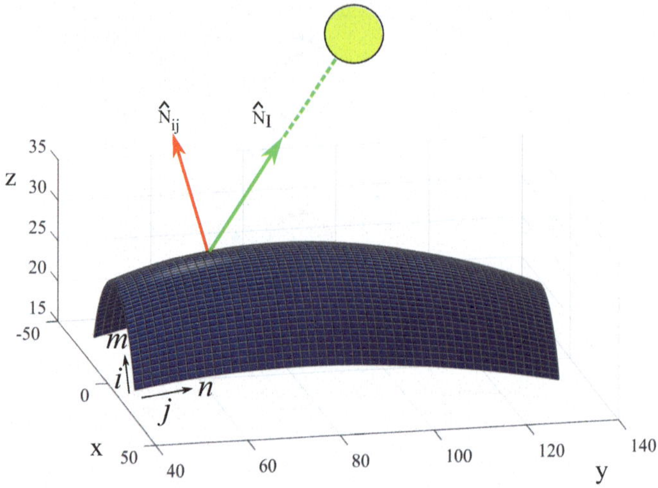

Fig. 17 Rectangular grids for the SA

produced. The entire panel can be discretized in several grids, as shown in Fig. 17. In the present case, it is divided into 10000 $[(m-1) \times (n-1) = 100 \times 100]$ small rectangular grids. Each grid is approximated as a small flat panel.

To estimate the total energy generated from the given panel, the entire duration for which sunshine was available can also be discretized. Therefore, the sunshine hour is divided into 100 small intervals for higher accuracy in energy calculations for reasonably accurate results. Figure 18 shows the analysis of an element at any time instance 'k'.

The length $(\vec{dl})$ and width $(\vec{db})$ of any grid can be represented in terms of position vector $(\vec{r})$ as:

$$\vec{dl}_{ij} = \vec{r}_{i,j+1} - \vec{r}_{i,j} \tag{18}$$

$$\vec{db}_{ij} = \vec{r}_{i+1,j} - \vec{r}_{i,j} \tag{19}$$

Since each grid is assumed to be flat, its surface normal $(\hat{\boldsymbol{N}}_{ij})$ can be calculated by taking the cross product:

$$\hat{\boldsymbol{N}}_{ij} \Rightarrow \hat{N}_{ij,x}\hat{\boldsymbol{i}} + \hat{N}_{ij,y}\hat{\boldsymbol{j}} + \hat{N}_{ij,z}\hat{\boldsymbol{k}} = \frac{\vec{dl}_{ij} \times \vec{db}_{ij}}{|\vec{dl}_{ij}||\vec{db}_{ij}|} \tag{20}$$

Moreover, the area of each grid (dA) can be obtained by taking dot product as:

$$dA_{ij} = |\vec{dl}_{ij}|.|\vec{db}_{ij}| \tag{21}$$

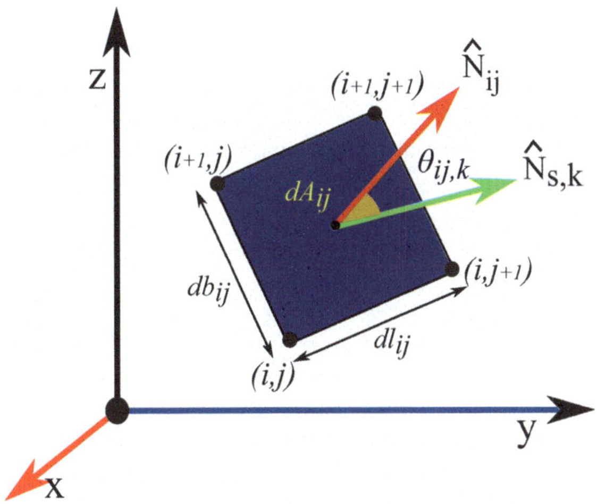

Fig. 18 Analysis of elementary A_{SA}

To calculate energy generated from each grid over a given duration, sun's ENU frame of reference need to be shifted to the system's local frame of reference through a rotation matrix $\boldsymbol{R_s}$. Assuming pitch angle to be zero and rolling angle as 90^0, it can be represented as:

$$\boldsymbol{R_s} = \begin{pmatrix} \cos\phi_y & -\sin\phi_y & 0 \\ \sin\phi_y & \cos\phi_y & 0 \\ 0 & 0 & 1 \end{pmatrix} \tag{22}$$

where ϕ_y is the yaw angle. The direction of solar irradiance to the system's frame is represented by:

$$\hat{\boldsymbol{N}}_{s,k} = \boldsymbol{R_s}\hat{\boldsymbol{N}}_{I,k} \tag{23}$$

Therefore, the angle between solar irradiance and surface normal can be estimated as:

$$\cos\theta_{ij,k} = \hat{\boldsymbol{N}}_{s,k} \cdot \hat{\boldsymbol{N}}_{ij} \tag{24}$$

Let $(dE_{ij,k})$ be the solar energy generated by each grid of area (dA_{ij}) in the duration of Δt_k. It can be represented in a mathematical form (neglecting negative flux) as:

$$dE_{ij,k} = \begin{cases} \eta_{SC} I_k dA_{ij} \cos\theta_{ij,k} \Delta t_k & \alpha \geq 0, -\frac{\pi}{2} \leq \theta_{ij,k} \leq \frac{\pi}{2} \\ 0 & \text{else} \end{cases} \tag{25}$$

I_k in the above equation is the solar radiation at time k. η_{sc} is the overall solar efficiency, which can be obtained through multiplying other efficiencies as given in the following equation.

$$\eta_{sc} = \eta_a \eta_c \eta_e \tag{26}$$

In the above equation, η_a is packing area efficiency. Similarly, η_c is solar cell conversion efficiency and η_e electrical component efficiency. Therefore, the total energy generated E_{sup} by the solar array can be obtained by integrating the mathematical equation developed for a grid over the entire solar array, as:

$$E_{\text{sup}} = \sum_{i=1}^{m} \sum_{j=1}^{n} \sum_{k=1}^{p} \eta_{SC} I_k dA_{ij} \cos\theta_{ij,k} \Delta t_k \tag{27}$$

The next subsection explains how the value of E_{sup} determined above is used to estimate the total propulsive power generated.

7 Effect of Temperature Rise on the Performance

The operational efficiency of the solar array is affected by the operating temperature. The temperature rise can reduce the performance of the panels significantly. The actual efficiency of the solar panels at a given temperature can be estimated using the the equation suggested by Skoplaki and Palyvos [21, 22]:

$$\eta_T = \eta_{\text{ref}} [1 - \beta (T - T_{\text{ref}})] \tag{28}$$

β in the above equation is temperature coefficient having a value of 0.0045 K^{-1}. By considering (T_{ref}) as 290 K, the reference efficiency (η_{ref}) of 15%, and the maximum temperature of the solar array as 360 K, up to 30% reduction in solar cell efficiency can be observed.

8 Case Study Considering Location as Mumbai

In the previous sections, we discussed the approach for calculating environmental parameters, viz., temperature, pressure, density, wind, solar irradiance, and solar energy generated from a solar array on a typical day of Mumbai. In this section, we will calculate all the parameters for the city of Mumbai. Once we have defined the geographic location, corresponding input parameters can be obtained, listed in Table 1.

The input parameters are of two types. The first corresponds to spot, which includes latitude, longitude, and altitude, and the second category includes day and time. With the defined set of input parameters, now as per the process described in

Table 1 Input parameters for Mumbai city

Input parameters	Value
Latitude	19.0760° N
Longitude	72.8777° E
Altitude	20 km
Day	15 July
Time	12 PM

Table 2 Output parameters for Mumbai city

Input	Value
Temperature	216 K+ΔT
Pressure	5529 Pa
Density	0.0889 kg/m^3
Wind speed	22 m/s
Solar irradiance	1307 W/m^2

Fig. 1 earlier, the corresponding output parameters, viz., temperature pressure, density, wind speed, and solar irradiance, can be calculated using the model discussed in the previous sections. The output parameters are listed in Table 2.

In Table 2, the temperature is given in terms of Δ T because 216 K is the standard temperature estimated by Eq. 1. However, that is based on the assumption of the ground temperature of 288 K. We have to add delta T as per the change in ground temperature with respect to the standard assumption for Mumbai on that specific day. Let us assume ground temperature on 15 July is 35 °C, i.e., 298 K. It gave ΔT of +10 °/K. This 10 °/K can be added to the standard calculation for a better estimate. For the estimation of solar power generated by a solar panel, the solar irradiance available is multiplied by the panel area, angle of incidence, and solar conversion efficiency. If we multiply the power output with the duration at which the power was available, we get the solar energy generated from the panel.

9 Summary

In this chapter, a novel energy model based on a numerical approach was presented to quickly estimate the energy generated from a solar array with a complex profile. Furthermore, various models for estimating environmental parameters, viz., temperature, pressure, and density, were also presented. The estimation of wind at any location through the described wind model was also discussed in detail. Some sample codes in MATLAB and Python were also provided to facilitate coupling with other analysis and design models.

References

1. A. Iqbal, et al., Renewable power for sustainable growth, 1E (Springer Nature, Part of the Lecture Notes inElectrical Engineering, vol. 723, 805 p). ISBN: 978-981-33-4080-0. https://doi.org/10.1007/978-981-33-4080-0
2. N. Fatema, et al., Intelligent Data-Analytics for Condition Monitoring: Smart Grid Applications, 1E (Academic Press, 2021). ISBN: 978-0-323-85510-5. https://doi.org/10.1016/C2020-0-02173-0
3. U.S. Standard Atmosphere, 1976. U.S. Government Printing Office, Washington, D.C
4. NOAA National Centers for Environmental Information (NCEI) https://www.ngdc.noaa.gov/
5. A.E. Hedin, Horizontal wind model (HWM)(1990). Planet. Space Sci. **40**(4), 556–557 (1992)
6. D.P. Drob, J.T. Emmert, G. Crowley, J.M. Picone, G.G. Shepherd, W. Skinner, P. Hays, R.J. Niciejewski, M. Larsen, C.Y. She, J.W. Meriwether, An empirical model of the Earth's horizontal wind fields: HWM07. J. Geophys. Res.: Space Phys. **113**(A12) (2008)
7. D.P. Drob, J.T. Emmert, J.W. Meriwether, J.J. Makela, E. Doornbos, M. Conde, G. Hernandez, J. Noto, K.A. Zawdie, S.E. McDonald, J.D. Huba, An update to the horizontal wind model (HWM): the quiet time thermosphere. Earth Space Sci. **2**(7), 301–319 (2015)
8. P. Alken, S. Maus, J. Emmert, D.P. Drob, Improved horizontal wind model HWM07 enables estimation of equatorial ionospheric electric fields from satellite magnetic measurements. Geophys. Res. Lett. **35**(11) (2008)
9. M.I. Alam, R.S. Pant, Multidisciplinary approach for solar area optimization of high altitude airships. Energy Convers. Manage. **164**, 301–310 (2018)
10. M.I. Alam, R.S. Pant, Multi-objective multidisciplinary design analyses and optimization of high altitude airships. Aerosp. Sci. Technol. **78**, 248–259 (2018)
11. A.K. Yadav, et al., *Soft Computing in Condition Monitoring and Diagnostics of Electrical and Mechanical Systems*, 1E (Springer Nature, Part of the Advances in Intelligent Systems and Computing, vol. 1096, p. 496, 2020). ISBN978-981-15-1532-3. https://doi.org/10.1007/978-981-15-1532-3
12. A. Iqbal, et al., Meta Heuristic and Evolutionary Computation: Algorithms and Applications, 1E (Springer Nature,Part of the Studies in Computational Intelligence, vol. 916, p. 849, 2020). ISBN 978-981-15-7571-6. https://doi.org/10.1007/978-981-15-7571-6
13. J.A. Alzubi, *AI and Machine Learning Paradigms for Health Monitoring System: Intelligent Data Analytics*, 1E (Springer Nature, Part of the Studies in Big Data, vol. 86, p. 513, 2020). ISBN: 978-981-33-4412-9. https://doi.org/10.1007/978-981-33-4412-9
14. S. Srivastava, et al., Applications of Artificial Intelligence Techniques in Engineering, vol. 1, 1E (Springer Nature, Part of the Advances in Intelligent Systems and Computing, vol. 698, 643 p, 2018). ISBN 978-981-13-1819-1. https://doi.org/10.1007/978-981-13-1819-1
15. S. Srivastava, et al., *Applications of Artificial Intelligence Techniques in Engineering*, vol. 2, 1E (Springer Nature, Part of the Advances in Intelligent Systems and Computing, vol. 697, 647 p, 2018). ISBN 978-981-13-1822-1. https://doi.org/10.1007/978-981-13-1822-1
16. H. Ran, R. Thomas, D. Mavris, A comprehensive global model of broadband direct solar radiation for solar cell simulation, in *45th AIAA Aerospace Sciences Meeting and Exhibit*, Jan 2007, p. 33
17. T. Pan, S. Wu, E. Dai, Y. Liu, Estimating the daily global solar radiation spatial distribution from diurnal temperature ranges over the Tibetan Plateau in China. Appl. Energy **107**, 384–393 (2013)
18. Q. Dai, X. Fang, A simple model to predict solar radiation under clear sky conditions. Adv. Space Res. **53**(8), 1239–1245 (2014)
19. R.C. Willson, S. Gulkis, M. Janssen, H.S. Hudson, G. Chapman, Observations of solar irradiance variability. Science **211**(4483), 700–702 (1981)

20. Q. Wang, J. Chen, G. Fu, D. Duan, H. Zhao, A methodology for optimisation design and analysis of stratosphere airship. Aeronaut. J. **113**(1146), 533–540 (2009)
21. E. Skoplaki, J.A. Palyvos, On the temperature dependence of photovoltaic module electrical performance: a review of efficiency/power correlations. Solar Energy **83**(5), 614–624 (2009)
22. E.J.A.P. Skoplaki, J.A. Palyvos, Operating temperature of photovoltaic modules: a survey of pertinent correlations. Renew. Energy **34**(1), 23–29 (2009)

Novel Application of Data-Driven Intelligent Approaches to Estimate Parameters of Photovoltaic Module for Condition Monitoring in Renewable Energy Systems

Omkar Singh, Amit Kumar Yadav, and Anjan Kumar Ray

Abstract Photovoltaic (PV) arrays do not have moving parts. So, these require comparatively less maintenance. However, PV arrays operate under outdoor conditions in severe environment and lead to undergo different faults. Therefore, PV arrays' fault diagnosis is necessary to make the PV energy systems more reliable. Due to varying environmental conditions and nonlinear PV characteristics, different artificial neural networks-based fault diagnosis has been proposed. But there are some concerns; e.g., fault diagnosis models are limited for mountainous region, and fault history is difficult to obtain using experimental analysis under outdoor condition. To address these concerns, this study proposes a new fault diagnostic techniques of PV module using extreme learning machine and multilayer feedforward neural network with Levenberg–Marquardt algorithm. For this, an experimental database of solar radiation, air and back surface module temperatures and electrical parameters of PV module are created by developing an experimental setup. This work is suitable for PV applications and researchers to estimate PV parameters for condition monitoring and would be useful for prior fault analysis of the PV module.

Keywords ELM · RBFNN · LM · PV · Fault

1 Introduction

Due to rise in pollution by conventional energy sources such as fossil fuels and nuclear, concern toward renewable energy (RE) sources is increasing around the world [1]. Solar energy is widely used RE source. Solar energy is safe, sustainable and eco-friendly with no greenhouse gas emissions [2–5]. In Europe, around 53

O. Singh
Electrical and Electronics Engineering, Raghu Engineering College (A), Visakhapatnam 531162, India

O. Singh · A. K. Yadav (✉) · A. K. Ray
Electrical and Electronics Engineering Department, National Institute of Technology Sikkim, Ravangla, Barfung Block, South Sikkim 737139, India
e-mail: amit1986.529@nitsikkim.ac.in

H. Malik et al. (eds.), *Intelligent Data Analytics for Power and Energy Systems*,
Lecture Notes in Electrical Engineering 802,
https://doi.org/10.1007/978-981-16-6081-8_21

Table 1 Failure classification [13]

Failure mode	Failure type	Cause
Order (0F)	Control and command signals	Human error and environment
First (1F)	Life cycle	Installation, manufacturing, design
Second (2F)	Excessive condition and constraints	Human error and environment

million tons of CO_2 emissions was reduced in 2012 by generating 100 GW from PV systems [6]. This gives rise to exponential growth of PV worldwide.

PV array in PV system is modeled to function under various climatic conditions like corrosion, wind, bad weather and variable solar radiation for generation of power [7]. Due to these conditions, various malfunctions and failures may arise, resulting in reduction in lifetime of the PV modules, loss of availability, safety hazards and lower total energy yield [8–12]. PV module failure classification is shown in Tables 1, 2, 3 and 4.

Table 2 PV failure by environment [14, 15]

Failure mode	Cause
Power loss	Dust, encapsulation degradation, nest of insects
Destruction of diodes, overvoltage, performance loss	Degradation of module due to light
Seal loss and deterioration of cells	Rust by passage of water
Destruction of module	Lightning storm
Destruction of cell and hotspot	Shading
Decrease of DC resistance to ground, insulation and adhesion loss, increase of leakage current, corrosion, hotspot	Penetration of moisture
Corrosion	Marine air

Table 3 Failure by human error [14, 15]

Failure mode	Cause
Installation	Stealing of module
Connection corrosion and water penetration	Missing of gland plugs on junction box
Shading and performance loss	Not correct inclination of module

Table 4 Life cycle failure [14, 15]

Failure mode	Cause
Accumulation of dust	Panel inaccessible
Fire, module destruction, short circuit	Poor isolation between inverter and module
Broken and snatching of module	Structure's weakness
Deformation of support	Poor mechanical strength of module supports
Module's mechanical stress	Improperly and inadequately installed support
Heating	No ventilation of module
Loss of performance	Less power production

2 Literature Review

The acquisition of real-time PV module data and environmental factors becomes an important task to comment upon the operation and analyze the performance of PV module. Stand-alone solar PV module is an effective solution for electrifying isolated zones and can similarly be beneficial for farming segment like solar water pump, solar heating, etc. In these remote areas, it is difficult to monitor and maintain the PV module performance. However, to do the monitoring of the module data acquisition systems are used which are costly and may not be cost effective for stand-alone PV systems. Several monitoring methods are proposed in [16, 17]. In the literature [14], the authors have used PLC for monitoring the PV module and also taking field measurements and uncertainty estimation. The authors in the literature [18] have proposed a PV smart monitoring system called Sentinella at PV level with wireless sensor network. A residential power line communication network for monitoring home-based PV module is given in the literature [19]. In the literature [20–22], the authors have developed an Arduino-based PV monitoring system with 3G connectivity for remote operation.

Multifold literature has been introduced for PV modeling. In the literature, a one-diode modeling of PV module is proposed. A two-diode model for PV modeling is proposed in [23]. In the literature [24], the authors provided a modeling of PV module for different irradiance levels. The literature [25, 26] provides a contrast among the exactness of one-diode and two-diode models for different environmental factors. The estimation of PV module parameters is very important to accurately obtain the PV module. Many authors have proposed different methods for estimation of PV parameters [27]. The working of the PV module is governed by some crucial PV module parameters like series resistance (R_s), shunt resistance (R_{sh}), reverse saturation current (I_0), photo generated current (I_{ph}) and ideality factor (η) [27–30]. Many literatures have proposed a four-parameter model [27] and five-parameter model [29]. As the builders do not deliver info about the PV module parameters, these must be calculated by the operator so as to approximate the performance and lifespan of the PV module. However, determining these parameters is not an easy task as they are extremely reliant on solar irradiance, module temperature and various climate settings [31]. In the literature [10], the authors have proposed many methods to determine these parameters. Many soft computing methods and bio-inspired methods are also proposed for measuring the PV parameter accurately. In the literature [18], the authors have used ABC algorithm and probabilistic neural network along with bee algorithm for estimation. The bio-inspired methods based on the GA [32], PSO [33], SA [34], HS [35], PS [36], DE [37] and ABC [38] are widely used in the modeling of PV systems because of their exactness. Though the authors have traditional methods like NR- [39] and LM [40]-based methods which provide good accuracy in parameter estimation, they are inefficient for fast varying weather condition. PV modules withstand severe outdoor conditions and factors; as a result, they degrade and their lifetime decreases. In [18], the authors presented an artificial neural network (ANN) method to determine the faults in the PV module. The authors in the literature [40] have proposed a method of detecting partial and permanent shading. In the literature [41], the authors use fuzzy logic to determine the fault signatures and parameters to comment upon the module health. An improved ML-based approach to identify the faults through estimation of PV parameters is given in the literature [42]. In the literature [43], thermography is performed with infrared camera for identifying the temperature of the PV module surface. In the literature [40], the author studied array current, array voltage and irradiance under all irradiation conditions using reference indicators at the commencement of the diagnosis process.

3 Methodology

3.1 Experimental Setup

The hardware arrangement is situated at 27° 17′ N 88° 21′ E in Ravangla, Sikkim, a northeastern state of India. The climate surroundings here are erratic, and solar irradiance and environmental temperature vary severely during the day. The Sova solar PV modules of rating 20 W, under investigation, are presented in Fig. 2. PV modules are kept at an angle of 27° 17′ [44, 45], for maximum yield. The data acquisition system can be used in any location. However, in areas where the climate surroundings are nearly similar throughout the day, the output of PV module is foreseeable. But at the location where the hardware setup is present, the climate varies suggestively all over the day.

Due to varying weather condition, the operation of the module becomes unpredictable. The ratings of the modules at STC, i.e., irradiance at 1000 (W/m^2) and module temperature at 298 K, are mentioned in Table 5. Along with the two modules, a sensor interface consisting of Arduino Uno board, OPT101, DHT11, ADY011 and ACS712 is used to obtain real-time module parameters. The sensor ratings are mentioned in Table 6. Block diagram of sensor interface and the hardware setup for

Table 5 PV module specifications mentioned in manufacturer datasheet

S. No.	Module datasheet specifications	Sova solar 20P
1	Rated max power (P_{mp})	20 W peak
2	Max voltage rated (V_{mp})	17.57 V
3	Max current rated (I_{mp})	1.19 A
4	Open ckt. voltage (V_{oc})	21.60 V
5	Short ckt. current (I_{sc})	1.29 A
6	Temp. coeff. of I_{sc} (α)	0.05% (°C)
7	Temp. coeff. of V_{oc} (β)	−0.32% (°C)
8	Temp. coeff. of P_{mp} (γ)	−0.43% (°C)
9	Number of cells in series (N_s)	32

Table 6 Sensor specifications and connected load

S. No.	Component	Specification
1	Light sensor (OPT101)	5 V
2	Temperature and humidity sensor (DHT11)	5 V
3	Voltage sensor (ADY011)	Less than 25 V
4	Current sensor (ACS712)	30 A
5	Resistive load 20 W module	502 Ω
6	Resistive load 74 W module	1970 Ω
7	Relay switch	12 A

the sensors is shown in Fig. 1. The sensor data obtained are then directly passed on to the computer using serial communication ports as shown in Fig. 2.

3.2 PV Parameter Estimation

3.2.1 PV Modeling

PV module modeling is mostly done by either one- or two-diode model methods that have been proposed in the literature. Though, frequently one-diode model is extensively implemented in the literature. One-diode model (ODM) consists of 5 parameters, while two-diode model (TDM) consists of 7 parameters. Though, TDM is more accurate than ODM. Also, TDM consists of extra parameters. Thus, in TDM the calculation intricacy rises; henceforth, one-diode model is mostly used.

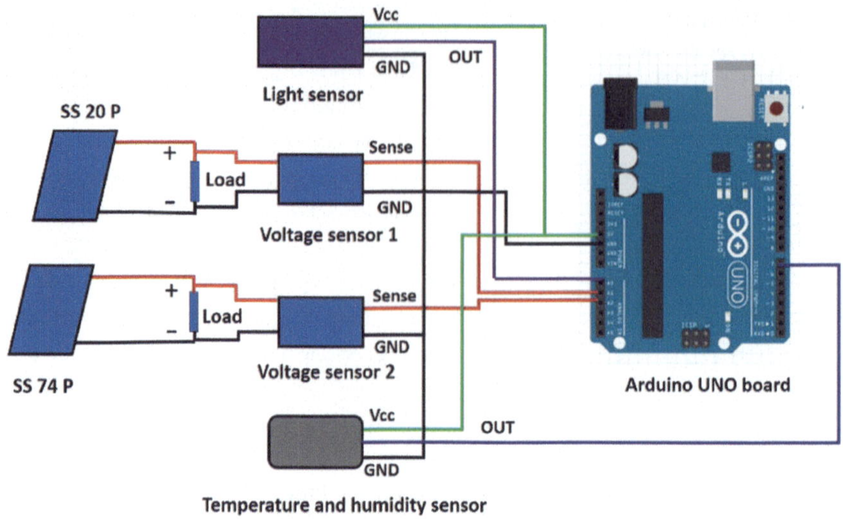

Fig. 1 Diagram of hardware interface and sensors

Fig. 2 Experimental setup located at Ravangla, Sikkim (27° 17′ N 88° 21′ E)

3.2.2 ODM PV Model

ODM is the simplest and widely implemented method for PV modeling. ODM consists of the following parameters, i.e., R_s, R_{sh}, I_0, I_{ph} and η as presented in Fig. 3. The equations describing ODM are given as follows:

$$I_L = I_{\text{ph}} - I_d - \frac{V_L + I_L R_s}{R_{\text{sh}}} \tag{1}$$

$$I_d = I_0 \exp \frac{V_L + I_L R_s}{\eta V_t} - 1 \tag{2}$$

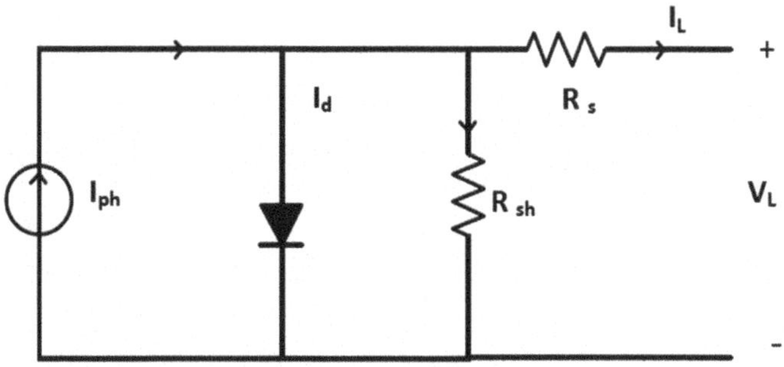

Fig. 3 One-diode model

$$V_T = \frac{N_s k T}{q} \tag{3}$$

$k =$ Boltzmann's constant.

$q =$ charge of electron.

$V_T =$ thermal voltage.

$m = 1.5$ for Si.

$N_s =$ no. of cells in series.

$T =$ module temperature in Kelvin.

Once real-time irradiance and temperature are extracted by the sensor. The module backside temperature is converted to cell temp., and the I_{ph}, V_t, I_o are calculated [29]. The efficacy, condition and longevity of the PV module are estimated; to do so, it is required to identify output voltage and current and also it is necessary to measure the module parameters [10]. Since the module operation is dependent on the module specifications and weather conditions. Though, giving a generalized model or method to obtain the module parameters and output is a difficult task. However, by using sensors and optimization techniques, these data are obtained. Since the I_{ph} is dependent the solar irradiance and module temperature [46]. The I_{ph} is obtained by:

$$I_{\mathrm{ph}} = I_{\mathrm{sc}} \frac{G}{G_{\mathrm{STC}}} \{1 + \alpha (T - T_{\mathrm{STC}})\} \tag{4}$$

$$I_o = \frac{I_{\mathrm{ph}}}{\exp \frac{V_{\mathrm{oc}}}{V_T} - 1} \tag{5}$$

where I_{sc} is short-circuit current at STC, G is solar irradiance, G_{STC} is solar irradiance at STC, α is temp. coeff. of short-circuit current and G is real-time irradiance; from equation, we can obtain [47]:

$$\frac{\mathrm{d}I_L}{\mathrm{d}V_L} = \frac{-\frac{I_o}{V_t}\exp\left(\frac{V_{\mathrm{mp}}+I_L R_s}{V_t}\right)}{1+\frac{I_o R_s}{V_t}\exp\left(\frac{V_{\mathrm{mp}}+I_L R_s}{V_t}\right)} \tag{6}$$

$$R_s = R_{s_o} + \frac{V_t}{I_{s_c}}\left(\frac{G T_{\mathrm{STC}}}{G_{\mathrm{STC}} T} - 1\right) \tag{7}$$

$$R_{\mathrm{sh}} = \frac{V_{\mathrm{mp}}\left(V_{\mathrm{mp}} + I_{\mathrm{mp}} R_s\right)}{V_{\mathrm{mp}} I_{\mathrm{ph}} - V_{\mathrm{mp}} I_o \exp\left(\frac{(V_{\mathrm{mp}}+I_{\mathrm{mp}} R_s)q}{N_s \eta k T}\right) + V_{m_p} I_o - P_{\mathrm{mp}}} \tag{8}$$

$$\eta = \frac{\beta T - V_{\mathrm{oc}} + V_{\mathrm{go}} N_s}{\frac{T\alpha}{I_L} - 3} \tag{9}$$

where V_{go} is the band gap energy for silicon PV cells.

In order to obtain accurate results, irradiance and temperature must be measured accurately. But while measuring the back temperature of PV module varies from the T. In [18], the authors proposed an equation which gives the T_{cell}:

$$T_{\mathrm{cell}} = T + \frac{G}{G_{STC}} \Delta T \tag{10}$$

where T_{cell} is temp. of cell, T is back temperature of the module and ΔT is 2 [18]. And also, we have the temperature coeff. of the PV module, i.e., α and β. From these coefficients, we can calculate the values of V_{oc} and I_{sc} as follows:

$$V_{\mathrm{oc}} = V_{\mathrm{oc,STC}} + m V_t \ln \frac{G}{G_{\mathrm{STC}}} + \beta(T - T_{\mathrm{STC}}) \tag{11}$$

$$I_{\mathrm{sc}} = I_{\mathrm{sc,STC}} \frac{G}{G_{\mathrm{STC}}} + \alpha(T - T_{\mathrm{STC}}) \tag{12}$$

3.3 Multilayer Feedforward Neural Network with Levenberg–Marquardt Algorithm

Neural network (NN) is processing device either a hardware or an algorithm whose working is inspired by brain. NN has the ability to learn by example. NN is well suited for real-time system due to its faster response. ANN is composed of highly interrelated processing element named as neurons, which can solve composite difficulties. ANN utilized 3 layers: 1 input, 1 hidden and 1 output (Fig. 4). ANN

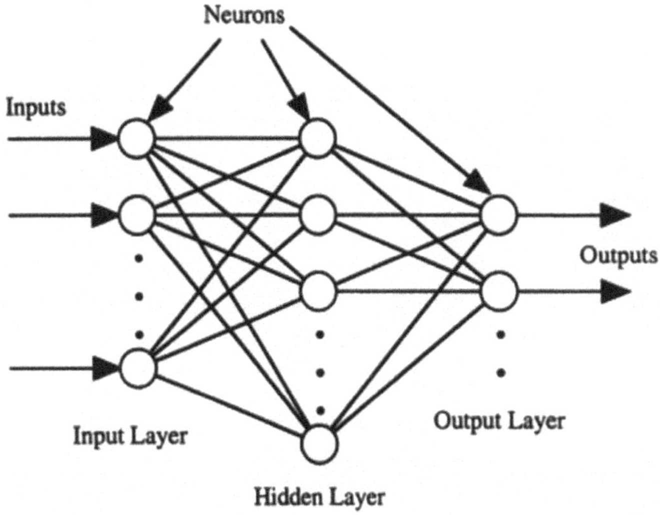

Fig. 4 Architecture of ANN

is used on various fields such as neurobiology, cognitive psychology, linguistics, philosophy, economics, engineering images, signal processing, physics, dynamical systems, statistical physics, mathematics, computer science and renewable energy [47–53]. The advantages of working with ANN are adaptive learning, pattern recognition, online applications and fault tolerance. The entire NN is trained based on the following steps [54]:

Step 0: Initialize the bias and weights (for easy calculation, it can be set to zero). Also, initialize the learning rate $\alpha(0 < \alpha \leq 1)$. For ease, α is set to 0.

Step 1: Until the final ending condition is 0, carry out steps 2–6.

Step 2: Carry out steps 3–5 until the final stopping condition is false.

Step 3: Similar activation function is applied to the input units in the input layer.

Step 4: First obtain the net input (y_{in}) and compute the output (y) of the network using the equations below:

$$y_{\text{in}} = b + \sum_{i=1}^{n} x_i w_i$$

where "n" is the number of input layer neurons. By applying activations over the net input, calculate the output:

$$y = f(y_{\text{in}}) = \left\{ \begin{array}{cc} 1 & \text{if } y_{\text{in}} > \theta \\ 0 & \text{if } -\theta \le y_{\text{in}} \le \theta \\ -1 & f\ y_{\text{in}} < -\theta \end{array} \right\}$$

Step 5: Weight and bias tuning: Compute the error between the calculated output and target output.

$$\begin{aligned} &\text{if } y \neq t, then \\ &\quad w_i(\text{new}) = w_i(\text{old}) + \alpha t x_i \\ &\quad b(\text{new}) = b(\text{old}) + \alpha t \\ &\text{else, we have} \\ &\quad w_i(\text{new}) = w_i(\text{old}) \\ &\quad b(\text{new}) = b(\text{old}) \end{aligned}$$

Step 6: Train the network until there is no weight variation. If the stopping condition for the network is not met, then start again from step 2.

3.3.1 Levenberg–Marquardt Algorithm (LMA)

The LMA comprises GNA (second-order algorithm) and GDA (second-order learning algorithm). The second-order algorithm minimizes faster than the first-order algorithms. The update rule of LMA is presented as:

$$w_{k+1} = w_k - \left(J_k^T J_k + \mu I\right)^{-1} . J_k e$$

where J_k is Jacobian matrix, μ is always positive and termed as combination coefficient, I is identity matrix.

3.4 Extreme Learning Machine (ELM)

To train the single hidden layer feedforward neural network (SLFNN) is done using the ELM method. Without any iterative tuning, the hidden nodes of the ELM are initiated as discussed in [15]. Assume that the SLFNN having M hidden neurons and activation function $h(x)$ is trained to estimate O distinct models (v_i, z_i) with zero error, where v_i denotes the input samples and $V_i = [v_{i1}, v_{i2}, \ldots, v_{in}]^T \varepsilon R^n$, and Z_i is the output of samples and $Z_i = [z_{i1}, z_{i2}, \ldots, z_{im}]^T \varepsilon R^m$. The inputs and biases in the ELM-based NN are generated randomly. The nonlinear SLFNNs are transformed into the following rapport:

$$I\gamma = U \tag{13}$$

where $I = \{i_{jk}\}\left\{\begin{array}{l} j = 1 \ldots O \\ k = 1 \ldots M \end{array}\right\}$ denotes the output matrix, $i_{jk} = h(w_k u_j + b_j)$ is the output of kth hidden neuron according to $v_i.w_k = [w_{k1}\, w_{k2}\, \ldots\, w_{jn}]\varepsilon$ which is the weight matrix linking kth hidden neuron and input neurons, and b_j signifies the bias of jth hidden neuron; $v_i.w_k$ shows the inner product of v_i and w_k; $\gamma = [\gamma_1 \ldots \gamma_k \ldots \gamma_m]_T$ $(k = 1 \ldots M)$ is the vector of weights of output vector, and $\gamma_k = [\gamma_{k1} \ldots \gamma_{k2} \ldots \gamma_{km}]^T$ represents the weight matrix linking the kth hidden neuron and output neurons; $Z = [z_1, z_2, \ldots, z_o]^T$ is the target matrix (anticipated outputs).

In order to determine the solution of the given linear system, the LS, the weights are used. The minimum norm of LS solution is given in Eq. 14

$$\hat{\Upsilon} = I^+ Z \tag{14}$$

where I^+ is the Moore–Penrose generalized inverse of matrix I. The minimum norm LS solution is unique and has the smallest norm among all the LS solutions.

3.5 ANN Estimation Models' Accuracy Evaluation

To evaluate AP prediction mode accuracy, mean bias error (MBE), MAE, RMSE and R^2 are used. These are given as follows:

$$\text{MAE} = \left(\frac{1}{N}\sum_{i=1}^{N}\left|\text{PV}_{i(\text{estimated})} - \text{PV}_{i(\text{measured})}\right|\right) \tag{15}$$

$$\text{MBE} = \frac{1}{N}\sum_{i=1}^{N}\left(\text{PV}_{i(\text{estimated})} - \text{PV}_{i(\text{measured})}\right) \tag{16}$$

$$\text{RMSE} = \sqrt{\frac{1}{N}\sum_{i=1}^{N}\left(\text{PV}_{i(\text{estimated})} - \text{PV}_{i(\text{measured})}\right)^2} \tag{17}$$

$$R^2 = 1 - \frac{(\text{RMSE})^2}{\text{var}\left(\text{PV}_{i(\text{estimated})}\right)} \tag{18}$$

where $\text{PV}_{i(\text{estimated})}$ is the estimated value of PV parameters, $\text{PV}_{i(\text{measured})}$ is measured PV parameter values and n is the number of samples.

The coefficient of determination R^2 measures how accurately the ANN models predict trends in measured values. By comparing the variance of the model data and model errors. For perfect prediction, $R^2 = 1$. The RMSE indicates root mean square of the errors. Both MAE and RMSE signify spread amount in the errors and in some sense represent the 1- and 2-norms of the errors. The MBE is the average prediction error representing the model error of an ANN model. MAE gives the average magnitude of prediction errors.

For estimating value of series and shunt resistance, ideality factor of PV module using ELM and FNN algorithm is presented in Fig. 5.

4 Results and Discussion

To estimate R_s,R_{sh} and η of PV module, G and T are taken as inputs. For estimation three Extreme learning machine (ELM1, ELM2, ELM3) and Feedforward Neural Network (FNN1, FNN2, FNN3). ELM1 and FNN1 are used to estimate series resistance, ELM2 and FNN2 are used to estimate shunt resistance, and ELM3 and FNN3 are used to estimate ideality factor of PV module. These models are trained with Levenberg–Marquardt algorithm. The performance plots (PP) of ELM1, ELM2, ELM3, FNN1, FNN2, FNN3 are shown in Figs. 6, 7, 8, 9, 10 and 11. As per the observations, the MSE is inversely proportional to the number of epochs. The comparison between measured and estimated values of R_s,R_{sh} and η is shown in Tables 1, 2 and 3, and error analysis is shown in Tables 7, 8, 9, 10, 11 and 12.

5 Conclusion

In the present study using R_s,R_{sh} and η of PV module are estimated using Multilayer Feedforward Neural Network and Extreme Learning Machine. The input variables for these models are solar irradiance and temperature. For this, an experimental setup is developed using sensors and microcontroller and these are used to measure solar irradiance, temperature, R_s,R_{sh} and η of 20 Wp PV module. Based on experimental reading, three multilayer feedforward neural networks and extreme learning machine are developed to estimate R_s,R_{sh} and η. The performance of developed models shows that MSE decreases with increase in epochs. These models are tested, and results show that for ELM models RMSE varies from 0.3474 to 0.6700 and for multilayer feedforward network RMSE varies from 0.1531 to 0.4585. This work is useful to estimate PV parameters proving useful for condition monitoring of PV module.

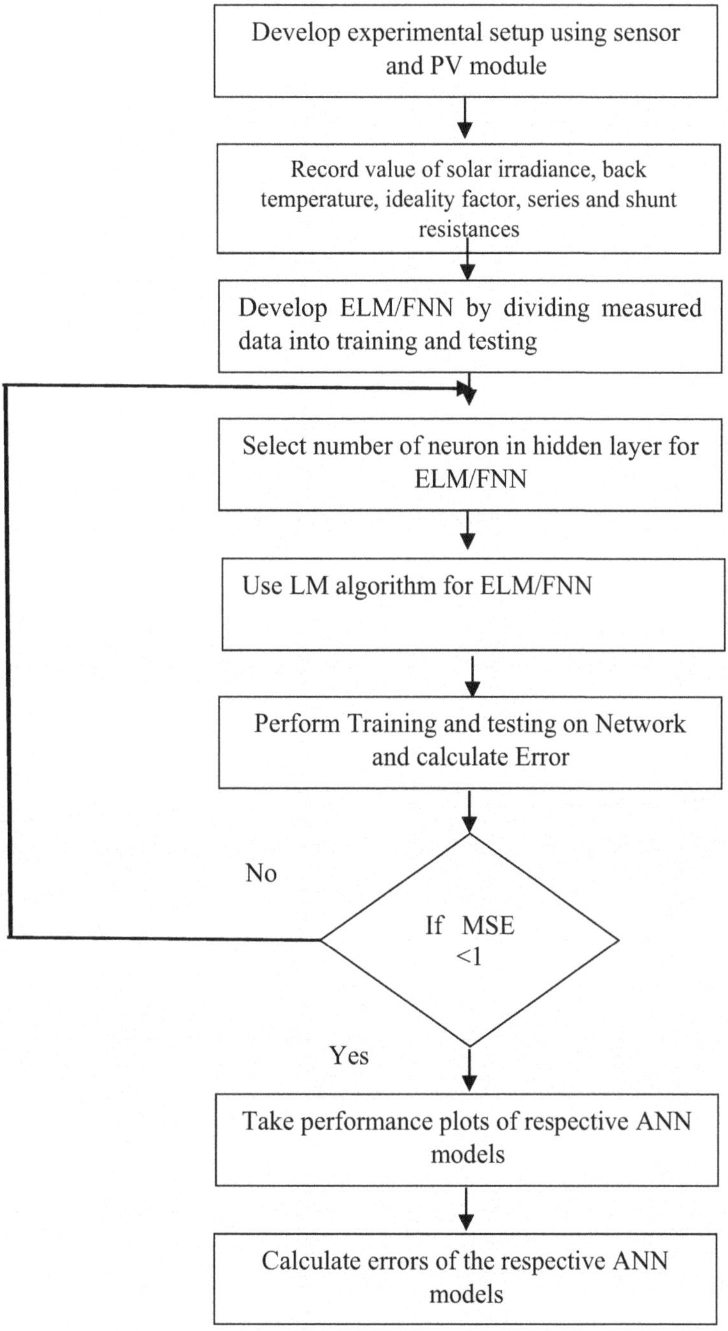

Fig. 5 Proposed algorithm for ANN-based PV Module parameter estimation

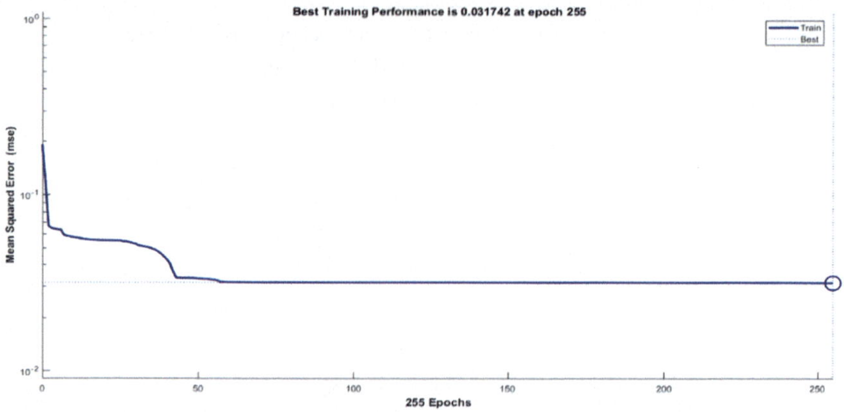

Fig. 6 PP of ELM1

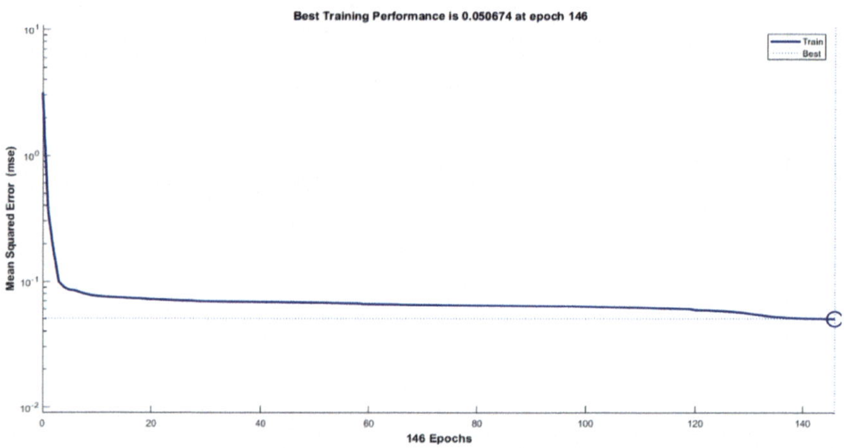

Fig. 7 PP of ELM2

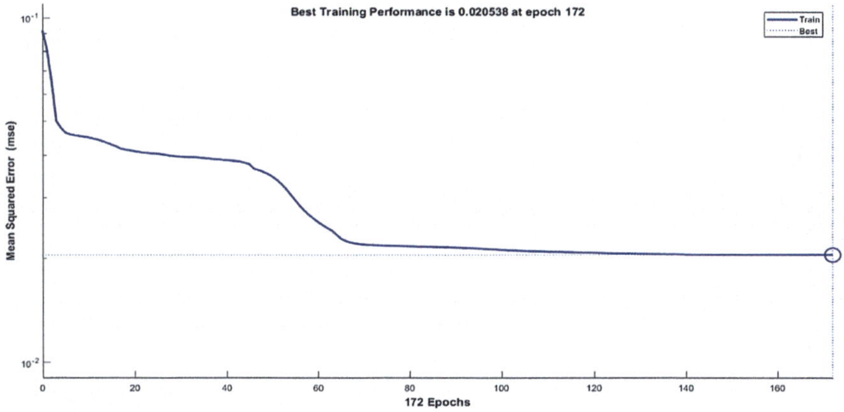

Fig. 8 PP of ELM3

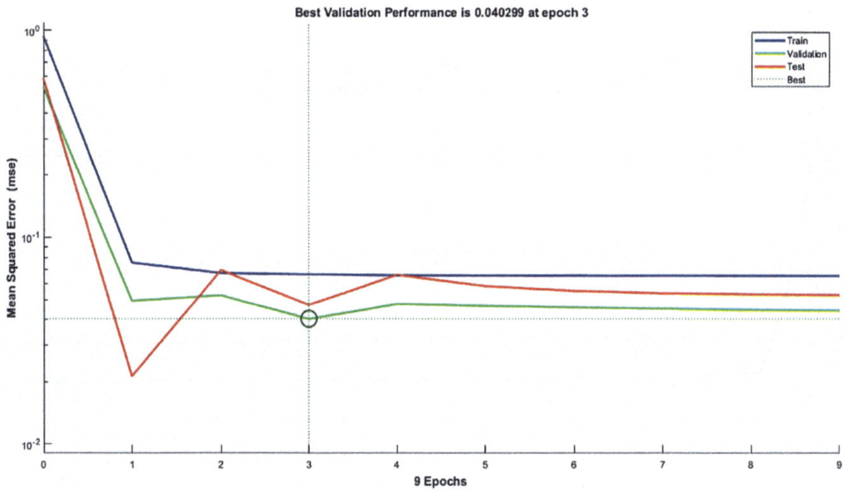

Fig. 9 PP of FNN1

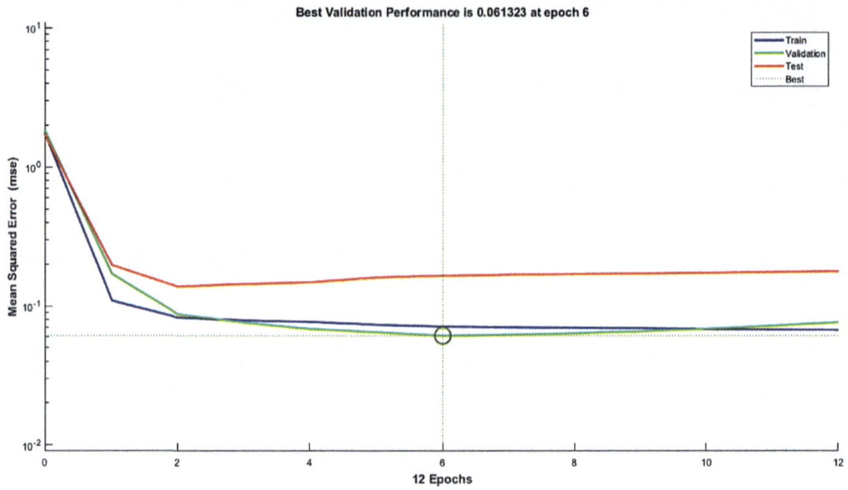

Fig. 10 PP of FNN2

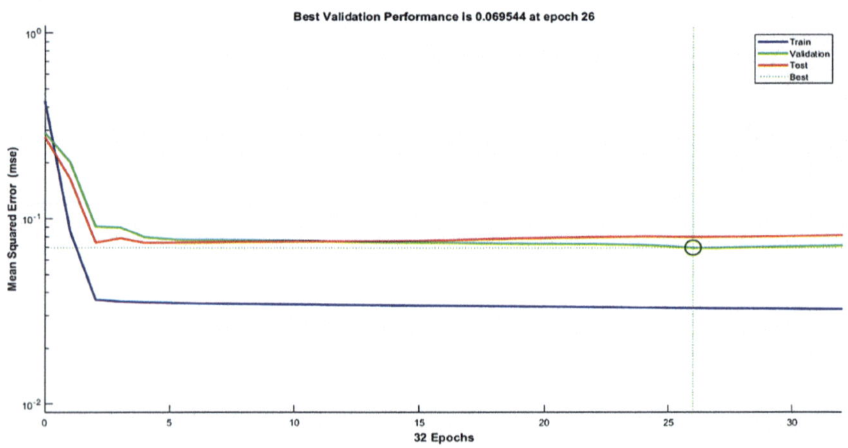

Fig. 11 PP of FNN3

Table 7 Comparison between measured and estimated values of series resistance

Measured	ELM1	Error	FNN1	Error
0.5750	0.8209	−0.2459	0.7950	−0.2200
0.4481	0.3421	0.1060	0.4902	−0.0421
0.4290	−0.4015	0.8305	0.5295	−0.1005
0.7870	0.8668	−0.0798	0.8100	−0.0231
0.7775	1.1077	−0.3302	0.9169	−0.1393
0.8047	1.1241	−0.3195	0.9310	−0.1263
0.9515	1.1241	−0.1726	0.9310	0.0206
0.9325	1.1077	−0.1752	0.9169	0.0156
0.6999	1.1077	−0.4078	0.9169	−0.2170
0.6789	0.4518	0.2271	0.3882	0.2907
0.1871	0.4518	−0.2646	0.3882	−0.2011

Table 8 Comparison between measured and estimated values of shunt resistance

Measured	ELM2	Error	FNN2	Error
0.2153	0.8113	−0.5960	0.3177	−0.1024
0.3260	0.7692	−0.4432	0.5872	−0.2612
0.3620	0.8622	−0.5002	0.6463	−0.2843
0.6481	0.9052	−0.2571	0.1500	0.4982
0.4737	0.7863	−0.3127	0.1794	0.2942
0.3851	0.7736	−0.3885	0.2041	0.1809
0.7287	0.7736	−0.0448	0.2041	0.5246
0.0377	0.7863	−0.7486	0.1794	−0.1418
0.1810	0.7863	−0.6053	0.1794	0.0015
0.1898	1.0341	−0.8443	0.2573	−0.0675
0.4636	1.0341	−0.5705	0.2573	0.2063

Table 9 Comparison between measured and estimated values of ideality factor

Measured	ELM3	Error	FNN3	Error
0.6317	1.1338	−0.5020	0.8358	−0.2040
0.5650	0.4531	0.1119	0.4624	0.1026
0.5650	−0.1185	0.6835	0.4855	0.0795
0.8258	1.6742	−0.8485	0.9164	−0.0906
0.8258	1.5755	−0.7497	0.9983	−0.1726
0.8334	1.4865	−0.6532	1.0039	−0.1706
1.0000	1.4865	−0.4865	1.0039	−0.0039
1.0000	1.5755	−0.5755	0.9983	0.0017
0.7502	1.5755	−0.8253	0.9983	−0.2481
0.7455	1.3401	−0.5946	0.5037	0.2418
0.3937	1.3401	−0.9464	0.5037	−0.1100

Table 10 Error analysis of ANN models for estimation of series resistance

Network	RMSE	MAE	MBE	R^2
ELM1	0.3474	0.2872	0.0756	1.2301
FNN1	0.1563	0.1269	0.0675	0.5483

Table 11 Error analysis of ANN models for estimation of shunt resistance

Network	RMSE	MAE	MBE	R^2
ELM2	0.5295	0.4828	0.4828	5.5189
FNN2	0.4585	0.3956	0.3956	−3.8892

Table 12 Error analysis of ANN models for estimation of ideality factor

Network	RMSE	MAE	MBE	R^2
ELM3	0.6700	0.6343	0.4897	−11.7671
FNN3	0.1531	0.1296	0.0522	0.3338

References

1. S.P. Europe, Global market outlook for solar power 2015–2019. *European Photovoltaic Industry Association*, Bruxelles, Tech. Rep, 2015
2. E. Garoudja, F. Harrou, Y. Sun, K. Kara, A. Chouder, S. Silvestre, Statistical fault detection in photovoltaic systems. Sol. Energy **150**, 485–499 (2017)
3. E. Garoudja, F. Harrou, Y. Sun, K. Kara, A. Chouder, S. Silvestre, Statistical fault detection in photovoltaic systems. Sol Energy **150**, 485–499 (2017)
4. S.R. Madeti, S. Singh, Monitoring system for photovoltaic plants: a review. Renew. Sustain. Energy Rev. **67**, 1180–1207 (2017)

5. P.T. Le, H.-L. Tsai, T.H. Lam, A wireless visualization monitoring, evaluation system for commercial photovoltaic modules solely in MATLAB/Simulink environment. Sol. Energy **140**, 1–11 (2016)
6. W. Chine, A. Mellit, A.M. Pavan, S.A. Kalogirou, Fault detection method for grid-connected photovoltaic plants. Renew. Energy **66**, 99–110 (2014)
7. B. Marion, R. Schaefer, H. Caine, G. Sanchez, Measured and modeled photovoltaic system energy losses from snow for Colorado and wisconsin locations. Sol. Energy **97**, 112–121 (2013)
8. S.R. Potnuru, D. Pattabiraman, S.I. Ganesan, N. Chilakapati, Positioning of PV panels for reduction in line losses and mismatch losses in PV array. Renew. Energy **78**, 264–275 (2015)
9. A.R. Reisi, M.H. Moradi, S. Jamasb, Classification and comparison of maximum power point tracking techniques for photovoltaic system: a review. Renew. Sustain. Energy Rev. **19**, 433–443 (2013)
10. R. Hariharan, M. Chakkarapani, G.S. Ilango, C. Nagamani, A method to detect photovoltaic array faults and partial shading in PV systems. IEEE J. Photovoltaics **6**(5), 1278–1285 (2016)
11. S. Daliento, A. Chouder, P. Guerriero, A.M. Pavan, A. Mellit, R. Moeini, P. Tricoli, Monitoring, diagnosis, and power forecasting for photovoltaic fields, 2017
12. G. Zwingelstein, *Diagnostic des défaillances: théorie et pratique pour les systèmes industriels*. Hermès, 1995
13. L. Bun, *Détection et localisation de défauts dans un système photovoltaïque*. PhD thesis, Université de Grenoble, 2011
14. A. Triki-Lahiani, A.B.-B. Abdelghani, I. Slama-Belkhodja, Fault detection and monitoring systems for photovoltaic installations: a review. Renew. Sustain. Energy Rev. **82**, 2680–2692 (2018)
15. G. Huang, G. Huang, S. Song, K. You, Trends in extreme learning machines: a review. Neural Netw. **61**, 32–48 (2015)
16. E. Garoudja, A. Chouder, K. Kara, S. Silvestre, An enhanced machine learning based approach for failures detection and diagnosis of PV systems. Energy Convers. Manage. **151**, 496–513 (2017)
17. I.A. Karim, Fault analysis and detection techniques of solar cells and PV modules, in *2015 International Conference on Electrical Engineering and Information Communication Technology (ICEEICT)*. IEEE, 2015, pp. 1–4
18. F.J. Sanchez-Pacheco, P.J. Sotorrío-Ruiz, J.R. Heredia-Larrubia, F. Pérez-Hidalgo, M.S. de Cardona, PLC-based PV plants smart monitoring system: field measurements and uncertainty estimation. IEEE Trans. Instrum. Meas. **63**(9), 2215–2222 (2014)
19. B. Andò, S. Baglio, A. Pistorio, G.M. Tina, C. Ventura, Sentinella: Smart monitoring of photovoltaic systems at panel level. IEEE Trans. Instrum. Meas. **64**(8), 2188–2199 (2015)
20. J. Han, J.-D. Jeong, I. Lee, S.-H. Kim, Low-cost monitoring of photovoltaic systems at panel level in residential homes based on power line communication. IEEE Trans. Consum. Electron. **63**(4), 435–441 (2017)
21. A. López-Vargas, M. Fuentes, M. Vivar, IoT application for real-time monitoring of solar home systems based on Arduino^{TM} with 3G connectivity. IEEE Sens. J. **19**(2), 679–691 (2018)
22. M. Fuentes, M. Vivar, J. Burgos, J. Aguilera, J. Vacas, Design of an accurate, low-cost autonomous data logger for PV system monitoring using Arduino^{TM} that complies with IEC standards. Sol. Energy Mater. Sol. Cells **130**, 529–543 (2014)
23. A. López-Vargas, M. Fuentes, M. Vivar, F.J. Muñoz-Rodríguez, Low-cost datalogger intended for remote monitoring of solar photovoltaic stand-alone systems based on Arduino. IEEE Sens. J. (2019)
24. W. Xiao, W.G. Dunford, A. Capel, A novel modeling method for photovoltaic cells, in *2004 IEEE 35th Annual Power Electronics Specialists Conference (IEEE Cat. No. 04CH37551)*, vol. 3. IEEE, 2004, pp. 1950–1956
25. M.G. Villalva, J.R. Gazoli, E. Ruppert Filho, Comprehensive approach to modeling and simulation of photovoltaic arrays. IEEE Trans. Power Electron. **24**(5), 1198–1208 (2009)
26. K. Ishaque, Z. Salam, H. Taheri, Simple, fast and accurate two-diode model for photovoltaic modules. Sol. Energy Mater. Sol. Cells **95**(2), 586–594 (2011)

27. N.M.A.A. Shannan, N.Z. Yahaya, B. Singh, Single-diode model and two-diode model of PV modules: a comparison, in *2013 IEEE International Conference on Control System, Computing and Engineering*. IEEE, 2013, pp. 210–214
28. M. De Blas, J. Torres, E. Prieto, A. Garcıa, Selecting a suitable model for characterizing photovoltaic devices. Renew. Energy **25**(3), 371–380 (2002)
29. A.N. Celik, N. Acikgoz, Modelling and experimental verification of the operating current of mono-crystalline photovoltaic modules using four-and five-parameter models. Appl. Energy **84**(1), 1–15 (2007)
30. V.L. Brano, A. Orioli, G. Ciulla, A. Di Gangi, An improved five-parameter model for photovoltaic modules. Sol. Energy Mater. Sol. Cells **94**(8), 1358–1370 (2010)
31. A.H. Fanney, B.P. Dougherty, M.W. Davis, Evaluating building integrated photovoltaic performance models, in *Proceedings of the 29th IEEE Photovoltaic Specialists Conference (PVSC)*, New Orleans, LA, USA. Citeseer, 2002, pp. 194–199
32. M. Zagrouba, A. Sellami, M. Bouaïcha, M. Ksouri, Identification of PV solar cells and modules parameters using the genetic algorithms: application to maximum power extraction. Sol. Energy **84**(5), 860–866 (2010)
33. K. Ishaque, Z. Salam, H. Taheri et al., Modeling and simulation of photovoltaic (PV) system during partial shading based on a two-diode model. Simul. Model. Pract. Theory **19**(7), 1613–1626 (2011)
34. K. El-Naggar, M. AlRashidi, M. AlHajri, A. Al-Othman, Simulated annealing algorithm for photovoltaic parameters identification. Sol. Energy **86**(1), 266–274 (2012)
35. A. Askarzadeh, A. Rezazadeh, Parameter identification for solar cell models using harmony search-based algorithms. Sol. Energy **86**(11), 3241–3249 (2012)
36. M. AlHajri, K. El-Naggar, M. AlRashidi, A. Al-Othman, Optimal extraction of solar cell parameters using pattern search. Renew. Energy **44**, 238–245 (2012)
37. K. Ishaque, Z. Salam, S. Mekhilef, A. Shamsudin, Parameter extraction of solar photovoltaic modules using penalty-based differential evolution. Appl. Energy **99**, 297–308 (2012)
38. E. Garoudja, K. Kara, A. Chouder, S. Silvestre, Parameters extraction of photovoltaic module for long-term prediction using artificial bee colony optimization, in *2015 3rd International Conference on Control, Engineering & Information Technology (CEIT)*. IEEE, 2015, pp. 1–6
39. A.M. Ameen, J. Pasupuleti, T. Khatib, Modeling of photovoltaic array output current based on actual performance using artificial neural networks. J. Renew. Sustain. Energy **7**(5), 053107 (2015)
40. S. Sivanandam, S. Deepa, *Principles of Soft Computing (with CD)* (Wiley, New York, 2007)
41. T. Ikegami, T. Maezono, F. Nakanishi, Y. Yamagata, K. Ebihara, Estimation of equivalent circuit parameters of PV module and its application to optimal operation of PV system. Sol. Energy Mater. Sol. Cells **67**(1–4), 389–395 (2001)
42. A. Belaout, F. Krim, A. Mellit, Neuro-fuzzy classifier for fault detection and classification in photovoltaic module, in *2016 8th International Conference on Modelling, Identification and Control (ICMIC)*. IEEE, 2016, pp. 144–149
43. Y. Chouay, M. Ouassaid, An intelligent method for fault diagnosis in photovoltaic systems, in *2017 International Conference on Electrical and Information Technologies (ICEIT)*. IEEE, 2017, pp. 1–5
44. Y. Hu, W. Cao, J. Ma, S.J. Finney, D. Li, Identifying PV module mismatch faults by a thermography-based temperature distribution analysis. IEEE Trans. Device Mater. Reliab. **14**(4), 951–960 (2014)
45. M. Kacira, M. Simsek, Y. Babur, S. Demirkol, Determining optimum tilt angles and orientations of photovoltaic panels in Sanliurfa, Turkey. Renew. Energy **29**(8), 1265–1275 (2004)
46. G. Ahmad, H. Hussein, H. El-Ghetany, Theoretical analysis and experimental verification of PV modules. Renew. Energy **28**(8), 1159–1168 (2003)
47. A.K. Yadav, H. Malik, S. Chandel, Selection of most relevant input parameters using WEKA for artificial neural network based solar radiation prediction models. Renew. Sustain. Energy Rev. **31**, 509–519 (2014)

48. G. Masson, M. Latour, M. Rekinger, I.-T. Theologitis, M. Papoutsi, Global market outlook for photovoltaics 2013–2017, *European Photovoltaic Industry Association*, 2013, pp. 12–32
49. A.K. Yadav, S. Chandel, Identification of relevant input variables for prediction of 1-minute time-step photovoltaic module power using artificial neural network and multiple linear regression models. Renew. Sustain. Energy Rev. **77**, 955–969 (2017)
50. A.K. Yadav, V. Sharma, H. Malik, S. Chandel, Daily array yield prediction of grid-interactive photovoltaic plant using relief attribute evaluator based radial basis function neural network. Renew. Sustain. Energy Rev. **81**, 2115–2127 (2018)
51. A.K. Yadav, S. Chandel, Solar radiation prediction using artificial neural network techniques: a review. Renew. Sustain. Energy Rev. **33**, 772–781 (2014)
52. A.K. Yadav, S. Chandel, Solar energy potential assessment of western Himalayan Indian state of Himachal Pradesh using J48 algorithm of WEKA in ANN based prediction model. Renew. Energy **75**, 675–693 (2015)
53. A.K. Yadav, H. Malik, S. Chandel, Application of rapid miner in ANN based prediction of solar radiation for assessment of solar energy resource potential of 76 sites in Northwestern India. Renew. Sustain. Energy Rev. **52**, 1093–1106 (2015)
54. P. Ramasamy, S. Chandel, A.K. Yadav, Wind speed prediction in the mountainous region of India using an artificial neural network model. Renew. Energy **80**, 338–347 (2015)

Intelligent Modeling and Analysis of a Transformerless Inverter System for Renewable Energy System

Sheikh Fareed Mohammad, Farhad Ilahi Bakhsh, Sheikh Suhail Mohammad, and Arshad Ali Hurra

Abstract The conventional sources of electrical energy are a significant cause of pollution and are being replaced by renewable energy sources (RESs) rapidly. Solar photovoltaic (SPV) system is the primary source among RESs. In this chapter, a single-phase two-stage transformerless SPV inverter system has been modeled and analyzed using MATLAB software. The first stage is a closed-loop DC-DC boost converter that regulates the solar arrays' DC output voltage. A full-bridge inverter that converts the DC power to sinusoidal AC power forms the second stage. The system has been tested at varying load and climatic conditions and can restore the DC voltage level from (90–140 V) the solar arrays to the 245 Vrms, 50 Hz AC output voltage. The proposed topology has been designed to meet the requirements of a typical household having a 2-kW rating with improved efficiency, reduced harmonics, and optimal use of selected components. The graphical abstract for the holistic work done in the chapter has been depicted in Fig. 1.

Keywords Solar photovoltaic systems (SPVs) · Single-phase full-bridge inverter · Pulse width modulation (PWM) · Maximum power point tracking (MPPT) · Dc-DC boost converter

1 Introduction

Photovoltaic (PV) systems as distributed generators are famous for being pollution-free, renewable, and applicable in different terrains. The market saw a significant decrease in the price of the PV panels. Since PV panels are static, a long lifetime (25 years with more than 80% power output) is inevitable with simple maintenance

S. F. Mohammad (✉) · A. A. Hurra
Department of Electrical Engineering, SSM College of Engineering, Baramulla (Jammu and Kashmir) 193121, India

F. I. Bakhsh · S. S. Mohammad
Department of Electrical Engineering, National Institute of Technology Srinagar, Hazratbal, Srinagar (Jammu and Kashmir) 190006, India
e-mail: farhad@nitsri.ac.in

H. Malik et al. (eds.), *Intelligent Data Analytics for Power and Energy Systems*, Lecture Notes in Electrical Engineering 802,
https://doi.org/10.1007/978-981-16-6081-8_22

[1]. For the conversion of DC power extracted from the solar PV panel (or arrays), a power converter is needed and fed to the AC load. In the late decade, a significant quantity of projects aims to employ the energy produced by PV systems as backup sources to support the already existing prime utility grid or be used in a standalone configuration. The production of sinusoidal output current that would meet the required standards is the main principle looked upon. This chapter focuses on modeling and analyzing a two-stage single-phase SPV inverter system that feeds power to an isolated load without the need for power storage devices such as batteries. It can be used for real-life application. Discussed below are some points showcasing the novelty of the work done in this chapter.

- Modeling and simulating a simplistic design of a single-phase two-stage SPV inverter system in MATLAB.
- Effective implementation of MPPT algorithm for a 2 KW solar system.
- Validation of its functioning at maximum power transfer scheme and at variable weather and load conditions.
- Neat and clean waveforms at the inverter output after effective filtration.

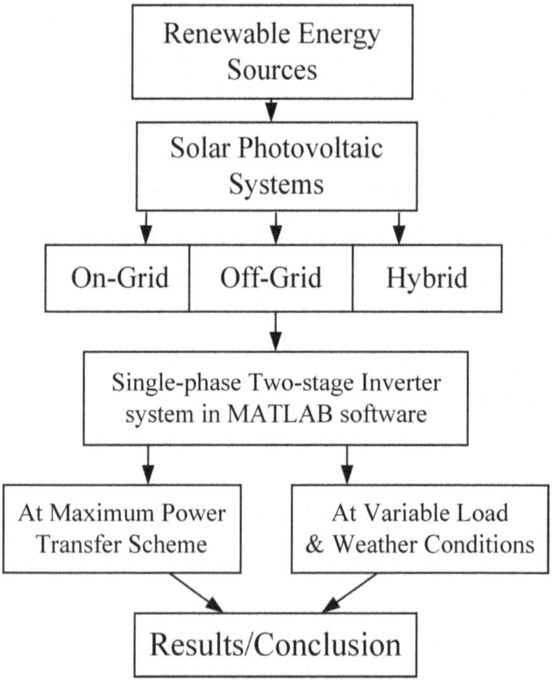

Fig. 1 Graphical abstract of the work done in the chapter

Fig. 2 PV cell

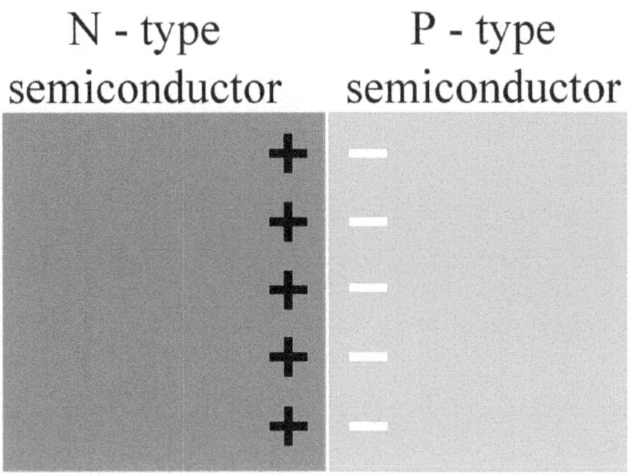

1.1 *Solar Photovoltaic (PV) Systems*

A solar photovoltaic (PV) system consists of several solar panels integrated with an inverter and other balance systems like charge controller, battery, etc., that use energy from the Sun to produce electricity [2]. Different applications of PV system demand different sizing. Although PV systems can operate both in on-grid and off-grid configurations, this work focuses on a system that is operating in standalone mode. An essential part and fundamental building block of a solar PV system is a solar cell/PV cell, which transforms the light energy into electrical energy.

1.1.1 The Photovoltaic Cell (PV)

A solar cell basically functions as an electrical tool transforming the solar energy into the form of electrical energy through a process known as the photovoltaic effect. A solar cell, as shown in Fig. 2, is a p-n junction diode. Electrical aspects—such as current, voltage, and resistance—differ upon being exposed to light (photons). PV panels or modules are a combination of these PV cells. The typical single-junction silicon-based solar cell generates a topmost of 0.5 to 0.6 V open-circuit voltage [3].

1.1.2 Working of a PV Cell

The basic principle of a PV cell relies on the photovoltaic effect, i.e., on exposure to the sunlight (photons), the voltage and electric current get induced into the material. The PV cell consists of three parts: n-type forms the top layer, the depletion layer, p-type forms the bottom layer, as shown in Fig. 3. The depletion layer consisting of neutral atoms apprehends the wavelength 350–1140 nm [3]. The photons emitted by the sunlight penetrate the depletion layer, thereby creating a free electron and a hole. Electrons which are free proceed toward top layer, and the holes proceed toward

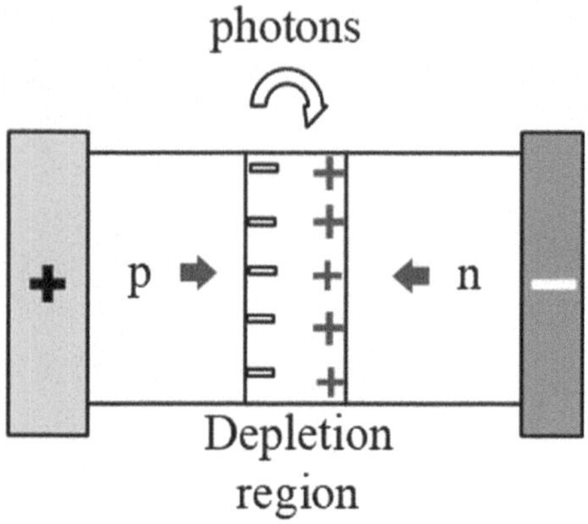

Fig. 3 Operation of a PV cell

bottom layer. Hence, by sustaining a conductor between the top and bottom layers, a path for the electrons to move toward the holes gets created, thereby constituting an electric current. Individual PV cell produces about 0.5–1 V [3]. Upon assembling these PV cells in array configuration, as shown in Fig. 4, required voltage and current can be obtained [4].

1.1.3 Types of Solar Photovoltaic Systems

Solar photovoltaic systems constitute mainly of three types which are used throughout the world; each type of system has its advantages and disadvantages and is used according to the different requirements. These include an on-grid solar PV system, off-grid solar PV system, and hybrid solar PV system.

On-Grid Solar PV system

As shown in Fig. 5, the on-grid solar PV system (grid-tied), whose apparatus consists of the solar panels, optimizer, grid-tie inverter, meter, and the power/utility grid. The electrical energy which is obtained from the solar panels is DC. An AC is used for powering most of our home appliances. An inverter is used to get AC, which then passes through the meter that feeds electricity into our home or the utility grid (if load demand at home is low), thereby selling off the access power and the owner gets paid in tariff. During nighttime or unfavorable weather conditions, the required electricity can be obtained from the utility grid. The on-grid solar system differs from the hybrid solar system because it does not allow electricity storage for later use, which comes as a significant system drawback.

Off-Grid Solar PV System

As shown in Fig. 6, an off-grid solar PV system (stand-alone) has battery storage instead of connectivity with the utility grid. Its apparatus also consists of solar panels, an off-grid inverter, and a battery bank (a generator can be an extra backup option). The DC coming from the solar panels is converted into AC using a solar inverter. After

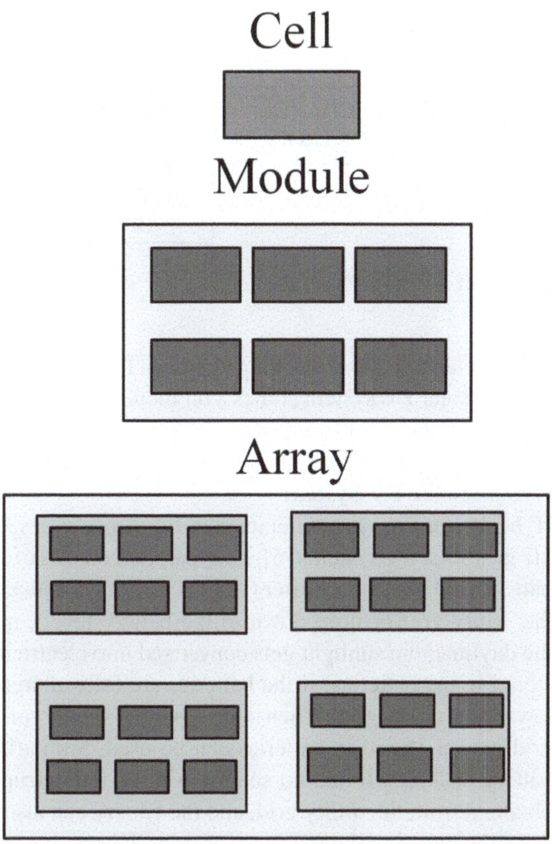

Fig. 4 Solar module and array

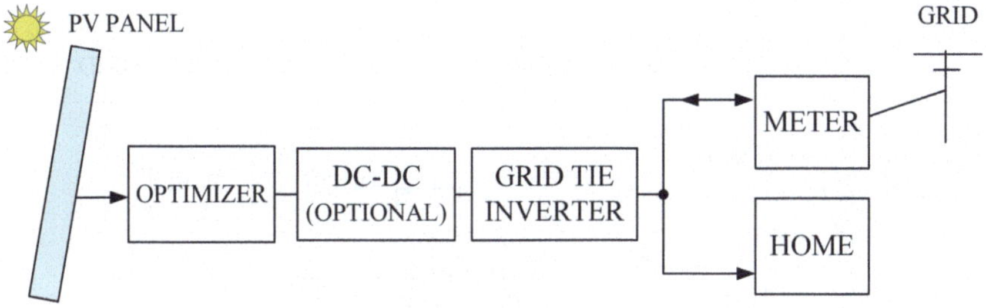

Fig. 5 On-grid solar PV system

meeting the home load demand, the excess amount of electricity left out gets stored in the batteries, which act as a power backup to be used during the nighttime or during unfavorable weather conditions when solar panels cannot generate electricity. The charge controller keeps charging and discharging batteries in check and disconnects solar panels when the latter is fully charged. This system is more beneficial for people

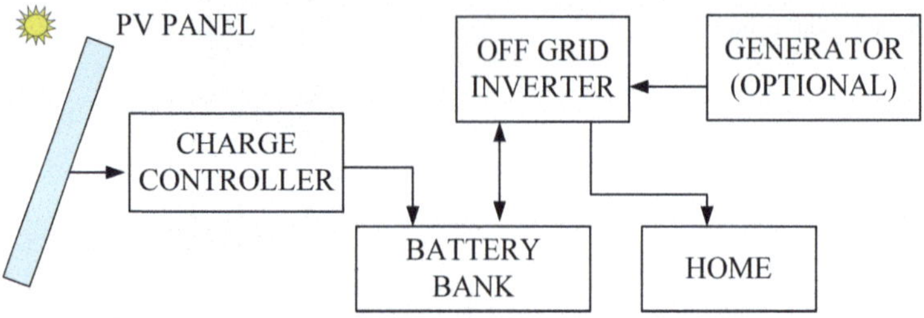

Fig. 6 Off-grid solar PV system

living in far-flung areas lacking approach to a fully functional utility grid. Cost of the off-grid solar PV system comes a bit more than the on-grid solar PV system because of the additional constrain of purchasing batteries.

Hybrid Solar PV System

Hybrid solar PV system depicted in Fig. 7 possesses advantages of duo on-grid and off-grid solar PV systems [5]. Its apparatus consists of solar panels, charge controller, battery bank, hybrid inverter (which can both produce synchronized power to match the utility grid or drop down to the battery level), meter, and utility grid. During the daytime, the sunlight gets convertsed into electricity. The excess amount will be stored in the batteries; if the batteries are fully charged, the power will be directed toward the utility grid. When there is neither support of solar energy nor the electricity grid, power stored in batteries can be used. Similarly, when there is no sufficient battery backup left and no solar power, an uninterrupted flow of electricity can be obtained from the utility grid, and the battery can also be charged. This system also costs the most in all three types of solar PV systems and is not commercially viable to all.

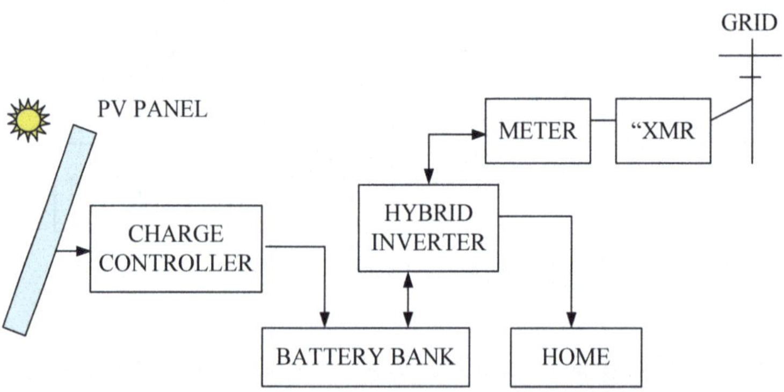

Fig. 7 Hybrid solar PV system

2 Single Phase Full-Bridge Inverter

The conversion of DC into AC is achieved with the help of an electrical circuit called an inverter. The necessity to convert is typical because most electrical appliances work and function upon an alternating input. A full-bridge or H-bridge, an inverter, is a circuit that converts DC to AC. It consists of four power MOSFETs or IGBTs switches, as shown in Fig. 8.

During the first half cycle, the switch Q1 and Q2, as shown in Fig. 10, close simultaneously (with the help of pulse width modulation, PWM technique) and through the second part of the pattern, switch Q4 along with Q3 close. During this process, as shown in Table 1, we obtain a square wave. A capacitor is placed on the input side to filter out the power instabilities, as demonstrated in Fig. 9. Additionally, an LC filter can be placed on the output of the inverter to filter out harmonics and smoothen the square waveform to a pure sine waveform [6].

One of the many essential aspects of inverter technology is its waveform at the output. There are two primary independent output AC waveforms produced by the inverter [7].

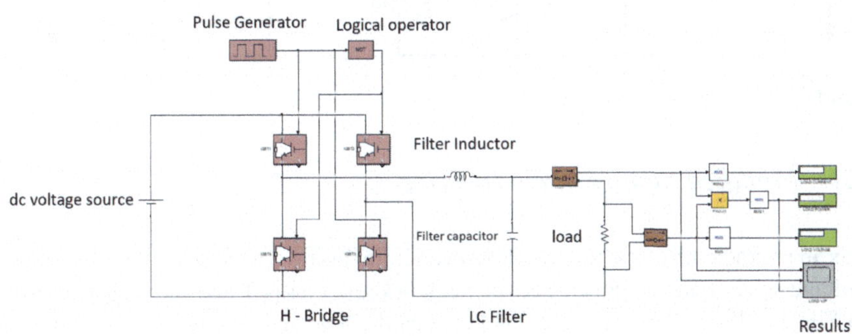

Fig. 8 Single-phase full-bridge inverter topology

Table 1 H-bridge switch states

Switch	State	Voltage across load
S1 and S2	ON	$+V_{DC}$
S3 and S4	ON	$-V_{DC}$
S1 and S3	ON	0
S2 and S4	ON	0

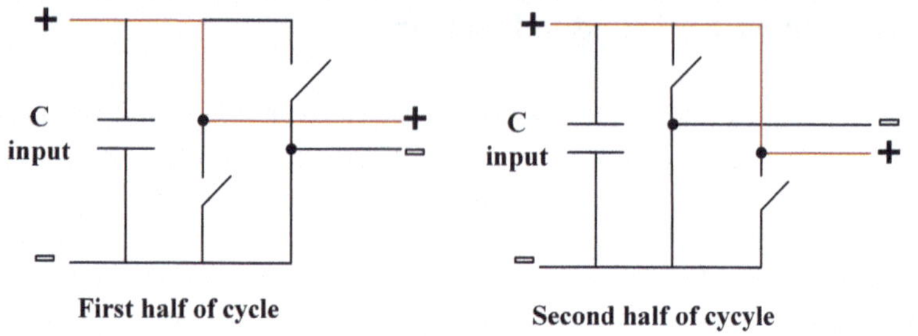

Fig. 9 Inverter cycle

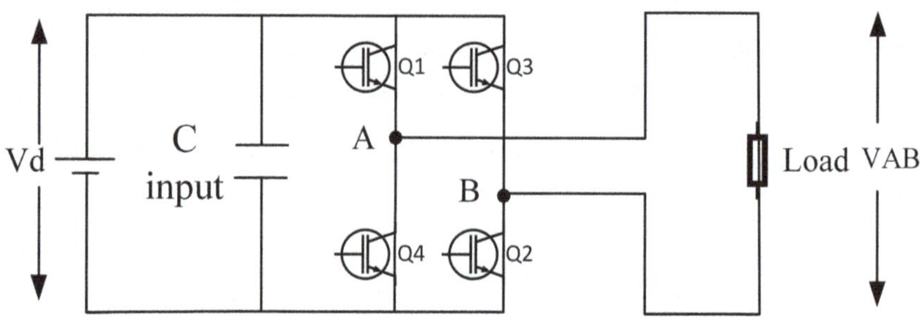

Fig. 10 Full-bridge inverter operation

2.1 *Modified Square Wave Inverter*

It is an upgradation of the square wave inverter. In this inverter system, three different voltage levels at the output side are present, $(+V_{DC})$, $(-V_{DC})$, and zero [1], as shown in Fig. 11. There also exists a delay between the +ve and −ve levels. This type of inverter is relatively cheap when compared with a pure sine wave inverter. It can be

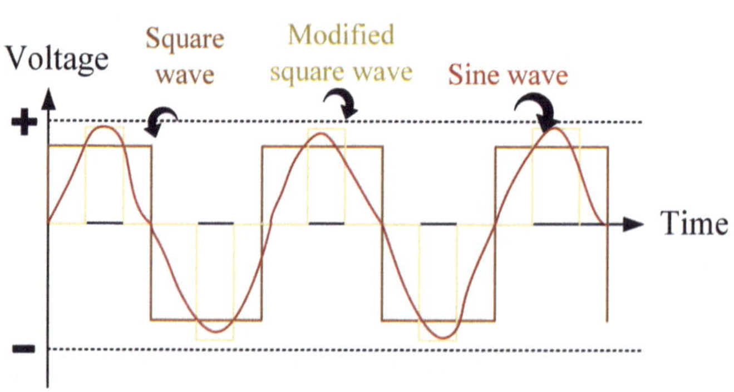

Fig. 11 Output waveforms of different inverters

used by most domestic electrical equipment, but its major drawback is that it has significant harmonic frequencies, which can cause specific loads to malfunction.

2.2 *Pure Sine Wave Inverter*

Since some equipment, such as light dimmers, medical equipment, and variable speed devices, may not work correctly with a modified sine wave Inverter, appliances of such kind must use a pure sine wave inverter which supplies sinusoidal ac output voltage similar to AC grid voltage. In a pure sine wave inverter, magnitude of output voltage and its frequency are kept in check by the PWM technique; hence, such an inverter is also known as a PWM inverter [8]. Generally, the output voltage signal of this inverter passes through a low-pass filter such as LC to reduce the THD and produce the final pure sine wave output [9].

3 Pulse Width Modulation (PWM)

PWM is a power electronic circuit controlling method which generates digital pulses as shown in Fig. 12 by changing pulse width (also called modulation); the controller/modulator (a device that breaks DC voltage into pulses) regulates power flow to the load. The inductor in an LC filter stocks charge during the "on" cycle and releases it, in accordance with the input or reference signal (low-frequency signal).

Figure 13 shows the generation of the PWM signal through a comparator. To produce an AC, a sine wave is required. Thus, we use a switching device such as IGBT to obtain a sine wave. When the signal is given to gate terminal of IGBT, the carrier and the reference signal are fed into the comparator to generate a sine wave. When the carrier signal rises, the width of the signal also increases and is known as pulse width modulation (PWM). By placing a passive LC filter, a perfect sine wave

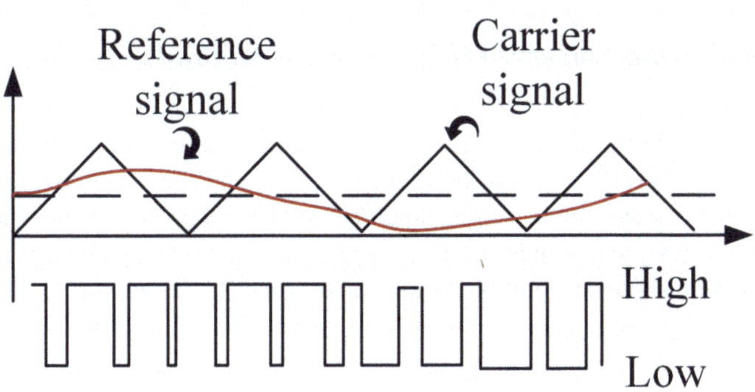

Fig. 12 Carrier and reference signals

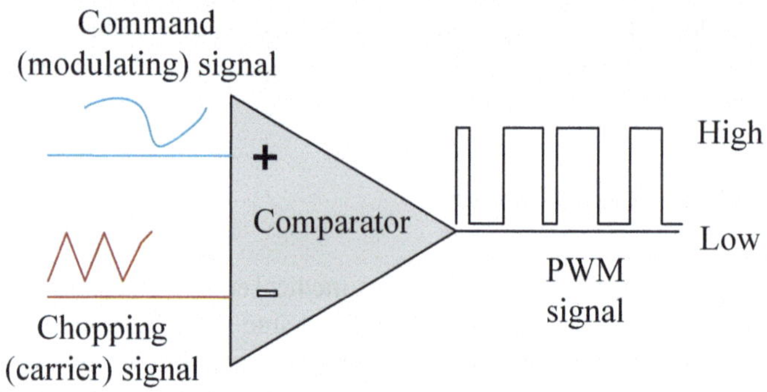

Fig. 13 PWM signal generation

gets generated. Sinusoidal pulse width modulation (SPWM) is widely employed in large-scale uses, such as motor control and inverter applications [10]. It is identified by constant amplitude pulses that have varying pulse widths. The choice of an SPWM technique is followed by the tolerable harmonic value at the output of an inverter [11]. H-bridge circuit consisting of IGBT switches, as shown in Fig. 10, is employed for both unipolar and bipolar inverter control schemes.

4 Maximum Power Point Tracking

Computations of maximum power point tracking such like perturb and observe, current sweep, constant voltage, fuzzy logic, incremental conductance method, and neural network are employed to operate PV at the highest power point, hence increasing system efficiency [12, 13]. This work has employed perturb and observe computation for implementing the maximum power point, as shown in Fig. 18 [14, 15].

4.1 Perturb and Observe (Hill Climbing Method)

This method is simple because it is easy to implement and is fast, with the disadvantage of power oscillations during perturbance around the maximum point. This drawback can be overcome by setting an error limit to end the recursion. The voltage sensor continuously calculates the PV voltage and measures power, as shown in Fig. 14. Maximum power point operation waveforms are shown in Fig. 15.

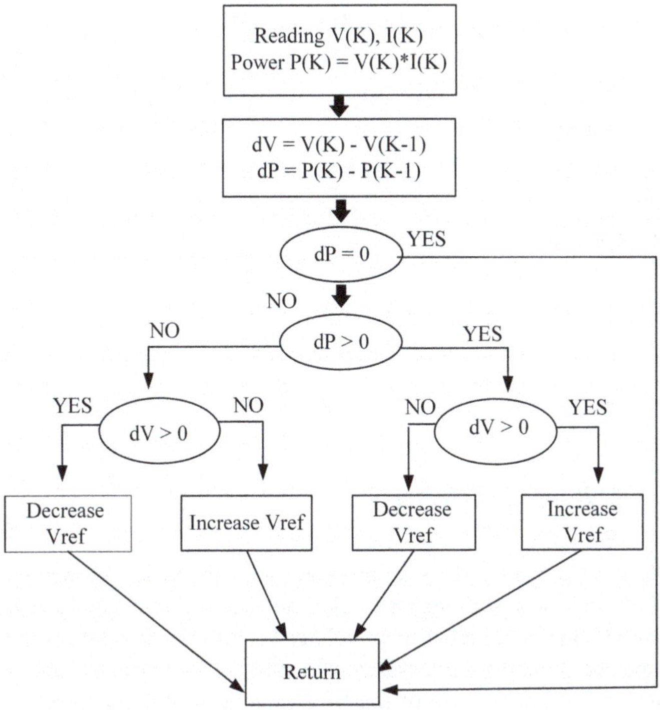

Fig. 14 Flowchart of perturb and observe method

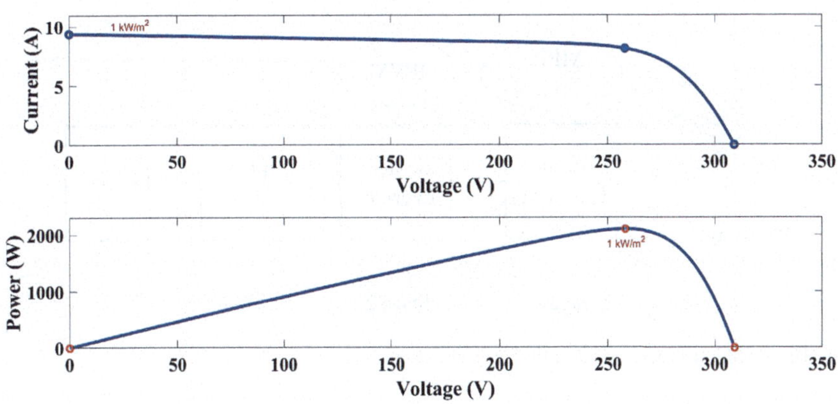

Fig. 15 MPPT tracking waveforms

5 Schematic Diagram of Inverter Topology

The complete block model of the inverter topology is depicted in Fig. 16. It consists of following main stages:

5.1 SPV Arrays

This system consists of (6) Soltech STH-350-WH solar panels connected in series and parallel (1 parallel string and six series-connected modules per string). The entire PV system specifications are listed in Table 2 at standard test conditions (STC).

5.2 DC-DC Boost Converter

The DC-DC boost converter plays an important act in the SPV system as it is responsible for increasing and sustaining an efficient and stable power supply in the system. The system operating with higher input voltage leads to lower input current required for a given power, reducing the losses considerably [1]. Its primary task is to adjust and improve output DC voltage of the SPV system. In this proposed topology, a boost converter operating in continuous conduction mode (CCM) has been designed

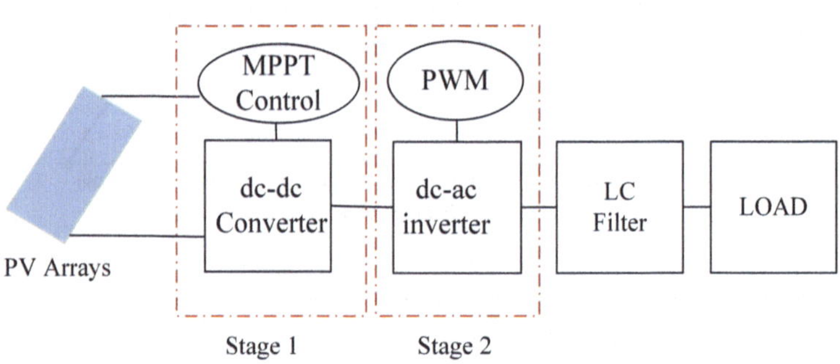

Fig. 16 Complete block model of the inverter system

Table 2 SPV system specifications at STC

Characteristics of parameters	Specifications
Maximum power	6 × 349 = 2094 W
Maximum power point voltage, Vmp	6 × 43 = 258 V
Maximum power point current, Imp	8.13 A
Voltage (open circuit), Voc	51.5 V
Current (short-circuit), Isc	9.4 A

Table 3 Specifications of the boost converter

Parameter	Value
Power rating (Pmax)	2 kW
Input voltage range (Vin)	(90–140) V
Output voltage	266 V
Max. load current (Pmax/Vout)	≈7.51 A
Switching frequency (fsw)	10 kHz
Vout ripple (ΔVout)	≤0.5%
Iin ripple (ΔIin)	≤20%

Table 4 Boost converter calculated parameters

Component	Value
IGBT switch	40 A, 500 V
Diode	60 A, 500 V
Boost inductance [8], L = Vip * (Vop − Vip)/fSW * ΔI * Vop =	5 mH
Output capacitor [8], C = Iop (Vop − Vip)/fSW * ΔV * Vop =	5000 uF
Minimum duty (Dmin) = (1-Vin-max/Vout)	0.661
Maximum duty (Dmax) = (1-Vin-min/Vout)	0.811

to step up the voltage of the SPV system to a higher constant voltage of 266 V. All selected components of the boost converter should meet the design specifications and the parameters tabulated in Table 3.

Table 4 represents the specifications of boost converter based on the considerations mentioned in Table 3.

5.3 H-Bridge Inverter

The H-bridge inverter is being modeled in accordance with the specifications given in Table 5. The choice of its components must meet these specifications.

Table 5 Specifications of the H-bridge inverter

Power rating	2 kW
Input voltage, (Vout of the boost converter)	266 V
Peak inverter current = $\sqrt{2} \times 2000/220$	≅12.8 A
Output AC voltage (r.m.s)	(240 ± 5%)
Frequency of AC output	(50 Hz ± 1%)
Total harmonic content (THD)	≤5%
PWM switching frequency	10 kHz

5.4 Output LC Filter

The output of the proposed full-bridge inverter is connected to a low-pass filter to generate a pure sinusoidal voltage at the output. The factors such as efficiency, voltage and current harmonics constraints, weight, volume, and cost must be considered when selecting the optimal filter design [16]. Also, the saturation of the inductor core should be avoided [17]. The LC filter has good performance in terms of current to voltage conversion and noise damping. Still, the filter capacitor may be insecure to line voltage harmonics, which result in large currents [18]. LC filter topology is illustrated in Fig. 17.

Equations for the LC filter design shown above (Fig. 17) are [19]:

$$fo = \omega 0/2\pi = 1/2\pi \times \sqrt{}\mathrm{LC}, \tag{1}$$

cutoff frequency for single-ended LC filter.

Fig. 17 LC filter design

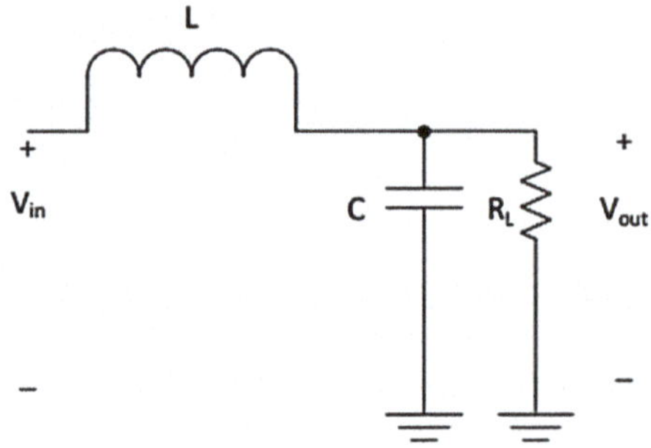

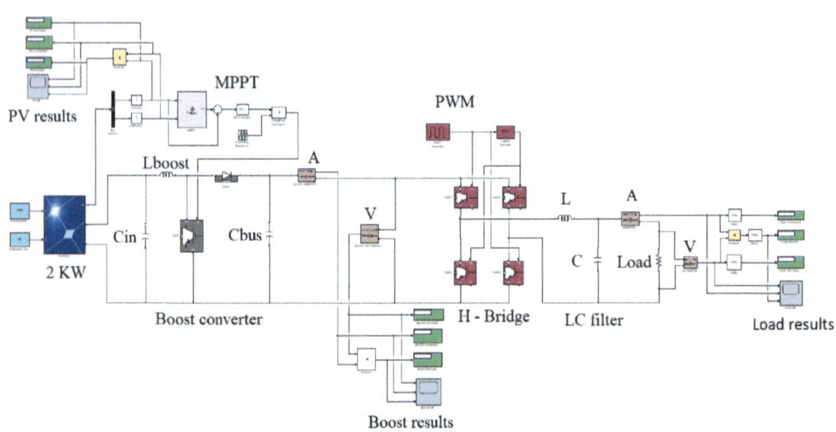

Fig. 18 Block model of the SPV inverter topology

Table 6 LC filter components

Component	Value
L	50 mH
C	20 mF

$$\omega 0 = 2\pi f o, \tag{2}$$

Transformation between frequency in hertz and radians.

$$Q = \mathrm{RL} \times \sqrt{C/L}, \tag{3}$$

Q stands for quality factor.

$$\zeta = 1/2Q = 1/2\mathrm{RL} \times \sqrt{C/L}, \tag{4}$$

Damping ratio.

It is often advisable to design a critically damped filter [20]. For the filter shown above,

$$Q = 0.707 = 1/\sqrt{2}.$$

$$L = \mathrm{RL} \times \sqrt{2}/\omega 0 \tag{5}$$

$$C = 1/\omega 0 \times \mathrm{RL} \times \sqrt{2}. \tag{6}$$

Table 6 depicts values of inductor and capacitor used in the LC filter of the proposed inverter topology.

5.5 *Harmonic Calculation*

Equations 7 and 8 are employed for total harmonic distortion of voltage and current [19].

$$\%\mathrm{THDv} = \sqrt{\sum_{h=2}^{\infty} \mathrm{V}^2\mathrm{h(rms)}/\mathrm{V1(rms)} \times 100\%} \tag{7}$$

$$\%\mathrm{THDi} = \sqrt{\sum_{h=2}^{\infty} \mathrm{I}_h^2\mathrm{(rms)}/\mathrm{I1(rms)} \times 100\%} \tag{8}$$

where Vh(rms) and Ih(rms) are the r.m.s values of hth harmonic of the voltage and current, respectively, V1(rms) and I1(rms) are the r.m.s values of actual voltage and current, respectively. Harmonic content for the Vinv and Iinv is analyzed using the fast Fourier transform (FFT) block, which is introduced in the (powergui block) of MATLAB–Simulink library.

6 Simulation and Validation of the Proposed SPV Inverter System

The whole Simulink block model of the standalone single-phase two-stage SPV inverter system is depicted in Fig. 18.

Detailed Simulink block of the closed-loop DC-DC boost converter is depicted in Fig. 19, where a PI controller is used to adjust its output at desired value (309 V).

Performance of the topology of the single-phase two-stage SPV system assessed considers the various operating conditions such as load changes and weather conditions (irradiance and temperature) to confirm the system's validity.

6.1 System Operating at Maximum Power Transfer Scheme

The system was tested at standard test weather conditions (1000 W/m^2 irradiance and 25 °C temperature) with the value of resistive load being equal to 33.33 ohms following the internal resistance of the PV panels. The voltage at the output of the boost converter is depicted in Fig. 20. Waveforms of different electrical quantities

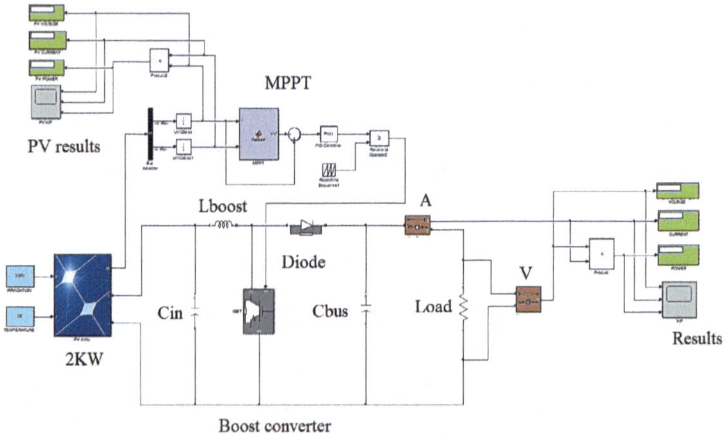

Fig. 19 Closed-loop model of DC-DC boost converter

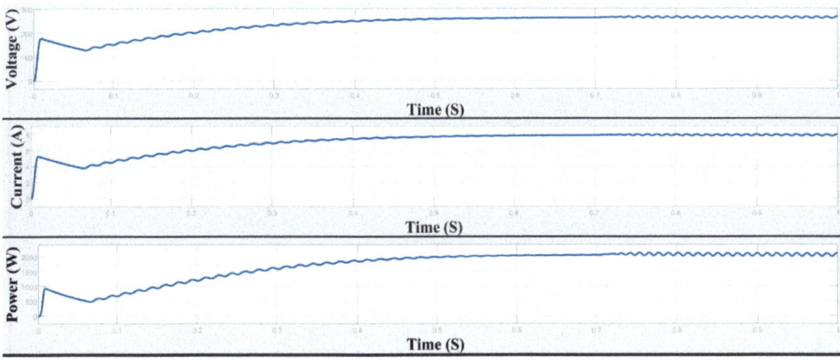

Fig. 20 Waveforms of boost converter output

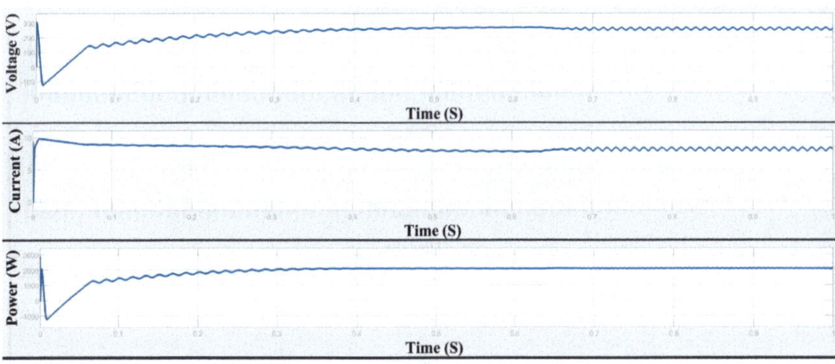

Fig. 21 Waveforms of PV panels

coming from the PV panels have been shown in Fig. 21. The waveforms at the inverter output before applying the filter have been shown in Fig. 22, while the waveforms after applying the filter have been shown in Fig. 23.

The results show that the system output voltage achieves a constant value after entering a steady state with a time delay of 0.5 s and equals 266 V. The system offers pretty good performance with neat and clean waveforms at different stages of the proposed SPV system topology, especially at the load end.

6.2 *System Operating on Variable Load, Temperature, and Irradiance Values*

As the system experiences various climatic and load change variations, its performance gets affected respectably; the system was exposed to multiple irradiances, temperature, and load changes at different time intervals so as to evaluate the overall

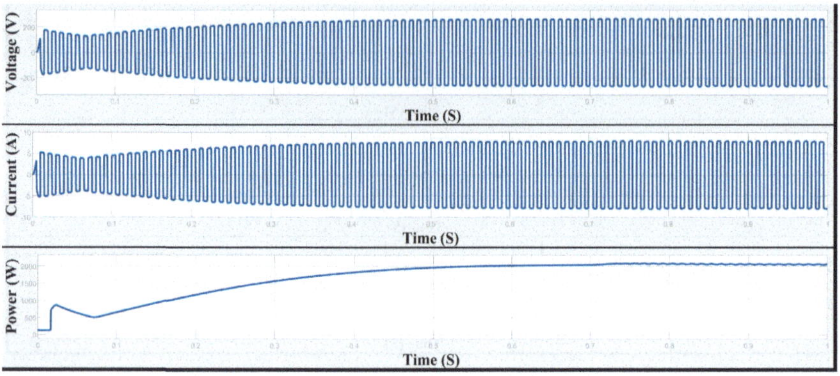

Fig. 22 Waveforms at the inverter output (without-filter)

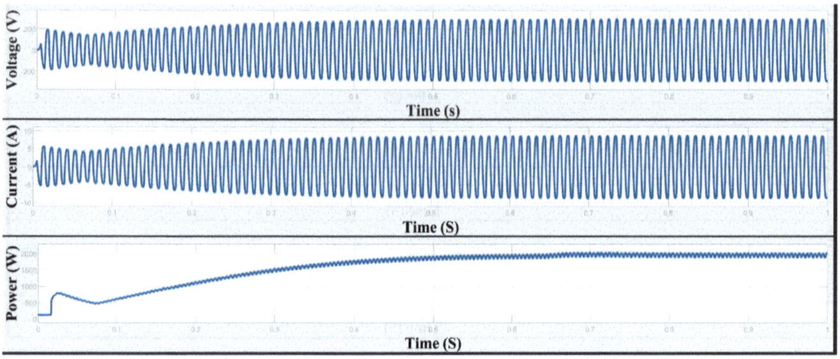

Fig. 23 Waveforms at the inverter output (with-filter)

outcome of the change in that variable on the system output parameters such as voltage, current, and power. The results thus acquired out of these simulations were acceptable and have been summarized in Table 7.

7 Conclusions

In this chapter, a single-phase two-stage SPV inverter system has been modeled and analyzed for a household with a 2-kW load. The proposed inverter model has been generated using MATLAB software. The system has been put to the test and evaluated under different load and climatic conditions to examine its efficiency and validity. The simulations showed that the proposed topology showcases a pretty good performance for all studied cases, and a decent sinusoidal output voltage of 240 Vrms, 50 Hz was obtained. The system was suitable enough to supply isolated

Table 7 Simulation results under different load and weather variations

S. No.	Irradiance (W/m^2)	Temp. (°C)	Resistive load (Ω)	Load voltage (V)	Load current (A)	Load power (W)
01	500	25	33.33	139.4	4.183	651.4
02	750	25	33.33	202.7	6.083	1378
03	1000	25	33.33	245.6	7.37	2022
04	1000	30	33.33	243.4	7.302	1985
05	1000	35	33.33	240.6	7.218	1940
06	1000	40	33.33	238.6	7.16	1909
07	1000	25	20	172.1	8.607	1695
08	1000	25	30	229.4	7.645	1963
09	1000	25	40	275	6.878	2117

load without energy storage units, thereby significantly lowering the system cost and maintenance.

References

1. N.H. Selman, J. Mahmood, Design and simulation of two stages single phase PV inverter operating in standalone mode without batteries. Int. J. Eng. Trends Technol. (2016). https://doi.org/10.14445/22315381/ijett-v37p217
2. A. Kumar, M. Andleeb, F.I. Bakhsh, Design and analysis of solar PV rooftop in Motihari. J. Phys. **1817**(1), 1–14 (2021). https://doi.org/10.1088/1742-6596/1817/1/012019
3. H. Malik, A. Iqbal, P. Joshi, S. Agrawal, F.I. Bakhsh, Metaheuristic and evolutionary computation: algorithms and applications, in *Studies in Computational Intelligence*, vol. 916 (Springer, Singapore, 2020). https://doi.org/10.1007/978-981-15-7571-6
4. M. Nisha, M. Andleeb, F.I. Bakhsh, Effect of partial shading on a PV array and its maximum power point tracking using particle swarm optimization. J. Phys. **1817**(1), 1–14 (2021). https://doi.org/10.1088/1742-6596/1817/1/012025
5. S.S. Mohammad, S.J. Iqbal, Improved dispatchability of solar photovoltaic system with battery energy storage, in *Proceedings of the 7th International Conference on Advances in Energy Research* (Springer, Singapore, 2021), pp. 1003–1013. https://doi.org/10.1007/978-981-15-5955-6_96
6. S. Islam, F.I. Bakhsh, Q.U. Islam, Modeling and analysis of the photovoltaic array feeding a SPWM inverter, in *Advances in Intelligent Systems and Computing, Applications of Artificial Intelligence Techniques in Engineering*, vol. 698 (Springer Book, Singapore, 2019), pp. 449–457. https://doi.org/10.1007/978-981-13-1819-1_42
7. S. Ahmad, F.I. Bakhsh, Transformer-based DC-to-AC grid-connected multi-level inverter topology, in *Renewable Power for Sustainable Growth. Lecture Notes in Electrical Engineering*, ed. by in A. Iqbal, H. Malik, A. Riyaz, K. Abdellah, S. Bayhan, vol. 723 (Springer, Singapore, 2021). https://doi.org/10.1007/978-981-33-4080-0_28
8. K. Rahman, A. Tariq, F.I. Bakhsh, Modeling and analysis of multilevel inverters using unipolar and bipolar switching schemes, in *IEEE-International Conference on Advances in Engineering, Science and Management (ICAESM)*, Tamil Nadu, India, 30–31 March, 2012, pp. 466–471. https://ieeexplore.ieee.org/document/6216159

9. M. Anwari, Recent trends in renewable energy and power electronics research. Penerbit UTM University, Malaysia, Tech. Rep., 2008. ISBN 978-983-52-0639-9
10. J.A. Lone, F.I. Bakhsh, Design and analysis of cascaded H bridge nine-level inverter in typhoon HIL. IOP Conf. Ser.: Mater. Sci. Eng. **804**(1), 1–8 (2020). https://doi.org/10.1088/1757-899X/804/1/012049
11. S. Ahmad, F.I. Bakhsh, M.S.J. Asghar, A novel DC-to-AC controlled multi-level inverter, in *International Conference on Multimedia, Signal Processing and Communication Technologies (IMPACT)*, AMU, India, Nov 2013, pp. 315–319. https://doi.org/10.1109/mspct.2013.6782144
12. A. Sinha, S.S. Sahu, MPPT control of standalone-PV system with battery as an energy storage element. Ph.D. dissertation, 2015. http://ethesis.nitrkl.ac.in/7858/
13. N. Sharma, F.I. Bakhsh, S. Mehta, Efficiency enhancement of a solar power plant using maximum power point tracking techniques, in *IEEE International Conference on Computational and Characterization Techniques in Engineering & Sciences (CCTES-18)*, Integral University, Lucknow, India, 14–15 Sept 2018, pp. 106–111. https://doi.org/10.1109/cctes.2018.8674146
14. S.S. Mohammad, S. Mishra, S.K. Sinha, V.K. Tayal, Integration of renewable energy resources for rural electrification, in *Proceeding of International Conference on Intelligent Communication, Control and Devices* (Springer, Singapore, 2017), pp. 545–551. https://doi.org/10.1007/978-981-10-1708-7_62
15. Vikas, A.A. Hurra, Modified MPPT based PI controller for standalone PV switched capacitor inverter. J. Emerg. Technol. Innov. Res. (2019). https://www.semanticscholar.org/paper/Modified-MPPT-Based-PI-Controller-For-Standalone-PV-Hurra/72a87c98ca4ce4c4b779765a73af938797905606
16. M.H. Rashid, *Power Electronics, Circuits, Devices, and Applications*, 3rd edn. (Pearson Education Inc., 2004). ISBN: 9780131011403. https://www.pearson.com/us/higher-education/product/Rashid-Power-Electronics-Circuits-Devices-and-Applications-3rd-Edition/9780131011403.html
17. A. Roshan, A DQ Rotating Frame Controller for Single Phase Full-Bridge Inverters Used in Small Distributed Generation Systems. Master thesis, Virginia Polytechnic Institute and State University, 2006. https://doi.org/10.1109/apex.2007.357582
18. H.B. Massawe, Grid Connected Photovoltaic Systems with Smart Grid functionality. Master thesis, Norwegian University, 2013. https://core.ac.uk/download/pdf/52105821.pdf
19. Texas Instruments Application Report "LC Filter design", SLAA701A-October- 2016-Revised Nov 2016. https://www.ti.com/lit/an/slaa701a/slaa701a.pdf?ts=1622458600856&ref_url=https%253A%252F%252Fwww.google.com%252F#:~:text=ABSTRACT,to%20as%20an%20LC%20filter
20. B.A. Johnson, Modeling and Analysis of a PV Grid Tied Smart Inverter's Support Functions, Master Thesis, The Faculty of California State University, San Luis Obispo, 2013. https://doi.org/10.15368/theses.2013.40

Intelligent Approach for Performance Investigation of Direct-Drive Generator-Based Wind Energy Conversion System Under Variable Speed Operation

Mohammed Aslam Husain, S. P. Singh, and Md. Tabrez

Abstract The performance of a wind energy conversion system (WECS) under employing a permanent magnet synchronous generator (PMSG) is investigated in this article under MATLAB/Simulink software environment. An intelligent approach for performance investigation of direct-drive generator-based system for conversion of wind energy under variable speed operation is presented here. A peak (max.) power point (location) tracking (MPPT) that is based on traditional tip speed control (TSC) technique and artificial intelligence relied MPPT estimation procedure is used to mine the maximum energy obtainable from the wind energy conversion system. The used MPPTs control strategies regulate the optimal value of active reference current which is maintained by the grid side converter's active current. Control strategy applied on converter (at the grid end) is used to regulate the overall power added to the grid in conjunction with converter on generator end so as to augment the per unit overall power from the generator.

Keywords PMSG · Field-oriented control · MPPT · Wind energy conversion

1 Introduction

Electricity produced by the wind turbines utilizes the wind energy to act as prime mover of the generator. Wind pressure on the blades of the wind turbine produces lift and exerts a rotating force. The rotational speed of the shaft ties with generator speed using the gear-box.

Modern wind turbines can work either with fixed speed or with variable speed mode. There are various benefits of variable speed wind turbine (VSWT) above fixed speed wind turbine and they are: increased annual energy production, reduced

---

M. A. Husain (✉) · S. P. Singh
Department of Electrical Engineering, REC, Ambedkar Nagar, Ambedkar Nagar, India

Md. Tabrez
Department of Electrical and Electronics Engineering, Motihari College of Engineering, Motihari, India

H. Malik et al. (eds.), *Intelligent Data Analytics for Power and Energy Systems*,
Lecture Notes in Electrical Engineering 802,
https://doi.org/10.1007/978-981-16-6081-8_23

mechanical pressures, improved quality of power, reduced pitch control complexity, and reduced acoustic noise. VSWT works at maximum-power coefficient along a varied wind speed range [1]. During the process, rotor speed continuously varies according to the speed of wind by diverse technique of-control so as to get maximum power point (MPP) at all speed of wind [2]. Maximum power point techniques (MPPTs) are used for maximizing the power output from the system. The concept of MPPT is applied in solar PV systems as well as wind generation systems [3–6].

Principally, DC generators, synchronous as well as asynchronous generators, are being deployed with a wind turbine in a wind energy conversion system [7]. Many researchers have advised the use of PMSGs for wind power extraction because of the self-excitation feature of PMSG. Moreover, the efficiency of these machines is higher as compared to induction machines [8].

At a particular velocity of wind, the turbine's power (mechanical output) is associated to its shaft speed. The shaft speed is controlled in order to maximize the wind power extraction. The terminal frequency of the PMSG is adjusted through the converters, which results in adjustment of shaft speed and finally extraction of maximum power [9–12]. The various important steps and contribution in this chapter are as follows:

- The continuous variation in wind speed makes it difficult to maximize the power yield. This is taken care by inserting a controller which works to maximize the power yield during varying wind speed conditions.
- The MPPT methods employed varies with turbine shape, but they can be broadly classified as power-signal feed.back (PSF), pertur.bation and obser.vation (P&O), and wind s.peed measurement (WSM) [13, 14]. In the P&O method (also known has hill climbing (HC) method), small perturbation is done in the speed of shaft (of turbine) and the resulting–change–in power pf turbine is observed. In the PSF method, the peak power is obtained using the prior experimental results and accordingly the control is done.
- In the last scheme, i.e., WSM, the wind along with the shaft speed is measured, and optimal tip-speed ratio is determined for the controller. The controller is adjusted prior to the installation of a new wind energy system [15–17].
- An intelligent approach for performance investigation of direct-drive generator-based WECS under variable speed operation is presented here. A MPPT based on traditional tip speed control (TSC) technique and artificial intelligence-based MPPT estimation technique is utilized to extract the peak power available from the wind energy conversion system. The used MPPTs control strategies regulate the optimal value of active reference current which is maintained by the grid side converter's active current.
- Control strategy applied on the converter (at the grid end) is used to regulate the overall power added to the grid in conjunction with converter on generator end so as to augment the per unit overall power from the generator.

2 Modeling of PMSG-Based Wind Turbine System

Generally, a VSWT energy system has two converters, one at the generator end and the other at grid end, and this is done to get the variable speed operation along with constant output frequency and output voltage. The wind energy conversion arrangement under consideration consists of a wind turbine, PMSG, and converter (at the generator end and at grid end) as shown in Fig. 1.

2.1 Wind Turbine Modeling

A wind turbine is generally modeled as aerodynamic model or mechanical model. The blades of a wind turbine are aerodynamic system of a wind turbine rotor. The wind has mechanical power associated with it. The turbine rotor absorbs the mechanical power from the wind and start rotating and thus converts the obtained kinetic power from the blowing air to power in mechanical form, P_T. This power, P_T, varies with the speed (V_W) of wind, with the angle (β), i.e., angle of turbine blade and the rotational speed (ω_T), this in turn depends on wind speed) of the turbine rotor [18, 19]. Mathematically mechanical power P_T and mechanical torque T_T of the wind turbine can be written as:

$$P_T = \frac{1}{2}\rho A C_p(\lambda, \beta) V_W^3 \tag{1}$$

$$T_T = \frac{1}{2\omega_T}\rho A C_p(\lambda, \beta) V_W^3 \tag{2}$$

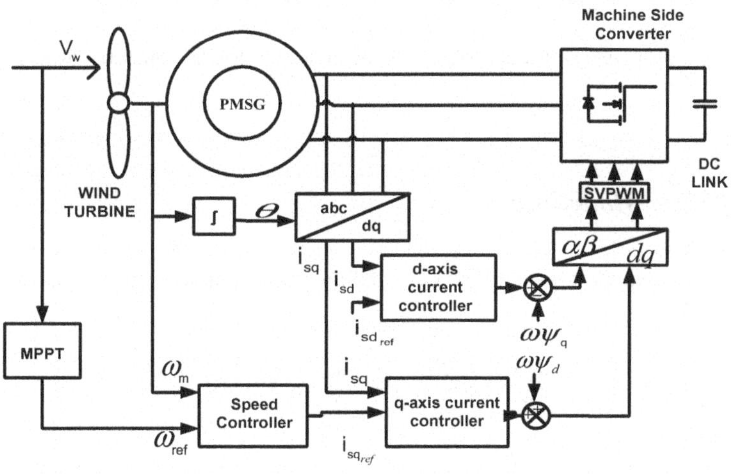

Fig. 1 Schematic diagram of FOC-based PMSG drive

Ration of tip speed can be expressed as below:

$$\lambda = \frac{\omega_T R}{V_w} \tag{3}$$

The speed of generator (ω_g) and the speed of rotor connected to turbine ω_T are same in modern PMSG-based wind system. The rotor efficiency (C_P) of a turbine is expressed by Eq. 4. It is expressed in terms of β and λ [9, 10].

$$C_p(\lambda, \beta) = 0 \cdot 22\left(\frac{116}{\phi} - 0.4\beta - 5\right)e^{\frac{-12.5}{\phi}} \tag{4}$$

$$\phi = \frac{1}{\left(\frac{1}{\lambda + 0.08\beta} - \frac{0.035}{\beta^3 + 1}\right)} \tag{5}$$

Neglecting the shaft and gearbox's moments of inertia, mechanical model of wind turbine train is a 2-mass scheme of the dynamics of rotor, having of a huge mass and another mass of small size, for the turbine's rotor inertia (J_τ) and inertia of the generator's rotor (J_g), respectively [11, 12]. The modelling of the cdrive train is described below given by Eqs. 6–8.

$$\frac{\mathrm{d}\omega_\tau}{\mathrm{d}t} = \frac{1}{J_\tau}[T_\tau - (K_s * \delta\theta + B * \delta\omega)] \tag{6}$$

$$\frac{\mathrm{d}(\delta\theta)}{\mathrm{d}t} = \delta\omega \tag{7}$$

$$\frac{\mathrm{d}\omega_g}{\mathrm{d}t} = \frac{1}{J_g}\left[\frac{1}{\eta_{\text{gear}}}(K_s * \delta\theta + B * \delta\omega)\right] - T_e \tag{8}$$

At this point, the single aggregate mass accounts for all parts (rotating) of the turbine.

2.2 *PMSG Modelling*

The electrical equation-based PMSG model in synchronously rotating frame can be given as [13]:

$$\frac{\mathrm{d}i_{\text{sd}}}{\mathrm{d}t} = -\frac{R_a}{L_d} i_{\text{sd}} + \omega \frac{L_q}{L_d} i_{\text{sq}} + \frac{1}{L_d} u_d \tag{9}$$

$$\frac{di_{sq}}{dt} = -\frac{R_a}{L_q} i_{sq} - \omega\left(\frac{L_d}{L_q} i_{sd} + \frac{1}{L_q}\psi_m\right) + \frac{1}{L_q} u_q \tag{10}$$

The nelectromagneticb torque:

$$T_e = \frac{3}{2} * p * \left[i_{\text{sq}} * \varphi_m + \left(L_d - L_q\right) * i_{\text{sq}} * i_{\text{sd}}\right] \tag{11}$$

If the kinductance of the d- and gq-axis is same

$$T_e = \frac{3}{2} p i_{\text{sq}} \psi_m \tag{12}$$

Thus, the model of PMSG is obtained using the above equations.

2.3 Field-Oriented Control (FOC)

The VSWTs are turning into the most appealing arrangement, because of the way that at almost all speeds, maximum power can be mined. Be that as it may, the measure of energy acquired from a WECS is influenced by not only the wind profile at the location, yet additionally on the methodology utilized for control of the WECS. A famous strategy for controlling of PMSG is the ffield-oriented control (FOC), that utilizes position of rotor and generator's speed [17].

FOC works as a closed loop strategy, and here, the torque is indirectly managed by the control of stator current. This control scheme is articulated in d-q axis frame. The obtained torque expression of a PMSG (surface mounted) in the dq frame is given as:

$$T_e = 1.5 p \psi_m i_{\text{sq}} \tag{13}$$

As the above equation shows that for constant Ψ_m, the torque can be managed by varying the stator current of q-axis stator. The torque can be regulated by keeping the desired value of torque angle.

3 Control of Wind Turbine Generator System

The output mechanical power P_T of the wind turbine is dependent on the TSR, λ (i.e., the proportion of the turbine shaft speed and the wind velocity). If there is a variation in wind speed or direction, the value of ω_t, i.e., the speed of turbine shaft (or ω_g, the speed of generator shaft speed), and wind turbine, the power of turbine P_T gets changed. Figure 2 represents variation of P_T with respect to ω_g.

Shaft speed that corresponds to the peak power can be calculated from the below given equation [13]:

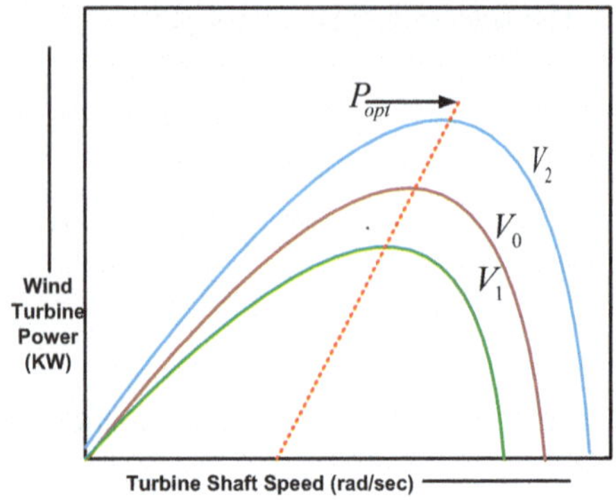

Fig. 2 Variation of output mechanical power of wind turbine w.r.t shaft speed

$$\omega_{P_{\max}} = \frac{V_w}{R}\left(\left(\frac{15-0.3\beta}{\pi}\right)\cos^{-1}\left(\frac{0.00184\beta(15-0.3\beta)}{\pi 0.5\rho^2 V_W^2(0.44-0.0167\beta)}\right)+3\right) \tag{14}$$

Since the wind turbine generally does not operate at optimal points (i.e., points at which low fC_P is there) for majority of the time, better operating points can be obtained by deploying an mMPPT.

3.1 *Artificial Intelligence-Based MPPT*

Tip-speed ratio (TSR) control is a traditional method of control in which slight perturbation in speed is applied to see the reflected change in power (Fig. 3), according. Nonetheless, in the neural network assessment-based technique, the neural network utilizes the speed and DFIG output, for the training purpose. The speed of rotor cannot vary unexpectedly due to larger system inertia, which prompts a continuous change in voltage output, using it the best speed control signal is gotten. Albeit the technique based on TSR has a moderately quicker response, yet huge transients in power are seen due to variation in speed and due to abrupt variation in the control

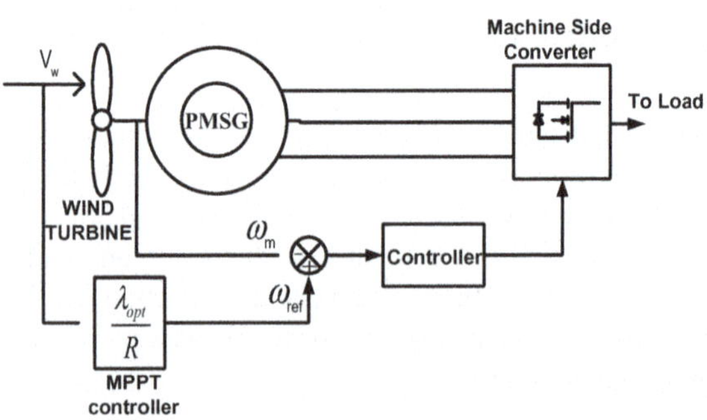

Fig. 3 TSR control of WECS

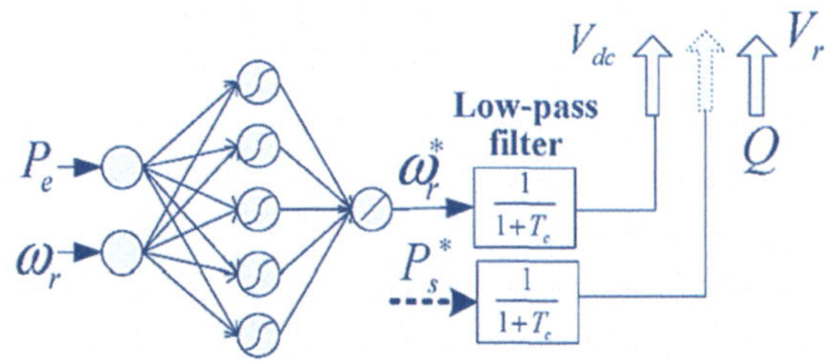

Fig. 4 ANN estimation-based MPPT control (where, active power is P_e and rotor speed is ω_r, the best possible control signal for speed is $\omega_r{}^*$ and the best possible control signal for power is $P_s{}^*$)

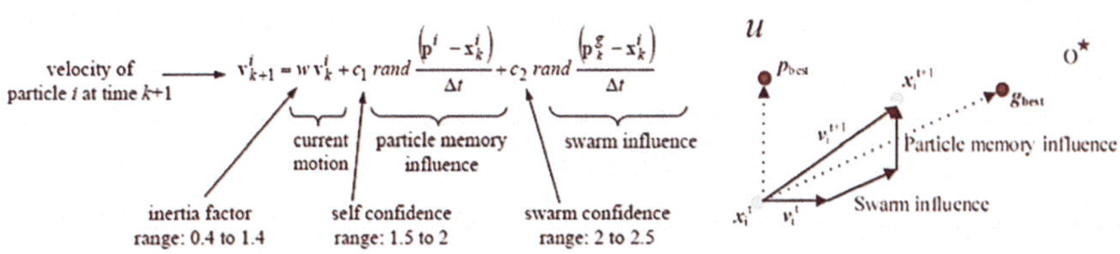

Fig. 5 Velocity of particles and their update process in PSO

signal. In case of ANN, GA and PSO type ANN estimation-based methods show a lower power transient as of the even transition in the control value of rotor speed [19–27]. Figure 4 shows the ANN guesstimated wind power MPPT control (where, active power is P_e and rotor speed is ω_r, the best possible control signal for speed is $\omega_r{}^*$ and the best possible control signal for power is $P_s{}^*$). Figure 5 shows the velocity of particles and their update process in PSO MPPT scheme.

4 Model Formulation and Results Demonstration

In this work, a grid-connected WECS based on PMSG is implemented and studied in the MATLAB/Simulink software environment. The active behavior of this presented system was studied for varying wind speed conditions under P&O, AI, and PSO MPPT in the proposed model. A similar and comparable results are found for each control method (Figs. 6, 7, 8, 9, and 10).

At $t = 3$ s, the wind speed was changed in a step of 3 m/s as represented in Fig. 4, i.e., the speed is changed from 8 to 11 m/s. Figure 8 represents the resultant change in the rotor speed (actual) for the corresponding change in wind speed. This is clear from the results, that a rapid tracking of optimal speed is achieved and hence MPPT operation takes place.

A comparison of various values obtained using controller and without using controller for the presented model is shown in Table 1. The results show that the use of controller results in higher extraction of power and thus validates it. This is also shown by the percentage increase of active power extracted.

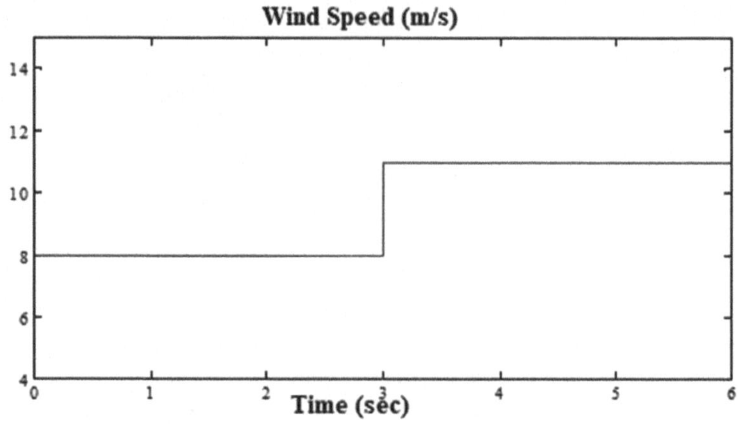

Fig. 6 Step variation in input wind speed

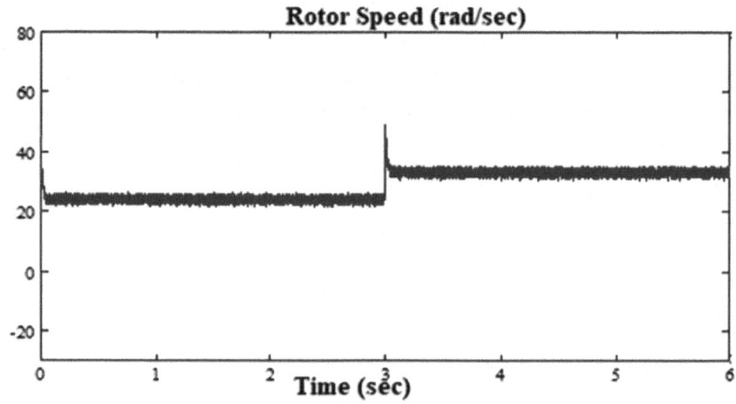

Fig. 7 Rotor speed (actual)

For varying wind speed, the resultant active and reactive powers are shown in Table 2. For all the above values, DC coupling voltage was taken as 500 V. Figure 11 shows the variation of active and reactive power w.r.t. wind speed. Figure 12 shows the tracking time of different MPPT method. Figure 13 gives the value of tracked power by different MPPT method.

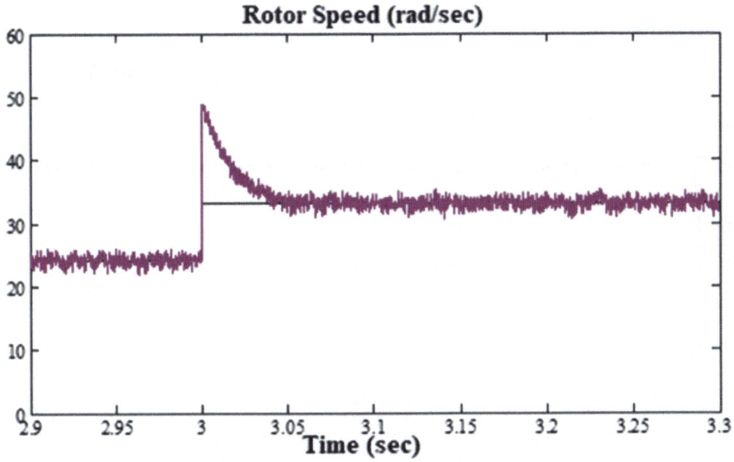

Fig. 8 Actual rotor speed and reference rotor speed

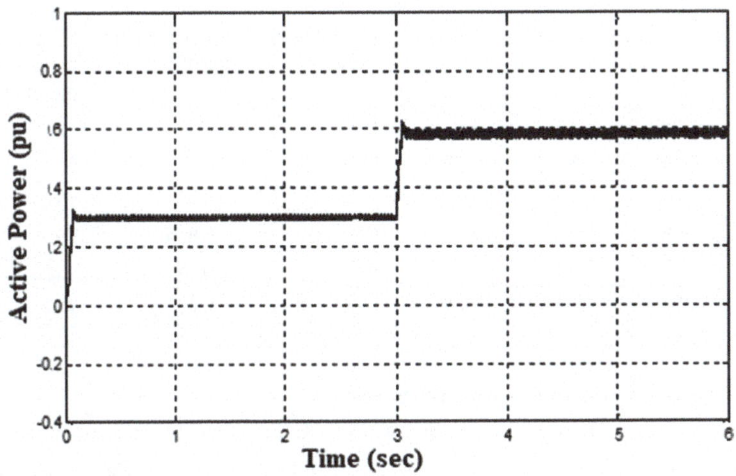

Fig. 9 Variation of active power injected to grid

5 Conclusion and Future Scope

An intelligent approach for performance investigation of direct-drive generator WECS under variable speed operation is shown here. A dynamic model that represents the proposed scheme is developed in MATLAB/Simulink environment. MPPT model based on TSC, ANN, GA, and PSO is discussed. A converter employed at the generator end adjusts the speed of turbine shaft and thus active power injected into the grid is achieved so as to extract maximum power for different wind speeds. The value of reference currents (for the closed-loop control) is selected in such a manner

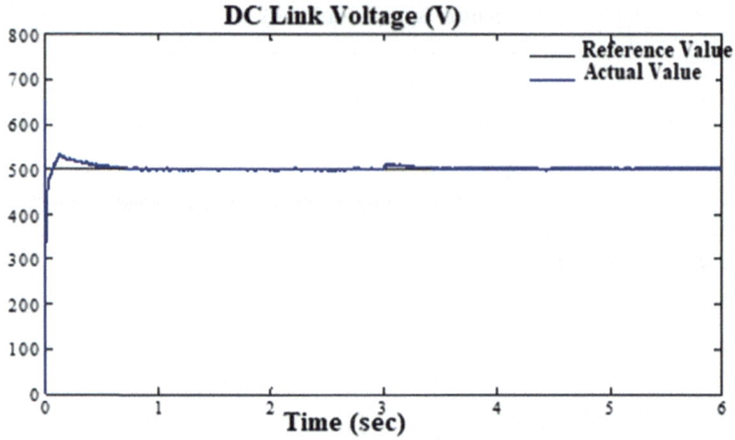

Fig. 10 DC link voltage

Table 1 Comparisons of active power injected to the grid

ω_m (m/s)	T_e (Nm)	T_m (Nm)	P at Grid	P at GS without MPPT	P at GS with MPPT	Percentage increase in power P (%)
6	55	64	0.11	0.0095	0.13	92
7	80	80	0.115	0.01	0.2	95
8	110	110	0.121	0.013	0.33	96
9	130	135	0.13	0.0156	0.425	96
10	170	170	0.14	0.019	0.55	96.5
11	200	200	0.145	0.022	0.62	96.4

Table 2 Comparisons of active and reactive power

ω_m (m/s)	ω_r (m/s)	T_e (Nm)	T_m (Nm)	DC (V)	Active power (pu)	Reactive power (pu)
6	20	55	64	500	0.11	0.001
7	22	80	80	500	0.115	0.005
8	24	110	110	500	0.121	0.01
9	28	130	135	500	0.13	0.015
10	30	170	170	500	0.14	0.03
11	34	200	200	500	0.145	0.045

so as to get power factor equal to unity. Different MATLAB simulation results were shown and it was observed that the MPPT operation was successful and the use of converter resulted in enhancing the overall system's performance.

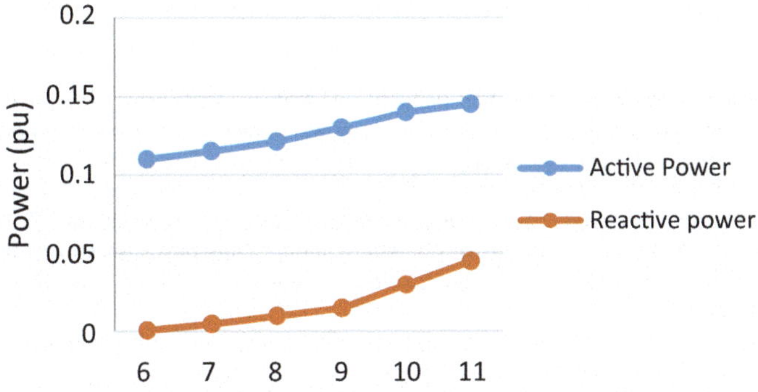

Fig. 11 Variation of active and reactive power w.r.t. wind speed

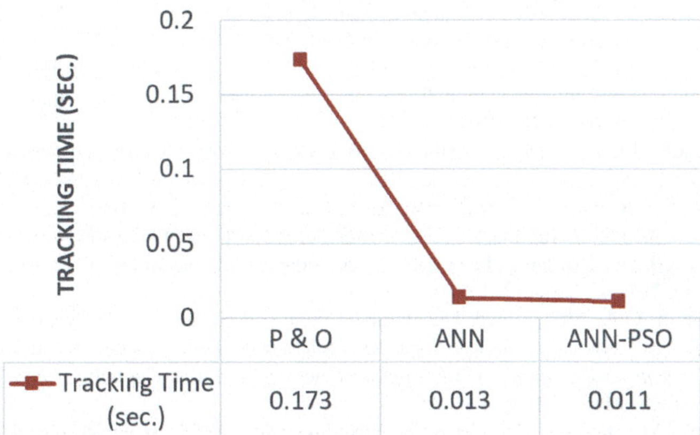

Fig. 12 Tracking time of different MPPT method

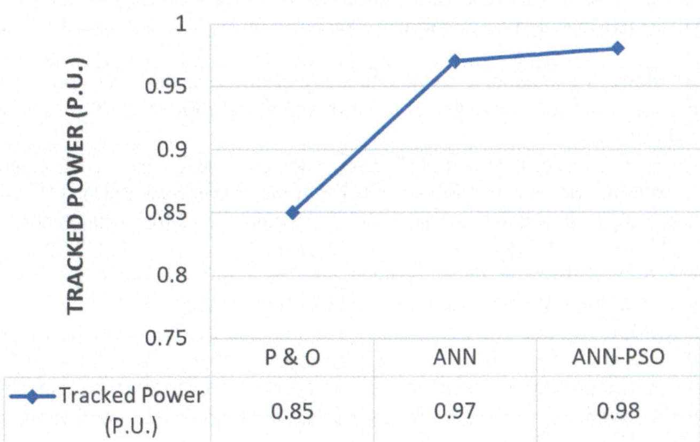

Fig. 13 Value of tracked power by different MPPT method

References

1. R. Sitharthan, T. Parthasarathy, S. Sheeba Rani, K.C. Ramya, An improved radial basis function neural network control strategy-based maximum power point tracking controller for wind power generation system. Trans. Inst. Meas. Control **41**(11), 3158–3170 (2019)
2. U. Yasin, *Tracking of Maximum Power from Variable Speed Wind Using Fuzzy Controller Based on Permanent Magnet Synchronous Generator* (Doctoral dissertation, ASTU) (2020)
3. M.A. Husain, A. Tariq, S. Hameed, M.S.B. Arif, A. Jain, Comparative assessment of maximum power point tracking procedures for photovoltaic systems. Green Energy & Environment **2**(1), 5–17 (2017)
4. A. Nouriani, H. Moradi, Variable speed wind turbine power control: a comparison between multiple MPPT based methods. Int. J. Dyn. Control 1–14 (2021)
5. M. Naseem, M.A. Husain, A.F. Minai, A.N. Khan, M. Amir, J. Dinesh Kumar, A. Iqbal, Assessment of meta-heuristic and classical methods for GMPPT of PV system. Trans. Electr. Electron. Mater. 1–18 (2021)
6. M. Hannachi, O. Elbeji, M. Benhamed, L. Sbita, Comparative study of four MPPT for a wind power system. Wind Eng. (2021). 0309524X21995946
7. R. Melício, V.M.F. Mendes, J.P.D.S. Catalão, Power converter topologies for wind energy conversion systems: integrated modeling, control strategy and performance simulation. Renew. Energy **35**(10), 2165–2174 (2010)
8. A. Hebala, O. Hebala, W.A. Ghoneim, H.A. Ashour, Multi-objective particle swarm optimization of wind turbine directly connected PMSG, in *2017 Nineteenth International Middle East Power Systems Conference (MEPCON)*. IEEE, Dec 2017, pp. 1075–1080
9. A. Jain, S. Shankar, V. Vanitha, Power generation using permanent magnet synchronous generator (PMSG) based variable speed wind energy conversion system (WECS): an overview. J. Green Eng. **7**(4), 477–504 (2017)
10. M.M. Amin, O.A. Mohammed, Development of a grid-connected wind generation system utilizing high frequency-based three-phase semicontrolled rectifier-current source inverter, in *2011 Twenty-Sixth Annual IEEE Applied Power Electronics Conference and Exposition (APEC)*. IEEE, Mar 2011, pp. 645–652
11. Y. Errami, M. Ouassaid, M. Maaroufi, Modeling and variable structure power control of PMSG based variable speed wind energy conversion system. J. Optoelectron. Adv. Mater. **15**(November–December 2013) (2013), 1248–1255
12. M.A. Husain, A. Tariq, Modeling of a standalone Wind-PV Hybrid generation system using MATLAB/SIMULINK and its performance analysis. Int. J. Sci. Eng. Res **4**(11), 1805–1811 (2013)
13. M. Tabrez, et al., A comparative simulation study of different sensorless permanent magnet synchronous motor drives using neural network and fuzzy logic. J. Intell. Fuzzy Syst. **35**(5), 5177–5184 (2018)
14. S. Muyeen, R. Takahashi, T. Murata, J. Tamura, A variable speed wind turbine control strategy to meet wind farm grid code requirements. IEEE Trans. Power Syst. **25**(1), 331–340 (2010)
15. M.A. Husain, A. Tariq, Modeling and study of a standalone PMSG wind generation system using MATLAB/SIMULINK. Univ. J. Electr. Electron. Eng. **2**(7), 270–277 (2014)
16. E. Spooner, A.C. Williamson, Direct coupled, permanent magnet generators for wind turbine applications. IEE Proc.-Electr. Power Appl. **143**(1), 1–8 (1996)
17. W.M. Lin, C.M. Hong, Intelligent approach to maximum power point tracking control strategy for variable-speed wind turbine generation system. Energy **35**(6), 2440–2447 (2010)
18. G. Yang, Y. Zhu, Application of a matrix converter for PMSG wind turbine generation system, in *The 2nd International Symposium on Power Electronics for Distributed Generation Systems*. IEEE, June 2010, pp. 185–189
19. R. Chedid, F. Mard, M. Basma, Intelligent control of a class of wind energy conversion system. IEEE Trans. Energy Conv. EC **14**, 1597–1604 (1999)
20. R. Tiwari, N. Ramesh Babu, Recent developments of control strategies for wind energy conversion system. Renew. Sustain. Energy Rev. **66**, 268–285 (2016)

21. A.K. Yadav, et al., Soft computing in condition monitoring and diagnostics of electrical and mechanical systems, in *Part of the Advances in Intelligent Systems and Computing*, vol. 1096. Springer Nature, 2020, pp. 496. ISBN 978-981-15-1532-3. https://doi.org/10.1007/978-981-15-1532-3
22. A. Iqbal, et al., Meta heuristic and evolutionary computation: algorithms and applications, in *Part of the Studies in Computational Intelligence*, vol. 916. Springer Nature, 2020, p. 849. ISBN 978-981-15-7571-6. https://doi.org/10.1007/978-981-15-7571-6
23. J.A. Alzubi, AI and machine learning paradigms for health monitoring system: intelligent data analytics, in *Part of the Studies in Big Data*, vol. 86. Springer Nature, 2020, p. 513. ISBN: 978-981-33-4412-9. https://doi.org/10.1007/978-981-33-4412-9
24. A. Iqbal et al., Renewable power for sustainable growth, in *Part of the Lecture Notes in Electrical Engineering*, vol. 723. Springer Nature, 2021, 805 p) ISBN: 978-981-33-4080-0. https://doi.org/10.1007/978-981-33-4080-0
25. N. Fatema, et al., *Intelligent Data-Analytics for Condition Monitoring: Smart Grid Applications* (Academic Press, 2021). ISBN: 978-0-323-85510-5. https://doi.org/10.1016/C2020-0-02173-0
26. S. Srivastava, et al., *Applications of Artificial Intelligence Techniques in Engineering*, vol. 1, Part of the Advances in Intelligent Systems and Computing, vol. 698, 643 p (Springer Nature, 2018). ISBN 978-981-13-1819-1. https://doi.org/10.1007/978-981-13-1819-1
27. S. Srivastava et al., Applications of Artificial Intelligence Techniques in Engineering, vol. 2, Part of the Advances in Intelligent Systems and Computing, vol. 697, 647 p (Springer Nature, 2018). ISBN 978-981-13-1822-1. https://doi.org/10.1007/978-981-13-1822-1

Performance Analysis of PV Module Using Pyramid Surface Texturing Approach

Rahul Kumar Bansal, Satyendra Singh, Alok Kumar Singh, and Md. Waseem Ahmad

Abstract Performance analysis of thin-film solar modules has been done using the pyramid texturing technique. To change the geometry of the surface of the solar cell through surface texturing technique, it increases the effective area of thin-film module. Significant improvement has been found by inserting a random pyramid structure. TCAD and PvSyst software is used to design and development of surface texturing and temperature-dependent loss minimization. Efficiency improvement of 3% has been achieved using this noble approach.

Keywords Efficiency improvement · Loss forecasting · PVsyst · Performance parameter · Thin-film solar PV module · TCAD · Surface texturing

1 Introduction

In 1954, first era of semiconductor silicon-based PV cells was imagined, with an efficiency of six percent, and used for space applications. Presently, the PV innovation is improving quickly since the late 1980s with a plan to improve efficiency and decrease cost. PV solar energy, solar board life is roughly 30 years [1]. Confirmed board outrageous warmth and cold, downpour, hail, storms is impervious to such hard conditions. Being a non-straight gadget, thusly its attributes change with the

R. K. Bansal (✉)
Department of Electrical Engineering, Dr. Radhakrishnan Institute of Technology, Jaipur, Rajasthan 302037, India

S. Singh
School of Electrical Skills, Bhartiya Skill Development University Jaipur, Rajasthan 302037, India

A. K. Singh
Department of Electrical Engineering, Nirwan University, Jaipur, Rajasthan 303301, India

Md. Waseem Ahmad
Department of Electrical Engineering, National Institute of Technology Karnataka, Surathkal 575025, India
e-mail: waseem@nitk.edu.in

H. Malik et al. (eds.), *Intelligent Data Analytics for Power and Energy Systems*,
Lecture Notes in Electrical Engineering 802,
https://doi.org/10.1007/978-981-16-6081-8_24

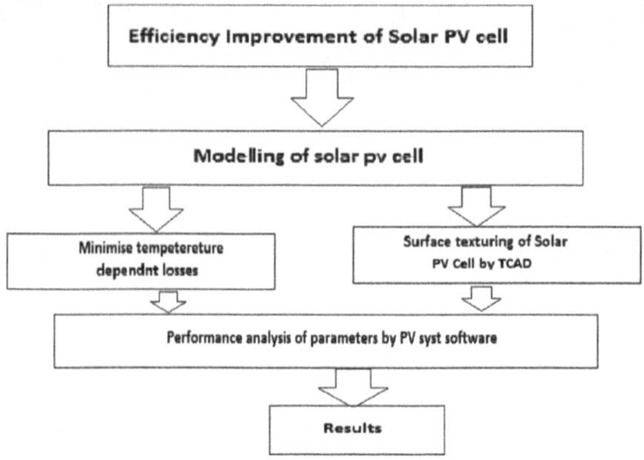

Fig. 1 Flowchart of work

change in solar light and temperature [2]. The procedure of the direct conversion of sunlight into power is called photovoltaic impact. Some semiconductor material like Si or Ge displays the property of photovoltaic impact which causes they retain the light photon and discharge electrons. A French physicist Edmund Becquerel in 1839 found that some material, when presented to light, delivered a little measure of electric flow. Later on, Albert Einstein depicted the photovoltaic impact and nature of light in 1905. Solar cells produce an electric field made by uncommonly treating a thin semiconductor wafer. When light strikes the solar cell's surface, electrons are ejected from the semiconductor material's particles. If the positive and negative sides of electrical channels switch, they structure electric circuit, and there will be a stream of electrons which result into electric flow [3, 4].

In the proposed work (as shown in Fig. 1),

- Mathematical modeling and design installation of solar photovoltaic arrangement with grid connection and losses have been analyzed for different manufacturing technologies. The performance characteristics of the thin-film module were clearly superior to those of the mono-Si and poly-Si modules for Jaipur. But these performance parameters have the drawback of the low spectral efficiency of thin-film modules.
- The second part of this research deals with the implementation of surface texturing to increase the active area of photovoltaic modules. Significant improvement was achieved by surface texturing of PV modules by inserting random pyramid structures [5, 6].

2 Intelligent Data Analytics for Performance Analysis of PV Module

2.1 *Data Analytics for Performance Analysis of PV Module*

There is much research work that has been done on the topic "intelligent data analytics for performance analysis of PV module." In this chapter, the author represents a comparative analysis of research work with different field publications through some graphs.

1. Scholarly works overtime:

This graph represents the publication data from the year of 1970–2021 in this field in different research sections (Fig. 2).

2. Field of study covered by most active institutes:

This graph represents the comparative study covered by most active institutes in different research fields [7] (Fig. 3).

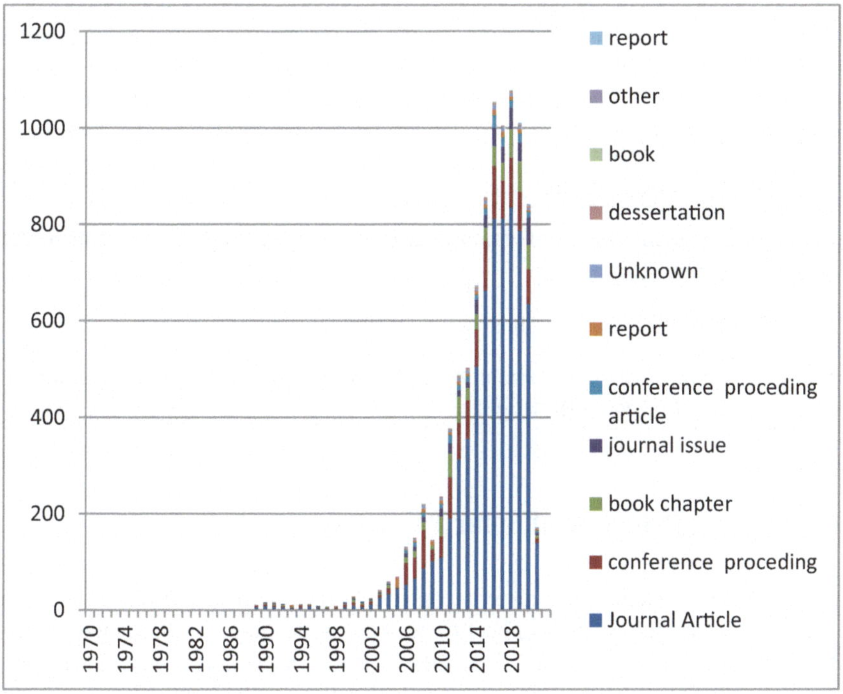

Fig. 2 Graph of scholarly works over time

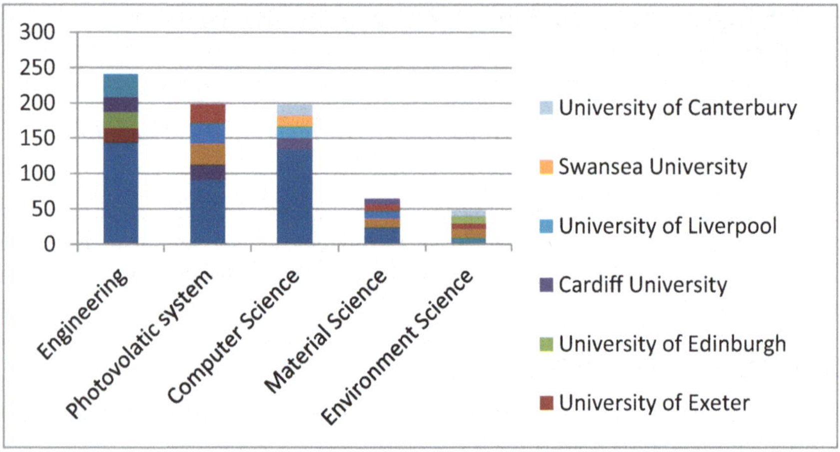

Fig. 3 Graph of the field of study covered by most active institutes

2.2 *Data Analysis for Application of Pyramid Surface Texturing Approach*

Much research work has been done on the topic "data analysis for application of pyramid surface texturing approach." In this chapter, the author represents a comparative analysis of research work with different field publications through some graphs.

1. Scholarly works overtime:

This graph represents the publication data from 1970 to 2021 in this field in the different research sections (Fig. 4).

2. Field of study covered by most active institutes:

This graph represents the comparative study covered by most active institutes in different fields of research [8] (Fig. 5).

3 Working Principle of Solar PV System

A solar PV cell is a device that directly converted sunlight into electricity. It operates on the basis of the photovoltaic effect [9]. Once the sun's shines reached at the surface of semiconductor material, which is doped with boron, phosphorus, etc. converts into electricity. A photovoltaic cell is a semiconductor diode. When a particle of sunlight strikes the surface of the solar cell, pair of electrons–holes are generated [10]. These

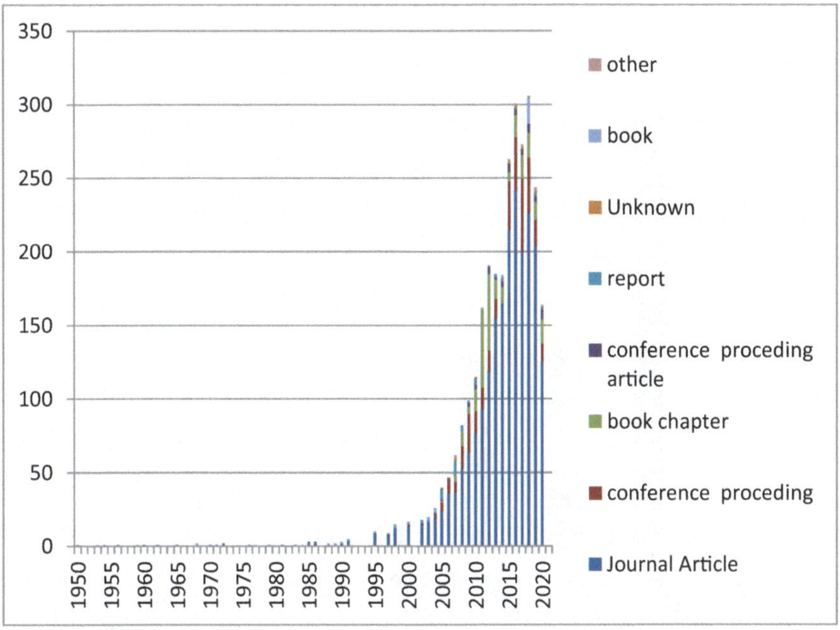

Fig. 4 Graph of scholarly works over time

electrons and holes are known as light-generated electrons and holes. These light-generated pairs can create electricity and start flowing current in the circuit. Figure 6 shows the working of solar photovoltaic cells [11].

4 Surface Texturing of Solar Cell

When the sun's rays are incident on the surface of the solar cell, it has been seen that a portion of energy reflected back and other portion is transmitted. Solar cell converts only transmitted portion of energy into electricity [12]. Generally, bare silicon reflects about 30% portion of incident rays and these reflected energies minimize the efficiency of solar PV cells. There are different techniques to increase the conversion rate of energy from sunlight into electricity. One of the techniques is to etch and text the front surfaces. This phenomenon is known as texturing the surface of the solar cell. This technique provides a chance to incident multiple times on the surface of solar cells that release more energy from sunlight [13, 14] (Fig. 7).

This technique increases the absorption rate of sunlight by increasing the effective area of thin-film solar PV cell [15]. This technique has also reduced the total cost of the same rating solar module. Since crystalline silicon is a semiconductor material that is not directly attached, the transmitted portion of sunlight is comparatively weak. To use lamination with single or multilayer exceeds the thickness of a few millimeter

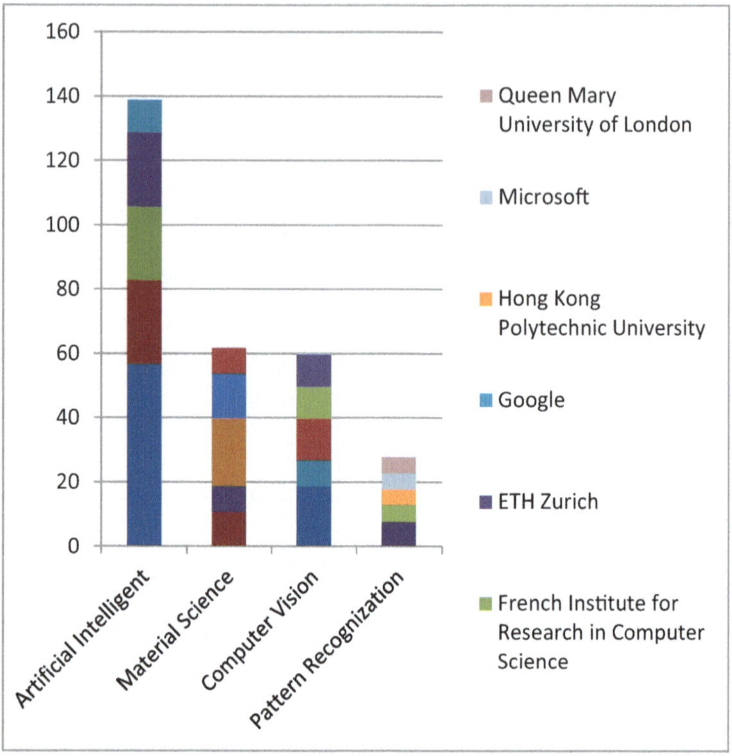

Fig. 5 Graph of the field of study covered by most active institutes

Fig. 6 PN junction solar cell

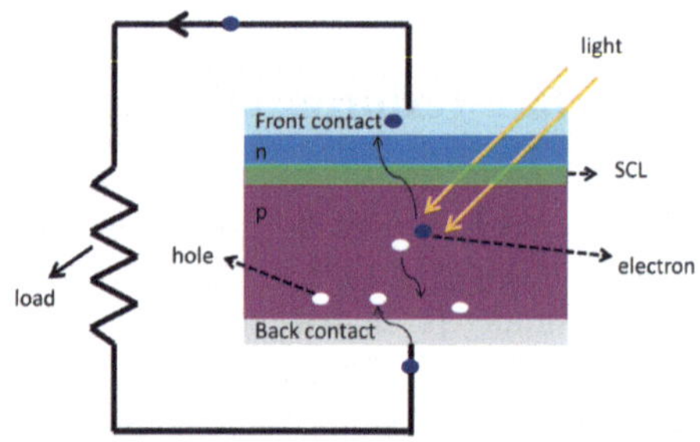

absorption rates of sunlight which increases the weight and cost of materials, and the quality recombines, resulting in resistance. The radiation efficiency is reduced. Textured surfaces can be made in a variety of ways. These methods are different for single-crystal silicon and polycrystalline silicon [16, 17] (Table 1).

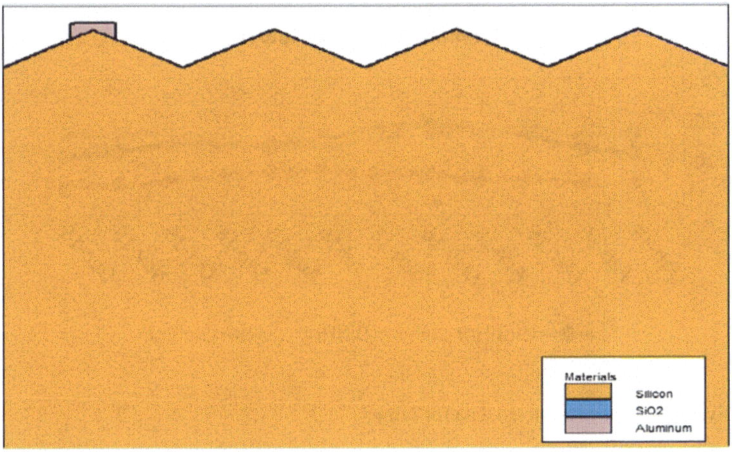

Fig. 7 Surface texturing on solar cell

Table 1 Data of surface texturing

Height of pyramid	5 μm
Thickness of Si	50 μm
Band gap (e-V)	1.12
Temperature	300 K

5 Simulation of Integrated PV System

Usually, PVsyst software is used for design and installation of solar PV arrangements for the 1 MWp system at Jaipur. Three different technologies have been selected for analyzing temperature-dependent losses. Comparative analysis system parameter and mathematical modeling of the system have been done using PVsyst and TCAD software. The results of the simulation process are more accurate as compared to other software [18, 19] (Fig. 8 and Table 2).

Using manufacturer datasheet and Meteonorm solar irradiation and geographical data, 1 MWp system was simulated for Jaipur. Comparative analysis of the active area of Jaipur is mentioned below. It can be noted that thin-film-based solar photovoltaic system required a comparatively larger area due to less spectral efficiency [20, 21] (Table 3).

Yield Simulation—Yield simulation is used to calculate hours per day of production for a power plant. Simulation of yield was done in PVsyst software for 1 MWp system to calculate a relative comparison of manufacturing technologies of solar PV system [22, 23] (Fig. 9).

Losses Simulation—Losses in solar photovoltaic systems have been classified into three clusters. Losses in photovoltaic systems are due to irradiance, losses due to

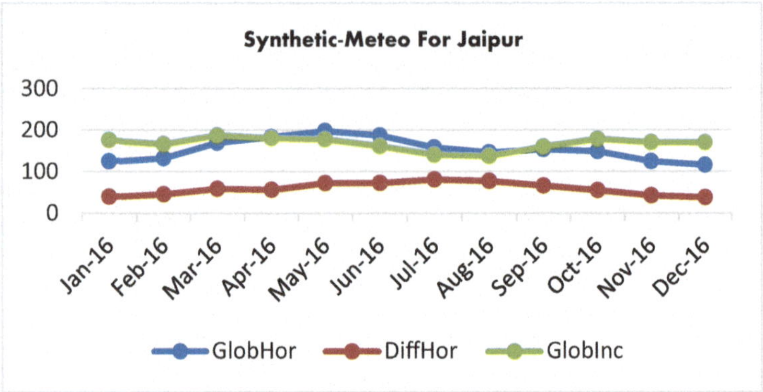

Fig. 8 Synthetic generated Meteo data for Jaipur

Table 2 Details of modules considered

Technology	Manufacturer	Model number
Poly-silicon	Tata Power	Tata-135
Thin film	Bosch Solar	BSM-EU1510
Mono-silicon	SunPower	BSM-EU 40123

Table 3 Comparative analysis of active area for Jaipur

Technology	Model number	Efficiency (%)	Active area of system (m^2)
Poly-silicon	Tata-135	14.6	6440
Mono-silicon	SunPower	17	5720
Thin film	BSM-EU1510	9.8	11,010

Fig. 9 Comparative analysis of yield for Jaipur

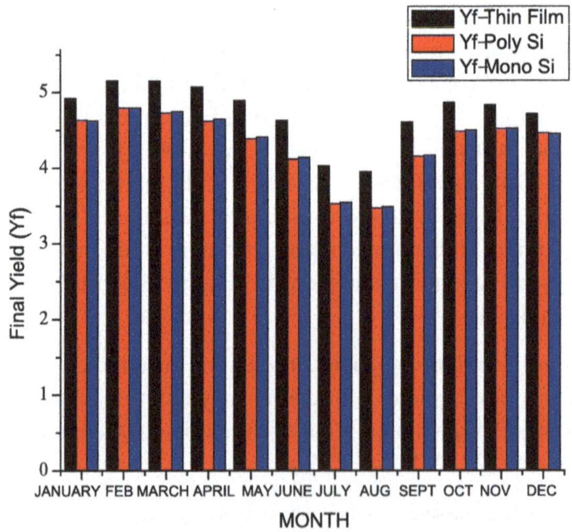

PV modules, inverter losses and losses due to soiling. Using PVsyst, temperature-dependent losses were calculated for different manufacturing technologies for the given system. It was evident that thin-film technologies perform better to counter losses due to the temperature and heating of solar panels (Fig. 10).

Performance Ratio—The final yield in hours/day divided by the reference yield is known as the performance ratio defined in hours/day of production from the solar photovoltaic system. The thin-film system's performance ratio was calculated using a PV system simulation and found to be comparatively better than poly-Si-based structures and mono-Si-based structures (Fig. 11).

Forecasting of Production—With PVsyst simulation tool, the annual production figures were computed for 1 MW system at Jaipur in terms of kWh/month. On

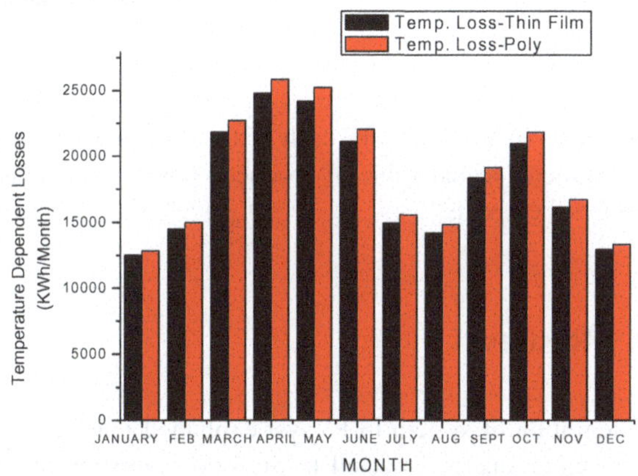

Fig. 10 Analysis of temperature-dependent losses

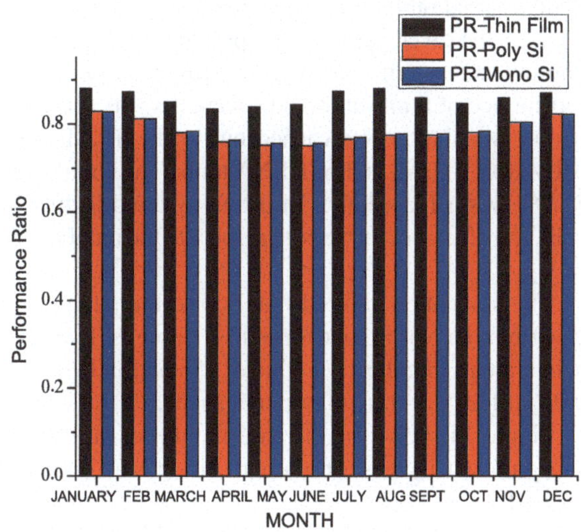

Fig. 11 Relative comparison of performance ratio

Fig. 12 Relative assessment of energy injected to grid

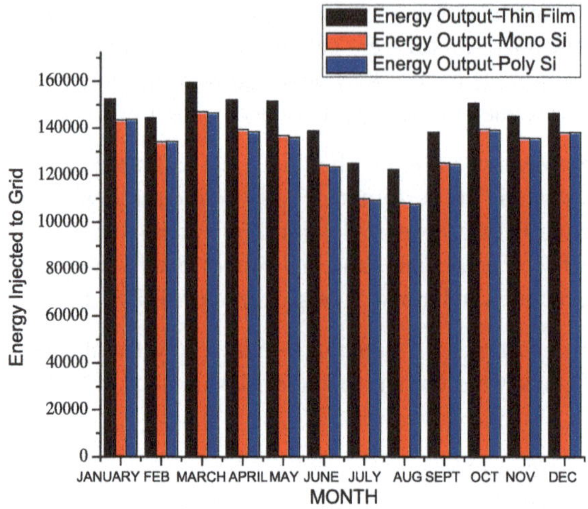

the basis of kWh/day, for each of the four production technologies, we also calculated the expected value of produced power from solar PV installations. Figure 12 shows the results.

6 Results

Integration of system in PVsyst explained that the overall operational parameters, i.e., yield, energy injected to the grid, temperature losses and performance ratio, was superior in the solar photovoltaic system based on thin films. Solar photovoltaic system based on thin films has low conversion efficiency, which costs more active area for the given system. The problem of countering the poor efficiency of solar photovoltaic systems based on thin film was addressed, and surface texturing of the top surface of thin-film-based solar cells was achieved in the TCAD tool. Pyramid texture simulation was carried out in the TCAD tool by designing solar cells in SILVACO ATLAS and ATHENA tools. The simulation showed that efficiency up to 11.7% was achieved after inserting pyramid texture in the top surface of the solar cell. The efficiency curve with respect to the number of pyramid textures is explained in Fig. 13.

The simulation showed that due to the addition of pyramid texture, the value of current increased four times in solar cells. Pyramid texture helped in increasing active area and collection of solar irradiation by multiple reflections in a solar cell with surface texturing of pyramid shape.

Figure 14 shows the increase in current in a solar cell with a pyramid texture. A significant increase in output current was observed in the simulated solar cell (Fig. 15).

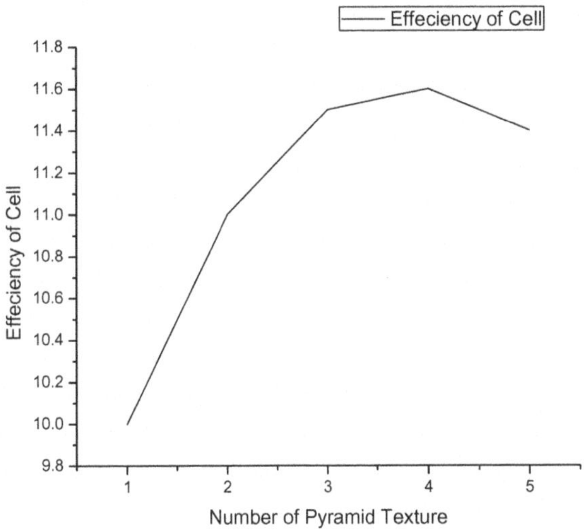

Fig. 13 Efficiency with respect to pyramid texture

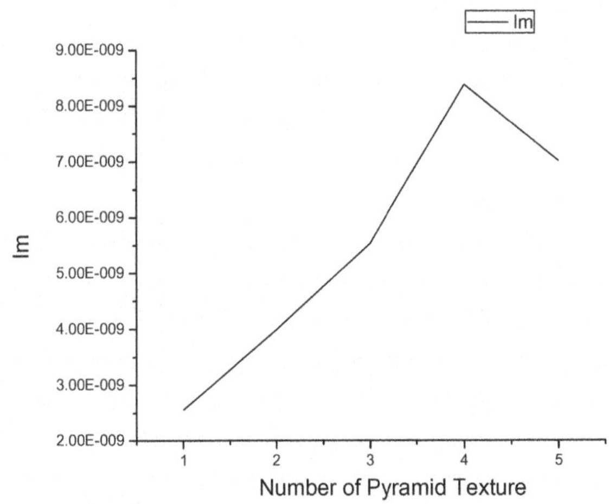

Fig. 14 Current with respect to pyramid texture

At constant voltage, current of pyramid textured solar cells increased significantly due to the increased active area of solar cells having pyramid texture. Simulation was carried out for a different number of the pyramids in order to calculate an optimum number of pyramids based on the area of the cell (Fig. 16).

Due to the increased efficiency, the relative area requirement for 1 MWp system for thin-film technology has shown a significant decrease. Figure 17 explains the close comparison between our conditions before and after surface texturing of modules. This decrease in the area was due to an increment achieved in the power output of panels. Figure 18 explains the relative comparison of power output by module before and after achieving surface texturing.

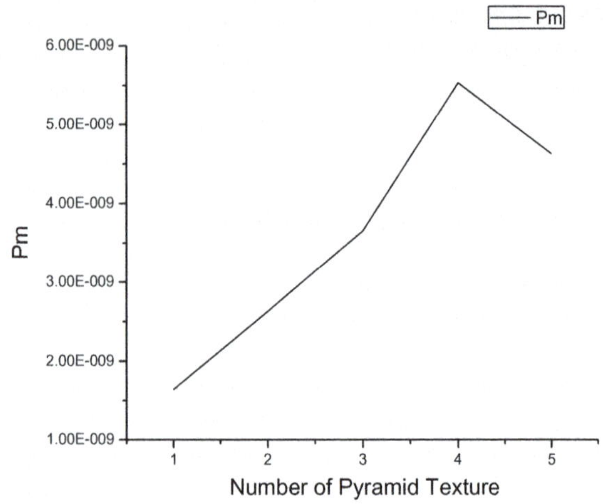

Fig. 15 Power with respect to pyramid texture

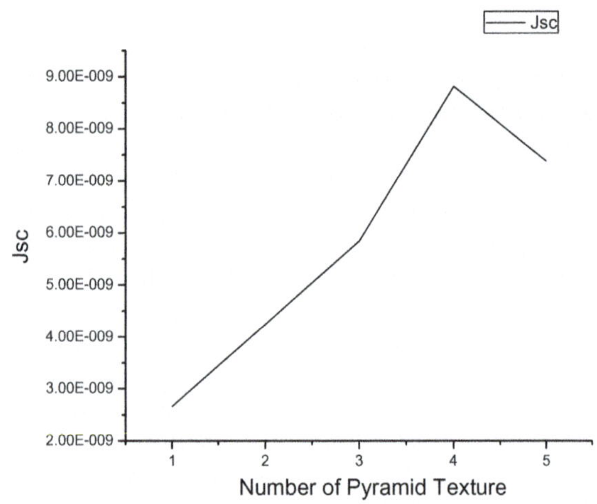

Fig. 16 Current density with respect to texture

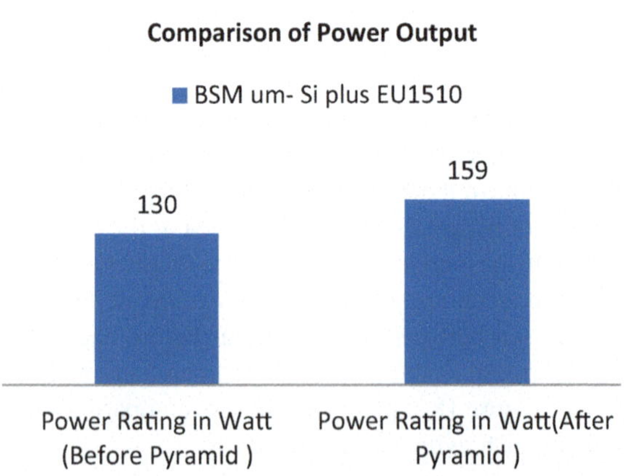

Fig. 17 Comparison of power output

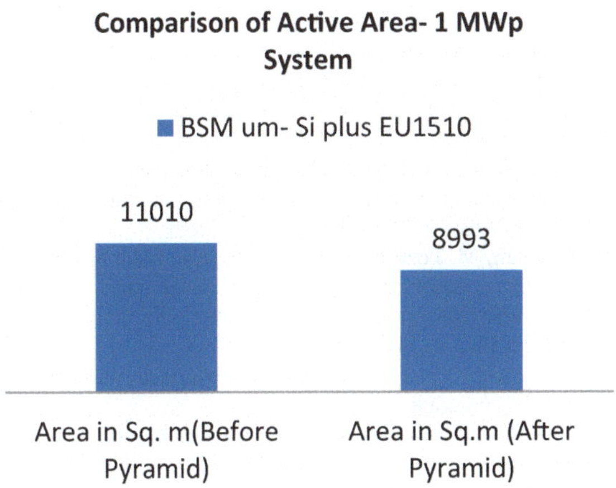

Fig. 18 Comparison of active area

7 Conclusion

In this chapter, the problem of temperature-dependent losses and low spectral efficiency of thin-film modules was addressed.

- The relative comparison of a performance parameter of 1 MWp grid-tied solar PV system was carried out on PVsyst software for different module manufacturing technologies for Jaipur, India.
- The analysis of results proved that thin-film technology performs best in terms of operating efficiency in high-temperature conditions. The value of energy injected into the grid was maximum for thin-film technology-based power plants.
- In the second part of the research, the problem of the low spectral efficiency of thin-film modules was addressed and for simulating the physical modeling of thin-film modules TCAD-based approach was used. Surface texturing on mesh structure of solar cell was achieved by SILVACO TCAD tool, and an increase in 3% efficiency was achieved.
- Analysis on PVsyst for improved module explained that there was significant increase in power and a significant reduction is required for the thin-film-based system after surface texturing.

References

1. A. Chaïb, M. Kesraoui, E. Kechadi, PV panel positioning using a robot manipulator, in *3rd International on Renewable and Sustainable Energy Conference (IRSEC)*. IEEE (2015)
2. S. Singh, M. Fozdar, A. Almutairi, S. Alyami, H. Malik, Strategic bidding in the presence of renewable sources for optimizing the profit of the power suppliers. IEEE Access. https://doi.org/10.1109/ACCESS.2021.3078288

3. A.K. Yadav, et al., Soft computing in condition monitoring and diagnostics of electrical and mechanical systems, in *Part of the Advances in Intelligent Systems and Computing*, vol. 1096, p. 496 (Springer Nature, 2020). https://doi.org/10.1007/978-981-15-1532-3
4. K. Masuko et al., Achievement of more than 25% conversion efficiency with crystalline silicon hetero-junction solar cell. IEEE J. Photovoltaic's **4**(6), 1433–1435 (2014)
5. S. Singh, M. Fozdar, Double-sided bidding strategy for power suppliers and large buyers with amalgamation of wind and solar based generation in a modern energy market. IET Gener. Transm. Distrib. **14**(6), 1031–1041 (2020)
6. S. Singh, M. Fozdar, Supplier's strategic bidding for profit maximization with solar power in a day-ahead market, in *Advances in Machine Learning and Computational Intelligence. Algorithms for Intelligent Systems* (Springer, Singapore, 2021). https://doi.org/10.1007/978-981-15-5243-4_74
7. https://www.lens.org/lens/search/scholar/analysis?q=Performance%20analysis%20of%20PV%20module
8. https://www.lens.org/lens/search/scholar/analysis?q=pyramid%20surface%20texturing%20Approach
9. S. Ahsan et al., Design and cost analysis of 1 kW photovoltaic system based on actual performance in Indian scenario. Perspect. Sci. **8**, 642–644 (2016)
10. A. Iqbal, et al., Meta Heuristic and Evolutionary Computation: Algorithms and Applications, in *Part of the Studies in Computational Intelligence*, vol. 916, p. 849 (Springer Nature, 2020), 978-981-15-7571-6. https://doi.org/10.1007/978-981-15-7571-6
11. O. Singh, S.K. Gupta, A review on recent Mppt techniques for photovoltaic system, in *2018 IEEMA Engineer Infinite Conference* (eTechNxT), 2018, pp. 1–6. https://doi.org/10.1109/ETECHNXT.2018.8385315
12. N. Kumar, P. Yadav, S.S. Chandel, Comparative analysis of four different solar photovoltaic technologies, in *2015 International Conference on Energy Economics and Environment (ICEEE)*, 2015, pp. 1–6. https://doi.org/10.1109/EnergyEconomics.2015.7235077
13. V. Sharma, A. Kumar, O.S. Sastry, S.S. Chandel, Performance assessment of different solar photovoltaic technologies under similar outdoor conditions. Energy **58**, 511–518 (2013). ISSN 0360-5442
14. I.J. Lee, U. Paik, J.G. Park, Solar cell implemented with silicon nanowires on pyramid-texture silicon surface. Solar Energy **91**, 256–262 (2013). ISSN 0038-092X
15. A. Iqbal, et al., Renewable Power for Sustainable Growth, in Part of the Lecture Notes in Electrical Engineering, vol. 723. Springer Nature, 2021, p. 805. 978-981-33-4080-0. https://doi.org/10.1007/978-981-33-4080-0
16. X. Tian, et al.: Pyramid size control and its effects on the performance of silicon heterojunction solar cells, in *China Semiconductor Technology International Conference*, Shanghai (2015)
17. S. Srivastava, et al., Applications of Artificial Intelligence Techniques in Engineering, vol. 1, in Part of the Advances in Intelligent Systems and Computing, vol. 698, p. 643 (Springer Nature, 2018). 978-981-13-1819-1. https://doi.org/10.1007/978-981-13-1819-1
18. W.S. Bennett, N.L. Niraj, B.F. Simeon, P.W. Thomas, Pyramidal surface textures for light trapping and antireflection in perovskite-on-silicon tandem solar cells. Opt. Express **22**, A1422–A1430 (2014)
19. E. Tarigan, Djuwari, F.D. Kartikasari, Techno-economic simulation of a grid-connected PV system design as specifically applied to residential in Surabaya, Indonesia. Energy Procedia **65**, 90–99 (2015)
20. Y.M. Irwan et al., Stand-alone photovoltaic (SAPV) system assessment using PVSYST software. Energy Procedia **79**, 596–603 (2015)
21. D. Kler, Y. Goswami, K.P.S. Rana, K. Vineet, A novel approach to parameter estimation of photovolataic using hybridized optimizer. J.: Energy Convers. Manage. **187**(1), 486–511 (2019)

22. A.A. Mona, E. Mahieddine, Performance analysis of rooftop PV systems in Abu Dhabi. Energy Procedia **42**, 689–697 (2013)
23. M.U. Hasan, S. Saha, M.E. Haque, Techno-economic analysis and performance evaluation of rooftop solar PV systems using actual data in Australian context, in *2020 IEEE Industry Applications Society Annual Meeting*, 2020, pp. 1–8. https://doi.org/10.1109/IAS44978.2020.9334784

Analysis and Modelling of Basic Wireless Power Transfer Compensation Topology: A Review

Bilal Alam, Noorul Islam, Ibra Subhan, and Mohammad Sarfraz

Abstract Wireless power transfer (WTP) emerged back in the early 1890s. This electrical technology captured major attention when a group of MIT researchers designed a functioning model of a WPT system that was efficiently able to transmit power and light a bulb of 60 W at a distance of 2 m. However, this concept was first put up by a Serbian scientist, Nikola Tesla, who introduced a system that could auspiciously transfer electrical power wirelessly. In this chapter, a detailed discussion on WPT and relevant methodologies and operation techniques is presented. A wireless power transfer system employs three distinct technologies: inductive, capacitive, and radiant. In this article, four fundamental compensation topologies of resonant WPTs are described, which are series–series (SS), parallel–parallel (PP), series–parallel (SP) and parallel–series (PS). Standard parameters and equation modelling have been discussed in this chapter. The world needs a wire-free system, and that is the hope of every other Tesla-influenced researcher. Every knowledge provided in the chapter is purely true to its deepest extent. The production and development of battery-powered devices face unparalleled technological difficulties because of the drawbacks of poor power density, exorbitant prices and bulky structure, etc. The wireless power transfer introduces a new energy acquisition method for electric vehicles as a novel energy pattern. This chapter also summarizes WPT approaches focusing on operating structures, mathematical and technological challenges focusing on the system for WPT.

Keywords Wireless power transfer · Tesla · Topologies · Electric vehicles · Plug-in charging

B. Alam · N. Islam · I. Subhan · M. Sarfraz (✉)
Department of Electrical Engineering, ZHCET, AMU, Aligarh 202001, India
e-mail: msarfraz@zhcet.ac.in

H. Malik et al. (eds.), *Intelligent Data Analytics for Power and Energy Systems*,
Lecture Notes in Electrical Engineering 802,
https://doi.org/10.1007/978-981-16-6081-8_25

1 Introduction

In the twentieth century, as we can see, pollution is increased significantly due to exhaust fumes (petrol, CNG, diesel and so many more) from conventional vehicles (CVs). The electric vehicle is a smart choice to strengthen our environment [1–3]. EVs are not a novel concept, having been discovered in the nineteenth century. Due to their moderate driving range, EVs are not widespread. But the idea of affordable EVs has not disappeared. Limited driving range, cost of EVs, efficiency has been an important obstacle that bound the acceptance of EVs on a broad scale. Moreover, with the implementation of efficient battery packs (Li-ion), rapid charging systems and relatively low cost, EVs can take over conventional vehicles soon [4, 5].

The prominence of EVs is boosting day by day due to certain advantages over conventional vehicles (CV). (1) EVs consume lower energy: EVs are more efficient than conventional vehicles. For EVs, approximately 86% of the electricity is needed to recharge the battery which powers the car motor. Conversely, only 20% of the fuel potential is absorbed as output power in the engine in CVs [6]. (2) Reduced fossil fuel reliance: EVs can reduce the fuel cost (low cost of electricity) compared to conventional vehicles' fuels [7]. (3) Environment-friendly: the use of EVs reduces fossil fuel consumption and results in a pollution-free environment [8]. (4) Recyclable: every battery is recyclable to minimize pollution while is also assisting us in creating a more environmentally sustainable climate. There are also drawbacks like wastage of fossil fuels, long refuelling time, more weight and space [9, 10].

This review article represents the background of the wireless power transfer system under Sect. 2. WPT has three distinct configurations for the wireless transmission of the energy; these configurations are inductive, capacitive and resonant. In the present scenario, the main concern is to achieve a larger driving range using the most efficient batteries, eventually improving the efficiency of EVs. In Sect. 3, Resonant WPT topologies are discussed under four categories based on series. A parallel combination of capacitors connections to transmitter and receiver under this section analysis of different topologies is outlined. Furthermore, standard parameters in wireless power transfer operation are described in Sect. 4.

2 Intelligent Data Analytic Related to Wireless Power Transfer

The wireless power transfer (WPT) system comprises two components: a transmitter and a receiver. These two parts are electrically insulated. The transmitter is connected to the mains of the supply, whereas the receiver is connected to the EV battery. Figure 1 consists of a basic circuit structure of the WPT. The transmitter and receiver consist of a coupling element as well as a power conversion device. The transmitter of the WPT consist of a coupling device that helps to generate an alternating field (maybe electric, magnetic and electromagnetic) [11, 12]. On the receiver side, the

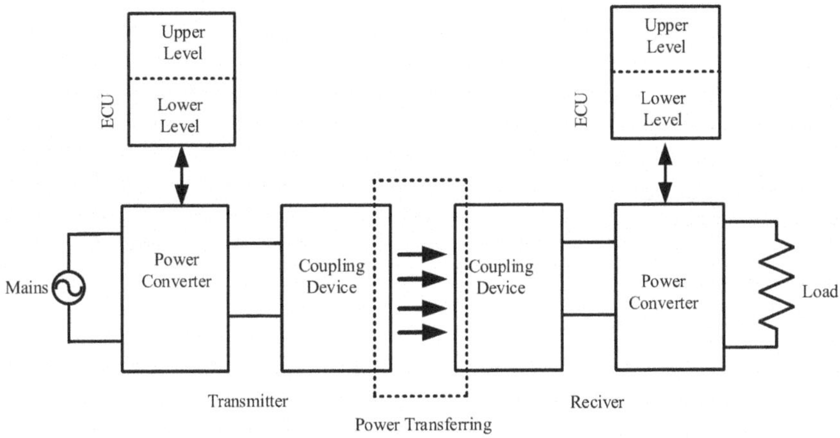

Fig. 1 Wireless power transfer structure

connection device is embedded by the transmitting device, generating an alternating field and functioning as a pickup, take the energy transported by the field [13].

The transmitter power conversion comprises a front-end power factor rectifier powered by a DC–DC converter (inverter) that supplies an AC voltage to the transmitter. The receiver's power conversion element absorbs energy from the receiving device and acts as a rectifier to power the battery. WPT consists of two electronic control units (ECU) that help control the WPT operation, i.e. upper and lower levels. The top levels manage the power segment, whereas the bottom level performs the power conversion [14].

2.1 WPTs Technology

A WPT structure comprises three different technologies inductive, capacitive and radiant. These topologies are in the field of the field concerned and are used for electricity transmission [15]. The common similarity between all three technologies is that the Tx device energized the surrounding space. The respective energy in IPT and CPT is contained in a given device volume.

$$W_e = \frac{1}{2}\varepsilon_0 E^2$$

$$W_m = \frac{1}{2}\mu_0 H^2$$

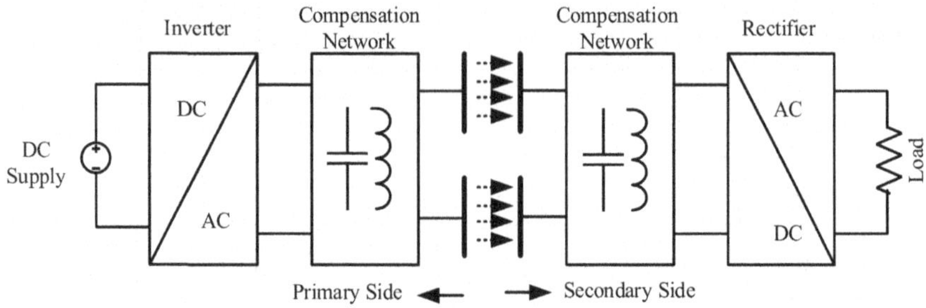

Fig. 2 Typical structure of CPT system

where E = Intensity of the electric field,

H = Intensity of the magnetic field.

μ_0 = Permeability of free space, ε_0 = Permittivity of free space.

2.1.1 Capacitive WPTs

Among all the PTS, presently, the CPT technology is being studied widely. The typical structure of the CPT is delineated in Fig. 2. The primary circuits of the CPT consist of a power electronics converter and the corresponding compensation network. Generally, the CPT system consists of four plates and is used to build a capacitor coupler. The primary and secondary (Tx and Rx) are connected with the metal plates. Two plates are used as a power transmitter on the transmitter side, and the remaining are used as a power receiver. The benefits of CPT technology lie in its low system cost, less weight and lower flow rate loss in surrounding metals. Still, it is only suitable for small-power, narrow distance applications such as IC, lightning, mobile charging and many more applications [14].

2.1.2 Inductive WPTs

Inductive WPTs operate on the concept of Faraday's law of electromagnetic induction. It consists of two sections, one connected to the primary is known as the transmitter (Tx), and it is connected to the main supply. The other section of the system is connected to the secondary, known as the receiver (Rx), which is connected to the load. When the electrical current flows through the wire. In this WPT system, an alternating voltage delivered to the primary Tx coil results in a different terminal voltage from the counterpart at about the equivalent frequency.

1. ***Operation of IPT***

The circuit diagram of an IPTs system is shown in Fig. 3. L_{PRI} and L_{SEC} are the self-inductances of the inductive coupling of the transmitter and receiver coil. Whereas K

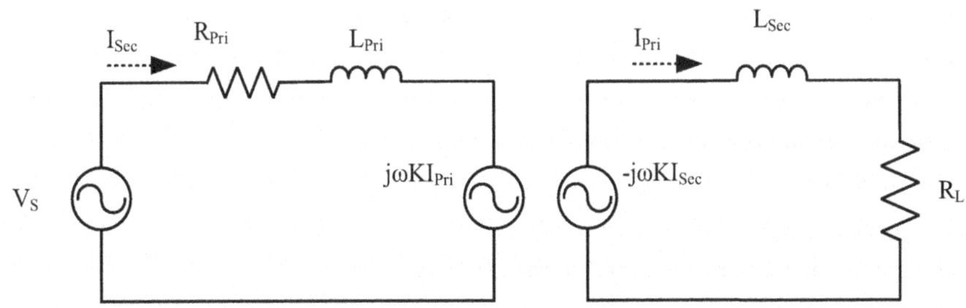

Fig. 3 Circuital scheme of IWPTs

represents the mutual inductance, R_{PRI} and R_{LXZ} resist the primary, secondary load resistance and load, respectively. V_S is the voltage that is inserted into the receiver side, and V_L is the voltage across the load side.

Applying KVL, voltage equations for transmitting and receiving side,

$$V_S = Z_T I_{\mathrm{Sec}} + j\omega K I_{\mathrm{pri}} \quad (1)$$

$$-j\omega K I_{\mathrm{Sec}} = Z_R I_{\mathrm{pri}} \quad (2)$$

Z_T and Z_R are the total impedance of the transmitter (Tx) and receiver (Rx) side, given by

$$Z_T = R_{\mathrm{Pri}} + j\omega L_{\mathrm{Pri}} \quad (3)$$

$$Z_R = R_L + j\omega L_{\mathrm{Sec}} \quad (4)$$

ω represents the angular frequency of the main supply V_S from Eqs. 3 and 4, the transmitting and receiving current are as follows

$$I_{\mathrm{Sec}} = \frac{V_S.Z_R}{Z_T.Z_R + \omega^2 K^2} \quad (5)$$

$$I_{\mathrm{Pri}} = -\frac{V_S.j\omega K}{Z_T.Z_R + \omega^2 K^2} \quad (6)$$

From Eqs. 4 and 5, we can find the complex power S_S and the power loss P_L across the load.

$$S_S = V_S.I_{\mathrm{Sec}} = \frac{|V_S^2|}{|Z_T.Z_R + \omega^2 K^2|^2} Z_R^*\left(Z_T.Z_R + \omega^2 K^2\right) \quad (7)$$

$$P_L = R_L |I_{\text{Sec}}|^2 = \frac{R_L.V_S^2\omega^2 K^2}{\left|Z_T.Z_R + \omega^2 K^2\right|^2} \quad (8)$$

Advantage and disadvantage of IPT System.
Advantages

- Quite easy method of power transfer.
- Efficiency is high as compared to the other systems.

Disadvantage

- Power transfer is done only for a small distance (few millimetres)
- Coupling factor and mutual inductance values are too low
- Due to resistive losses, heating of the IPT system happens.

2.1.3 Resonant WPTs

In a resonant WPT system, to reduce the level of current and voltage required to power the WPT device, a capacitor is connected to the compensation network. At resonant conditions, the shrinking magnetic field of the inductor induces an electrical current in its windings, as well as the capacitor is charged. The discharge current produces an alternating current in the inductor that provides the magnetic field [16].

The resonant coupling works on the resonance concept between the transmitter and receiver coils by the massachusetts institute of technology (MIT) the resonant coupling is also known as resonant inductive coupling. There are many topologies if we connect the capacitor in a series–parallel combination to the Tx and Rx. Each configuration has benefits and drawbacks, which are explained in the later section.

By inserting the capacitor in series, efficiency can be increased. For example, in Fig. 4, the resonant coupling WPT system's circuit schematic is illustrated. Wherein the capacitor (C_{Sec}) is inserted in series with the inductor in receiving circuit. Due to the insertion of the capacitor in the receiver, part impedance is given by

$$Z_R = R_L + j\left(\omega L_{\text{Sec}} - \frac{1}{\omega C_{\text{Sec}}}\right) \quad (9)$$

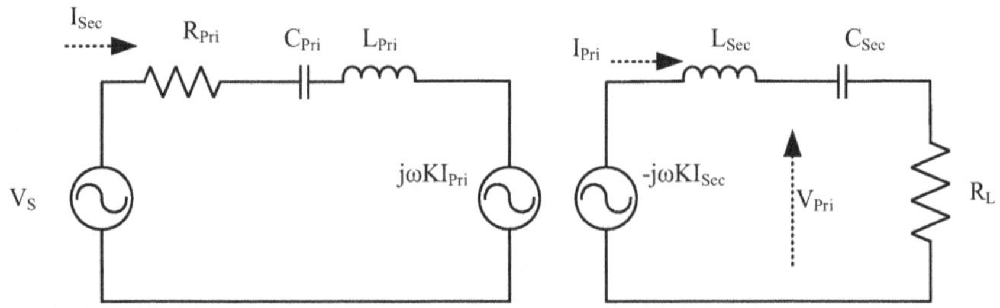

Fig. 4 Circuit diagram of IWPT system

C_{Pri} is the capacitor that is connected to the transmitter side. The effect on the transmitter impedance section is given by.

$$Z_T = R_{\text{Pri}} + j\left(\omega L_{\text{Pri}} - \frac{1}{\omega C_{\text{Pri}}}\right) \quad (10)$$

Under the resonant condition,

$$\omega^2 C_{\text{Pri}} = \frac{1}{L_{\text{Pri}}} \quad (11)$$

$$\omega^2 C_{\text{Sec}} = \frac{1}{L_{\text{Sec}}} \quad (12)$$

At this resonant frequency, the Tx and Rx section impedance is minimum and equal to

$$Z_T = R_{\text{Pri}} \quad (13)$$

$$Z_R = R_L \quad (14)$$

3 Resonant Topologies of WPTs

The transmitter (Tx) and receiver (Rx) are separated by a certain distance in the WPT system. The critical factors are mutual inductance and coupling coefficient that affect the WPT capability. Compensation for leaking inductance is required as mutual inductance increases. With the help of a resonant circuit, the performance of the WPT system is enhanced. For compensating circuit, to enhance the performance of the wireless power system, capacitors are connected in series and parallel to the transmitter (Tx) and receiver (Rx). There are four basic fundamental topologies of resonant WPTs, and these are: aeries–series (SS), parallel–parallel (PP), series–parallel (SP) and the last one is parallel–series (PS) [17].

In WPTs, high-value leakage inductance is linked with both Tx and Rx coil. A large amount of primary current is required for better power transfer. Losses are generated due to the large current flowing in the primary coils, which results from poor efficiency. Some topologies are implemented to enhance the WPT system's efficiency [7, 18].

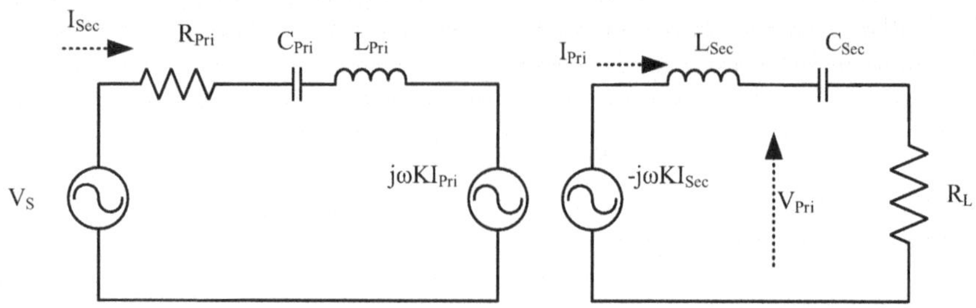

Fig. 5 Series–series (SS) topologies

3.1 Analysis of Compensation Topologies Structure

One of the most challenging tasks is determining topologies; before selecting the compensation topologies, we need to know the equational modelling, advantage and disadvantage. After choosing the right topology for the specific work to meet an application-specific requirement must be selected. The topologies of SS, SP, PS and PP are clarified in resect section [19].

3.1.1 Series–Series (SS) Topologies

Figure 5 delineates the circuit diagram of the SS topologies. The equation modelling is discussed in Sect. 2. Equations (9)–(14) express the figure of merit.

The efficiency of the RCWPT can be obtained by Eqs. (13) and (14).

$$\eta_{R\text{Sec}} = \frac{\omega^2 K^2}{\omega^2 K^2 + R_L R_{\text{Pri}}} \tag{15}$$

where R_{Sec} stands for resonant for Rx side.

3.1.2 Series–Parallel (SP) Topologies

Figure 6 depicts the SP topology's schematic diagram, where the Tx having a capacitor connected in series, whereas in Rx, a capacitor is connected in parallel. Some basic calculations of the SP topology.

$$Z_{T,\text{SP}} = R_{\text{Pri}} + j\omega L_{\text{Pri}} - j\frac{1}{\omega C_{\text{pri}}} \tag{16}$$

$$Z_{R,\text{SP}} = j\omega L_{\text{Sec}} + \frac{1}{\frac{1}{R_L} + j\omega C_R} \tag{17}$$

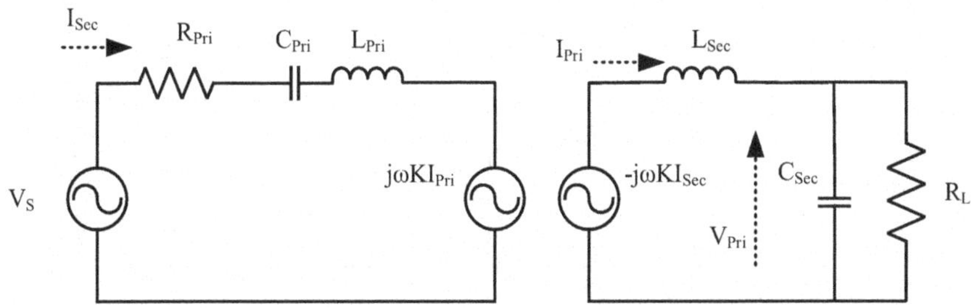

Fig. 6 Series–parallel (SP) topology

The Rx side impedance is entirely resistive and is determined by,

$$L_{\mathrm{Sec,SP}} = \frac{C_{\mathrm{Sec}} R_L^2}{1 + \omega^2 C_{\mathrm{Sec}}^2 R_L^2} \tag{18}$$

By Eqs. (16) and (17), we get $Z_{R,\mathrm{SP}}$

$$Z_{R,\mathrm{SP}} = \frac{R_L}{1 + \omega^2 C_{\mathrm{Sec}}^2 R_L^2} + j\frac{\omega L_{\mathrm{Sec}}}{1 + \omega^2 C_{\mathrm{Sec}}^2 R_L^2} \tag{19}$$

Under Eq. (18), we get $Z_{R,\mathrm{SP}}$,

$$Z_{R,\mathrm{SP}} = \frac{R_L}{1 + \omega^2 C_{\mathrm{Sec}}^2 R_L^2} \tag{20}$$

Equation (18) is directly proportional to the load (R_L) in some applications where the load is not constant. As in the case of electric vehicle charging, reaching Eq. (20) means that (i) the supply frequency needs to be continuously turned to preserve the resonance state of the Rx section. (ii) If the resonant state in the receiving side is retained. This cannot be achieved on the Tx side, where the resonant frequency is set and determined by the first Eq. (11) and (12).

The quality factor can be achieved by

$$Q = \frac{1}{\omega C_{\mathrm{Sec}}} \tag{21}$$

3.1.3 Parallel–Series (SP) Topology

Figure 7 shows the topology of PS by using the theorem of superposition, Tx current $I_{T,\mathrm{PS},}$ and the source current $I_{S;\mathrm{PS}}$ can be represented as

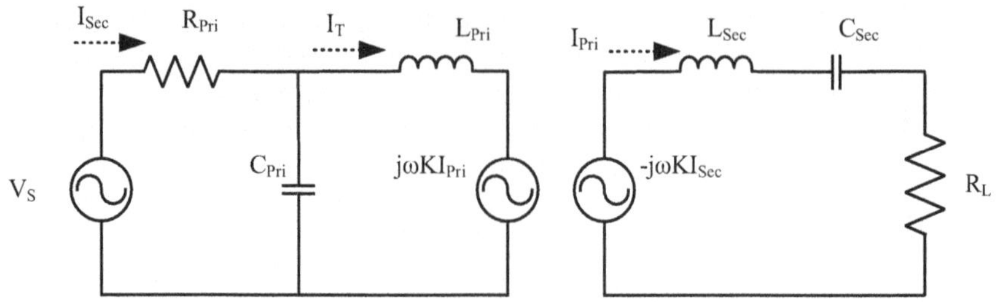

Fig. 7 Parallel–series (PS) topology

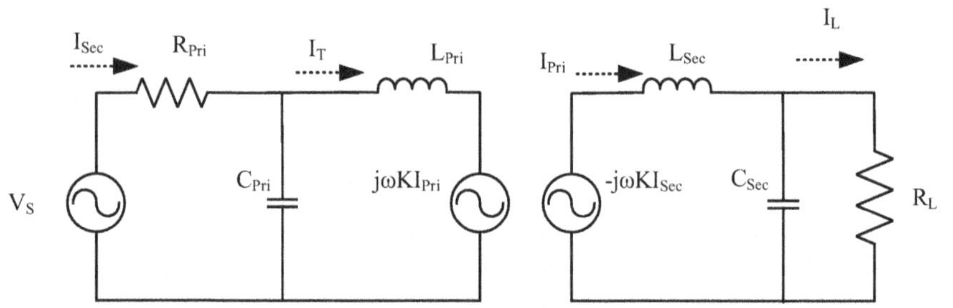

Fig. 8 Parallel–series (PP) topology

$$I_{T,\mathrm{PS}} = \frac{-j\left(\frac{V_s}{\omega L_{\mathrm{Pri}}}\right)(R_{\mathrm{Pri}} + j\omega L_{\mathrm{Pri}})Z_R}{\left[Z_R R_{\mathrm{Pri}} + \omega^2 K^2\left(1 + \omega^2 C_T^2 R_{\mathrm{Pri}}^2\right)\right] + j\omega L_{\mathrm{pri}}} \tag{22}$$

$$I_{\mathrm{Pri,SP}} = \frac{j(\omega M)^2\left(\frac{V_S}{\omega L_{\mathrm{Pri}}}\right)(R_{\mathrm{Pri}} + j\omega L_{\mathrm{Pri}})\left(1 - j\omega C_{\mathrm{Pri}} R_{\mathrm{pri}}\right)}{\left[Z_R R_{\mathrm{Pri}} + \omega^2 K^2\left(1 + \omega^2 C_T^2 R_{\mathrm{Pri}}^2\right)\right] + j\omega L_{\mathrm{pri}}} \tag{23}$$

Figure 7 Parallel–series (PS) topology and received current $I_{\mathrm{Pri,PS}}$ is expressed as

$$I_{\mathrm{Pri,PS}} = \frac{-(\omega K)\left(\frac{V_S}{\omega L_T}\right)(R_{\mathrm{Pri}} + j\omega L_{\mathrm{Pri}})}{\left[Z_R R_S + \omega^2 K^2\left(1 + \omega^2 C_T^2 R_S^2\right)\right] + j\omega L_{\mathrm{Pri}}} \tag{24}$$

$$\eta_{\mathrm{PS}} = \frac{\frac{Q_{\mathrm{Pri}}+1}{Q_T}}{Q_T + \left(\frac{1}{Q_T}\right)\left(1 + k^2 Q_{\mathrm{Sec}}\left(1 + \frac{Q_T}{Q_T}\right)\right)} \tag{25}$$

3.1.4 Parallel–Parallel (PP) Topology

PP topology consists of a capacitor on both Tx and Rx in parallel as seen in Fig. 8. I_L is given by.

$$I_L = \frac{I_R}{|1 + \omega C_{\text{Sec}} R_L|} \tag{26}$$

The efficiency of the PP topology can be calculated by

$$\eta_{\text{PP}} = \frac{\frac{Q_T^2+1)}{Q_T}}{Q_T + \left(\frac{1}{Q_T}\right)\left(1 + \frac{k^2\left(Q_R^2+1\right)^2}{Q_T \cdot Q_R}\right)} \tag{27}$$

where Q_T and Q_R are a quality factor of Tx and Rx and k is the coupling factor.

$$Q_T = \frac{\omega L_{\text{Pri}}}{R_{\text{Pri}}} \tag{28}$$

$$Q_R = \frac{\omega L_{\text{Sec}}}{R_L} \tag{29}$$

$$k = \frac{K}{\sqrt{L_{\text{Pri}} . L_{\text{Sec}}}} \tag{30}$$

4 Standards in Wireless Power Transfer Operation

In the twentieth century, there has been a rise in interest in the energy transfer between coils at a variable voltage at some distance (few mm, cm distance). The following companies are focusing on developing the WPTs system with EVs for a longer driving range.

4.1 Firms Working for WPTs

4.1.1 KAIST and OLEV Technology

KAIST (Korean Advanced Institute of Science and Technology) established a system to pass power to buses wirelessly. In [20], two parallel conductors are switched on when the buses are driving in the segment. This protects the people around the power line with a magnetic field when the bus is static. And due to the static and driving

Table 1 Specification of EV: KAIST

Quantity	Specification
System type	(EV) single-phase
Input current	200 A
Operating frequency	20 kHz
Efficiency	80%
Power transmitting distance	35 cm

mode, the efficiency is increased. The receiver is connected to the bus (located 20 cm above the road), and the transmitter is placed under the surface (which is 15 cm) of the road, making an effective distance of 35 cm. The range is attributed to picking up modules must be wide, and many modules must be required to make it possible to pass 100 Kw. The detail of KAIST is given in Table 1.

4.1.2 Qualcomm Halo

Qualcomm Halo WEVC (Wireless electric vehicle charging) software was developed over many years of effort and hard work. From the University of Auckland, A proven leader in wireless power transfer and the cycle of technology and economic growth [3, 21]. The technical details are shown in Table 2.

Maker Status:

- "Qualcomm and Delta Motorsport Announce Integration of Wireless Electric Vehicle Charging into Delta E-4 Coupé Electric Vehicles for London Trial" [22].
- "Qualcomm Becomes Founding Technology Partner of FIA Formula Championship" [22].

Table 2 Specifications of Qualcomm Halo

Quantity	Specification
Input voltage	240 V AC
Output voltage	300 V DC
Output power	3Kw
Coil pad separation	(180 ± 30) mm
Misalignment	$\pm$ 150 mm
Efficiency	85%
Parameters of coil pad	$(800 \times 400 \times 30)$ mm
Output specification	(250–300) Volt in 10 steps
Operated frequency	20 kHz

Table 3 Specifications of WiTricity Corp (WiT-3300)

Quantity	Specifications
Frequency	145 kHz
Tolerance to longitudinal positional	±20 cm *X*-axis ±10 cm on *Y*-axis
Output power	300 watts-3.3 kW (DC)
Output voltage	(350–400) Volt-DC at 3.3 kW, 18 cm displacement
Enclosure for source module	(50 × 50 × 3.75) cm; 12.5 kg
Enclosure for capture module	(50 × 50 × 3.75) cm; 12.5 kg
Assembly RF amplifier	(22 × 33 × 13) cm; 4.2 kg
Assembly of on-vehicle rectifier	(20 × 28 × 7) cm; 3.6 kg

4.1.3 WiTricity Corp

WiTricity Corp is one of the reputed manufactured MNCs in the world. The company was founded in 2007 to sell the technologies developed in 2006 by a team of MIT physicists. The wireless parking/charging system for electric trucks was the WIT-3300 implementation package. The following are technical requirements shown in Table 3 [23, 24].

5 Conclusion

WPT has proven itself as a possible technology that could be utilized to solve power cables (plug-in charging) in various applications. However, it still needs to overcome many technical challenges to become one of the next extraordinary technology. This review article gives a general background of the WPT technologies (capacitive, inductive and resonant). Equation modelling of the basic four compensation topology has been discussed; the technical aspects discussed are efficiency and power quality. From a sustainable perspective, the wireless power transfer system are faster, more dependable, have a minimal environmental impact and safer than plug-in charging. Standards and EMF restriction could reduce the effect on humans.

References

1. F. Musavi, M. Edington, W. Eberle, Wireless power transfer: a survey of EV battery charging technologies, in *2012 IEEE Energy Convers. Congr. Expo. ECCE 2012*, 2012, pp. 1804–1810. https://doi.org/10.1109/ECCE.2012.6342593
2. D. Patil, M.K. McDonough, J.M. Miller, B. Fahimi, P.T. Balsara, Wireless power transfer for vehicular applications: overview and challenges. IEEE Trans. Transp. Electrif. **4**(1), 3–37 (2017). https://doi.org/10.1109/TTE.2017.2780627
3. A. Ahmad, M.S. Alam, R. Chabaan, A comprehensive review of wireless charging technologies for electric vehicles. IEEE Trans. Transp. Electrif. **4**(1), 38–63 (2017). https://doi.org/10.1109/TTE.2017.2771619
4. J.A.A. Hasmat Malik, N. Fatema, *AI and Machine Learning Paradigms for Health Monitoring System: Intelligent Data Analytics* (2021)
5. S.S. Mohammed, K. Ramasamy, T. Shanmuganantham, Wireless power transmission—a next generation power transmission system. Int. J. Comput. Appl. **1**(13), 102–105 (2010). https://doi.org/10.5120/274-434
6. P.L. Kempker, N.M. van Dijk, W. Scheinhardt, H. Van Den Berg, J. Hurink, Smart charging of electric vehicles. Int. Ser. Oper. Res. Manag. Sci. **248**, 387–404 (2017). https://doi.org/10.1007/978-3-319-47766-4_14
7. J.L. Villa, J. Sallán, J.F. Sanz Osorio, A. Llombart, High-misalignment tolerant compensation topology for ICPT systems. IEEE Trans. Ind. Electron. **59**(2), 945–951 (2012). https://doi.org/10.1109/TIE.2011.2161055
8. K. Aditya, S.S. Williamson, Comparative study of series-series and series-parallel topology for long track EV charging application, in *2014 IEEE Transp. Electrif. Conf. Expo Components, Syst. Power Electron.—From Technol. to Bus. Public Policy, ITEC 2014*, 2014. https://doi.org/10.1109/itec.2014.6861793
9. H. Malik, A. Iqbal, P. Joshi, S. Agrawal, I.B. Farhad, *Metaheuristic and Evolutionary Computation: Algorithms and Applications*, vol. 916, 2021
10. C.C. Mi, G. Buja, S.Y. Choi, C.T. Rim, Modern advances in wireless power transfer systems for roadway powered electric vehicles. IEEE Trans. Ind. Electron. **63**(10), 6533–6545 (2016). https://doi.org/10.1109/TIE.2016.2574993
11. S.B. Atif Iqbal, H. Malik, A. Riyaz, K. Abdellah, *Renewable Power for Sustainable Growth*, 2020
12. A. Ahmad, R.C.H. Motors, Comparative Analysis of Power Pad for Wireless Charging of Electric Vehicles, 2019, pp. 1–7. https://doi.org/10.4271/2019-01-0865.Abstract
13. C.L.A.P.H.M. Budhia, A generalized coupling model for Capacitive Power Transfer systems. https://doi.org/10.1109/IECON.2010.5675014
14. X.C.L.C.Z. Juan, Analysis of power transfer characteristic of capacitive power transfer system and inductively coupled power transfer system. https://doi.org/10.1109/MEC.2011.6025703
15. M. Bertoluzzo, M.K. Naik, G. Buja, Preliminary investigation on contactless energy transfer for electric vehicle battery recharging, in *2012 IEEE 7th Int. Conf. Ind. Inf. Syst. ICIIS 2012*, 2012. https://doi.org/10.1109/ICIInfS.2012.6304787
16. W.C. Brown, The history of power transmission by radio waves. IEEE Trans. Microw. Theory Tech. **32**(9), 1230–1242 (1984). https://doi.org/10.1109/TMTT.1984.1132833
17. R.J. Parise, Model To Predict Performance of All Electric, no. 20011, 2002, pp. 1–6
18. N. Jamal, S. Saat, Y. Yusmarnita, T. Zaid, M.S.M. Isa, A.A.M. Isa, Investigations on capacitor compensation topologies effects of different inductive coupling links configurations. Int. J. Power Electron. Drive Syst. **6**(2), 274–281 (2015). https://doi.org/10.11591/ijpeds.v6.i2.pp274-281
19. G. B., K.N.M.M. Bertoluzzo, Characteristic Evaluation of Wireless Battery Chargers for Electric Vehicle, 2013, pp. 1–6
20. S. Anthony, World's First Road-Powered Electric Vehicle Network Switches on in South Korea—ExtremeTech

21. A. Ahmad, M.S. Alam, Vatsala, Y. Varshney, R.H. Khan, A state of the Art review on Wireless Power Transfer a step towards sustainable mobility, 2018. https://doi.org/10.1109/INDICON.2017.8488032
22. A.M. Ahmed, O.O. Khalifa, Wireless power transfer for electric vehicle charging. AIP Conf. Proc. **2306** (2020). https://doi.org/10.1063/5.0032383
23. http://www.witricity.com/pages/ev-charging-system
24. M.S.A. Khan, Wajahat, A. Ahmad, F. Ahmad, A Comprehensive Review of Fast Charging Infrastructure for Electric Vehicles. https://doi.org/10.1080/23080477.2018.1437323

Comprehensive Data Analysis of Power and Energy System: A Review of Microgrid Applications and Status

Gaurav Kumar and Shafqat Nabi Mughal

Abstract A microgrid is an independent power generation unit. It is able of providing the electricity to a local load like school, college, street light, hospital, etc., using the renewable energy sources and battery banks. Microcontrollers are playing an important role to operate the microgrid automatically. The microgrids mode of operation is determined by controllers. We use different algorithms to control the operation of a microgrid. Microgrid may be used in two different modes of operations. One is grid connected mode which injects power into the grid and another is islanding or off-grid mode of operation. This chapter discusses important data analytics related to microgrid besides power and energy systems. In addition, a comprehensive review on the emergence of microgrid is done along with its importance. This chapter also reviews different types of controllers used in microgrid systems. The current status of microgrid in India is also reviewed.

Keywords Microgrid · Controller design · Droop control of inverter · Voltage control mode · P-Q mode · Data analytics

Nomenclature

AC	Alternating Current
DC	Direct Current
RER	Renewable Energy Resources
DG	Distributed Generation
VSI	Voltage Source Inverter
PV	Photovoltaics
MPPT	Maximum Power Point Tracker

G. Kumar · S. N. Mughal (✉)
BGSB University, Rajouri, Jammu and Kashmir, India
e-mail: snmughal@bgsbu.ac.in

G. Kumar
e-mail: gaurav@bgsbu.ac.in

H. Malik et al. (eds.), *Intelligent Data Analytics for Power and Energy Systems*,
Lecture Notes in Electrical Engineering 802,
https://doi.org/10.1007/978-981-16-6081-8_26

PLL	Phased Locked Loop
DER	Distributed Energy Resources
ADB	Asian Development Bank
SCADA	Supervisory Control & Data Acquisition
LAN	Local Area Network
WAN	Wide Area Network
PWM	Pulse Width Modulation

1 Introduction

According to the IEEE statement, an assortment of connected AC/DC stacks and disseminated energy assets with certain electrical qualities that capacity as a solitary controlled unit comparable to the network and may associate and withdraw from the framework to work in both lattice associated and island modes is known as microgrid. Figure 1 shows the fundamental design of microgrid. With the expansion in populace and industrialization, world energy request is likewise developing. It is difficult to satisfy the heap need for conventional energy assets. Nuclear power age has a great deal of burdens.

The majority of the age is with coal. Coal is restricted in amount. Subsequently the world is moving from nuclear power to environmentally friendly renewable energy resources (RER). RER are the best choice to full fill the energy utilization request. The worldwide energy measurable yearbook 2018 expressed those absolute energy

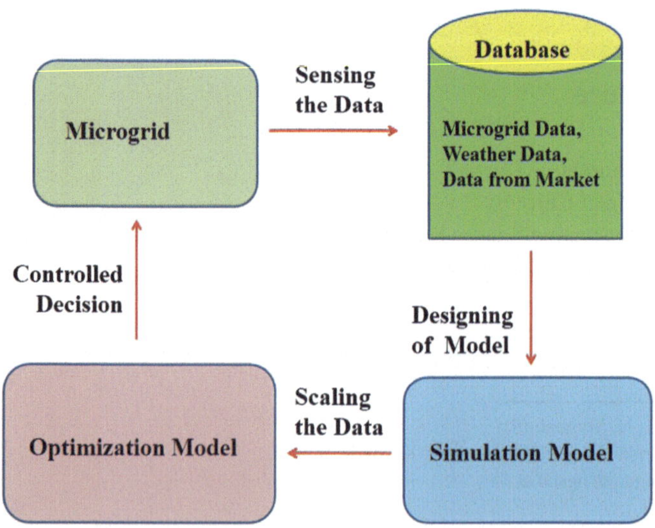

Fig. 1 Typical microgrid structure

utilization increments by 2.3% in 2017 as a correlation with 2016 [1]. China has been the world's biggest energy consumer since 2009 [2], expressed that in light of speedy industrialization and urbanization, the interest of energy is expanding step by step, still the hole between energy organic market isn't satisfied [2, 3]. The utilization of energy is additionally expanding in Asian nations like India, Russia, Malaysia, and so on. In Japan, energy utilization has been expanding since 2013. Because of energy proficiency enhancements in the US, the energy utilization is fixed just as the majority of the provincial local area are immature and have an absence of power access. As per [4, 5], energy prerequisite of the non-industrial nations is expanding. The energy emergency issue is the impediment in the improvement of a country. The energy emergency can be overwhelmed with the assistance of the environmentally friendly power Sources, elective on the planet [6, 7]. The most elevated limit of hydro is introduced in China followed by South America, Focal Asia, and Europe, and so forth.

The microgrid controllers play a critical role in the microgrid's automated operation and control, both in grid linked and islanded modes. These controller includes a number of characteristics that allow them to properly coordinate dispersed energy resources based on their power-generating capability to service essential and non-critical loads [8]. These controllers incorporate the control capacities that characterize the microgrid as a framework that can oversee itself, work self-ruling, and associate with and disengage from the fundamental dissemination framework for the trade of force and the inventory of auxiliary administrations. They also incorporate the elements of the microgrid energy the executive's framework (MEMS).

The main source of microgrids is renewable energy system (RESs). These RESs are associated in equal and feed capacity to utility lattice known as the Distributed Generators (DG) as demonstrated in Fig. 2. We need a voltage source inverter to connect the DGs to the AC Burden and utility network since the voltage and frequency of DGs differs from that of the utility grid or the burden needed [9]. The voltage source

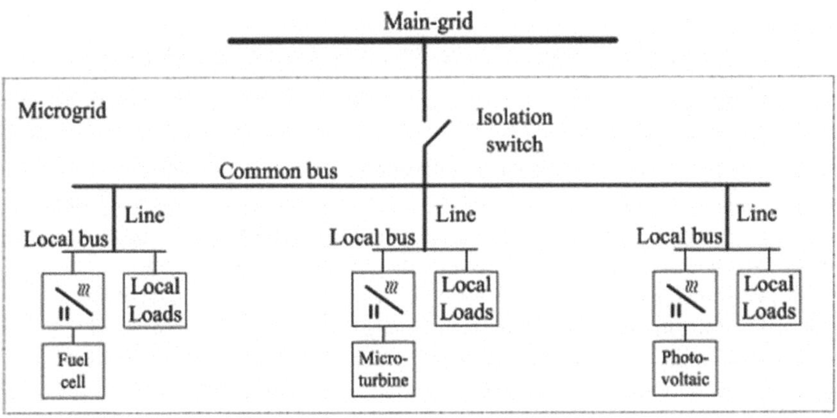

Fig. 2 Structure of a microgrid

inverter controls the voltage, frequency, and active power output. The AC loads are taken care of [4]. Activity and control procedure of an Inverter may differ as indicated by the sort of method of activity of microgrid. On the off chance that the VSI is in Lattice associated mode, at that point VSI will be constrained by the network voltage and frequency utilizing the PLL. There is no fixed procedure to control the microgrid in associated and islanding method of activity. The significant focus point of the distributed generation on the controlling of DG units center around the essential control systems because of their power-sharing difficulties [10–14]. Based on the above discussion, some of the important points which were felt as research gaps in the successful implementation of microgrids are as follows:

- The key sources of energy generations which supply microgrids need to be identified and investigated in detail.
- The synchronization between the different energy supplies as per the load demands needs to be handled carefully.
- The controlling of microgrids needs to be assisted with protective devices and coordination mechanisms for its successful operation need to be devised.
- For the fast deployment of microgrids, the economic aspects of various RES needs to be further investigated.

In the light of these shortcoming, this chapter explains the emergence of microgrid and its importance using RES. A successful microgrid needs to be controlled and this chapter is centered on the various methodologies to control the method of activity of microgrids and to control the voltage source inverter to take care of the capacity to the utility framework and neighborhood AC Loads. Further, this research work provides the important statistics related with microgrids besides power and energy systems. This work will help the utilities to understand microgrid operations and its importance to provide the uninterrupted power supply with better power quality in the future.

2 Intelligent Data Analytics Related to Power and Energy Systems

Because of the growth of pervasive computing devices, the use of big data in the energy business is gaining fresh interest. The application of data analytics in power and energy systems can help to enhance the industry's control, monitoring, and efficiency. Mechanistic models can be supplemented with data-based models to determine operational parameters. In the form of research publications and patents, a huge quantity of information about power and energy systems is being created. Table 1 depicts data analytics based on published research publications, whereas Table 2 depicts data analytics based on patents registered throughout the world for power and energy systems.

Table 1 Data analytics based on research articles dated 22-06-2021 (power and energy systems)

Works in set	Works cited by scholarly	Scholarly citations	Prominent Authors	Prominent institutions	Source
1,086,760	817,998	24,595,869	(i) Frede Blaabjerg (ii) Bhim Singh (iii) Joseph M Guerroro	(i) Max Planck Society (ii) Chinese Academy of Sciences (iii) Harvard University	[15]

Table 2 Data analytics based on patent information dated 22-06-2021 (power and energy systems)

Works cited by patents	Citing patents	Patent citations	Prominent inventors	Prominent applicants	Source
48,473	129,998	190,606	(i) Camphausen Ray (ii) Majumder Ritwik (iii) Willert Erin	(i) Siemens (ii) Bristol-Myers Squibb (iii) Gen Electric	[16]

The majority of traditional analytical techniques in power systems are model-based. In contrast, techniques created in the IT industry are frequently based only on data and they make use of artificial intelligence and internet of things [17–19]. In many energy benchmark situations, a combination of data and model-based methods appears to be the most successful. Data is collected by a centralized unit using modern measuring infrastructure such as smart meters and other specialized devices, which then feeds the data back to the centralized unit. Various measuring devices are put at various stages based on the demands; all of these devices create a huge amount of bidirectional data that travels to the centralizing unit and consumer premises, resulting in a vast volume of data in various forms. Data technology may aid power system analysis and optimization by utilizing historical data and past knowledge of power networks, therefore improving the economic efficiency and security of the power system.

3 Intelligent Data Analytics Related with Microgrids

Due to the unidirectional power flow, the rising usage of distributed generators in response to the need for green energy resources is upsetting the hegemony of large-scale centralized power plants, making the old centralized control technique less effective. Microgrids have the ability to solve long-standing energy challenges in developing nations such as India, Pakistan, and others. A large amount of researchers are focusing on the development and design on microgrids. A large amount of data is

Table 3 Data analytics based on research articles dated 22-06-2021 (microgrid)

Works in set	Works cited by Scholarly	Scholarly citations	Prominent authors	Prominent Institutions	Source
37,926	23,432	463,873	(i) Joseph M Guerrero (ii) Frede Blaabjerg (iii) Juan C Vasquez	(i) Aalborg University (ii) Electric Power Research Institute (iii) North China Electric Power University	[20]

Table 4 Data analytics based on patent information dated 22-06-2021 (microgrid)

Works cited by patents	Citing patents	Patent citations	Prominent inventors	Prominent applicants	Source
1,248	2,071	2,893	(i) Camphausen Ray (ii) Majumder Ritwik (iii) Willert Erin	(i) Siemens (ii) Bristol-Myers Squibb (iii) Gen Electric	[21]

getting generated related with microgrids in the form of research articles and patents. Table 3 shows the data analytics based on research articles published while Table 4 shows the data analytics based on the patents filed across world for microgrids.

The microgrid market was worth USD 8.29 billion in 2019 and is expected to grow to USD 25.45 billion by 2026, with a CAGR of 21.5% from 2021 to 2026 [22–24]. During the projection period, North America is anticipated to be the largest market, with the bulk of demand coming from nations such as the United States and Canada. Microgrid technology is being used by major countries to deliver power to distant areas. InfraCo Asia and Electricite de France (EDF) created a microgrid project for Kha Laing village in Myanmar, which was authorized by the Myanmar government in 2019. Lanka Electricity Company Ltd. (LECO) inked a contract with the Dimo and Dhybrid GmbH consortia in June 2020 to build Sri Lanka's first-ever national grid-tied Renewable Energy-based Microgrid on the University of Moratuwa's campus. The initiative will begin on a new paradigm in the field of distributed generation in Sri Lanka, thanks to a funding from the Asian Development Bank (ADB).

Maharashtra Energy Development Agency released a tender for a 29.4 KW microgrid project in Maharashtra, India, in 2019. The project is expected to cost more than INR 6,140,000 and will supply power to more than three communities. In Behlolpur, Bihar, Tata Power Delhi Distribution Ltd launched its second microgrid project in 2019. (India). The project will offer a safe and consistent power supply to about 1200 people living in 220 homes. In addition, in 2019, Tata Power signed a contract with the Bihar government to build 16 microgrid projects along the Nepal border by 2021.

India was designated a 100% electrified country by Indian Prime Minister Narendra Modi in 2018. This assertion, however, is predicated on a definition that many people may disagree with, and it is frequently contested by the beneficiaries themselves. When 10% of all houses and government offices have electricity, a hamlet is deemed 100% electrified. While this is a significant step toward linking a town to a central grid, it does not necessarily imply that the whole population has access to power. Microgrids can help with this. Microgrids provide a way for communities that are sporadically linked to the grid, or that have intermittent power supply and frequent blackouts, to continue with their daily activities. Local residents enjoy the reliable electricity they provide and are willing to pay for it.

3.1 Optimization and Data Analytics Approaches for Microgrid Design

Future microgrid is expected to be a combination of interconnected network with small scale microgrid and self-sustained microgrid. Optimization approaches are useful in designing of microgrid. Optimization can be used with the help of database. To use the optimization technique, data is required from different sources. Data is being collected from microgrid, weather station, customers and from market. The data collection is done in real time. Computers take the decision of microgrid operation according to the optimization algorithm. In designing of microgrid, data and communication layers are also required for optimization [25, 26]. Cost of microgrid can be reduced with the help of optimization using the load data and microgrid data in real time. Data is stored in database. Database may be near to the microgrid or may be online. To use the database, communication is also required. For communication, we use wide area networks (WAN), local area networks (LAN), optical fiber, SCADA [27]. Data center contains business with the servers, virtual computers, cloud storage and more facility for rent.

4 Importance of Microgrid

Microgrids are very necessary for any country because the populations are expending. Microgrids are fundamental for developing countries like India on account of its high populace. A particularly tremendous populace has remained without a central supply grid. A central supply grid required the generation, transmission and distribution network, which is exorbitant just as upkeep cost is high. The power bill is likewise expensive. The principle purpose for the microgrid idea is to lessen the non-conventional energy assets. Microgrids are using the sustainable power sources known as distributed generators. Existing utility network are exceptionally stacked. Cost of power may differ as indicated by the region. There are certain policies are made by the agencies for the installation of microgrid [28, 29]. The age costs in a microgrid rely upon area, limit, establishment costs, and so on, thus the cost per kWh from a microgrid is difficult to quantify. Despite the fact that microgrids offer advantages, including as bringing down transmission and circulation costs because of the way that they might be introduced close to the load center. Assuming we utilize renewable power, we can reduce the power cost. Power isn't modest. For example, according to a recent Stanford analysis, the average cost of grid power in Gujarat towns is $0.06/kWh; nonetheless, the microgrid energy cost would be doing $0.38/kWh. So we can say that environmentally friendly power sources with battery banks are the acceptable substitution of the energy from thermal and nuclear power. A few groups imagine that the advantages exceed at the extra expense. Microgrid will be helpful for ranchers; additionally, they can utilize the sunlight-based siphons as opposed to utilizing the matrix supply. Limited scope organizations can be controlled with the miniature frameworks. Be that as it may, the miniature network itself has numerous difficulties. We will examine every one of the difficulties in the forthcoming areas. For instance, individuals didn't acknowledge the new innovation without any problem. Sustainable power introduced limit is appeared in Fig. 3 for the top sustainable power maker nations. China is at the top situation with 153 GW of age followed by Joined State, Germany, Italy, Spain, Japan and India.

They are afraid to use it. So this may be one challenge for use of microgrids. Scalability, power theft, and an ultimate extension of the microgrid are all issues from a commercial standpoint. States do not have a plan and fund for microgrid implementation. Renewable Generation Growth will be 98 GW by 2022. To provide the electricity in low amounts near to end-users or load centers. The combination of distributed generators makes the microgrids. Increase energy efficiency dramatically, minimize carbon emissions, enhance grid resiliency, and demand some additional transmission and distribution capacity.

Distributed generators are also called decentralized generation is or on location generation appeared in Fig. 4. On location generation implies generating sources close to the load centers, rather than long transmission of energy. Different advantages

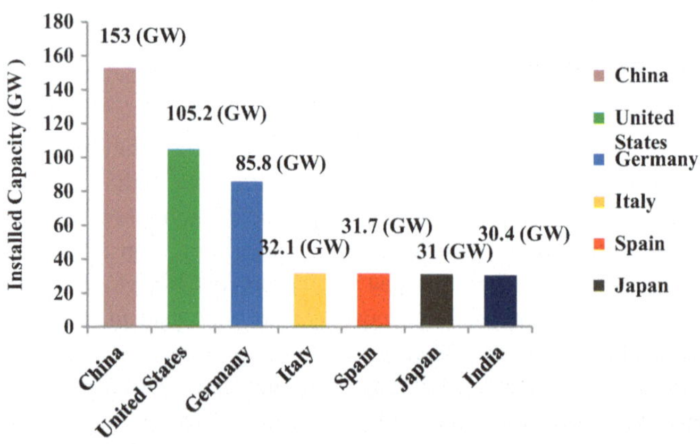

Fig. 3 Top nations' installed renewable energy capacity

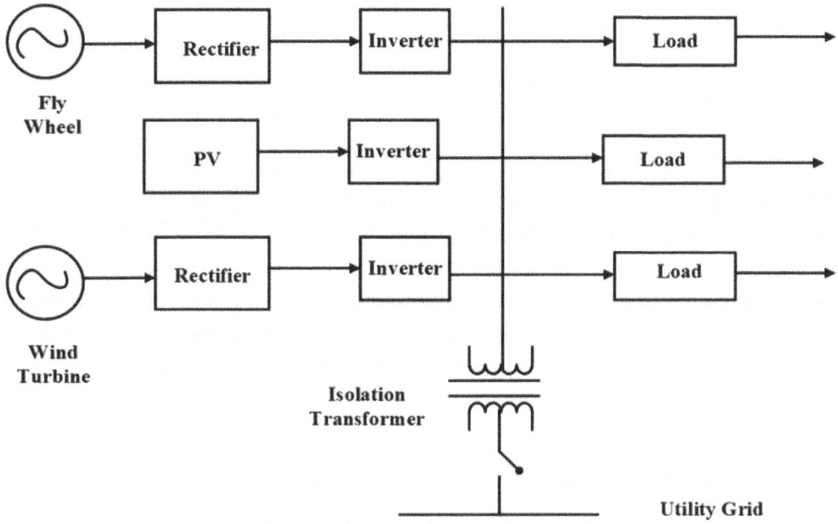

Fig. 4 Block diagram of distributed generators in a microgrid

of on location generation are and there is no transmission and distribution cost. Yet, to create the power close to the load community is additionally difficult. Because monetary growth outpaces power supply growth in some parts of the country, and various districts have capacity constraints in delivering power where and when it is needed, it is vital to empower local power transmission options.

The vast bulk of electricity in the United States is nowadays generated from interconnected sources (such as coal, gaseous petroleum, nuclear, and massive hydropower), which convey a great deal of control over long distances. While this strategy has certain advantages, it also has some drawbacks. In the United States, 6% of generated electricity is lost due to transmission breakdowns, primarily due to the long distances the power must travel. Building new transmission lines (or revamping old ones) may be expensive and time-consuming, with numerous challenges and delays. Millions of people might be left without electricity if the electric grid falls down. Distributed generators avoid the great majority of these problems by delivering more modest power systems near to where they are needed. Neighborhood renewables, for example, sunlight-based and wind; squander to energy; and joined warmth energy and force (otherwise called cogeneration), which includes recuperating heat from a customary force plant to warm designs or maybe water. A variety of commercial sectors and administrative barriers have slowed the expansion of distributed generation in the United States. However, as more people see the value of a well-designed distributed generating system, these barriers are dissipating, and their innovation costs are falling. The continued development of distributed generation, backed by strategy shifts, may help businesses and networks overcome system flaws and become more resilient.

5 Control Approaches in Microgrid Systems

5.1 Centralized Method

Complex core processing units that regulate the whole system characterize centralized control. From afar, it employs a communication system and a control distribution system. A focal framework, which mimics a human mind, as well as sensors and control sensors, make up the communication system (voltage, frequency and current). The data from the sensors is sent to the focal framework. When all social event information is complete, the focused framework computes the control variables for each control hardware and provides them to each regulator [30–33].

5.2 Decentralized Approach

The circulation framework will be parted into astute little networks, known as microgrids, and each of them will be associated with the primary grid in a decentralized style. A microgrid has its own controller and own sources of power [34]. In linked mode, Fig. 5 depicts the fundamental layout of a typical microgrid. As a result, the distribution system may regulate itself in pieces. The crucial element is that each component may function independently [32]. Figure 5 shows the grid connected voltage source inverter.

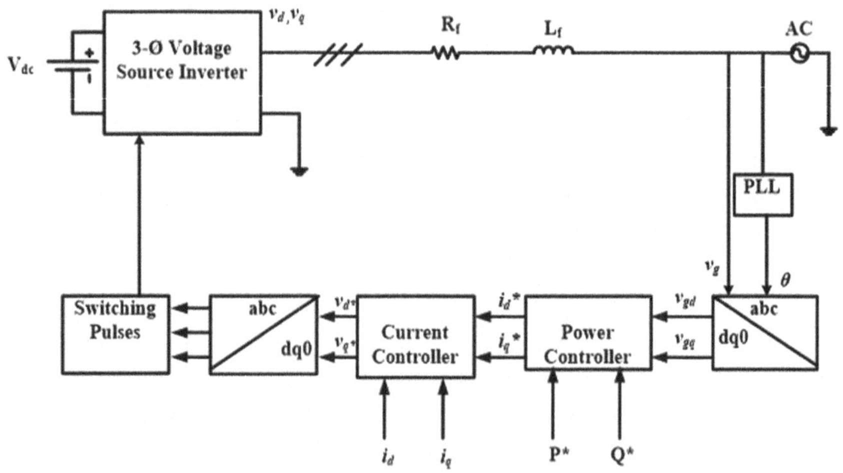

Fig. 5 VSI controller block diagram based on *P-Q* theory

5.3 P-Q Control in Grid Connected Mode

Microsources, as described in the introduction, require certain interfaces (inverters) to connect to the network. An inverter can be operated in one of two ways. Figure 5 shows the *P*-*Q* control of inverter: To offer exact active and reactive power set points, inverters are employed. Inverters are regulated to take control of the load using the specified frequency information along with voltage. According to the load requirement, the voltage source inverter active and reactive power output will be specified [34, 35]. The energy executives are the major item in grid connected mode. In connected mode, the MPPT is utilized to calculate active and reactive power references. The PI controller uses these *P* and *Q* values to generate PWM pulses.

Many control studies have been conducted till date for the operation of renewable energy sources in *P*-*Q* mode or, alternatively, consistent current control mode [8], on the basis that in this mode, the current moving from inverter to lattice is constant according to the inventory of active power from solar and wind-powered systems [31–33]. In associated method of activity DGs communicate the power from DC to AC loads. Figure 5 depicts a square chart of a voltage source inverter with *P*-*Q* control. The *dq*0 reference frame transformation, which employs PLL to identify the *d*- and *q*-axis components of the ac side currents, provides the foundation for these control approaches. Without PLL, we can't predict the utility grid's phase and frequency, hence, it's a crucial part of this control system. A PID regulator is employed in this control system to regulate the active and reactive power of the inverter. The utility grid controls the inverter's voltage and frequency. The only two elements to consider are active and reactive power. From a revolving to a stationary reference frame, both voltage and current are swapped. This common reference frame is used to translate the voltage source inverter. The common reference frame is *dq*0, which rotates at utility grid frequency.

5.3.1 PCM Control

The power control management system will not be able to modify the voltage because of the network related way of activity; as a result, a coherent selection of controls is the sorts of current regulations.

5.3.2 VCM Control

Controlling a microgrid in island mode using *P*-*Q* control mode is not a smart idea due to the reference voltage and the down to earth on workability of regulating the load demand [4]. The control activity in island mode differs from that in grid linked mode. The framework is now disconnected from the upstream organization, and the utility grid no longer covers the VSI's voltage and frequency. As a result, the voltage of surrounding loads must be directed efficiently, which necessitates the use of a VCM control to regulate the output of the VSI in response to load variations [34, 36, 37].

5.4 V-f Control or Droop Control in Islanding Mode

In island mode, the system's principal purpose controlling frequency along with voltage [28, 35, 36, 38–53]. The inverter can generally function in one of two modes when connecting energy sources to a utility network, as shown in Fig. 6, grid connected mode or isolated mode. The two modes of controller are voltage control mode and power control mode.

Voltage and frequency regulation in an islanding microgrid will be handled by voltage source inverters using the voltage-frequency method or the voltage-frequency method plus the voltage-frequency method plus the voltage-frequency method plus the voltage-frequency method plus the voltage-frequency method plus the voltage-frequency method plus the voltage-(V-f control mode). Communication-less mode is also known as V-f control mode. In this circumstance, at least one VSI should be set to V-f control mode in order to keep the system voltage and frequency within safe limits [13, 35]. The significance of this control method is that it eliminates the need for communication between parallel linked inverters by relying just on droop control. The whole load power is shared by inverters. Figure 6 depicts the block diagram of a VSI with V-f control. After measuring the voltage and current, p and q are computed. The low pass filter with cut-off frequency (ω_c) is used to pass calculated p and q. The P and Q are computed using following equations.

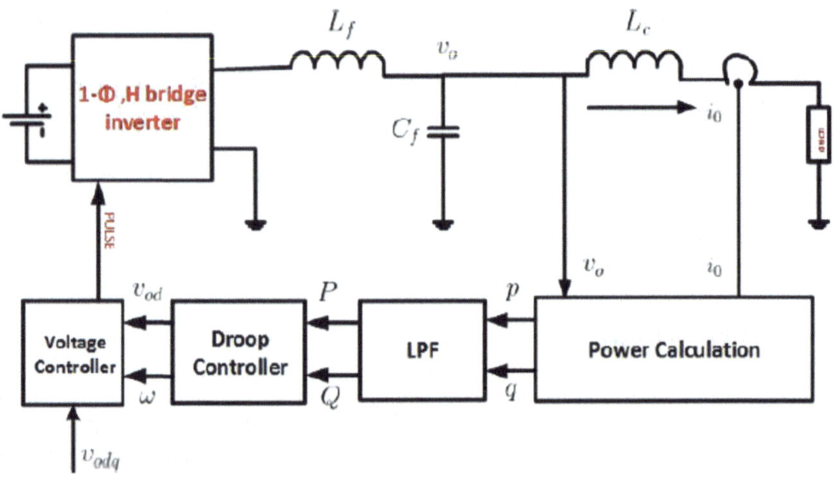

Fig. 6 Block diagram of VSI in island mode (V-f mode)

$$P = \frac{\omega_c}{S + \omega_c} p \tag{1}$$

$$Q = \frac{\omega_c}{S + \omega_c} q \tag{2}$$

True power sharing across inverters may be achieved via inverter frequency drooping artificially. The droop gain determines the phase, and the frequency is determined by integrating the frequency. This causes negative feedback, comparable to the governor drop in a steam turbine and the inertia characteristics of conventional generators. When a generator's power consumption rises, the voltage rotation slows and the rotation angle lowers. The nominal frequency set-point is represented by ω_n in the following equations, where the angle of the inverter voltage varies in response to real power flow in the intended negative sense and with a gain determined by the droop.

$$\omega = \omega_n - m_p P \tag{3}$$

$$V_{\mathrm{od}}^* = V_n - n_q Q \tag{4}$$

$$V_{\mathrm{oq}}^* = 0 \tag{5}$$

To redistribute the reactive power across the parallel connected multiple inverters as seen in (4) and (5), a droop co-efficient is introduced in the voltage magnitude. The d-axis output voltage has a nominal set value of V_n, whereas the quadrature axis voltage is 0. According to the control approach, the output voltage magnitude reference is aligned to the d-axis of the inverter reference frame, and the q-axis reference is set to zero. As a result, the voltage source inverter generates ω_n and V_{od} switching pulses. For the provided range of frequency and voltage magnitude, the droop coefficients m_p and n_q are determined using Eqs. (6) and (7).

$$m_p = \frac{\omega_{\max} - \omega_{\min}}{P_{\max}} \tag{6}$$

$$n_q = \frac{V_{\mathrm{od}\max} - V_{\mathrm{od}\min}}{Q_{\max}} \tag{7}$$

Then voltage controller which contains PI controller generates the modulation index and phase angle accordingly. PWM pulses are constructed from the comparison of reference signal and carrier signal for Voltage Source Inverter.

5.5 Tertiary Control of Microgrid

The last control level in the functioning of a microgrid is tertiary control, which incorporates economic factors. Optimal control is another term for it. Power flow between the microgrid and the main grid is managed by tertiary control [54]. The amplitude and frequency of the distributed generator's voltages may be adjusted in grid linked mode to manage power flow from microgrid to main grid. The reference voltages for distributed generators are generated by measuring the local power at the point of common connection. The gossiping algorithm [55–61], according to the tertiary control, may be used to optimize the cost of linked operations.

5.6 Different Stochastic Control Approaches Used in a Microgrid

A lot of research is going on, based on the stochastic modeling of microgrids. According to these modeling, microgrid performance can be evaluated and have the ability of specific operation, planning and control functions. Different type of stochastic control approaches is discussed below.

1. State preparation Model

To create a stochastic model for a microgrid, the Stochastic Hybrid System (SHS) can be utilized [62]. In a microgrid, the SHS model can capture probabilistic events like as device failure and continuous/discrete mode behavior. The term "discrete mode" refers to the state of equipment such as solar panels (on/off/shutdown), wind turbines (on/off/shutdown), and the mode of operation of a microgrid, which can be linked or islanding.

2. State Prediction

This Based on real-time observations, this approach is used to forecast the state of power system dynamics such as bus voltages and phase angles for the whole system [63]. There are three different types of measures that are commonly used:

- Analog measurements: includes magnitudes of bus voltages, active/reactive power flows/injections
- Discrete measurement: includes the position of switches and circuit breakers
- Pseudo measurement: includes estimated power generation and load prediction.

Based on the techniques discussed above the stochastic model can be modeled. These techniques are useful in the forecasting the microgrid generation and load status. The forecasting calculations are done with the help of local measurements and data from the database. Data analytics and optimization is very useful tool to predict the performance of microgrid and to reduce the energy cost.

6 Status of Microgrid in India

Microgrid is defined as the controllable framework comprising of backs, loads, storage system that can work with and without a network a grid connected mode. India is one of the quickly developing nations and most of populace is living in country networks [37, 64]. With the increment in populace, energy utilization rate increments and most of power request is finished by power grid. An enormous populace of India has absence of power choices and no power by any means. Numerous impetuses programs are controlled by the public authority for the arranging of microgrid. In the most recent few years, more power was delivered from the backs as an examination with the customary fuel sources. The Indian government has been zeroing in on the organization of microgrid with the main utility grid.

Table 5 shows the total production of electricity from various sources in India [65–67]. Figure 7 shows the biggest introduced limit of wind power followed by sun based, biomass, and hydropower. The greater part of the miniature lattices set-ups in the nation is worked and packed in northern and eastern parts. Likewise, the biomass and waste to energy programs are grown as of late. Microgrids and

Table 5 Installed capacity in India up to November 2020

S. No.	Source of generation	Installed capacity (MW)	Percentage (%)
1	Thermal power	231,320.72	61.82
2	Nuclear power	6780.00	1.81
3	Hydro energy	45,699.22	12.21
4	Renewable energy	90,399.11	24.16
Total		374,199.04	100

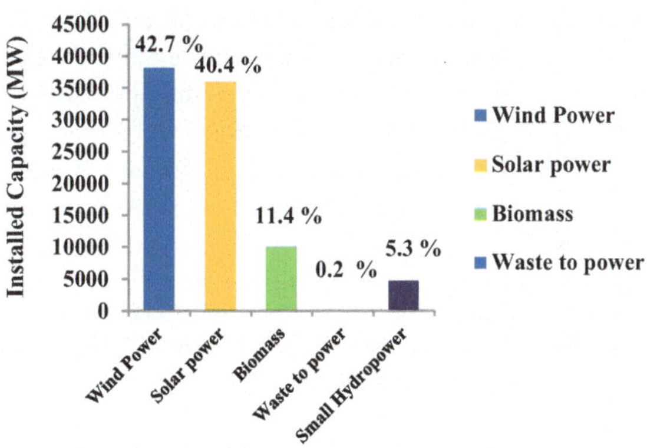

Fig. 7 As of September 30, 2020, installed grid interactive Renewable Power Capacity (excluding major hydropower).

minigrids can be urgent and cost-saving choices to the augmentation of principle power framework. Numerous endeavors ought to be made by the public authority for the advancement of backs. The primary hotspots for the microgrid projects are sun powered, wind, minihydro, and biomass. A portion of the major microgrids projects are microgrids in Dharnai town in Bihar, Sagar island Microgrid in Sundarbans, husk power-based microgrid in Bihar, Orissa environmentally friendly power improvement office (oreda) projects in Orissa, Sikkim environmentally friendly power improvement organization (Sreda) projects, and so forth. In the Dharnai town, there is no power association with the fundamental framework. The microgrid comprising of spv utilizes 280 sun oriented boards to create power for in excess of 400 houses. The microgrid framework involving RESs likewise comprises of 60 sunlight-based boards [68, 69]. There are three kinds of microgrids, like smaller than usual, little, medium, and enormous disseminated age innovation having limits of 0.001–0.005, 0.005–5, 5–50, and 50–300 MW, individually. The service of force and sustainable power announced that sunlight-based PV created 28,181 MW of force in India up to Walk 2019 and target is 175 GW by 2022 [70–80].

7 Conclusion

We can make small microgrids and we can connect them to each other. It may be a part of smart grid also. Many renewable energy resources are coupled to three-phase voltage source inverters, which are controlled in constant current mode and feed electricity into the grid in connected mode of microgrid at the point of common connection. The distribution network has become more puzzling as a result of the use of sustainable power assets in the power industry and microgrid, as well as their use as DERs. In this manner, the distribution network's control conduct should be precise. This study demonstrates how microgrids and their control methodologies satisfy distribution system assumptions. Microgrids must be capable of operating in both island and main grid linked modes. The importance of capacity devices in the strength of microgrids is not allowed due to supporting DGs. Following islanding, the primary goal of a microgrid is to protect basic burdens as they become necessary, while non-essential burdens are organically eliminated when load fluctuates. In general, microgrid control tactics may be divided into two categories: communication methods and communication techniques without communication. In this research work, various data analytics related with microgrids besides power and energy systems are also discussed. Moreover, current microgrid status of India is also described.

Acknowledgements The research work is carried by funding from NPIU in the form of research grant with sanction no. BGSBU/TEQIP-III/RGS/015. The authors are thankful to the BGSB University for assisting the necessary help.

References

1. A. Belke, F. Dobnik, C. Dreger, Energy consumption and economic growth: new insights into the cointegration relationship. Energy Economics **33**(5), 782–789 (2011)
2. I. Ozturk, A. Acaravci, CO_2 emissions, energy consumption and economic growth in Turkey. Renew. Sustain. Energy Rev. **14**(9), 3220–3225 (2010)
3. M.F. Tahir, C. Haoyongm, K. Mehmood, N. Ali, J.A. Bhutto, Integrated energy system modeling of China for 2020 by incorporating demand response, heat pump and thermal storage. IEEE Access **7**, 40095–40108 (2019)
4. A. Chauhan, S. Upadhyay, A. Srivastav, M.T. Khan, An intelligent load management technique for remotely located areas powered by grid connected solar/diesel generator based system, in *2020 International Conference on Power Electronics & IoT Applications in Renewable Energy and its Control (PARC)* (2020), pp. 320–324
5. A.K. Yadav, et al., *Soft Computing in Condition Monitoring and Diagnostics of Electrical and Mechanical Systems*, 1E (Springer Nature, Part of the Advances in Intelligent Systems and Computing, 2020), p. 496
6. A. Iqbal, et al., *Meta Heuristic and Evolutionary Computation: Algorithms and Applications*, 1E (Springer Nature, Part of the Studies in Computational Intelligence, 2020), p. 849
7. J.A. Alzubi, *AI and Machine Learning Paradigms for Health Monitoring System: Intelligent Data Analytics*, 1E (Springer Nature, Part of the Studies in Big Data, 2020), p. 513
8. G. Kumar, S.N. Mughal, S. Banoth, Design and development of an AC/DC microgrid for isolated power system and grid connected system. Think India J. **22**(4), 7031–7035 (2019)
9. S.H. Hu, C.Y. Kuo, T.L. Lee, J.M. Guerrero, Droop-controlled inverters with seamless transition between islanding and grid-connected operations, in *2011 IEEE Energy Conversion Congress and Exposition* (2011), pp. 2196–2201
10. W.R. Issa, A. Abusara, S.M. Sharkh, Impedance interaction between islanded parallel voltage source inverters and the distribution network, in *IET Conference Publications*, vol. 2014, no. 628 CP (2014)
11. J.M. Guerrero, et al., Steady-state invariant-frequency control of parallel redundant uninterruptible power supplies, in *IEEE 2002 28th Annual Conference of the Industrial Electronics Society, IECON 02*, (2002), pp. 274–277
12. M.M. Ramadan, P. Koutb, A. Magdy, Droop control for parallel-connected three-phase UPS units with different ratings. Menoufia J. Electron. Eng. Res. **29**(1), 13–20 (2020)
13. M.C. Chandorkar, D.M. Divan, R. Adapa, Control of parallel connected inverters in standalone AC supply systems. IEEE Trans. Ind. Appl. **29**(1), 136–143 (1993)
14. D. Wu, F. Tang, J.M. Guerrero, J.C. Vasquez, Autonomous control of distributed generation and storage to coordinate P/Q sharing in islanded microgrids-An approach beyond droop control, in *2014 IEEE International Energy Conference (ENERGYCON)*, (2014), pp. 983–988
15. https://www.lens.org/lens/search/scholar/analysis?q=Power%20and%20Energy%20System&preview=true
16. https://www.lens.org/lens/search/patent/analysis?scholarQueryId=810849a4-8d70-40aa-a5ea2ef46d53c478
17. S. Srivastava, et al., *Applications of Artificial Intelligence Techniques in Engineering*, vol. 1, 1E (Springer Nature, Part of the Advances in Intelligent Systems and Computing, 2018), vol. 698, p. 643
18. S. Srivastava, et al., *Applications of Artificial Intelligence Techniques in Engineering*, vol. 2, 1E (Springer Nature, Part of the Advances in Intelligent Systems and Computing, 2018), vol. 697, p. 647
19. P. Nandankar, R. Thaker, S.N. Mughal, M. Saidireddy, J.E.A.L. Kostka, M.V.A. Nag, An Iot based healthcare data analytics using fog and cloud computing. Turk. J. Physiother. Rehabil. **32**(3), 3401–3409 (2021)
20. https://www.lens.org/lens/search/scholar/analysis?q=microgrid&n=50&preview=false
21. https://www.lens.org/lens/search/patent/analysis?n=50&preview=false&scholarQueryId=810849a4-8d70-40aa-a5ea-2ef46d53c478

22. Mordor Intelligence, Microgrid Market—Growth, Trends, COVID-19 Impact, and Forecasts (2021–2026). Available at https://www.mordorintelligence.com/industry-reports/micro-grid-market
23. A. Iqbal, et al., *Renewable Power for Sustainable Growth*, 1E (Springer Nature, Part of the Lecture Notes in Electrical Engineering), vol. 723, p. 805
24. N. Fatema, et al., *Intelligent Data-Analytics for Condition Monitoring: Smart Grid Applications*, 1E (Academic Press, 2021)
25. H. Qiu, W. Gu, X. Xu, G. Pan, P. Liu, Z. Wu, L. Wang, A historical-correlation-driven robust optimization approach for microgrid dispatch. IEEE Trans. Smart Grid (2020)
26. R. Rahmani, I. Moser, A.L. Cricenti, Modelling and optimization of microgrid configuration for green data centres: a metaheuristic approach. Future Gener Comput Syst **108**, 742–750 (2020)
27. A. Bani-Ahmed, L. Weber, A. Nasiri, H. Hosseini, Microgrid communications: state of the art and future trends, in *2014 International Conference on Renewable Energy Research and Application (ICRERA)*, (2014), pp. 780–785
28. A. Ali, W. Li, R. Hussain, X. He, B.W. Williams, A.H. Memon, Overview of current microgrid policies, incentives and barriers in the European Union, United States and China. Sustainability **9**(7), 1146 (2017)
29. H. Van Pham, J.L. Rueda, I. Erlich, Probabilistic evaluation of voltage and reactive power control methods of wind generators in distribution networks. IET Renew. Power Gener. **9**(3), 195–206 (2015)
30. H. Hatta, H. Kobayashi, Demonstration study on centralized voltage control system for distribution line with sudden voltage fluctuations (2008), pp. 70–70
31. S.K. Mishra, S. Sahoo, B. Sahoo, S.K. Jena, Energy-efficient service allocation techniques in cloud: a survey. IETE Tech. Rev. **37**(4), 339–352 (2020)
32. P. Borazjani, N.I.A. Wahab, H.B. Hizam, A.B.C. Soh, A review on microgrid control techniques, in *2014 IEEE Innovative Smart Grid Technologies-Asia (ISGT ASIA)*, (2014), pp. 749–753
33. H. Hatta, H. Kobayashi, A study of centralized voltage control method for distribution system with distributed generation, in *19th International Conference on Electricity Distribution (CIRED)* (2017)
34. J.P. Lopes, C.L. Moreira, A.G. Madureira, Defining control strategies for microgrids islanded operation. IEEE Trans. Power Syst. **21**(2), 916–924 (2006)
35. S. Xiaofeng, L. Qingqiu, T. Yanjun, Z. Chen, An improved control method of power electronic converters in low voltage micro-grid, in *2011 International Conference on Electrical Machines and Systems* (2016), pp. 1–6
36. I.R.E. Series, Microgrids and active distribution networks. The Institution of Engineering and Technology (2009)
37. N.H. Saad, A.A. El-Sattar, A.E. Mansour, A novel control strategy for grid connected hybrid renewable energy systems using improved particle swarm optimization. Ain Shams Eng. J. **9**(4), 2195–2214 (2018)
38. T.L. Vandoorn, B. Renders, L. Degroote, B. Meersman, L. Vandevelde, Active load control in islanded microgrids based on the grid voltage. IEEE Trans. Smart Grid **2**(1), 139–151 (2010)
39. W. Bai, K. Lee, Distributed generation system control strategies in microgrid operation. IFAC Proc. Vol. **47**(3), 11938–11943 (2014)
40. E. Khazaei, A. Jamaledini, Optimal Operation of Islanded Microgrid Operation Based on the JAYA Optimization Algorithm (2019)
41. T.S. Basso, R. DeBlasio, IEEE 1547 series of standards: interconnection issues. IEEE Trans. Power Electron. **19**(5), 1159–1162 (2004)
42. X. Tang, Z. Qi, Energy storage control in renewable energy based microgrid, in *2012 IEEE Power and Energy Society General Meeting* (2012), pp. 1–6
43. S. Bahramirad, W. Reder, Islanding applications of energy storage system, in *2012 IEEE Power and Energy Society General Meeting* (2012), pp. 1–5
44. T. Hiyama, T. Funabashi, Multi-agent based operation and control of isolated power system with dispersed power sources including new energy storage device, in *ICREPQ* (2004)

45. U. Kwhannet, N. Sinsuphun, U. Leeton, T. Kulworawanichpong, Impact of energy storage in micro-grid systems with DGs, in *2010 International Conference on Power System Technology* (2010), pp. 1–6
46. L. Wang, C. Singh, Environmental/economic power dispatch using a fuzzified multi-objective particle swarm optimization algorithm. Electr. Power Syst. Res. **77**(12), 1654–1664 (2007)
47. R. Ahshan, M.T. Iqbal, G.K. Mann, J.E. Quaicoe, Micro-grid system based on renewable power generation units, in *CCECE* (2010), pp. 1–4
48. T. Hammons, L.L. Lai, International practices in distributed generation development worldwide, in *2007 IEEE Power Engineering Society General Meeting* (2007), pp. 1–5
49. M.E. Baran, I.M. El-Markabi, A multiagent-based dispatching scheme for distributed generators for voltage support on distribution feeders. IEEE Trans. Power Syst. **22**(1), 52–59 (2007)
50. R. Roche, B. Blunier, A. Miraoui, V. Hilaire, A. Koukam, Multi-agent systems for grid energy management: a short review, in *IECON 2010–36th Annual Conference on IEEE Industrial Electronics Society* (2010), pp. 3341–3346
51. T. Nagata, Y. Tao, H. Fujita, An autonomous agent for power system restoration, in *IEEE Power Engineering Society General Meeting* (2004), pp. 1069–1074
52. Y.J. Reddy, S. Dash, A. Ramsesh, Y.P. Kumar, K.P. Raju, Monitoring and control of real time simulated microgrid with renewable energy sources, in *2012 IEEE Fifth Power India Conference* (2012), pp. 1–6
53. J. Dang, R.G. Harley, Islanded microgrids black start procedures with wind power integration, in *2013 IEEE Power & Energy Society General Meeting* (2013), pp. 1–5
54. J.M. Guerrero, J.C. Vásquez, J. Matas, M. Castilla, L.G.D. Vicuña, M. Castilla, Hierarchical control of droop-controlled AC and DC microgrids—a general approach toward standardization. IEEE Trans. Ind. Electron. **58**, 158–172 (2011)
55. K.D. Brabandere, K. Vanthournout, J. Driesen, G. Deconinck, R. Belmans, Control of microgrids, in *Proceedings of IEEE Power & Energy Society General Meeting* (2007), pp. 1–7
56. A. Pantoja, N. Quijano, A population dynamics approach for the dispatch of distributed generators. IEEE Trans. Ind. Electron. **58**, 4559–4567 (2011)
57. K. Vanthournout, K.D. Brabandere, E. Haesen, J. Driesen, G. Deconinck, R. Belmans, Agora: distributed tertiary control of distributed resources, in *Proceedings of 15th Power Systems Computation Conference* (2005), pp. 1–7
58. T. Tanabe, et al., Optimized operation and stabilization of microgrids with multiple energy resources, in *Proceedings of 7th International Conference on Power Electron* (2007), pp. 74–78
59. E. Barklund, N. Pogaku, M. Prodanović, C. Hernandez-Aramburo, T.C. Green, Energy management in autonomous microgrid using stability-constrained droop control of inverters. IEEE Trans. Power Electron. **23**, 2346–2352 (2008)
60. C.A. Hernandez-Aramburo, T.C. Green, N. Mugniot, Fuel consumption minimization of a microgrid. IEEE Trans. Ind. Appl. **41**, 673–681 (2005)
61. K. Vanthournout, A Semantic Overlay Network Based Robust Data Infrastructure, Applied to the Electric Power Grid, Ph.D. dissertation, Facultiet Ingenieurswetenschappen, Leuven, Belgium (2006)
62. M. Strelec, K. Macek, A. Abate, Modeling and simulation of a microgrid as a stochastic hybrid system, in *Proceedings of the 3rd IEEE PES International Conference and Exhibition on Innovative Smart Grid Technologies (ISGT Europe)*, Berlin, Germany, (2012), pp. 1–9
63. A. Monticelli, F.F. Wu, Network observability: identification of observable islands and measurement placement. IEEE Trans. Power App. Syst. **104**, 1035–1041 (1985)
64. P. INDIA, Census of India 2011 Provisional Population Totals. Office of the Registrar General and Census Commissioner, New Delhi (2011)
65. M.A. Hossain, H.R. Pota, W. Issa, M.J. Hossain, Overview of AC microgrid controls with inverter-interfaced generations. Energies **10**(9), 1300 (2017)
66. S. Upadhyay, M.P. Sharma, Selection of a suitable energy management strategy for a hybrid energy system in a remote rural area of India. Energy **94**, 352–366 (2016)

67. V.M. Balijepalli, S.A. Khaparde, R.P. Gupta, Y. Pradeep, SmartGrid initiatives and power market in India, in *IEEE PES General Meeting* (2010), pp. 1–7
68. T.L. Vandoorn, B. Meersman, L. Degroote, B. Renders, L. Vandevelde, A control strategy for islanded microgrids with dc-link voltage control. IEEE Trans. Power Delivery **26**(2), 703–713 (2011)
69. S.N. Mughal, Y.R. Sood, R.K. Jarial, A proposal for evaluating the clearness index based on extra terrestrial radiations at a given site, in *Intelligent Techniques and Applications in Science and Technology. ICIMSAT 2019. Learning and Analytics in Intelligent Systems*, ed. by S. Dawn, V. Balas, A. Esposito, S. Gope, vol. 12 (Springer, Cham, 2020), pp. 524–531
70. S.N. Mughal, S. Ahmed, G. Kumar, Uninterruptible power supply design using solar-wind hybrid combination with minimal use of power supplied from the grid. Dogo Rangsang Res. J. **10**(07) (2020)
71. S.N. Mughal, Y.R. Sood, R.K. Jarial, A review on Solar Photovoltaic technology and future trends. Int. J. Sci. Res. Comput. Sci. Eng. Inf. Technol. **4**(1) (2018)
72. S.N. Mughal, Y.R. Sood, R.K. Jarial, A neural network-based time-series model for predicting global solar radiations. IETE J. Res. 1–13 (2021)
73. S.N. Mughal, Y.R. Sood, R.K. Jarial, A proposal on techno-financial design aspects of photovoltaic system for the twin districts of Rajouri and Poonch (Jammu & Kashmir). EAI Endorsed Trans. Energy Web **8**(31), 1–11 (2020)
74. S.N. Mughal, et al., Design and techno-financial analysis of solar photovoltaic plant for school of engineering and technology at BGSB University, Rajouri (J&K), in *Applications of Computing, Automation and Wireless Systems in Electrical Engineering*. Lecture Notes in Electrical Engineering, ed. by S. Mishra, Y. Sood, A. Tomar, vol. 553 (2019), pp. 231–243
75. H. Malik, A.K. Yadav, A novel hybrid approach based on relief algorithm and fuzzy reinforcement learning approach for predicting wind speed. Sustain. Energy Technol. Assessm. **43** (2020). https://doi.org/10.1016/j.seta.2020.100920
76. W. Ahmad et al., Development of wide area monitoring system for smart grid application. J. Intell. Fuzzy Syst. (2021). https://doi.org/10.3233/JIFS-189752
77. A. Azeem et al., k-NN and ANN based deterministic and probabilistic wind speed forecasting intelligent approach. J. Intell. Fuzzy Syst. **35**(5), 5021–5031 (2018). https://doi.org/10.3233/JIFS-169786
78. P. Arora et al., Wind energy forecasting model for northern-western region of India using decision tree and MLP neural network approach. Interdiscip. Environ. Rev. **19**(1), 13–30 (2018). https://doi.org/10.1504/IER.2018.089766
79. T. Khursheed et al., Multi-step ahead time-series wind speed forecasting for smart-grid application. J. Intell. Fuzzy Syst. (2021). https://doi.org/10.3233/JIFS-189736
80. M.A. Alotaibib et al., A new hybrid model combining EMD and neural network for multi-step ahead load forecasting. J. Intell. Fuzzy Syst. (2021). https://doi.org/10.3233/JIFS-189775

Role of Blockchain in IoT Enabled Power and Energy Related Healthcare System-Platform for the Development of IoT Security

Vishal Sharma and Niranjan Lal

Abstract In the concept of IoT technology, various physical objects, things, machines and even peoples can communicate with each other through the internet. Predictably, millions of devices would be connected and shared in the coming future. In this technology, there are various types of sensors attached to these devices. These sensors can use different types of connections like RFID, Zigbee, Bluetooth, Wi-Fi and the internet via wireless and wired network configurations. There are many IoT enabled applications like smart home, intelligent traffic, smart energy, innovative healthcare, smart agriculture, etc., and these applications enhance the existing infrastructure a lot. There are many challenges with this technology, such as data privacy and security, user authentication, attacks countermeasure, easy deployment and self-maintenance. IoT enabled healthcare system is one of the applications which can be boost by integrated blockchain technology. Although blockchain technology is based on expensive computation, it involves the overhead of high bandwidth and delay. Therefore, it is not entirely fit for resource-constrained IoT devices suggested smart healthcare system (Dwivedi et al. in Sensors, 2019; Dorri et al. in Blockchain in Internet of Things: Challenges and Solutions). The key objective of this chapter is to investigate and proposed a modified power and energy-efficient architecture to integrate the IoT and the blockchain in the application of digital healthcare information system to eliminate the overhead of the blockchain technology security and privacy benefits. This integration might be beneficial not only for the patients but also for the doctors, medical caregivers, insurance companies, pharmacies and hospitals. In this chapter, we have discussed the power and energy-efficient architecture of IoT enabled healthcare system using blockchain and its application in numerous fields of the healthcare system (Srivastava in A Light and Secure Healthcare Blockchain for IoT Medical Devices. IEEE, CCECE, Edmonton, AB, Canada, 2019). In the last section, we discussed various benefits, challenges, opportunities, various security challenges in IoT enabled healthcare with a Blockchain system with the solution and currently ongoing research along with future directions.

V. Sharma · N. Lal (✉)
CSE, SET, Mody University, Lachamangarh, Sikar, Rajasthan, India

H. Malik et al. (eds.), *Intelligent Data Analytics for Power and Energy Systems*,
Lecture Notes in Electrical Engineering 802,
https://doi.org/10.1007/978-981-16-6081-8_27

Keywords Blockchain · IoT · Healthcare · Security · Privacy · Patient monitoring · Power · Energy · E-health

Nomenclature

IoT	Internet of Things
HER	Electronics Health Record
ABE	Attribute Based Encryption
BC	Blockchain
Wi-Fi	Wireless-Fidelity
RPM	Remote Patient Monitoring
RFID	Radio Frequency Identification
ARX	Addition-Rotation-XoR

1 Introduction

Internet of things (IoT): IoT could be introducing such that it is a network of various physical objects or even persons are considered as things. These so-called things are embedded with different types of software and sensors. These things can be communicated with each other or exchange and collect data without any human intervention, as shown in Fig. 1. These sensors might be a temperature sensor, pressure sensors, motion sensors, light sensors, moisture sensors, etc., and these can make things intelligent and "IoT" concept introduced by Kevin Ashton in his presentation to Proctor and Gamble [1, 2].

As shown in Fig. 2, As reported by Global Data, the IoT market was $130bn, and it would expect to reach more than $320bn by 2023 [1, 3]. Storing health record digitally, research data related to biomedical, health data about Insurance, Drug Supply data, Medical Education data, Remote Patient Monitoring and many more interrelated data. As mentioned earlier, blockchain technology that can integrate with IoT

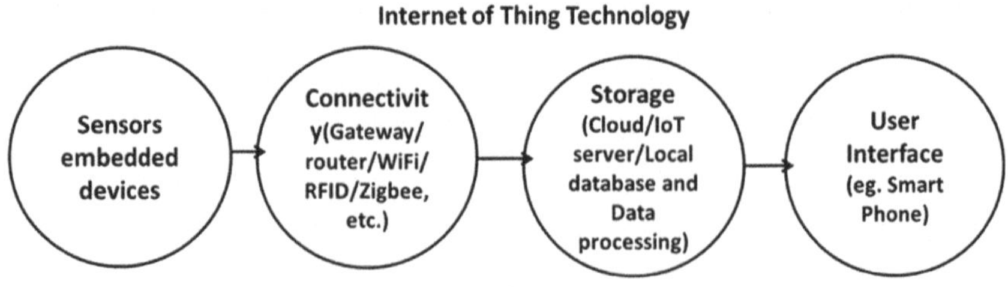

Fig. 1 IoT structure

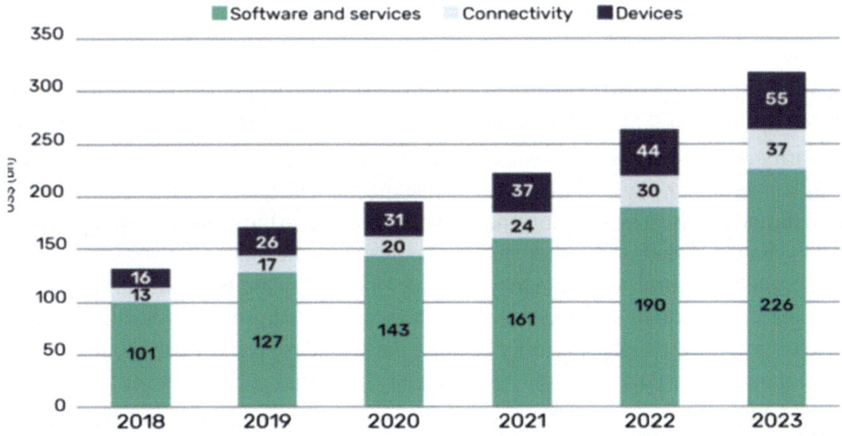

Fig. 2 Global IoT market analysis

would be an outstanding solution. Because of its features as immutability, decentralization, transparency, cryptographic security and interoperability. Satoshi Nakamoto firstly introduced this technology in his white chapter for cryptocurrency as bitcoin. Nowadays, academician, government and industries are showing great interest in research in this direction. Although blockchain technology is based on expensive computation, it involves the overhead of high bandwidth and delay. Therefore, it is not entirely fit for resource-constrained IoT devices suggested smart healthcare system [4, 5].

Healthcare is one of the patients monitoring field remotely for the cure and care of patients. Because these technologies come with severe risks of privacy and data transfer security and the logs of the data transactions in the block which could reveal the privacy of the user, these types of privacy and security problems of healthcare data could outcomes a delay in treatment progress and even death of the patient [6]. The most important contributions of this chapter can be summarized as follows:

- Although, A lot of possible solutions can come out of getting integration of a blockchain and IoT security system.
- The implementation of decentralized mechanism of a blockchain on a whole IoT network can help in several ways.
- At each and every levels, it will confirm the proper security by guaranteeing the privacy preservation and data protection.
- In addition, if scalability issues of blockchain network might be resolve it would provide an effective functioning in the system.
- Integration of the Latest technology IoT and blockchain with Healthcare system will open new challenges for research and development for the organizations.

2 Literature Review

Due to a shortage of proper and sufficient infrastructure, healthcare providers cannot access this critical information when they want to need it. Whether there is a central authority or not, the possibility of tampering with the data remains. If data is stored within a particular physical machine, then someone who has access to it can modify the data and misuse it or even corrupt the data. To maintain the privacy of patient's data and exchange of data with other situated bodies in the healthcare ecosystem, provenance, access control, data integrity and interoperability are very crucial. Recently, with the innovation of new technologies, the security and privacy of healthcare data have been given the highest priority because these types of privacy and security problems of healthcare data could outcomes a delay in treatment progress and even death of the patient.

The authors of [7] have discussed several literature reviews for blockchain-enabled EHR. They only focused on the privacy and security context. They also discussed numerous research challenges of the integration of blockchain in smart healthcare. The authors of [8] have discussed the various solutions for improving the current difficulties of building EHR using blockchain. They also optimized different matrices for performance majors. Authors of [9–11] have discussed the challenges and benefits of a blockchain-enabled pharmaceutical supply chain to improve drug governance.

The authors of [12] have discussed a lightweight consensus mechanism and decentralized approach at different three platforms to control remote patient monitoring. They have also done some performance analysis for verifying the tool. The authors of [13] have discussed permissioned blockchain architecture to enhance the security and privacy of remote patient monitoring systems. They have also discussed the integration of machine learning with this approach. The authors of [6] have addressed a blockchain architecture for securely monitoring the RPM. They have also discussed some lightweight cryptographically techniques like ARX and ring signature to enhance security and privacy. Authors of [14] have discussed permissioned blockchain architecture by using smart contact to monitor the real-time remote patient monitoring. Authors of [15] have discussed a patient-centric agent-based architecture that comprises a lightweight communication protocol to impose data security of different real-time patient monitoring system segments.

Blockchain is based on the principle of the linked list. So, therefore, a blockchain is a series of blocks, and each block is connected to its previous block by some link. Every block in a blockchain is made up of a hash value, the set of transactions, timestamp and Nonce value. A blockchain has the previous blocks' hash values and makes a sequence of cryptographically secured blocks [16]. Satoshi Nakamoto first introduced this concept in 2009 in its white chapter [17]. This concept was raising for digital currency, Bitcoin. This technology was developed for eliminating the necessity for a central authority to provide the trust for directing the transactions among the entities. Therefore, blockchain is a distributed, decentralized and trusted digital ledger of commerce [18]. This technology works in a node-to-node network, and every block in this network is immutable, tamper-proof and transparent. This technology plays a significant role in securing an IoT network because of its properties (Immutability, resilience, auditability, transparency, data encryption) [16].

The book by Yadav et al. [19] addresses the issues related to condition monitoring, fault diagnosis in smart buildings and combines the latest technologies. Jafar et al. [20] cover the soft computing approaches in engineering and the healthcare system for the new research area and researchers. They have integrated the Artificial Intelligence, Machine Learning and Data-analytics approaches. A book presented by A. Iqbal et al. [21, 22] covers renewable energy for sustainable growth with various renewable energy and management applications, and Fatema et al. [23] presented the condition monitoring using Data-Analytics and machine learning. Srivastava et al., in their book volume 1 [24] and volume 2 [25], discussed various emerging techniques applicable in industries, engineering and other applications.

In [26], they discussed different use cases and the applicability of blockchain technology in several industries. In [27], authors discussed other issues of blockchain technology adoption in healthcare, the different challenges in healthcare, and different possible solutions. In [28], they discussed detailed design choices implementation of blockchain technology by other researchers in various situations. In [29], they analyzed 39 research studies, and they provide a comprehensive statistic on different platforms wherein blockchain technology has been applied to improve the healthcare system.

This is the critical field of the healthcare system that involves collecting biomedical data from sensors or from the IoT devices remotely to observe the status of patient's vital signs outside the traditional healthcare environments such as the healthcare places, viz. hospital, clinic, etc. Hence, tampering with remotely collected biomedical data is possible when storing, sharing and retrieving these data. In [12], they have discussed a lightweight consensus mechanism and decentralized approach at different three platforms to control remote patient monitoring. They have also done some performance analysis for verifying the mechanism.

In [6], they have discussed permissioned blockchain architecture to enhance the privacy and security of remote patient monitoring systems. They have also discussed the integration of machine learning with this approach. In [14], they have addressed a blockchain architecture for securely monitoring the RPM. They have also addressed some lightweight cryptographically techniques like ARX and ring signature to enhance security and privacy. In [15], they have discussed permissioned blockchain architecture using smart contact to monitor the real-time RPM. In [30], they have discussed a patient-centric agent-based architecture using lightweight protocol for data security of different real-time patient monitoring system segments.

3 Research Gap and Suggestions

Technology's development and innovation have made our life easy everywhere and in every field like engineering, management, agriculture, in our real-time activities and the healthcare system. The Healthcare system is a significant sector nowadays, and the vast amount of data is produced by the healthcare sector.

3.1 Research Gap

However, as with most emerging technologies, several characteristics are associated with blockchain and IoT problems that make them technically challenging, associated with our research problems. The whole world taking advantages of increasing technologies in each field, but it is found that the health care data is increasing day by day in the healthcare organizations throughout the world, so there will be a problem with data storage, privacy and security. We need a proper approach that can secure the data for further development in companies and businesses analysis. Based on the existing prototypes and developed applications, we have identified below research gaps in the current systems that motivate us to work in blockchain, IoT and the healthcare system.

- Integration of blockchain with the healthcare system is not proper.
- No efficient, lightweight cryptography algorithm is used with blockchain for data security in a resource-constrained platform.
- Data privacy is higher end in distributed ledger transaction environment to optimize the resources efficiently.
- An efficient blockchain distributed system required for scalability that consumes less computation power in execution.
- Need to improve the security and privacy with optimum resources using IoT.
- Need to design a smart contract-based blockchain application for smart healthcare user's data privacy and identity verification with the integration of IoT.
- How can doctor access the patient's health data in the absence of the patient with the help of IoT, blockchain securely.

3.2 Suggestions

- The use of blockchain technology will be the right choice to secure the data privacy and security
- IoT can solve the storage problem, decentralized issues, power issue and energy saving
- The integration of blockchain technology with IoT for healthcare systems will develop a better system
- The use of blockchain and IoT can be grasped to support healthcare services and ecosystems.

4 Layered Architecture of IoT

As shown in Fig. 3, the layered architecture of IoT is classified into three layers, i.e., application layer, network layer and perception layer. Next, we briefly discussed each layer [31, 32].

Application Layer: This layer is the topmost layer of the architecture. The primary responsibility of this layer is to provide the required services or information to meet the essential need of the user. This layer offers different services, backup and storage. For estimating the future state of the device, it provides analysis services [31].

Network Layer: This layer provides an integrated or heterogeneous infrastructure by connecting numerous networks (Wi-Fi, Bluetooth, Zigbee, RFID, etc.). It provides the route and transmits the data acquired from the perception layer by secured channels to central control. It also provides the services like data processing, decision making and computing. This layer is also having devices like cloud, hub, gateways, switches, etc. [32].

Perception Layer: This is the bottom layer of the architecture and is responsible for interacting with various sensors, actuator, RFID and other physical devices. That's why this layer is also known as the sensor layer. This layer is also responsible for data sensing, measuring and sometimes processing of these data. This layer is connected things to the internet [32].

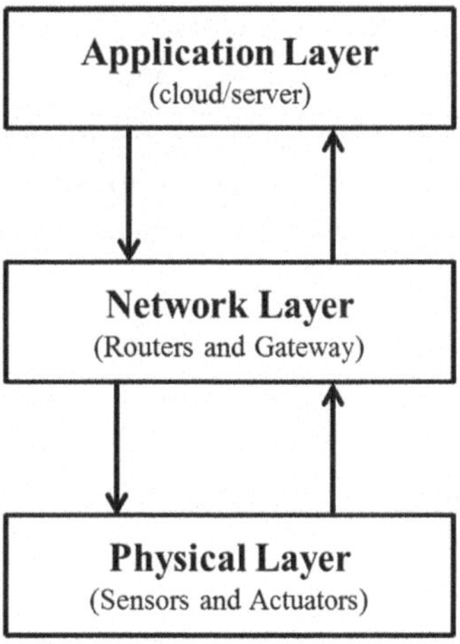

Fig. 3 IoT layered architecture

4.1 Application Areas of IoT

IoT has many application areas. According to the many present technological solutions, various IoT application sectors are available. Some of the sectors are defined here, as have shown in Fig. 4. IoT in Smart Home: A smart home might be possible by utilizing the internet, remotely detect the presence of human and activities, self-controlled appliances and air conditioner controlling, based on user desire and convenience [1]. IoT in Smart City: A smart city might utilize internet-enabled devices to collect and analyze data. Governments use this information to improve the public utilities, infrastructure, surveillance and other facilities in a town, which would increase the life quality in cities [33].

IoT in Smart Industry: Innovative industrial IoT solutions use real-time data from various IoT devices, boost process efficiency, reduce human intervention and energy costs. These acquired data can minimize idle time due to maintenance, absence of assets and shortage of human resources in the industry [1].

IoT in Smart Energy: A smart grid and smart meter can achieve smart energy. The production costs can be effectively managed and controlled by this and save energy; compared to the conventional grid, and it consistently proves more reliable to achieve these goals. Furthermore, intelligent meters by modern energy meter make consumption of the power measurable and help in monitor and control electrical devices [1].

IoT in Smart Agriculture: Using IoT devices to collect real-time data to monitor crop field, the farmers make smart decisions, automate the irrigation system and watch the livestock [33].

IoT in Smart Transportation: Smart transportation systems are improved applications whose purpose is to provide state-of-the-art services relating to different transport, manage traffic modes and empower several users to be better up-to-date and

Fig. 4 IoT application areas

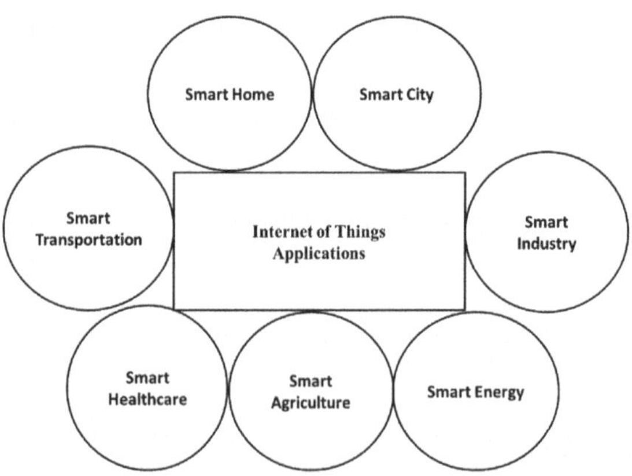

make safer, more coordinated and more imaginative use of transportation networks [1].

IoT in Smart Healthcare: This is one of the demanding IoT technology applications that has been identified in the healthcare sector through the digital concept of health [26]. Enhance the quality, patient's health data could be exploited through IoT support and improved patient care, safety and security and it could also increase the patient's life expectancy. There is a massive possibility in innovative medical devices for different purposes that can monitor various vibrant and valuable human body functions such as blood pressure, respiratory rate, heart rate movement monitoring, etc. Remote patient's health monitoring is also one of the exciting outlooks in this scenario that might be applied to support the products and devices of IoT. It could be enabled the prophecy of the patient body's vital signs and prevention of assured diseases and hazardous states of life [1, 33].

4.2 Challenges of IoT

In the IoT implementation, various devices communicate with each other and store information and data without human intervention. With this, there are some challenges to the application of the Internet of Things, in this section briefly addresses the most common challenges of the IoT [1].

Security: Security presents a significant in IoT to connect various things and devices. Different vendors manufacture these devices; these might be running other operating principles and standards [1]. Due to this heterogeneity, Most IoT devices are vulnerable to cyber-attacks [30]. These types of attacks can manipulate the data and personal information of the users stored on this platform. Thus, it is technically complex to preserve a constant level of security athwart all manufactured IoT devices [33].

Privacy: In the IoT context, a growing number of devices would also increase the chances of sensitive information leaking through unauthorized handling [1]. Hence, these approaches are necessarily to be developed across a broad range of expectations [33].

Scalability: Scalability refers to the term in IoT platform to add new devices and services in the network without affecting the existing quality of services of the network. But due to heterogeneity in the network and devices, adding new devices and services to the existing infrastructure is difficult [1].

Interoperability: Caused of the nonexistence of standardization and mostly the devices' heterogeneity in IoT platform, interoperability is also becoming a big problem [1]. It is studied due to the proposed distributed architecture of the IoT platform security, and privacy breaches of the entire IoT platform have upstretched in its different things. Consequently, securely accessing the basic functionalities related to the platform Security and Privacy becomes the main module in this. Lots of existing technologies like machine learning, fog computing, edge computing and blockchain might be a solution for the privacy and security of IoT network [34]. Blockchain can be a viable solution for a heterogonous network of IoT devices [35]. Blockchain technology is beneficial in forming a trusted and secure configuration for heterogeneous IoT devices [36, 37].

5 Blockchain Architecture

Blockchain technology works in a node-to-node network. Every block in this network is immutable, tamper-proof and transparent because of its properties like immutability, transparency, auditability, data encryption and operational resilience. Figure 5 shows the steps needed to build a blockchain network. The first block in the chain of blocks is called a generic block. Other blocks in the network are called miner block. Miner nodes can add the newly constructed block to the chain by solving a mathematical puzzle named as proof of work called consensus algorithm [38]. Thus, all the participating nodes can build a trusted network over the untrusted participants [39].

Types of Blockchain: Based on the nodes' permission, blockchain is classified into three types.

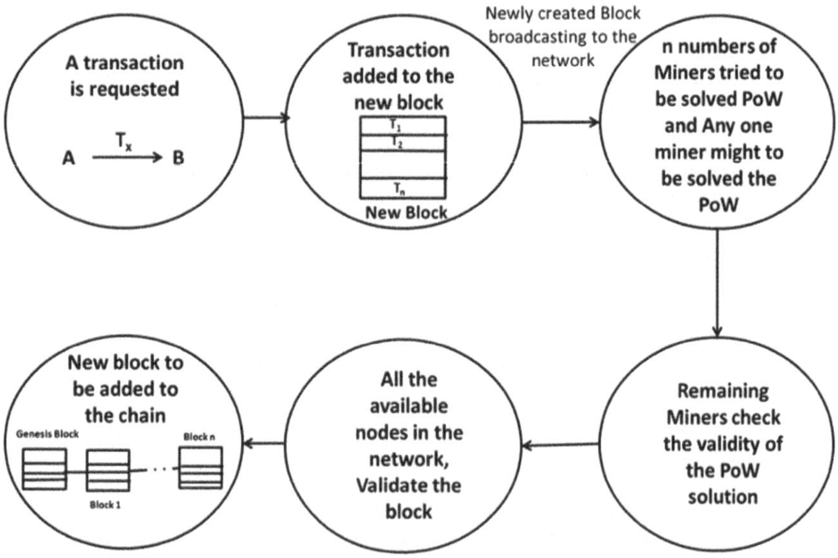

Fig. 5 Blockchain Architecture

Public Blockchain: It is a type of permissionless blockchain. It gives anyone the right to access it. Anyone can also check the overall history of the blockchain along with making any transactions through it [39, 40]. Example—Bitcoin.

Private Blockchain: It is a type of permissioned blockchain. This type of blockchain shared only among the trusted participants. The overall control of the network is kept centralized in it [39, 40]. Example; Ethereum.

Consortium blockchain: it is a semi-private blockchain. It has a controlled user group, and it provides most transaction privacy [39, 40].

6 Use of Blockchain in Numerous Fields of the Healthcare System

Nowadays, technologies are integrating with other areas, and there are many blockchains in healthcare systems. Some we have covered in this section:

Digital health record: DHR is the most popular field of blockchain in Healthcare. This field is generally used for digitally creating, managed and storing the patient's personal and health-related data. Hence, tampering in the database of collected biomedical data is possible when storing, sharing and retrieving these data by the central authorities or malicious person. In [7], they discussed several literature reviews for Blockchain-enabled EHR. They only focused on the privacy and security context. They also discussed numerous research challenges. In [8], they discussed the various solutions for improving the current difficulties to build HER by using blockchain. They also optimized various matrices for performance majors. In [41], they discussed scalable architecture for sharing electronic health records with two different blockchain using a multi-channel hyperledger blockchain. In [42], they discussed a proposed hybrid architecture of edge nodes and blockchain to enable E-Health Records. They also presented two different schemes for user's signatures authentication without exposing the user's sensitive information. They encrypted HER data on the edge node by attribute-based multi-signature and a multi-authority attribute-based encryption (ABE) scheme.

Research data related to biomedical: Historical data of different diseases are stored in a database for finding the cure and medicines of generic illnesses. These data can be exchanged between various healthcare stakeholders. In [43, 44], they discussed that blockchain technology could help remove the fabrication of data and the barring of unwanted clinical research results in clinical trials [45, 46]. They also discussed that, because of the anonymization and immutability properties of blockchain. The patients easily might permit consent for their healthcare data to be used for clinical trials. Therefore, blockchain might be reform the biomedical research.

Health data related to Insurance: Digital health record can be used to process insurance claim in the healthcare system. Several research chapters [45–48] find that claim processing for insurance is a favorable area for applying blockchain technology in the Healthcare area.

Drug Supply data: This is a healthy supply chain for medicines and pharmacy equipment, mainly in the drug or pharmaceutical industries. The delivery of forged or cheap medications can have terrible consequences for the patients, which is a common problem in the pharma industry. In [9–11], they discussed blockchain-enabled pharmaceutical supply chain challenges and benefits to improve drug governance.

Medical Education data: In this methodology, in online education, we don't know the validity and accountability of the knowledge acquired, created and shared and also the mediators those providing certification and credentialing of healthcare professionals. In [44], they discussed health professional education systems based on values, based on competence,= and should offer credential services without any central authority.

7 IoT-Blockchain Integration

IoT can exploit the blockchain for ensuring the integrity of the sensor's data. Therefore, blockchain is a very efficient way by which sensors can exchange data without interfering with a third party for establishing trust. But, the integrity of IoT and blockchain depends on the memory capabilities and power consumption in various applications [49]. IoT can take the advantages of blockchain technology in different ways like Trust building, Acceleration of data exchange, cost reduction and security improvement.

Although integration of BC with IoT is not effortless and it is necessary to pay attention to some serious issues related to this.

- Cryptographic mathematical puzzle-solving or mining (Say PoW) is computationally exhaustive, which is unsuitable for most resource-constrained IoT devices.
- As the number of nodes increases in the blockchain networks, the networks of blockchain scales poorly as most IoT networks have large numbers of nodes in the network.
- Due to transaction confirmation over the extensive network, the throughput of the network decreases and the latency increases. Which may be unacceptable for the limited bandwidth's IoT network.

Therefore, blockchain technology may thus have to be implemented in selected IoT network. The IoT-blockchain architecture application must be such that it can handle massive traffic generated by the network. Also, it must provide data security and attack and threat–resilient [50]. The following features of blockchain like decentralized, trustless, autonomous, anonymity and immutability make it suitable to be applied in numerous applications like Smart Healthcare, Smart Home, Smart City, etc. Ethereum (permissioned) blockchain can remove trusted centralized authority or

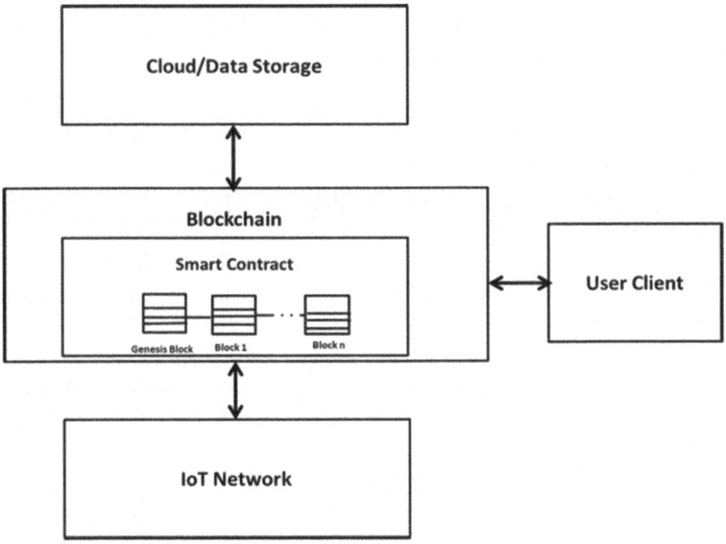

Fig. 6 IoT and blockchain amalgamation architecture

human control by using smart contracts as executed the rules automatically based on current situations and the rules are publicly available, thereby endorsing transparency [51].

So, this chapter investigates a blockchain integrated architecture with IoT for remote patient monitoring that conveys energy and power-efficient lightweight and decentralized structure for security and privacy preservation of patient's health data, suitable for IoT devices, which holds the advantages of blockchain while overcoming the above-discussed issues in integrating blockchain in IoT networks [4, 6]. Figure 6 shows the integration of IoT and blockchain.

A smart contract is a computer program that operates the many different logics that the network members settled. It runs on the blockchain and provides a way of interfacing to interact with the data. A smart contract program is available to each member present on the network [50]. As the transactions are added to the blockchain, in the exact way participating members add the smart contracts to the blockchain thus, these are included in the blocks [52]. All the changes are recorded in the next block; even the transactions updated the smart contract states after the change had been made. This process makes smart contracts immutable in the same manner as in transactions [53].

8 Role of Blockchain in IoT Enabled Energy and Power-Efficient Healthcare System

Every day a massive amount of critical data has produces by the healthcare sector generates. These data often remain dispersed and disorganized across various systems [54, 55]. Due to a shortage of proper and sufficient infrastructure, healthcare providers cannot access this vital information when they want to need it. Whether there is a central authority or not, the possibility of tampering with the data remains. If data is stored within a particular physical machine, then someone who has access to it can modify the data and misuse it or even corrupt the data [4, 55, 56]. To maintain the privacy of patient's data and exchange of data with other situated bodies in the healthcare ecosystem, provenance, access control, data integrity and interoperability are very crucial [57]. Energy and power-efficient BIoT healthcare architecture is shown in Fig. 7 [58].

With the recent immense popularity and growth of advanced technologies, integration with the healthcare system is the main domain to help medical practitioners. The gathered information turned into data delivers better performance when used effectively by algorithms. The timely availability of crucial medical data can save people's lives in emergencies. Thus, collected data has become an essential concern for the healthcare community regarding user's data security and privacy.

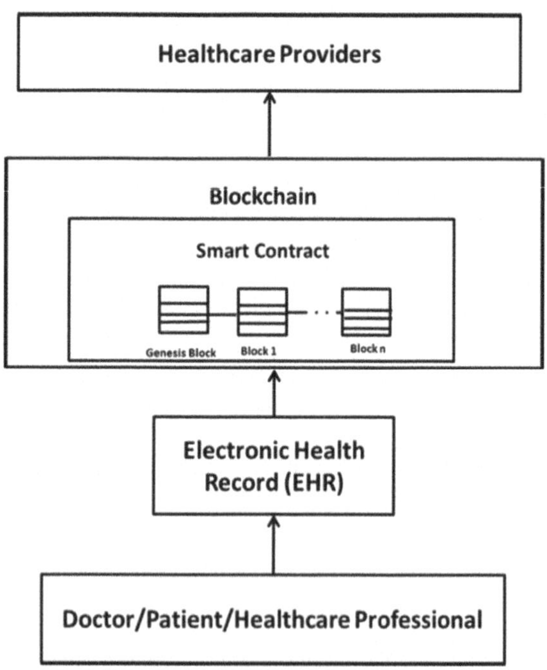

Fig. 7 Blockchain in IoT enabled healthcare system

In this proposed architecture, three stakeholders (Patient, Doctor, Healthcare Professional) of the overall system architecture. They will communicate with each other through the proposed blockchain network. The patient's health data would be store in an off-chain database (EHR) in a cloud network. The hash value of the data would be store in blockchain. The lightweight cryptographic algorithm and the digital signature scheme would be used for security and privacy preservation, and authentication. To solve the severe challenges of Healthcare, blockchain has been suggested as a solution to be secured sharing of health-related data and obedience to the law of data privacy.

9 Conclusion

From the literature review, we have concluded that the traditional blockchain technology is not compatible in a resource-constrained environment. In our research, we will find the solution for integrating blockchain with IoT and healthcare system to secure sensor's data and privacy. Blockchain has become a hot topic in academic research today. This chapter includes a systemic review of the importance of blockchain in IoT enabled power and the energy-efficient healthcare system. This chapter expresses that blockchain technology has numerous healthcare uses cases like storing health record digitally, research data related to biomedical, health data about Insurance, Drug Supply data, Medical Education data and Remote Patient Monitoring. In our study, we have given maximum attention to the blockchain-based remote patient monitoring system. We have discussed several limitations and the challenges of the blockchain-based remote patient monitoring system and current approaches to developing this power and energy-efficient architecture application. In future work, we simulate over proposed architecture and analyze it for performance security of healthcare data using blockchain and IoT. Our approach also motivates the researcher to work in this latest research area and opens the way to work on challenges discussed.

References

1. N. Sandro, Petar, Diego, Luigi, Internet of Things (IoT): opportunities, issues and challenges towards a smart and sustainable future. J. Cleaner Prod. **274** (2020)
2. Ahanger et al., Internet of Things: a comprehensive study of security issues and defense mechanisms. IEEE Access **7**, 11020–11028 (2019)
3. M. Froese, Global IoT market, Wind power Eng. & Dev. Accessed on 26 June 2021
4. Dwivedi, et al., A decentralized privacy-preserving healthcare blockchain for IoT. Sensors (2019)
5. A. Dorri, S. Kanhere, R. Jurdak, Blockchain in Internet of Things: Challenges and Solutions, arxiv
6. G. Srivastava, A Light and Secure Healthcare Blockchain for IoT Medical Devices. IEEE, CCECE, Edmonton, AB, Canada, 2019, pp. 1–5

7. Shuyun, Debiao, Li, Kumar, Khan, Kim, Applications of blockchain in ensuring the security and privacy of electronic health record systems: a survey. Comput. Secur. **97** (2020)
8. Tanwar, et al., Blockchain-based electronic healthcare record system for healthcare 4.0 applications. JISA **50** (2020)
9. Ahmadi, Benjelloun, M. El Kik, Sharma, Chi, Zhou, Drug Governance: IoT-based Blockchain Implementation in the Pharmaceutical Supply Chain. MobiSecServ, Miami Beach, FL, USA, 2020, pp. 1–8
10. Premkumar, Srimathi, Application of Blockchain and IoT towards Pharmaceutical Industry, ICACCS, Coimbatore, India, 2020, pp. 729–733
11. Tseng, Liao, Chong, Liao, Governance on the Drug Supply Chain via Gcoin Blockchain. IJPRPH (2018)
12. M.A. Uddin, A decentralized patient agent controlled blockchain for remote patient monitoring. WiMob, Barcelona, Spain, 2019, pp. 1–8
13. J. Hathaliya, Blockchain-Based Remote Patient Monitoring in Healthcare 4.0, IEEE IACC, Tiruchirappalli, India, 2019, pp. 87–91
14. Griggs, Healthcare Blockchain System Using Smart Contracts for Secure Automated Remote Patient Monitoring. J. Med. Syst. **42**, 130 (2018)
15. Uddin, Stranieri, Gondal, Balasubramanian, Continuous patient monitoring with a patient centric agent, a block arch. IEEE Access **6**, 32700–32726 (2018)
16. V. Sharma, N. Lal, A Detail dominant approach for IoT and blockchain with their research challenges, ICONC3, Sikar, India, 2020, pp. 1–6
17. S. Nakamoto, Bitcoin: A Peer-to-Peer Electronic Cash System. Cryptography Mailing list at https://metzdowd.com (2009)
18. M. Jena, et al., Blockchain technology: a survey on applications and security privacy Challenges. IoT **8** (2019)
19. A.K. Yadav, *Soft Computing in Condition Monitoring and Diagnostics of Electrical and Mechanical Systems*, 1E, Springer Nature, Part of the Adv. in Intell. Syst. Comput., vol. 1096, 2020, p. 496
20. J.A. Alzubi, *AI and Machine Learning Paradigms for Health Monitoring System: Intelligent Data Analytics*, 1E, Springer Nature, Part of the Studies in Big Data, vol. 86, 2020, p. 513
21. A. Iqbal, *Meta Heuristic and Evolutionary Computation: Algorithms and Applications*, 1E, Springer Nature, Part of the Studies in Compu. Intell., vol. 916, 2020, p. 849
22. A. Iqbal, *Renewable Power for Sustainable Growth*, 1E, Springer Nature, Part of the Lecture Notes in Electrical and Engineering, vol. 723, 2020, p. 805
23. N. Fatema, *Intelligent Data-Analytics for Condition Monitoring: Smart Grid Applications*, 1E (Academic Press, 2021)
24. S. Srivastava, *Applications of Artificial Intelligence Techniques in Engineering*, vol. 1, 1E, Springer Nature, Part of the Adv. in Intell. Syst. and Comp., vol. 698, 2018, p. 643
25. S. Srivastava, *Applications of Artificial Intelligence Techniques in Engineering*, vol. 2, 1E, Springer Nature, Part of the Adv. in Intell. Syst. Comput., vol. 697, 2018, p. 647
26. Jaoude, Saade, Blockchain applications–usage in different domains. IEEE Access **7**, 45360–45381
27. Agbo, Mahmoud, Eklund, Blockchain technology in healthcare: a systematic review. Healthcare **7**(2), 56
28. O'Donoghue, Design choices and trade-offs in health care blockchain implementations: systematic review. J. Med. Internet Res. **21** (5), e12426
29. Hasselgren, Kralevska, Gligoroski, Pedersen, Faxvaag, Blockchain in healthcare and health sciences—a scoping review. Int. J. Med. Inform. **134**
30. Jemal, Kornegay,Security assessment of blockchains in heterogenous IoT networks: invited presentation. Annual CISS, MD, USA, 2019, pp. 1–4
31. Karie, IoT Threat Detection Advances Challenges and Future Directions, The Workshop on ETSecIoT, Sydney, pp. 22–29, 2020
32. K. Kaur, A Survey on Internet of Things—Architecture, Applications, and Future Trends, in *ICSCCC-2018*, 2018, pp. 581–583

33. Goswami, Padhya, Patel, Internet of Things: Applications, Challenges and Research Issues. I-SMAC, Palladam, India, 2019, pp. 47–50
34. V. Hassija et al., A survey on IoT security: application areas, security threats, and solution architectures. IEEE Access **7**, 82721–82743 (2019)
35. Q. Zhu, et al., *Applications of Distributed Ledger Technologies to the Internet of Things: A Survey* (ACM, 2019), pp. 1–34
36. J. Raju, et al., Cyber Security in IoT Devices using Blockchain Concept (2020)
37. A. Dorri, et al., Blockchain in Internet of Things: Challenges and Solutions. Arxiv, 2016
38. V. Sharma, N. Lal, A novel comparison of consensus algorithms in blockchain. AAMS **19** (2020)
39. M. Daniel, O. Benedict, Blockchain mechanisms for IoT security. Internet of Things **1**, 1–13 (2018)
40. M. Singh, et al., Blockchain: a game changer for securing IoT data, in *IEEE, (WF-IoT)*, 2018, pp. 51–55
41. Fernandes, Rocha, A.F.D. Conceição, Horita, Scalable architecture for sharing EHR using the Hyperledger Blockchain, IEEE ICSA-C, Salvador, Brazil, pp. 130–138, 2020
42. H. Guo, Attribute-based multi-signature and encryption for EHR management: a blockchain-based solution. IEEE ICBC, Toronto, ON, Canada, 2020, pp. 1–5
43. Radanović, Likić, Opportunities for use of blockchain technology in medicine. Appl. Health Econ. Health Policy **16**, 583–590 (2018)
44. Funk, Riddell, Ankel, Cabrera, Blockchain technology: a data framework to improve validity, trust, and accountability of information exchange in health professions education. Acad. Med. 1791–1794 (2018)
45. Saldamli, Reddy, Bojja, Gururaja, Doddaveerappa, Tawalbeh,Health Care Insurance Fraud Detection Using Blockchain, SDS, Paris, France, 2020, pp. 145–152
46. W. Liu, A Blockchain-Based System for Anti-Fraud of Healthcare Insurance. IEEE: ICCC, Chengdu, China, 2019, pp. 1264–1268
47. J. Gera,Blockchain Technology for Fraudulent Practices in Insurance Claim Process. ICCES, Coimbatore, India, pp. 1068–1075, 2020
48. Zhou, MIStore: A Blockchain-Based Medical Insurance Storage System. J. Med. Syst. **42**, 149 (2018)
49. Thakore, Vaghashiya, Patel, Doshi, Blockchain-based IoT: a survey. Proc. Comput. Sci. **155**, 704–709 (2019)
50. Mehedi et al., Blockchain-based security management of IoT infrastructure with Ethereum transactions. Iran J. Comput. Sci. **2**, 189–195 (2019)
51. Hang, Kim, Design and implementation of an integrated IoT blockchain platform for sensing data integrity. Sensors **19**, 2228 (2019)
52. A. Dorri, et al., Blockchain for IoT Security and Privacy: The Case Study of a Smart Home, 2019
53. Panarello, et al., Blockchain and IoT integration: a systematic survey. Sensors (2018)
54. Hathaliya, Tanwar, An exhaustive survey on security and privacy issues in Healthcare 4.0. Comput. Commun. **153**, 311–335 (2020)
55. Srivastava, et al., A Light and Secure Healthcare Blockchain for IoT Medical Devices. IEEE CCECE, Edmonton, Canada, 2019, pp. 1–5
56. Y. Sun, Security and privacy for the internet of medical things enabled healthcare systems: a survey. IEEE Access **7**, 183339–183355 (2019)
57. William, Christian, Blockchain technology for healthcare: facilitating the transition to patient-driven interoperability. Comput. Str. Biotech. **16**, 224–230 (2018)
58. S. Khezr et al., Blockchain technology in healthcare: a comprehensive review and directions for future research. Appl. Sci. **9**, 1736 (2019)

Security Challenges in Internet of Things (IoT) Integrated Power and Energy (PaE) Systems

Kamaldeep, Manisha Malik, and Maitreyee Dutta

Abstract The power and energy (PaE) systems are undergoing a major transformation to ensure cleaner energy with sustainable economic growth. The focus of this transformation is the introduction of Internet of Things (IoT) which has enabled real-time control and monitoring, fault location detection, smart metering, building automaton and smart grid with smart sensors and programmable hardware. This integration of IoT and power systems offers several opportunities of development and growth in economy as well as to human lives. However, there are certain security and privacy challenges with the deployment of IoT for power systems. The advancements in technology have enabled the attackers to launch attacks on IoT-enabled systems. Thus, this review paper provides comprehension into the security issues and challenges in integrating IoT to power and energy systems.

Keywords Internet of Things · 6LowPAN · RPL · Network security · Smart metering · Power and energy systems

1 Introduction

The Internet of Things (IoT) has dominated and revolutionized many sectors of development like home automaton, smart transportation and cities and intelligent health care. The exponential growth of IoT is fueled by the widespread adoption of the Internet and the continuous depreciation in the cost and size of sensor technology during the last few decades. Other influencing factors include the availability of cheap Internet connections, development of built-in sensors in devices, the mobile insurgency, and a multitude of companies designing required IoT applications and software. By 2030, the IoT will see an explosive growth, where the total number of Internet connected smart devices will be 125 million [28].

Kamaldeep (✉) · M. Malik · M. Dutta
National Institute of Technical Teachers Training and Research, Chandigarh, India
e-mail: manisha.cse@nitttrchd.ac.in

H. Malik et al. (eds.), *Intelligent Data Analytics for Power and Energy Systems*,
Lecture Notes in Electrical Engineering 802,
https://doi.org/10.1007/978-981-16-6081-8_28

Since the last decade, IoT has even dominated the energy sector. The transformation of electric power and energy (PaE) systems into sustainable and cleaner systems began in 2015 with the climate change agreement [9]. The aim was to improve the reliability and efficiency of power distribution and reduce carbon emission. Additionally, the transformation is driven by the low cost of batteries and semiconductors used for developing sensing materials and the increased ease of access to the Internet technology. IoT provides Internet connectivity to everyday objects, and all physical things like goods, people, machines, appliances, and vehicles are things in IoT. One of the early incarnations of consumer applications of IoT in power sector is the smart meters and smart thermostats [33]. Later, the IoT solutions have entered the industrial domain where the major application in power sector is the maintenance and condition monitoring of assets [13, 19, 20, 22].

The IoT technology is proving to be a strong enabler of smart energy management systems that are both Internet-enabled as well as intelligent. With a goal to limit energy costs, reduce carbon emission, maximize revenue, and optimize power flow and real-time feedback, power and energy systems are being developed with IoT technology. In IoT, the data generated from energy resources will be sensed, gathered, communicated, and analyzed over the Internet. This will enable power engineers to avoid any excess power generation, and if generated, they can supply it to the power grid. Hence, we can say that with IoT, power systems will become smarter, i.e., more cost effective, sustainable, efficient, reliable, and secure.

The consolidation of IoT with PaE systems provides numerous opportunities to build smarter cities and a smarted world. Both the technologies have great social and financial impact, and thus, ensuring the security and privacy of IoT-enabled PaE systems becomes crucial. The security concern can be seconded by the increased number of attacks on IoT devices in the last decade. For instance, in 2019, the hackers hit the Norsk Hydro project with ransomware costing the company over $40 million in losses [7]. In 2013, the attack on retailer target was initiated through the automated electric heating system causing the company losses of $290 million [37].

Through this paper, we will highlight the security challenges with IoT-enabled PaE systems and enumerate the various possibilities associated while addressing those challenges (Fig. 1). We set our domain outright and talk about security challenges in IoT, in general which are applicable to each domain of PaE systems also. The three aspects taken into consideration for securing the IoT-integrated PaE systems are shown in Fig. 2.

To summarize, the contributions that our review brings to securing the IoT-enabled PaE systems are the following:

- A detailed review of the research proposals in IoT-integrated PaE security;
- A breakdown of different security aspects and solutions proposed;
- A summary of current literature analyzing the advantages and disadvantages of the different proposals;
- A walk through of open issues and challenges to achieve robust IoT-integrated PaE systems.

The remainder of the paper is broken down as follows: In Sect. 2, we review related literature of IoT standards and security services. Section 3 discusses the security challenges and presents recommended solutions for securing IoT-integrated PaE systems. Section 4 concludes the work and provides scope for future research.

2 Related Work

2.1 IoT Standards

IoT uses different technologies implementing various standards to enable the communication of a single device with the Internet or other similar devices. The variability among IoT devices utilizing different standards gives rise to a new challenge, interoperability [29]. In IoT-integrated systems, two types of standards exist: communication and network protocol standards and regulatory standards to ensure data privacy and security. One of these standardized communication protocols designed by the IEEE and IETF [12] is listed in Fig. 3. The efforts of these working groups

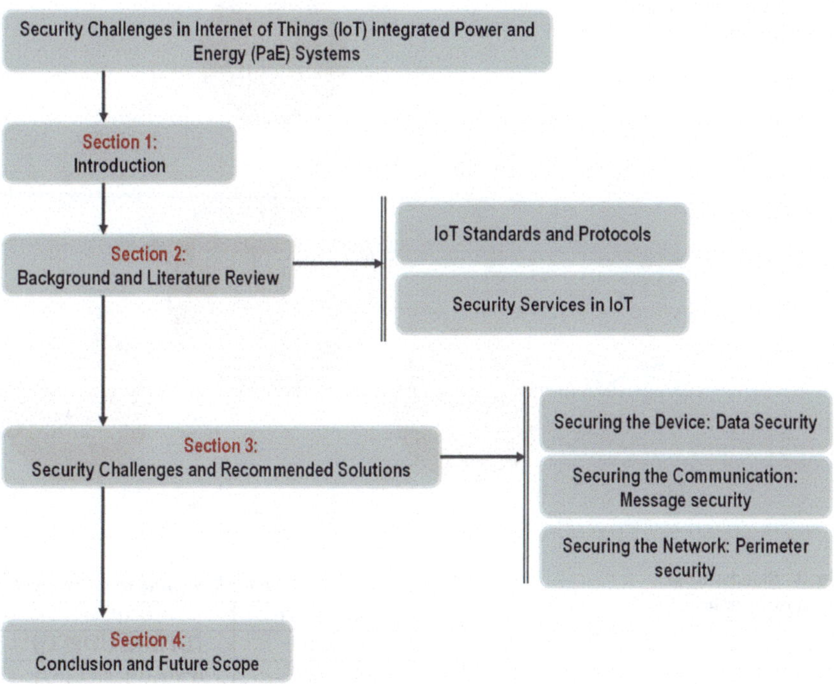

Fig. 1 Holistic organization of this chapter

play a pivotal role in realizing futuristic IoT applications including smart grid and other industrial power applications.

As given in Fig. 3, the Constrained Application Protocol (CoAP) [39] runs at the application layer. Developed by a working group, the Constrained RESTful Environments (CoRE), CoAP was designed with aim of minimum energy and computation overhead and support for multicasting within the network. It is said to be alternative of the modern HTTP protocol with request and response packet exchanges between the CoAP client and the CoAP server. At the transport layer, the User Datagram Protocol (UDP) [32] is implemented in IoT as it does not incur the overhead for a handshake before actual communication between entities and is thus suitable for IoT. The IPv6 protocol at the network layer functions the same way as in the conventional Internet. However, the routing protocol, i.e., the IPv6 Routing Protocol for Low Power and Lossy Networks (RPL) developed by the Routing over Low power and Lossy networks (ROLL) group, is a new routing protocol developed by IETF in RFC 6550 [45]. A novel layer called as the adaptation layer implements the 6LoWPAN protocol [30] to allow the transmission of bigger IPv6 packets over small packet size compatible IEEE 802.15.4 networks. It details mechanisms for compressing, fragmenting, and

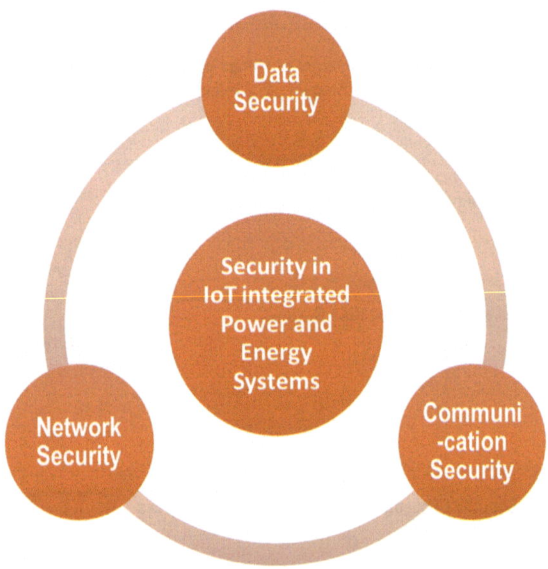

Fig. 2 Security aspects in IoT-integrated power and energy systems

Fig. 3 Standardized protocol stack in IoT

Layer in IoT Stack	Relevant IoT Protocol
Application	CoAP[41]
Transport	TCP[34], UDP[33]
Network	IPv6[10], RPL[47]
Adaptation	6LoWPAN[31]
Data Link	IEEE 802.15.4[2]
Physical	

reassembling of headers to allow routing of IPv6 packets over IoT networks. The last layers data link and physical define mechanisms of medium access and encodings of the packets. IEEE 802.15.4 [2] protocol can also be used with higher protocols like WirelessHART and Zigbee.

2.2 *Security Services for IoT*

The Internet Protocol Version 6 (IPv6) enables the connection of every possible physical object to the Internet. This seamless connection gives rise to numerous IoT applications, smart PaE systems management being one of them. Since the objects are globally accessible and involve sensitive applications, real deployment of security in IoT is pivotal requirement. Some of the security services essential for IoT are as follows:

- Confidentiality: Confidentiality ensures that the data flow between sender and receiver cannot be intercepted and secret contents are unrevealed to the attacker. This holds true for data in transit with end-to-end message security as well as for data stored on IoT device should be concealed from unauthorized access. This is realized by encryption/decryption.
- Availability: Availability ensures that the services offered by IoT devices are always available and run efficiently. To ensure availability, detection of intrusions and other malicious activities are essential. Intrusion detection and prevention systems (IDPS) and firewalls are deployed to ensure availability.
- Data and Message Integrity: The secret message contents being exchanged between two entities cannot be undetectably changed. It applied to data stored on the IoT devices. Integrity check algorithms like hash functions are implemented to ensure data and message integrity.
- Authentication: Authentication ensures source integrity in a sense that entities involved in communication are able to verify each other with what both of them are claiming to be. A number of such authentication schemes exist [41].
- Non-Repudiation: Lastly, non-repudiation ensures that the entities involved in communication cannot deny of being in the communication partly or fully.

Thus, in order to provide multi-dimensional security, data security, communication security as well as network security needs to be ensured as we proceed to discuss in Sects. 3.1, 3.2 and 3.3. A review of security issues in smart grid is discussed in [16, 21, 43].

3 Security Challenges and Recommended Solutions

3.1 Securing the Device: Data Security

The first and the most important aspect is to secure the sensitive data stored on an IoT device. This data can be the sensed data, cryptographic credentials like pre-shared symmetric keys, public key, authentication certificates, and even secure identities. Since guarding and protecting the IoT devices physically is practically impossible, hardware-based protection mechanisms like trusted platform module (TPM) or smart cards are often used. Other software-based secure storage solutions like Codo [3] also exist.

The PaE IoT hardware or infrastructure damage causes the device disconnection from the network. For instance, PaE systems may be sabotaged by clipping connection to the network or by damaging the smart metering units [4]. This is true for power stations as well. In the year 2014, an assault on California power station whose power lines for electricity go from residences to businesses in Silicon Valley knocked out 17 giant transformers. Such a terrorist attack can also aim nuclear power plants which will have devastating effects on human health and the environment. Therefore, such attacks can be minimized by reducing the accessibility to critical PaE systems IoT infrastructure.

Ideally, the data stored in IoT-based PaE systems devices should be saved in encrypted format accompanied by its hash. However, such an approach involves steps of decryption, data integrity check followed by encryption and integrity protected again to transmit over the channel in case when any remote hosts ask for the data. This way cryptographic operations are run twice which eventually consume a lot of resources.

When discussing data security, the location of storing the data and cryptographic method of encryption are of utmost important. For the location, with recent advancements, flash memories have gained researchers' attention. Research works have focused on storing data on flash memory and then transmitting it directly for secure communication. For instance, data can be stored in format compatible with the security protocol working at the lower layers so that it may be exchanged over the network without the need of any further cryptographic processing.

The rapid generation of newer data at the IoT-integrated PaE systems device end clearly indicates that it cannot be stored on the device itself and needs to be communicated to an infrastructure that can conceptually support infinite storage is always available for real time processing and analysis. The infrastructure in this case is the cloud computing.

3.2 Securing the Communication: Message Security

Communication between devices can be secured by using the security services listed above. The standard Internet security solutions can be implemented at the different layers of the standardized IoT stack presented in Table 1. For PaE systems devices implementing the standardized IoT stack for communication, the protocols and respective security mechanisms proposed in the literature are as follows:

- **IEEE 802.15.4 Security (Link Layer)**: The IEEE 802.15.4-2011 standard provides security at the link layer by efficiently utilizing the hardware supporting the symmetric cryptography. The most common symmetric algorithm implemented on IoT devices at link layer is the advanced encryption standard (AES). The standard incorporates multiple security modes like counter (CTR) mode, cipher block chaining (CBC) modes and counter with cipher block chaining (CCM) modes of AES. AES not only ensures confidentiality by providing encryption of messages at the link layer but also ensures data authenticity and integration when implemented using the CBC mode. In addition, the radio chips implementing IEEE 802.15.4 allow storage of upto 255 access control entries. However, management of keys and initialization vector values in ACL entries remains an area of open research.
- **Internet Protocol Security (IPSec) and RPL Security (Network and Routing Layer)**: In IoT, the IPSec framework enables the encryption and authentication at the network layer just like conventional Internet. However, the application of IPSec in IoT and its resource implications have been analyzed in related literature [34]. The RPL specification [45] includes mechanisms to secure RPL control message as well as supports some basic security modes. However, security in RPL is mostly dependent on the underlying application area since each application is mapped to an objective function which optimizes the target application scenario. For instance, RFC 5867 [14] is for building automaton, and RFC 5548 [18] is for modern low-power applications.
- **Security in 6LoWPAN (Adaptation Layer)**: The raw specification of 6LoWPAN does not define any security mechanisms, but relevant research on the security vulnerabilities and consideration of applying network layer security like IPSec onto the adaptation layer is also underway. However, the resource constraints of 6LoWPAN networks must be kept in mind while applying network layer security mechanisms that work in tandem with adaptation layer security. Such works have been analyzed in [36, 38]. Another approach includes the optimization of 6LoWPAN headers by compression to function the same as encapsulating payload (ESP) and authentication header (AH) of the IPSec [11, 34].
- **Datagram Transport Layer Security (DTLS) (Application and Transport Layer)**: DTLS is a transport layer security (TLS) mutant with methods to cater to unreliable nature of the UDP protocol. In order to bind with DTLS, CoAP defines security modes with pre-shared key, authenticated key, raw public key, and no security mode. The optimization in DTLS has been done in [35] with the compression of DTLS headers to reduce the number of bytes transmitted, thus minimizing communication overhead.

Regardless of the security service providers by different IoT layers, a few challenges still exist that we proceed to discuss next. To be able to implement any security protocol or algorithm at the IoT layer, key management is a central issue that needs attention. Since keys are the core of cryptography, appropriate key management solutions must be designed. A comprehensive review of such solutions has been presented in our earlier work in [26]. Another area of research is to develop cryptographic solutions that require zero configuration on the PaE systems integrated with IoT [4]. One such identity-based encryption protocol for smart grid is proposed in [42]. In other words, it means that the devices will be able to configure themselves as a system without depending on any central smart meter systems. However, the authors in [4] observed that success of such proposals will depend on standards being developed and diligently followed by the original equipment manufacturers (OEM). To achieve strong hardware-based authentication and key management, another approach to protect the device is to incorporate physically unclonable functions (PUF). PUF are physical security primitives providing digital fingerprints based on physical properties of a device which cannot be replicated even if the same manufacturing process is repeated.

3.3 Securing the Network: Perimeter Security

Even with communication security in place, an IoT network is susceptible to a large number of attacks against availability services of the network. Such attacks disrupt the network by interrupting the routing mechanisms or are mainly denial of service (DoS). A DoS attack consumes network services by overloading the bandwidth which results in legitimate traffic being dropped or slowed. For instance, in PaE systems, the real-time sensor measurement data can be delayed or slowed in a DoS attack which may result in inaccurate forecasts, delayed transmission status, and even entire failure of network of sensing and actuating devices. A DoS attack can be implemented at multiple layers of the IP stack: transport, network, data link, and even application layer [27]. These attacks are usually flooding-based, but they can also be triggered with a few packets [25]. DoS attacks have the capability to damage the PaE systems and hence solutions like the firewall, intrusion detection systems (IDS), traffic filtering, and air gapping.

The most economical approach to prevent DoS on PaE systems is traffic filtering using a network or host firewall. In traffic filtering, aim is to distinguish between DoS and legitimate traffic with distributed infrastructure and then redirect the malicious traffic [46]. However, this approach becomes difficult to implement when we have large inflow of traffic. In addition, this technique will only prevent from DoS attack originating from outside of the PaE systems network.

Another approach is to implement an IDS on PaE systems networks. IDS can be classified as signature-based or anomaly-based. Since signature IDS are based on known rules or signatures, they cannot detect zero day attacks. However, they utilize less resources of the network and are thus lightweight to implement on resource

Table 1 Summary of security challenges and recommended solutions for IoT-integrated PaE systems

Security challenge	Recommended solutions and relevant references
Data security	• Secure data aggregation and transmission [1, 44] • Secure cloud and Fog storage solutions [8] • Restricted access to critical PaE infrastructure
Communication security	• Proposals for effective key management in IoT [26] • Development of lightweight cryptographic primitives [35] • Ensuring authentication and non-repudiation [10, 17]
Network security	• Traffic filtering [46] • Ensuring availability by developing intrusion detection and prevention systems for PaE [6, 31, 40] • Intrusion response systems [15] • Risk assessment

constrained IoT-based PaE systems. On the other hand, anomaly IDS can detect zero day attacks by finding anomalies in the network traffic. However, in this case, there is trade-off between detection performance and network utilization to implement anomaly IDS on IoT-based PaE systems. Thus, hybrid IDS may be proposed for smart PaE systems that incorporates the benefits of both signature- and anomaly-based IDS. In such an IDS, resource-intensive anomaly IDS tasks can be delegated to a resource-rich device in the PaE systems network, and lightweight rules can be deployed on the constrained IoT devices in the PaE systems network [15].

Apart from the two approaches discussed, two comparatively expensive approaches are the air gapped network and the big pipes. In an air gapped network, a secure network is physically isolated from an insecure network [5]. This approach incurs huge cost to build separate infrastructures for the PaE systems. Big pipes, on the other hand, are scaled bandwidth network connections that have the ability to absorb large volume of traffic to mitigate the DoS attack.

In situations when intrusion cannot be prevented or detected, appropriate logging procedures must be in place for audit control and forensic analysis. These systems are called intrusion response systems, and these have been extensively surveyed for IoT in [15]. Some other applications of artificial intelligence in power systems have been discussed extensively in [23, 24].

A summary of security challenges and recommended solutions for IoT-integrated PaE systems have been listed in Table 1.

4 Conclusion and Future Scope

The security challenges of integrating IoT with power and energy systems were presented in this paper. The digitization of current power systems with IoT can help in optimized energy consumption, reduced carbon emission, improvement in efficiency and reliability, and robustness of power networks. The role of IoT standard protocols for communication and security was presented, wherein a detailed analysis of the three aspects, namely security, data, communication and network security, was provided.

For the continuous growth of IoT-integrated PaE systems, it is important to develop robust security solutions to cater to the increased vulnerabilities. Some of the recommended solutions were presented in the paper. The client server and centralized nature of IoT make IoT-based PaE susceptible to number attacks and are easy to compromise. Therefore, as a potential future direction, blockchain can be integrated with IoT-based PaE to ensure robustness in security and privacy. Blockchain provides a robust decentralized service with zero intervention of the third party, and with secure distributed ledger technology, millions of IoT devices can be synchronized. In IoT-based PaE, the implementation of blockchain will propel the IoT efficacy by providing an architecture for securing the distributed PaE systems. Research is also foreseen on quantum cryptographic solutions as quantum computers are developed. Finally, security algorithms built for IoT-based PaE should aim to consume minimum energy to realize the ultimate objective of green IoT.

References

1. A.A. Abd El-Latif, B. Abd-El-Atty, I. Mehmood, K. Muhammad, S.E. Venegas-Andraca, J. Peng, Quantum-inspired blockchain-based cybersecurity: securing smart edge utilities in iot-based smart cities. Inf. Process. Manage. **58**(4), 102, 549 (2021). https://doi.org/10.1016/j.ipm.2021.102549. https://www.sciencedirect.com/science/article/pii/S0306457321000546
2. I.S. Association, Ieee 802.15.4-2015/cor 1-2018—ieee standard for low-rate wireless networks corrigendum 1 (2015). Available: https://standards.ieee.org/standard/8021542015-Cor1-2018.html
3. I.E. Bagci, M.R. Pourmirza, S. Raza, U. Roedig, T. Voigt, Codo: confidential data storage for wireless sensor networks, in *2012 IEEE 9th International Conference on Mobile Ad-Hoc and Sensor Systems (MASS 2012)*. IEEE (2012), pp. 1–6
4. T. Baumeister, Literature review on smart grid cyber security. Collaborative Software Development Laboratory at the University of Hawaii (2010)
5. G. Bedi, G.K. Venayagamoorthy, R. Singh, R.R. Brooks, K.C. Wang, Review of internet of things (iot) in electric power and energy systems. IEEE Internet of Things J. **5**(2), 847–870 (2018). https://doi.org/10.1109/JIOT.2018.2802704
6. R. Berthier, W.H. Sanders, Specification-based intrusion detection for advanced metering infrastructures, in *2011 IEEE 17th Pacific Rim International Symposium on Dependable Computing* (2011), pp. 184–193. https://doi.org/10.1109/PRDC.2011.30
7. B. Briggs, Hackers Hit Norsk Hydro with Ransomware. The Company Responded with Transparency (2019). https://news.microsoft.com/transform/hackers-hit-norsk-hydro-ransomware-company-responded-transparency/

8. A. Caudhari, R. Bansode, Securing iot devices generated data using homomorphic encryption, in *Intelligent Computing and Networking*. ed. by V.E. Balas, V.B. Semwal, A. Khandare, M. Patil (Springer Singapore, Singapore, 2021), pp. 219–226
9. CMP11, P.U.C.C.C.C.: Paris Agreement (2015). http://ec.europa.eu/clima/policies/international/negotiations/paris/index.htm
10. M.M. Fouda, Z.M. Fadlullah, N. Kato, R. Lu, X.S. Shen, A lightweight message authentication scheme for smart grid communications. IEEE Trans. Smart Grid **2**(4), 675–685 (2011). https://doi.org/10.1109/TSG.2011.2160661
11. J. Granjal, E. Monteiro, J.S. Silva, Network-layer security for the internet of things using tinyos and blip. Int. J. Commun. Syst. **27**(10), 1938–1963 (2014)
12. J. Granjal, E. Monteiro, J.S. Silva, Security in the integration of low-power wireless sensor networks with the internet: a survey. Ad Hoc Networks **24**, 264–287 (2015)
13. A. Iqbal, H. Malik, A. Riyaz, K. Abdellah, S. Bayhan, *Renewable Power for Sustainable Growth* (Springer, Berlin, 2021)
14. J.E. Martocci, P.D. Mil, N. Riou, W. Vermeylen, Building automation routing requirements in low-power and lossy networks. RFC 5867, RFC Editor (2010). https://datatracker.ietf.org/doc/html/rfc5867
15. Kamaldeep, M. Dutta, J. Granjal, Towards a secure internet of things: a comprehensive study of second line defense mechanisms. IEEE Access **8**, 127, 272–127, 312 (2020). https://doi.org/10.1109/ACCESS.2020.3005643
16. N. Komninos, E. Philippou, A. Pitsillides, Survey in smart grid and smart home security: issues, challenges and countermeasures. IEEE Commun. Surv. Tutor. **16**(4), 1933–1954 (2014). https://doi.org/10.1109/COMST.2014.2320093
17. X. Lu, W. Wang, J. Ma, Authentication and integrity in the smart grid: an empirical study in substation automation systems. Int. J. Distrib. Sens. Networks **8**(6), 175,262 (2012). https://doi.org/10.1155/2012/175262
18. E.M. Dohler, E.T. Watteyne, E.T. Winter, E.D. Barthel, Routing requirements for urban low-power and lossy networks. RFC 5548, RFC Editor (2009). https://datatracker.ietf.org/doc/html/rfc5548
19. H. Malik, *Metaheuristic and Evolutionary Computation* (Springer Nature, 2020)
20. H. Malik, N. Fatema, J.A. Alzubi, *AI and Machine Learning Paradigms for Health Monitoring System: Intelligent Data Analytics* (Springer, Berlin, 2021)
21. H. Malik, N. Fatema, A. Iqbal, *Intelligent Data-Analytics for Condition Monitoring: Smart Grid Applications* (Elsevier, Amsterdam, 2021)
22. H. Malik, A. Iqbal, A.K. Yadav, Soft computing in condition monitoring and diagnostics of electrical and mechanical systems, in *Novel Methods for Condition Monitoring and Diagnostics* (Springer, Berlin, 2020), p. 499
23. H. Malik, S. Srivastava, Y.R. Sood, A. Ahmad, *Applications of Artificial Intelligence Techniques in Engineering: SIGMA 2018, Volume 1*, vol. 698 (Springer, Berlin, 2018)
24. H. Malik, S. Srivastava, Y.R. Sood, A. Ahmad, *Applications of Artificial Intelligence Techniques in Engineering: SIGMA 2018, Volume 2*, vol. 697 (Springer, Berlin, 2019)
25. M. Malik, M. Dutta, Implementation of single-packet hybrid ip traceback for ipv4 and ipv6 networks. IET Inf. Secur. **12**(1), 1–6 (2018)
26. M. Malik, M. Dutta, J. Granjal, A survey of key bootstrapping protocols based on public key cryptography in the internet of things. IEEE Access **7**, 27443–27464 (2019). https://doi.org/10.1109/ACCESS.2019.2900957
27. M. Malik, M. Dutta, et al., Defending ddos in the insecure internet of things: a survey, in *Artificial Intelligence and Evolutionary Computations in Engineering Systems* (Springer, Berlin, 2018), pp. 223–233
28. I. Markit, The internet of things: a movement, not a market. Tech. rep., London (2018). https://cdn.ihs.com/www/pdf/IoT_ebook.pdf
29. A. Meddeb, Internet of things standards: who stands out from the crowd? IEEE Commun. Mag. **54**(7), 40–47 (2016). https://doi.org/10.1109/MCOM.2016.7514162

30. G. Montenegro, N. Kushalnagar, J. Hui, D. Culler, Transmission of ipv6 packets over ieee 802.15.4 networks. RFC 4944, IETF (2007). http://www.rfc-editor.org/rfc/rfc4944.txt
31. M. Panthi, Identification of disturbances in power system and DDoS attacks using machine learning. IOP Conf. Ser.: Mater. Sci. Eng. **1022**, 012, 096 (2021). https://doi.org/10.1088/1757-899x/1022/1/012096
32. J. Postel, User datagram protocol. STD 6, IETF (1980). http://www.rfc-editor.org/rfc/rfc768.txt. http://www.rfc-editor.org/rfc/rfc768.txt
33. A. Ramamurthy, P. Jain, The Internet of Things in the Power Sector: Opportunities in Asia and the Pacific (issue 48). Asian Development Bank (2017)
34. S. Raza, S. Duquennoy, J. Hoglund, U. Roedig, T. Voigt, Secure communication for the internet of things-a comparison of link-layer security and ipsec for 6lowpan. Secur. Commun. Networks **7**(12), 2654–2668 (2014). https://doi.org/10.1002/sec.406
35. S. Raza, H. Shafagh, K. Hewage, R. Hummen, T. Voigt, Lithe: lightweight secure coap for the internet of things. IEEE Sens. J. **13**(10), 3711–3720 (2013). https://doi.org/10.1109/JSEN.2013.2277656
36. R. Riaz, K.H. Kim, H.F. Ahmed, Security analysis survey and framework design for ip connected lowpans, in *2009 International Symposium on Autonomous Decentralized Systems* (2009), pp. 1–6. https://doi.org/10.1109/ISADS.2009.5207373
37. K. Security, Target hackers broke in via hvac company (2014). https://krebsonsecurity.com/2014/02/target-hackers-broke-in-via-hvac-company/
38. Z. Shelby, C. Bormann, *6LoWPAN: The Wireless Embedded Internet*, vol. 43 (Wiley, New York, 2011)
39. Z. Shelby, K. Hartke, C. Bormann, The constrained application protocol (coap). RFC 7252, RFC Editor (2014). http://www.rfc-editor.org/rfc/rfc7252.txt
40. V.K. Singh, M. Govindarasu, *Cyber Kill Chain-Based Hybrid Intrusion Detection System for Smart Grid* (Springer International Publishing, Cham, 2021), pp. 571–599
41. R.E. Smith, *Authentication: From Passwords to Public Keys* (Addison-Wesley Longman Publishing Co., Inc., 2001)
42. H.K.H. So, S.H.M. Kwok, E.Y. Lam, K.S. Lui, Zero-configuration identity-based signcryption scheme for smart grid, in *2010 First IEEE International Conference on Smart Grid Communications* (2010), pp. 321–326. https://doi.org/10.1109/SMARTGRID.2010.5622061
43. S. Sridhar, A. Hahn, M. Govindarasu, Cyber-physical system security for the electric power grid. Proc. IEEE **100**(1), 210–224 (2012). https://doi.org/10.1109/JPROC.2011.2165269
44. A. Ullah, M. Azeem, H. Ashraf, A.A. Alaboudi, M. Humayun, N. Jhanjhi, Secure healthcare data aggregation and transmission in iot—a survey. IEEE Access **9**, 16849–16865 (2021). https://doi.org/10.1109/ACCESS.2021.3052850
45. T. Winter, P. Thubert, A. Brandt, J. Hui, R. Kelsey, P. Levis, K. Pister, R. Struik, J. Vasseur, R. Alexander, Rpl: Ipv6 routing protocol for low-power and lossy networks. RFC 6550, RFC Editor (2012). http://www.rfc-editor.org/rfc/rfc6550.txt
46. S.T. Zargar, J. Joshi, D. Tipper, A survey of defense mechanisms against distributed denial of service (ddos) flooding attacks. IEEE Commun. Surv. Tutor. **15**(4), 2046–2069 (2013)

Intelligent Analysis of a Hybrid Energy System with Telecom Load

Mohd. Khursheed Siddiqui, Asim Rahman Ansari, Khadim Moin Siddiqui, Mohammad Sami, and Munish Kumar

Abstract Thousands of people throughout the world are facing lack of power supply. In developing countries such as India, lack of power is the major problem in the sustainable development of the rural areas. Renewable energy sources (RES) are widely used for supplying power to remote areas. Hybrid energy system is the intelligent option to provide a reliable and cost-effective power supply in the rural area where grid access is limited. This case study examines the practical financial possibilities for grid-connected solar PV system for a telecom load in Lucknow. The telecom load is examined using the hybrid optimization model for electric renewable (HOMER) software for the minimum cost and optimal design. Load profile in HOMER is selected according to the telecom load pattern considered for the case study. RES taken into account for study comprises of solar PV with battery storage. The meteorological solar irradiance data for SPV in Lucknow, Uttar Pradesh, India, (latitude of 26.84^0 North and longitude of 80.94^0 East) has been considered from NASA surface metrology and Websites of solar energy. The cost of energy (COE) and net present cost (NPC) are analysed in HOMER. Furthermore, the sensitivity parameter such as change in solar irradiation is also considered in the model. The analysis gives the performance of the hybrid system and proposes an optimum configuration for the telecom load.

Keywords Solar PV system · Battery · Hybrid system homer · Techno-economic

Mohd. K. Siddiqui (✉) · M. Sami · M. Kumar
Integral University, Lucknow, India

A. R. Ansari
FET, MJP Rohilkhand University, Bareilly, India

K. M. Siddiqui
SRMGPC, Lucknow, India

H. Malik et al. (eds.), *Intelligent Data Analytics for Power and Energy Systems*,
Lecture Notes in Electrical Engineering 802,
https://doi.org/10.1007/978-981-16-6081-8_29

1 Introduction

Variable capacity challenging terrain and the additional cost of conventional fuel energy, grids and availability are either not possible or costlier in remote areas [1–3]. A feasible solution to power station sites is to utilize non-conventional resources [3–5]. The conventional energy resources contain a lot of potential in the field of research and growth [6, 7]. For varying natural conditions, the sustainable load demand has always been the challenging task. For fulfilling the demand, one need to use other energy resources with the integration of converters.

In India, the use of SPV system [8, 9] for the power supply may give a pleasing substitute from economic point of view for power sector in place of using conventional diesel generators in near future [10, 11]. Furthermore, utilization of renewable energy resources (RES) such as solar PV, wind generation and biomass will help to reduce the carbon emission level in country hybrid energy system based on SPV/DG/battery can install as per on-site demand with less fuel consumption thus giving an attractive cost-mitigating solution [8, 9]. Hybrid energy system with RES also provides reduced sound and carbon pollution which gives an additional benefit in some regions [12].

Definitely, power technologies are growing fast day by day, but power sector is still facing challenge to supply sustainable power to remote areas [13–15]. As the RES is becoming popular, now the power companies are looking toward RES as a solution of providing sustainable power supply to remote areas [16, 17]. Hence, based on the load demand of telecom tower and availability of RES, a hybrid energy system (HES) is proposed in this study. The presented HES comprises of grid, SPV panels, battery storage and advance power electronic converters. Modelling and simulation of each component of the proposed HES has been done by using MATLAB/Simulink [18]. The adjustment of the size of the system as well as the coast IC analysis has been done in HOMER.

2 Telecom Site Location

The telecom sites can be categorized on the basis of topology location, operation and expenditure. Out of these, the fixed base site is constructed once for a particular location. This chapter deals with fixed type base stations. From economic viewpoint, the base stations can be categorized on the basis of grid site reliability as follows:

(a) **Reliable grid site:** For such a site, power from grid is the cheapest option available which could be used if available to supply the base station. Here, in urgent situation battery backup is used.

(b) **Untrustworthy grid location:** Such a site may lose the connection from grid for many hours that may be dangerous for the telecom load and for the associated equipments.

(c) **Off-grid site:** Such a site does not have grid connection and are usually located remote areas.

The solutions for aforesaid sites are SPV/DG, with deep discharge batteries or two generators [17, 19, 20]. In this chapter, hybrid grid/SPV/batteries-based hybrid system is evaluated in regions like India, where the fuel cost is high. The important point is that the generators is the best option in countries where the diesel is cheap, but on the other hand, the regions where diesel is costlier, hybrid solar or battery–generator cycle could be best options. Generally, the battery–generator cycles are economically better than the hybrid solar-system.

3 Solar PV Scheme Design and Configurations

In various applications, loads are fed from a RES which is much better than others because of reliability and quality concerns. This chapter discusses the combination of only solar PV and battery system along with grid connected arrangement. The coupling of the solar PV system may be DC type, AC type or both as per the requirement. Coupling with DC bus is the simplest way [12–14]. While the AC coupling is the most complex because of synchronization issue becomes more burdensome and the inverter cost has to be incorporated [12–14]. As per the system requirement, the SPV system can be designed many ways. Every design has its own merit and limitation [21–23].

Any of the above two configurations can be considered for system design. The first one may be grid/SPV system without battery (Fig. 1a) and another one is standalone SPV system with battery storage (Fig. 1b) [12–14]. Figure 1a shows the configuration one where the SPV panels are connected to inverters, which is feeding power to the load through the bus bar. Figure 1b shows the connection diagram for second configuration which is able to supply and to store energy, respectively, during high power demand configuration one is used to supply the load. But during low power demand or no load demand, the battery stored energy via charge controller. These batteries are utilized as a backup during semi-shiny day or night hours (Table 1).

3.1 Selection of PV Module

As huge power supply is needed, having small area so mono crystalline silicon module can be selected. The price and efficiency play a vital role in choosing PV System. High cost is incurred for PV system installation, which is around 60% of overall system installation [15]. Efficiency of PV module and the material affect the PV panel cost. The silicon solar cell has very high cost. Mono crystalline silicon cell is used in this design. Technology is the main source for improvisation in efficiency of solar PV cell. Among all the semiconductor materials, silicon solar cell gives

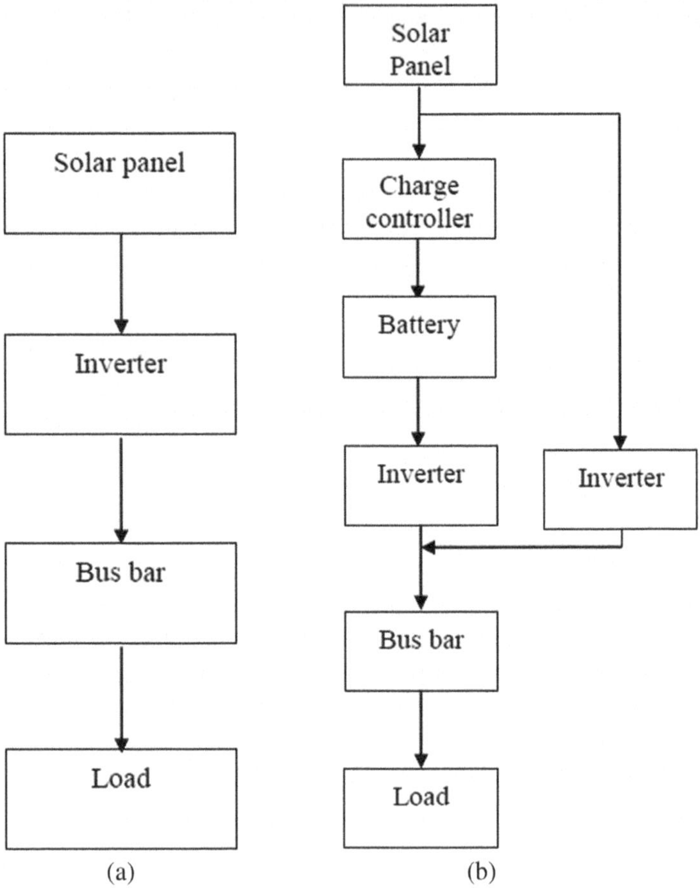

Fig. 1 **a** and **b** Possible design of solar SPV

Table 1 Inverter/rectifier selection quantity

Quantity (with units)	Inverter	Rectifier
House of operation (hrs/year)	20	3698
Energy in (kwh/year)	0	8051
Energy out (kwh/year)	0	6843
Losses (kwh/year)	0	1208

maximum efficiency, and in areas of high temperature, low temperature coefficient PV panels are selected for better output (Fig. 2).

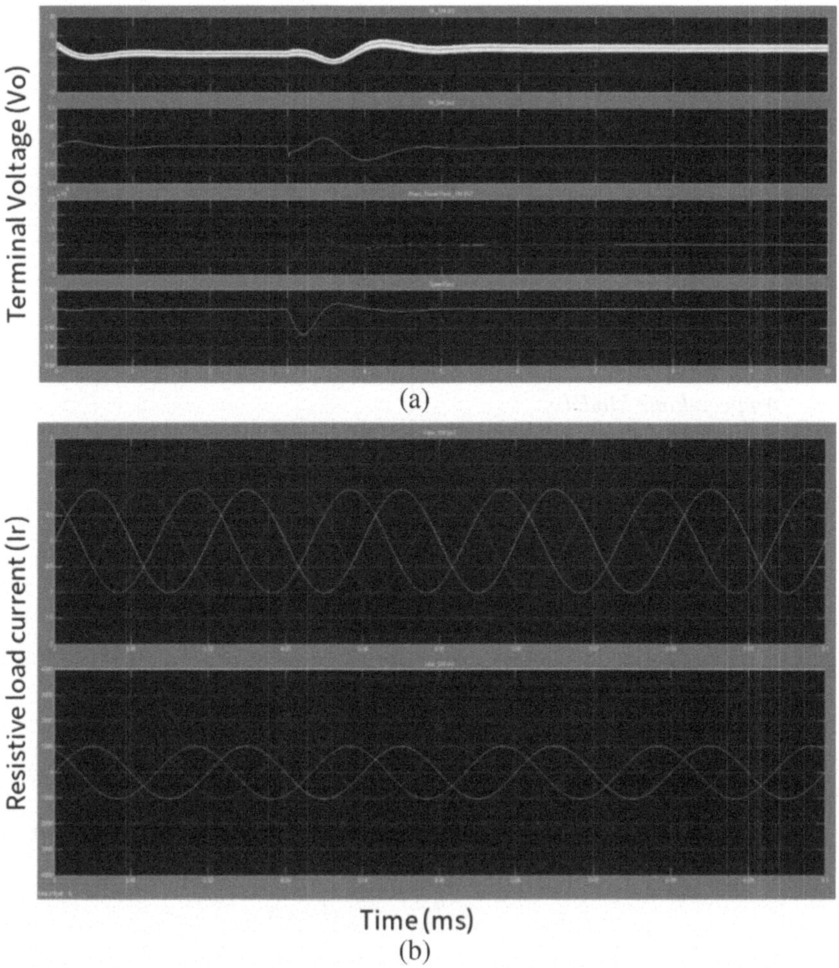

Fig. 2 **a** Three-phase terminal voltages. **b** Three-phase resistive load current

3.2 Inverter Selection

The converters are employed for maintaining the bidirectional power flow in between AC and DC buses. The convertor selected in this study is of 0.75 kW capacity. The installation, replacement and operation–maintenance cost are $388, $300 and $50, respectively. Life of inverter is assumed as 15 years and efficiency 90%. Rectifier of 100% ratted capacity with an efficiency of 85% has been chosen (Table 2).

Table 2 Inverter/rectifier selection

Quantity (with unit)	Inverter	Rectifier
Capacity (kW)	3.00	3.00
Mean output (kW)	0.00	0.78
Min. output (kW)	0.00	0.00
Max. output (kW)	0.00	3.00
Capacity factor (%)	0.00	26.0

3.3 Battery Model

(i) ***Ampere-hour Model:***
The relation between battery capacity and charge current is given as [4, 8].

$$C_{\max}(i) = \frac{S_{\max o}pAT}{1 - e^{-pT} + A\left(pT - 1 + e^{-pT}\right)}. \quad (1)$$

where
$C_{\max}(i)$ = maximum battery capacity (Ah).
p = rate constant/h.
A = available charge capacity/total capacity.
The above data can be collected from constant current charges/discharges method.

(ii) ***Voltage base Model***
The voltage equation is given as;

$$V_f = V_j + rN + \frac{\lambda N}{(\xi - N)} \quad (2)$$

where
V_f = the full internal charging voltage attained by the battery post initial transient.
r = this parameter gives initial linier relation between the battery voltage and state of charge (SoC).
λ, ξ = These parameters show the variation in battery voltage during continuous discharged/charged state.
N = normalized maximum capacity at given current.

(iii) ***Lifetime Model***
The lifetime model can be represented by exponential curve that is fitted to normally available as follows.

$$f_c = \tau + \zeta e^{-\sigma Rc} + \gamma e^{-\kappa Rc} \quad (3)$$

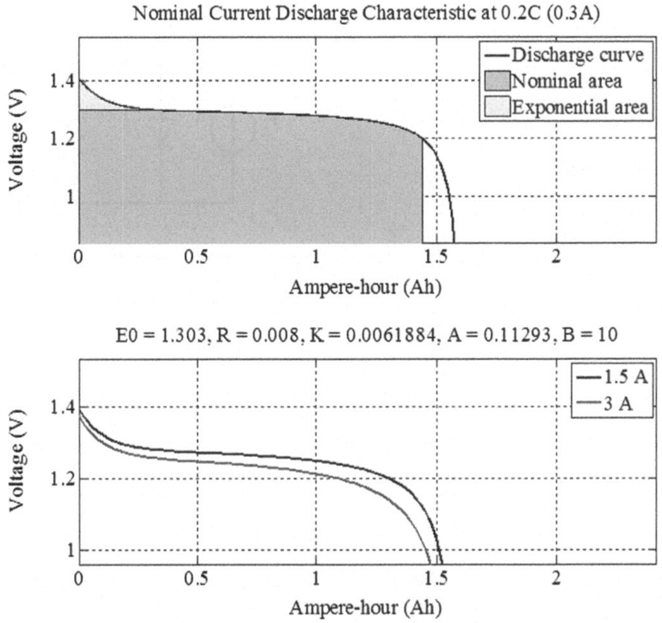

Fig. 3 Battery *V–I* characteristic

Here

f_c, denotes to failure of cycle.

$\tau, \zeta, \sigma, \kappa, \gamma$, are curve fitting constants.

R_c is the value of frictional DoD also known as range of cycle (Fig. 3).

Battery charging/discharging characteristics are obtained in MATLAB/Simulink environment and are depicted in Fig. 5.

4 Modelling of PV System

The model of SPV cell based on standard equivalent circuit available in literature [1, 12, 13] is depicted in Fig. 4.

The *V–I* characteristics at 30° C temperature and at different irradiance condition *S* is shown in Fig. 5

This is clear from Fig. 7 that the I_{sc} linearly related with the insolation, whereas the V_{oc} change nonlinearly with change in insolation [2, 21, 16, 24].

The effect of ambient temperature variation on the *V–I* characteristic is shown in Fig. 6. The value of I_{sc} at given temperature *T* can be calculated as;

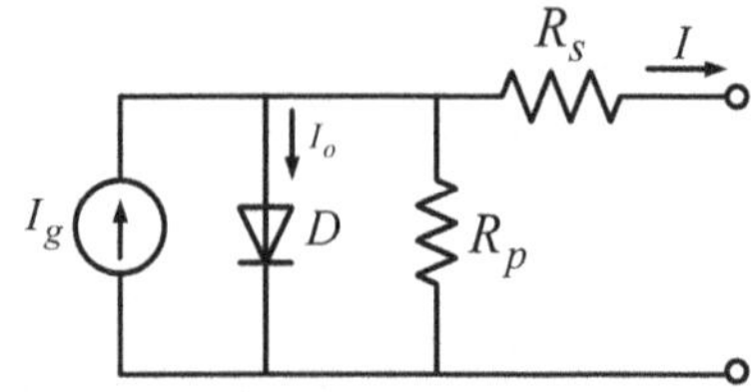

Fig. 4 Equivalent circuit of SPV cell

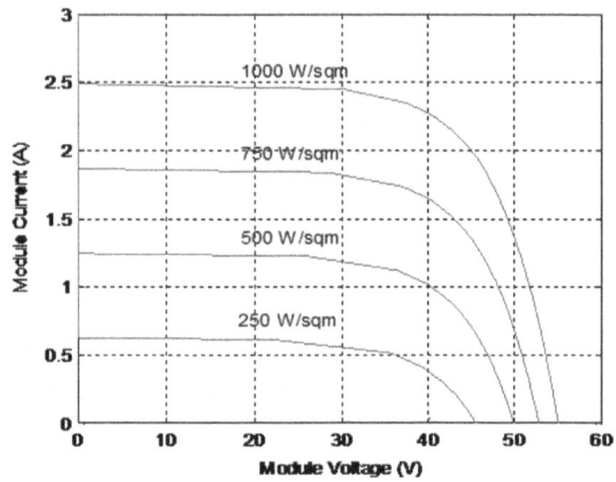

Fig. 5 SPV module *I–V* characteristics at different insolation

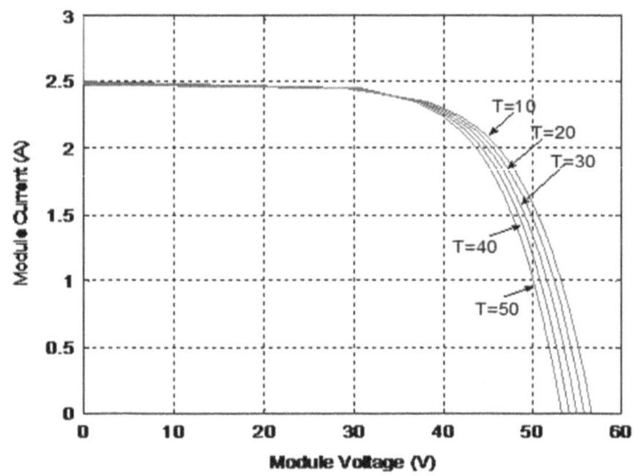

Fig. 6 *I–V* characteristics of a SPV module at different temperature (1000 W/m^2 isolation)

$$I_{sc(t)} = I_{sc(t0)}[1 + \delta(T - T_0)] \tag{4}$$

where T is cell temperature (°K*or* °C).

δ denote to coefficient of temperature.

$T_0 = 25°$ C (slandered ambient temperature).

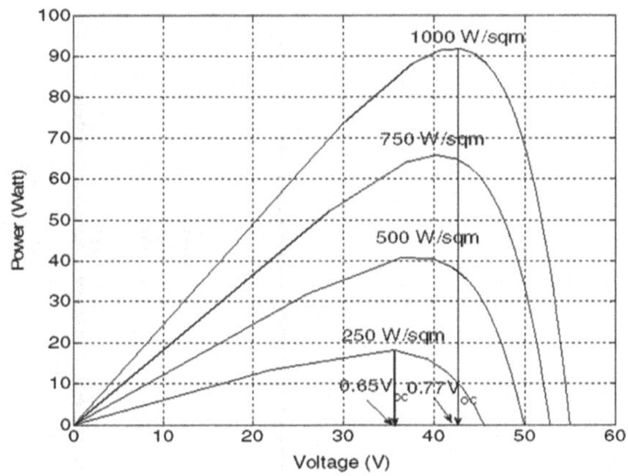

Fig. 7 Standard l *P–V* characteristic of a SPV module at different isolations (at 30° C ambient temperature)

The modelling of the solar PV module at different temperature variations is done in MATLAB/Simulink environment and the result is presented in Fig. 6.

The *P–V* characteristics at different solar irradiations are shown in Fig. 7. From the characteristics, it is clear that the open-circuit voltage at maximum power point is lies between 65 and 77% approximately.

The effect of temperature variation on the *P–V* curve is examined and the characteristic is presented in Fig. 8. It can be observed that the reduction in the maximum power (P_{max}) because of temperature variation is very small as compared to irradiation change [17].

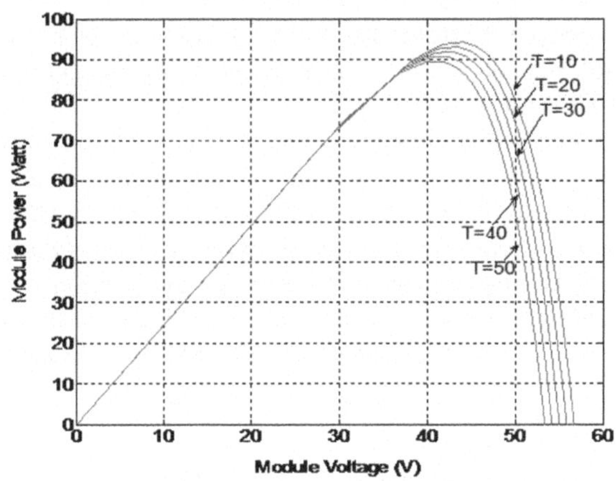

Fig. 8 *P–V* characteristic of a SPV module at different ambient temperature (at 1000 W/m^2 isolation level)

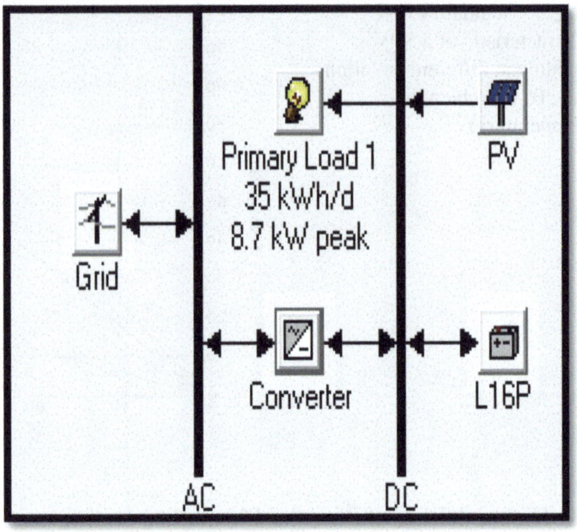

Fig. 9 Connection diagram of hybrid solar PV system in HOMER

5 Modelling of Hybrid SPV/Grid System in HOMER

The simulation of complete HES is done by employing HOMER software [19, 20]. Generally, PV system is taken as a main supply source for remote on-grid telecom site to feed DC bus. Solar PV system along with the storage battery is used to feed DC bus and the power from grid is converted to DC with the help of converter as shown in Fig. 9.

For designing and simulation of SPV system, HOMER software has been employed. HOMER has various inbuilt mathematical models for small units like SPV module, converter, inverter, etc. HOMER stands for hybrid optimization model for electric renewable [22].

HOMER also helps in sizing and to meet the sensitivity analysis for obtaining the best suitable structure of HES [18, 19]. As per the load requirement of telecom site and the availability of renewable energy sources, HES is comprises of SPV panels, battery storage and advanced power electronics devices. For the presented study, SPV is adopted as the primary energy source to supply the power to load and battery backup is also used for any emergency. Sizing of the system is realized by HOMER and modelling has been completed in MATAB (Simulink).

6 Sizing Selection of the System

System modelling and its implementation have been done in HOMER environment [13–15]. The proposed hybrid system comprises of solar PV source, conventional battery source and converter for AC and DC buses.

6.1 PV System

HOMER software is adopted for designing and simulation of PV system. HOMER has different built-in mathematical models to realize various component like SPV module, battery bank and inverter, etc. (Fig. 10).

For this case, SPV is considered as primary energy source supplying power to the telecom load whereas electric grid and a battery bank has been used for backup purpose. The output power daily profile for SPV array is depicted in Fig. 11.

To design the above-mentioned systems, many constraints have to be considered, i.e. region, horizon, the size of system, etc. To develop a HES, various parts have to be chosen like PV array, battery storage and inverter, series parallel combination of modules, etc. HOMER facilitates many consumers to connect different parts for developing an optimum PV system and also give simulation result. Using HOMER software, similar HES for this case study has been developed whose component detail is given in Table 3.

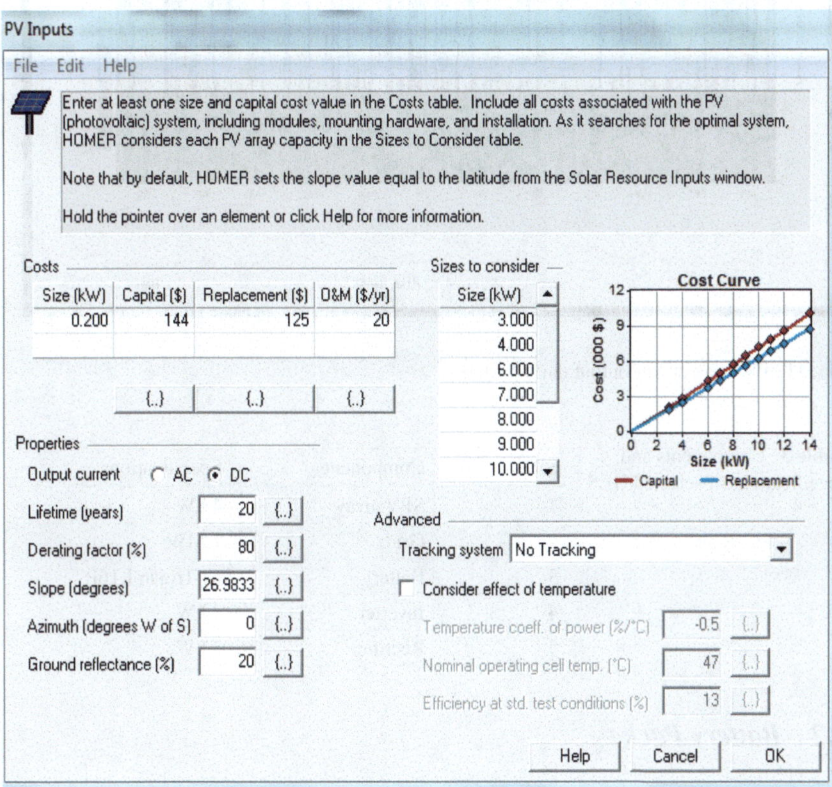

Fig. 10 PV array in HOMER software

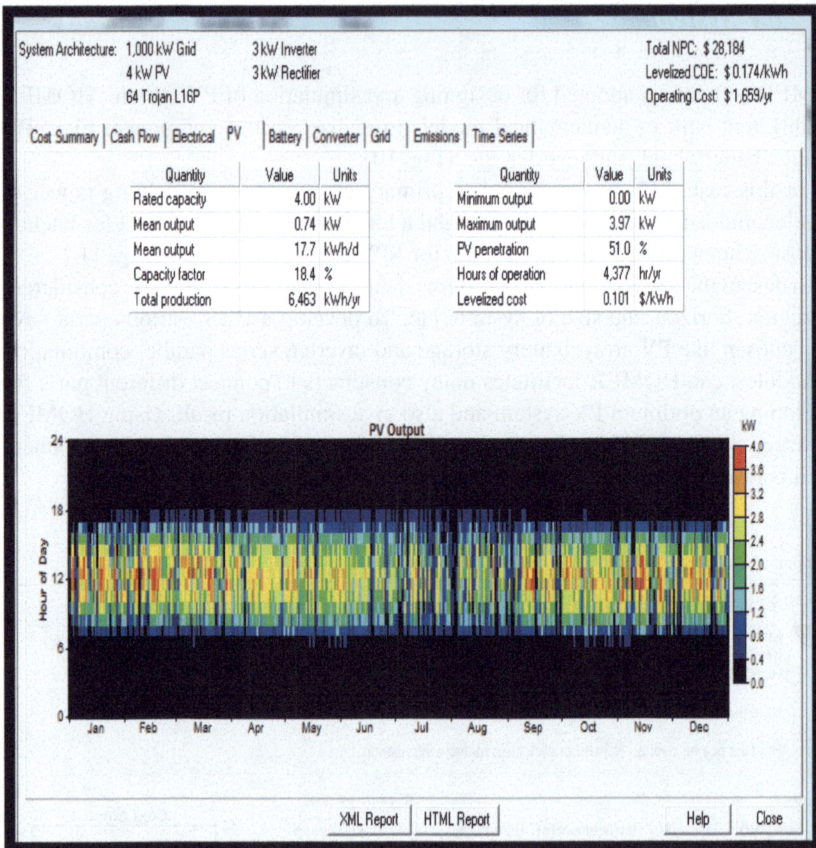

Fig. 11 PV array power output daily profile

Table 3 Components and specifications of selected HES

S.N.	Components	Specifications
1	SPV array	4 kW
2	Grid	1 MW
3	Battery	64 Trojan L16P
4	Inverter	3 kW
5	Rectifier	3 kW

6.2 *Battery Pack*

For tackling the irregular character of RES, the battery bank is employed. A selected RES relies upon many factors such as climate conditions. The diesel sets are provided for irregular supply of energy [17, 19, 20]. The HOMER software is developed by

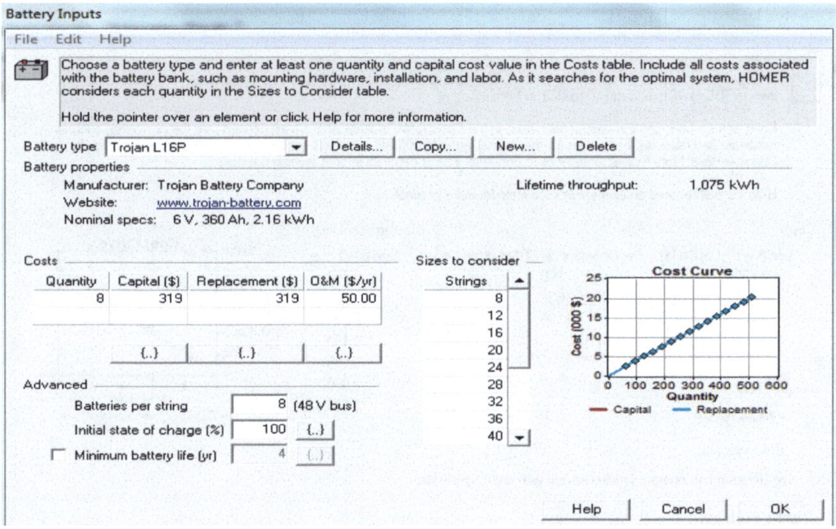

Fig. 12 Battery at HOMER software

National Energy Laboratory (NREL) for the model optimization in the USA. The economic and sensitivity analysis is done by the proposed simulation model. The needed energy optimization has been done by HOMER in every year in the USA [18]. By varying the proposed model parameters, the sensitivity analysis is done with the satisfaction of all restrictions. In the integrated model analysis, the cost and revenue have been considered for the life span of the project. The analysis has been done by considering the parameters such as fuel costs, maintenance cost and cost buying from the grid. The annualized operation, maintenance costs and annualized replacement cost for all the components [1] come in the category of total annualized cost. The capital recovery factor covers the present value into a series of equal annual cash flows over the entire lifetime of the projects. HOMER comprises of many inbuilt model of battery bank, and for this case study, the battery model 64 Trojan L16P has been selected having nominal voltage of 6 V and capacities 360 Ah with lifetime throughput of 1075 kWh. To obtain 48 V DC bus, 8 batteries per string is selected. Capital cost per battery is chosen as $40 (Fig. 12).

6.3 AC to DC Converter/Inverter

A DC to AC inverter is used for converting DC supply to AC supply to feed the DC loads. In this case study, a 3 kW ratting inverter is utilized, and the capital, replacement and O & M cost are $1659, respectively. Inverter lifetime is 15 year, while the efficiency is 90%. Similarly, the rectifier capacity with respect to inverter, efficiency is 95% and 85%, respectively (Fig. 13).

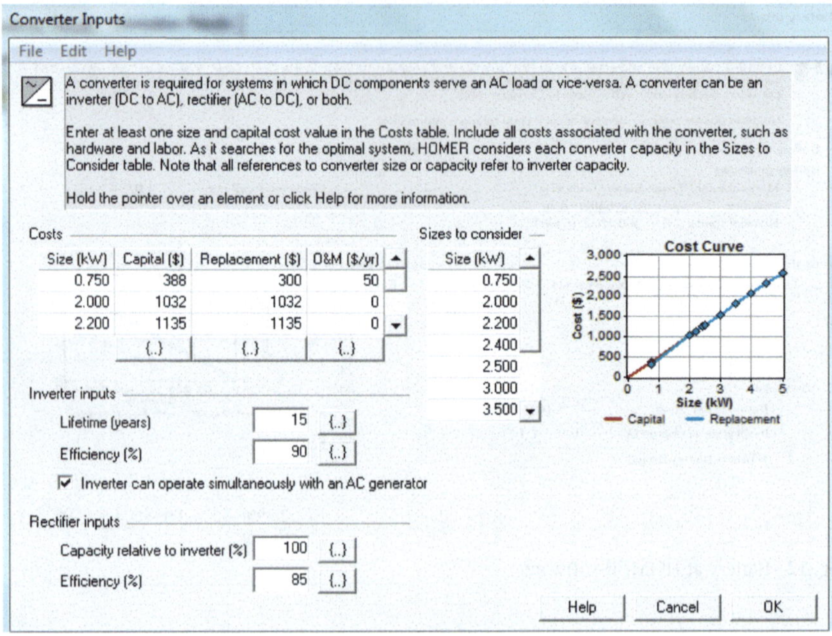

Fig. 13 Converter at HOMER software

The global horizontal radiation variation over the year is shown in Fig. 14a, whereas the variation of isolation during one year in Lucknow is depicted in Fig. 14b. The average isolation for month of January for one week shown in Fig. 14c.

Load profile is selected according to the telecommunication load pattern, having peak load and average load of 8.71 kW and 1.44 kW, respectively. The average energy variation obtained during a is 34.7 kWh/day with the load factor 0.166. The load variation curve is shown in Fig. 15.

7 Simulation Results

7.1 Techno-coast Analysis

The techno-coast analysis is carried out in HOMER and given in Table 7 is discussed:

Initial capital cost is \$6976 (Fig. 16).

O & M Cost is \$ 18,417.

Replacement cost \$3645.

The total net present cost (NPC) is \$28,184.

Levelized cost of energy is \$0.174/kWh.

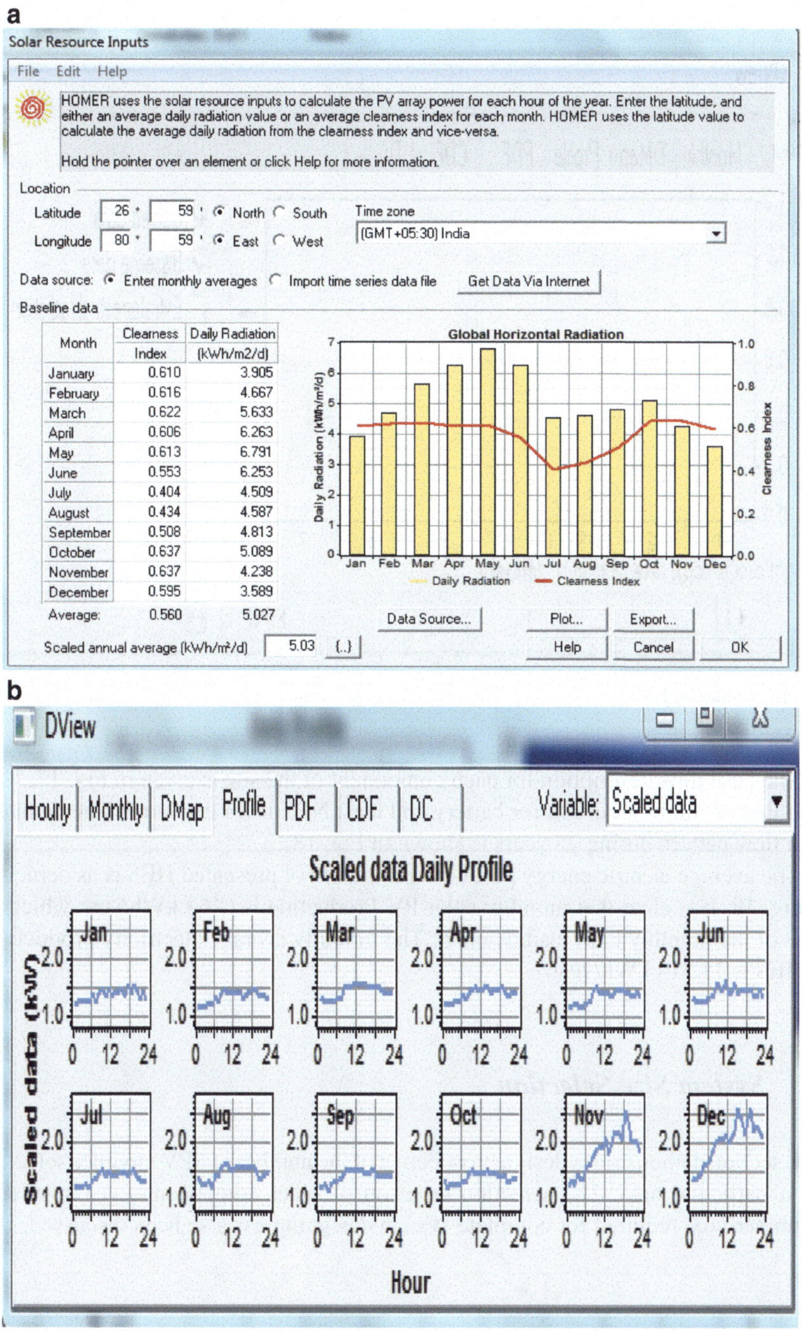

Fig. 14 **a** Solar irradiance profile in HOMER for Lucknow. **b** Average isolation per day profile for one year. **c** Daily average isolation profile per week month of January in Lucknow

c

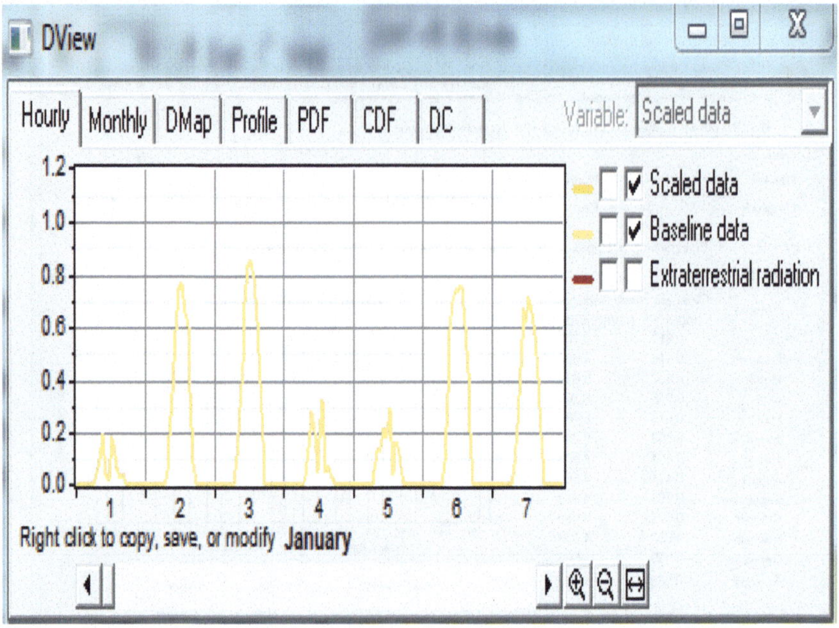

Fig. 14 (continued)

The cash flow description for each component of the site is given in Fig. 17. It is seen that the highest NPC is for battery and least NPC is for converter. The detailed cash flow pattern during 25 years is shown in Fig. 18.

The average electric energy produced per month of presented HES is as depicted in Fig. 19. It is clear that monthly solar PV Production is 6463 kWh/year which is 51% of the monthly total load demand. The monthly average electricity production of HES is 14,514 kWh/year.

7.2 System Size Selection

This section demonstrates design prospective of the number of SPV modules selected for a particular area. The selection criterion of other components, i.e. inverters, combiner box, required for complete system designing has also been discussed.

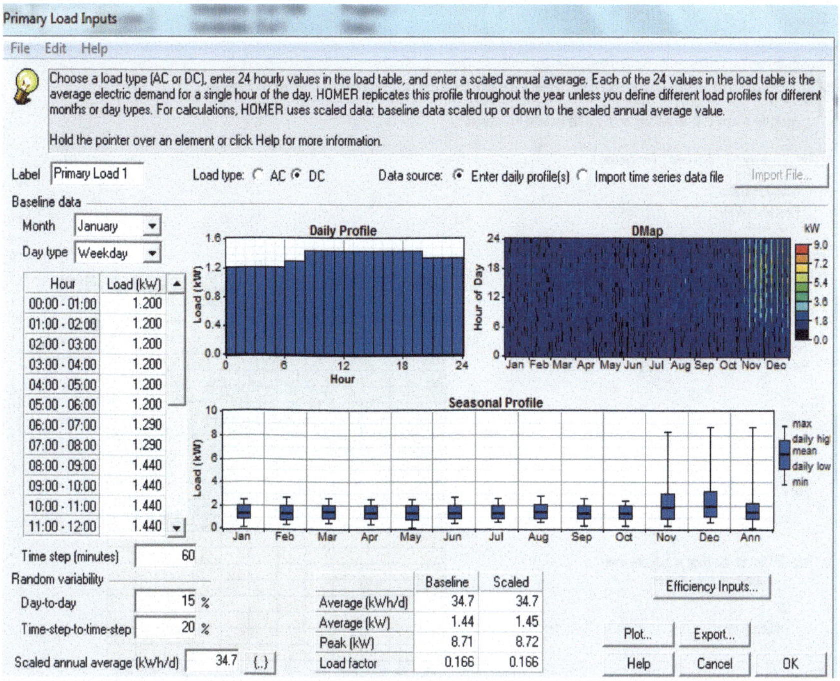

Fig. 15 Load profile window at HOMER

7.3 *Number of Module Selection*

Modules calculation for modules installed on both roof top and/or facade by using the method mentioned below,

Total module required = All the area under usage/area of one SPV module. Table 4 gives the total module required for this case study (Table 5).

The HES considered in this chapter is SPV with grid connection; it is clear from Fig. 20 that the breakeven point is obtained at 3.08 km. Furthermore, it has been examined that for of grid system, the installation cost remains constant, whereas it increases with increase in distance (Table 6).

8 Result Demonstration and Validation

HOMER software execute several simulations in order to obtain the optimal HES. To obtain the optimum hybrid energy system. The best optimum HES obtained in this case study is having an SPV array of 4 kW, a 3 kW converter, 64 L16P Trojan battery with a 1000 kW grid connected to supply the primary load. Overall net present cost (NPC), initial capital cost (ICC) and cost of energy (COE) obtained

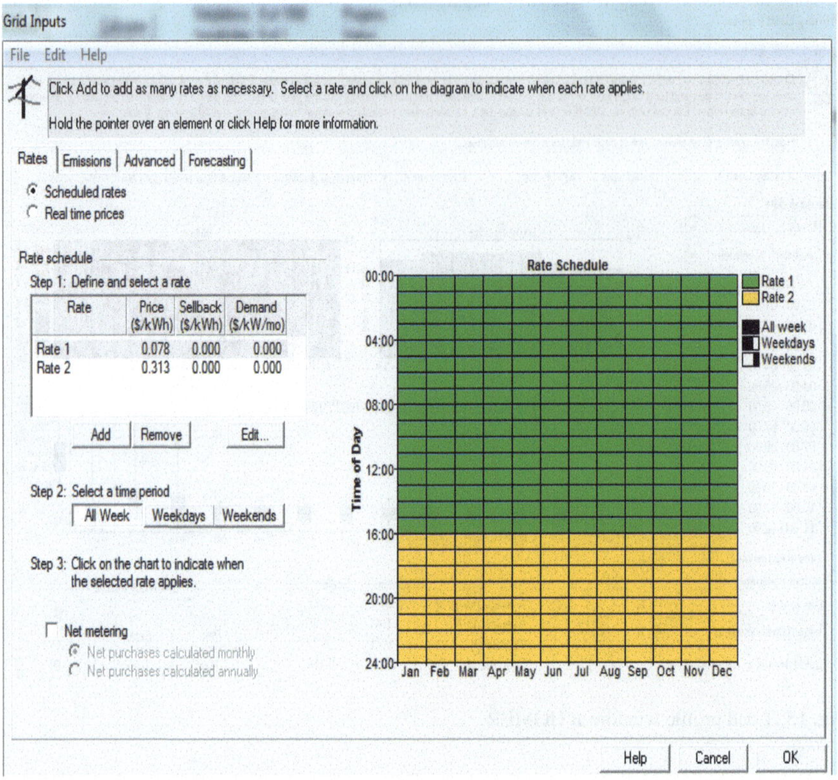

Fig. 16 Grid inputs to the system

from the simulation is \$28,184 (Rs. 2,118,309), \$6976 (Rs. 524,316.16) and 0.174\$ (Rs. 13.077)/kWh, respectively (Tables 7 and 8).

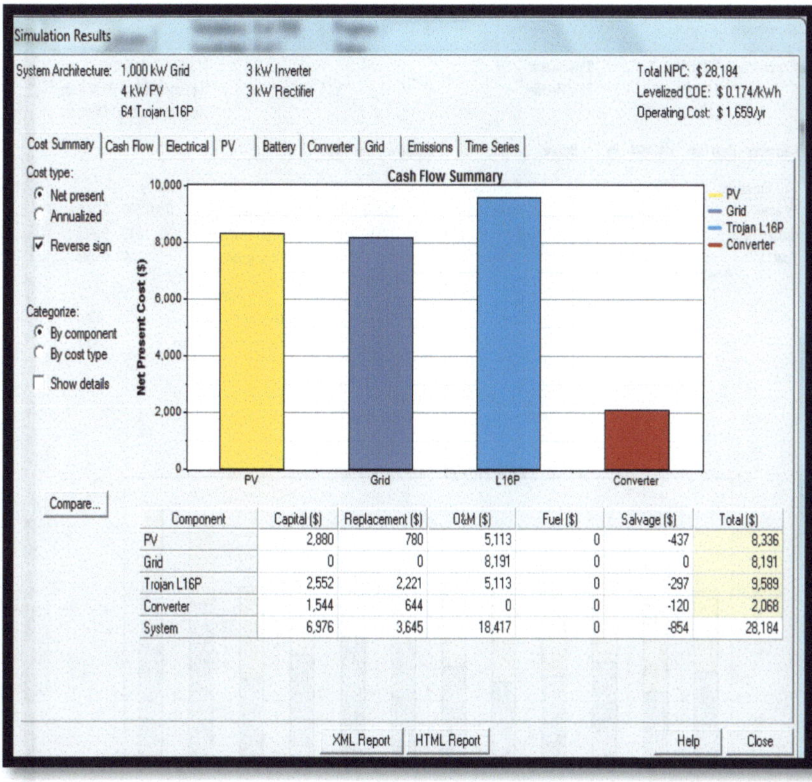

Fig. 17 Cash flow pattern for all the components of selected HES

Fig. 18 Exact cash flow for the proposed HES

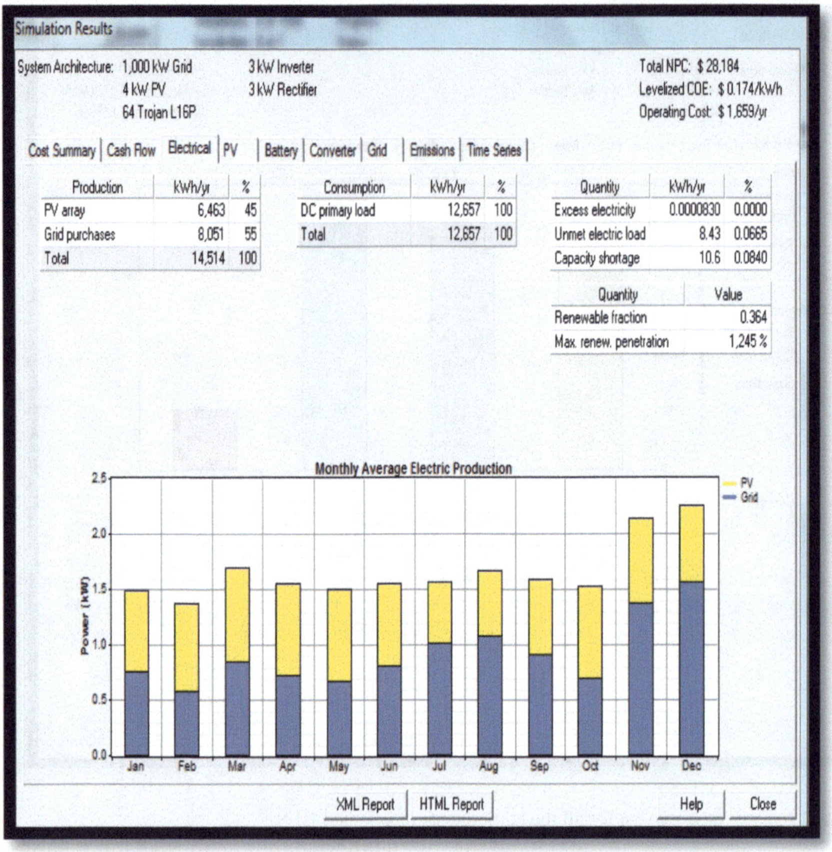

Fig. 19 Monthly energy produce by SPV

Table 4 Approximate calculation of modules for rooftop and facade area PV

Panel installation location	Usable area (m^2)	No. of modules
Roof	408	252
Façade	380.50	240
Total	788.50	492

Table 5 Quantity of SPV modules

Quantity	Value with unit
Rated capacity	4.00 kW
Capacity factor	18.4%
Mean output	0.738 kW
Mean output	17.7 kWh/d
Total production	6,463 kWh/year
PV penetration	51.0%
Min. output	0.00 kW
Max. output	3.97 kW
Levelized cost	0.101 $/kWh
Hours of operation	4377 h/year

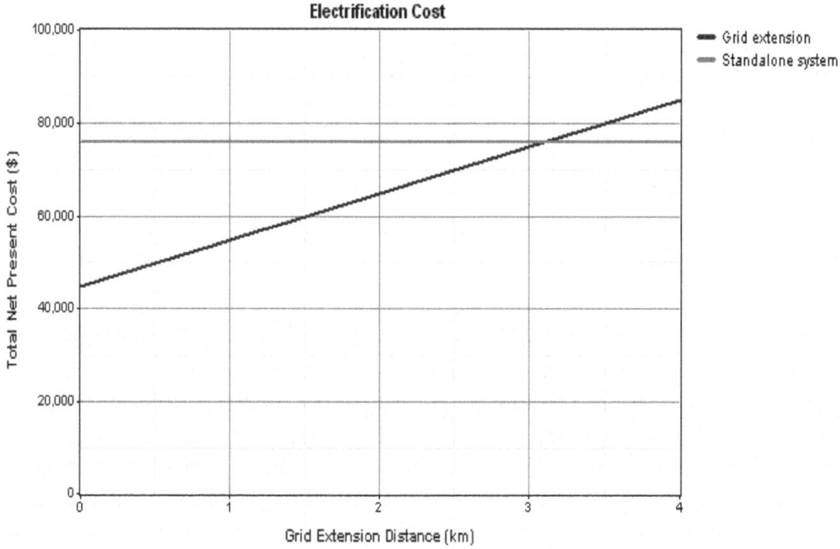

Fig. 20 Electrification cost for on-grid and off grid system

Table 6 Price outline

Present net full price	$28,184
Levelized energy price	$0.174/kWh
Operating price	$1659/year

Table 7 Net present costs

Component	Capital ($)	Replacement ($)	O&M ($)	Fuel ($)	Salvage ($)	Total ($)
PV	2880	780	5113	0	−437	8336
Grid	0	0	8191	0	0	8191
Battery	2552	2221	5113	0	−297	9589
Converter	1544	644	0	0	−120	2068
System	6976	3645	18417	0	−854	28,184

Table 8 Annualized costs

Component	Capital $/year	Replacement ($/year)	O&M $/year	Fuel $/year	Salvage $/year	Total $/year
PV	225	61	400	0	−34	652
Grid	0	0	641	0	0	641
Battery	200	174	400	0	−23	750
Converter	121	50	0	0	−9	162
System	546	285	1441	0	−67	2205

9 Conclusions

This chapter deals with SPV/battery/grid-based hybrid system as an optimum solution for supplying power to the telecommunication load in Lucknow, India. Techno-economic aspects are also examined in detail using HOMER software. Mostly, telecom sites in Lucknow are grid connected. Though still, there are some base stations that are totally dependent on the DG. Thus, there is scope of implementing renewable energy system. The comparison of cost among different solutions based on grid, solar PV and battery and their combination are checked for different loading conditions, and it is concluded that the capital cost of the PV is maximum while the operating and maintenance cost is minimum as compared to other components. Moreover, the energy production of the PV is highest that is 51% of the total production of the hybrid system. Further, it is hoped that in future, the installation cost will be decreased because of fast technology development. Finally, from the economic analysis of this case study, it has been concluded that hybrid power system could be an intelligent solution for supply power to telecommunication load in Lucknow.

References

1. M. Khursheed, M.A. Mallick, A. Iqbal, Modeling and analysis of the photovoltaic array feeding a single phase SPWM inverter. Int. J. Adv. Res. Sci. Eng. **06**(11), 403–413 (2017)

2. H. Doukas, K.D. Patlitzianas, A.G. Kagiannas, J. Psarras, Renewable energy sources and rationale use of energy development in the countries of GCC: myth or reality? Renew. Energy **31**(6), 755–770 (2006)
3. N.R. Deevela, B. Singh, T.C. Kandpal, Load profile of telecom towers and potential renewable energy power supply configurations, in *EEE International Conference on Power Electronics, Drives and Energy Systems (PEDES)* (Chennai, India, 2019)
4. M. Khursheed, M.A. Mallick, A. Iqbal, Tuning of controllers for a boost converter used to interface battery source to BTS load of a telecommunication site, in *International Conference on Renewable Power (ICRP-2020)*, LNEE (BGSB University, Rajouri, J&K, India, 2020)
5. M. Khursheed, M.A Mallick, A. Iqbal, Performance analysis of closed loop control of diesel generator power supply for base transceiver (BTS) load. Int. J. Innovative Technol. Explor. Eng. (IJTEE) **8**(9), 2483–2495 (2019) ISSN: 2278–3075 (Online)
6. H. Abu-Rub, A. Iqbal, S. Amin, G. Abdelkadar, N. Salem, S. Mansoor, T. Mirza, S. Ahmed, Feasibility analysis of solar photo-voltaic array cladding on commercial towers in Doha, Qatar—a case study. Int. J. Sustain. Energy **29**(2), 76–86 (2010)
7. K. Sopian, O.M. Yusof, R. MohdAzharAbd, Performance of a photovoltaic diesel hybrid system in Malaysia. ISESCO Sci. Technol. Vision, **1**, 37–39 [Retrieved] 15 March 2012
8. M.Z.C. Wanik, A. Bousselham, A. Elrayyah, Real-time simulation modeling for PV-battery based microgrid system, in *IEEE International Conference on Power System Technology (POWERCON)* (Wollongong, NSW, Australia, 2016)
9. A. Iqbala, M. Saad Bin Arif, S.Md. Ayobb, K. Siddiqui, Analysis of a solar PV/Battery/DG set-based hybrid system for a typical telecom load: a case study. Int. J. Sustain. Energy **36**(3), 259–276 (2015). https://doi.org/10.1080/14786451.2015.1017497
10. J.S. Goud, R. Kalpana, Optimal sizing of hybrid power supply system for telecommunication BTS load to ensure reliable power at lower cost, in *EEE International Conference on Technological Advancements in Power and Energy (TAP Energy)* (Kollam, India, 2018)
11. R. Atallah, P. Singh, Design and simulation of an integrated PV solar electric system, in *IEEE Global Humanitarian Technology Conference (GHTC)* (San Jose, CA, USA, 2017)
12. T. Lambert, P. Gilman, P. Lilienthal, *Integration of Alternative Sources of Energy-Micropower System Modeling with HOMER* (John Wiley & Sons, 2006)
13. T. Bano, K.V.S. Rao, The effect of solar PV module price and capital cost on the levelized electricity cost of the solar PV power plant in the context of India, in *International Conference on Power and Energy Systems: Towards Sustainable Energy (PESTSE)* (IEEE Xplore, Bengaluru, India, 2016)
14. R. Rajbongshi, D. Borgohain, S. Mahapatra, Optimization of PV-biomass-diesel and grid base hybrid energy systems for rural electrification by using HOMER. Energy **126**, 461–474 (2017). https://doi.org/10.1016/j.energy.2017.03.056
15. S. Avikal, R. Singhal, R. Sajwan, Selection of best power supply source for telecom towers in remote areas. Int. J. Math. Eng. Manage. Sci. **5**(5), 913–925 (2020)
16. B. Panajotovic, B. Odadzic, Design and intelligent control of hybrid power system in telecommunication, in *IEEE Xplore, 5th IEEE Mediterranean Electrotechnical Conference* 2010
17. A. Iqbal et al., *Renewable Power for Sustainable Growth*, vol 723 1E (Springer Nature, Part of the Lecture Notes in Electrical Engineering, 2021), p. 805. ISBN: 978–981–33–4080–0. https://doi.org/10.1007/978-981-33-4080-0
18. S. Srivastava et al., *Applications of Artificial Intelligence Techniques in Engineering*, vol 2–697 1E (Springer Nature, Part of the Advances in Intelligent Systems and Computing, 2018), p. 647, ISBN 978–981–13–1822–1. https://doi.org/10.1007/978-981-13-1822-1
19. N. Fatema et al., *Intelligent Data-Analytics for Condition Monitoring: Smart Grid Applications*, 1E (Academic Press, 2021), ISBN: 978–0–323–85510–5. https://doi.org/10.1016/C2020-0-02173-0
20. Srivastava et al., *Applications of Artificial Intelligence Techniques in Engineering*, vol 1–698 1E (Springer Nature, Part of the Advances in Intelligent Systems and Computing, 2018), p. 643, ISBN: 978–981–13–1819–1. https://doi.org/10.1007/978-981-13-1819-1

21. Alternative Energy Foundation, Combating Qatar's CO2 Emissions, Carbon World 2009 Addresses Reduction. Available: http://www.alternativeenergyfoundation.org/releases/2008/combating-qatars-co2-emissions-carbon-world-2009-addresses-reduction/
22. A. Iqbal et al., *Meta Heuristic and Evolutionary Computation: Algorithms and Applications*, vol 916 1E (Springer Nature, Part of the Studies in Computational Intelligence, 2020), p. 849. ISBN 978–981–15–7571–6. https://doi.org/10.1007/978-981-15-7571-6
23. J.A. Alzubi, *AI and Machine Learning Paradigms for Health Monitoring System: Intelligent Data Analytics*, vol 86 1E (Springer Nature, Part of the Studies in Big Data, 2020), p. 513. ISBN: 978–981–33–4412–9. https://doi.org/10.1007/978-981-33-4412-9
24. A.K. Yadav et al., *Soft Computing in Condition Monitoring and Diagnostics of Electrical and Mechanical Systems*, vol 1096 1E (Springer Nature, Part of the Advances in Intelligent Systems and Computing, 2020), p. 496. ISBN 978–981–15–1532–3. https://doi.org/10.1007/978-981-15-1532-3

Comparative Analysis of Nature-inspired Optimization Techniques for Data Analytics

Rajesh Ranjan, Pradeep Kumar, and Amit Kumar Yadav

Abstract Many scientific and engineering issues can be solved with the right tools, expressed as optimization problems with complex nonlinear constraints. Highly nonlinear issues frequently necessitate advanced optimization techniques, and standard algorithms may struggle to handle them. Because of their versatility and effectiveness, nature-inspired algorithms are getting more popular. This study provides an in-depth examination of a number of current nature-inspired algorithms, with MATLAB implementation for solving sphere function. The various algorithms used for the comparison are social group optimization, particle swarm optimization (PSO), simulated annealing, genetic algorithm (GA), cuckoo search, artificial bee colony (ABC), farmland fertility optimization, firefly optimization, water cycle optimization (WCO), teaching–learning-based optimization (TLBO), and antlion optimization. Through experimental results, we will show the best optimization technique for a given number of iteration and for a given function (here, it is sphere function).

Keywords SGO · PSO · SA · GA · CS · ABC · FFO · FO · WCO · TLBO · ALO

1 Introduction

In many cases, optimization is critical and important, from engineering aspect to daily life activity and to business scheduling. Optimization of a problem/objective means to get the best possible solution from the set of solutions of a given problem/function [1]. In general, to optimize, a problem requires sophisticated tools and sometimes very cumbersome. This may lead to inaccurate or wrong results. In recent years, nature-inspired algorithm is widely used to get optimized results for complex objectives. These algorithms or optimization techniques are based on natural behaviour of various species such as swarm of birds, school of fish, and social behaviour of human. These algorithms do not involve complex mathematical calculations instead

R. Ranjan · P. Kumar · A. K. Yadav (✉)
Electrical and Electronics Engineering Department, National Institute of Technology Sikkim, Barfung Block, South Sikkim, Ravangla, Sikkim 737139, India
e-mail: amit1986.529@nitsikkim.ac.in

H. Malik et al. (eds.), *Intelligent Data Analytics for Power and Energy Systems*, Lecture Notes in Electrical Engineering 802,
https://doi.org/10.1007/978-981-16-6081-8_30

update their solution as per the natural behaviour of species. The characteristics of nature-inspired algorithms are population-based having multiple agents, and each agent is a possible solution, evolution of population is generally iterative, movement of agents in search space are quasi-deterministic, all algorithms carry out both local and global searches, and selection of the best possible solution is done on the basis of 'survival of fittest' [2, 3].

In this study, we will see different optimization algorithm and its implementation on our objective. Optimization is a process that compares many solutions iteratively until a solution satisfying the objective is found. Today's optimization algorithms are of two categories (1) deterministic algorithm and (2) stochastic algorithm. In this chapter, we will be dealing with stochastic algorithms. The objective proposed in this chapter is to make a comparative analysis of some nature-inspired algorithms (social group organization, particle swarm optimization, simulated annealing, genetic algorithm, cuckoo search algorithm, artificial bee colony optimization, farmland fertility, firefly optimization, water cycle optimization, teaching–learning-based optimization, and antlion optimization). The problem taken in this chapter to make analysis is to minimize the sphere function.

2 Methodology

2.1 Social Group Optimization (SGO)

Each person (better to say a candidate for viable solution) in SGO is given some level of knowledge and problem-solving abilities. This is comparable to the 'fitness'. The best solution is the best person. The best person tries to deliver knowledge to each member of the group that will raise the group's overall knowledge level. The best individual in the group is the one who has the maximum knowledge and has ability to solve the problem. In the learning phase, knowledge of each person in the group is enhanced through interaction among themselves, as well as the group's best member at the time [4]. The following are the stages for implementing SGO [5–13]:

Step 1: Initializing parameters and enumeration of the problem.

Take the population size of (N), number of generations be (g), number of variables (D), and boundary of variables (ub, lb). Define the optimization problem as: minimize $f(X)$. Subject to = (×1,×2,×3,...,×D), so that X = (×1,× 2,×3, ...,×D), where $f(X)$ is the objective function and X is a vector for design variables with upper and lower bounds.

Step 2: Make a population start. To generate random population, nos. of parameters and population size are fed from user. The population size in SGO refers to the total number of people, while the characteristics refer to the total number of qualities. The following is a list of the people who make up this population:

Population = $[x_{1,1}x_{1,2}x_{1,3}\ldots\ldots\ldots\ldots x_{1,D};\ \ldots\ldots;\ \ldots\ldots;\ \ldots\ldots;\ \ldots\ldots.;\ x_{N,1}x_{N,2}\ldots\ldots\ldots.x_{N,D}]$ Determine the fitness value of the population $f(X)$.

Step 3: Phase of improvement, then using Eq. (1), compute gbestg, which is the best answer for that iteration. Similarly, to the improving phase, each person gains information from the best in their group, i.e. gbest.

for $i = 1{:}\,N$
for $j = 1{:}\,D$

$$X_{\text{new}(i,j)} = C * X_{\text{old}(i,j)} + r * \left(\text{gbest}(j) - X_{\text{old}(i,j)}\right).$$

end
end.

The self-introspection factor has a value of c. For a given situation, the value of c can be determined empirically. After a thorough examination of our examined problems, we have set it to 0.2 in this work, and r is a random value, r U. (0, 1). Accept X new if it provides a better value for the function.

Step 4: Acquiring phase As previously stated, in the acquiring phase, a member of a social group interacts with the best individual in the group, i.e. the gbest of the group, as well as other members of the group at random, in order to acquire knowledge. In the 'acquiring phase', the mathematical expression is defined.

Step 5: Criteria for termination. If the maximum generation number is reached, stop the experiment; otherwise, continue Steps 3–4 [5].

2.2 Particle Swarm Optimization (PSO)

James Kennedy and Russell C Eberhart created it in 1995. Intelligent optimization algorithm is inspired by social behaviour of birds and fishes. Belongs to a class of optimization problem called meta-heuristic. In swarm intelligence, agents are called 'particle'. Particles have to fetch a level of 'information' from the swarm. At the beginning, no particle has any idea about the 'information'. But, with collective behaviour like communication with each other, learning from the information helps them to find the information [14–21].

- Swarm: A set of particles who has memory.
- Every particle is a set of potential solution
- Particle position $X\_((i, j)) = (X\_((i, 1)), X\_((i, 2)) \ldots. X\_((i, n))) \in \Re\hat{}n$
- Particle velocity $w\_((i, j)) = (w\_((i, 1)), w\_((i, 2)) \ldots.$ w_ $((i, n))) \in \Re\hat{}n$
- Each particle remembers its own best position
- Swarm has a global best position.
- Communicating with each other about their best position, they explore the search space to get the global result/solution.

Steps to implement PSO:

Step 1: Initialize the population, the search space's dimension and maximum iteration, and the fitness function's global best cost.

Step 2: Define Max and Min of the search space and initialize position and velocity of particle.

Step 3: Find cost/fitness function and update personal and global best cost.

Step 4: Update velocity and each particle as per the formula:

$$V_{(i,j)}(t+1) = \omega V_{(i,j)}(t+1) + c_1 r_1 \left(x^{lb}_{(i,j)}(t) - x_{(i,j)}(t)\right) + c_2 r_2 \left(x^{\text{Global}}(t) - x_{(i,j)}(t)\right) \quad (1)$$

$$x_{(i,j)}(t+1) = x_{(i,j)}(t) + V_{(i,j)}(t+1) \quad (2)$$

where ω is inertia coefficient. c_1 and c_2 are acceleration coefficients.

Step 5: Go to Step 3 and repeat until termination condition [14].

2.3 Simulated Annealing (SA)

Among various optimization methods, this is the most extensively used heuristic approaches for handling optimization problems. The annealing process in metallurgy was used as inspiration for SA. In the process of annealing, molecular configurations of metal particles are changed so that the potential energy of the mass is minimized by slowly cooling metals after they have been exposed to intense heat.

Classical approach of any optimization method is to iteratively compare the outputs of objective functions running with the current and surrounding points in the domain, keeping the adjacent points as the base solution for the next iteration if the adjacent points produce a better result than the current one. Or else, the algorithm is terminated without looking for better results in a larger region. As a result, the algorithm is vulnerable to becoming stuck in local minima or maxima. The main idea of the metropolis criterion is that it should be run at random to extra search the neighbourhood of the candidate solution to prevent being trapped at local extreme points. Take a minimization problem and define an objective function to be more specific.

$f(x_i)$ corresponds to the $x_i = x_1, x_2, ..., x_n$ argument set with n_R.

Take x_{i+1} as a new candidate only if $f(x_{i+1}) < f(x_i)$. Otherwise, write w = exp $[f(x_{i+1}) - f(x_i)/T_c]$, where T_c is the current temperature parameter and s is a random number such that $0 < s < 1$.Then, if the relation of $w > s$ is true, you also accept the x_{i+1} as a new candidate, else reject and go back to previous step and generate another solution.

Steps to implement SA:

```
Create random initial solution Y
E_old = cost(Y);
For (temp=temp_max ; temp>=temp_min ; temp=next_temp(temp))
  {
      for (i=0; i<=i_max ; i++)
        {
          Successor_func(Y); // Randomized function
          E_new  = cost(Y);
          Delta=  E_new  - E_old;
          If (Delta>0)
              If(random()>=exp(-Delta/k*temp))
                Undo_func(Y);
              Else
                E_old = E_new ;
           Else
             E_old = E_new ;
        }
  } (chaturvedi, 2007)
```

2.4 Genetic Algorithm (GA)

A population of guesses is used to start GA. These are frequently distributed at random throughout the search space. To lead the population towards convergence at the global optimum, a common algorithm employs three operators: selection, crossover, and mutation. Although an increasing number of GAS utilize 'real-valued' encodings, or encodings that have been selected to mirror in some way the inherent data structure of the problem, these first guesses are typically maintained as binary encodings of the genuine variables. The three primary operators then process this initial population. Selection tries to put pressure on the population in the same way that natural selection does in biological systems. Individuals who do poorly are weeded out, while those who perform better, or are fitter, have a better chance to pass on the information they carry to their successors. Crossover permits solutions to exchange information in a manner akin to that of a sexually reproducing organism. One method is to take two people who have been promoted by the selection operator, pick a single locus inside the binary strings at random, and swap all the information to the right of this locus between them. Mutation is a method of changing the value of individual bits in strings at random. In most cases, mutation is used sparingly. A new population will be generated after selection, crossover, and mutation have been applied to the first population, and the generational counter will be incremented by one. This selection, crossover, and mutation process is repeated until a certain number of generations have passed, or a convergence requirement has been satisfied [6–13, 22].

Steps to implement GA:

(1) Randomly generate population *P*.
(2) Calculate fitness of the population generated.
(3) Repeat the below process until convergence is achieved:

(a) Parents are taken from general population.
(b) Crossover and population generation
(c) Generate a new population and perform mutations on it.
(d) Determine the fitness of a new population

2.5 Cuckoo Search

Xin-she Yang and Suash Deb created the cuckoo search (CS) optimization algorithm in 2009. It was taken from cuckoo's, obligate brood parasitism, in which they lay their eggs in the nests of some other host birds (of different kind). Some cuckoo species have evolved to the point where female parasitic cuckoos are highly skilled at imitating the colour and design of a few specific host species' eggs. The chance of eggs being abandoned is reduced and raises the likelihood of them developing. The intruding cuckoos can cause direct conflict with host birds. Whenever a host bird realizes that the eggs are not theirs, it will either discard the foreign eggs or leave its nest and create a new one at different location [23].

Mathematical model of cuckoo search algorithm (C.S.)

Three different operators defined for C.S.

1. Levy light—Generate a New Candidate Solution by Perturbing Current X with

$$X_{\text{new}} = X + R * C.$$

C = change of position

$$C = 0.01 * S * (X - \text{best})$$

2. R = *random number with normal distribution* $\in (0, 1)$
3. S= random step size and generated by levy distribution as:

$$v = R; \quad u = R * \sigma_u$$

$$S = u/|v|^{1/\beta} \quad \text{where } \beta = 1.5; \; u \sim N\left(0, \sigma_u^2\right) \quad \text{and} \quad v \sim N\left(0, \sigma_v^2\right).$$

$$\sigma_v = 1 \quad \text{and} \quad \sigma_u = \frac{\Gamma(1+\beta)\cdot \sin(\Pi \cdot \beta/2)}{\Gamma(1+\beta)/2 \cdot \beta \cdot 2^{(\beta-1)/2}} = 0.6966.$$

$$X_{\text{new}} = X + R * 0.01 * S * (X - \text{best})$$

best = global best solution of the nest.

2.6 Artificial Bee Colony

Karaboga's artificial bee colony (ABC) algorithm is a swarm-based meta-heuristic technique for numerical problem optimization that he created in 2007 [24]. The sophisticated foraging activities of honeybees served as inspiration. The method is based on a model for honeybee colony foraging behaviour developed by Tereshko and Loengarov in 2005 [24]. The model is made up of three main elements: working and unemployed foraging bees, as well as food sources. First two elements, i.e. employed and unemployed foraging bees, look for huge food sources near to their hive, that is the third component. This model also identifies two important behaviours that are needed for self-organization and collective intelligence: forager recruitment to rich food sources, which results in positive feedback, and forager desertion of low food sources, which results in negative feedback. The artificial bees then randomly find a population of initial solution vectors and iteratively enhance them using the following strategies: migrating to better solutions via a neighbour search mechanism while leaving inferior answers.

The general steps followed by this algorithm is:

1. Initialization
2. Repeat

   Phase 2.1: Employed bees
   Phase 2.2: Onlooker bees
   Phase 2.3: Scout bees
   Phase 2.4: Write down the best solution.

3. Until the termination criteria are satisfied [24].

2.7 Farmland Fertility

This is a population-based nature-inspired optimization technique. Farmers partition their farmland into different sections depending on soil quality. Soil quality of each section is different, and it can be changed by adding some specific materials. Each section of farmland has memory unit. These memory units record best quality of soil on adding specific material till date (called as local memory). A series of memory units keep the record of the best quality soil in all sections of farmland (called global memory). Local memory saves the best solution from each section's last visit, while global memory saves the best solution from all sections up to this point. Farmers employ both memory and experience to improve the soil quality of areas on their next visit. Farmland fertility is divided into six stages [25]:

1. Initial Values: Production of population

   Total population in search space, $N = K * n$
   $K =$ nos. of sections; $n =$ nos. of solutions in a section
   Random production of search space: $x_{i,j} = L_j + \text{rand}(0, 1) * \left(U_j - L_j\right)$.

2. Determining soil quality in each part.
3. Update memories.

   Number of best local memory $M_{\text{local}} = \text{round}(t * n) \quad 0.1 < t < 1.$
   Number of best global memory $M_{\text{global}} = \text{round}(t * N) \quad 0.1 < t < 1.$

4. Change in soil quality for different section of farmland, $h = a$*rand (); $a =$ nos. between 0 and 1

$$X_{\text{new}} = h * \left(X_{ij} - X_{\text{global}}\right) + X_{ij}$$

5. Soil's combination is done to improve the quality of existing solution.
6. Final condition, at this stage, existing solutions are found as per the objective function in the entire search space [25].

2.8 Firefly Optimization

Fireflies are winged beetles or insects that create light and flicker at night. Bioluminescence is a light that emits no infrared or ultraviolet wavelengths and is produced chemically in the lower abdomen. The flash light is specifically used to lure partners or prey. The flash light also served as a warning system for the fireflies, reminding them of prospective predators. Yang created the firefly algorithm. It is a meta-heuristic algorithm inspired by fireflies' flashing behaviour and bioluminescent communication.

The firefly algorithm was created based on the following assumptions [26]:

(1) Because fireflies are unisexual, they will be attracted to each other regardless of their gender.
(2) Because their appeal is proportional to their brightness, the less light firefly will be drawn to the brighter firefly. However, as the distance between the two fireflies was extended, the two fireflies' attraction decreased.
(3) If the fireflies' brightness is equal, the fireflies will move at random [26].

2.9 Water Cycle Optimization

Nature-inspired the proposed WCA, which is inspired from real-world observations of the water cycle and how rivers and streams flow down the hill to fall into the sea. To further comprehend this, consider the following explanation as how rivers are formed and water flow to the sea. When water flows down the hill from one location to different location, it forms a river or stream. As a result, the majority of rivers originate high in the mountains, where snow and ancient glaciers melt. Rivers are continuously flowing downhill. Rain water is harvested, and along with other streams, they go downwards, and finally enters the sea. The proposed method, like prior meta-heuristic algorithms, begins with a population known as raindrops. We will start by presuming we are in the midst of rain or precipitation. As a sea, we select the best individual (best raindrop). Then, a few good raindrops are taken to make a river, while the other raindrops form streams that run into rivers and the sea [27].

The amount of water that each river receives from the streams is determined by the amount of flow, which will be detailed in the following subsections. The amount of water in a stream that goes into rivers and/or the sea differs from the amount of water in other streams. Rivers also flow to the sea, which is at the mountain's lowest point. The values of variables in the problem must be organized to form an array so that an optimization problem using population dependent meta-heuristic approach can be solved. Such an array is referred to as a 'chromosome' or a 'particle position" in GA and PSO terminology, respectively. As a result, a single solution is referred to as a 'raindrop' in the suggested method. A raindrop is an array of 1 * Nvar in an Nvar dimensional optimization problem. For continuous and discrete situations, each decision variable values can be expressed as a floating-point number (actual value) or a predetermined set. The fitness of a raindrop is determined by evaluating the cost function (C), that is written as:

$$\text{Population} = \begin{bmatrix} \text{Stream1} \\ \text{Stream2} \\ . \\ . \\ \text{Stream}N \end{bmatrix} = \begin{pmatrix} x_{11} & \cdots & x_{1N} \\ \vdots & \ddots & \vdots \\ x_{N1} & \cdots & x_{NN} \end{pmatrix}$$

$$\text{Total Population} = \begin{bmatrix} \text{Sea} \\ \text{River1} \\ \text{River2} \\ \text{River3} \\ \text{Stream4} \\ \text{Stream5} \\ . \\ . \\ \text{Stream}N\text{pop} \end{bmatrix} = \begin{pmatrix} x_{11} & \cdots & x_{1,N} \\ \vdots & \ddots & \vdots \\ x_{N\text{pop},1} & \cdots & x_{N\text{pop},N} \end{pmatrix}$$

Fitness $= f(\text{Stream}, i) = f(\times\ 1, \times\ 2, \ldots, \times\ N)$ and $i = 1, 2, 3, \ldots N$.

The raindrop having the lowest value among the others is referred to as a sea. In actuality, N_{sr} is the sum of the user parameter,number of rivers and a single sea. Rest of the population is computed (raindrops generate streams that run to rivers or may directly enter the sea). $N\text{sr}$ = Number of Rivers + Sea

$$N_{\text{stream}} = N_{\text{pop}} - N_{\text{sr}}.$$

Updated positions of streams and rivers are evaluated as:

$$X_{\text{Stream}}^{i+1} = X_{\text{Stream}}^{i} + \text{rand} * C * \left(X_{\text{River}}^{i} - X_{\text{Stream}}^{i}\right)$$

$$X_{\text{Stream}}^{i+1} = X_{\text{Stream}}^{i} + \text{rand} * C * \left(X_{\text{Sea}}^{i} - X_{\text{Stream}}^{i}\right)$$

$$X_{\text{River}}^{i+1} = X_{\text{River}}^{i} + \text{rand} * C * \left(X_{\text{Sea}}^{i} - X_{\text{River}}^{i}\right)$$

Evaporation Condition

Evaporation can cause delay in the convergence. Evaporation is a natural phenomenon, and this can be observed in rivers also. Plants discharge water into rivers and lakes during photosynthesis. Water evaporates into the atmosphere and condenses at the higher altitudes of atmosphere to form water droplets. These water droplets enter the earth's atmosphere as rain.

$$\text{If } \|X_{\text{Sea}}^{i} = X_{\text{River}}^{i}\| < d_{\max} \quad i = 1, 2, 3, \ldots N_{\text{sr}-1}.$$

End of the Evaporation and raining process.

Where $d_{\max}$ is a very small no. but not zero [27].

2.9.1 Teaching–learning-based Optimization (TLBO)

TLBO algorithm is inspired by the teaching–learning process. It depends on teaching of a teacher and learner's output in the class. This algorithm explains two primary ways of learning: (i) learning from a teacher (called as the teacher phase) and (ii) learning from other students (called as the student phase or learner phase). In this optimization technique, learner's population is studied, and different subjects are provided to the learners. These subjects can be considered as different design variables for the optimization problem, with a learner's outcome being comparable to the optimization problem's fitness value.

For the overall population, the instructor is believed to be the best option. The design variables are the parameters that go into an objective function of considered optimization problem and objective function's best will give optimum solution. The teaching–learning-based optimization (TLBO) algorithm is a population-based algorithm that mimics the classroom's environment of teacher—student/learner. The standard parameters required for this algorithm such as population size and generation number, with no algorithm-specific control parameters required. The 'teaching phase' and the 'learning phase' are the two components of the TLBO's operation [28].

2.9.2 Antlion Optimization (ALO)

The antlion optimizer (ALO), a newly developed artificial intelligence-based optimization technique, was introduced by Mirjalili. It is a intelligence of swarm-based strategy that was inspired by the larval behaviour of antlions. The programme uses particles to simulate antlions and ants in order to execute an optimization process based on the mathematical and logical characteristics of antlion hunting behaviour. In the real world, antlions hunt ants with underground traps. The ALO builds on this fact by focusing on interactions between antlion and ant particles. Ants are transported over the search space, while antlions are permitted to hunt them with their traps to replicate these interactions. The antlions become fitter by utilizing the traps at this point. The motions of ants are simulated in the ALO utilizing the 'random walk' approach. The ants' shifting positions are used as parameters for objective function (s) find the best values [29].

3 Results and Discussions

For comparative analysis, MATLAB code is used [Appendix] to minimize the objective function $x_1^2 + x_2^2$ with boundary condition $-5 \leq x_1, x_2 \leq 5$. Total number of iterations are 100 and number of variables = 2. The plot of objective function vs number of iterations for social group optimization, PSO, simulated annealing, GA, cuckoo search, ABC, farmland fertility, firefly optimization, water cycle optimization, teaching–learning-based optimization, and antlion optimization is shown in Figs. 1, 2, 3, 4, 5, 6, 7, 8, 9, 10, and 11. The comparative analysis is shown in Table 1 which shows GA gives better result.

Farmland Fertility: Iteration 100: Best Cost = 1.2415e−32.

4 Conclusions

This chapter presents the comparative analysis of different nature-inspired optimization techniques. The objective of the chapter is to minimize sphere function using the various optimization techniques mentioned below and compare their results for a certain number of iterations (taken 100). The lower bounds and upper bounds considered are [ub lb] = [−5 5], and the number of variables considered is 2. Table 1 shows the result of different optimization techniques. It is clear from the results that the order of convergence of above techniques can be stated as: GA > farmland fertility > TLBO > PSO > ALO > SA > ABC > CSO > firefly optimization > SGO > WCO. This order may change on varying any of the parameters such as objective function, nos. of variables taken, nos. of iterations considered, lower bound, and upper bound.

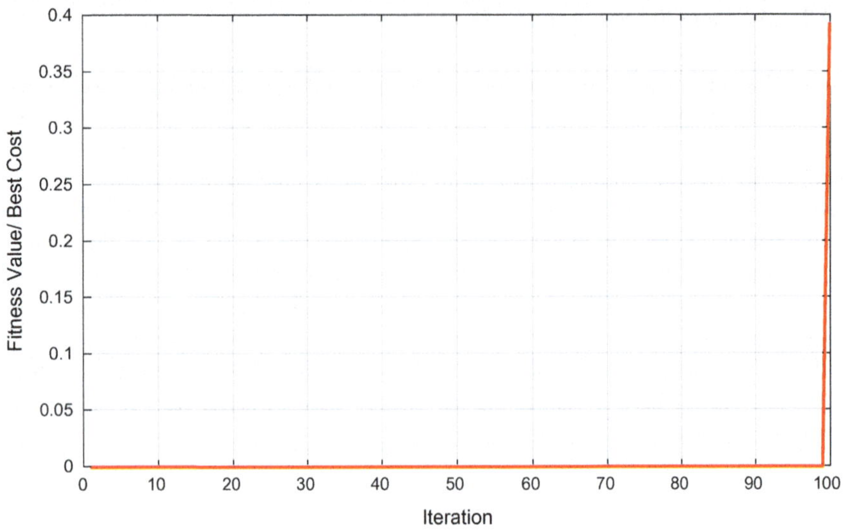

Fig. 1 Plot of number of iteration versus min. value of the objective function (best cost)

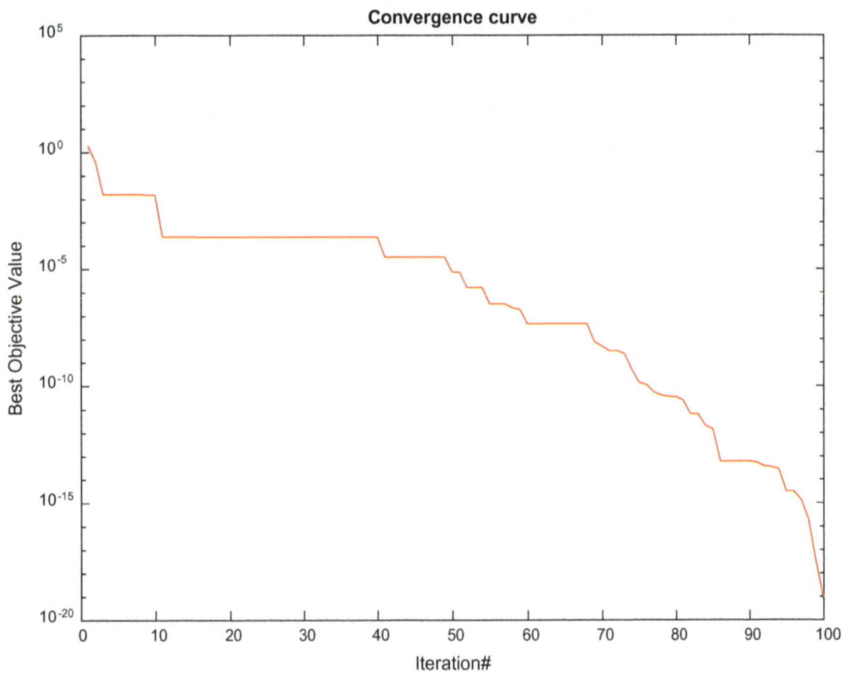

Fig. 2 Plot between nos. of iteration versus best objective value for PSO

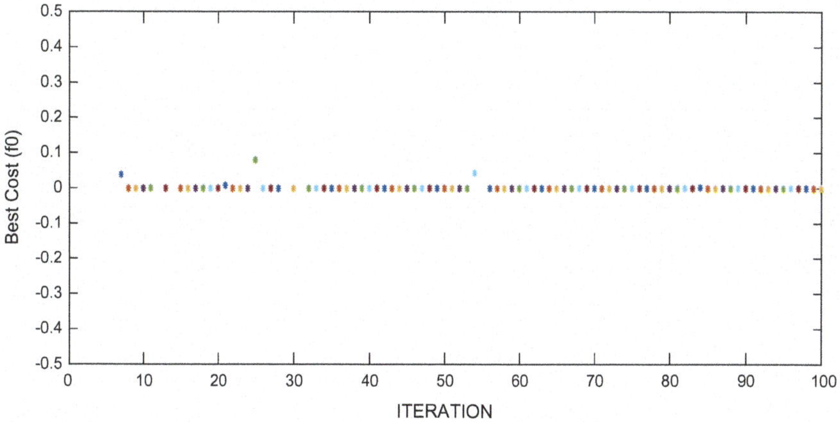

Fig. 3 Plot of nos. of iteration versus best cost for simulated annealing

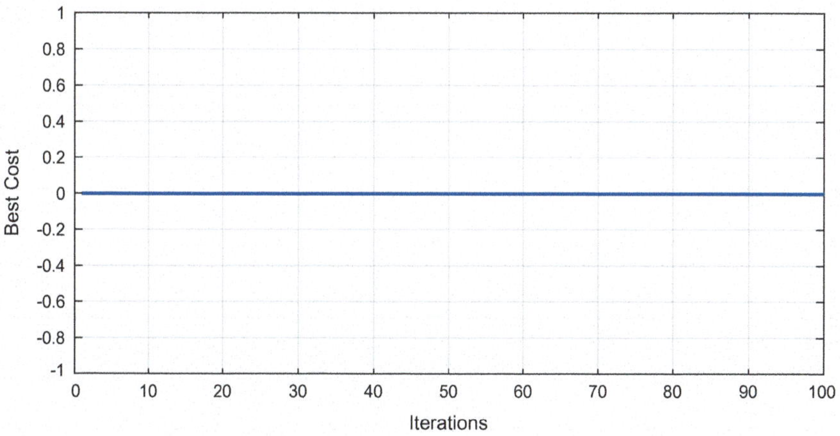

Fig. 4 Plot of nos. of iteration versus best cost for genetic algorithm

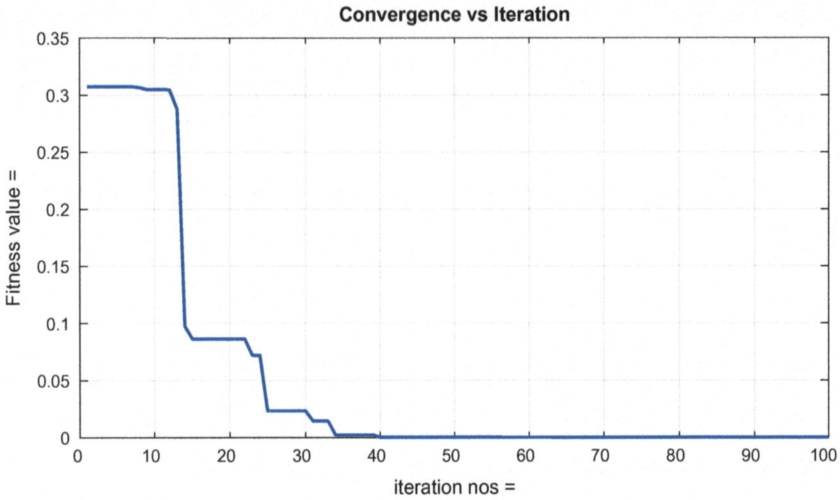

Fig. 5 Plot of nos. of iteration versus best objective value/fitness value for cuckoo search optimization

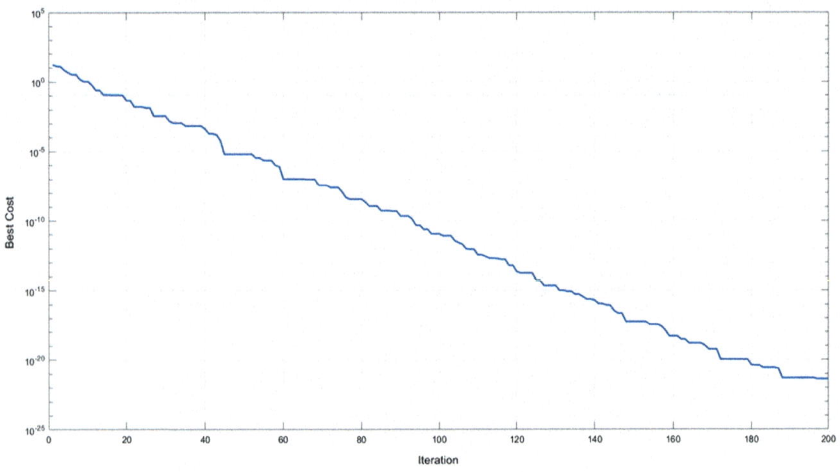

Fig. 6 Plot of nos. of iteration versus best cost for artificial bee colony

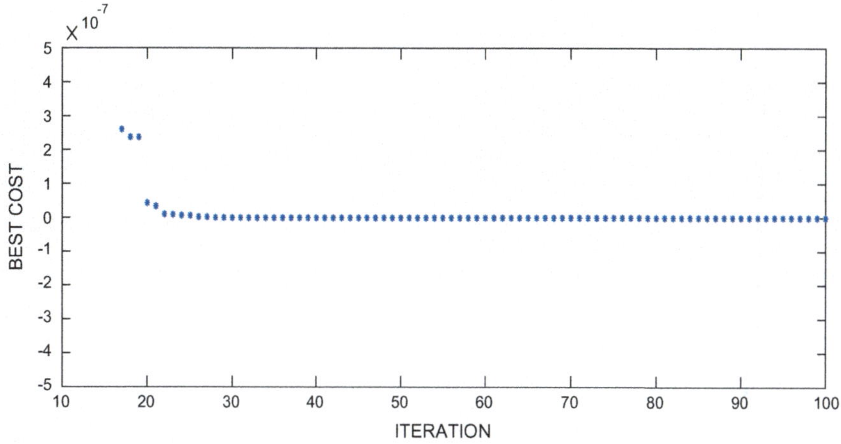

Fig. 7 Plot of nos. of iteration versus best cost for farmland fertility

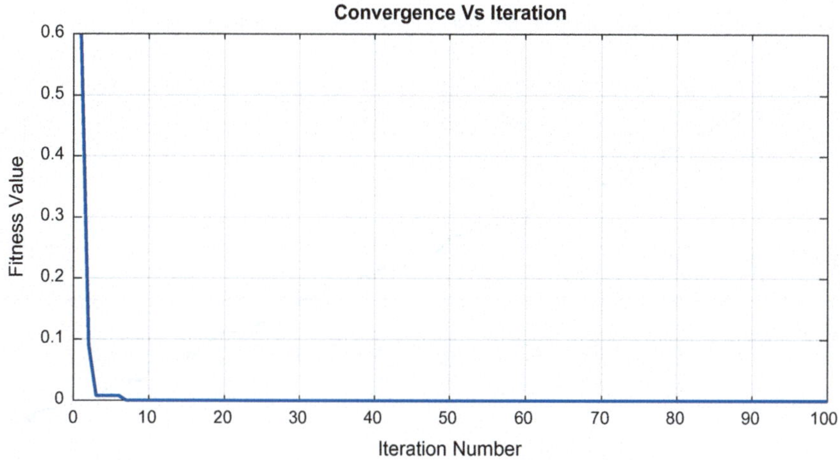

Fig. 8 Plot of nos. of iteration versus fitness value for firefly optimization

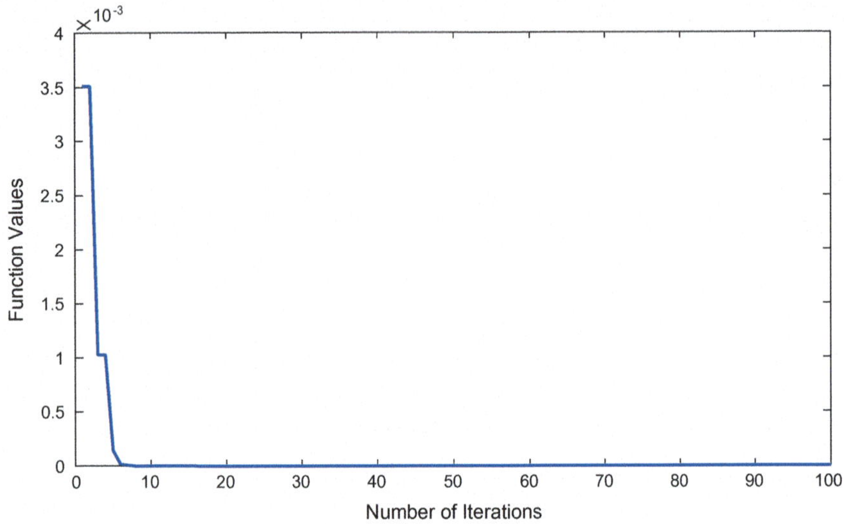

Fig. 9 Plot of nos. of iteration versus function (fitness value) for water cycle optimization

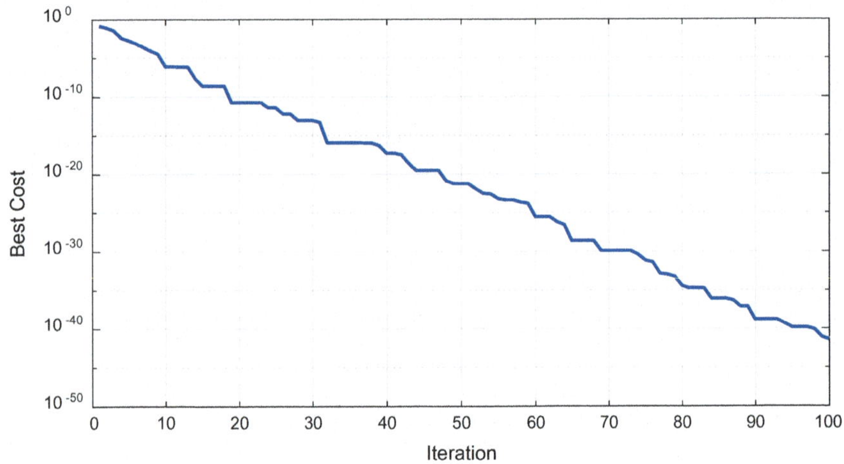

Fig. 10 Plot of nos. of iteration versus best cost for TLBO

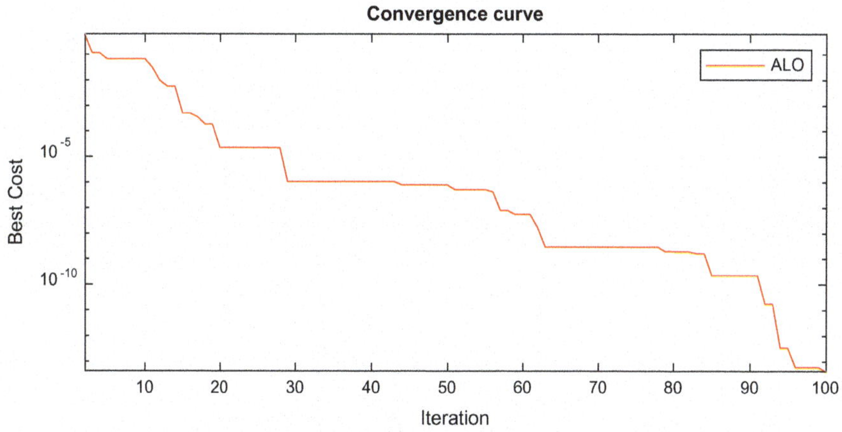

Fig. 11 Plot of nos. of iteration versus best cost for ALO

Table 1 Comparative analysis

Optimization technique	Best cost/minimum value of objective function
1. Social group optimization (SGO)	0.3161
2. Particle swarm optimization (PSO)	7.00e−18
Iteration# 80 swarm.GBEST.O = 6.693e−12; Iteration# 81 Swarm.GBEST.O = 4.9089e−12; Iteration# 82 Swarm.GBEST.O = 4.2091e−12; Iteration# 83 Swarm.GBEST.O = 4.0776e−12; Iteration# 84 Swarm.GBEST.O = 4.0456e−12; Iteration# 85 Swarm.GBEST.O = 3.1056e−12; Iteration# 86 Swarm.GBEST.O = 2.7705e−12;	
Iteration# 87 Swarm.GBEST.O = 3.2228e−13; Iteration# 88 Swarm.GBEST.O = 3.2228e−13; Iteration# 89 Swarm.GBEST.O = 3.0616e−13; Iteration# 90 Swarm.GBEST.O = 2.7096e−13; Iteration# 91 Swarm.GBEST.O = 2.7096e−13; Iteration# 92 Swarm.GBEST.O = 1.1127e−13; Iteration# 93 Swarm.GBEST.O = 2.3508e−14; Iteration# 94 Swarm.GBEST.O = 2.7326e−15; Iteration# 95 Swarm.GBEST.O = 1.552e−15; Iteration# 96 Swarm.GBEST.O = 1.4921e−15; Iteration# 97 Swarm.GBEST.O = 1.49e−15; Iteration# 98 Swarm.GBEST.O = 5.9217e−16; Iteration# 99 Swarm.GBEST.O = 2.2015e−17; Iteration# 100 Swarm.GBEST.O = 7.0031e−18	
3. Simulated annealing (SA)	3.03e−13
ITER NO = 80: BEST COST = 3.19570202990689e−20; ITER NO = 81: BEST COST = 0.0412685216914903; ITER NO = 82: BEST COST = 3.19570202990689e−20; ITER NO = 83: BEST COST = 1.09344423849242e−13; ITER NO = 84 : BEST COST = 3.19570202990689e−20; ITER NO = 85: BEST COST = 3.19570235297056e−20; ITER NO = 86: BEST COST = 3.66171992620219e−07; ITER NO = 87: BEST COST = 3.19570235297057e−20; ITER NO = 88: BEST COST = 3.1957023529706e−20; ITER NO = 89: BEST COST = 3.1957023529706e−20; ITER NO = 90: BEST COST = 3.1957023529706e−20; ITER NO = 91: BEST COST = 3.19570235294698e−20; ITER NO = 92: BEST COST = 6.86870888496198e−07; ITER NO = 93: BEST COST = 3.19570235294698e−20; ITER NO = 94: BEST COST = 5.31384316422321e−09; ITER NO = 95: BEST COST = 3.19570235294698e−20; ITER NO = 96: BEST COST = 3.17074554505465e−20; ITER NO = 97: BEST COST = 3.17074554505465e−20; ITER NO = 98: BEST COST = 3.17074554505465e−20; ITER NO = 99: BEST COST = 3.17074554505465e−20; ITER NO = 100: BEST COST = 3.03052836849092e−13	
4. Genetic algorithm (GA)	0

(continued)

Table 1 (continued)

Optimization technique											Best cost/minimum value of objective function
Iteration 80: Best cost = 0; Iteration 81: Best cost = 0; Iteration 82: Best cost = 0; Iteration 83: Best cost = 0; Iteration 84: Best cost = 0; Iteration 85: Best Cost = 0; Iteration 86: Best cost = 0; Iteration 87: Best cost = 0; Iteration 88: Best cost = 0; Iteration 89: Best cost = 0; Iteration 90: Best cost = 0; Iteration 91: Best Cost = 0; Iteration 92: Best cost = 0; Iteration 93: Best cost = 0; Iteration 94: Best cost = 0; Iteration 95: Best cost = 0; Iteration 96: Best cost = 0; Iteration 97: Best Cost = 0; Iteration 98: Best Cost = 0; Iteration 99: Best cost = 0; Iteration 100: Best cost = 0											
5. Cuckoo search (CSO)											3.94e−07
Iteration 80: Best cost = 0.00014179; Iteration 81: Best cost = 0.00014084; Iteration 82: Best cost = 0.00014084; Iteration 83: Best cost = 0.00014084; Iteration 84: Best cost = 0.00014084; Iteration 85: Best cost = 0.00014084; Iteration 86: Best cost = 0.00014084; Iteration 87: Best cost = 0.00014084; Iteration 88: Best cost = 3.941e−07; Iteration 89: Best cost = 3.941e−07; Iteration 90: Best cost = 3.941e−07; Iteration 91: Best cost = 3.941e−07; Iteration 92B: Best cost = 3.941e2207; Iteration93: Best Cost = 3.941e−07; Iteration 94: Best cost = 3.941e−07; Iteration 95: Best cost = 3.941e−07; Iteration 96: Best cost = 3.941e−07; Iteration 97: Best cost=3.941e−07; Iteration 98: Best Cost = 3.941e−07; Iteration 99: Best cost = 3.941e−07; Iteration 100: Best cost = 3.941e−07											
6. Artificial bee colony (ABC)											4.22e−11
7. Farmland fertility											1.80e−31
Iteration 80: Best cost = 2.7093e−25; Iteration 81: Best cost = 2.7093e−25; Iteration 82: Best cost = 2.5752e−25; Iteration 83: Best cost = 2.0042e−26; Iteration 84: Best cost = 1.1124e−26; Iteration 85: Best cost = 2.5731e−27; Iteration 86: Best cost = 2.3939e−27; Iteration 87: Best cost = 1.7761e−27; Iteration 88: Best cost = 1.3358e−27; Iteration 89: Best cost = 3.7965e−28; Iteration 90: Best cost = 2.7475e−29; Iteration 91: Best cost = 2.7475e−29; Iteration 92: Best cost = 2.7475e−29; Iteration 93: Best cost = 2.7475e−29; Iteration 94: Best cost = 1.888e−29; Iteration 95: Best cost = 1.888e−29; Iteration 96: Best cost = 1.888e−29; Iteration 97: Best cost = 5.5421e−31; Iteration 98: Best cost = 5.5421e−31; Iteration 99: Best cost = 5.2761e−31; Iteration 100: Best cost = 1.8002e−31											
8. Firefly optimization											0.0003
Iter. No.	80	81	82	83	84	85	86	87	88	89	
Best cost	0.0003	0.0003	0.0003	0.0003	0.0003	0.0003	0.0003	0.0003	0.0003	0.0003	

(continued)

Table 1 (continued)

Optimization technique											Best cost/minimum value of objective function
Iter. no.	90	91	92	93	94	95	96	97	98	99	
Best cost	0.0003	0.0003	0.0003	0.0003	0.0003	0.0003	0.0003	0.0003	0.0003	0.0003	
Iter. no. 100: Best cost 0.0003											
9. Water cycle optimization (WCO)											15.5103
Iteration: 80 Fmin = 17.9506; Iteration: 81 Fmin = 17.6907; Iteration: 82 Fmin = 17.5029; Iteration: 83 Fmin = 17.4787; Iteration: 84 Fmin = 17.4245; Iteration: 85 Fmin = 17.3528; Iteration: 86 Fmin = 17.3123; Iteration: 87 Fmin = 17.3122; Iteration: 88 Fmin = 17.2675; Iteration: 89 Fmin = 17.2126; Iteration: 90 Fmin = 17.2039; Iteration: 91 Fmin = 16.8348; Iteration: 92 Fmin = 16.7586; Iteration: 93 Fmin = 16.4633; Iteration: 94 Fmin = 15.6368; Iteration: 95 Fmin = 15.6045; Iteration: 96 Fmin = 15.5738; Iteration: 97 Fmin = 15.5531; Iteration: 98 Fmin = 15.5331; Iteration: 99 Fmin = 15.5249; Iteration: 100 Fmin = 15.5103											
10. Teaching learning based optimization (TLBO)											2.05e−21
Iteration 80: Best cost = 9.2328e−17; Iteration 81: Best cost = 6.4121e−17; Iteration 82: Best cost = 3.77e−17; Iteration 83: Best cost = 2.3461e−17; Iteration 84: Best cost = 9.3755e−18; Iteration 85: Best cost = 9.3755e−18; Iteration 86: Best cost = 3.7343e−18; Iteration 87: Best cost = 1.6645e−18; Iteration 88: Best cost = 1.3241e−18; Iteration 89: Best cost = 5.2904e−19; Iteration 90: Best cost = 4.0033e−19; Iteration 91: Best Cost = 1.5796e−19; Iteration 92: Best cost = 8.7141e−20; Iteration 93: Best cost = 8.7015e−20; Iteration 94: Best cost = 6.2848e−20; Iteration 95: Best cost = 2.3365e−20; Iteration 96: Best cost = 1.3163e−20; Iteration 97: Best cost = 1.163e−20; Iteration 98: Best cost = 2.0506e−21; Iteration 99: Best cost = 2.0506e−21; Iteration 100: Best cost = 2.0506e−21											
11. Ant lion optimization (ALO)											4.22e−14

References

1. M. Gen, R. Cheng, *Genetic Algorithms and Engineering Design* (John Wiley & Sons, 1997)
2. D.K. Chaturvedi, in *Soft Computing Techniques and its Applications in Electrical Engineering*, Studies in Computational Intelligence, vol 103 (Springer, India, 2007), ISBN 978-3-540-77480-8
3. X.-S. Yang, in *Nature-Inspired Algorithms and Applied Optimization*, Studies in Computational Intelligence, vol 744 (Springer, London 2017), ISBN 978-3-319-67668-5
4. S. Satapathy, A .Naik, Social group optimization (SGO): a new population evolutionary optimization technique. Complex Intell. Syst. **2**, 173–203 (2016)
5. X. Feng, Y. Wang, H. Yu, F. Luo, A novel intelligence algorithm based on the social group optimization behaviors. IEEE Trans. Syst. Man Cybern. Syst. **48**(1), 65–76 (2018). https://doi.org/10.1109/TSMC.2016.2586973
6. A.K. Yadav et al., in *Soft Computing in Condition Monitoring and Diagnostics of Electrical and Mechanical Systems*, Part of the Advances in Intelligent Systems and Computing, vol 1096 (Springer Nature, 2020), p. 496, ISBN 978-981-15-1532-3. https://doi.org/10.1007/978-981-15-1532-3.
7. A. Iqbal et al., in *Meta Heuristic and Evolutionary Computation: Algorithms and Applications*, Part of the Studies in Computational Intelligence, vol 916 (Springer Nature, 2020), p. 849, ISBN 978-981-15-7571-6. https://doi.org/10.1007/978-981-15-7571-6
8. J.A. Alzubi, in *AI and Machine Learning Paradigms for Health Monitoring System: Intelligent Data Analytics*, Part of the Studies in Big Data, vol 86 (Springer Nature, 2020), p. 513, ISBN: 978-981-33-4412-9. https://doi.org/10.1007/978-981-33-4412-9
9. A. Iqbal et al., in *Renewable Power for Sustainable Growth*, Part of the Lecture Notes in Electrical Engineering, vol 723 (Springer Nature, 2021), p. 805, ISBN: 978–981–33–4080–0. https://doi.org/10.1007/978-981-33-4080-0
10. N. Fatema et al., *Intelligent Data-Analytics for Condition Monitoring: Smart Grid Applications* (Academic Press, 2021), ISBN: 978–0–323–85510–5. https://doi.org/10.1016/C2020-0-02173-0
11. S. Srivastava et al., in *Applications of Artificial Intelligence Techniques in Engineering*, vol 1 Part of the Advances in Intelligent Systems and Computing, vol 698 (Springer Nature, 2018), p. 643, ISBN 978–981–13–1819–1. https://doi.org/10.1007/978-981-13-1819-1
12. S. Srivastava et al., in *Applications of Artificial Intelligence Techniques in Engineering*, vol 2 Part of the Advances in Intelligent Systems and Computing, vol 697 (Springer Nature, 2018), p. 647, ISBN 978–981–13–1822–1. https://doi.org/10.1007/978-981-13-1822-1
13. H. Malik et al., Digital transformation through advances in artificial intelligence and machine learning. J. Intell. Fuzzy Syst. (2021). https://doi.org/10.3233/JIFS-189787
14. Y. Zhang, S. Wang, G. Ji, A comprehensive survey on particle swarm optimization algorithm and its applications. Math. Prob. Eng. **2015**(931256), 38 (2015)
15. P. Vinoop et al., PSO-NN-based hybrid model for long-term wind speed prediction: a study on 67 cities of India, in *Applications of Artificial Intelligence Techniques in Engineering*, Advances in Intelligent Systems and Computing, vol 697 (2018), pp. 319–327. https://doi.org/10.1007/978-981-13-1822-1_29
16. T. Mahto et al., Fractional order control and simulation of wind-biomass isolated hybrid power system using particle swarm optimization, in *Applications of Artificial Intelligence Techniques in Engineering*, Advances in Intelligent Systems and Computing, vol 698 (2018), pp. 277–287. https://doi.org/10.1007/978-981-13-1819-1_28
17. T. Mahto et al., Load frequency control of a solar-diesel based isolated hybrid power system by fractional order control using particle swarm optimization. J. Intell. Fuzzy Syst. **35**(5), 5055–5061 (2018). https://doi.org/10.3233/JIFS-169789
18. N. Fatema et al., Data-driven occupancy detection hybrid model using particle swarm optimization based artificial neural network, in *Metaheuristic and Evolutionary Computation: Algorithms and Applications*, Studies in Computational Intelligence (Springer Nature, 2020), pp. 283–297. https://doi.org/10.1007/978-981-15-7571-6_13

19. N. Fatema et al., Deterministic and probabilistic occupancy detection with a novel heuristic optimization and back-propagation (BP) based algorithm. J. Intell. Fuzzy Syst. (2021). https://doi.org/10.3233/JIFS-189748
20. N. Fatema et al., Metaheurestic algorithm based hybrid model for identification of building sale prices, in *Metaheuristic and Evolutionary Computation: Algorithms and Applications*, Studies in Computational Intelligence (Springer Nature, 2020), pp. 689–704. https://doi.org/10.1007/978-981-15-7571-6_32.
21. A.F. Minai et al., Metaheuristics Paradigms for renewable energy systems: advances in optimization algorithms, in *Metaheuristic and Evolutionary Computation: Algorithms and Applications*, Studies in Computational Intelligence (Springer Nature, 2020), pp. 35–61. https://doi.org/10.1007/978-981-15-7571-6_2
22. D.E. Goldberg, *Genetic Algorithm in Search, Optimization and Machine Learning* (Pearson Education, India, 1989)
23. A.S. Joshi, O. Kulkarni, Dr. G.M. Kakandikar, Dr. V.M Nandedkar, Cuckoo search optimization—a review. ICAAMM-2016 (2017)
24. D. Karaboga, B. Basturk, A powerful and efficient algorithm for numerical function optimization: artificial bee colony (ABC) algorithm. J. Glob. Optim. **39**(3), 459–471 (2007)
25. H. Shayanfar, F.S. Gharehchopogh, Farmland fertility: a new metaheuristic algorithm for solving continuous optimization problems. Appl. Soft Comput. **71**(2018), 728–746 (2018)
26. A. Kavousi, H. Samet, F. Marzbani, A new hybrid modified firefly algorithm and support vector regression model for accurate short term load forecasting. Expert Syst. Appl. **41**(13), 6047–6056 (2014)
27. A. Sadollah, H. Eskandar, A. Bahreininejad, J.H. Kim, Water cycle algorithm for solving multi objective optimization problems. Appl. Soft Comput. **27**, 279–298 (2014)
28. X. Chen, B. Xu, K. Yu, W. Du, Teaching-learning-based optimization with learning enthusiasm mechanism and its application in chemical engineering. J. Appl. Math. **2018**(1806947), 19 (2018)
29. S. Kumar, A. Kumar, A brief review on ant lion optimization algorithm, in *2018 International Conference on Advances in Computing, Communication Control and Networking (ICACCCN)*, (IEEE, 2018), pp. 236–240. https://doi.org/10.1109/ICACCCN.2018.8748862

Appendix
MATLAB Codes

H. Malik et al. (eds.), *Intelligent Data Analytics for Power and Energy Systems*,
Lecture Notes in Electrical Engineering 802,
https://doi.org/10.1007/978-981-16-6081-8

1.SGO

```
clc
clear all;
D=2;
lb=[-5 -5];
ub=[5 5];
N= 20;
c=0.25;
max_iter = 100;

for i=1:N;
   for j =1:D
       pos(i,j) = lb(:,j) + rand.*(ub(:,j)-lb(:,j));
   end
end
fx=fun(pos);


%% IMPROVING PHASE
for iter=1:max_iter
   [best, bestind] = min(fx);
   Xbest = pos(bestind,:);
end
   %%
 for j=1:size(pos,1)
   X=pos(j,:);
   Xnew = c.*X + rand(size(X)).*(Xbest-X);

 end
   %% CHECK BOUND
   for kk=1:size(Xnew,2)
      if Xnew(kk)>ub(kk)
         Xnew(kk)= ub(kk);
      elseif Xnew(kk)<lb(kk)
         Xnew(kk)=lb(kk);
      end
   end
fnew = fun(Xnew);
if fnew<fx(j,:)
   pos(j,:)=Xnew;
   fx(j,:)=fnew;
end
[bestG indG] = min(fx);
gbest = pos(indG,:);
%%
for i=1:N
   X = pos(i,:);
   p2c = ceil(rand*D);
```

2.PSO

```

clear
close all
clc

% Problem preparation
problem.nVar = 2;
problem.ub = 5 * ones(1, 2);
problem.lb = -5 * ones(1, 2);
problem.fobj = @ObjectiveFunction;

% PSO parameters
noP = 20;
maxIter = 100;
visFlag = 1; % set this to 0 if you do not want visualization

RunNo  = 30;
BestSolutions_PSO = zeros(1 , RunNo);


 [ GBEST   , cgcurve ] = PSO( noP , maxIter, problem , visFlag ) ;

 disp('Best solution found')
 GBEST.X
 disp('Best objective value')
 GBEST.O



function [ GBEST ,  cgCurve ] = PSO ( noP, maxIter,  problem, dataVis )

% Define the details of the objective function
nVar = problem.nVar;
ub = problem.ub;
lb = problem.lb;
fobj = problem.fobj;

% Extra variables for data visualization
average_objective = zeros(1, maxIter);
cgCurve = zeros(1, maxIter);
FirstP_D1 = zeros(1 , maxIter);
position_history  =  zeros(noP  , maxIter  , nVar );
```

```
    partner = ceil(rand*N);
    while(partner==i)
       partner = ceil(rand*N);
    end
    Xp = pos(partner,:);
    fp = fun(Xp);
    %%
    if fx(i,:)<fp
       Xnew = X+rand(size(X)).*(X-
Xp)+rand(size(X)).*(gbest - X);
    else
       Xnew = X-rand(size(X)).*(X-
Xp)+rand(size(X)).*(gbest - X);
    end
%%%
for kk=1:size(Xnew,2)
    if Xnew(kk)>ub(kk)
       Xnew(kk) = ub(kk);
    elseif Xnew(kk)<lb(kk)
       Xnew(kk) = lb(kk);
    end
end
%%
fnew = fun(Xnew);
if fnew<fx(i,:)
    pos(i,:)=Xnew;
    fx(i,:)=fnew;
end
end

%%
[optival, optind] = min(fx);
 BestFx(iter) = optival;
 BestX(iter,:)=pos(optind,:);

 %show iteration information
 disp(['Iteration' num2str(iter) ': Best Cost = '
num2str(BestFx(iter))]);

 plot (BestFx, 'r', 'LineWidth', 2);
 %xlabel('Iteration');
 %ylabel('Fitness Value;);
 %grid on;

 %%%% Objective function

 function out =fun(X)
x1=X(:,1);
x2=X(:,2);
out = x1.^2+x2.^2;
```

```
% Define the PSO's paramters
wMax = 0.9;
wMin = 0.2;
c1 = 2;
c2 = 2;
vMax = (ub - lb) .* 0.2;
vMin  = -vMax;

% The PSO algorithm

% Initialize the particles
for k = 1 : noP
    Swarm.Particles(k).X = (ub -lb) .*
rand(1,nVar) + lb;
    Swarm.Particles(k).V = zeros(1, nVar);
    Swarm.Particles(k).PBEST.X =
zeros(1,nVar);
    Swarm.Particles(k).PBEST.O = inf;

    Swarm.GBEST.X = zeros(1,nVar);
    Swarm.GBEST.O = inf;
end


% Main loop
for t = 1 : maxIter

    % Calcualte the objective value
    for k = 1 : noP

       currentX = Swarm.Particles(k).X;
       position_history(k , t , : ) = currentX;


       Swarm.Particles(k).O                =
fobj(currentX);
       average_objective(t)                =
average_objective(t)                       +
Swarm.Particles(k).O;

       % Update the PBEST
       if        Swarm.Particles(k).O      <
Swarm.Particles(k).PBEST.O
          Swarm.Particles(k).PBEST.X       =
currentX;
          Swarm.Particles(k).PBEST.O       =
Swarm.Particles(k).O;
       end
```

end	% Update the GBEST
	if Swarm.Particles(k).O < Swarm.GBEST.O
	Swarm.GBEST.X = currentX;
	Swarm.GBEST.O = Swarm.Particles(k).O;
	end
	end
	
	% Update the X and V vectors
	w = wMax - t .* ((wMax - wMin) / maxIter);
	
	FirstP_D1(t) = Swarm.Particles(1).X(1);
	
	for k = 1 : noP
	Swarm.Particles(k).V = w .* Swarm.Particles(k).V + c1 .* rand(1,nVar) .* (Swarm.Particles(k).PBEST.X - Swarm.Particles(k).X) ...
	+ c2 .* rand(1,nVar) .* (Swarm.GBEST.X - Swarm.Particles(k).X);
	
	% Check velocities
	index1 = find(Swarm.Particles(k).V > vMax);
	index2 = find(Swarm.Particles(k).V < vMin);
	
	Swarm.Particles(k).V(index1) = vMax(index1);
	Swarm.Particles(k).V(index2) = vMin(index2);
	
	Swarm.Particles(k).X = Swarm.Particles(k).X + Swarm.Particles(k).V;
	
	% Check positions
	index1 = find(Swarm.Particles(k).X > ub);
	index2 = find(Swarm.Particles(k).X < lb);
	
	Swarm.Particles(k).X(index1) = ub(index1);
	Swarm.Particles(k).X(index2) = lb(index2);

	``
	    end
	
	    if dataVis == 1
	        outmsg = ['Iteration# ', num2str(t) , ' Swarm.GBEST.O = ' , num2str(Swarm.GBEST.O)];
	        disp(outmsg);
	    end
	
	    cgCurve(t) = Swarm.GBEST.O;
	    average_objective(t) = average_objective(t) / noP;
	
	    fileName = ['Resluts after iteration # ' , num2str(t)];
	    save(fileName)
	end
	
	GBEST = Swarm.GBEST;
	
	if dataVis == 1
	    iterations = 1: maxIter;
	
	%% Draw the landscape
	% figure
	%
	% x = -50 : 1 : 50;
	% y = -50 : 1 : 50;
	%
	% [x_new , y_new] = meshgrid(x,y);
	%
	% for k1 = 1: size(x_new, 1)
	% for k2 = 1 : size(x_new , 2)
	% X = [x_new(k1,k2) , y_new(k1, k2)];
	% z(k1,k2) = ObjectiveFunction(X);
	% end
	% end
	%
	% subplot(1,5,1)
	% surfc(x_new , y_new , z);
	% title('Search landscape')
	% xlabel('x_1')
	% ylabel('x_2')
	% zlabel('Objective value')
	% shading interp
	% camproj perspective
	% box on
	% set(gca,'FontName','Times')
	``

```
%

%% Visualize the cgcurve
   subplot(1,5,2);
   semilogy(iterations , cgCurve, 'r');
   title('Convergence curve')
   xlabel('Iteration#')
   ylabel('Weight')

%% Visualize the average objectives
%    subplot(1,5,3)
%                semilogy(iterations ,
average_objective , 'g')
%         title('Average objecitves of all
particles')
%    xlabel('Iteration#')
%    ylabel('Average objective')
%
 %% Visualize the fluctuations
%    subplot(1,5,4)
%    plot(iterations , FirstP_D1, 'k');
%    title('First dimention in first Particle')
%    xlabel('Iteration#')
%    ylabel('Value of the first dimension')
%
 %% Visualize the search history
%    subplot(1,5,5)
%    hold on
%    for p = 1 : noP
%       for t = 1 : maxIter
%          x = position_history(p, t , 1);
%          y = position_history(p, t , 2);
%              myColor = [0+t/maxIter  0 1-
t/maxIter ];
%          plot(x , y , '.' , 'color' , myColor );
%       end
%    end
%    contour(x_new , y_new , z);
%                plot(Swarm.GBEST.X(1)   ,
Swarm.GBEST.X(2) , 'og');
%    xlim([lb(1) , ub(1)])
%    ylim([lb(2) , ub(2) ])
%    title('search history')
%    xlabel('x')
%    ylabel('y')
%    box on
%
%      set(gcf , 'position' , [128          372
1634       259])
%
```

<table>
<tr><td></td><td><pre>%

End</pre></td></tr>
<tr><td>3.SA
<pre>
clear
clc
f=@(x)(x(1)^2+x(2)^2)

x0=[0 0];
l=[-5,-5];
u=[5,5];
Mmax=100;

TolFun=1e-4;

x=x0;fx=feval(f,x);f0=fx;

for m=0:Mmax

   T=m/Mmax;
   mu=10^(T*100);

   for k=0:10

      dx=mu_inv(2*rand(size(x))-1,mu).*(u-l);
      x1=x+dx;

      x1=(x1 < l).*l+(l <= x1).*(x1 <= u).*x1+(u < x1).*u;

      fx1=feval(f,x1);df=fx1-fx;

      if (df < 0 || rand < exp(-T*df/(abs(fx)+eps)/TolFun))==1
         x=x1;fx=fx1;
      end

      if fx1 < f0 ==1
      end

   end
      x0=x1;f0=fx1;
      plot(m,f0,'*')
      hold on

end
</pre></td><td>4.GA
<pre>
clc;
clear;
close all;
%% Problem Definition
problem.CostFunction = @(x) MinOne(x);
problem.nVar = 2;

%% GA Parameters considered
params.MaxIt = 100;
params.nPop = 100;
params.beta = 1;
params.pC = 1;
params.mu = 0.02;
%% Run GA
out = RunGA(problem, params);

%% Results

figure;
plot(out.bestcost, 'LineWidth', 2);
xlabel('Iterations');
ylabel('Best Cost');
grid on;

function out = RunGA(problem, params)
   % Problem
   CostFunction = problem.CostFunction;
   nVar = problem.nVar;

   % Params
   MaxIt = params.MaxIt;
   nPop = params.nPop;
   beta = params.beta;
   pC = params.pC;
   nC = round(pC*nPop/2)*2;
   mu = params.mu;

   % Template for Empty Individuals
   empty_individual.Position = [];
   empty_individual.Cost = [];

   % Best Solution Ever Found
   bestsol.Cost = inf;
</pre></td></tr>
</table>

```function x=mu_inv(y,mu)```
 	
```x=(((1+mu).^abs(y)-1)/mu).*sign(y);```
```end```
	```% Initialization```
	```pop = repmat(empty_individual, nPop, 1);```
	```for i = 1:nPop```
	
	```% Generate Random Solution```
	```pop(i).Position = randi([0, 1], 1, nVar);```
	
	```% Evaluate Solution```
	```pop(i).Cost = CostFunction(pop(i).Position);```
	
	```% Compare Solution to Best Solution Ever Found```
	```if pop(i).Cost < bestsol.Cost```
	```bestsol = pop(i);```
	```end```
	```end```
	
	```% Best Cost of Iterations```
	```bestcost = nan(MaxIt, 1);```
	```% Main Loop```
	```for it = 1:MaxIt```
	
	```% Selection Probabilities```
	```c = [pop.Cost];```
	```avgc = mean(c);```
	```if avgc ~= 0```
	```c = c/avgc;```
	```end```
	```probs = exp(-beta*c);```
	
	```% Initialize Offsprings Population```
	```popc = repmat(empty_individual, nC/2, 2);```
	
	```% Crossover```
	```for k = 1:nC/2```
	
	```% Select Parents```
	```p1 = pop(RouletteWheelSelection(probs));```
	```p2 = pop(RouletteWheelSelection(probs));```
	
	```% Perform Crossover```
	```[popc(k, 1).Position, popc(k, 2).Position] = ...```
	```MyCrossover(p1.Position,```

	p2.Position);
	
	end
	
	% Convert popc to Single-Column Matrix
	popc = popc(:);
	
	% Mutation
	for l = 1:nC
	
	% Perform Mutation
	popc(l).Position = Mutate(popc(l).Position, mu);
	
	% Evaluation
	popc(l).Cost = CostFunction(popc(l).Position);
	
	% Compare Solution to Best Solution Ever Found
	if popc(l).Cost < bestsol.Cost
	bestsol = popc(l);
	end
	
	end
	
	% Merge and Sort Populations
	pop = SortPopulation([pop; popc]);
	
	% Remove Extra Individuals
	pop = pop(1:nPop);
	
	% Update Best Cost of Iteration
	bestcost(it) = bestsol.Cost;
	% Display Itertion Information
	disp(['Iteration ' num2str(it) ': Best Cost = ' num2str(bestcost(it))]);
	end
	
	% Results
	out.pop = pop;
	out.bestsol = bestsol;
	out.bestcost = bestcost;
	
	end

	function [y1, y2] = DoublePointCrossover(x1, x2)
	
	nVar = numel(x1);
	
	q = randperm(nVar);
	j1 = min(q(1), q(2));
	j2 = max(q(1), q(2));
	
	y1 = [x1(1:j1) x2(j1+1:j2) x1(j2+1:end)];
	y2 = [x2(1:j1) x1(j1+1:j2) x2(j2+1:end)];
	
	end
	
	function z = MinOne(x)
	% z = sum(x);
	z = sum(x.^2); %SPHERE FUNCTION
	end
	
	function y = Mutate(x, mu)
	flag = (rand(size(x)) < mu);
	y = x;
	y(flag) = 1 - x(flag);
	end
	
	function [y1, y2] = MyCrossover(x1, x2)
	m = randi([1, 3]);
	switch m
	case 1
	[y1, y2] = SinglePointCrossover(x1, x2);
	
	case 2
	[y1, y2] = DoublePointCrossover(x1, x2);
	
	otherwise
	[y1, y2] = UniformCrossover(x1, x2);
	end
	end
	
	function [y1, y2] = SinglePointCrossover(x1, x2)
	nVar = numel(x1);
	j = randi([1, nVar-1]);
	y1 = [x1(1:j) x2(j+1:end)];
	y2 = [x2(1:j) x1(j+1:end)];

```
end
function pop = SortPopulation(pop)
   [~, so] = sort([pop.Cost]);
   pop = pop(so);
end

function [y1, y2] = UniformCrossover(x1, x2)
   alpha = randi([0, 1], size(x1));
   y1 = alpha.*x1 + (1-alpha).*x2;
   y2 = alpha.*x2 + (1-alpha).*x1;
end
```

5.CSO

```
format short
clear all
clc

% % input parameters
D=2;
lb=[-5 -5];
ub=[5 5];
N=20;
n=N;
Pa=0.25;
max_iter=100;

% % generate initial population
for i=1:N
   for j=1:D
      nest(i,j) = lb(:,j)+rand.*(ub(:,j)-lb(:,j));
   end
end
fx=fns(nest);
beta=1.5;
sigma = (gamma(1+beta)*sin(pi*beta/2)/(gamma((1+beta)/2)*beta*2^((beta-1)/2)))^(1/beta);

% % main loop

for iter = 1:max_iter
   [fnv,indf] = min(fx);
   best = nest(indf,:);
   for j= 1:N
      s=nest(j,:);
      X=s;
% % %      levy flight
      u=randn(size(s))*sigma;
      v=randn(size(s));
      step=u./abs(v).^(1/beta);
```

6.ABC

```
clc;
clear;
close all;

%% Problem Definition

CostFunction=@(x) Sphere(x);        % Cost Function

nVar=5;             % Number of Decision Variables

VarSize=[1 nVar];   % Decision Variables Matrix Size

VarMin=-10;         % Decision Variables Lower Bound
VarMax= 10;         % Decision Variables Upper Bound

%% ABC Settings

MaxIt=200;          % Maximum Number of Iterations

nPop=100;           % Population Size (Colony Size)

nOnlooker=nPop;     % Number of Onlooker Bees

L=round(0.6*nVar*nPop); % Abandonment Limit Parameter (Trial Limit)

a=1;                % Acceleration Coefficient Upper Bound
```

```
Xnew=X+randn(size(s)).*0.01.*step.*(X-
best);
%%%       check bounds
for kk=1:size(Xnew,2)
   if Xnew(kk)>ub(kk)
      Xnew(kk)=ub(kk);
   elseif Xnew(kk)<lb(kk)
      Xnew(kk)=lb(kk);
   end
end
fnew=fns(Xnew);
if fnew<fx(j,:)
   nest(j,:)=Xnew;
   fx(j,:)=fnew;
end
   end

[fmin,k1]=min(fx);
best=nest(k1,:);
k=rand(size(nest))<Pa;
stepsize=rand*(nest(randperm(n),:)-
nest(randperm(n),:));
new_nest=nest+stepsize.*k;

for ii=1:size(nest,1)
   s=new_nest(ii,:);
   for kk=1:size(s,2)
      if s(kk)>ub(kk)
         s(kk)=ub(kk);
      elseif s(kk)<lb(kk)
         s(kk)=lb(kk);
      end
   end
   new_nest(ii,:)=s;

   fnew=fns(s);
   if fnew<fx(ii,:)
      nest(ii,:)=s;
      fx(ii,:)=fnew;
   end
end
[optval,optind]=min(fx);
BestFx(iter)=optval;
BestX(iter,:)=nest(optind,:);

disp(['Iteration' num2str(iter) 'Best Cost='
num2str(BestFx(iter))]);

plot(BestFx, 'LineWidth',2);
```

```
%% Initialization

% Empty Bee Structure
empty_bee.Position=[];
empty_bee.Cost=[];

% Initialize Population Array
pop=repmat(empty_bee,nPop,1);

% Initialize Best Solution Ever Found
BestSol.Cost=inf;

% Create Initial Population
for i=1:nPop

pop(i).Position=unifrnd(VarMin,VarMax,V
arSize);

pop(i).Cost=CostFunction(pop(i).Position);
   if pop(i).Cost<=BestSol.Cost
      BestSol=pop(i);
   end
end

% Abandonment Counter
C=zeros(nPop,1);

% Array to Hold Best Cost Values
BestCost=zeros(MaxIt,1);

%% ABC Main Loop

for it=1:MaxIt

   % Recruited Bees
   for i=1:nPop

      % Choose k randomly, not equal to i
      K=[1:i-1 i+1:nPop];
      k=K(randi([1 numel(K)]));

      % Define Acceleration Coeff.
      phi=a*unifrnd(-1,+1,VarSize);

      % New Bee Position

newbee.Position=pop(i).Position+phi.*(pop
(i).Position-pop(k).Position);
```

```
xlabel('iteration nos =');
ylabel('Fitness value =');
title('Convergence vs Iteration');
grid on

end
%% Objective function
function out = fns(X)
x1 = X(:,1);
x2 = X(:,2);
out = x1.^2+x2.^2;
end
```

```
        % Evaluation

newbee.Cost=CostFunction(newbee.Positio
n);

        % Comparision
        if newbee.Cost<=pop(i).Cost
           pop(i)=newbee;
        else
           C(i)=C(i)+1;
        end

    end

    % Calculate Fitness Values and Selection
Probabilities
    F=zeros(nPop,1);
    MeanCost = mean([pop.Cost]);
    for i=1:nPop
        F(i) = exp(-pop(i).Cost/MeanCost); %
Convert Cost to Fitness
    end
    P=F/sum(F);

    % Onlooker Bees
    for m=1:nOnlooker

        % Select Source Site
        i=RouletteWheelSelection(P);

        % Choose k randomly, not equal to i
        K=[1:i-1 i+1:nPop];
        k=K(randi([1 numel(K)]));

        % Define Acceleration Coeff.
        phi=a*unifrnd(-1,+1,VarSize);

        % New Bee Position

newbee.Position=pop(i).Position+phi.*(pop
(i).Position-pop(k).Position);

        % Evaluation

newbee.Cost=CostFunction(newbee.Positio
n);

        % Comparision
        if newbee.Cost<=pop(i).Cost
           pop(i)=newbee;
```

	else
	C(i)=C(i)+1;
	end
	
	end
	
	% Scout Bees
	for i=1:nPop
	if C(i)>=L
	
	pop(i).Position=unifrnd(VarMin,VarMax,VarSize);
	
	pop(i).Cost=CostFunction(pop(i).Position);
	C(i)=0;
	end
	end
	
	% Update Best Solution Ever Found
	for i=1:nPop
	if pop(i).Cost<=BestSol.Cost
	BestSol=pop(i);
	end
	end
	
	% Store Best Cost Ever Found
	BestCost(it)=BestSol.Cost;
	
	% Display Iteration Information
	disp(['Iteration ' num2str(it) ': Best Cost = ' num2str(BestCost(it))]);
	
	end
	
	%% Results
	
	figure;
	%plot(BestCost,'LineWidth',2);
	semilogy(BestCost,'LineWidth',2);
	xlabel('Iteration');
	ylabel('Best Cost');
	grid on;

7.FARMLAND FERTILITY VarMin=-5; VarMax=5; Nvar=2; FunNumber=1; CostFunction=@(x) (x(1)^2+x(2)^2);	**8.Firefly** clc; clear all; close all; %% Input parameters

VarSize=[1 Nvar];
MaxIt=100;
k=2;
n=50;
NPop=k*n;
alpha=0.6;
betha=0.4;
W=1;
Q=0.5;
EmptyFarmland.Position=[];
EmptyFarmland.Cost=inf;
pop=repmat(EmptyFarmland,NPop,1);
 	
for i=1:NPop
 	
pop(i).Position=unifrnd(VarMin,VarMax,VarSize);
 	
pop(i).Cost=CostFunction(pop(i).Position);
 	
end
RandIdx=randsample(NPop,NPop);
Section=cell(1,k);
for s=1:k
aj=n*(s-1)+1:n*s;
Section{s}.Pop=pop(RandIdx(aj));
Section{s}.LocalMem=[];
end
 	
BestSol.Cost=inf;
BestSol.Position=[];
BestCost=zeros(MaxIt,1);
 	
for It=1:MaxIt
 	
FitSection=inf(1,k);
for s=1:k
 	
FitSection(s)=mean([Section{s}.Pop.Cost]);
end
 	
t=0.02;
CLocal=round(t*n);
 	
Section=UpdateLocalMem(Section,k,CLocal);
CGlobal=round(t*NPop);
 	
GlobalMem=repmat(EmptyFarmland,CGlob
	D=2; %Dimension
	lb=[-5 -5]; %Lower bound of the variable
	ub=[5 5]; %Upper bound of the variable
	N=20; %Population Size
	alpha=1; %Randomness Strength 0-1
	beta0=1; %Attractiveness constant
	gamma =0.01; %Absorbtion co-efficient
	theta = 0.97; %Randomness Reduction Factor
	iter_max=100; %Maximum number of iteration
	
	%%Generate initial population randomly
	
	for i=1:N
	for j=1:D
	pop(i,j)=lb(:,j)+ rand.*(ub(:,j)-lb(:,j));
	end
	end
	
	%% Evaluate objective function
	
	fx = fun(pop);
	
	alpha = alpha*theta;
	scale = abs(ub-lb);
	
	%% Firefly Algorithm Main Loop
	
	for iter=1:iter_max
	% Two loops over all the n fireflies
	for i=1:N
	for j=1:N
	fx(i)=fun(pop(i,:));
	%Brighter/more attractive
	if fx(i)<fx(j)
	pop(i,:)=pop(i,:);%Since its minimization(lesser value will remain same)
	else fx(i)>fx(j)
	Xi = pop(i,:);
	Xj = pop(j,:);
	r = sqrt(sum((Xi-Xj).^2));
	beta = beta0*exp(-gamma*r.^2);
	steps = alpha.*(rand(1,D)-0.5).*scale;

```
al,1);

[GlobalMem,PopMain,CostMain]=FindGlobalSoultion(Section,Nvar,n,CGlobal,GlobalMem);

 [~,idx]=max(FitSection);
 %
 for s=1:k

    if (s==idx)
       for i=1:n
          h=alpha*unifrnd(-1,+1,VarSize);
          Xij=Section{s}.Pop(i).Position;
          XGlobal=GlobalMem(randi([1 CGlobal],1)).Position;
          Xnew=h.*(Xij-XGlobal)+Xij;
          Xnew=max(Xnew,VarMin);
          Xnew=min(Xnew,VarMax);
          FitNew=CostFunction(Xnew);

if(FitNew<=Section{s}.Pop(i).Cost)

Section{s}.Pop(i).Position=Xnew;
             Section{s}.Pop(i).Cost=FitNew;
          end
       end
    else
       for i=1:n
          h=betha*unifrnd(-1,1,VarSize);
          Xij=Section{s}.Pop(i).Position;
          Xuj=PopMain(randi([1 NPop],1),:);
          Xnew=h.*(Xij-Xuj)+Xij;
          Xnew=max(Xnew,VarMin);
          Xnew=min(Xnew,VarMax);
          FitNew=CostFunction(Xnew);

if(FitNew<=Section{s}.Pop(i).Cost)

Section{s}.Pop(i).Position=Xnew;
             Section{s}.Pop(i).Cost=FitNew;
          end
       end
    end
 end


Section=UpdateLocalMem(Section,k,CLocal);
  for s=1:k
```

```
Xnew = Xi + beta*(Xi-Xj) +steps;

%% Check If Xnew lies withhin limit
for k=1:size(Xnew,2)
if Xnew(k)>ub(k)
Xnew(k)=ub(k);
elseif Xnew(k)<lb(k)
Xnew(k)=lb(k);
end
end
%% Perform Greedy Selection
fnew =fun(Xnew);
if fnew< fx(i) %%Since its minimization we use '<' sign
fx(i)=fnew;
pop(i,:)=Xnew;
end

end
end
end
%% Memorise the BEST
[opt_value,opt_index] = min(fx);
BestFx(iter)=opt_value;
BestX(iter,:)=pop(opt_index,:);

%% Show Iteration Information
%
% disp('Iteration no. ')
t1(iter,1)= iter;
% disp('value')
t1(iter,2)=(BestFx(iter))
% disp(['Iteration ' t1 ': Best Value =' t2]);

%% Plotting the Result
plot(BestFx, 'LineWidth',2);
xlabel('Iteration Number');
ylabel('Fitness Value');
title('Convergence Vs Iteration');
grid on
end
```

```
    for i=1:n

        if(Q>rand)
            W=W*0.1;
            Xij=Section{s}.Pop(i).Position;
            b=randi([1 CGlobal],1);
            XGlobal=GlobalMem(b).Position;
            Xnew=unifrnd(-
1,+1,VarSize).*(Xij-XGlobal)+Xij;
            Xnew=max(Xnew,VarMin);
            Xnew=min(Xnew,VarMax);
            FitNew=CostFunction(Xnew);

if(FitNew<=Section{s}.Pop(i).Cost)

Section{s}.Pop(i).Position=Xnew;
              Section{s}.Pop(i).Cost=FitNew;
           end

        else

            Xij=Section{s}.Pop(i).Position;
            b=randi([1 CLocal],1);
            idxlocal=Section{s}.LocalMem(b);

XLocal=Section{s}.Pop(idxlocal).Position;
            Xnew=unifrnd(-
1,+1,VarSize).*(Xij-XLocal)+Xij;
            Xnew=max(Xnew,VarMin);
            Xnew=min(Xnew,VarMax);
            FitNew=CostFunction(Xnew);

if(FitNew<=Section{s}.Pop(i).Cost)
            Section{s}.Pop(i).Position=Xnew;
              Section{s}.Pop(i).Cost=FitNew;
            end

        end

    end

  end

GlobalFind=FindGlobalSoultion(Section,Nv
ar,n,CGlobal,GlobalMem);
  if(GlobalFind(1).Cost<BestSol.Cost)
    BestSol.Cost=GlobalFind(1).Cost;

BestSol.Position=GlobalFind(1).Position;
  end
```

<table>
<tr><td>

```
   BestCost(It)=BestSol.Cost;

   disp(['Iteration ' num2str(It) ': Best Cost =
' num2str(BestCost(It))]);

end
minresult=BestCost(end);
   plot(BestCost,'*');

function
[GlobalMem,PopMain,CostMain]=FindGlo
balSoultion(Section,Nvar,n,CGlobal,Global
Mem)

PopMain=[];
CostMain=[];
for s=1:numel(Section)
 PopSection=[Section{s}.Pop.Position];

PopMain=[PopMain;reshape(PopSection,n,
Nvar)];
 CostMain=[CostMain
[Section{s}.Pop(1:end).Cost]];
end

[~,idx]=sort(CostMain);
for i=1:CGlobal

GlobalMem(i).Position=PopMain(idx(i),:);
   GlobalMem(i).Cost=CostMain(idx(i));
end
end

function
Section=UpdateLocalMem(Section,k,CLoca
l)

  for s=1:k
     [~,idx]=sort([Section{s}.Pop.Cost]);
     for i=1:CLocal

Section{s}.LocalMem=idx(1:CLocal);
     end
  end
end
```

</td><td></td></tr>
<tr><td>

9.WCO

```
close all
clear all
```

</td><td>

10.TLBO

```
 clc;
clear all;
```

</td></tr>
</table>

```
clc

global nvars

nvars=100;
objective_function=@sphere;
LB=ones(1,nvars)*-5;
UB=ones(1,nvars)*5;
Npop=50;
Nsr=2;
max_it=100;
dmax=1e-16;

[Xmin,Fmin,NFEs,Elapsed_Time]=WCA(objective_function,LB,UB,nvars,Npop,Nsr,dmax,max_it)

function z = sphere(x);

global nvars
z=0;
 for i=1:nvars
  zz=x(i)^2;
   z=zz+z;
 end


end
function
[Xmin,Fmin,NFEs,Elapsed_Time]=WCA(objective_function,LB,UB,nvars,Npop,Nsr,dmax,max_it)


% INPUTS:

% objective_function:        sphere function
% LB:                        Lower bound of a problem
% UB:                        Upper bound of a problem
% nvars:                     Number of design variables
% Npop                       Population size
% Nsr                        Number of rivers + sea
% dmax                       Evporation condition constant
% max_it:                    Maximum number of iterations
```

```
close all;
%% Problem Definition
% Cost Function
CostFunction = @(x) Sphere(x);
nVar = 10;            % Number of Unknown Variables
VarSize = [1 nVar]; % Unknown Variables Matrix Size
VarMin = -10;        % Unknown Variables Lower Bound
VarMax =  10;        % Unknown Variables Upper Bound
%% TLBO Parameters
MaxIt = 5;            % Maximum Number of Iterations
nPop = 50;           % Population Size
%% Initialization
% Empty Structure for Individuals
empty_individual.Position = [];
empty_individual.Cost = [];
% Initialize Population Array
pop = repmat(empty_individual, nPop, 1);
% Initialize Best Solution
BestSol.Cost = inf;
% Initialize Population Members
for i=1:nPop
   pop(i).Position = unifrnd(VarMin, VarMax, VarSize);
   pop(i).Cost = CostFunction(pop(i).Position);

   if pop(i).Cost < BestSol.Cost
      BestSol = pop(i);
   end
end
% Initialize Best Cost Record
BestCosts = zeros(MaxIt,1);
%% TLBO Main Loop
for it=1:MaxIt

   % Calculate Population Mean
   Mean = 0;
   for i=1:nPop
      Mean = Mean + pop(i).Position;
   end
   Mean = Mean/nPop;

   % Select Teacher
   Teacher = pop(1);
   for i=2:nPop
```

```
% OUTPUTS:

% Xmin:                    Global optimum
solution
% Fmin:                    Cost of global
optimum solution
% NFEs:                Number of function
evaluations
% Elapsed_Time             Elasped time for
solving an optimization problem
%% Default Values for WCA
format long g
if (nargin <5 || isempty(Npop)), Npop=50;
end
if (nargin <6 || isempty(Nsr)), Nsr=4; end
if (nargin <7 || isempty(dmax)), dmax=1e-
16; end
if (nargin <8 || isempty(max_it)),
max_it=1000; end
%% ---------------------------------------------
-------------------------
% Create initial population and form sea,
rivers, and streams
tic
N_stream=Npop-Nsr;

ind.position=[];
ind.cost=[];

pop=repmat(ind,Npop,1);

for i=1:Npop
   pop(i).position=LB+(UB-
LB).*rand(1,nvars);

pop(i).cost=objective_function(pop(i).positi
on);
end

[~, index]=sort([pop.cost]);
%------------ Forming Sea --------------------
--------------------------
sea=pop(index(1));
%------------Forming Rivers -----------------
----------------------------
river=repmat(ind,Nsr-1,1);
for i=1:Nsr-1
   river(i)=pop(index(1+i));
end
```

```
        if pop(i).Cost < Teacher.Cost
            Teacher = pop(i);
        end
    end

    % Teacher Phase
    for i=1:nPop
        % Create Empty Solution
        newsol = empty_individual;

        % Teaching Factor
        TF = randi([1 2]);

        % Teaching (moving towards teacher)
        newsol.Position = pop(i).Position ...
            + rand(VarSize).*(Teacher.Position
- TF*Mean);

        % Clipping
        newsol.Position                    =
max(newsol.Position, VarMin);
        newsol.Position                    =
min(newsol.Position, VarMax);

        % Evaluation
        newsol.Cost                        =
CostFunction(newsol.Position);

        % Comparision
        if newsol.Cost<pop(i).Cost
            pop(i) = newsol;
            if pop(i).Cost < BestSol.Cost
                BestSol = pop(i);
            end
        end
    end

    % Learner Phase
    for i=1:nPop

        A = 1:nPop;
        A(i)=[];
        j = A(randi(nPop-1));

        Step      =      pop(i).Position      -
pop(j).Position;
        if pop(j).Cost < pop(i).Cost
            Step = -Step;
        end
```

```
%------------ Forming Streams-----------------
-----------------------------
stream=repmat(ind,N_stream,1);
for i=1:N_stream
   stream(i)=pop(index(Nsr+i));
end
%--------- Designate streams to rivers and
sea ------------------------
cs=[sea.cost;[river.cost]';stream(1).cost];
f=0;
if length(unique(cs))~=1
   CN=cs-max(cs);
else
   CN=cs;
   f=1;
end

NS=round(abs(CN/sum(CN))*N_stream);
if f~=1
   NS(end)=[];
end
NS=sort(NS,'descend');
% ------------------------ Modification on NS
-----------------------
i=Nsr;
while sum(NS)>N_stream
   if NS(i)>1
      NS(i)=NS(i)-1;
   else
      i=i-1;
   end
end

i=1;
while sum(NS)<N_stream
   NS(i)=NS(i)+1;
end

if find(NS==0)
   index=find(NS==0);
   for i=1:size(index,1)
      while NS(index(i))==0

NS(index(i))=NS(index(i))+round(NS(i)/6);
         NS(i)=NS(i)-round(NS(i)/6);
      end
   end
end

NS=sort(NS,'descend');
```

```
      % Create Empty Solution
      newsol = empty_individual;

      % Teaching (moving towards teacher)
      newsol.Position = pop(i).Position +
rand(VarSize).*Step;

      % Clipping
      newsol.Position =
max(newsol.Position, VarMin);
      newsol.Position =
min(newsol.Position, VarMax);

      % Evaluation
      newsol.Cost =
CostFunction(newsol.Position);

      % Comparision
      if newsol.Cost<pop(i).Cost
         pop(i) = newsol;
         if pop(i).Cost < BestSol.Cost
            BestSol = pop(i);
         end
      end
   end

   % Store Record for Current Iteration
   BestCosts(it) = BestSol.Cost;

   % Show Iteration Information
   disp(['Iteration ' num2str(it) ': Best Cost =
' num2str(BestCosts(it))]);

end
%% Results
figure;
%plot(BestCosts, 'LineWidth', 2);
semilogy(BestCosts, 'LineWidth', 2);
xlabel('Iteration');
ylabel('Best Cost');
grid on;
Objective Function:-

function z = Sphere(x)
   z = sum(x.^2);
end
```

```
NB=NS(2:end);
%%
%----------- Main Loop for WCA -------------
-------------------------------
disp('******************* Water Cycle
Algorithm
(WCA)*******************');
disp('*Iterations     Function Values *');
disp('*******************************
**********************************
**');
FF=zeros(max_it,1);
for i=1:max_it
   %---------- Moving stream to sea-----------
---------------------------
   for j=1:NS(1)

stream(j).position=stream(j).position+2.*ran
d(1).*(sea.position-stream(j).position);


stream(j).position=min(stream(j).position,U
B);

stream(j).position=max(stream(j).position,L
B);


stream(j).cost=objective_function(stream(j).
position);

       if stream(j).cost<sea.cost
          new_sea=stream(j);
          stream(j)=sea;
          sea=new_sea;
       end
   end
   %---------- Moving Streams to rivers-------
---------------------------
   for k=1:Nsr-1
       for j=1:NB(k)

stream(j+sum(NS(1:k))).position=stream(j+
sum(NS(1:k))).position+2.*rand(1,nvars).*(r
iver(k).position-
stream(j+sum(NS(1:k))).position);


stream(j+sum(NS(1:k))).position=min(strea
m(j+sum(NS(1:k))).position,UB);
```

11. ALO

MATLAB FILE 1

```
function
[Elite_antlion_fitness,Elite_antlion_position
,Convergence_curve]=ALO(N,Max_iter,lb,
ub,dim,fobj)

% Initialize the positions of antlions and
ants
antlion_position=initialization(N,dim,ub,lb)
;
ant_position=initialization(N,dim,ub,lb);

% Initialize variables to save the position of
elite, sorted antlions,
% convergence curve, antlions fitness, and
ants fitness
Sorted_antlions=zeros(N,dim);
Elite_antlion_position=zeros(1,dim);
Elite_antlion_fitness=inf;
Convergence_curve=zeros(1,Max_iter);
antlions_fitness=zeros(1,N);
ants_fitness=zeros(1,N);

% Calculate the fitness of initial antlions
and sort them
for i=1:size(antlion_position,1)

antlions_fitness(1,i)=fobj(antlion_position(i
,:));
end

[sorted_antlion_fitness,sorted_indexes]=sor
t(antlions_fitness);

for newindex=1:N

Sorted_antlions(newindex,:)=antlion_positi
on(sorted_indexes(newindex),:);
end

Elite_antlion_position=Sorted_antlions(1,:);
Elite_antlion_fitness=sorted_antlion_fitness
(1);

% Main loop start from the second iteration
since the first iteration
```

```
stream(j+sum(NS(1:k))).position=max(strea
m(j+sum(NS(1:k))).position,LB);


stream(j+sum(NS(1:k))).cost=objective_fun
ction(stream(j+sum(NS(1:k))).position);

        if
stream(j+sum(NS(1:k))).cost<river(k).cost

new_river=stream(j+sum(NS(1:k)));
            stream(j+sum(NS(1:k)))=river(k);
            river(k)=new_river;

            if river(k).cost<sea.cost
               new_sea=river(k);
               river(k)=sea;
               sea=new_sea;
            end
         end
      end
   end
   %---------- Moving rivers to Sea -----------
---------------------------

   for j=1:Nsr-1

river(j).position=river(j).position+2.*rand(1,
nvars).*(sea.position-river(j).position);


river(j).position=min(river(j).position,UB);

river(j).position=max(river(j).position,LB);


river(j).cost=objective_function(river(j).posi
tion);

      if river(j).cost<sea.cost
         new_sea=river(j);
         river(j)=sea;
         sea=new_sea;
      end
   end
   %-------------- Evaporation condition and
raining process--------------
   % Check the evaporation condition for
rivers and sea
   for k=1:Nsr-1
```

```
% was dedicated to calculating the fitness
of antlions
Current_iter=2;
while Current_iter<Max_iter+1

   % This for loop simulate random walks
   for i=1:size(ant_position,1)
      % Select ant lions based on their
fitness (the better anlion the higher chance
of catching ant)

Rolette_index=RouletteWheelSelection(1./s
orted_antlion_fitness);
      if Rolette_index==-1
         Rolette_index=1;
      end

      % RA is the random walk around the
selected antlion by rolette wheel

RA=Random_walk_around_antlion(dim,M
ax_iter,lb,ub,
Sorted_antlions(Rolette_index,:),Current_it
er);

      % RA is the random walk around the
elite (best antlion so far)

[RE]=Random_walk_around_antlion(dim,
Max_iter,lb,ub,
Elite_antlion_position(1,:),Current_iter);

      ant_position(i,:)=
(RA(Current_iter,:)+RE(Current_iter,:))/2;
% Equation (2.13) in the paper
   end

   for i=1:size(ant_position,1)

      % Boundar checking (bring back the
antlions of ants inside search
      % space if they go beyoud the
boundaries
      Flag4ub=ant_position(i,:)>ub;
      Flag4lb=ant_position(i,:)<lb;

ant_position(i,:)=(ant_position(i,:).*(~(Flag
4ub+Flag4lb)))+ub.*Flag4ub+lb.*Flag4lb;
```

```
        if ((norm(river(k).position-sea.position)<dmax) || rand<0.1)
            for j=1:NB(k)
stream(j+sum(NS(1:k))).position=LB+rand(1,nvars).*(UB-LB);
            end
        end
    end
    % Check the evaporation condition for streams and sea
    for j=1:NS(1)
        if ((norm(stream(j).position-sea.position)<dmax))
stream(j).position=LB+rand(1,nvars).*(UB-LB);
        end
    end
    %---------------------------------------------------------------------
    dmax=dmax-(dmax/max_it);

    disp(['Iteration: ',num2str(i),' Fmin= ',num2str(sea.cost)]);
    FF(i)=sea.cost;
end
%% Results and Plot
toc;
Elapsed_Time=toc;
plot(FF,'LineWidth',2);
xlabel('Number of Iterations');
ylabel('Function Values');
NFEs=Npop*max_it;
Xmin=sea.position;
Fmin=objective_function(Xmin);
end
```

```
ants_fitness(1,i)=fobj(ant_position(i,:));

    end

    % Update antlion positions and fitnesses based of the ants (if an ant
    % becomes fitter than an antlion we assume it was cought by the antlion
    % and the antlion update goes to its position to build the trap)
double_population=[Sorted_antlions;ant_position];
    double_fitness=[sorted_antlion_fitness ants_fitness];

    [double_fitness_sorted I]=sort(double_fitness);
double_sorted_population=double_population(I,:);

antlions_fitness=double_fitness_sorted(1:N);
Sorted_antlions=double_sorted_population(1:N,:);

    % Update the position of elite if any antlinons becomes fitter than it
    if antlions_fitness(1)<Elite_antlion_fitness
Elite_antlion_position=Sorted_antlions(1,:);
Elite_antlion_fitness=antlions_fitness(1);
    end

    % Keep the elite in the population
Sorted_antlions(1,:)=Elite_antlion_position;
    antlions_fitness(1)=Elite_antlion_fitness;

    % Update the convergence curve
Convergence_curve(Current_iter)=Elite_antlion_fitness;

    % Display the iteration and best optimum
```

	obtained so far
	if mod(Current_iter,50)==0
	display(['At iteration ', num2str(Current_iter), ' the elite fitness is ', num2str(Elite_antlion_fitness)]);
	end
	
	Current_iter=Current_iter+1;
	End
	
	MATLAB FILE 2
	
	function func_plot(func_name)
	
	[lb,ub,dim,fobj]=Get_Functions_details(func_name);
	
	switch func_name
	case 'F1'
	x=-100:2:100; y=x; %[-100,100]
	surfc(x,y,f,'LineStyle','none');
	
	end
	
	MATLAB FILE 3
	function [lb,ub,dim,fobj] = Get_Functions_details(F)
	
	switch F
	case 'F1'
	fobj = @F1;
	lb=-100;
	ub=100;
	dim=10;
	end
	
	end
	function o = F1(x)
	o=sum(x.^2);
	end
	
	MATLAB FILE 4
	% This function creates the first random population
	
	function X=initialization(SearchAgents_no,dim,ub,lb)

	Boundary_no= size(ub,2); % numnber of boundaries
	
	% If the boundaries of all variables are equal and user enter a signle
	% number for both ub and lb
	if Boundary_no==1
	X=rand(SearchAgents_no,dim).*(ub-lb)+lb;
	end
	
	% If each variable has a different lb and ub
	if Boundary_no>1
	for i=1:dim
	ub_i=ub(i);
	lb_i=lb(i);
	
	X(:,i)=rand(SearchAgents_no,1).*(ub_i-lb_i)+lb_i;
	end
	end
	
	MATLAB FILE 5
	clear all
	clc
	
	SearchAgents_no=40; % Number of search agents
	
	Function_name='F1'; % Name of the test function that can be from F1 to F23 (Table 1,2,3 in the paper)
	
	Max_iteration=500; % Maximum numbef of iterations
	
	% Load details of the selected benchmark function
	[lb,ub,dim,fobj]=Get_Functions_details(Function_name);
	
	[Best_score,Best_pos,cg_curve]=ALO(SearchAgents_no,Max_iteration,lb,ub,dim,fobj);
	
	figure('Position',[500 500 660 290])
	%Draw search space
	subplot(1,2,1);
	func_plot(Function_name);
	title('Test function')
	xlabel('x_1');

	ylabel('x_2');
	zlabel([Function_name,'(x_1 , x_2)'])
	grid off
	
	%Draw objective space
	subplot(1,2,2);
	semilogy(cg_curve,'Color','r')
	title('Convergence curve')
	xlabel('Iteration');
	ylabel('Best score obtained so far');
	
	axis tight
	grid off
	box on
	legend('ALO')
	
	display(['The best solution obtained by ALO is : ', num2str(Best_pos)]);
	display(['The best optimal value of the objective funciton found by ALO is : ', num2str(Best_score)]);
	
	MATLAB FILE 6
	
	% This function creates random walks
	
	function [RWs]=Random_walk_around_antlion(Dim ,max_iter,lb, ub,antlion,current_iter)
	if size(lb,1) ==1 && size(lb,2)==1 %Check if the bounds are scalar
	  lb=ones(1,Dim)*lb;
	  ub=ones(1,Dim)*ub;
	end
	
	if size(lb,1) > size(lb,2) %Check if boundary vectors are horizontal or vertical
	  lb=lb';
	  ub=ub';
	end
	
	I=1; % I is the ratio in Equations (2.10) and (2.11)
	
	if current_iter>max_iter/10
	  I=1+100*(current_iter/max_iter);
	end

```
if current_iter>max_iter/2
   I=1+1000*(current_iter/max_iter);
end

if current_iter>max_iter*(3/4)
   I=1+10000*(current_iter/max_iter);
end

if current_iter>max_iter*(0.9)
   I=1+100000*(current_iter/max_iter);
end

if current_iter>max_iter*(0.95)
   I=1+1000000*(current_iter/max_iter);
end


% Dicrease boundaries to converge towards antlion
lb=lb/(I); % Equation (2.10) in the paper
ub=ub/(I); % Equation (2.11) in the paper

% Move the interval of [lb ub] around the antlion [lb+anlion ub+antlion]
if rand<0.5
   lb=lb+antlion; % Equation (2.8) in the paper
else
   lb=-lb+antlion;
end

if rand>=0.5
   ub=ub+antlion; % Equation (2.9) in the paper
else
   ub=-ub+antlion;
end

% This function creates n random walks and normalize accroding to lb and ub
% vectors
for i=1:Dim
   X = [0 cumsum(2*(rand(max_iter,1)>0.5)-1)']; % Equation (2.1) in the paper
   %[a b]--->[c d]
   a=min(X);
   b=max(X);
   c=lb(i);
   d=ub(i);
```

	``
	  X_norm=((X-a).*(d-c))./(b-a)+c; %
	Equation (2.7) in the paper
	  RWs(:,i)=X_norm;
	end
	
	MATLAB FILE 7
	
	function choice =
	RouletteWheelSelection(weights)
	 accumulation = cumsum(weights);
	 p = rand() * accumulation(end);
	 chosen_index = -1;
	 for index = 1 : length(accumulation)
	   if (accumulation(index) > p)
	     chosen_index = index;
	     break;
	   end
	 end
	 choice = chosen_index;
	``

The manufacturer's authorised representative in the EU is Springer
Nature Customer Service Centre GmbH, Europaplatz 3, 69115 Heidelberg,
Germany. If you have any concerns regarding our products, please
contact ProductSafety@springernature.com

Printed and bound by CPI Group (UK) Ltd, Croydon, CR0 4YY
06/10/2026
02225252-0018